Advanced Structured Materials

Volume 253

Series Editors

Common engineering materials are reaching their limits in many applications, and new developments are required to meet the increasing demands on engineering materials. The performance of materials can be improved by combining different materials to achieve better properties than with a single constituent, or by shaping the material or constituents into a specific structure. The interaction between material and structure can occur at different length scales, such as the micro, meso, or macro scale, and offers potential applications in very different fields.

This book series addresses the fundamental relationships between materials and their structure on overall properties (e.g., mechanical, thermal, chemical, electrical, or magnetic properties, etc.). Experimental data and procedures are presented, as well as methods for modeling structures and materials using numerical and analytical approaches. In addition, the series shows how these materials engineering and design processes are implemented and how new technologies can be used to optimize materials and processes.

Advanced Structured Materials is indexed in Google Scholar and Scopus.

Holm Altenbach · Avetik Varazdat Sahakyan
Editors

Contact Mechanics and Relevant Problems

Springer

Editors
Holm Altenbach
Faculty of Mechanical Engineering
Otto von Guericke University Magdeburg
Magdeburg, Sachsen-Anhalt, Germany

Avetik Varazdat Sahakyan
Institute of Mechanics
National Academy of Sciences
of Armenia
Yerevan, Armenia

ISSN 1869-8433 ISSN 1869-8441 (electronic)
Advanced Structured Materials
ISBN 978-3-032-20868-2 ISBN 978-3-032-20869-9 (eBook)
https://doi.org/10.1007/978-3-032-20869-9

This Springer imprint is published by the registered company Springer Nature Switzerland AG
The registered company address is: Gewerbestrasse 11, 6330 Cham, Switzerland

Preface

The book is a collection of 40 articles by eminent 83 scientists from 5 countries (mainly from Armenia and the Russian Federation), who participated and presented papers at the IX International Conference "Current Problems of Continuum Mechanics", held on September 22-26, 2025, in Tsaghkadzor, Armenia. The first session of the conference was dedicated to the memory of the renowned scientist in the field of contact interaction mechanics and mathematical methods for solving integral equations - Corresponding Member of the National Academy of Sciences of Armenia, Doctor of Physical and Mathematical Sciences, Professor Suren Manukovich Mkhitaryan.

The topics of the articles are wide enough:

- General Problems,
- Contact Mechanics,
- Mechanics of Growing Solids,
- Mechanics of Technological Processes,
- Mechanics of Additive Manufacturing,
- Thin-walled Structures,
- Elasticity,
- Viscoelasticity,
- Plasticity,
- Creep,
- Tribology,
- Fracture Mechanics,
- Coupled Fields in Solids,
- Composites,
- Soil Mechanics,
- Fluid Mechanics,
- Wave Motion
- Robotics,
- Optimal Design,
- Control Theory,

- Mechanics of Nano-systems,
- Structural Mechanics,
- Biomechanics,
- Hydraulics and Hydraulic Devices, and
- Experimental Mechanics.

The purpose of the book is to raise interest in the conference and expand the geography of its participants. It also presents works by authors that have often only been published in Russian.

Magdeburg, Germany *Holm Altenbach*
Yerevan, Armenia *Avetik V. Sahakyan*
December 2025

Contents

List of Contributors

Mher L. Aghalovyan
Armenian State University of Economics, Nalbandyan St. 128, Yerevan, 0025 &
Institute of Mechanics, National Academy of Sciences of the Republic of Armenia,
24b Marshal Baghramyan Ave, Yerevan, 0019, Republic of Armenia,
e-mail: mheraghalovyan@yahoo.com

Karo L. Aghayan
Institute of Mechanics, National Academy of Sciences of the Republic of Armenia,
24b Marshal Baghramyan Ave, Yerevan, 0019, Republic of Armenia,
e-mail: karo.aghayan@gmail.com

Holm Altenbach
Institut für Werkstoffe, Technologien und Mechanik, Fakultät für Maschinenbau,
Otto-von-Guericke-Universität Magdeburg, Universitätsplatz 2, 39106 Magdeburg,
Germany,
e-mail: holm.altenbach@ovgu.de

Harutyun A. Amirjanyan
Institute of Mechanics, National Academy of Sciences of the Republic of Armenia,
24b Marshal Baghramyan Ave, Yerevan, 0019, Republic of Armenia,
e-mail: amirjanyan@gmail.com

Alexander R. Arutyunyan
St. Petersburg State University, 7/9 Universitetskaya Emb., St. Petersburg, 199034,
Russian Federation,
e-mail: a.arutyunyan@spbu.ru

Vahan V. Avetisyan
Institute of Mechanics, National Academy of Sciences of the Republic of Armenia,
24b Marshal Baghramyan Ave, Yerevan, 0019, Republic of Armenia,
e-mail: vanavet@yahoo.com

Emil A. Barseghyan
Yerevan State University, 1 Alec Manukyan, Yerevan, 0025, Republic of Armenia,
e-mail: emil.barsegyann@gmail.com

Vanya R. Barseghyan
Institute of Mechanics, National Academy of Sciences of the Republic of Armenia, 24b Marshal Baghramyan Ave, Yerevan, 0019, & Faculty of Mathematics and Mechanics, Yerevan State University, 1 Alec Manukyan, Yerevan, 0025, Republic of Armenia,
e-mail: barseghyan@sci.am

Alexei A. Bondarchuk
Vorovich Institute of Mathematics, Mechanics and Computer Science, Southern Federal University, Milchakova Street 8a, Rostov-on-Don, 344090, Russian Federation, e-mail: melchior@list.ru

Alexander V. Chentsov
Moscow State University of Civil Engineering, Moscow, Russian Federation,
e-mail: chentsov@ipmnet.ru

Illia Chernomorets
School of Computer Science and Mathematics, Keele, Staffordshire, ST5 5BG, UK,
e-mail: i.chernomorets@keele.ac.uk

Artavazd Z. Darbinyan
Institute of Mechanics, National Academy of Sciences of the Republic of Armenia, 24b Marshal Baghramyan Ave, Yerevan, 0019, Republic of Armenia,
e-mail: darbinyan_1954@mail.ru

Lilit L. Dashtoyan
Institute of Mechanics of the National Academy of Sciences of Armenia, 24/2 M.Baghramyan, Yerevan 0019, Republic of Armenia,
e-mail: lilit.dashtoyan@sci.am

Dmitry A. Filippov
Yaroslav-the-Wise Novgorod State University, 41 Bolshaya St., Petersburg Str., Veliky Novgorod, Russian Federation,
e-mail: Dmitry.Filippov@novsu.ru

Tigran A. Galichyan
Institute of Mechanics, National Academy of Sciences of the Republic of Armenia, 24b Marshal Baghramyan Ave, Yerevan, 0019, Republic of Armenia,
e-mail: Galichyantigran@gmail.com

David V. Gandilyan
Russian Academy of Sciences, Ishlinsky Institute for Problems in Mechanics RAS, Prospekt Vernadskogo, Moscow, 119526, Russian Federation,
e-mail: david.ghandilyan@mail.ru

Gnun Z. Gevorgyan
Institute of Mechanics, National Academy of Sciences of the Republic of Armenia, 24b Marshal Baghramyan Ave, Yerevan, 0019, Republic of Armenia,
e-mail: gnungev@gmail.com

Hrant A. Gevorgyan
Institute of Mechanics of the National Academy of Sciences of the Republic of Armenia, Baghramyan Ave. 24B, Yerevan, Armenia,
e-mail: hrgevorkian@mail.ru

Gurgen R. Ghulghazaryan
Khachatur Abovian Armenian State Pedagogical University, 17 Tigran Mets Ave, Yerevan, 0010, Republic of Armenia,
e-mail: ghulghazaryangurgen08@aspu.am

Lusine G. Ghulghazaryan
Institute of Mechanics, National Academy of Sciences of the Republic of Armenia, 24b Marshal Baghramyan Ave, Yerevan, 0019, & Khachatur Abovian Armenian State Pedagogical University, 17 Tigran Mets Ave, Yerevan, 0010, Republic of Armenia,
e-mail: ghulghazaryanlusine08@aspu.am

Aram H. Grigoryan
Institute of Mechanics, National Academy of Sciences of the Republic of Armenia, 24b Marshal Baghramyan Ave, Yerevan, 0019, Republic of Armenia,
e-mail: grigoryanaram4@gmail.com

David H. Grigoryan
Institute of Mechanics, National Academy of Sciences of the Republic of Armenia, 24b Marshal Baghramyan Ave, Yerevan, 0019, Republic of Armenia,
e-mail: davit-grigoryan@yandex.com

Marine S. Grigoryan
Institute of Mechanics, National Academy of Sciences of the Republic of Armenia, 24b Marshal Baghramyan Ave, Yerevan, 0019 & National University of Architecture and Construction of Armenia, 105 Teryan Street, Yerevan, 0009, Republic of Armenia,
e-mail: marinegrigoryan17@gmail.com

Sarkis G. Hairoyan
Yerevan State University, 1 Alec Manukyan, Yerevan, 0025, Republic of Armenia,
e-mail: hairoyan@ysu.am

Lusine V. Hakobyan
Institute of Mechanics of the National Academy of Sciences of Armenia, 24/2 M.Baghramyan, Yerevan 0019, Republic of Armenia, e-mail: lusine.vahram.hakobyan@gmail.com

Vahram N. Hakobyan
Institute of Mechanics, National Academy of Sciences of the Republic of Republic of Armenia, 24b Marshal Baghramyan Ave, Yerevan, 0019, Republic of Armenia,
e-mail: vhakobyan@sci.am

Levon A. Harutyunyan
Institute of Mechanics, National Academy of Sciences of the Republic of Armenia, 24b Marshal Baghramyan Ave, Yerevan, 0019, Republic of Armenia,
e-mail: arutyunyanlevon@gmail.com

Samvel H. Jilavyan
Faculty of Mathematics and Mechanics, Yerevan State University, Alex Manoogian 1, 0025, Yerevan, Republic of Armenia,
e-mail: samjilavyan@ysu.am

Eghine G. Kanetsyan
Institute of Mechanics, National Academy of Sciences of the Republic of Armenia, 24b Marshal Baghramyan Ave, Yerevan, 0019 & National University of Architecture and Construction of Armenia, 105 Teryan Stree, Yerevan, 0009, Republic of Armenia, Republic of Armenia,
e-mail: ekanetsyan@gmail.com

Julius Kaplunov
School of Computer Science and Mathematics, Keele, Staffordshire, ST5 5BG, UK,
e-mail: j.kaplunov@keele.ac.uk

Koryun A. Karapetyan
Institute of Mechanics, National Academy of Sciences of the Republic of Armenia, 24b Marshal Baghramyan Ave, Yerevan, 0019, Republic of Armenia,
e-mail: koryun.a.karapetyan@gmail.com

Eugene A. Kazakov
Vorovich Institute of Mathematics, Mechanics and Computer Science, Southern Federal University, Milchakova Street 8a, Rostov-on-Don, 344090, Russian Federation,
e-mail: Eugene.A.KazakoV@yandex.ru

Kirill E. Kazakov
Ishlinsky Institute for Problems in Mechanics of the Russian Academy of Sciences, Vernadskogo Ave 101 Bldg 1, Moscow, 119526 & Bauman Moscow State Technical University, 2nd Baumanskaya str., 5, p. 1, Moscow, Russian Federation,
e-mail: kazakov@ipmnet.ru, kazakov-ke@bmstu.ru

Sergey O. Kirichenko
St. Petersburg State University, 7/9 Universitetskaya Emb., St. Petersburg, 199034, Russian Federation,
e-mail: sergey.kirichenko@spbu.ru

Alexey V. Krylov
Bauman Moscow State Technical University, 2nd Baumanskaya str., 5, building 1, Moscow, 105005, Russian Federation,
e-mail: kav1982@bmstu.ru

Vladimir K. Kudymov
St. Petersburg State University, 7/9 Universitetskaya Emb., St. Petersburg, 199034, Russian Federation,
e-mail: v.k.kudymov@gmail.com

Konstantin K. Kukudzhanov
Ishlinsky Institute for Problems in Mechanics of the Russian Academy of Sciences, Vernadskogo Ave 101 Bldg 1, Moscow, 119526, Russian Federation,
e-mail: kkukudzhanov@yandex.ru

Konstantin V. Kukudzhanov
Ishlinsky Institute for Problems in Mechanics of the Russian Academy of Sciences, Vernadskogo Ave 101 Bldg 1, Moscow, 119526, Russian Federation,
e-mail: kconstantin@ipmnet.ru

Svetlana P. Kurdina
Bauman Moscow State Technical University, 2nd Baumanskaya str., 5, p. 1, Moscow, Russian Federation,
e-mail: kurdinasp@bmstu.ru

Dmitrii Legatiuk
Chair of Mathematics, Universität Erfurt, Nordhäuser Str. 63, 99089 Erfurt, Germany,
e-mail: dmitrii.legatiuk@uni-erfurt.de

Ekaterina M. Lokhova
National Research Moscow State University of Civil Engineering, Yaroslavskoye Hwy, 26, Moscow, 129337, Russian Federation,
e-mail: elm97@mail.ru

Irina S. Maksimova
Peoples' Friendship University of Russia named after Patrice Lumumba, Mikluho-Maklaya st., 6, Moscow, 117198, Russian Federation,
e-mail: irismax@yandex.ru

Yeghish S. Manukyan
Institute of Mechanics, National Academy of Sciences of the Republic of Armenia, 24b Marshal Baghramyan Ave, Yerevan, 0019, Republic of Armenia,
e-mail: exishe.manukyan@gmail.com

Stella R. Martirosyan
Institute of Mechanics, National Academy of Sciences of the Republic of Armenia, 24b Marshal Baghramyan Ave, Yerevan, 0019, Republic of Armenia,
e-mail: mechinsstella@mail.ru

Aram G. Matevosyan
Yerevan State University, 1 Alec Manukyan, Yerevan, 0025, Republic of Armenia,
e-mail: matevosaram@gmail.com, amatevosyan@ysu.am

Marine A. Mikilyan
Institute of Mechanics, National Academy of Sciences of the Republic of Armenia, 24/2 Marshal Baghramyan ave., 0019, Yerevan, Republic of Armenia,
e-mail: marine.mikilyan@rau.am

Manuk H. Mkrtchyan
Institute of Mechanics, National Academy of Sciences of the Republic of Armenia, 24b Marshal Baghramyan Ave, 0019, Yerevan, Republic of Armenia,
e-mail: mkmanuk@yandex.ru

Mher M. Mkrtchyan
Institute of Mechanics, National Academy of Sciences of the Republic of Armenia, 24b Marshal Baghramyan Ave, Yerevan, 0019, Republic of Armenia,
e-mail: mmher628@gmail.com

Mushegh S. Mkrtchyan
Institute of Mechanics, National Academy of Sciences of the Republic of Armenia, 24b Marshal Baghramyan Ave, Yerevan, 0019 & National University of Architecture and Construction of Armenia, 105 Teryan Stree, Yerevan, 0009, Republic of Armenia,
e-mail: mmushegh195@gmail.com

Oleg V. Mkrtychev
National Research Moscow State University of Civil Engineering, Yaroslavskoye Hwy, 26, Moscow, 129337, Russian Federation,
e-mail: mkrtychev@yandex.ru

Pavel E. Morozov
St. Petersburg State University, 7/9 Universitetskaya Emb., St. Petersburg, 199034, Russian Federation,
e-mail: comitcont@yandex.ru

V. E. Nazaikinskii
Ishlinsky Institute for Problems in Mechanics RAS, Vernadsky Ave. 101-1, Moscow, 119526 Russian Federation,
e-mail: nazaikinskii@googlemail.com

Heikki Orelma
Tampere Institute of Mathematics, Hepolamminkatu 51, 33720 Tampere, Finland,
e-mail: heikki.orelma@proton.me

Danila Prikazchikov
School of Computer Science and Mathematics, Keele, Staffordshire, ST5 5BG, UK,
e-mail: d.prikazchikov@keele.ac.uk

Andrey A. Reshetov
National Research Moscow State University of Civil Engineering, Yaroslavskoye Hwy, 26, Moscow, 129337, Russian Federation,
e-mail: andrew331@bk.ru

Avetik V. Sahakyan
Institute of Mechanics, National Academy of Sciences of the Republic of Armenia, 24b Marshal Baghramyan Ave, Yerevan, 0019, Republic of Armenia,
e-mail: avetik.sahakyan@sci.am

Sona V. Sahakyan
Institute of Mechanics, National Academy of Sciences of the Republic of Armenia, 24b Marshal Baghramyan Ave, Yerevan, 0019, Republic of Armenia,
e-mail: sonasahakyan1999@gmail.com

Regina R. Saitova
St. Petersburg State University, 7/9 Universitetskaya Emb., St. Petersburg, 199034, Russian Federation,
e-mail: rigastr@yandex.ru

Stanislav O. Sapel'nikov
Yaroslav-the-Wise Novgorod State University, 41 Bolshaya St., Petersburg Str., Veliky Novgorod, Russian Federation,
e-mail: Stanislav.Sapelnikov@novsu.ru

Sergey B. Sapozhkov
Yaroslav-the-Wise Novgorod State University, 41 Bolshaya St., Petersburg Str., Veliky Novgorod, Russian Federation,
e-mail: ssb@novsu.ru

Azat M. Sargsyan
Institute of Mechanics, National Academy of Sciences of the Republic of Armenia, 24b Marshal Baghramyan Ave, Yerevan, 0019, Republic of Armenia,
e-mail: azat-sargsyan@mail.ru

Karen S. Sargsyan
Institute of Mechanics, National Academy of Sciences of the Republic of Armenia, 24b Marshal Baghramyan Ave, Yerevan, 0019, Republic of Armenia,
e-mail: sarkaren1999@gmail.com,

Boris N. Semenov
St. Petersburg State University, 7/9 Universitetskaya Emb., St. Petersburg, 199034, Russian Federation,
e-mail: b.semenov@spbu.ru

Suren P. Seyranyan
Institute of Mechanics, National Academy of Sciences of the Republic of Armenia, 24b Marshal Baghramyan Ave, Yerevan, 0019, Republic of Armenia,
e-mail: surenparuir@gmail.com

Artur O. Shakhverdov
Bauman Moscow State Technical University, 2nd Baumanskaya str., 5, building 1, Moscow, 105005, Russian Federation,
e-mail: zimin@bmstu.ru

Yuri V. Sidorov
St. Petersburg State University, 7/9 Universitetskaya Emb., St. Petersburg, 199034, Russian Federation,
e-mail: arutalr@rambler.ru

Mihran G. Stakyan
Institute of Mechanics, National Academy of Sciences of the Republic of Armenia, 24b Marshal Baghramyan Ave, Yerevan, 0019, Republic of Armenia,
e-mail: stakyan.mihran@yandex.ru

Yuri V. Sudenkov
St. Petersburg State University, 7/9 Universitetskaya Emb., St. Petersburg, 199034, Russian Federation,
e-mail: y.sudenkov@yandex.ru

Mezhlum A. Sumbatyan
Vorovich Institute of Mathematics, Mechanics and Computer Science, Southern Federal University, Milchakova Street 8a, Rostov-on-Don, 344090, Russian Federation,
e-mail: masumbatyan@sfedu.ru

Timur A. Tagirov
National Research Moscow State University of Civil Engineering, Yaroslavskoye Hwy, 26, Moscow, 129337, Russian Federation,
e-mail: tagirovv_timur@mail.ru

Evgeny A. Titarenko
Southern Federal University, Milchakova Street 8a, Rostov-on-Don, 344090, Russian Federation,
e-mail: evtitarenko@sfedu.ru

Mher S. Torosyan
Institute of Mechanics, National Academy of Sciences of the Republic of Armenia, 24b Marshal Baghramyan Ave, Yerevan, 0019, Republic of Armenia,
e-mail: mhertorosyan@list.ru

Artur B. Tovmasyan
Institute of Mechanics, National Academy of Sciences of the Republic of Armenia, 24b Marshal Baghramyan Ave, Yerevan, 0019, Republic of Armenia,
e-mail: tovmasyanarthur01@gmail.com

Konstantin B. Ustinov
Russian Academy of Sciences, Ishlinsky Institute for Problems in Mechanics RAS, Prospekt Vernadskogo, Moscow, 119526, Russian Federation,
e-mail: ustinov@ipmnet.ru

Sona S. Valesyan
Institute of Mechanics, National Academy of Sciences of the Republic of Armenia, 24b Marshal Baghramyan Ave, Yerevan, 0019, Republic of Armenia,
e-mail: svalesyan@yahoo.com

Anna V. Vardanyan
St. Petersburg State University, 7/9 Universitetskaya Emb., St. Petersburg, 199034, Russian Federation,
e-mail: vardanyan_a@yahoo.com

Tatevik G. Vasilyan
Institute of Mechanics, National Academy of Sciences of the Republic of Armenia, 24b Marshal Baghramyan Ave, Yerevan, 0019, Republic of Armenia,
e-mail: tat.vasilyan@gmail.com

Elizaveta S. Yalych
Baltic State Technical University, 1st Krasnoarmeyskaya St., Building 1, St. Petersburg, 190005, Russian Federation,
e-mail: ialych_es@voenmeh.ru

Tatevik V. Zakaryan
Institute of Mechanics, National Academy of Sciences of the Republic of Armenia, 24b Marshal Baghramyan Ave, Yerevan, 0019, Republic of Armenia,
e-mail: zaqaryantatevik@mail.ru

Vahe G. Zakaryan
Institute of Mechanics, National Academy of Sciences of the Republic of Armenia, 24b Marshal Baghramyan Ave, Yerevan, 0019, Republic of Armenia,
e-mail: vahezaqaryan@mail.ru

Elena G. Zemtsova
St. Petersburg State University, 7/9 Universitetskaya Emb., St. Petersburg, 199034, Russian Federation,
e-mail: ezimtsova@yandex.ru

Boris A. Zimin
St. Petersburg State University, 7/9 Universitetskaya Emb., St. Petersburg, 199034, Russian Federation, e-mail: bazimin@mail.ru

Vladimir N. Zimin
Bauman Moscow State Technical University, 2nd Baumanskaya str., 5, building 1, Moscow, 105005, Russian Federation,
e-mail: zimin@bmstu.ru

Chapter 1
On Mixed Spatial Problems for Plates with a Plane of Elastic Symmetry

Mher L. Aghalovyan, Tatevik V. Zakaryan, and Artur B. Tovmasyan

Abstract The spatial problem for plates having a plane of elastic symmetry is solved. In particular, crystals of the monoclinic system (13 independent elastic constants) have such property. It is assumed that the face surface of the plate is subjected by normal and tangential loads, and the base of the plate is rigidly fixed. For solving such problems, hypotheses of classical and existing refined theories of plates and shells are not applicable.

Key words: Plate, Anisotropy, 3D problems, Asymptotic solution

1.1 Statement of the Problem, Basic Equations and Relationships

Let the plate occupies the area (Fig. 1.1)

$$D = \{(x, y, z) : 0 \leq x \leq a, 0 \leq y \leq b, -h \leq z \leq h, \quad 2h \ll l, \quad l = \min(a, b)\}$$

It is required to find in this area the solutions to the equilibrium equations

Mher L. Aghalovyan
Armenian State University of Economics, Nalbandyan St. 128, Yerevan, 0025 & Institute of Mechanics, National Academy of Sciences of the Republic of Armenia, 24b Marshal Baghramyan Ave, Yerevan, 0019, Republic of Armenia, e-mail: mheraghalovyan@yahoo.com

Tatevik V. Zakaryan (corresponding author) · Artur B. Tovmasyan
Institute of Mechanics, National Academy of Sciences of the Republic of Armenia, 24b Marshal Baghramyan Ave, Yerevan, 0019, Republic of Armenia, e-mail: zaqaryantatevik@mail.ru, tovmasyanarthur01@gmail.com

H. Altenbach and A. V. Sahakyan, *Contact Mechanics and Relevant Problems*, Advanced Structured Materials 253, https://doi.org/10.1007/978-3-032-20869-9_1

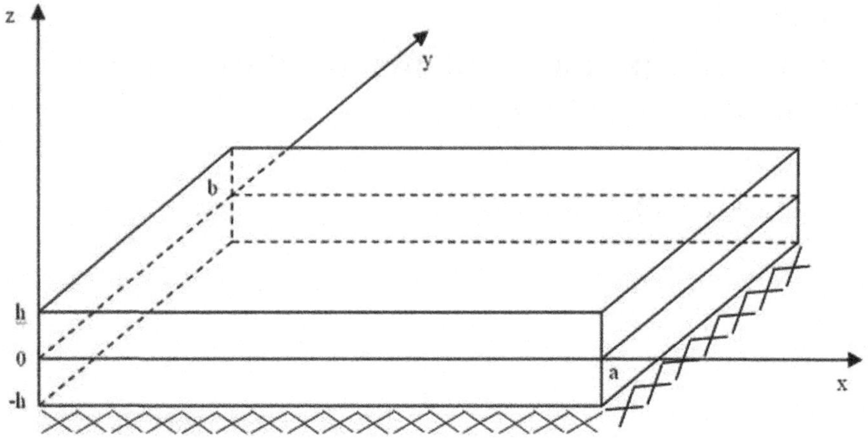

Fig. 1.1: The structure of the plate.

$$\begin{aligned}
&\frac{\partial \sigma_{xx}}{\partial x}+\frac{\partial \sigma_{xy}}{\partial y}+\frac{\partial \sigma_{xz}}{\partial z}+F_x(x,y)=0,\\
&\frac{\partial \sigma_{xy}}{\partial x}+\frac{\partial \sigma_{yy}}{\partial y}+\frac{\partial \sigma_{yz}}{\partial z}+F_y(x,y)=0,\\
&\frac{\partial \sigma_{xz}}{\partial x}+\frac{\partial \sigma_{yz}}{\partial y}+\frac{\partial \sigma_{zz}}{\partial z}+F_z(x,y)=0,
\end{aligned} \tag{1.1}$$

where F_x, F_y, F_z are volumetric forces and relations of elasticity [3, 5]

$$\begin{aligned}
&\frac{\partial u}{\partial x}=a_{11}\sigma_{xx}+a_{12}\sigma_{yy}+a_{13}\sigma_{zz}+a_{16}\sigma_{xy},\\
&\frac{\partial v}{\partial y}=a_{12}\sigma_{xx}+a_{22}\sigma_{yy}+a_{23}\sigma_{zz}+a_{26}\sigma_{xy},\\
&\frac{\partial w}{\partial z}=a_{13}\sigma_{xx}+a_{23}\sigma_{yy}+a_{33}\sigma_{zz}+a_{36}\sigma_{xy},\\
&\frac{\partial v}{\partial z}+\frac{\partial w}{\partial y}=a_{44}\sigma_{yz}+a_{45}\sigma_{xz},\\
&\frac{\partial u}{\partial z}+\frac{\partial w}{\partial x}=a_{45}\sigma_{yz}+a_{55}\sigma_{xz},\\
&\frac{\partial u}{\partial y}+\frac{\partial v}{\partial x}=a_{16}\sigma_{xx}+a_{26}\sigma_{yy}+a_{36}\sigma_{zz}+a_{66}\sigma_{xy}
\end{aligned} \tag{1.2}$$

at the following boundary conditions
at $z = h$:

$$\sigma_{xz}(x,y,h)=\sigma_{xz}^{+}(x,y),\ \sigma_{yz}(x,y,h)=\sigma_{yz}^{+}(x,y),\ \sigma_{zz}(x,y,h)=-\sigma_{zz}^{+}(x,y), \tag{1.3}$$

at $z = -h$:

$$u(x, y, -h) = 0, \quad v(x, y, -h) = 0, \quad w(x, y, -h) = 0. \tag{1.4}$$

The boundary conditions on the lateral surfaces have not been defined concretely yet; they determine the appearance of a boundary layer, which, as usual, can be considered separately.

1.2 Asymptotic Solution of the Problem

For solving the set problem, in Eqs. (1.1) and relations (1.2) we pass to dimensionless coordinates and displacements

$$\xi = \frac{x}{l}, \quad \eta = \frac{y}{l}, \quad \zeta = \frac{z}{h}, \quad \mathrm{U} = \frac{u}{l}, \quad \mathrm{V} = \frac{v}{l}, \quad \mathrm{W} = \frac{w}{l}. \tag{1.5}$$

As a result, we obtain the system singularly perturbed by a small parameter $\varepsilon = h/l$

$$\begin{aligned}
&\frac{\partial\sigma_{11}}{\partial\xi} + \frac{\partial\sigma_{12}}{\partial\eta} + \varepsilon^{-1}\frac{\partial\sigma_{13}}{\partial\zeta} + lF_x(x, y) = 0,\\
&\frac{\partial\sigma_{12}}{\partial\xi} + \frac{\partial\sigma_{22}}{\partial\eta} + \varepsilon^{-1}\frac{\partial\sigma_{23}}{\partial\zeta} + lF_y(x, y) = 0,\\
&\frac{\partial\sigma_{13}}{\partial\xi} + \frac{\partial\sigma_{23}}{\partial\eta} + \varepsilon^{-1}\frac{\partial\sigma_{33}}{\partial\zeta} + lF_z(x, y) = 0,\\
&\frac{\partial U}{\partial\xi} = a_{12}\sigma_{11} + a_{12}\sigma_{22} + a_{13}\sigma_{33} + a_{16}\sigma_{12},\\
&\frac{\partial V}{\partial\eta} = a_{12}\sigma_{11} + a_{22}\sigma_{22} + a_{23}\sigma_{33} + a_{26}\sigma_{12},\\
&\varepsilon^{-1}\frac{\partial W}{\partial\zeta} = a_{13}\sigma_{13} + a_{23}\sigma_{22} + a_{33}\sigma_{33} + a_{36}\sigma_{12},\\
&\frac{\partial W}{\partial\eta} + \varepsilon^{-1}\frac{\partial V}{\partial\zeta} = a_{44}\sigma_{23} + a_{45}\sigma_{13},\\
&\frac{\partial W}{\partial\xi} + \varepsilon^{-1}\frac{\partial U}{\partial\zeta} = a_{45}\sigma_{23} + a_{55}\sigma_{13},\\
&\frac{\partial V}{\partial\xi} + \frac{\partial U}{\partial\eta} = a_{16}\sigma_{11} + a_{26}\sigma_{22} + a_{36}\sigma_{33} + a_{66}\sigma_{12}.
\end{aligned} \tag{1.6}$$

It is easy to verify, that the known asymptotics for stresses and displacements of the classical theory of plates leads to the contradictory system [1]. For solving the similar classes of problems (second and mixed boundary value problems) the fundamentally new asymptotics has been established [1, 2]. The solution of the singularly perturbed system (1.9) is the sum of the solutions of the external problem (I^{out}) and the boundary layer (I_b): $I = I^{\text{out}} + I_b$. The solution to the external problem we will seek out in the form [1, 2]

$$\sigma_{ij}^{\text{out}} = \varepsilon^{-1+s}\sigma_{ij}^{(s)}\,(\xi,\eta,\zeta)\,,\quad \text{i}, j = 1,2,3,\quad s = \overline{0,N},$$

$$\left(U^{\text{out}}, V^{\text{out}}, W^{\text{out}}\right) = \varepsilon^{s}\left(U^{(s)}, V^{(s)}, W^{(s)}\right) \tag{1.7}$$

The asymptotics (1.7) fundamentally differs from the asymptotics of the same value in the classical theory plates. For example, in the bending problems according to classical theory of plates, we have [2, 4]

$$I^{\text{out}} = \varepsilon^{q_i+s} I^{(S)}, s = \overline{0,N},$$

where $q_i = -2$ for $\sigma_{xx}, \sigma_{xy}, \sigma_{yy}$; $q_i = -1$ for σ_{xz}, σ_{yz};

$$q_i = 0 \text{ for } \sigma_{zz};\;\; q_i = -2 \text{ for } U, V;\;\; q_i = -3 \text{ for } W. \tag{1.8}$$

The immediate check can convince, that the solution obtained by asymptotics (1.8), cannot satisfy the boundary conditions (1.3) and (1.4). Physically, this is also obvious, since in the classical theory of plates it is considered that the normal displacement W does not depend on the coordinate z. In this case, it is impossible to satisfy conditions (1.3) and (1.4) relatively σ_{zz} and W.

By substituting (1.7) into (1.6) and equating the coefficients of the same powers of ε in each equation, we will obtain the following consistent system for determining unknown functions $\sigma_{ij}^{(s)}, U^{(s)}, V^{(s)}, W^{(s)}$

$$
\begin{aligned}
&\frac{\partial \sigma_{11}^{(s-1)}}{\partial \xi}+\frac{\partial \sigma_{12}^{(s-1)}}{\partial \eta}+\frac{\partial \sigma_{13}^{(s)}}{\partial \zeta}+F_x^{(s)}(x,y)=0,\\
&\frac{\partial \sigma_{12}^{(s-1)}}{\partial \xi}+\frac{\partial \sigma_{22}^{(s-1)}}{\partial \eta}+\frac{\partial \sigma_{23}^{(s)}}{\partial \zeta}+F_y^{(s)}(x,y)=0,\\
&\frac{\partial \sigma_{13}^{(s-1)}}{\partial \xi}+\frac{\partial \sigma_{23}^{(s-1)}}{\partial \eta}+\frac{\partial \sigma_{33}^{(s)}}{\partial \zeta}+F_z^{(s)}(x,y)=0,\\
&\frac{\partial U^{(s-1)}}{\partial \xi}=a_{11}\sigma_{11}^{(s)}+a_{12}\sigma_{22}^{(s)}+a_{13}\sigma_{33}^{(s)}+a_{16}\sigma_{12}^{(s)},\\
&\frac{\partial V^{(s-1)}}{\partial \eta}=a_{12}\sigma_{11}^{(s)}+a_{22}\sigma_{22}^{(s)}+a_{23}\sigma_{33}^{(s)}+a_{26}\sigma_{12}^{(s)},\\
&\frac{\partial W^{(s)}}{\partial \zeta}=a_{13}\sigma_{11}^{(s)}+a_{23}\sigma_{22}^{(s)}+a_{33}\sigma_{33}^{(s)}+a_{36}\sigma_{12}^{(s)},\\
&\frac{\partial W^{(s-1)}}{\partial \eta}+\frac{\partial V^{(s)}}{\partial \zeta}=a_{44}\sigma_{23}^{(s)}+a_{45}\sigma_{13}^{(s)},\\
&\frac{\partial W^{(s-1)}}{\partial \xi}+\frac{\partial U^{(s)}}{\partial \zeta}=a_{45}\sigma_{23}^{(s)}+a_{55}\sigma_{13}^{(s)},\\
&\frac{\partial V^{(s-1)}}{\partial \xi}+\frac{\partial U^{(s-1)}}{\partial \eta}=a_{16}\sigma_{11}^{(s)}+a_{26}\sigma_{22}^{(s)}+a_{36}\sigma_{33}^{(s)}+a_{66}\sigma_{12}^{(s)},\\
&F_x^{(0)}(x,y)=\frac{h^2}{l}F_x,\ \ F_x^{(s)}=0,\ \ s\neq 0,\ \ Q^{(m)}=0,\ \ m<0.
\end{aligned}
\tag{1.9}
$$

From the elasticity relations of the system (1.9), all stresses are expressed through the components of the displacement vector according to the formulas

$$
\begin{aligned}
\sigma_{13}^{(s)}&=\frac{1}{\Delta_1}\left(a_{44}\frac{\partial U^{(s)}}{\partial \zeta}-a_{45}\frac{\partial V^{(s)}}{\partial \zeta}+\sigma_{13*}^{(s)}\right),\\
\sigma_{23}^{(s)}&=\frac{1}{\Delta_1}\left(-a_{45}\frac{\partial U^{(s)}}{\partial \zeta}+a_{55}\frac{\partial V^{(s)}}{\partial \zeta}+\sigma_{23*}^{(s)}\right),\\
\sigma_{33}^{(s)}&=\frac{1}{\Delta}\left(A_{33}\frac{\partial W^{(s)}}{\partial \zeta}+\sigma_{33*}^{(s)}\right),\\
\sigma_{11}^{(s)}&=\frac{1}{\Delta}\left(A_{31}\frac{\partial W^{(s)}}{\partial \zeta}+\sigma_{11*}^{(s)}\right),\\
\sigma_{22}^{(s)}&=\frac{1}{\Delta}\left(A_{32}\frac{\partial W^{(s)}}{\partial \zeta}+\sigma_{22*}^{(s)}\right),\\
\sigma_{12}^{(s)}&=\frac{1}{\Delta}\left(A_{34}\frac{\partial W^{(s)}}{\partial \zeta}+\sigma_{12*}^{(s)}\right),
\end{aligned}
\tag{1.10}
$$

where

$$
\begin{aligned}
&\sigma_{13*}^{(s)} = a_{44}\frac{\partial W^{(s-1)}}{\partial \xi} - a_{45}\frac{\partial W^{(s-1)}}{\partial \eta},\\
&\sigma_{23*}^{(s)} = -a_{45}\frac{\partial W^{(s-1)}}{\partial \xi} + a_{55}\frac{\partial W^{(s-1)}}{\partial \eta},\\
&\sigma_{11*}^{(s)} = A_{11}\frac{\partial U^{(s-1)}}{\partial \xi} - A_{21}\frac{\partial V^{(s-1)}}{\partial \eta} - A_{41}\left(\frac{\partial V^{(s-1)}}{\partial \xi} + \frac{\partial U^{(s-1)}}{\partial \eta}\right),\\
&\sigma_{22*}^{(s)} = -A_{12}\frac{\partial U^{(s-1)}}{\partial \xi} + A_{22}\frac{\partial V^{(s-1)}}{\partial \eta} + A_{42}\left(\frac{\partial V^{(s-1)}}{\partial \xi} + \frac{\partial U^{(s-1)}}{\partial \eta}\right),\\
&\sigma_{33*}^{(s)} = A_{13}\frac{\partial U^{(s-1)}}{\partial \xi} - A_{23}\frac{\partial V^{(s-1)}}{\partial \eta} - A_{43}\left(\frac{\partial V^{(s-1)}}{\partial \xi} + \frac{\partial U^{(s-1)}}{\partial \eta}\right),\\
&\sigma_{12*}^{(s)} = -A_{14}\frac{\partial U^{(s-1)}}{\partial \xi} + A_{24}\frac{\partial V^{(s-1)}}{\partial \eta} + A_{44}\left(\frac{\partial V^{(s-1)}}{\partial \xi} + \frac{\partial U^{(s-1)}}{\partial \eta}\right),
\end{aligned}
\tag{1.11}
$$

$$
\sigma_{ij}^{(m)} = 0, \quad U^{(m)} = V^{(m)} = W^{(m)} = 0 \quad \text{at } m < 0,
$$

$$
\Delta = \begin{vmatrix} a_{11} & a_{12} & a_{13} & a_{16} \\ a_{12} & a_{22} & a_{23} & a_{26} \\ a_{13} & a_{23} & a_{33} & a_{36} \\ a_{16} & a_{26} & a_{36} & a_{66} \end{vmatrix}, \qquad \Delta_1 = a_{44}a_{55} - a_{45}^2,
$$

A_{ij} is the corresponding minor of the determinant Δ, where i is the number of the row , j is the number of column of the element of determinant.

By substituting the values of $\sigma_{13}^{(s)}, \sigma_{23}^{(s)}, \sigma_{33}^{(s)}$ in the first three equations of the system (1.10), for determining $U^{(s)}, V^{(s)}, W^{(s)}$ we will obtain

$$
\begin{aligned}
&\frac{\partial^2 U^{(s)}}{\partial \zeta^2} = \bar{F}_x^{(s)}, \quad \frac{\partial^2 V^{(s)}}{\partial \zeta^2} = \bar{F}_y^{(s)}, \quad \frac{\partial^2 W^{(s)}}{\partial \zeta^2} = \bar{F}_z^{(s)},\\
&U^{(s)} = \int_0^{\zeta} \mathrm{d}\,\zeta \int_0^{\zeta} \bar{F}_x^{(s)} \mathrm{d}\,\zeta + C_1^{(s)}(\xi,\eta)\,\zeta + C_2^{(s)}(\xi,\eta),\\
&V^{(s)} = \int_0^{\zeta} \mathrm{d}\,\zeta \int_0^{\zeta} \bar{F}_y^{(s)} \mathrm{d}\,\zeta + C_3^{(s)}(\xi,\eta)\,\zeta + C_4^{(s)}(\xi,\eta),\\
&W^{(s)} = \int_0^{\zeta} \mathrm{d}\,\zeta \int_0^{\zeta} \bar{F}_z^{(s)} \mathrm{d}\,\zeta + C_5^{(s)}(\xi,\eta)\,\zeta + C_6^{(s)}(\xi,\eta),
\end{aligned}
\tag{1.12}
$$

$$\overline{F}_x^{(s)} = -a_{55}\left(F_x^{(s)} + \frac{\partial\sigma_{11}^{(s-1)}}{\partial\xi} + \frac{\partial\sigma_{12}^{(s-1)}}{\partial\eta} + \frac{1}{\Delta_1}\frac{\partial\sigma_{13^*}^{(s)}}{\partial\zeta}\right)$$

$$+ a_{45}\left(F_y^{(s)} + \frac{\partial\sigma_{12}^{(s-1)}}{\partial\xi} + \frac{\partial\sigma_{22}^{(s-1)}}{\partial\eta} + \frac{1}{\Delta_1}\frac{\partial\sigma_{23^*}^{(s)}}{\partial\zeta}\right),$$

$$\overline{F}_y^{(s)} = a_{45}\left(F_x^{(s)} + \frac{\partial\sigma_{11}^{(s-1)}}{\partial\xi} + \frac{\partial\sigma_{12}^{(s-1)}}{\partial\eta} + \frac{1}{\Delta_1}\frac{\partial\sigma_{13^*}^{(s)}}{\partial\zeta}\right)$$

$$- a_{44}\left(F_y^{(s)} + \frac{\partial\sigma_{12}^{(s-1)}}{\partial\xi} + \frac{\partial\sigma_{22}^{(s-1)}}{\partial\eta} + \frac{1}{\Delta_1}\frac{\partial\sigma_{23^*}^{(s)}}{\partial\zeta}\right), \tag{1.13}$$

$$\bar{F}_z^{(s)} = -\left[\frac{\Delta}{A_{33}}\left(F_z^{(s)} + \frac{\partial\sigma_{13}^{(s-1)}}{\partial\xi} + \frac{\partial\sigma_{23}^{(s-1)}}{\partial\eta}\right) + \frac{\partial\sigma_{33*}^{(s)}}{A_{33}\,\partial\zeta}\right],$$

$$\sigma_{13}^{(s)} = \frac{1}{\Delta_1}\left(a_{44}C_1^{(s)} - a_{45}C_3^{(s)}\right) + d_1^{(s)},$$

$$\sigma_{23}^{(s)} = \frac{1}{\Delta_1}\left(-a_{45}C_1^{(s)} + a_{55}C_3^{(s)}\right) + d_2^{(s)},$$

$$\sigma_{33}^{(s)} = \frac{A_{33}}{\Delta}\left(\int_0^{\zeta} \bar{F}_z^{(s)}\mathrm{d}\,\zeta + C_5^{(s)}\right) + \frac{\sigma_{33*}^{(s)}}{\Delta}.$$

Having satisfied the boundary conditions (1.3), (1.4), we obtain the following system of algebraic equations with respect to $C_j^{(s)}(\xi,\eta)$, $j = 1, 2, \ldots, 6$

$$\frac{1}{\Delta_1}\left(a_{44}C_1^{(s)} - a_{45}C_3^{(s)}\right) + d_1^{(s)} = \sigma_{xz}^{+(s)},$$

$$\frac{1}{\Delta_1}\left(-a_{45}C_1^{(s)} + a_{55}C_3^{(s)}\right) + d_2^{(s)} = \sigma_{yz}^{+(s)},$$

$$\frac{A_{33}}{\Delta}\left(\int_0^{\zeta} \bar{F}_z^{(s)}\mathrm{d}\zeta + C_5^{(s)}\right) + \frac{\sigma_{33*}^{(s)}}{\Delta} = -\sigma_{zz}^{+(s)},$$

$$-\int_{-1}^{0}\mathrm{d}\zeta\int_0^{\zeta} \bar{F}_x^{(s)}\mathrm{d}\zeta - C_1^{(s)}(\xi,\eta) + C_2^{(s)}(\xi,\eta) = 0, \tag{1.14}$$

$$-\int_{-1}^{0}\mathrm{d}\zeta\int_0^{\zeta} \bar{F}_y^{(s)}\mathrm{d}\zeta - C_3^{(s)}(\xi,\eta) + C_4^{(s)}(\xi,\eta) = 0,$$

$$\int_{-1}^{0}\mathrm{d}\zeta\int_0^{\zeta} \bar{F}_z^{(s)}\mathrm{d}\zeta - C_5^{(s)}(\xi,\eta) + C_6^{(s)}(\xi,\eta) = 0,$$

where from

$$
\begin{aligned}
&C_1^{(s)} = a_{55}\sigma_{xz}^{+(s)} + a_{45}\sigma_{yz}^{+(s)} - \left(a_{55}d_1^{(s)} + a_{45}d_2^{(s)}\right)_{\zeta=1},\\
&C_3^{(s)} = a_{45}\sigma_{xz}^{+(s)} + a_{44}\sigma_{yz}^{+(s)} - \left(a_{45}d_1^{(s)} + a_{44}d_2^{(s)}\right)_{\zeta=1},\\
&\sigma_{xz}^{+(0)} = \sigma_{xz}^{+}, \quad \sigma_{xz}^{+(s)} = 0, \quad s \neq 0,\\
&C_5^{(s)} = -\frac{\Delta\sigma_{zz}^{+(s)}}{A_{33}} - \frac{\sigma_{33*}^{(s)}}{A_{33}} - \int_0^1 \bar{F}_z^{(s)}\,\mathrm{d}\,\zeta,\\
&C_2^{(s)} = \int_{-1}^{0} \mathrm{d}\,\zeta \int_0^{\zeta} \bar{F}_x^{(s)}\,\mathrm{d}\,\zeta + C_1^{(s)},\\
&C_4^{(s)} = \int_{-1}^{0} \mathrm{d}\,\zeta \int_0^{\zeta} \bar{F}_y^{(s)}\,\mathrm{d}\,\zeta + C_3^{(s)}, \quad C_6^{(s)} = \int_{-1}^{0} \mathrm{d}\,\zeta \int_0^{\zeta} \bar{F}_z^{(s)}\,\mathrm{d}\,\zeta + C_5^{(s)},\\
&d_1^{(s)} = \frac{1}{\Delta_1}\left(a_{44}\int_0^{\zeta} \bar{F}_x^{(s)}\,\mathrm{d}\,\zeta - a_{45}\int_0^{\zeta} \bar{F}_y^{(s)}\,\mathrm{d}\,\zeta + \sigma_{13*}^{(s)}\right),\\
&d_2^{(s)} = \frac{1}{\Delta_1}\left(-a_{45}\int_0^{\zeta} \bar{F}_x^{(s)}\,\mathrm{d}\,\zeta + a_{55}\int_0^{\zeta} \bar{F}_y^{(s)}\,\mathrm{d}\,\zeta + \sigma_{23*}^{(s)}\right).
\end{aligned}
\tag{1.15}
$$

Using the values of the functions $C_j^{(s)}\ (\xi, \eta)$ by formulas (1.10) and (1.11) will be determined all components of the stress tensor and displacement vectors. If the functions σ_{xz}^{+}, σ_{yz}^{+}, σ_{zz}^{+} are algebraic polynomials of degree m, the iterative process terminates at the approximation m+1. As a result we will have mathematically precise solution to the external problem.

The found solution satisfies to the boundary conditions (1.3) and (1.4) and, as a rule, will not satisfy the conditions at the end sections of the plate. The arising discrepancy is resolved by solving for the boundary layer. The magnitudes of the boundary layer decreases exponentially with distance from the side surface towards the inside of the plate. The solution for the boundary layer and its conjugation (merging) with the solution to the external problem is carried out in the manner described in [1, 2]. Therefore, the solution we found to the external problem is a mathematically exact solution to the spatial layer.

1.3 On Mathematically Exact Solutions

If the functions σ^+_{xz}, σ^+_{yz}, σ^+_{zz} are the algebraic polynomials the corresponding to them solution is mathematically exact. Let us illustrate what has been said with two examples.

a) The stress-strain state of a plate caused by the weight of the plate. For this case

$$F_x = 0,\;\; F_y = 0,\;\; F_z = -\rho g,\;\; \sigma^+_{xz} = 0,\;\; \sigma^+_{yz} = 0,\;\; \sigma^+_{zz} = 0. \tag{1.16}$$

The iteration terminates at the initial approximation. According to formulas (1.15) we will have

$$C_1^{(0)} = C_2^{(0)} = C_3^{(0)} = C_4^{(0)} = 0,$$

$$C_5^{(0)} = -\frac{\rho g}{A_{33}},\;\; C_6^{(0)} = -\frac{3\rho g}{2A_{33}}, \tag{1.17}$$

$$\sigma_{13}^{(0)} = 0,\;\; \sigma_{23}^{(0)} = 0,\;\; \sigma_{12}^{(0)} = 0,\;\; U^{(0)} = 0,\;\; V^{(0)} = 0,$$

$$\sigma_{33}^{(0)} = \frac{\rho g h^2}{\Delta l}(\zeta - 1),\quad W^{(0)} = \frac{\rho g h^2}{2A_{33}l}\left(\zeta^2 - 2\zeta - 3\right),$$

$$\sigma_{xx} = 0,\;\; \sigma_{yy} = 0,\;\; \sigma_{xy} = 0,\;\; \sigma_{yz} = 0,\;\; \sigma_{xz} = 0,\quad u = 0,\;\; v = 0,$$

$$\sigma_{zz} = \frac{\rho g h}{\Delta}(\zeta - 1),\quad w = \frac{\rho g h^2}{2A_{33}}\left(\zeta^2 - 2\zeta - 3\right),$$

b) Let

$$F_x = 0,\quad F_y = 0,\quad F_z = 0, \tag{1.18}$$

$$\sigma^+_{xz} = 0,\;\; \sigma^+_{yz} = 0,\;\; \sigma^+_{zz} = a_0 + a_1\xi + a_2\eta.$$

Then at $s = 0$

$$\sigma_{13}^{(0)} = 0,\; \sigma_{23}^{(0)} = 0, \sigma_{33}^{(0)} = -\left(a_0 + a_1\xi + a_2\eta\right),$$

$$\sigma_{11}^{(0)} = -\frac{A_{31}}{A_{33}}\left(a_0 + a_1\xi + a_2\eta\right),\quad \sigma_{22}^{(0)} = -\frac{A_{32}}{A_{33}}\left(a_0 + a_1\xi + a_2\eta\right), \tag{1.19}$$

$$\sigma_{12}^{(0)} = -\frac{A_{34}}{A_{33}}\left(a_0 + a_1\xi + a_2\eta\right),$$

$$W^{(0)} = -\frac{\Delta}{A_{33}}(\zeta + 1)\left(a_0 + a_1\xi + a_2\eta\right),\;\; U^{(0)} = 0,\;\; V^{(0)} = 0.$$

At $s = 1$

$$\bar{F}_x^{(1)} = \frac{1}{A_{33}}\left[a_{55}\left(A_{31}a_1 + A_{34}a_2 + \frac{\Delta}{\Delta_1}\left(a_{44}a_1 + a_{45}a_2\right)\right) - \right.$$

$$\left. -a_{45}\left(A_{34}a_1 + A_{32}a_2 + \frac{\Delta}{\Delta_1}\left(a_{45}a_1 + a_{55}a_2\right)\right)\right],$$

$$\bar{F}_y^{(1)} = \frac{1}{A_{33}}\left[a_{45}\left(A_{31}a_1 + A_{34}a_2 + \frac{\Delta}{\Delta_1}(a_{44}a_1 + a_{45}a_2)\right) - \right.$$
$$\left. -a_{44}\left(A_{34}a_1 + A_{32}a_2 + \frac{\Delta}{\Delta_1}(a_{45}a_1 + a_{55}a_2)\right)\right],$$
$$\bar{F}_z^{(1)} = 0, \tag{1.20}$$
$$C_1^{(1)} = -\bar{F}_x^{(1)} + \frac{2\Delta a_1}{A_{33}}, \quad C_2^{(1)} = \frac{\bar{F}_x^{(1)}}{2} + C_1^{(1)},$$
$$C_3^{(1)} = -\bar{F}_y^{(1)} + \frac{2\Delta a_2}{A_{33}}, \quad C_4^{(1)} = \frac{\bar{F}_y^{(1)}}{2} + C_3^{(1)},$$
$$C_5^{(1)} = 0, \quad C_6^{(1)} = 0.$$

According to formulas (1.10), (1.11), (1.13)

$$W^{(1)} = 0, \quad \sigma_{11}^{(1)} = 0, \quad \sigma_{22}^{(1)} = 0, \quad \sigma_{12}^{(1)} = 0, \quad \sigma_{33}^{(1)} = 0,$$
$$U^{(1)} = \bar{F}_x^{(1)}\frac{\zeta^2}{2} + C_1^{(1)}\zeta + C_2^{(1)}, \quad V^{(1)} = \bar{F}_y^{(1)}\frac{\zeta^2}{2} + C_3^{(1)}\zeta + C_4^{(1)},$$
$$\sigma_{13}^{(1)} = \frac{1}{\Delta_1}\left(a_{44}\frac{\partial U^{(1)}}{\partial \zeta} - a_{45}\frac{\partial V^{(1)}}{\partial \zeta}\right), \quad \sigma_{23}^{(1)} = \frac{1}{\Delta_1}\left(-a_{45}\frac{\partial U^{(1)}}{\partial \zeta} + a_{55}\frac{\partial V^{(1)}}{\partial \zeta}\right).$$

As a result we will obtain exact solution

$$w = lW^{(0)}, \quad u = hU^{(1)}, \quad v = hV^{(1)},$$
$$\sigma_{xx} = \varepsilon^{-1}\sigma_{11}^{(0)}, \quad \sigma_{yy} = \varepsilon^{-1}\sigma_{22}^{(0)}, \quad \sigma_{zz} = \varepsilon^{-1}\sigma_{33}^{(0)}, \quad \sigma_{xy} = \varepsilon^{-1}\sigma_{12}^{(0)}, \tag{1.21}$$
$$\sigma_{xz} = \sigma_{13}^{(1)}, \quad \sigma_{yz} = \sigma_{23}^{(1)}.$$

From formulas (1.19) and (1.21) it follows that the nonuniformity of the normal load σ_{zz}^{+} leads to the appearance of tangential displacements and transverse shear stresses, however, they are an order of magnitude smaller.

1.4 Solution of the Second Mixed Boundary Value Problem

The asymptotic method also allows to solve the spatial problem of a plate when on its facial surface normal loads act and to the surface tangential displacements are imparted at $z = h$:

$$u\,(\mathrm{x,y,h}) = u^{+}\,(\mathrm{x,y})\,, v\,(\mathrm{x,y,h}) = v^{+}\,(\mathrm{x,y})\,, \tag{1.22}$$
$$\sigma_{zz}\,(\mathrm{x,y,h}) = -\sigma_{zz}^{+}\,(x,y)\,.$$

It is assumed that the lower surface of the plate is rigidly fixed (conditions (1.4)). Using formulas (1.13), satisfying conditions (1.4) and (1.22) with respect to u,v, we obtain the system

$$
\begin{aligned}
&\int_0^1 d\zeta \int_0^\zeta \bar{F}_x^{(s)} d\zeta + C_1^{(s)}(\xi,\eta) + C_2^{(s)}(\eta) = u^{+(s)},\\
&\int_0^1 d\zeta \int_0^\zeta \bar{F}_y^{(s)} d\zeta + C_3^{(s)}(\xi,\eta) + C_4^{(s)}(\xi,\eta) = v^{+(s)},\\
&-\int_{-1}^0 d\zeta \int_0^\zeta \bar{F}_x^{(s)} d\zeta - C_1^{(s)}(\xi,\eta) + C_2^{(s)}(\xi,\eta) = 0,\\
&-\int_{-1}^0 d\zeta \int_0^\zeta \bar{F}_y^{(s)} d\zeta - C_3^{(s)}(\xi\eta) + C_4^{(s)}(\xi\eta) = 0,\\
&u^{+(0)} = \frac{u^+}{l},\quad u^{+(s)} = 0,\quad s \neq 0;\quad (u,v)\,.
\end{aligned}
\tag{1.23}
$$

Having solved this system, we determine the values of $C_1^{(s)}, C_2^{(s)}, C_3^{(s)}, C_4^{(s)}$

$$
\begin{aligned}
&C_1^{(s)}(\xi,\eta) = \frac{1}{2}\left(u^{+(s)} - \int_{-1}^1 d\zeta \int_0^\zeta \bar{F}_x^{(s)} d\zeta\right),\\
&C_2^{(s)}(\xi,\eta) = \frac{1}{2}\left(u^{+(s)} + \int_{-1}^0 d\zeta \int_0^\zeta \bar{F}_x^{(s)} d\zeta - \int_0^1 d\zeta \int_0^\zeta \bar{F}_x^{(s)} d\zeta\right),\\
&C_3^{(s)}(\xi,\eta) = \frac{1}{2}\left(v^{+(s)} - \int_{-1}^1 d\zeta \int_0^\zeta \bar{F}_y^{(s)} d\zeta\right),\\
&C_4^{(s)}(\xi,\eta) = \frac{1}{2}\left(v^{+(s)} + \int_{-1}^0 d\zeta \int_0^\zeta \bar{F}_y^{(s)} d\zeta - \int_0^1 d\zeta \int_0^\zeta \bar{F}_y^{(s)} d\zeta\right).
\end{aligned}
\tag{1.24}
$$

Having satisfied condition (1.22) with respect to σ_{zz} and condition (1.4) with respect to w, we obtain the system

$$
\begin{aligned}
&\frac{A_{33}}{\Delta}\left(\int\limits_0^1 \bar{F}_z^{(s)}\mathrm{d}\zeta + C_5^{(s)}\right) + \frac{\sigma_{33*}^{(s)}(\zeta=1)}{\Delta} = -\sigma_{zz}^{+(s)}, \\
&-\int\limits_{-1}^{0}\mathrm{d}\zeta\int\limits_0^{\zeta}\bar{F}_z^{(s)}\mathrm{d}\zeta - C_5^{(s)}(\xi,\eta) + C_6^{(s)}(\xi,\eta) = 0, \\
&\sigma_{zz}^{+(0)} = \sigma_{zz}^{+},\quad \sigma_{zz}^{+(s)} = 0,\ s \neq 0.
\end{aligned}
\tag{1.25}
$$

from where its follows

$$
\begin{aligned}
C_5^{(s)} &= -\frac{1}{A_{33}}\left(\Delta\sigma_{zz}^{+(s)} + \sigma_{33*}^{(s)}(\zeta=1)\right) - \int\limits_0^1 \bar{F}_z^{(s)}\mathrm{d}\zeta, \\
C_6^{(s)} &= \int\limits_{-1}^{0}\mathrm{d}\zeta\int\limits_0^{\zeta}\bar{F}_z^{(s)}\mathrm{d}\zeta + C_5^{(s)}.
\end{aligned}
\tag{1.26}
$$

As before, if u^+, v^+, σ_{zz}^+ are algebraic polynomials, the iteration breaks and we obtain the mathematically exact solution. For example at

a)

$$
F_x = 0,\quad F_y = 0,\quad F_z = 0,\quad u^+ = 0,\quad v^+ = 0,\quad \sigma_{zz}^+ = a_0 + a_1\xi + a_2\eta. \tag{1.27}
$$

We will have at $s = 0$

$$
\begin{aligned}
&C_1^{(0)} = 0,\quad C_2^{(0)} = 0,\quad C_3^{(0)} = 0,\quad C_4^{(0)} = 0, \\
&C_5^{(0)} = C_6^{(0)} = -\frac{\Delta}{A_{33}}(,a_0 + a_1\xi + a_2\eta), \\
&\sigma_{13}^{(0)} = 0,\quad \sigma_{23}^{(0)} = 0,\qquad \sigma_{33}^{(0)} = -(a_0 + a_1\xi + a_2\eta), \\
&\sigma_{11}^{(0)} = -\frac{A_{31}}{A_{33}}(a_0 + a_1\xi + a_2\eta),\qquad \sigma_{22}^{(0)} = -\frac{A_{32}}{A_{33}}(a_0 + a_1\xi + a_2\eta), \\
&\sigma_{12}^{(0)} = -\frac{A_{34}}{A_{33}}(a_0 + a_1\xi + a_2\eta), \\
&W^{(0)} = -\frac{\Delta}{A_{33}}(1+\zeta)(a_0 + a_1\xi + a_2\eta),\quad U^{(0)} = 0,\quad V^{(0)} = 0
\end{aligned}
\tag{1.28}
$$

and at $s \geq 1$, $\sigma_{ij}^{(s)} = 0$, $U^{(s)} = 0$, $V^{(s)} = 0$, $W^{(s)} = 0$.
Consequently

$$
\begin{aligned}
&\sigma_{xx} = \varepsilon^{-1}\sigma_{11}^{(0)},\quad \sigma_{yy} = \varepsilon^{-1}\sigma_{22}^{(0)},\quad \sigma_{zz} = \varepsilon^{-1}\sigma_{33}^{(0)}, \\
&\sigma_{xy} = \varepsilon^{-1}\sigma_{12}^{(0)},\quad \sigma_{xz} = 0,\quad \sigma_{yz} = 0, \\
&w = lW^{(0)},\qquad u = 0,\qquad v = 0.
\end{aligned}
\tag{1.29}
$$

Let us also consider the case

b)

$$F_x = 0,\ \ F_y = 0,\ \ F_z = 0,\quad v^+ = 0,\ \ \sigma^+_{zz} = 0,\ \ u^+ = l\,(a_0 + a_1\xi + a_2\eta)\,. \quad (1.30)$$

Then at $s = 0$

$$\begin{gathered}
C_1^{(0)} = C_2^{(0)} = \frac{1}{2}\,(a_0 + a_1\xi + a_2\eta)\,,\ C_3^{(0)} = 0,\ C_4^{(0)} = 0,\ \ C_5^{(0)} = 0,\ \ C_6^{(0)} = 0,\\
\sigma_{13}^{(0)} = \frac{a_{44}}{2\Delta_1}\,(a_0 + a_1\xi + a_2\eta)\,,\quad \sigma_{23}^{(0)} = -\frac{a_{45}}{2\Delta_1}\,(a_0 + a_1\xi + a_2\eta)\,,\\
\sigma_{33}^{(0)} = 0,\ \sigma_{11}^{(0)} = 0,\ \sigma_{22}^{(0)} = 0,\ \sigma_{12}^{(0)} = 0,\ \ U^{(0)} = \frac{u^+}{l},\ \ W^{(0)} = 0,\ \ V^{(0)} = 0.
\end{gathered} \quad (1.31)$$

At $s = 1$

$$\begin{gathered}
\sigma_{11*}^{(1)} = A_{11}a_1 - A_{41}a_2,\quad \sigma_{22*}^{(1)} = -A_{12}a_1 + A_{42}a_2,\\
\sigma_{33*}^{(1)} = A_{13}a_1 - A_{43}a_2,\quad \sigma_{12*}^{(1)} = -A_{14}a_1 + A_{44}a_2,\\
\bar{F}_z^{(1)} = \frac{\Delta}{2A_{33}\Delta_1}\,({}_{44}a_1 - {}_{45}a_2)\,,\\
C_1^{(1)} = 0,\ \ C_2^{(1)} = 0,\quad C_3^{(1)} = 0,\ C_4^{(1)} = 0,\\
C_5^{(1)} = -\frac{\sigma_{33*}^{(1)}}{A_{33}} - \bar{F}_z^{(1)},\ \ C_6^{(1)} = -\frac{3}{2}\bar{F}_z^{(1)} - \frac{\sigma_{33*}^{(1)}}{A_{33}},\\
\sigma_{11}^{(1)} = \frac{1}{\Delta}\left(A_{31}\left(\bar{F}_z^{(1)}\zeta + C_5^{(1)}\right) + \sigma_{11*}^{(1)}\right),\\
\sigma_{22}^{(1)} = \frac{1}{\Delta}\left(A_{32}\left(\bar{F}_z^{(1)}\zeta + C_5^{(1)}\right) + \sigma_{22*}^{(1)}\right),\\
\sigma_{33}^{(1)} = \frac{1}{\Delta}\left(A_{33}\left(\bar{F}_z^{(1)}\zeta + C_5^{(1)}\right) + \sigma_{33*}^{(1)}\right),\\
\sigma_{12}^{(1)} = \frac{1}{\Delta}\left(A_{34}\left(\bar{F}_z^{(1)}\zeta + C_5^{(1)}\right) + \sigma_{12*}^{(1)}\right),\\
\sigma_{13}^{(1)} = 0,\ \ \sigma_{23}^{(1)} = 0,\\
U^{(1)} = 0,\ \ V^{(1)} = 0,\quad W^{(1)} = \bar{F}_z^{(1)}\frac{\zeta^2}{2} + C_5^{(1)}\zeta + C_6^{(1)}.
\end{gathered} \quad (1.32)$$

Consequently we will have

$$\begin{gathered}
w = hW^{(1)},\ \ u = lU^{(0)},\ \ v = 0,\\
\sigma_{xx} = \sigma_{11}^{(1)},\quad \sigma_{yy} = \sigma_{22}^{(1)},\quad \sigma_{zz} = \sigma_{33}^{(1)},\quad \sigma_{xy} = \sigma_{12}^{(1)}.
\end{gathered} \quad (1.33)$$

As follows from formula (1.29), in case a), the main role normal displacements, normal stresses will be play, as well as tangential stress. The tangential displacements and the transverse shear stresses are negligibly small. In case b), according to (1.33), the main role play the tangential displacement u, the normal displacement is an order of magnitude smaller, and the other tangential displacement is negligibly small. From the stresses, in unlike case a), the transverse shear stresses play the main role, the other stresses are an order of magnitude smaller.

1.5 Conclusions

The spatial 3D problem for anisotropic plates with a plane of elastic symmetry is solved. In particular, this property have the crystals of the monoclinic system. For solving such problems, the hypotheses of the classical theory of plates, as well as known refined theories are not applicable. For solving the problem the asymptotic method for solving singularly perturbed differential equations proved to be effective. The solution consists of the solution of the external problem and the boundary layer. Asymptotics, which is fundamentally different from the asymptotic the stress tensor components and displacement vectors of the classical theory, was established, which made it possible to find the solutions to the problem. The cases are established when the asymptotic solution becomes mathematically exact. It is shown that the unevenness of the acting normal load leads to the appearance of tangential displacements and transverse stresses, but they are an order of magnitude smaller. It is shown that the unevenness of the acting normal load leads to the appearance of tangential displacements and transverse stresses, but they are an order of magnitude smaller. The examples of illustrative character are given.

References

[1] Agalovyan LA (2009) Asimptoticheskaya teoriya deformiruyemykh tonkostennykh sistem (in Russ., Asymptotic theory of deformable thin-walled systems). In: Collection Mechanics 2009: Proceedings of the International School – Conference of Young Scientists. Yerevan. Publishing House of National Yerevan State University of Architecture and Construction, pp. 5-35

[2] Aghalovyan LA (2015) Asymptotic Theory of Anisotropic Plates and Shells. World Scientific Publishing, Singapore

[3] Dieulesaint E, Royer D (1980) Elastic Waves in Solids: Applications to Signal Processing. Wiley, New York, Chichester

[4] Goldenweiser AL (1962) Postroyeniye priblizhennoy teorii izgiba plastinki metodom asimptoticheskogo integrirovaniya uravneniy teorii uprugosti (in Russ., Construction of the approximate theory of plate bending by the method of asymptotic integration of the equations of elasticity theory). Journal of Applied Mathematics and Mechanics 26(4):668-686

[5] Lekhnitsky SG (1963) Theory of Elasticity of an Anisotropic Body. Holden-Day, San Francisco

Chapter 2
Plane Problem of the Theory of Elasticity for a Compound Plane Weakened by a Lune Opening

Karo L. Aghayan, Levon A. Harutyunyan, and Vahe G. Zakaryan

Abstract This study investigates a compound elastic plane containing a plane with an opening formed by two partially overlapping circular arcs and two luneshaped inserts of a dissimilar material, with shared vertices. By employing a modified bipolar coordinate system, the contact problem for the compound plane weakened by an opening is formulated and subsequently reduced to a system of two Fredholm integral equations of the second kind, which is amenable to an iterative solution through the method of successive approximation. The solution procedure leads to a characteristic transcendental equation parameterized by the physical and geometric properties of the investigated problem. Under specific physical conditions, this equation serves as an implicit finiteness criterion, defining the set of geometric parameters required to preclude stress singularities at the material contact vertices. The domain of geometric parameters is mapped, revealing distinct regions of low stress and singularity partitioned by boundary curves.

Key words: Plane problem, Compound body, Bipolar coordinates, Lune, Fourier transform, Singular integrals, Fredholm equation

2.1 Introduction

Extensive research has been devoted to the governing singularities and stress state factors in compound elastic bodies in the form of an elastic plane containing stress discontinuities—such as cracks, overlays, inclusions, and acute-angled cavities. In this regard, reference is made here to [5, 7, 8, 13, 16], which put forward various approaches to solving issues related to determining the singular behavior of elastic

Karo L. Aghayan · Levon A. Harutyunyan · Vahe G. Zakaryan (corresponding author)
Institute of Mechanics, National Academy of Sciences of the Republic of Armenia, 24b Marshal Baghramyan Ave, Yerevan, 0019, Republic of Armenia, e-mail: karo.aghayan@gmail.com, arutyunyanlevon@gmail.com, vahezaqaryan@mail.ru

H. Altenbach and A. V. Sahakyan, *Contact Mechanics and Relevant Problems*,
Advanced Structured Materials 253, https://doi.org/10.1007/978-3-032-20869-9_2

stresses near the end points of stress discontinuities. The formulation and solution of various problems of the stress-strain state in compound elastic bodies, and a precise definition of the degree of singularity of destructive stresses at acute-angled points of contact surfaces is addressed with particular rigor in the following studies [1, 3, 4, 7, 9–11]. The present study extends investigation of the plane problems formulated for the interior domains of circular lunes, as examined in [3, 9, 10]. The present work builds upon related analyses for external domains, specifically studies of non-homogeneous planes with circular segment-shaped openings [3, 10, 11].

We derive a solution for the contact problem of elasticity theory for a piecewise-homogeneous elastic plane composed of a plane with an opening in the form of two partially overlapping circular arcs and two lune-shaped inserts with shared vertices. Using the integral Fourier transform in bipolar coordinates, the problem is reformulated as a system of two Riemann-Hilbert boundary value problems [6, 15, 16]. To solve the above, a system of Fredholm integral equations of the second kind was used, by the method of successive approximations. We obtain a transcendental equation that delineates the domain of the finite and infinite stresses at the vertices of contact surfaces, according to the geometric and physical parameters of the problem. The problem was implemented numerically for pertinent specific cases. The domain of geometric parameters is mapped, revealing distinct regions of finite and infinite stresses at the vertices of the contact surfaces.

2.2 Problem Formulation and Governing Equations

This study investigates the contact problem of the theory of elasticity for a compound plane weakened by an opening. The opening geometry, defined within a bipolar coordinate system, is bounded by symmetric circular arcs β_0. The compound plane consists of two parts: the outer, infinite part defined by the inequality $\beta \leq |\beta_0|$, and the inner, finite part in the form of a pair of symmetrical lunes bounded by two circular arcs $\beta_0 < |\beta| < \beta_1$ (Fig. 2.1). Exploiting the symmetry of the domain about the axis Oy, the study is confined to the right half-plane, defined in the bipolar coordinate system by the inequality $0 \leq \beta < \infty$. The outer part (hereinafter called the Half-Plane) has elastic characteristics μ_1 (shear modulus) and ν_1 (Poisson's ratio), and the inner part (hereinafter called the Lune) has elastic characteristics μ_2, ν_2. The Lune and the Half-Plane come into full contact along the contact line $\beta = \beta_0$. External loads, in the form of normal and shear stresses, are determined using the functions $\varphi_1(\alpha)$ and $\varphi_2(\alpha)$ on the line $\beta = \beta_1$.

The problem thus reduces to determining the stress-strain state of the investigated compound body, and looking into the nature of the distribution of contact stresses near the end points of the material separation line. The solution employs a biharmonic stress function approach in conjunction with a modified bipolar coordinate system. The connection between rectangular coordinates and a modified bipolar coordinate system is determined by formulas [15]

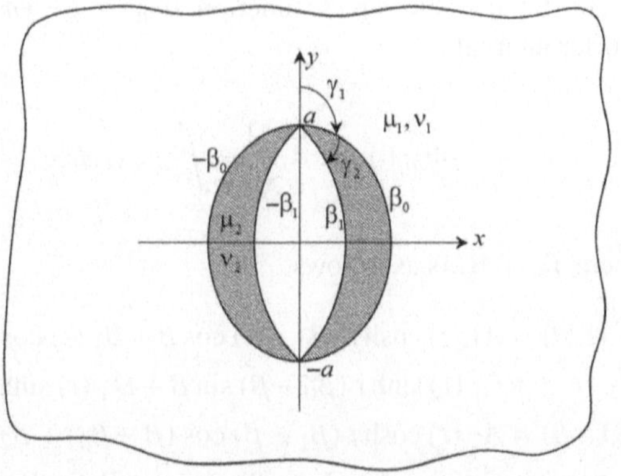

Fig. 2.1 Schematic problem geometry

$$gx = \sin\beta, gy = \sinh\alpha, ag = \cosh\alpha - \cos\beta, \tag{2.1}$$

where a is a dimensional parameter. The solution is constructed from stress functions $\Phi_m(\alpha, \beta)$ $(m = 1, 2)$, where each function satisfies the biharmonic equation within its respective domain

$$\left[\frac{\partial^4}{\partial\alpha^4} + 2\frac{\partial^4}{\partial\alpha^2\partial\beta^2} + \frac{\partial^4}{\partial\beta^4} - 2\frac{\partial^2}{\partial\alpha^2} + 2\frac{\partial^2}{\partial\beta^2} + 1\right](g\Phi_m(\alpha, \beta)) = 0 \tag{2.2}$$

Stress and displacements in modified bipolar coordinates are given as stress functions $\Phi_m(\alpha, \beta)$ by the following formulas [16]

$$a\sigma_\alpha^{(m)}(\alpha, \beta) = \left[(\cosh\alpha - \cos\beta)\frac{\partial^2}{\partial\beta^2} - \sinh\alpha\frac{\partial}{\partial\alpha} - \sin\beta\frac{\partial}{\partial\beta} + \cosh\alpha\right](g\Phi_m(\alpha, \beta)),$$

$$a\sigma_\beta^{(m)}(\alpha, \beta) = \left[(\cosh\alpha - \cos\beta)\frac{\partial^2}{\partial\alpha^2} - \sinh\alpha\frac{\partial}{\partial\alpha} - \sin\beta\frac{\partial}{\partial\beta} + \cos\beta\right](g\Phi_m(\alpha, \beta)),$$

$$a\tau_{\alpha\beta}^{(m)}(\alpha, \beta) = -(\cosh\alpha - \cos\beta)\frac{\partial^2}{\partial\alpha\partial\beta}(g\Phi_m(\alpha, \beta)),$$

$$u_m(\alpha, \beta) = \frac{g}{2\mu_m}\left[(1 - 2\nu_m)\frac{\partial\Phi_m(\alpha, \beta)}{\partial\alpha} - \frac{\partial\Psi_m(\alpha, \beta)}{\partial\beta}\right],$$

$$\mathrm{v}_m(\alpha, \beta) = \frac{g}{2\mu_m}\left[(1 - 2\nu_m)\frac{\partial\Phi_m(\alpha, \beta)}{\partial\beta} + \frac{\partial\Psi_m(\alpha, \beta)}{\partial\alpha}\right], \quad m = 1, 2, \tag{2.3}$$

where μ_m and $\nu_m (m = 1, 2)$ are the elastic characteristics of the materials of the domains; the relationship between $g\Psi_m(\alpha, \beta)$ and $g\Phi_m(\alpha, \beta)$ is presented by the formula

$$\frac{\partial^2(g\Psi_m(\alpha, \beta))}{\partial\alpha\partial\beta} = (1 - \nu_m)\left(\frac{\partial^2}{\partial\alpha^2} - \frac{\partial^2}{\partial\beta^2} - 1\right)(g\Phi_m(\alpha, \beta)). \tag{2.4}$$

The biharmonic stress function $\Phi_m(\alpha,\beta)$ $(m=1,2)$ is defined through the Fourier intergal

$$g\Phi_m(\alpha,\beta)=\frac{1}{\sqrt{2\pi}}\int\limits_{-\infty}^{\infty} f_m(t,\beta)e^{-it\alpha}dt \quad (m=1,2), \tag{2.5}$$

where $f_m(t,\beta)$ is as follows

$$\begin{aligned} f_1(t,\beta) &= A_1(t)\cosh t(\beta_0-\beta)\cos\beta + B_1(t)\cosh t\beta\cos(\beta_0-\beta) \\ &\quad + C_1(t)\sinh t(\beta_0-\beta)\sin\beta + D_1(t)\sinh t\beta\sin(\beta_0-\beta), \\ f_2(t,\beta) &= A_2(t)\cosh t(\beta_1-\beta)\cos(\beta-\beta_0) + B_2(t)\cosh t(\beta-\beta_0)\cos(\beta_1-\beta) \\ &\quad + C_2(t)\sinh t(\beta_1-\beta)\sin(\beta-\beta_0) + D_2(t)\sinh t(\beta-\beta_0)\sin(\beta_1-\beta). \end{aligned} \tag{2.6}$$

Unknown integration functions $A_m(t)$, $B_m(t)$, $C_m(t)$ and $D_m(t)$ $(m=1,2)$ are determined based on the following boundary and contact terms

$$\begin{aligned} &\mathrm{V}_1(\alpha,0)=\tau_{\alpha\beta}^{(1)}(\alpha,0)=0, \\ &(g\Phi_2(\alpha,\beta))|_{\beta=\beta_1}=\varphi_1(\alpha); \quad \left.\frac{\partial(g\Phi_2(\alpha,\beta))}{\partial\beta}\right|_{\beta=\beta_1}=\varphi_2(\alpha), \end{aligned} \tag{2.7}$$

$$\begin{aligned} &g\Phi_1(\alpha,\beta)=g\Phi_2(\alpha,\beta); \quad \frac{\partial(g\Phi_1(\alpha,\beta))}{\partial\beta}=\frac{\partial(g\Phi_2(\alpha,\beta))}{\partial\beta} \quad \text{at } \beta=\beta_0, \\ &U_1(\alpha,\beta_0)=U_2(\alpha,\beta_0); \quad \mathrm{V}_1(\alpha,\beta_0)=\mathrm{V}_2(\alpha,\beta_0). \end{aligned} \tag{2.8}$$

It is assumed that the functions $\varphi_m(\alpha)$ $(m=1,2)$ meet the Fourier integral expansion terms. Based on the boundary terms (2.7) and the first two terms (2.8), the following expressions are derived for the unknown integration functions $A_m(t)$, $B_m(t)$, $C_m(t)$ and $D_m(t)$ $(m=1,2)$ through the unknown contact stresses $P_m(t)$ $(m=1,2)$

$$\begin{aligned} A_1(t) &= P_1(t)\frac{2t\sin\gamma_1}{\Delta_1(t)} - P_2(t)\frac{\sinh 2t\gamma_1\sin\gamma_1}{\Delta_1(t)\Delta_3(t)}, \\ B_1(t) &= P_1(t)\frac{2sht\gamma_1}{\Delta_1(t)} + P_2(t)\frac{\sinh t\gamma_1\sin 2\gamma_1}{\Delta_1(t)\Delta_3(t)}, \\ C_1(t) &= -P_1(t)\frac{2\sin\gamma_1}{\Delta_1(t)} - P_2(t)\frac{\sin 2\gamma_1\sin\gamma_1}{\Delta_1(t)\Delta_3(t)}, \\ D_1(t) &= P_1(t)\frac{2t\sinh t\gamma_1}{\Delta_1(t)} - P_2(t)\frac{\sinh 2t\gamma_1\sinh t\gamma_1}{\Delta_1(t)\Delta_3(t)}, \\ A_2(t) &= P_1(t)\frac{\cosh t\gamma_2}{\Delta_4(t)} - \bar{\varphi}_1(t)\frac{\cos\gamma_2}{\Delta_4(t)}, \\ B_2(t) &= -P_1(t)\frac{\cos\gamma_2}{\Delta_4(t)} + \bar{\varphi}_1(t)\frac{\cosh t\gamma_2}{\Delta_4(t)}, \end{aligned}$$

$$C_2(t) = P_1(t)\left[\frac{t\cosh t\gamma_2}{\Delta_2(t)} + \frac{(t^2+1)\sinh t\gamma_2 \sin\gamma_2}{2\Delta_2(t)\Delta_4(t)}\right] + P_2(t)\frac{\sinh t\gamma_2}{\Delta_2(t)}$$
$$-\bar{\varphi}_1(t)\left[\frac{t\cos\gamma_2}{\Delta_2(t)} + \frac{(t^2+1)\sinh 2t\gamma_2 \sin 2\gamma_2}{2\Delta_2(t)\Delta_4(t)}\right] + \bar{\varphi}_2(t)\frac{t\sin\gamma_2}{\Delta_2(t)}, \tag{2.9}$$

$$D_2(t) = -P_1(t)\left[\frac{t\cos\gamma_2}{\Delta_2(t)} + \frac{(t^2+1)\sinh 2t\gamma_2 \sin\gamma_2}{2\Delta_2(t)\Delta_4(t)}\right] - P_2(t)\frac{t\sin\gamma_2}{\Delta_2(t)}$$
$$+\bar{\varphi}_1(t)\left[\frac{t\cosh t\gamma_2}{\Delta_2(t)} + \frac{(t^2+1)\sinh t\gamma_2 \sin 2\gamma_2}{2\Delta_2(t)\Delta_4(t)}\right] - \bar{\varphi}_2(t)\frac{\sinh t\gamma_2}{\Delta_2(t)},$$

$$\bar{\varphi}_m(t) = \frac{1}{\sqrt{2\pi}}\int_{-\infty}^{\infty} \varphi_m(\alpha)e^{it\alpha}d\alpha \quad (m = 1, 2),$$

$$\Delta_1(t) = \sinh 2t\gamma_1 + t\sin 2\gamma_1, \quad \Delta_2(t) = \sinh^2 t\gamma_2 - t^2\sin^2\gamma_2,$$
$$\Delta_3(t) = \sinh^2 t\gamma_1 + \sin^2\gamma_1, \quad \Delta_4(t) = \sinh^2 t\gamma_2 + \sin^2\gamma_2$$
$$\gamma_1 = \beta_0, \quad \gamma_2 = \beta_1 - \beta_0.$$

If the displacement equality terms are met, i.e. the last two terms in (2.8), the following system of defining singular integral equations is formed for determining unknown functions $P_m(t)$

$$M_1(t)\mathrm{X}_1(t) + M_2(t)\mathrm{X}_2(t) + N_1(t) = \int_{-\infty}^{\infty} \frac{e^{-(t-\tau)\beta_0}}{\sinh(t-\tau)\pi}\mathrm{X}_1(\tau)d\tau,$$
$$M_2(-t)\mathrm{X}_1(t) + M_1(-t)\mathrm{X}_2(t) + N_2(t) = \int_{-\infty}^{\infty} \frac{e^{(t-\tau)\beta_0}}{\sinh(\tau-t)\pi}\mathrm{X}_2(\tau)d\tau, \tag{2.10}$$

where

$$\mathrm{X}_1(t) = (h + a_{21}(t))P_1(t) + a_{22}(t)P_2(t) + a_{23}(t)\bar{\varphi}_1(t) + a_{24}(t)\bar{\varphi}_2(t),$$
$$\mathrm{X}_2(t) = (h - a_{21}(t))P_1(t) - a_{22}(t)P_2(t) - a_{23}(t)\bar{\varphi}_1(t) - a_{24}(t)\bar{\varphi}_2(t),$$
$$M_1(t) = \frac{1}{2ha_{22}(t)}\left[a_{11}(t)a_{22}(t) + (h + a_{12}(t))(h - a_{21}(t))\right] - t,$$
$$M_2(t) = \frac{1}{2ha_{22}(t)}\left[a_{11}(t)a_{22}(t) - (h + a_{12}(t))(h - a_{21}(t))\right],$$
$$N_1(t) = M_3(t)\bar{\varphi}_1(t) + M_4(t)\bar{\varphi}_2(t),$$
$$N_2(t) = M_3(-t)\bar{\varphi}_1(t) + M_4(-t)\bar{\varphi}_2(t),$$
$$M_3(t) = a_{13}(t) - \frac{a_{23}(t)}{a_{22}(t)}(h + a_{12}(t)),$$
$$M_4(t) = a_{14}(t) - \frac{a_{24}(t)}{a_{22}(t)}(h + a_{12}(t)),$$

$$
\begin{aligned}
a_{11}(t) &= h_2 \frac{t^2+1}{\Delta_1(t)} \sin 2\gamma_1 + t \frac{t^2+1}{\Delta_2(t)} \sin^2 \gamma_2, \\
a_{12}(t) &= \frac{h_2}{\Delta_1(t)} (\cosh 2t\gamma_1 + \cos 2\gamma_1) + \frac{1}{2\Delta_2(t)} (\sinh 2t\gamma_2 - t \sin 2\gamma_2), \\
a_{13}(t) &= -\frac{t^2+1}{\Delta_2(t)} \sinh t\gamma_2 \sin \gamma_2, \\
a_{14}(t) &= \frac{1}{\Delta_2(t)} (t \cosh t\gamma_2 \sin \gamma_2 - \sinh t\gamma_2 \cos \gamma_2), \\
a_{21}(t) &= \frac{h_2}{\Delta_1(t)} (\cosh 2t\gamma_1 - \cos 2\gamma_1) + \frac{1}{2\Delta_2(t)} (\sinh 2t\gamma_2 + t \sin 2\gamma_2), \\
a_{22}(t) &= \frac{h_2}{\Delta_1(t)} \sin 2\gamma_1 + \frac{t}{\Delta_2(t)} \sin^2 \gamma_2, \\
a_{23}(t) &= -\frac{1}{\Delta_2(t)} (t \cosh t\gamma_2 \sin \gamma_2 + \sinh t\gamma_2 \cos \gamma_2), \\
a_{24}(t) &= \frac{1}{\Delta_2(t)} \sinh t\gamma_2 \sin \gamma_2, \\
h_1 &= \frac{1-\mu}{1+\mu}, \quad h_2 = \frac{\mu(1-\nu_1)}{1-\nu_2}, \quad \mu = \frac{\mu_2}{\mu_1}, \quad h = \frac{h_1}{2} + h_2 - 1.
\end{aligned}
\tag{2.11}
$$

Thus, system (2.10) is the governing equation set; its solution is equivalent to solving the formulated problem.

2.3 Solution for the Governing Equation Set

Singular integral equations (2.10) are expressed as follows

$$
\begin{aligned}
M_1(t) Y_1(t) + M_4(t) Y_2(t) + K_1(t) &= -\frac{1}{\pi} \int_{-\infty}^{\infty} \frac{\theta'(\tau) Y_1(\tau)}{\theta(\tau) - \theta(t)} d\tau, \\
M_4(-t) Y_1(t) + M_1(-t) Y_2(t) + K_2(t) &= \frac{1}{\pi} \int_{-\infty}^{\infty} \frac{\theta'(\tau) Y_2(\tau)}{\theta(\tau) - \theta(t)} d\tau,
\end{aligned}
\tag{2.12}
$$

where

$$
\begin{aligned}
&Y_1(t) = e^{t(\beta_1 - \pi)} X_1(t), \quad Y_2(t) = e^{-t(\beta_0 + \pi)} X_2(t), \\
&K_1(t) = e^{t(\beta_0 - \pi)} N_1(t), \quad K_2(t) = e^{-t(\beta_0 + \pi)} N_2(t), \\
&M_4(t) = e^{2t\beta_0} M_2(t), \quad \theta(t) = e^{2t\pi}.
\end{aligned}
\tag{2.13}
$$

Integral equations (2.12) are reformulated as a Riemann - Hilbert boundary value problem [4, 9, 12, 14].

A new piecewise holomorphic function is introduced

$$R_m(z) = \frac{1}{2\pi i}\int_{-\infty}^{\infty} \frac{\theta'(\tau) Y_m(\tau)}{\theta(\tau) - \theta(t)} d\tau \quad (m = 1, 2), \tag{2.14}$$

which is required to vanish at infinity. For any point of the contour t, the following applies according to the Sokhotski–Plemelj theorem

$$\begin{aligned} R_m^+(t) - R_m^-(t) &= Y_m(t), \\ R_m^+(t) + R_m^-(t) &= \frac{1}{\pi i}\int_{-\infty}^{\infty} \frac{\theta'(\tau) Y_m(\tau)}{\theta(\tau) - \theta(t)} d\tau \quad (m = 1, 2). \end{aligned} \tag{2.15}$$

By introducing these expressions into the given system of singular integral equations (2.12), the following Riemann-Hilbert boundary value problems are obtained:

$$\begin{aligned} R_1^+(t) &= c_{11}(t) R_1^-(t) + c_{12}(t) R_2^-(t) + T_1(t), \\ R_1^+(t) &= c_{21}(t) R_1^-(t) + c_{22}(t) R_2^-(t) + T_2(t), \end{aligned} \tag{2.16}$$

where

$$\begin{aligned} & c_{11}(t) = 1 + \frac{2i(M_1(-t) + i)}{\Delta_3(t)}, \quad c_{22}(t) = 1 - \frac{2i(M_1(t) + i)}{\Delta_3(t)}, \\ & c_{12}(t) = \frac{2iM_4(t)}{\Delta_3(t)}, \quad c_{21}(t) = -\frac{2iM_4(-t)}{\Delta_3(t)}, \\ & T_1(t) = \frac{1}{\Delta_3(t)}[M_4(t) K_2(t) - (M_1(-t) - i) K_1(t)], \\ & T_2(t) = \frac{1}{\Delta_3(t)}[M_4(-t) K_1(t) - (M_1(t) + i) K_2(t)], \\ & \Delta_3(t) = (M_1(t) + i)(M_1(-t) - i) - M_4(t) M_4(-t). \end{aligned} \tag{2.17}$$

Denoting the coefficient matrix of system (2.16) by $G(t)$, and the vectors with components $R_m(t)$ and $T_m(t)$ $(m = 1, 2)$ by $R(t)$ and $T(t)$, the boundary value problems (2.16) can be represented in matrix form

$$R^+(t) = G(t) R^-(t) + T(t) \tag{2.18}$$

In the general case, equation (2.18) is solved by regularization [14], which transforms it into a Fredholm integral equation of the second kind. From both parts of the equation, singular integrals, interpreted in the sense of the Cauchy principal value, are eliminated, in compliance with certain rules

$$R^+(t) = \frac{1}{\pi i}\int_{-\infty}^{\infty} \frac{\theta'(\tau) R^+(\tau)}{\theta(\tau) - \theta(t)} d\tau, \quad R^-(t) = -\frac{1}{\pi i}\int_{-\infty}^{\infty} \frac{\theta'(\tau) R^-(\tau)}{\theta(\tau) - \theta(t)} d\tau \tag{2.19}$$

By introducing $R^+(t)$ from equation (2.18) into the first equation (2.19), the following is obtained

$$G(t)\,R^-(t) + T(t) = \frac{1}{\pi i}\int_{-\infty}^{\infty} \frac{\theta'(\tau)}{\theta(\tau)-\theta(t)}\left[G(\tau)\,R^-(\tau) + T(\tau)\right]d\tau. \tag{2.20}$$

Multiplying this equation by $[G(t)]^{-1}$ and adding to the second equation (2.19), the following is obtained:

$$\begin{aligned} R^-(t) &- \frac{1}{2\pi i}\int_{-\infty}^{\infty} \frac{\theta'(\tau)}{\theta(\tau)-\theta(t)}\left\{[G(t)]^{-1}G(\tau) - E\right\} R^-(\tau)d\tau \\ &= \frac{[G(t)]^{-1}}{2}\left[\frac{1}{\pi i}\int_{-\infty}^{\infty} \frac{\theta'(\tau)\,T(\tau)}{\theta(\tau)-\theta(t)}d\tau - T(t)\right], \end{aligned} \tag{2.21}$$

where E - is the identity matrix.

The integral on the left of (2.21) is proper, since the integrand no longer has a singularity at the point $\tau = t$. The formulation thus results in a system of Fredholm integral equations of the second kind. This equation is solvable, since on the contour the elements of the matrix $G(t)$ and the free terms included in equation (2.18) satisfy the Hölder condition, and the determinant of the matrix on the contour is not zero

$$\det G(t) = \frac{t+i}{t-i} \neq 0 \text{ on the contour} \tag{2.22}$$

The index of the boundary value problem in question is -1, and the homogeneous problem has a sole zero solution. In this case, the Fredholm equation (2.21) is always equivalent to the original equation [4, 6, 9, 14]. The transformed and simplified vector equation (2.21) is reduced to the following system of Fredholm integral equations of the second kind with respect to the functions $P_m(t)$ $(m = 1, 2)$

$$\begin{aligned} P_1(t) &= \frac{\Delta_1(t)\,\Delta_2(t)}{(t^2+1)\,\Delta(t)}\left\{\int_{-\infty}^{\infty}\left[H_{11}(t,\tau)\,P_1(\tau) + H_{12}(t,\tau)\,P_2(\tau)\right]d\tau + F_1(t)\right\}, \\ P_2(t) &= \frac{\Delta_1(t)\,\Delta_2(t)}{(t^2+1)\,\Delta(t)}\left\{\int_{-\infty}^{\infty}\left[H_{21}(t,\tau)\,P_1(\tau) + H_{22}(t,\tau)\,P_2(\tau)\right]d\tau + F_2(t)\right\}, \end{aligned} \tag{2.23}$$

where

$$H_{11}(t,\tau)=\frac{\sinh(t-\tau)\beta_0}{\sinh(t-\tau)\pi}$$
$$\times\left[h(t-\tau)(a_{21}(\tau)-a_{12}(t))-a_{12}(t)a_{11}(\tau)-\left(t^2+1\right)a_{22}(t)a_{12}(\tau)\right]$$
$$+\frac{\cosh(t-\tau)\beta_0}{\sinh(t-\tau)\pi}\left[h\left(\left(t^2+1\right)a_{22}(t)-a_{11}(\tau)\right)+(t-\tau)\left(a_{12}(t)a_{21}(\tau)-h^2\right)\right],$$

$$H_{12}(t,\tau)=\frac{\sinh(t-\tau)\beta_0}{\sinh(t-\tau)\pi}\left[a_{11}(t)(h+a_{11}(\tau))+\left(t^2+1\right)\left(a_{21}(\tau)-h^2\right)\right]$$
$$+\frac{\cosh(t-\tau)\beta_0}{\sinh(t-\tau)\pi}\left[h\left(t^2+1\right)(a_{21}(\tau)-a_{21}(t))-(t-\tau)a_{11}(t)a_{21}(\tau)\right],$$
$$H_{21}(t,\tau)=\frac{\sinh(t-\tau)\beta_0}{\sinh(t-\tau)\pi}$$
$$\times\left[h^2+h(t-\tau)a_{22}(\tau)-a_{12}(t)a_{12}(\tau)-\left(t^2+1\right)a_{22}(t)a_{22}(\tau)\right]$$
$$+\frac{\cosh(t-\tau)\beta_0}{\sinh(t-\tau)\pi}\left[h(a_{12}(t)-a_{12}(\tau))-(t-\tau)a_{22}(\tau)\right],$$
$$H_{22}(t,\tau)=\frac{\sinh(t-\tau)\beta_0}{\sinh(t-\tau)\pi}\left[a_{11}(t)a_{12}(\tau)-\left(t^2+1\right)a_{21}(t)a_{22}(\tau)\right]$$
$$+\frac{\cosh(t-\tau)\beta_0}{\sinh(t-\tau)\pi}\left[h\left(\left(t^2+1\right)a_{22}(\tau)-a_{11}(t)\right)-(t-\tau)a_{11}(t)a_{22}(\tau)\right],\tag{2.24}$$

$$F_1(t)=\int_{-\infty}^{\infty}\left\{\frac{\sinh(t-\tau)\beta_0}{\sinh(t-\tau)\pi}\left[(h(t-\tau)\right.\right.$$
$$-(t^2+1)a_{22}(t)(a_{23}(\tau)\bar{\varphi}_1(\tau)+a_{24}(\tau)\bar{\varphi}_2(\tau))$$
$$-a_{12}(t)(a_{13}(\tau)\bar{\varphi}_1(\tau)+a_{14}(\tau)\bar{\varphi}_2(\tau))]$$
$$+\frac{\cosh(t-\tau)\beta_0}{\sinh(t-\tau)\pi}\left[(t-\tau)a_{12}(t)(a_{23}(\tau)\bar{\varphi}_1(\tau)+a_{24}(\tau)\bar{\varphi}_2(\tau))\right.$$
$$\left.\left.-h(a_{13}(\tau)\bar{\varphi}_1(\tau)+a_{24}(\tau)\bar{\varphi}_2(\tau))\right]\right\}d\tau$$
$$+\left(h\left(t^2+1\right)-ta_{11}(t)\right)(a_{23}(t)\bar{\varphi}_1(t)+a_{24}(t)\bar{\varphi}_2(t))$$
$$+ta_{21}(t)(a_{13}(t)\bar{\varphi}_1(t)+a_{14}(t)\bar{\varphi}_2(t)),$$

$$F_2(t) = \int_{-\infty}^{\infty} \left\{ \frac{\sinh(t-\tau)\beta_0}{\sinh(t-\tau)\pi} \left[-(t^2+1)a_{21}(t)\,(a23(t)\bar{\varphi}_1(\tau) \right. \right.$$
$$+ a_{21}(\tau)\bar{\varphi}_2(\tau) + a_{11}(t)\,(a_{13}(\tau)\,\bar{\varphi}_1(\tau) + a_{14}(\tau)\,\bar{\varphi}_2(\tau))\,]$$
$$+ \frac{\cosh(t-\tau)\beta_0}{\sinh(t-\tau)\pi} \left[\left(h\left(t^2+1\right) - (t-\tau)\,a_{11}(t) \right) (a_{23}(\tau)\bar{\varphi}_1(\tau)) \right.$$
$$\left. \left. + a_{14}(\tau)\bar{\varphi}_2(\tau) \right] \right\} d\tau$$
$$- h\left(t^2+1\right)(a_{13}(t)\,\bar{\varphi}_1(t) + a_{14}(t)\,\bar{\varphi}_2(t))$$
$$+ \left(t^2+1\right)(ta_{21}(t) - a_{11}(t)) + 2th\,(a_{23}(t)\,\bar{\varphi}_1(t) + a_{24}(t)\,\bar{\varphi}_2(t)),$$

$$\Delta(t) = h_2\left[\sinh 2t\,(\gamma_1+\gamma_2) + t\sin 2\,(\gamma_1+\gamma_2)\right] - h_1h_2\sinh 2t\gamma_1\Delta_2(t)$$
$$- \Delta_1(t)\left[\frac{h_1^2}{4}\Delta_2(t) - h_1\sinh^2 t\gamma_2 + h_2 - 1 \right]. \tag{2.25}$$

Taking into account the obtained values for contact stresses, the following is obtained:

$$\tau_{\alpha\beta}^{(m)}(\alpha,\beta_0) = \frac{i\,(\cosh\alpha - \cos\beta_0)}{a\sqrt{2\pi}} \int_{-\infty}^{\infty} P_2(t)\,e^{-it\alpha}dt,$$
$$\sigma_\beta^{(m)}(\alpha,\beta_0) = -\frac{1}{a\sqrt{2\pi}} \int_{-\infty}^{\infty} \left\{ \left[t^2(\cosh\alpha - \cos\beta_0) + it\sinh\alpha - \cos\beta_0 \right] P_1(t) \right.$$
$$\left. + \sin\beta_0 P_2(t) \right\} e^{-it\alpha}dt. \tag{2.26}$$

In the general case, any of the three stresses (2.3), according to (2.5), (2.9) and (2.23), can be expressed by an improper integral of the form

$$\tau(x,y) = \int_{-\infty}^{\infty} \frac{\bar{\tau}(x,y,t)}{\Delta(t)} \left(1+\frac{y}{a}\right)^{-(1+it)} \left(1-\frac{y}{a}\right)^{-1+it} dt \tag{2.27}$$

with the possibility of its approximate calculation using various numerical methods, once $\bar{\tau}(x,y,t)$ is determined. To investigate the behavior of stresses in the vicinity of the edge of the contact $y = \pm a$ surface (i.e. at points $\alpha = \pm\infty$), the integral (2.27) along the real axis is supplemented by the integral along the right (at $y < 0$ or $\alpha < 0$) or left (at $y > 0$ or $\alpha > 0$) semicircle of radius $R \to \infty$ with the center at the origin [2, 4, 6, 7]. As the radius of the semicircle increases, sequence R can be selected so that the integral (2.27) over the semicircle tends to zero. By applying the residue theorem, (2.27) can be represented as an infinite series

$$\tau(x,y) = 2\pi i \left\{ \frac{\bar{\tau}(x,y,t_1)}{\Delta'(t_1)} \frac{\left(1+\frac{y}{a}\right)^{-(1+it_1)}}{\left(1-\frac{y}{a}\right)^{1-it_1}} + \sum_{k=2,3} \left[f(x,y,t_k), t_k \right] \right\}, \tag{2.28}$$

where $f(x, y, t_k)$ is the integrand in (2.27), and $t_k = \xi_k - i\eta_k$ are the roots of equation $\Delta(t) = 0$ arranged as per ascending positive values η_k.

The transcendental equation $\Delta(t) = 0$ has four parameters γ_1 and γ_2 and combinations of characteristics of elastic materials h_1 and h_2. Consequently, the behavior of the stress field near the edge $y = a$ $(\alpha = \infty)$ is governed by the magnitude of the imaginary unit of the first simple root $t_1 = \xi_1 - i\eta_1$ of the equation (2.28). If $\eta_1 > 1$, then the stress state near the edge is zero; if $\eta_1 < 1$, the phenomenon of stress concentration occurs. If $\eta_1 = 1$, the stresses at the edges of the contact surface are finite and generally non-zero.

From equation $\Delta(t) = 0$, based on the physical parameters of the problem, it is possible to determine the domains of finite and infinite stresses at the edge of the contact surface and the boundary between these domains. The above boundary is determined by the following function

$$h_2 \sin 2(\gamma_1 + \gamma_2) - \left(h_1 \sin^2 \gamma_2 + h_2 - 1\right) \sin 2\gamma_1 = 0. \tag{2.29}$$

2.4 Specific Cases

According to the designations of parameters h_1 and h_2, equation (2.29), and consequently the shape of the curves of the dependence of geometric parameters γ_1 and γ_2, will be determined by the ratio of the shear moduli $\mu = \mu_2/\mu_1$ and Poisson's ratios ν_1 and ν_2. These curves represent the boundary between the finite and infinite stress domains calculation at the edge of the contact surface. For any pair γ_1, γ_2 from the domain bounded by the coordinate axes and the curve, the stresses at the contact vertex will be finite. Figures 2.2, 2.3 and 2.4 show curves corresponding to

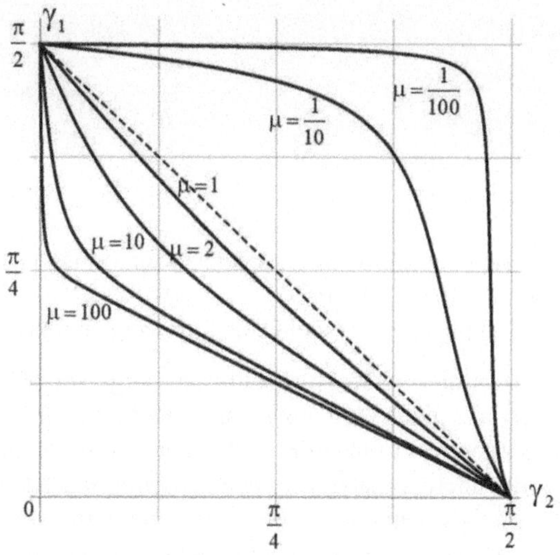

Fig. 2.2 The boundary of the low stress domain at $\nu_1 = 0.2,\ \nu_2 = 0.35$.

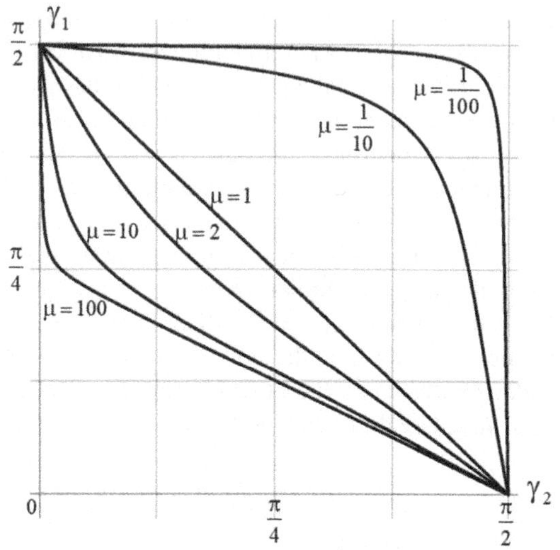

Fig. 2.3 Boundary at $\nu_1 = \nu_2 = 0.25$.

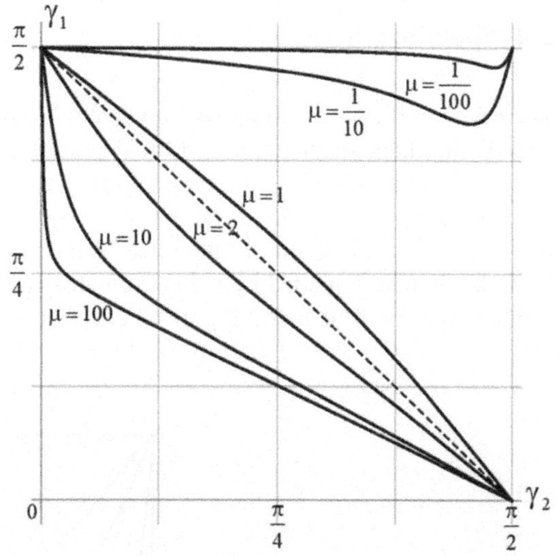

Fig. 2.4 Boundary at $\nu_1 = 0.35$, $\nu_2 = 0.2$.

different values of the parameter μ for values $\nu_1 = 0.2$, $\nu_2 = 0.35$, $\nu_1 = \nu_2 = 0.25$ and $\nu_1 = 0.35$, $\nu_2 = 0.2$.

Graphical results yields the following conclusions:

1. Regardless of Poisson's ratios, the larger the shear modulus of the internal material μ_2 at a given modulus μ_1, the smaller the finite stress domain.
2. Regardless of the ratio of the shear moduli μ, an increase in the Poisson's ratio of the internal material ν_2 at a certain ν_1 leads to a decrease in the domain of finite stresses, while an increase in the Poisson's ratio of the external material ν_1 at a certain ν_2 leads to an increase in this domain.

Figure 2.3 shows that, in the case of a homogeneous plane with a notch, i.e. when $\nu_1 = \nu_2, \mu = 1$, the stresses at the vertices $y = \pm a(\alpha = \pm$ inf will be finite only if the total angle $\gamma_1 + \gamma_2$ does not exceed $\frac{\pi}{2}$. For a non-homogeneous plane, the stress singularities at the notch vertices for angles $\gamma_1 + \gamma_2 > \frac{\pi}{2}$ can be mitigated - or entirely suppressed - through appropriate selection of the constituent material mechanical properties.

2.5 Conclusion

A plane problem for a piecewise-homogeneous plane with circular material boundary lines is solved using a modified system of bipolar coordinates and the integral Fourier transform. The plane is composed of a plane with an opening formed by two partially overlapping circular arcs and two lune-shaped inserts with shared vertices, forming an opening between them. On the free boundary of the lunes, normal and shear stresses are specified, and on the line of conjugation of the lunes with the plane, the full contact terms are met. The solution is obtained by solving a system of two Fredholm integral equations of the second kind. A transcendental equation is derived which, for given physical parameters, defines the functional relationship between geometric parameters that ensures a bounded stress state at the material contact vertex. In the plane of geometric parameters, this equation is a curve separating the domains of change in angles γ_1 and γ_2, when the stresses at the ends of the material separation line have zero or infinite values. The graphical results demonstrate the following: while a homogeneous plane with an opening of $\gamma_1 + \gamma_2 > \frac{\pi}{2}$ exhibits stresses at the vertex tending to infinity, the introduction of material heterogeneity can make this vertex low stress.

References

[1] Aghayan KL, Amirjanyan HA, Zakaryan VG (2016) Anti-plane contact problem for a compound elastic half-space with a finite crack. Proceedings of 8th Internatonal Conference Contemporary Problems of Architecture and Construction, pp 50-53. Yerevan, Armenia,

[2] Aghayan KL, Zakaryan VG (2025) Surface electro-elastic shear waves in piezo-electrical half-space with semi-infinite electrode. In: Altenbach H (ed) Current Developments in Solid Mechanics and Their Applications. Advanced Structured Materials, Vol 223, pp 1-16. Springer, Cham, DOI 10.1007/978-3-031-90022-8_1

[3] Alexanyan RK, Harutyunyan LA, Sedrakyan AM (2024) Ploskaya zadacha dlya vneshnosti krugovogo segmenta (in Russ., Plane problem for the appearance of the circular segment). Slovak International Scientific Journal 80:34-38, DOI 10.5281/zenodo.10651872

[4] Bogy DV (1971) Two edge-bonded elastic wedges of different materials and wedge angles under surface tractions. Journal of Applied Mechanics 38(2):377-386, DOI 10.1115/1.3408786
[5] Cherepanov GP (1983) Mekhanika razrusheniya kompozitsionnykh materialov (in Russ., Fracture Mechanics of Composite Materials). Nauka, Moscow
[6] Chibrikova LE (1956) O krayevoy zadache Rimana dlya avtomorfnykh funktsiy (in Russ., On Riemann boundary value problem for automorphic functions). Uchenyye zapiski Kazanskogo gosudarstvennogo universiteta 116(4):59-110
[7] Chobanyan KS (1987) Nagruzheniya v sostavnykh uprugikh tel (in Russ., Loading in composite elastic bodies). Publishing House AS ASSR, Yerevan
[8] Hakobyan VN (2014) Smeshannyye granichnyye zadachi vzaimodeystviy sploshnykh deformiruyemykh tel s kontsentratorami napryazheniy razlichnykh tipov (in Russ., Mixed boundary value problems of interactions of solid deformable bodies by concentrators of stresses of various types). Gitutyun, NAS, Yerevan
[9] Harutyunyan LA (1976) Ploskaya zadacha teorii uprugosti dlya sostavnoy oblasti, obrazovannoy iz dvukh lunochek (in Russ., Plane problem of elasticity theory for the composite region, formed from two holes. Izvestiya AS ASSR, Mechanics 29(1):51-66
[10] Harutyunyan LA, Edoyan VA (2017) Ploskaya zadacha teorii uprugosti dlya vneshnosti krugovoy lunochki (in Russ., Plane problem of elasticity theory for the appearance of a circular hole). Materialy V International Conference, pp 31-32. Tsaghkadzor, Armenia
[11] Harutyunyan LA, Zakaryan VG (2025) Ploskaya zadacha dlya sostavnoy ploskosti s konechnym otverstiyemPloskaya zadacha dlya sostavnoy ploskosti s konechnym otverstiyem (in Russ., Plane problem for composite plane with finite opening). Materialy IX International Conference, pp 31-35. Tsaghkadzor, Armenia
[12] Muskhelishvily NI (2008) Singular Integral Equations: Boundary Problems of Function Theory and Their Application to Mathematical Physics, Dover Books on Physics, 2nd ed. Dover
[13] Popov GY (1982) Kontsentratsiya uprugikh napryazheniy vozle shtampov, razrezov, tonkikh vklyucheniy i podklyucheniy (in Russ., Concentration of elastic stresses near stamps, cuts, thin inclusions and connections). Nauka, Moscow
[14] Vekua NP (1967) Systems of Singular Integral Equations. Noordhoff, Groningen
[15] Uflyand YS (1950) Bipolyarnyye koordinaty v teorii uprugosti (in Russ., Bipolar coordinates in elasticity). GTTI, Moscow
[16] Uflyand YS Integral'nyye preobrazovaniya v zadachakh teorii uprugosti (in Russ., Integral transformations in the problems of elasticity theory). Nauka, Leningrad

Chapter 3
Analysis of Torsional Vibrations of a Round Stamp Coupled to a Composite Half-Space with an Interphase Round Rigid Inclusion

Harutyun A. Amirjanyan, Aram H. Grigoryan, Avetik V. Sahakyan, and Mezhlum A. Sumbatyan

Abstract An axisymmetric contact problem of torsional vibrations of a circular disk attached to a two-layer half-space with a thin rigid circular inclusion in the interface plane is considered. The problem is reduced to solving a system of integral equations with a kernel in the form of a discontinuous Weber-Schafheitlin integral. The problem is solved by the mechanical quadrature method based on the quadrature rule for the integrals with mentioned kernel. A detailed numerical analysis of the problem is carried out and some specific properties of the influence of some external parameters on the vibrations of both the disk and the inclusion are revealed.

Key words: Torsion vibrations, Contact, Elasticity, Inclusion, Quadrature rule, Weber-Schafheitlin integral, Two-layer half-space, Axisymmetric problem

3.1 Introduction

In the solid mechanics, contact problems take a special place, since contact is the main method of applying loads to a deformable body, and the stress concentration in the contact zone can lead to the destruction of the material. Analytical solutions can only be obtained for a very limited class of contact problems, so it is important to develop numerical and numerical-analytical methods for solving them.

Harutyun A. Amirjanyan · Aram H. Grigoryan · Avetik V. Sahakyan (corresponding author)
Institute of Mechanics, National Academy of Sciences of the Republic of Armenia, 24b Marshal Baghramyan Ave, Yerevan, 0019, Republic of Armenia, e-mail: amirjanyan@gmail.com, grigoryanaram4@gmail.com, avetik.sahakyan@sci.am

Mezhlum A. Sumbatyan
Vorovich Institute of Mathematics, Mechanics and Computer Science, Southern Federal University, Milchakova Street 8a, 344090, Rostov-on-Don, Russian Federation,
e-mail: masumbatyan@sfedu.ru

H. Altenbach and A. V. Sahakyan, *Contact Mechanics and Relevant Problems*,
Advanced Structured Materials 253, https://doi.org/10.1007/978-3-032-20869-9_3

The problem of forced torsional vibrations of an elastic half-space was first solved in [11]. Further, a similar problems were considered for layered and continuously inhomogeneous foundations. The problem of torsion of a two-layer half-space by a rigid circular punch was considered in [6], where the asymptotic method of "large λ" [15] was used to construct the solution, where λ is the ratio of the layer thickness to the punch radius. Among many other works in this area, we note the works [1–3, 5, 8–10, 13, 14]. To solve these problems, various approximate methods were used, which are effective for sufficiently large values λ .

In this paper, we consider the contact problem of torsional vibrations of a circular disk attached to a two-layer half-space with a rigid circular inclusion in the interface plane. The problem is reduced to solving a system of integral equations with a kernel in the form of a discontinuous Weber-Schaffheitlin integral [4]. The solution is constructed using the mechanical quadrature method based on the quadrature rule for integrals with the mentioned kernel.

3.2 Statement of the Problem and Derivation of the Basic Equations

Let us have a compound half-space composed of a layer of thickness H, the material of which has a shear modulus G_1, and a half-space with a shear modulus G_2. In the interface plane there is a round thin rigid inclusion of radius$b$, attached to the matrix. Let us introduce a cylindrical coordinate system (r, φ, z) with the axisOz coinciding with the axis Oz of the Cartesian coordinate system $Oxyz$ and a common origin located on the free surface of the compound half-space above the center of the rigid inclusion (Fig. 3.1).

Let, further, on its upper edge $z = 0$ the layer be coupled to an absolutely rigid circular disk of radius a with its center on the axis Oz, to which harmonically oscillating in time torque $Me^{-i\omega t}$ is applied counterclockwise. It is obvious that some harmonic oscillations of the outer disk and the inner inclusion with an oscillation

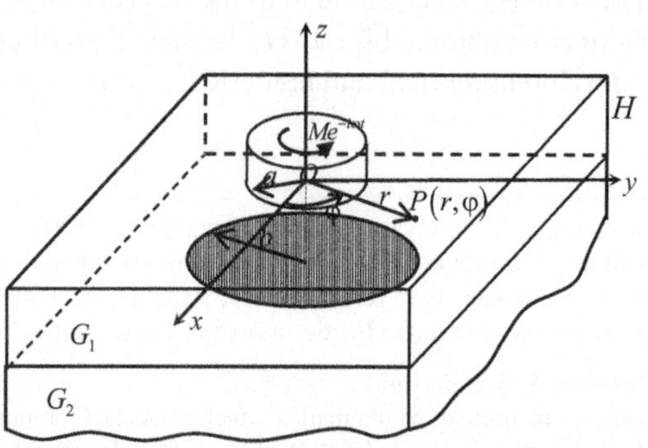

Fig. 3.1 Scheme of the stated problem

frequency ω take place, but at the same time, a phase shift of these oscillations in relation to the oscillations of the torque is possible.

The aim of the present work is to investigate the dependence of the amplitude values and phase shifts of the oscillations of the outer disk and the inner inclusion on such external parameters as the ratio of the linear velocity of the disk boundary points to the velocity of transverse elastic waves in the layer material, the ratio of the shear moduli, the ratio of the velocities of transverse elastic waves in the materials of the layer and the half-space, the ratio of the radii of the disk and the inclusion.

Let us derive the main equations for the problem. Since we are considering a torsion problem with axial symmetry, the displacement components may be written as follows

$$u_r = u_r(r, z, t) = 0, \quad u_z = u_z(r, z, t) = 0, \quad u_\varphi = u_\varphi(r, z, t) \neq 0 \tag{3.1}$$

where u_r, u_z, u_φ are radial, vertical and circular displacements. Furthermore, to determine $u_\varphi(r, z, t)$, we have the following differential equation

$$\frac{\partial^2 u_\varphi}{\partial r^2} + \frac{1}{r}\frac{\partial u_\varphi}{\partial r} + \frac{\partial^2 u_\varphi}{\partial z^2} - \frac{u_\varphi}{r^2} - \frac{1}{c_2^2}\frac{\partial^2 u_\varphi}{\partial t^2} = 0\,, \quad c_2^2 = \frac{G}{\rho} \tag{3.2}$$

The problem under consideration involves steady-state torsional oscillations with a frequency ω overall, i.e., all components of the stress-strain state depend on time through the factor $e^{-\imath\omega t}$. Having taken this factor into account in the inertial term, we will omit it below and deal with the amplitude values of all quantities, retaining their notations.

Next, we reduce the solution of the contact problem described above to the solution of an integral equation. To this end, let us first construct the solution to the following auxiliary mixed boundary value problem

$$\begin{cases} \dfrac{\partial^2 u_\varphi^{(1)}}{\partial r^2} + \dfrac{1}{r}\dfrac{\partial u_\varphi^{(1)}}{\partial r} + \dfrac{\partial^2 u_\varphi^{(1)}}{\partial z^2} - \dfrac{u_\varphi^{(1)}}{r^2} + \left(\dfrac{\omega}{c_2^{(1)}}\right)^2 u_\varphi^{(1)} = 0, \\ \tau_{z\varphi}^{(1)}\Big|_{z=0} = G_1 \dfrac{\partial u_\varphi^{(1)}}{\partial z}\Big|_{z=0} = \tau\,(r) \end{cases} \tag{3.3}$$

with $0 < r < \infty$, $-H < z < 0$ and for $-\infty < z < -H$

$$\begin{cases} \dfrac{\partial^2 u_\varphi^{(2)}}{\partial r^2} + \dfrac{1}{r}\dfrac{\partial u_\varphi^{(2)}}{\partial r} + \dfrac{\partial^2 u_\varphi^{(2)}}{\partial z^2} - \dfrac{u_\varphi^{(2)}}{r^2} + \left(\dfrac{\omega}{c_2^{(2)}}\right)^2 u_\varphi^{(2)} = 0, \\ u_\varphi^{(2)}, \tau_{z\varphi}^{(2)} \to 0, \qquad z \to -\infty, \\ u_\varphi^{(1)}\Big|_{z=-H} - u_\varphi^{(2)}\Big|_{z=-H} = 0; \quad \tau_{z\varphi}^{(1)}\Big|_{z=-H} - \tau_{z\varphi}^{(2)}\Big|_{z=-H} = g_\tau\,(r)\,, \end{cases} \tag{3.4}$$

where $\tau(r)$ and $g_\tau(r)$ are considered to be some known finite functions defined on intervals $(0,a)$ and $(0,b)$, respectively, $c_2^{(i)}$ $(i=1,2)$ are velocities of elastic transverse waves in the layer and the half-space. We will solve the boundary value problems (3.3) - (3.4) using the Hankel integral transform method, for which we introduce the Hankel transforms into consideration, as follows

$$\left\{\overline{u_\varphi^{(i)}}(\lambda,z);\ \overline{\tau_{z\varphi}^{(i)}}(\lambda,z)\right\}=\int\limits_0^\infty\left\{u_\varphi^{(i)}(r,z);\ \tau_{z\varphi}^{(i)}(r,z)\right\}J_1(\lambda r)\,rdr\quad(i=1,2),\tag{3.5}$$

where $J_1(x)$ is the Bessel function of the first kind, and λ $(0<\lambda<\infty)$ is a Hankel spectral parameter. Taking into account the known formula, for example see [16],

$$\int\limits_0^\infty\left(\frac{d^2f}{dr^2}+\frac{1}{r}\frac{df}{dr}-\frac{n^2}{r^2}f\right)J_n(pr)rdr=-p^2\bar{f}(p)$$

the two-dimensional mixed boundary value problem (3.3) - (3.4) is transformed to the following boundary value problem with respect to the Hankel transforms

$$\frac{d^2\overline{u_\varphi^{(i)}}}{dz^2}-\left(\lambda^2-\left(\frac{\omega}{c_2^{(i)}}\right)^2\right)\overline{u_\varphi^{(i)}}=0\quad\begin{cases}l-H<z<0 & \text{for } i=1,\\ -\infty<z<-H & \text{for } i=2,\end{cases}\tag{3.6}$$

$$\begin{gathered}G_1\left.\frac{\partial\overline{u_\varphi^{(1)}}}{\partial z}\right|_{z=0}=\bar{\tau}(\lambda);\quad\left.\overline{u_\varphi^{(1)}}\right|_{z=-H}-\left.\overline{u_\varphi^{(2)}}\right|_{z=-H}=0,\\ G_1\left.\frac{\partial\overline{u_\varphi^{(1)}}}{\partial z}\right|_{z=-H}-G_2\left.\frac{\partial\overline{u_\varphi^{(2)}}}{\partial z}\right|_{z=-H}=\overline{g_\tau}(\lambda).\end{gathered}\tag{3.7}$$

The solution to the boundary value problem (3.6) - (3.7) has the following form

$$\begin{aligned}\overline{u_\varphi^{(1)}}(\lambda,z)&=-\frac{\operatorname{ch}\gamma_1 z}{\Delta}\frac{\bar{g}_\tau}{G_1}+\frac{\gamma_1\operatorname{ch}\gamma_1(z+H)+G_*\gamma_2\operatorname{sh}\gamma_1(z+H)}{\Delta}\frac{\bar{\tau}}{G_1\gamma_1},\\ \overline{u_\varphi^{(2)}}(\lambda,z)&=-\frac{e^{\gamma_2(z+H)}\operatorname{ch}\gamma_1 H}{\Delta}\frac{\bar{g}_\tau}{G_1}+\frac{e^{\gamma_2(z+H)}}{\Delta}\frac{\bar{\tau}}{G_1}.\end{aligned}\tag{3.8}$$

where $G_*=G_2/G_1$,

$$\Delta=\gamma_1\operatorname{sh}\gamma_1 H+G_*\gamma_2\operatorname{ch}\gamma_1 H,$$

$$\bar{\tau}(\lambda)=\int\limits_0^a\tau(r)J_1(\lambda r)\,rdr,\quad\bar{g}_\tau(\lambda)=\int\limits_0^b g_\tau(r)J_1(\lambda r)\,rdr,$$

and

$$\gamma_j(\lambda) = \begin{cases} \sqrt{\lambda^2 - \left(\frac{\omega}{c_2^{(j)}}\right)^2}, & \frac{\omega}{c_2^{(j)}} \le \lambda < \infty, \\ -i\sqrt{\left(\frac{\omega}{c_2^{(j)}}\right)^2 - \lambda^2}, & 0 \le \lambda < \frac{\omega}{c_2^{(j)}}, \end{cases} \quad (j = 1, 2)$$

Applying the inverse Hankel transform to (3.8), in particular, for points on the free surface of the layer and points on the interface plane between the materials, we obtain

$$\begin{aligned} u_\varphi^{(1)}(r,0) &= \frac{1}{G_1}\int_0^a K_{11}(s,r)\,\tau(s)s\,ds - \frac{1}{G_1}\int_0^b K(s,r)\,g_\tau(s)s\,ds, \\ u_\varphi^{(2)}(r,-H) &= \frac{1}{G_1}\int_0^a K(s,r)\,\tau(s)s\,ds - \frac{1}{G_1}\int_0^b K_{22}(s,r,z)\,g_\tau(s)s\,ds, \end{aligned} \tag{3.9}$$

where

$$\begin{aligned} K(s,r) &= \int_0^\infty \frac{\lambda J_1(\lambda r)\,J_1(\lambda s)}{\Delta}\,d\lambda, \\ K_{11}(s,r) &= \int_0^\infty \frac{\gamma_1 \operatorname{ch}\gamma_1 H + G_* \gamma_2 \operatorname{sh}\gamma_1 H}{\Delta}\frac{\lambda}{\gamma_1} J_1(\lambda r)\,J_1(\lambda s)\,d\lambda, \\ K_{22}(s,r) &= \int_0^\infty \frac{\operatorname{ch}\gamma_1 H}{\Delta}\lambda J_1(\lambda r)\,J_1(\lambda s)\,d\lambda. \end{aligned} \tag{3.10}$$

Under the disc and the inclusion, we have the following conditions

$$\begin{aligned} & u_\varphi^{(1)}(r,0) = \delta_1 r \quad (0 < r < a), \\ & u_\varphi^{(1)}(r,-H) = u_\varphi^{(2)}(r,-H) = \delta_2 r \quad (0 < r < b), \end{aligned} \tag{3.11}$$

where δ_1 and δ_2 are amplitude values of rotation angles of the disc and the inclusion. Conditions (3.11) lead to the following equations

$$\begin{aligned} & \int_0^a K_{11}(s,r)\,\tau(s)s\,ds - \int_0^b K(s,r)\,g_\tau(s)s\,ds = G_1\delta_1 r \qquad (0 < r < a), \\ & \int_0^a K(s,r)\,\tau(s)s\,ds - \int_0^b K_{22}(s,r)\,g_\tau(s)s\,ds = G_1\delta_2 r \qquad (0 < r < b). \end{aligned} \tag{3.12}$$

In addition to the system of equations (3.12), the conditions of dynamic equilibrium of the disc:

$$\int_0^a \tau(r)\, r^2 dr = \frac{M}{2\pi} \tag{3.13}$$

and of the inner inclusion

$$\int_0^b g_\tau(s) s^2 ds = 0 \tag{3.14}$$

should also be satisfied. The system of equations (3.12) - (3.14) is the governing system of equations for the problem, allowing us to find both the distribution of contact tangential stresses under the disk and the jump in contact stresses at the inclusion, as well as the angles of rotation of the disk and the inclusion.

3.3 Algorithm for Solving the Governing System

Let us pass to dimensionless quantities and reduce the integration intervals to the interval (0, 1), then we obtain:

$$\begin{aligned}
&\int_0^1 Q_{11}(\xi,\eta)\,\varphi_1(\xi)\xi d\xi - b_*^2 \int_0^1 Q(b_*\xi,\eta)\,\varphi_2(\xi)\xi d\xi = \delta_1\eta \quad (0<\eta<1),\\
&\int_0^1 Q(\xi,b_*\eta)\,\varphi_1(\xi)\,\xi d\xi - b_*^2 \int_0^1 Q_{22}(b_*\xi,b_*\eta)\,\varphi_2(\xi)\xi d\xi = \delta_2 b_*^2\eta \quad (0<\eta<1),
\end{aligned} \tag{3.15}$$

where

$$\varphi_1(\xi) = \frac{\tau(a\xi)}{G_1}; \quad \varphi_2(\xi) = \frac{g_\tau(b\xi)}{G_1}; \; k = \frac{a\omega}{c_2^{(1)}}; \quad \gamma_{1*} = \begin{cases} \sqrt{\alpha^2-k^2}, \\ -i\sqrt{k^2-\alpha^2}, \end{cases}$$

$$\gamma_{2*} = \begin{cases} \sqrt{\alpha^2-\theta^2k^2} \\ -i\sqrt{\theta^2k^2-\alpha^2} \end{cases}, \quad \theta = \frac{c_2^{(1)}}{c_2^{(2)}}, \quad H_* = \frac{H}{a}, \quad b_* = \frac{b}{a},$$

$$Q_{11}(\xi,\eta)=\int\limits_0^\infty \frac{\gamma_{1*}\operatorname{ch}\gamma_{1*}H_*+G_*\gamma_{2*}\operatorname{sh}\gamma_{1*}H_*}{\gamma_{1*}\operatorname{sh}\gamma_{1*}H_*+G_*\gamma_{2*}\operatorname{ch}\gamma_{1*}H_*}\frac{\alpha}{\gamma_{1*}}J_1(\alpha\xi)J_1(\alpha\eta)\,d\alpha,$$

$$Q(\xi,\eta)=\int\limits_0^\infty \frac{\alpha J_1(\alpha\xi)J_1(\alpha\eta)}{\gamma_{1*}\operatorname{sh}\gamma_{1*}H_*+G_*\gamma_{2*}\operatorname{ch}\gamma_{1*}H_*}\,d\alpha, \tag{3.16}$$

$$Q_{22}(\xi,\eta)=\int\limits_0^\infty \frac{\operatorname{ch}\gamma_{1*}H_*}{\gamma_{1*}\operatorname{sh}\gamma_{1*}H_*+G_*\gamma_{2*}\operatorname{ch}\gamma_{1*}H_*}\alpha J_1(\alpha\xi)J_1(\alpha\eta)\,d\lambda.$$

By evaluating the behavior of the integrands of these kernels at infinity, we find that the middle kernel is regular, while the first and third kernels have a singularity in the form of a discontinuous Weber-Schafheitlin integral $L_1(\xi,\eta)$, and they can be represented as:

$$\begin{aligned} Q_{11}(\xi,\eta) &= L_1(\xi,\eta)+R_{11}(\xi,\eta),\\ Q_{22}(b_*\xi,b_*\eta) &= \frac{1}{(1+G_*)\,b_*}L_1(\xi,\eta)+R_{22}(b_*\xi,b_*\eta), \end{aligned} \tag{3.17}$$

where

$$L_1(\xi,\eta)=\int\limits_0^\infty J_1(\alpha\eta)J_1(\alpha\xi)\,d\alpha,$$

$$R_{11}(\xi,\eta)=\int\limits_0^\infty \left(\frac{\gamma_{1*}+G_*\gamma_{2*}\tanh\gamma_{1*}H_*}{\gamma_{1*}\tanh\gamma_{1*}H_*+G_*\gamma_{2*}}\frac{\alpha}{\gamma_{1*}}-1\right)J_1(\alpha\xi)J_1(\alpha\eta)\,d\alpha,$$

$$R_{22}(\xi,\eta)=\int\limits_0^\infty \left(\frac{\alpha}{\gamma_{1*}\tanh\gamma_{1*}H_*+G_*\gamma_{2*}}-\frac{1}{1+G_*}\right)J_1(\alpha\xi)J_1(\alpha\eta)\,d\alpha.$$

Finally, we write the system of equations in the following form

$$\int_0^1 \left[L_1(\xi,\eta) + R_{11}(\xi,\eta)\right] \varphi_1(\xi)\xi d\xi - b_*^2 \int_0^1 Q(b_*\xi,\eta)\,\varphi_2(\xi)\xi d\xi = \delta_1\eta \qquad (0<\eta<1),$$
$$\int_0^1 Q(\xi,b_*\eta)\,\varphi_1(\xi)\xi d\xi - \int_0^1 \left[\frac{b_*}{1+G_*}L_1(\xi,\eta) + b_*^2 R_{22}(b_*\xi,b_*\eta)\right]\varphi_2(\xi)\xi d\xi = \delta_2 b_*^2\eta. \tag{3.18}$$

Condition (3.13) is transformed to the following one

$$\int_0^1 \varphi_1(\xi)\,\xi^2 d\xi = \frac{M}{2\pi G_1 a^3} = M_* \tag{3.19}$$

and condition (3.14) takes the following form

$$\int_0^1 \varphi_2(\xi)\xi^2 d\xi = 0. \tag{3.20}$$

Taking into account that the contact stresses under the disc and the stress jump on the inclusion have a square root singularity, let us seek the solution to the system (3.18) - (3.20) in the following form

$$\varphi_k(\xi) = \frac{\varphi_{*k}(\xi)}{\sqrt{1-\xi^2}} \qquad (k=1,2), \tag{3.21}$$

where $\varphi_{*k}(\xi)$ $(k=1,2)$ are smooth functions bounded on the segment $[0,1]$. Let us use the quadrature rules [7], which are a special case of a more general quadrature rule obtained in [12],

$$\int_0^1 L_1(r,\rho)\frac{\rho}{\sqrt{1-\rho^2}} f(\rho)\,d\rho = \sum_{i=1}^{n} w_i f(x_i) T(r,i),$$

$$\int_0^1 Q(r,\rho)\frac{\rho}{\sqrt{1-\rho^2}} f(\rho)\,d\rho = \sum_{i=1}^{n} w_i \frac{Q(r,x_i) f(x_i)}{x_i}, \tag{3.22}$$

$$\int_0^1 \frac{\rho^2}{\sqrt{1-\rho^2}} f(\rho)\,d\rho = \sum_{i=1}^{n} w_i f(x_i),$$

where

$$L_1(r,\rho) = \int_0^\infty J_1(rt) J_1(\rho t)\,dt,$$

$$w_i = \frac{2(4n+1)}{(2n+3)(4n^2-1)\,x_i P_{n-1}^{(2,\frac{1}{2})}(\xi_i)\, P_{n-1}^{(1,-\frac{1}{2})}(\xi_i)},$$

$$x_i = \sqrt{\frac{1-\xi_i}{2}}, \quad P_n^{(1,-\frac{1}{2})}(\xi_i) = 0,$$

$$T(r,i) = r\sum_{m=0}^{n-1} \frac{(4m+3)}{2}\left(\frac{\Gamma\left(m+\frac{3}{2}\right)}{\Gamma(m+2)}\right)^2 P_m^{(1,-\frac{1}{2})}(\xi_i)\, P_m^{(1,-\frac{1}{2})}\left(1-2r^2\right).$$

Using the quadrature rules (3.22), the system of equations (3.18) and the conditions (3.19) - (3.20) can be written in the form

$$\sum_{i=1}^{n} w_i \varphi_{*1}(x_i)\left[T(\eta,i) + \frac{1}{x_i} R_{11}(x_i,\eta)\right] - b_*^2 \sum_{i=1}^{n} w_i \varphi_{*2}(x_i) \frac{Q(b_* x_i,\eta)}{x_i} = \delta_1 \eta,$$

$$\sum_{i=1}^{n} w_i \varphi_{*1}(x_i) \frac{Q(x_i,b_*\eta)}{x_i}$$

$$-\sum_{i=1}^{n} w_i \varphi_{*2}(x_i)\left[\frac{b_*}{1+G_*} T(\eta,i) + \frac{b_*^2}{x_i} R_{22}(b_* x_i, b_*\eta)\right] = \delta_2 b_*^2 \eta,$$

$$\sum_{i=1}^{n} w_i \varphi_{*1}(x_i) = M_*, \qquad \sum_{i=1}^{n} w_i \varphi_{*2}(x_i) = 0.$$

By choosing the nodes of the quadrature formula as just the same collocation points, we obtain the following closed system of linear algebraic equations with respect to $2n$ values of $\varphi_{*k}(x_i)$ $(k = 1,2;\ i = 1,n)$, as well as the angles of rotation of the disc δ_1 and the inclusion δ_2

$$\begin{gathered}
\sum_{i=1}^{n} w_i \varphi_{*1}(x_i)\left[T(x_k, i) + \frac{1}{x_i} R_{11}(x_i, x_k)\right] + b_*^2 \sum_{i=1}^{n} w_i \varphi_{*2}(x_i) \frac{K_{12}(x_i, x_k)}{x_i} \\
- \delta_1 x_k = 0, \\
\sum_{i=1}^{n} w_i \varphi_{*1}(x_i) \frac{K_{21}(x_i, x_k)}{x_i} - \frac{b_*^2}{1+G_*} \sum_{i=1}^{n} w_i \varphi_{*2}(x_i)\left[T(x_k, i) + \frac{1+G_*}{x_i} R_{22}(x_i, x_k)\right] \\
- \delta_2 b_*^2 x_k = 0, \\
\sum_{i=1}^{n} w_i \varphi_{*1}(x_i) = M_*, \qquad \sum_{i=1}^{n} w_i \varphi_{*2}(x_i) = 0 \qquad (k = 1, \ldots, n).
\end{gathered} \tag{3.23}$$

As soon as the coefficients $\varphi_{*k}(x_i)$ $(k = 1, 2;\ i = 1, n)$ are determined, the sought functions are reconstructed using the formulas

$$\varphi_k(\xi) = \frac{\xi}{2\sqrt{1-\xi^2}} \sum_{j=1}^{n} w_j \varphi_{*k}(x_j) \sum_{m=0}^{n-1} \frac{(4m+3)(2m+1)}{(m+1)} P_m^{\left(1,-\frac{1}{2}\right)}\left(1-2\xi^2\right) P_m^{\left(1,-\frac{1}{2}\right)}$$

with $k = 1, 2$.

It should be noted that the presence of functions $\gamma_{j*}(\alpha)$ $(j = 1, 2)$ in the regular kernels of system (3.15) that take both real and imaginary values leads to complex solutions of system (3.23). By representing complex numbers δ_k $(k = 1, 2)$ in exponential form, we can see that the moduli of these numbers represent the amplitudes of the angles of rotation of the disk and the inclusion, while the arguments represent the phase shift of their oscillations with respect to the oscillations of the torque.

3.4 Numerical Analysis

Let us perform a numerical analysis on dependence of the amplitude values of rotation angles of the disk and the inclusion, and the phase shifts of their oscillations relatively the oscillations of the torque $Me^{-\imath\omega t}$ applied to the disk on various external parameters of the problem. Let us assume the layer thickness to be constant and equal to the radius of the disk ($H_* = 1$), and the dimensionless amplitude of the torque also be considered constant and equal to the unit value ($M_* = 1$). For the remaining external parameters, we consider the following ranges of variation: $0 \le k \le 1$, $1 \le \theta \le 3$, $0.2 \le G_* \le 3$, $0.2 \le b_* \le 4$.

Figures 3.2 and 3.3 show the graphs of the disk rotation angle $|\delta_1|$ and of the inclusion rotation angle $|\delta_2|$ as a function of the torque oscillation frequency k for different values of the parameters G_*, θ, b_*. In Figs. 3.2 and 3.3, the curves corresponding to the sequence of values of the varying parameter are arranged from the top to the bottom. In the case of intersecting curves, follow their arrangement in the vicinity of $k = 1$. Based on the graphs in Figs. 3.2 and 3.3, the following conclusions can be drawn:

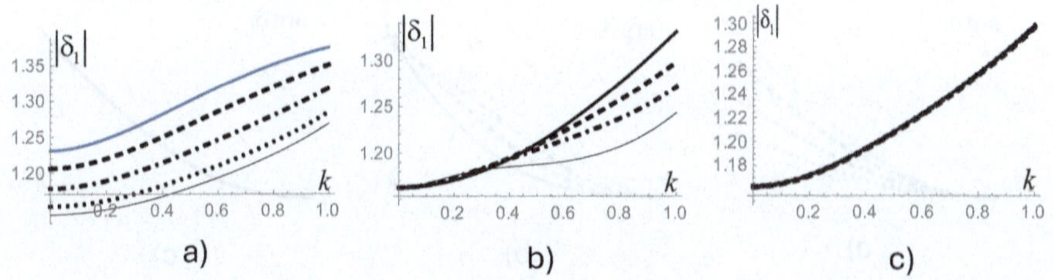

$G_* = \{0.25, 0.5, 1, 2, 3\}$, $\theta = 1.5$, $b_* = 2$

$\theta = \{1, 1.5, 2, 3\}$, $G_* = 1.5$, $b_* = 2$

$b_* = \{0.25, 0.5, 1, 2, 3\}$, $\theta = 1.5$, $G_* = 1.5$

Fig. 3.2: Graphs of the disk rotation angle versus frequency k.

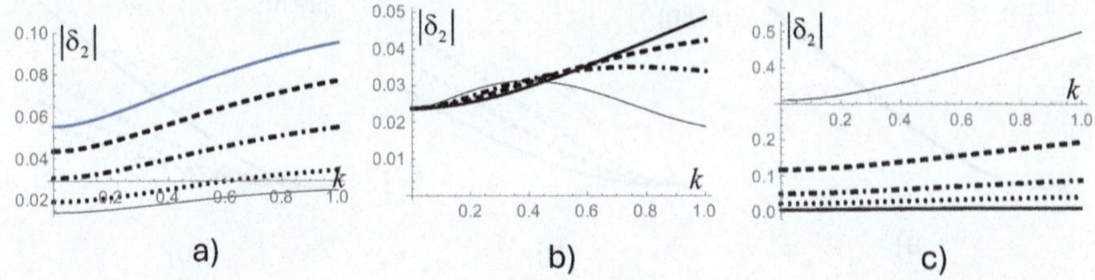

$G_* = \{0.25, 0.5, 1, 2, 3\}$, $\theta = 1.5$, $b_* = 2$

$\theta = \{1, 1.5, 2, 3\}$, $G_* = 1.5$, $b_* = 2$

$b_* = \{0.25, 0.5, 1, 2, 3\}$, $\theta = 1.5$, $G_* = 1.5$

Fig. 3.3: The inclusion rotation angle versus frequency k.

1. Increasing the ratio of the shear modulus of the half-space to the shear modulus of the layer, while keeping the ratio of the velocities of the transverse elastic waves in the layer to the velocity of these waves in the half-space constant, leads to a decrease in the angles of rotation of both the disk and the inclusion, and in the latter case – to a significantly more substantial decrease;
2. The influence of the ratio of the velocities of the transverse elastic waves in the layer and the half-space, with the same value of the ratio of elastic moduli, is more significant at high frequencies of torque oscillations, and in this case, the conclusion in point 1 remains valid;
3. The magnitude of the inclusion radius has practically no effect on the dependence of the disc's angle of rotation on the frequency of torque oscillations, while the angle of rotation of inclusion, along with an increase of its radius, decreases quite sharply, tending towards zero and smoothing out its dependence on the frequency of torque oscillations.

Figures 3.4 and 3.5 show the graphs of the phase shift of the disk oscillations $\arg \delta_1$ and of the inclusion oscillations $\arg \delta_2$ as a function of the torque oscillation frequency k, for different values of the parameters G_*, θ, b_*. Particularly noteworthy, among the graphs in Figs. 3.4 and 3.5, are the graphs in Figs. 3.4c and 3.5a,

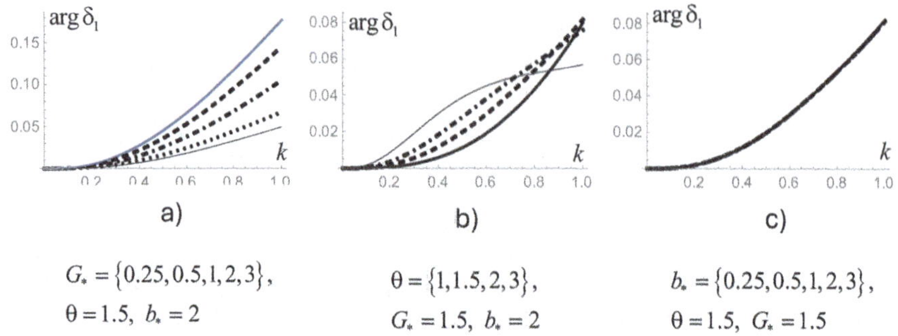

Fig. 3.4: The phase shift of the disk oscillations versus frequency k.

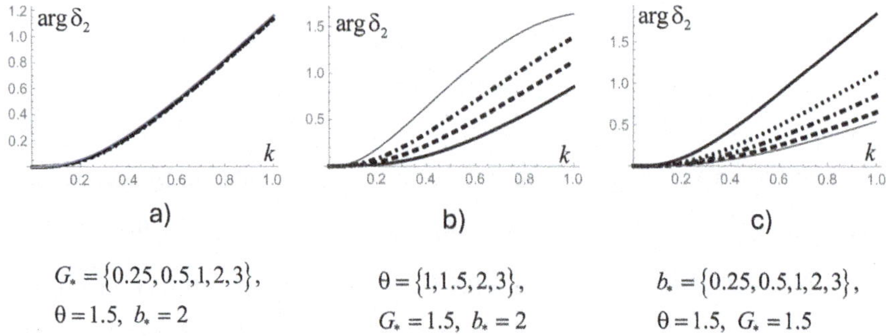

Fig. 3.5: The phase shift of the inclusion oscillations versus frequency k.

which indicate the practical independence of the phase shift of the disk oscillations versus the inclusion radius and the phase shift of the inclusion oscillations versus the ratio of the shear moduli. Taking into account that, according to Figs. 3.2c and 3.4c, the radius of the inclusion practically does not affect the disk's oscillations, let us investigate the dependence of the oscillations of the disk and the inclusion on the radius of the inclusion at different values of other parameters.

Figures 3.6 and 3.7 show the graphs of the rotation angle of the disk $|\delta_1|$ and of the inclusion $|\delta_2|$ versus the inclusion radius b_* with the frequency of the torque oscillations $k = 0.5$ and different values of parameters G_* and θ.

Figure 3.8(a,b) and 3.9(a,b) show the graphs of the phase shift of the disk oscillations $\arg \delta_1$ and of the inclusion oscillations $\arg \delta_2$ as a function of the inclusion radius b_*, with the frequency of the torque oscillations $k = 0.5$ and different values of parameters G_* and θ.

From Figs. 3.6 - 3.9, we observe that changing the radius of the inclusion does not affect the phase deviation of the disk oscillations and practically does not affect the amplitude of its oscillations. The seemingly strong dependence of the disk oscillation amplitude on the inclusion radius in the vicinity of the value $b_* = 1$ shown in Fig. 3.7a is a consequence of a more detailed scale on the ordinate axis.

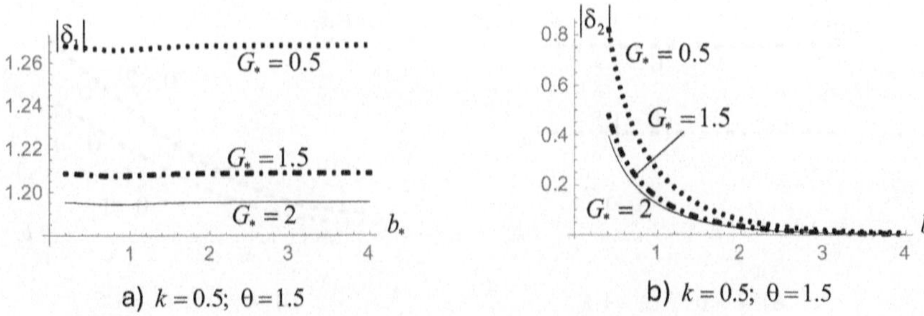

Fig. 3.6: Rotation angle of the disk and of the inclusion versus radius b_*.

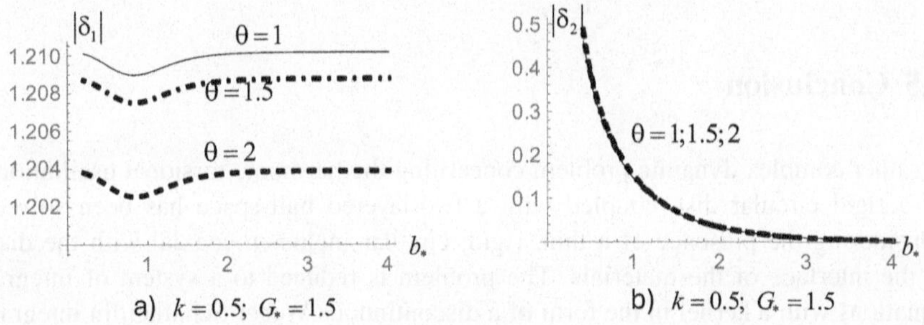

Fig. 3.7: Rotation angle of the disk and of the inclusion versus radius b_*, for another combination of parameters.

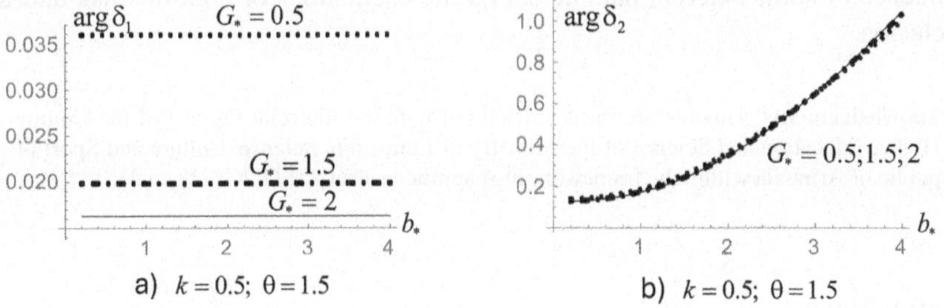

Fig. 3.8: The phase shift of oscillations of the disk and of the inclusion versus radius b_*.

It is also very interesting that the parameter θ, at a certain value G_*, does not influence the dependence of the amplitude of the inclusion oscillations on its radius, and that the parameter G_*, at a certain value θ, does not influence the dependence of the phase shift of the inclusion oscillations on its radius.

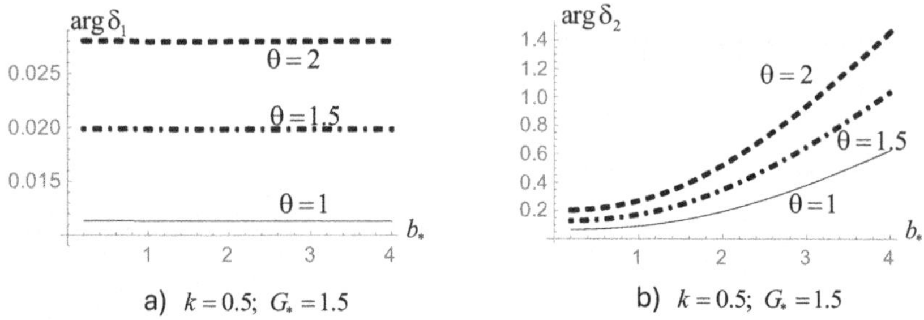

Fig. 3.9: The phase shift of oscillations of the disk and of the inclusion versus radius b_*, for another combination of parameters.

3.5 Conclusion

A rather complex dynamic problem concerning the harmonic torsional oscillations of a rigid circular disk coupled with a two-layered half-space has been solved, considering the presence of a thin, rigid, circular inclusion coaxial with the disk at the interface of the materials. The problem is reduced to a system of integral equations with a kernel in the form of a discontinuous Weber-Schafheitlin integral. To calculate the integrals with such a kernel, we use a specific quadrature formula. This formula allowed us, by using the method of mechanical quadratures, to conduct a fairly detailed numerical analysis of the problem and to reveal the features of the influence of some external parameters on the oscillations of both the disk and the inclusion.

Acknowledgements This research was carried out with the financial support of the Committee of Higher Education and Science of the Ministry of Education, Science, Culture and Sport of the Republic of Armenia within the framework of scientific project 25RG-2C221.

References

[1] Aizikovich SM (1978) Krucheniye kruglym shtampom neodnorodnogo poluprostranstva (in Russ., Torsion of a non- homogeneous half-space by a round stamp). In: Raschet obolochek i plastin. Rostov-on-Don, RISI, pp. 156-169

[2] Aizikovich SM, Vasiliev AS (2013) A bilateral asymptotic method of solving the integral equation of the contact problem of the torsion of an elastic half-space inhomogeneous in depth. Journal of Applied Mathematics and Mechanics 77(1):91-97, DOI 10.1016/j.jappmathmech.2013.04.011

[3] Bufler H (1963) Die Torsion der inhomogenen dicken Platte. ZAMM - Journal of Applied Mathematics and Mechanics / Zeitschrift für Angewandte Mathe-

matik und Mechanik 43(9):389-401, DOI 10.1002/zamm.19630430902

[4] Copson ET (1933) On the discontinuous integral of Weber and Schafheitlin. The Quarterly Journal of Mathematics os-4(1):134-139, DOI 10.1093/qmath/os-4.1.134

[5] Dhaliwal RS, Singh BM, Vrbik J, Khan SM (1984) Torsional oscillations by an annular disk on a semi-infinite cylinder embedded in an elastic half-space. Engineering Fracture Mechanics 20(5-6):719–727, DOI 10.1016/0013-7944(84)90081-X

[6] Grilitski DV (1961) Krucheniye dvukhsloynoy uprugoy sredy (in Russ., Torsion of a two-layered elastic media). Prikladnaya Mechanika 7(1):89-95

[7] Hakobyan VV, Gasparyan AV, Sahakyan AV (2025) Mechanical quadratures for solution of the problem of torsion of an inhomogeneously aging viscoelastic layer by a round rigid washer. In: Altenbach H (ed) Current Developments in Solid Mechanics and Their Applications, Advanced Structured Materials, Vol 223, pp. 609-620, DOI 10.1007/978-3-031-90022-8_36

[8] Kanetsyan EG, Mkrtchyan MS, Mkhitaryan SM (2022) Ob osesimmetrichnom kruchenii uprugogo sloya posredstvom dvukh absolyutno zhestkikh krugovykh tsilindricheskikh shtampov (in Russ., On axisymmetric torsion of an elastic layer by means of two absolutely rigid circular cylindrical punches). Reports NAS RA, Mecanics 122(4):265-276, DOI 0.54503/0321-1339-2022.122.4-265

[9] Mukherjee S (1984) Torsion problem of a circular die in a non-homogeneous elastic layer perfectly bonded to a non-homogeneous elastic layer rigidly fixed at the other end. Indian Journal of Theoretical Physics 32(1):39-46

[10] Nikishin VS (1976) Osesimmetrichnyye kontaktnyye zadachi teorii uprugosti dlya neodnorodnykh sred (in Russ., Axisymmetric contact problems of elasticity theory for inhomogeneous media). In: Soobshcheniya po prikladnoy matematike (3):51-103

[11] Reissner E., Sagoci HF (1944) Forced torsional oscillations of an elastic half-space, I. Journal of Applied Physics 15(9):652–654, DOI 10.1063/1.1707489

[12] Sahakyan AV (2025) Kvadraturnaya formula dlya integrala s yadrom, porozhdennym integralom Vebera – Shafkheytlina, v primenenii k resheniyu osesimmetrichnoy kontaktnoy zadachi teorii uprugosti (in Russ., Quadrature formula for an integral with a kernel generated by the Weber–Schafheitlin integral, applied to solving an axisymmetric contact problem of elasticity theory). Reports NAS RA, Mechanics 78(3-4): in print

[13] Singh BM, Danyluk HT, Selvadurai APS (1989) The Reissner-Sagoci problem for a half-space with a surface constraint. Journal of Applied Mathematics and Physics (ZAMP) 40:762–768, DOI 10.1007/BF00945876

[14] Vasilyev AS, Sadyrin EV, Fedotov IA (2013) Kontaktnaya zadacha o kruchenii kruglym shtampom transversal'no-izotropnogo uprugogo polupostranstva c neodnorodnymi transversal'no-izotropnym pokrytiem (in Russ., Contact problem on torsion of transversely isotropic elastic half-space with inhomogeneous transversely isotropic coating by round die). Vestnik of Don State Technical University 13(1-2):25-34

[15] Vorovich II, Alexandrov VM, Babeshko VA (1974) Neklassicheskiye smeshannyye zadachi teorii uprugosti (in Russ., Non-classical mixed problems of the theory of elasticity). Nauka, Moscow

[16] Zhislin GM (2012) Integral'nyye preobrazovaniya v zadachakh matematicheskoy fiziki (in Russ., Integral transformations in problems of mathematical physics. Electronic teaching and methodological manual). Nizhny Novgorod

Chapter 4
Mechanical Properties of Ultrahigh Molecular Weight Polyethylene After Natural and Deformation Aging

Alexander R. Arutyunyan and Anna V. Vardanyan

Abstract The effect of natural and deformation aging on changes of mechanical properties under tension and creep conditions of ultrahigh molecular weight polyethylene (UHMWPE) specimens were investigated. The aging of specimens was carried out according to the following programs: natural aging of specimens in laboratory conditions for 11 months (program 1) and natural aging of specimens in laboratory conditions for 5 months, deformation aging of specimens in fatigue experiments, followed by natural aging of specimens in laboratory conditions for 6 months (program 2). Aging according to programs 1 and 2 did not lead to a change to ultimate tensile strength of the UHMWPE specimens. However, there is a significant decrease in deformation at the fracture moment. For UHMWPE specimens after aging according to programs 1 and 2, a decrease of the creep fracture time compared to specimens without aging is observed. At the same time, the reduction of the creep fracture time is more significant for specimens after aging according to program 2. Thus, the deformation aging leads to a more significant degradation of the properties of ultrahigh molecular weight polyethylene.

Key words: Polymer materials, Ultrahigh molecular weight polyethylene, Natural aging, Deformation aging, Mechanical properties, Tensile experiments, Creep experiments

4.1 Introduction

In recent years, various polymer materials have become increasingly widespread and developed. They are being implemented in almost all spheres of life, industry and construction, in particular, in aircraft, rocket, automotive, shipbuilding, railway

Alexander R. Arutyunyan (corresponding author) · Anna V. Vardanyan
St. Petersburg State University, 7/9 Universitetskaya Emb., St. Petersburg, 199034, Russian Federation, e-mail: a.arutyunyan@spbu.ru, vardanyan_a@yahoo.com

H. Altenbach and A. V. Sahakyan, *Contact Mechanics and Relevant Problems*,
Advanced Structured Materials 253, https://doi.org/10.1007/978-3-032-20869-9_4

transport, in the manufacture of agricultural machinery and sports equipment. The advantages of such materials compared to metals are lightness, heat resistance, corrosion resistance and resistance to aggressive environments such as alkalis or concentrated acids. However, polymer materials are less stable than metals, so the aging and fracture processes in these materials are more intense.

Because these materials are used in critical areas of engineering practice, this makes their long-term aging, creep and fatigue characteristics of paramount importance. At the same time, the physic-chemical characteristics of polymer materials change after prolonged use, which is largely due to the aging process [1, 3–5, 7, 9–12, 14]. The mechanical characteristics of polymers directly depend on their structure. Various external factors also have a great influence on the mechanical properties of polymer materials: temperature, pressure, environment, duration and frequency of loading and the type of stress state. Therefore, it is necessary to study the aging processes of these materials. The effect of aging on the mechanical properties of ultrahigh molecular weight polyethylene of various grades was considered, in particular, in [2, 6, 8, 13, 15].

Specimens of ultra-high molecular weight polyethylene (UHMWPE) suitable for use in medical prostheses, were investigated in [2]. Specimens were irradiated in air using a Co source to a total dose of 10 Mrad. The effects of a 10 Mrad dose on percentage crystallinity, tensile behavior and dynamic mechanical response were established 3 days after irradiation, and irradiated specimens were then retested at intervals over a one year period. During the ageing period, specimens were stored in polyethylene bags in darkness. Crystallinity in UHMWPE was found to increase progressively during this period with consequent changes in mechanical behavior. Upon irradiation, UHMWPE shows an increase in ultimate tensile strength and a fall in elongation at break. Elongation at break has been shown to be particularly sensitive to radiation dosage, and reflects the reduced ability of the polymer to undergo plastic deformation and flow with increasing radiation induced crosslinking.

In [6], were compared the mechanical behavior of naturally (shelf) aged and accelerated aged tibial inserts using a previously validated miniature specimen testing technique known as the small punch test. Tibial inserts made of GUR 1120 and sterilized with 25 to 40 kGy of gamma radiation (in air) in 1988, 1993 and 1997 were obtained; a subset of the 1997 implants were subjected to 4 weeks of accelerated aging in air at 80°C. To determine the spatial variation of mechanical properties within each insert, miniature specimens were machined from the surface and subsurface regions of the inserts. Aging significantly affected to elastic modulus, initial load, ultimate load, ultimate displacement and work to failure. The accelerated aging lead to mechanical degrade the surface of GUR 1120 tibial components to an extent comparable to that seen after 10 years of natural aging. Test specimens showed a fracture morphology consistent with the decreased ductility and toughness UHMWPE undergoes a spatially no uniform change towards a less ductile (more brittle) mechanical behavior after gamma irradiation in air and shelf aging.

In [8] were tested the hypothesis that the oxidation resistance of Vitamin E-blended UHMWPE would be influenced by trace doses of antioxidant, resin, and radiation treatment. Trace concentrations (less 500 ppm w/w%) of Vitamin E were blended separately with GUR 1020 and 1050 resins and molded into disks. From each disk, three groups of 10 mm thick blocks were machined:

1. no irradiation (control);
2. 30 kGy of gamma irradiation in nitrogen; and
3. 75 kGy of gamma irradiation in air.

Specimens were subjected to three aging protocols

a) no aging (control);
b) two weeks and
c) four weeks of accelerated aging at 70°C and 5 atm oxygen.

The minimum concentrations of Vitamin E needed to stabilize UHMWPE during accelerated tests was determined, which depended upon the method of radiation processing. For the 30 and 75 kGy irradiated materials, the addition of 125 ppm or more Vitamin E was sufficient to maintain baseline mechanical and chemical properties through two weeks of accelerated aging. For these groups, the addition of 375 ppm or 500 ppm, respectively, was necessary to maintain baseline mechanical and chemical properties throughout the four-week accelerated aging period. UHMWPE resin molecular weight did not have an effect on oxidation behavior. The results of this experiment therefore supported the hypotheses that trace concentrations of Vitamin E, coupled with radiation treatment - but not resin grade - influence the mechanical and oxidative degradation behavior of UHMWPE.

[13] is devoted to assess the importance and extent of mechanical properties alterations of ultrahigh molecular weight polyethylene's of two sources, Hercules 1900 and RCH 1000C. UHMWPE specimens were subjected to gamma-ray irradiation and aging in serum and argon environments. The effect of irradiation is to introduce crosslinks and to increase the degree of crystallinity resulting in changes in the tensile properties in the direction of higher stiffness and reduced ductility. Aging for six months produced similar changes with subtle differences. The increase in elastic modulus and yield point stress with irradiation are consistent with the increase in the degree of crystallinity brought about by the irradiation. At the same time as the yield point stress is raised, the elongation at the yield point is reduced with irradiation. The very drastic reduction in the elongation at fracture with high irradiation dose levels indicates a very undesirable effect of loss of ductility at doses exceeding 2.5 Mrad.

The performance of ultra-highmolecularweight polyethylene (UHMWPE) barrier nets in marine environments are investigated by Fourier transform infrared spectroscopy, thermogravimetry, scanning electron microscopy and tensile experiments in [15]. The chemical, morphological, thermal stability and strength changes after aging in salt spray, hygrothermal and ultraviolet (UV) environments are characterized. An environmental spectrum is designed to simulate a real service environment and predict the service life of UHMWPE. The results show that UV energy can activate UHMWPE molecules and lead to chain breaking, which lowers the breaking strength more efficiently than salt spray. In a hygrothermal environment, the UHMPE fibers bond into clumps, which causes a slight increase in breaking strength after the initial rapid decrease with aging time. The degradation mechanism after aging in the environmental spectrum is the same to that in the real service environment. The acceleration ratio of the environmental spectrum increases with increasing aging time, which may be caused by the cross-linking and degradation of macromolecular chains in the material. The environmental spectrum given by work [15] can be used to evaluate performance and predict the service life of UHMWPE barrier nets in marine environments.

In this paper, experimental studies of the natural and deformation aging of ultrahigh molecular weight polyethylene specimens in tensile and creep experiments have been carried out.

4.2 Materials and Research Methods

In the tensile, creep and fatigue experiments, specimens in the form of double-sided blades made of ultrahigh molecular weight polyethylene (PE 1000) were used. The dimensions and geometry of the specimens are shown on Fig. 4.1. The dimensions are given in millimeters. The thickness of the specimens was about 5 mm. The experiments on tension and creep were performed on a tearing test machine TINIUS OLSEN H10K-T. Fatigue experiments were carried out on a desktop fatigue servo hydraulic test machine Si-Plan SH-B.

The study of the effect of aging on the mechanical properties of ultrahigh molecular weight polyethylene specimens was carried out according to the following programs:

Program 1. Natural aging of specimens at laboratory conditions for 11 months.

Program 2. Natural aging of specimens at laboratory conditions for 5 months; deformation aging of specimens in fatigue experiments with a frequency of 10 Hz, cycle asymmetry coefficient equal to 0 at stress of 20 MPa (0.9 of the ultimate

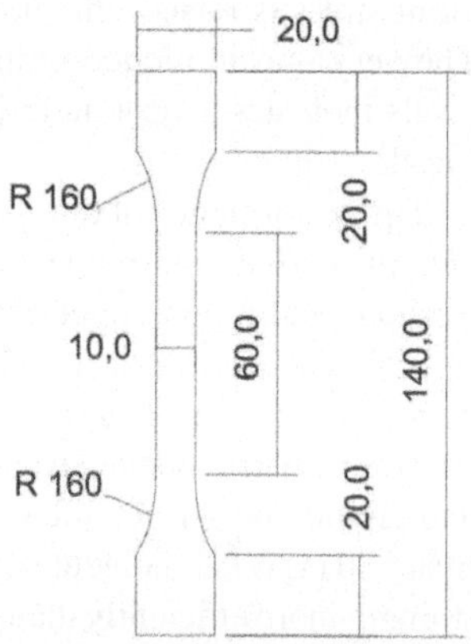

Fig. 4.1 Geometric dimensions of ultrahigh molecular weight polyethylene specimens (in mm).

tensile strength) up to 500 000 cycles, followed by natural aging of specimens at laboratory conditions for 6 months.

4.3 Experimental Studies of the Aging of Ultrahigh Molecular Weight Polyethylene Specimens in Tensile Experiments

To study the effect of considered aging programs on changes of mechanical properties of ultrahigh molecular weight polyethylene specimens were conducted tensile experiments at a rate of 5 mm/min.

The obtained tension stress-strain diagrams for specimens without aging (curve 1), after natural aging according to program 1 (curve 2) and after deformation and natural aging according to program 2 (curve 3) are shown on Fig. 4.2. From Fig. 4.2 it can be seen, that aging according to programs 1 and 2 did not lead to a change of the ultimate tensile strength of the UHMWPE specimens. However, a significant decrease in deformation at fracture moment is observed. Thus, for specimens without aging, it is equal to 363 %, and for specimens after aging according to programs 1 and 2 it is equal to 290 % and 226 %, respectively.

4.4 Experimental Studies of the Aging of Ultrahigh Molecular Weight Polyethylene Specimens in Creep Experiments

The creep experiments were carried out at room temperature at various stress levels. The obtained creep curves for UHMWPE specimens without aging (curves 1), after natural aging according to program 1 (curves 2) and after deformation and natural aging according to program 2 (curves 3) for stress levels of 18.75 MPa (a), 16.25 MPa (b) and 15 MPa (c) are shown on Fig. 4.3. From Fig. 4.3 it can be seen, that for UHMWPE specimens after natural aging according to program 1, a decrease in the time to creep fracture is observed. For specimens after deformation and natural

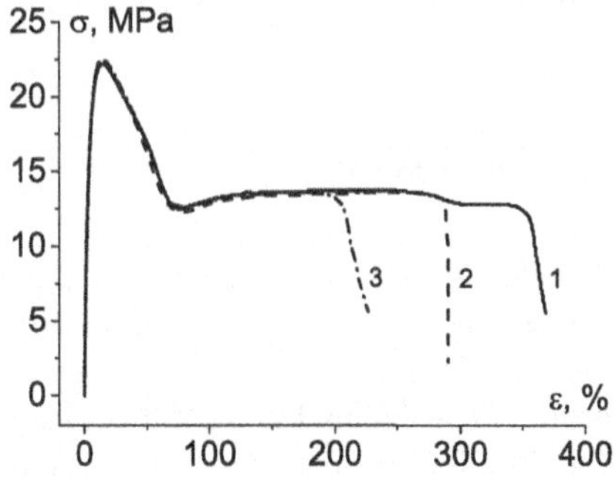

Fig. 4.2 Tensile stress-strain diagrams of ultrahigh molecular weight polyethylene specimens without aging (curve 1), after natural aging according to program 1 (curve 2) and after deformation and natural aging according to program 2 (curve 3).

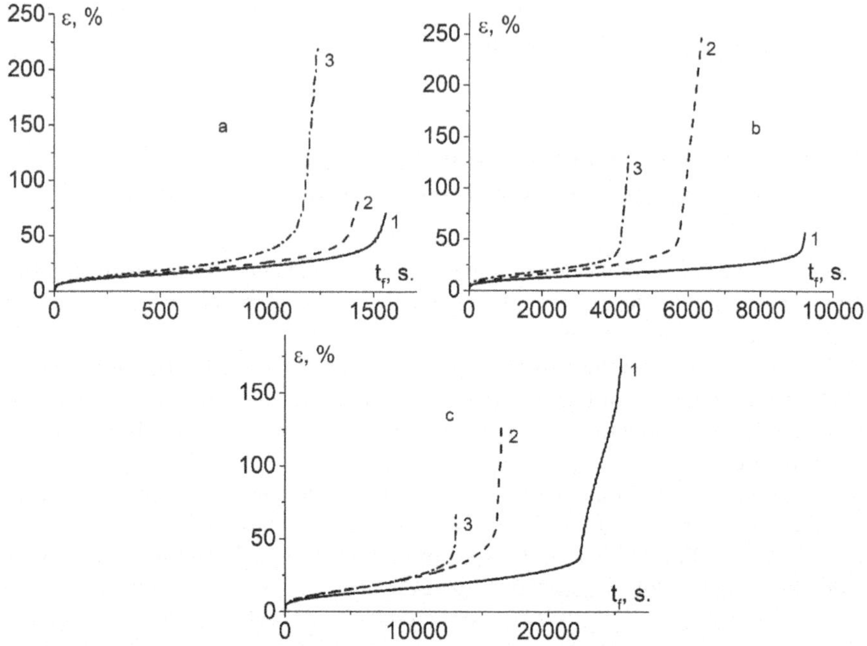

Fig. 4.3: Creep curves of ultrahigh molecular weight polyethylene specimens without aging (curves 1), after natural aging according to program 1 (curves 2) and after deformation and natural aging according to program 2 (curves 3) for stress levels of 18.75 MPa (a), 16.25 MPa (b) and 15 MPa (c) (t_f is the fracture time).

aging according to program 2, this decrease is even more significant. Thus, for a stress level equal to 18.75 MPa, there is a decrease in the time to creep fracture by 1.1 and 1.25 times for specimens aged according to programs 1 and 2, respectively compared to non-aged specimens. For stresses of 16.25 MPa and 15 MPa, also a decrease in creep fracture times by 1.5 and 2 times, respectively, for programs 1 and 2 is observed. Thus, the deformation aging leads to a significantly greater reduction in the time to creep fracture, as well as a more significant degradation of the properties of ultrahigh molecular weight polyethylene.

4.5 Conclusion

The effect of natural and deformation aging on the change of the mechanical properties of specimens made of ultrahigh molecular weight polyethylene (PE1000) in tensile and creep experiments under various programs was investigated. Aging according to programs 1 and 2 did not lead to a change of the ultimate tensile strength of the UHMWPE specimens. However, there is a significant decrease in deformation at the fracture moment. Thus, for specimens without aging, it is equal to 363 %, and for specimens after aging according to programs 1 and 2 it is equal to 290 % and 226 %, respectively. For UHMWPE specimens after aging according to programs 1 and 2, there is a decrease in the time to creep fracture compared to specimens without aging. At the same time, this decrease is more significant for specimens after aging according to program 2. For a stress level equal to 18.75 MPa, a decrease in the time to creep fracture by 1.1 and 1.25 times is observed for specimens aged according to programs 1 and 2, respectively. For stresses of 16.25 MPa and 15 MPa, there is also a 1.5 times decrease in the time to fracture for program 1 and a 2 times decrease for program 2. Thus, the deformation aging leads to a significantly greater reduction in the time to creep fracture, as well as a more significant degradation of the properties of ultrahigh molecular weight polyethylene.

Acknowledgements The work was carried out with the financial support of the Russian Science Foundation (grant No. 25-21-00376).

References

[1] Arutyunyan RA (2004) Problema deformacionnogo stareniya i dlitel'nogo razrusheniya v mekhanike materialov (in Russ., The Problem of Deformation Aging and Long-Term Fracture in the Mechanics of Materials). St. Petersburg State University Publishing House, St. Petersburg

[2] Birkinshaw C, Buggy M, Daly S, O'Neill M (1988) Mechanism of Ageing in Irradiated Polymers. Polymer Degradation and Stability 22(3):285-294 DOI 10.1016/0141-3910(88)90016-X

[3] Bochkarev RN, Filatov IS. (1990) Starenie materialov na osnove polivinikhlorida v usloviyakh kholodnogo klimata (in Russ., Aging of Materials Based on Polyvinyl Chloride in a Cold Climate. Nauka Publ. Siberian branch, Novosibirsk

[4] de Bruijn JCM (1992) The Failure Behavior of High Density Polyethylene with an Embrittled Surface Layer due to Weathering. Delft University press, Delft

[5] Crawford R, Martin P (2020) Plastics Engineering, 4th ed. Butterworth-Heinemann DOI doi.org/10.1016/C2015-0-00550-0

[6] Edidin AA, Jewett CW, Kalinowski A, Kwarteng K, Kurtz SM (2000) Degradation of mechanical behavior in UHMWPE after natural and accelerated aging. Biomaterials 21(14):1451-1460 DOI 10.1016/s0142-9612(00)00021-1

[7] Filatov IS (1983) Klimaticheskaya ustoichivost' polimernykh materialov. (in Russ., Climatic Stability of Polymer Materials). Nauka, Moscow

[8] Kurtz SM, Dumbleton J, Siskey RS, Wang A, Manley M (2009) Trace concentrations of vitamin E protect radiation crosslinked UHMWPE from oxidative degradation. Journal of Biomedical Materials Research. Part A. 90(2):549-563 DOI 10.1002/jbm.a.32122.

[9] McKeen L (2021) Effect of Long Term Thermal Exposure on Plastics and Elastomers, 2nd ed. Elsevier DOI 10.1016/C2020-0-02194-8
[10] Mills N, Jenkins M, Kukureka S (2020) Plastics. Microstructure and Engineering Applications, 4th ed., Butterworth-Heinemann
[11] Nadai A (1963) Theory of Flow and Fracture of Solids, vol. 2. McGraw-Hill, New York
[12] Narain R. (ed.) (2020) Polymer Science and Nanotechnology. Fundamentals and Applications. Elsevier DOI 10.1016/C2018-0-01134-2
[13] Roe R-J, Grood ES, Shastri R, Gosselin CA, Noyes FR (1981) Effect of radiation sterilization and aging on ultrahigh molecular weight polyethylene. Journal of Biomedical Materials Research 15(2):209-230 DOI 10.1002/jbm.820150209
[14] Struik LCE (1978) Physical Aging in Amorphous Polymers and Other Materials. Elsevier, Amsterdam, Oxford, New York
[15] Zhang W, Jing X, Bai Y, Shan X, Qi X, Yan M, Cui Z (2022) Study on the Aging Behavior of an Ultra-High MolecularWeight Polyethylene Fiber Barrier Net in a Marine Environment. Materials 15(16):5599 DOI 10.3390/ma15165599

Chapter 5
Optimization of Rotational Motions of a Two-Link Electromechanical Manipulator Considering Heat Losses Drives

Vahan V. Avetisyan

Abstract The problem of optimizing the rotational motions of a two-link electromechanical manipulator is considered according to the criterion of minimizing the total heat losses in the drives. Diagrams are constructed in the plane of generalized coordinates of the manipulator (the joint angles). These diagrams make it possible to determine the type of final configuration, the directions of link rotations, and the control methods that ensure the prescribed positioning of the gripper within the workspace according to the considered optimality criterion.

Key words: Two-link electromechanical manipulator, Energy-optimal control, Terminal configuration, Link rotation direction

5.1 Introduction

Electromechanical two-link robotic manipulators are widely used in modern automated control systems, where the issues of energy efficiency, precision, and compliance with technological constraints are of particular importance [7, 11]. They can operate either autonomously or as part of multi-link robotic assemblies, in which the two links perform the main portion of motion during various technological operations. One of the approaches to the rational design of control regimes is their optimization according to a chosen performance criterion, such as the duration of transport operations or energy expenditure. An essential aspect in developing optimal control algorithms is accounting for the structural and geometric characteristics of the manipulator. For a planar two-link manipulator, each position of the end effector corresponds to two possible configurations that differ by the sign of the angle between the links. Consequently, the quality of the end-effector motion toward a

Vahan V. Avetisyan
Institute of Mechanics, National Academy of Sciences of the Republic of Armenia, 24b Marshal Baghramyan Ave, Yerevan, 0019, Republic of Armenia, e-mail: vanavet@yahoo.com

H. Altenbach and A. V. Sahakyan, *Contact Mechanics and Relevant Problems*,
Advanced Structured Materials 253, https://doi.org/10.1007/978-3-032-20869-9_5

desired position depends both on the type of final configuration and on the control strategy ensuring its realization. In [4, 5], optimal and suboptimal control laws were derived for a two-link manipulator with massless links in a two-point positioning problem with a payload. It was shown that the transition time to the terminal state significantly depends on the configuration type, and the problem of selecting the optimal configuration was addressed. In [2], a graph–analytical approach was developed for constructing suboptimal time-efficient program controls that enable two-link manipulators to move from an initial rest configuration to any desired final rest configuration. In [8], using the method of parametric optimization, a suboptimal control law minimizing a quadratic performance functional was obtained for the planar two-link manipulator, taking into account the possible configurations corresponding to the given initial and terminal positions of the end effector. Various optimization methods for robot control, including that of two-link manipulators, as well as approaches to the design of their structural parameters, are discussed in [3, 6, 9, 10, 12]. The model of mechanical plane two – link manipulator with statically balanced second link is considered in [1]. Assuming that the manipulator design allows clockwise and counterclockwise rotation of the links, it was established that the manipulator can be brought to the same final configuration by various combinations of rotations of the links. For each of the two types of final configurations, the graphic – analytical procedure solved the problem of choosing the directions of rotations of the manipulator links and determining the control method for which the transition time reaches a minimum value. The optimal type of final configuration was found by a straightforward computation.

In the present study, an electromechanical model of a two-link manipulator with a statically balanced second link is considered. The input voltages and currents in the rotor circuits of the electric drives are subject to constraints. The optimality criterion is defined as the total heat losses in the motor windings. Diagrams are constructed in the plane of the manipulator's generalized joint angles, which make it possible, for given coordinates of the terminal end – effector position, to determine the type of final configuration, the directions of link rotation (clockwise or counterclockwise), and the optimal control strategy that ensures the transfer of the end effector from the initial rest configuration to the prescribed terminal configuration with minimal heat losses in the electric drives.

5.2 System Description

We will consider a two – link manipulator with straight links of equal length (Fig. 5.1) and the joint O_2. The joint O_1 connects the first link G_1 with a stationary base. At the end of the second link G_2, at the point O_3, a gripper that grasps a transported object (a load) is fixed. The control is performed by two independent drives D_1 and D_2. Each drive contains a direct current electric motor with separate excitation, as well as a gearhead. The stators of the electric motors and bodies of the gearheads are fixed on the stationary base and on the link G_1, respectively. The rotational axis of the rotor of the electric motor of goes through its center of mass and coincides with the axis of the joint O_2. The friction is not taken into account. Let as introduce the following natation: φ_1 is the angle between the line O_1O_2 and the stationary line O_1X (the angle of rotation of the first link G_1); φ_2 is the angle between lines O_2C and O_1O_2 (the angle of rotation of the second link G_2 with respect to the first link, measured in the counter – clockwise direction), where C is the center of mass of the second link; $a = |O_1O_2| = |O_2O_3|$, $L = |O_2C|$; M is the mass of the second link; m is the mass of the rotor of motor's drive D_2; I_1, I_2 is the moments of inertia of the links G_1, G_2 in reference to the joint axes O_1, O_2; $J^{(i)}$ is the moment of inertia of the rotor of the armature of the electric motor of D_i reference to the rotational axis; L_i, R_i is the coefficient of inductance and the winding impedance of motor's rotor of D_i; k_i is the coefficient of proportionality between the current and the moment μ_i of electromagnetic forces produced by the motor; j_i is the current in the winding of the armature of D_i; u_i is the control electric voltage, which is fed at the motor of D_i; n_i be the gear ratio of the reduction box of the drive D_i $(i = 1, 2)$.

The equations of motion of the considered electromechanical system is as follows:

$$\frac{d}{dt}\frac{\partial T}{\partial \dot{\varphi}_i} - \frac{\partial T}{\partial \varphi_i} = n_i k_i \mu_i, \tag{5.1}$$

$$L_i \frac{dj_i}{dt} - R_i j_i + n_i k_i \dot{\varphi}_i = u_i, \tag{5.2}$$

$$|u_i| \le U_i, \quad |j_i| \le j_i^0, \; i = 1, 2, \tag{5.3}$$

where

$$T = \frac{(A_{11} + 2MaL\cos\varphi_2)\dot{\varphi}_1^2}{2} + \frac{(A_{12} + 2MaL\cos\varphi_2)\dot{\varphi}_1\dot{\varphi}_2 + A_{22}\dot{\varphi}_1^2}{2},$$

$$A_{11} = I_1 + I_2 + J^{(1)}n_1^2 + J^{(2)} + (M + m)a^2,$$

$$A_{12} = I_2 + J^{(2)}n_2,$$

$$A_{22} = I_2 + J^{(2)}n_2^2.$$

Equations (5.1) are Lagrange equations, which describe the motion of the manipulator under the effect of the moments of the electromagnetic forces $\mu_i = k_i j_i$ $(i = 1, 2)$

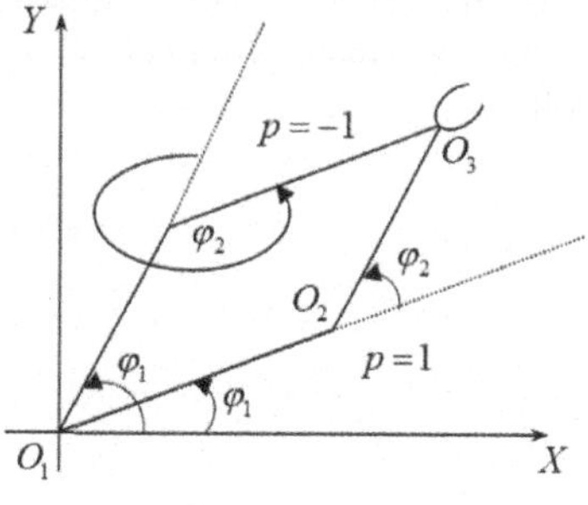

Fig. 5.1 Two configurations of a two-link manipulator corresponding to the gripper position O_3.

applied to the rotors of motors. Equations (5.2) describe the balance of the electric voltages in the circuits of armature. Inequalities (5.3) define the restrictions on the controlling voltages and currents in the circuits of the motor armatures, which must be satisfied during the motion. The detailed development and analysis of Eqs. (5.1) and (5.2) are presented in [7]. Let us pass on to the following dimensionless variables in the (5.1) – (5.3)

$$\begin{aligned} &t' = \frac{t}{\bar{T}}, \quad u_i' = \frac{u_i}{U_i}, \quad k_i' = \frac{k_i n_i}{U_i \bar{T}}, \quad R_i' = \frac{R_i A_{ii}}{k_i n_i U_i \bar{T}^2}, \\ &j_i' = j_i \left(k_i n_i U_i A_{ii}^{-2} \bar{T}^5\right)^{1/2}, \quad L_i' = \frac{L_i A_{ii}}{k_i n_i U_i \bar{T}^3}, \quad i = 1, 2. \end{aligned} \tag{5.4}$$

and drop the primes for simplicity. We obtain

$$\begin{aligned} &\ddot{\varphi}_1 + A_{12} A_{11}^{-1} \ddot{\varphi}_2 + MaLA_{11}^{-1} [2\ddot{\varphi}_1 \cos\varphi_2 + 2\ddot{\varphi}_2 \cos\varphi_2 \\ &\qquad - \dot{\varphi}_2 (2\dot{\varphi}_1 - \dot{\varphi}_2) \sin\varphi_2] = k_1^{1/2} j_1, \\ &\ddot{\varphi}_2 + A_{12} A_{22}^{-1} \ddot{\varphi}_1 + MaLA_{22}^{-1} (2\ddot{\varphi}_1 \cos\varphi_2 + \dot{\varphi}_1^2 \sin\varphi_2 = k_2^{1/2} j_2, \end{aligned} \tag{5.5}$$

$$L_i \frac{dj_i}{dt} - R_i j_i + k_i^{1/2} \dot{\varphi}_i = k_i^{-1/2} u_i, \quad i = 1, 2, \tag{5.6}$$

$$|u_i| \le 1, |j_i| \le j_i^0, \quad i = 1, 2. \tag{5.7}$$

By U_i, j_i^0, $\bar{T}$ in (5.4) we denote the parameters, which have the dimension of the voltage, current and time, respectively, where U_i is the maximally admissible voltage and current for the drive's motors, and $\bar{T}$ is some characteristic time of an operation, performed by the steady state of currents in the circuits of motor's rotors ($\tau_i = L_i / R_i$ in the initial dimensional variables) are much less than the time of a transportation operation, and the center of mass C of the link G_i is on the axis of the joint O_2, i.e., $L = 0$, after the elimination of the variables j_i, system (5.5)–(5.7) can be represented in the following from:

$$\begin{aligned} &R_1 \ddot{\varphi}_1 + R_1 A_{12} A_{11}^{-1} \ddot{\varphi}_2 + k_1 \dot{\varphi}_1 = u_1, \\ &R_2 A_{12} A_{22}^{-1} \ddot{\varphi}_1 + R_2 \ddot{\varphi}_2 + k_2 \dot{\varphi}_2 = u_2, \end{aligned} \tag{5.8}$$

$$|u_i| \le 1, \quad |u_i - k_i \dot{\varphi}_i| \le j_i^0 k_i^{1/2} R_i = \eta_i, \quad i = 1, 2. \tag{5.9}$$

System (5.8), (5.9) will be regarded as the initial conditions, which after the rotation of the coordinate system O_1XY by the angle between the first link and the O_1X is as follows

$$\begin{aligned} &\varphi_1(0) = 0, \quad \dot{\varphi}_1(0) = 0, \\ &\varphi_2(0) = \varphi_2^0, \quad \dot{\varphi}_2(0) = 0. \end{aligned} \tag{5.10}$$

5.3 The Statement of the Control Problem

In plane – parallel motion of a two–link manipulator the Cartesian coordinates (x, y) of the gripper O_3 are uniquely determined by generalized coordinates φ_1, φ_2

$$\begin{aligned} x &= a\left[\cos\varphi_1 + \cos(\varphi_1 + \varphi_2)\right], \\ y &= a\left[\sin\varphi_1 + \sin(\varphi_1 + \varphi_2)\right]. \end{aligned} \tag{5.11}$$

Let us suppose that the reading of the angles φ_1, φ_2 is performed in the counter – clockwise direction from the corresponding lines O_1X and O_1O_2 (Fig. 5.1).

It follows from (5.11) that two configurations correspond to each position of the grip. Let us denote them by

$$\{\varphi_1(x, y), \varphi_2(x, y)\}^{(p)}, \quad p = \pm 1.$$

Furthermore, we shall drop the arguments (x, y) of the function $\varphi_i(x, y)$ for simplicity. There is the following connection between the angles φ_i^{T}, $i = 1, 2$, which correspond to the different values of p

$$\begin{aligned} \varphi_1^{\mathrm{T}}\big|_{p=-1} &= \varphi_1^{\mathrm{T}}\big|_{p=1} + \varphi_2^{\mathrm{T}}\big|_{p=1}, \\ \varphi_2^{\mathrm{T}}\big|_{p=-1} &= 2\pi - \varphi_2^{\mathrm{T}}\big|_{p=1}. \end{aligned} \tag{5.12}$$

Let us suppose that the considered manipulator admits a change of the angles $\Delta\varphi_i$, $\Delta\varphi_i = \varphi_i - \varphi_i^0$ in the domain $\Omega = \{\Delta\varphi_i\colon\ |\Delta\varphi_i| \le 2\pi,\ i = 1, 2\}$ while moving. This assumption allows us to move the gripper from an arbitrary initial position into the prescribed one in the operation zone $D = \left\{(x, y)\colon x^2 + y^2 \le 4a^2\right\}$. We note that pairs of generalized coordinates (φ_1, φ_2), $(\varphi_1 - 2\pi, \varphi_2)$, $(\varphi_1, \varphi_2 - 2\pi)$, $(\varphi_1 - 2\pi, \varphi_2 - 2\pi)$ determine the same configuration of the manipulator. However, different parametric representations of this configuration contain information about the way to reach this particular configuration by the manipulator from the initial position ($\varphi_1^0 = 0, \varphi_2^0 = 0$). Thus, for instance, the pair (φ_1, φ_2), $\varphi_1 > 0$, $\varphi_2 > 0$ means that the first and the second links were turned counter – clockwise by the angles φ_1 and φ_2, respectively. While the configuration $(\varphi_1 - 2\pi, \varphi_2)$, means that for reaching the same configuration the first link was turned clockwise by the angle φ_2 counter – clockwise.

In accordance with this, any configuration can by represented by one the following four forms:

$$\begin{aligned} &\{\varphi_1^{\mathrm{T}}, \varphi_2^{\mathrm{T}}\} = (\varphi_1^{\mathrm{T}}, \varphi_2^{\mathrm{T}})_m, \quad m = 1, \ldots, 4, \\ &(\varphi_1^{\mathrm{T}}, \varphi_2^{\mathrm{T}})_1 = (\varphi_1^{\mathrm{T}}, \varphi_2^{\mathrm{T}}), \\ &(\varphi_1^{\mathrm{T}}, \varphi_2^{\mathrm{T}})_2 = (\varphi_1^{\mathrm{T}} - 2\pi, \varphi_2^{\mathrm{T}}), \\ &(\varphi_1^{\mathrm{T}}, \varphi_2^{\mathrm{T}})_3 = (\varphi_1^{\mathrm{T}}, \varphi_2^{\mathrm{T}} - 2\pi), \\ &(\varphi_1^{\mathrm{T}}, \varphi_2^{\mathrm{T}})_4 = (\varphi_1^{\mathrm{T}} - 2\pi, \varphi_2^{\mathrm{T}} - 2\pi), \\ &0 \le \varphi_1^{\mathrm{T}}, \varphi_2^{\mathrm{T}} \le 2\pi. \end{aligned} \tag{5.13}$$

The four different ways of representation in (5.13), which correspond to a combination of rotations in the clockwise and counter – clockwise directions, and realize the given transition on the plane $(\varphi_1^{\mathrm{T}}, \varphi_2^{\mathrm{T}})$, and define four different points of the form $(\varphi_1^{\mathrm{T}}, \varphi_2^{\mathrm{T}})_m$, $m = 1, \dots, 4$.

Assume that for system (5.8), (5.9) we define the quality criterion Q, which can characterize the property of optimality of the movement in some sense, for instance, in the sense of maximum speed, or minimum energy expenditures. It follows from the aforesaid that the movement quality of system (5.8), (5.9) depends not only on the way of giving the controls u_i, that are included in (5.8), (5.9)), but on the values of m and p, which determine the movement direction and the type of the final configuration of the considered manipulator, as well. Therefore, we have

$$\begin{aligned} &Q = Q((\varphi_1^{\mathrm{T}}, \varphi_2^{\mathrm{T}})_m^{(p)}, u), \quad (\varphi_1^{\mathrm{T}}, \varphi_2^{\mathrm{T}})_m^{(p)} \in \Phi, \\ &u = (u_1, u_2), \quad p = \pm 1, \ m = 1, 2, 3, 4. \end{aligned} \tag{5.14}$$

It is worth noting that the transition of the manipulator from the state (5.10) $(\varphi_1(0), \varphi_2(0)) = (0, \varphi_2^0)$ to the final state $(\varphi_1(T), \varphi_2(T)) = (\varphi_1^{\mathrm{T}}, \varphi_2^{\mathrm{T}})$ is equivalent to the transition from the zero state with the configuration $\{0, 0\}$ to the state with the configuration $\{\varphi_1^{\mathrm{T}}, \varphi_2^{\mathrm{T}} - \varphi_1^0\}$. Therefore, without loss of generality, we can restrict our considerations by the configuration $\{\varphi_1^0, \varphi_2^0\} = \{0, 0\}$.

Consider the following problem of optimal control of the manipulator. Let, at the initial instant, the manipulator gripper be in the position $\{x^0, y^0\}$, which is defined by the initial configuration $\{\varphi_1^0, \varphi_2^0\} = \{0, 0\}$:

$$\begin{aligned} &x(0) = a\left[\cos\varphi_1^0 + \cos(\varphi_1^0 + \varphi_2^0)\right] = x^0, \quad \dot{x}(0) = 0, \\ &y(0) = a\left[\sin\varphi_1^0 + \sin(\varphi_1^0 + \varphi_2^0)\right] = y^0, \quad \dot{y}(0) = 0, \quad \varphi_1^0 = \varphi_2^0 = 0. \end{aligned} \tag{5.15}$$

It is required to choose the type $p = \pm 1$ of the final configuration, the directions $m = 1, \dots, 4$ of the link rotations and the method $u = (u_1, u_2)$ of the control of system (5.8), (5.9) that provide

$$Q = \min_p \min_m \min_u Q\left[(\varphi_1^{\mathrm{T}}, \varphi_2^{\mathrm{T}})_m^{(p)}, u\right], \tag{5.16}$$

when moving the gripper in the prescribed position in the operation zone with the zero speed at the final moment of the process

$$\begin{aligned} &x(T) = a\left[\cos\varphi_1^{\mathrm{T}} + \cos(\varphi_1^{\mathrm{T}} + \varphi_2^{\mathrm{T}})\right] = x^{\mathrm{T}}, \quad \dot{x}(T) = 0, \\ &y(T) = a\left[\sin\varphi_1^{\mathrm{T}} + \sin(\varphi_1^{\mathrm{T}} + \varphi_2^{\mathrm{T}})\right] = y^{\mathrm{T}}, \quad \dot{y}(T) = 0. \end{aligned} \tag{5.17}$$

5.4 The Diagram Algorithm for Determining Optimal Motion

Consider the case where the optimality criterion Q represents the total heat losses in manipulator's drives for the fixed time T, i.e.,

$$Q = \sum_{i=1}^{2} \int_{0}^{T} j_i^2 R_i \, dt. \tag{5.18}$$

After the change of variables by formulas (5.4) and denoting

$$Q' = Q_1' + \gamma Q_2' = Q T^2 / A_{11}$$

(omitting the primes), the functional Q will have the from

$$Q = \int_{0}^{T} \frac{(u_1 - k_1 \dot{\varphi}_1)^2}{k_1 R_1} dt + \gamma \int_{0}^{T} \frac{(u_2 - k_2 \dot{\varphi}_2)^2}{k_2 R_2} dt \tag{5.19}$$

with

$$\gamma = \frac{A_{22}}{A_{11}}.$$

We find the first internal minimum in (5.16) as the result of solving the problem of construction for the optimal with respect to the heat losses (5.19) control u_i for system (5.8), (5.9) under the boundary conditions

$$\begin{aligned} &\varphi_i(0) = 0, \quad \dot{\varphi}_i(0) = 0, \\ &\varphi_i(T) = \varphi_i^{\mathrm{T}}, \quad \dot{\varphi}_i(T) = 0, \quad i = 1, 2. \end{aligned} \tag{5.20}$$

Let us the variables in (5.8), (5.9), (5.19), (5.20)

$$\begin{aligned} v_i &= (u_i - k_i \dot{\varphi}_i)/(k_i R_i)^{1/2}, \quad i = 1, 2, \\ \theta_1 &= (k_1^{-1} R_1)^{1/2} \varphi_1 + (k_1^{-1} R_1)^{1/2} A_{12} A_{11}^{-1} \varphi_2, \\ \theta_2 &= (k_2^{-1} R_2)^{1/2} A_{12} A_{22}^{-1} \varphi_1 + (k_2^{-1} R_2)^{1/2} \varphi_2. \end{aligned} \tag{5.21}$$

When using variables (5.21), the equations of motion, the functional and the boundary conditions have more simple form

$$\begin{aligned}
&\ddot{\theta}_i = v_i, \quad Q = \int_0^T (v_1^2 + \gamma v_2^2)\,dt,\\
&\left|v_i + A_i\dot{\theta}_1 + B_i\dot{\theta}_2\right| \le (k_i R_i)^{-1/2} = \eta_i^{(1)}, \quad |v_i| \le j_i^0 R_i^{1/2} = \eta_i^{(2)},\\
&\theta_i(0) = 0, \ \ \dot{\theta}_i(0) = 0, \ \ \theta_i(T) = \alpha_i\varphi_1^{\mathrm{T}} + \beta_i\varphi_2^{\mathrm{T}} = \theta_i^{\mathrm{T}}, \ \ \dot{\theta}_i(T) = 0, \ \ i = 1,2,\\
&\alpha_1 = (k_1^{-1}R_1)^{1/2}, \ \ \beta_1 = (k_1^{-1}R_1)^{1/2}A_{12}A_{11}^{-1},\\
&\alpha_2 = (k_2^{-1}R_2)^{1/2}A_{12}A_{22}^{-1}, \ \ \beta_2 = (k_2^{-1}R_2)^{1/2},\\
&A_1 = k_1R_1^{-1}A_{11}A_{22}(A_{11}A_{22} - A_{12}^2)^{-1},\\
&B_1 = -(k_1R_1^{-1}k_2R_2^{-1})^{1/2}A_{12}A_{22}(A_{11}A_{22} - A_{12}^2)^{-1},\\
&A_2 = k_2R_2^{-1}A_{11}A_{12}(A_{11}A_{22} - A_{12}^2)^{-1},\\
&B_2 = -(k_1R_1^{-1}k_2R_2^{-1})^{1/2}A_{11}A_{12}(A_{11}A_{22} - A_{12}^2)^{-1}.
\end{aligned} \tag{5.22}$$

Applying Pontryagin's maximum principle to problem (5.22) (not taking into account the restrictions on the controls v_i, $i = 1,2$), we obtain the following expression for the optimal trajectories, controls, and the values of the functional, which correspond to them:

$$\begin{aligned}
&\theta_i = -2\theta_i^{\mathrm{T}}T^{-3}t^3 + 3\theta_i^{\mathrm{T}}T^{-2}t^2,\\
&v_i = -12\theta_i^{\mathrm{T}}T^{-3}t + 6\theta_i^{\mathrm{T}}T^{-2}, \quad i = 1,2,\\
&Q = 12T^{-3}\left[(\theta_1^{\mathrm{T}})^2 + \gamma(\theta_2^{\mathrm{T}})^2\right].
\end{aligned} \tag{5.23}$$

Solving system of inequalities (5.22) with respect to v_i, we obtain

$$\begin{aligned}
v_i^{(1)} &= \max\left\{-\eta_i^{(1)}, -\eta_i^{(2)} - A_i\dot{\theta}_1 - B_i\dot{\theta}_2\right\} \le v_i\\
&\le \min\left\{\eta_i^{(1)}, \eta_i^{(2)} - A_i\dot{\theta}_1 - B_i\dot{\theta}_2\right\} = v_i^{(2)}.
\end{aligned} \tag{5.24}$$

It follows from (5.24) that the interval $\left[v_i^{(1)}, v_i^{(2)}\right]$ is not empty, i.e., $v_i^{(1)} \le v_i^{(2)}$ if and only if the condition

$$\left|A_i\dot{\theta}_1 + B_i\dot{\theta}_2\right| \le \eta_i^{(1)} + \eta_i^{(2)}, \quad i = 1,2. \tag{5.25}$$

is satisfied. The optimal control must satisfy conditions (5.24) for all $t \in [0,T]$. Consider the case where the inequality $\eta_i^{(1)} \le \eta_i^{(2)}$ in (5.28) is satisfied, which is the case for many industrial manipulators, where

$$-\eta_i^{(1)} \le v_i(t) + A_i\dot{\theta}_1 + B_i\dot{\theta}_2 \le \eta_i^{(1)}, \quad i = 1,2. \tag{5.26}$$

Let us substitute the expression for the optimal control $v_i(t)$ and trajectory $\theta_i(t)$ (5.23) in (5.26), and then go to variables φ_i by means of (5.22). After simple transformations (5.26) will have the from

$$|H_i(t,T)| \le \eta_i^{(1)}, \quad H_i(t,T) = E_i t^2 + 2F_i t + C_i, \tag{5.27}$$

where

$$E_i = -6T^{-3}\varphi_i^{\mathrm{T}},$$
$$F_i = -3T^{-2}\left[(2\alpha_1 T^{-1} + i - 2)\varphi_1^{\mathrm{T}} + (2\beta_1 T^{-1} - i + 1)\varphi_2^{\mathrm{T}}\right],$$
$$C_i = 6T^{-2}(\alpha_i\varphi_1^{\mathrm{T}} + \beta_i\varphi_2^{\mathrm{T}}), \quad i = 1, 2.$$

It is necessary and sufficient for the validity of inequality (5.27) on the interval $[0, T]$ that the following relations hold

$$|H_i(0,T)| \le \eta_i^{(1)}, \ |H_i(t_i^*)| = \eta_i^{(1)}, \ 0 \le t_i^* \le T, \ t_i^* = -F_i E_i^{-1}, \ |H_i(T,T)| \le \eta_i^{(1)}, \tag{5.28}$$

where

$$H_i(0,T) = C_i = 6T^{-2}(\alpha_i\varphi_1^{\mathrm{T}} + \beta_i\varphi_2^{\mathrm{T}}),$$
$$H_i(T,T) = E_i T^2 + 2F_i T + C_i$$
$$= 2T^{-1}\left[(3T^{-1}\alpha_i - 2T^{-1}\alpha_1 - i + 2)\varphi_1^{\mathrm{T}}\right.$$
$$\left. + (3T^{-1}\beta_i - 2T^{-1}\beta_1 + i - 1)\varphi_2^{\mathrm{T}} - 3\varphi_1^{\mathrm{T}}\right],$$
$$H_i(t_i^*, T) = C_i - F_i^2 E_i^{-1} = 6T^{-2}(\alpha_i\varphi_1^{\mathrm{T}} + \beta_i\varphi_2^{\mathrm{T}})$$
$$+ (6T^{-3}\varphi_i^{\mathrm{T}})^{-1}\left\{3T^{-2}\left[(2T^{-1}\alpha_i + i - 2)\varphi_1^{\mathrm{T}} + (2T^{-1}\beta_1 - i + 1)\varphi_2^{\mathrm{T}}\right]\right\}^2.$$

Let us also pass on to the variables φ_i in the expression for $v_i(t)$ in (5.23). The optimal control $u_i(t)$ will be have the form

$$u_i(t) = \varphi_i(t) - 12T^{-3}(\gamma_i\varphi_i^{\mathrm{T}} + \delta_i\varphi_2^{\mathrm{T}})t + 6T^2(\gamma_i\varphi_1^{\mathrm{T}} + \delta_i\varphi_2^{\mathrm{T}}), \quad i = 1, 2,$$
$$\gamma_1 = k_1^{-1}R_1, \ \delta_1 = k_2^{-1}R_1A_{12}A_{11}, \ \gamma_2 = k_1^{-1}R_2A_{12}A_{22}, \ \delta_2 = k_2^{-1}R_2. \tag{5.29}$$

Thus, control (5.29) provides the minimum of the functional (5.19) under restrictions (5.9) and boundary conditions (5.20) only for those finite values of φ_i^{T}, $i = 1, 2.$ that satisfy inequality (5.28).

Instead of the repeated computation of functional (5.16), when determining the second and third minima (with respect to m and p) for every point $\{\varphi_1, \varphi_2\} = \{\varphi_1^{\mathrm{T}}, \varphi_2^{\mathrm{T}}\}$, we suggest the following diagram method, which allows us to determine Q depending on the location of the point in the configuration plane $(\varphi_1^{\mathrm{T}}, \varphi_2^{\mathrm{T}})$. Let us introduce new coordinate systems on the plane $(\varphi_1^{\mathrm{T}}, \varphi_2^{\mathrm{T}})$ with origins K_m at the points

$$K_1 = (0, 0), \quad K_2 = (2\pi, 0), \quad K_3 = (0, 2\pi), \quad K_4 = (2\pi, 2\pi),$$

which corresponds to transformations of the coordinates. Under this transition, the rectangular domain Ω maps on the rectangular domain

$$\Phi = \left\{(\varphi_1^{\mathrm{T}}, \varphi_2^{\mathrm{T}}) : 0 \le \varphi_1^{\mathrm{T}}, \varphi_2^{\mathrm{T}} \le 2\pi\right\}.$$

The points $(\varphi_1^{\mathrm{T}}, \varphi_2^{\mathrm{T}})_m \in \Omega$, $m = 1, \dots, 4$ can be regarded as one point $(\varphi_1^{\mathrm{T}}, \varphi_2^{\mathrm{T}}) \in \Phi$, however, with respect to the coordinate systems $K_m\varphi_1^{\mathrm{T}}\varphi_2^{\mathrm{T}}$, $m = 1, \dots, 4$ respectively. Thus, for example, instead of the point $(\varphi_1^{\mathrm{T}}, \varphi_2^{\mathrm{T}})_2 = (\varphi_1^{\mathrm{T}} - 2\pi, \varphi_2^{\mathrm{T}})$, $0 \le \varphi_1^{\mathrm{T}}, \varphi_2^{\mathrm{T}} \le 2\pi$, which is in the second quadrant of the plane $O_1\varphi_1^{\mathrm{T}}\varphi_2^{\mathrm{T}}$, we can consider the point $(\varphi_1^{\mathrm{T}}, \varphi_2^{\mathrm{T}}) \in \Phi$, but relative to the coordinate system $K_2\varphi_1^{\mathrm{T}}\varphi_2^{\mathrm{T}}$ with the origin at the point $K_2 = (2\pi, 0)$. There is the following correspondence between the domains of Ω and Φ. The distinct points in the domain Ω that are the inverse images of the point $\{\varphi_1^{\mathrm{T}}, \varphi_2^{\mathrm{T}}\}$ correspond to the fixed configuration $(\varphi_1^{\mathrm{T}}, \varphi_2^{\mathrm{T}}) \in \Phi$. And conversely, every point $(\varphi_1^{\mathrm{T}}, \varphi_2^{\mathrm{T}}) \in \Phi$ is the image of the points $(\varphi_1^{\mathrm{T}}, \varphi_2^{\mathrm{T}})_m \in \Omega$, $m = 1, \dots, 4$, which determine the same configuration $\left\{\varphi_1^{\mathrm{T}}, \varphi_2^{\mathrm{T}}\right\}$ for any m. Thus, we can perform our studies in the domain Φ. At every point of the domain Φ due to (5.13), (5.14), (5.22), (5.23) the four values of the quality functional are determined

$$Q_1 = 12T^{-3}\bar{Q}_1, \quad Q_2 = 12T^{-3}\bar{Q}_2, \quad Q_3 = 12T^{-3}\bar{Q}_3, \quad Q_4 = 12T^{-3}\bar{Q}_4, \tag{5.30}$$

where

$$\begin{aligned}
\bar{Q}_1 &= \left[(\alpha_1\varphi_1^{\mathrm{T}} + \beta_1\varphi_2^{\mathrm{T}})^2 + \gamma(\alpha_2\varphi_1^{\mathrm{T}} + \beta_2\varphi_2^{\mathrm{T}})^2\right], \\
\bar{Q}_2 &= \left\{\left[\alpha_1(\varphi_1^{\mathrm{T}} - 2\pi) + \beta_1\varphi_2^{\mathrm{T}}\right]^2 + \gamma\left[\alpha_2(\varphi_1^{\mathrm{T}} - 2\pi) + \beta_2\varphi_2^{\mathrm{T}}\right]^2\right\}, \\
\bar{Q}_3 &= \left\{\left[\alpha_1\varphi_1^{\mathrm{T}} + \beta_1(\varphi_2^{\mathrm{T}} - 2\pi)\right]^2 + \gamma\left[\alpha_2\varphi_1^{\mathrm{T}} + \beta_2(\varphi_2^{\mathrm{T}} - 2\pi)\right]^2\right\}, \\
\bar{Q}_4 &= \left\{\left[\alpha_1(\varphi_1^{\mathrm{T}} - 2\pi) + \beta_1(\varphi_2^{\mathrm{T}} - 2\pi)\right]^2 + \gamma\left[\alpha_2(\varphi_1^{\mathrm{T}} - 2\pi) + \beta_2(\varphi_2^{\mathrm{T}} - 2\pi)\right]^2\right\}.
\end{aligned} \tag{5.31}$$

Since all the functionals $Q_i = 12T^{-3}\bar{Q}_i$, $i = 1, \dots, 4$ contain a common positive factor $12T^{-3}$, their optimization is equivalent to the optimization of the $\bar{Q}_i = \bar{Q}_i(\varphi_1^{\mathrm{T}}, \varphi_2^{\mathrm{T}})$, $i = 1, \dots, 4$. The optimization of functionals (5.31) with respect to the origin $K_m\varphi_1^{\mathrm{T}}\varphi_2^{\mathrm{T}}$, $m = 1, \dots, 4$ essentially depends on the values of the problem parameters and requires sufficiently complicated calculations. Therefore, to illustrate the proposed methodology, we consider the manipulator with the following dimensional parameters [7]

$$\begin{aligned}
&I_1 = 21.8\,\mathrm{kg}\cdot\mathrm{m}^2, \quad I_2 = 5.9\,\mathrm{kg}\cdot\mathrm{m}^2, \\
&J^{(1)} = 9.1\times10^{-4}\mathrm{kg}\cdot\mathrm{m}^2, \quad J^{(2)} = 2.45\times10^{-4}\mathrm{kg}\cdot\mathrm{m}^2, \\
&M = 53\,\mathrm{kg}, \quad m = 5.3\,\mathrm{kg}, \quad n_1 = 180, \quad n_2 = 163, \quad a = 1\,\mathrm{m}, \\
&R_1 = 3.68\,\Omega, \quad R_2 = 3.6\,\Omega, \\
&L_1 = 3.68\times10^{-2}\,\mathrm{H}, \quad L_2 = 3.6\times10^{-2}\,\mathrm{H}, \\
&k_1 = 0.316\,\mathrm{N}\cdot\mathrm{m/A}, \quad k_2 = 0.233\,\mathrm{N}\cdot\mathrm{m/A}, \\
&U_1 = U_2 = U = 110\,\mathrm{V}, \quad j_1^0 = 2.9\mathrm{A}, \quad j_2^0 = 2\mathrm{A}.
\end{aligned} \tag{5.32}$$

For the manipulator with parameters (5.32) as the unit of the measurement of the time in formula (5.4), we take the value

$$\bar{T} = n_2 k_2 U_2^{-1} \approx 0.345\,\mathrm{s},$$

which is equal to the time of rotation of the link G_2 by the angle equal to one radian when moving with the stationary angular velocity

$$\dot{\varphi}_2 = (n_2 k_2)^{-1} U_2 \approx 2.9\,\mathrm{s}^{-1}.$$

Using the dimensionless variables in accordance with (5.4) and performing the necessary calculations, we find that $\eta_i^{(1)} < \eta_i^{(2)}$. This means that for optimal control (5.23) the final points $(\varphi_1^{\mathrm{T}}, \varphi_2^{\mathrm{T}})$ must satisfy inequalities (5.28). After substituting the values of parameters (5.32) and the time T, which were made dimensionless by formula (5.4), in the expressions for functionals $\bar{Q}_m$, $m = 1, \ldots, 4$ (5.31) and minimizing them with respect to the origin K_m, $m = 1, \ldots, 4$, we obtain the diagram depicted in Fig. 5.2. This diagram allows us to determine the optimal movement of the manipulator. For the given initial and final values φ_2^0 and φ_1^{T}, φ_2^{T}, using the diagram, one can determine the directions of rotation of the links, which is optimal in the sense of the minimum of the quality functional, and performs the movement of the manipulator to the final position (the diagram can be shifted in a parallel way along the axis φ_2^{T} depending on the value of $\varphi_2^{\mathrm{T}} \neq 0$). The directions of rotations are determined by the index of the functional, which corresponds to the domain in which the point $(\varphi_1^{\mathrm{T}}, \varphi_2^{\mathrm{T}})$ is located. For instance, if the point $(\varphi_1^{\mathrm{T}}, \varphi_2^{\mathrm{T}})$ belongs to the domain denoted by $\bar{Q}_2$, then the first and the second links must be rotated by the angles $2\pi - \varphi_1^{\mathrm{T}}$ clockwise and φ_2^{T} counter – clockwise, respectively.

For complete solution of the problem, i.e., for finding the minimum of functional (5.16) with respect to the parameter p, we must take into account the following fact. The configurations $\{\varphi_1^{\mathrm{T}}, \varphi_2^{\mathrm{T}}\}^{(1)}$ and $\{\varphi_1^{\mathrm{T}}, \varphi_2^{\mathrm{T}}\}^{(2)}$, which correspond to the same position of manipulator's gripper, can belong to the domain with one and the same index of the functional $\bar{Q}_m$, $m = 1, \ldots, 4$. In both cases, we determine the optimal

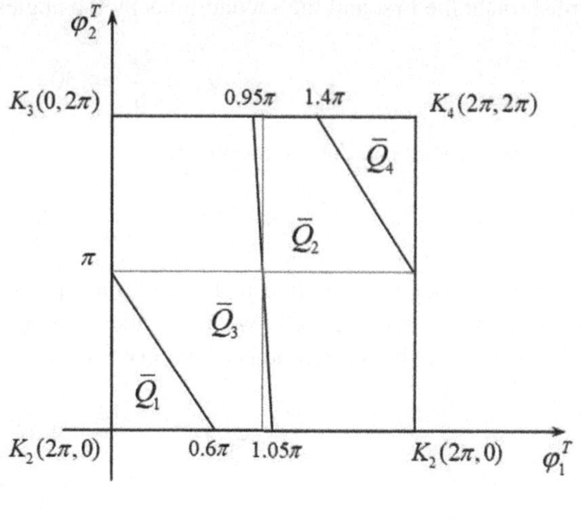

Fig. 5.2 Diagram for determining the optimal directions of rotation of the manipulator links.

values $p = 1$ or $p = -1$ by the simple calculation of the values of the functional by formulas (5.31).

5.5 Example

Let us give the following initial and final coordinates of manipulator's gripper

$$x^0 = 2\,\text{m}, \quad y^0 = 0, \quad x^{\text{T}} = -1.37\,\text{m}, \quad y^{\text{T}} = -0.37\,\text{m}.$$

In the angular variables, following initial and final conditions (5.11), (5.12) correspond to them

$$\begin{aligned} &\varphi_1^0 = \varphi_2^0 = 0, \quad \varphi_1^{\text{T}} = 150^0, \quad \varphi_2^{\text{T}} = 90^0 \quad (p = 1), \\ &\varphi_1^0 = \varphi_2^0 = 0, \quad \varphi_1^{\text{T}} = 240^0, \quad \varphi_2^{\text{T}} = 270^0 \quad (p = -1). \end{aligned} \tag{5.33}$$

In radians, the final angles (5.33) have the following values

$$\begin{aligned} &\varphi_1^{\text{T}} = \frac{5\pi}{6}, \quad \varphi_2^{\text{T}} = \frac{\pi}{2}, \qquad (p = 1), \\ &\varphi_1^{\text{T}} = \frac{4\pi}{3}, \quad \varphi_2^{\text{T}} = \frac{3\pi}{2}, \qquad (p = -1). \end{aligned} \tag{5.34}$$

The verification shows that points (5.34) satisfy inequality (5.28), i.e., they are admissible. From the diagram (Fig. 5.2) constructed for manipulator, whose dynamics is described by equations (5.8) and parameters are defined in (5.32), we find that final point (5.34) for $p = 1$ belongs to the domain, denoted by $\bar{Q}_3$ and point (5.34) for $p = -1$ belongs to the domain, denoted by $\bar{Q}_2$. This means that for the movement of the manipulator from the initial configuration $\varphi_1^0 = 0$, $\varphi_2^0 = 0$ to the final configuration (5.34) ($p = 1$) with the minimal heat losses, in accordance with (5.13), we must rotate the first and the second links by the angles

$$\varphi_1^{\text{T}} = \frac{5\pi}{6} = 150^0$$

counter – clockwise, the second the angles

$$\varphi_2^{\text{T}} = 2\pi - \frac{\pi}{9} = 270^0$$

counter – clockwise relative to the first one, respectively, using optimal rule of control (5.29), see Fig. 5.3. We move the manipulator to the final configuration (5.34) ($p = -1$) by the rotations of the first and second links by the angles

$$\varphi_1^{\text{T}} = 2\pi - \frac{4\pi}{3} = 120^0$$

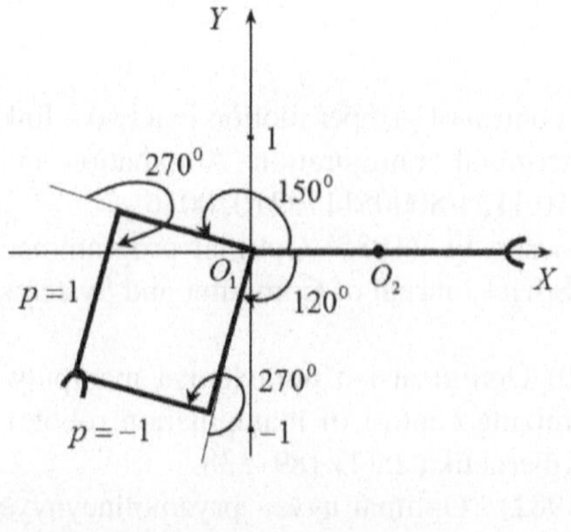

Fig. 5.3: Optimal choice of link rotation directions for configurations $p = 1$ and $p = -1$.

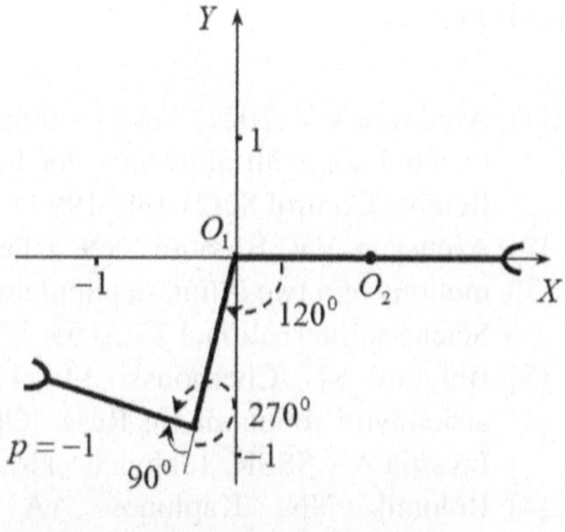

Fig. 5.4: Comparative analysis of the directions of rotation of links for the configuration $p = -1$.

clockwise and

$$\varphi_2^{\mathrm{T}} = \frac{3\pi}{2} = 270^0$$

counter–clockwise relative to the first link, respectively (Fig. 5.3).

If we use the initial variables and calculate the values of the functionals $\bar{Q}_2, \bar{Q}_3$, then we obtain $\bar{Q}_2 = 0.97$ and $\bar{Q}_3 = 1.49$. It means that for the movement of the gripper to the given position, we need to select the final configuration that correspond to the value $p = -1$.

Let us fix the configuration that corresponds to the value $p = -1$ and confirm the importance of choice of the directions of rotations, for manipulator's links by a numerical example. Figure 5.4 shows the final configuration of the manipulator for $p = -1$. At first glance, the rotation of the second link relative to the first by the angle of

$$\varphi_2^{\mathrm{T}} = \frac{\pi}{2} = 90^0$$

clockwise, accompanied by a rotation of the first link by the angle of

$$\varphi_1^{\mathrm{T}} = 2\pi - \frac{4\pi}{3} = 120^0$$

clockwise, seems to be optimal here. However, as the optimization of functionals (5.31) demonstrates, the value of the functional $\bar{Q}_1$, which corresponds to the case considered, appears to be considerably greater than the value of

$$\bar{Q}_3 : \bar{Q}_1 \approx 1.37 > 0.97 \approx \bar{Q}_3.$$

Thus, the example demonstrates that the choice of the direction and the final configuration is of considerable importance in minimizing the heat losses in manipulator's drives during their motion.

References

[1] Avetisyan VV (2021) Time – optimal control of gripper motion in a two – link manipulator with allowance for the terminal configuration. Automation and Remote Control 82(2):189–199 DOI 10.1134/S0005117921020016

[2] Avetisyan VV, Bolotnik NN, Chernousko FL (1985) Optimal programmed motions of a two – link manipulator. Soviet Journal of Computer and Systems Sciences International 23(5) 65–73

[3] Bolotnik NN, Chernousko FL (1990) Optimizatsiya upravleniya manipulyatsionnymi robotami (in Russ., Optimizing control of manipulation robots). Izvestia AN SSSR. Tekhnicheskaya Kibernetika 28(1):189–238

[4] Bolotnik NN, Kaplunov AA (1982) Optimal'nyye pryamolineynyye peremeshcheniya gruza pri pomoshchi dvuzvennogo manipulyatora (in Russ., Optimal linear motions of cargo by a two – link manipulator). Izvestia AN SSSR. Tekhnicheskaya Kibernetika 20(1) 68–74

[5] Bolotnik NN, Kaplunov AA (1983) Optimizatsiya upravleniya i konfiguratsiy dvuzvennogo manipulyatora (in Russ., Optimizing the control and configurations of a two – link manipulator). Izvestia AN SSSR. Tekhnicheskaya Kibernetika 21(4) 123–131

[6] Chernousko FL (1994) Computational Optimal Control. In: Bulirsch, R., Kraft, D. (eds) Computational Optimal Control. ISNM International Series of Numerical Mathematics, vol 115. Birkhäuser, Basel. DOI 10.26516/978-5-9624-2309-8.2024.1-224

[7] Chernousko FL, Bolotnik NN, Gradetsky VG (1994) Manipulation Robots: Dynamics, Control and Optimization. CRC Press, Boca Raton

[8] Demydyuk MV, Hoshovs'ka N (2019) Parametric optimization of the transport operations of a two-manipulator. Journal of Mathematial Sciences 238(2):174–188 DOI 10.1007/s10958-019-04227-8

[9] Galicki M (2020) Energy optimal control of mobile manipulators subject to compensation of external disturbance forces. Mechanism and Machine Theory 167:104550 DOI 10.1016/j.mechmachtheory.2021.104550

[10] Hussain S, Saad M, Baimagambetov A, Saeed K (2025) A robust and energy – efficient trajectory planning framework for high degree of freedom robots, arXiv:2503.11716 DOI 10.48550/arXiv.2503.11716

[11] Khalil W, Dombre E (2002) Modeling, Identification and Control of Robots, CRC Press, Boca Raton

[12] Oršanský P, Ftorek B, Vittek J (2020) Energy optimal trajectories for electro – mechanical systems. IOP Conference Series: Materials Science and Engineering, 776(1):012113 DOI 10.1088/1757-899X/776/1/012113

Chapter 6
On a Control Problem of a Linear Non-Stationary Dynamic System with Integral Constraints on the State Vector

Vanya R. Barseghyan and Emil A. Barseghyan

Abstract This paper investigates a control problem for a linear nonstationary dynamic system subject to multiple integral-type constraints. A control problem is formulated for a linear time-varying dynamic system where integral constraints are imposed on the components of the state vector over two distinct intervals of the system's operational timeline. Conditions for the complete controllability of the described linear dynamic system under integral constraints are obtained. The condition of existence of a solution to this control problem is derived, and the solution itself is constructed. A constructive approach is proposed for solving the control problem for a linear dynamic system with integral constraints. The control function is obtained such that the system trajectory, originating from a prescribed initial state, attaining a final state while simultaneously satisfying all given integral conditions. As an application, the control of a material point moving within a vertical plane under the influence of reactive force and gravitational forces, subject to integral conditions is considered. Explicit analytical expressions for the control functions and corresponding phase trajectories are obtained, and their graphical representations are provided.

Key words: Dynamic system, Non-local control problem, Integral condition, Existence of the solution, Completely controllable, Control function

---

Vanya R. Barseghyan (corresponding author)
Institute of Mechanics, National Academy of Sciences of the Republic of Armenia, 24b Marshal Baghramyan Ave, Yerevan, 0019, & Faculty of Mathematics and Mechanics, Yerevan State University, Yerevan, Republic of Armenia, e-mail: barseghyan@sci.am

Emil A. Barseghyan
Yerevan State University, 1 Alec Manukyan, Yerevan, 0025, Republic of Armenia, e-mail: emil.barsegyann@gmail.com

H. Altenbach and A. V. Sahakyan, *Contact Mechanics and Relevant Problems*,
Advanced Structured Materials 253, https://doi.org/10.1007/978-3-032-20869-9_6

6.1 Introduction

Mathematical modeling of various dynamic processes leads to nonlocal control problems, which are described by both partial differential equations and ordinary differential equations. A characteristic feature of problems with nonlocal conditions is the presence of constraints that link the values of state parameters at different intermediate points or over intervals with integral conditions alongside classical boundary conditions. The necessity to address nonlocal conditions arises from the need to develop mathematical models and investigate the corresponding problems that meet the demands of modern natural science.

The variety of integral-type nonlocal models is examined, in particular, in reference [13], where their physical justifications are also discussed. Nonlocal problems formulated as integral conditions arise in the mathematical modeling of phenomena across diverse fields, particularly in situations where the boundary of the domain where the process occurs is inaccessible for direct measurement [19]. It is noteworthy that problems with nonlocal conditions have been studied for both partial differential equations [2, 15, 20] and ordinary differential equations [1, 3, 5]. In [14], a nonlocal problem with an integral condition for a fourth-order equation is considered, and the existence of a unique solution is proven.

Important applied nonlocal control problems with multipoint intermediate conditions relate to systems which are described by partial differential equations, examined in [4, 8–12, 16, 17], and for systems governed by ordinary differential equations, studied in [6, 7]. Specifically, [16] considers a boundary value problem for the string oscillation equation with a velocity specified at a certain moment in time and constructs its solution. In turn, [17] investigates the classical solution of a mixed problem with an integral condition for the one-dimensional wave equation. In papers [4, 8–12] problems of boundary control and optimal control of string vibrations are examined. For these problems, alongside classical boundary (initial and final) conditions, the values of the deflection function or the velocities of points at intermediate times are prescribed. In papers [6, 7] are dedicated to the spatial motion of mechanical systems (a quadcopter, a two-link manipulator) and the construction of control laws that ensure movements with specified values for part of the phase vector coordinates at intermediate time instants. The control functions and corresponding phase trajectories, which account for the intermediate values of part of the coordinates at given times, have been constructed.

It should be noted, that in a case, when integral conditions are subjected on several time intervals of a dynamic system's operation, they are potentially significant, yet they have not been thoroughly investigated. This paper considers a nonlocal control problem for a linear time-varying dynamic system subject to boundary conditions and integral constraints imposed on the components of the state vector over two distinct parts of the system's operating time interval. Conditions for the complete controllability of the dynamic system under these integral constraints are formulated. The conditions are obtained under which a solution to this control problem exists and proceeds to construct the solution explicitly. A constructive approach for solving both the control and optimal control problems for a linear dynamic system with such integral conditions is proposed. The control functions are constructed, under whose influences the system's phase trajectories satisfy all stipulated conditions of the problem. As an application, the problem of controlling a material point moving in a vertical plane under the influence of a reactive force and gravitational force, subject to integral conditions, is considered. Explicit analytical expressions for the control functions and the corresponding state coordinates are obtained, and their plots are presented.

6.2 Formulation of the Problem

Let us consider a controllable process, dynamic of which is described by linear non-stationary differential equation,

$$\dot{x}(t) = A(t)x(t) + B(t)u(t) + f(t), \quad t \in [t_0, T], \tag{6.1}$$

where $x(t) \in R^n$ is a vector-column, $A(t)-(n\times n),\ B(t)-(n\times r),\ f(t)-(n\times 1)$ are matrices, components of which are continuous on $t_0 \leq t \leq T,\ u(t) - (r \times 1)(r \leq n)$ is a vector-column, whose elements are almost everywhere continuous and bounded on time interval $[t_0, T]$.

Let initial and final states of the system (6.1),

$$x(t_0) = x_0,\ x(T) = x_T, \tag{6.2}$$

be known and intermediate times $0 \leq t_0 < t_1 < t_2 < t_3 < t_4 < T$.

It is considered that state vector $x(t)$ of the system (6.1) satisfies to the following integral constraints,

$$\int_{t_1}^{t_2} F_1(t)x(t)dt = \alpha_1, \tag{6.3}$$

$$\int_{t_3}^{t_4} F_2(t)x(t)dt = \alpha_2, \tag{6.4}$$

where $F_1(t) - (q_1 \times n),\ F_2(t) - (q_2 \times n)$ are matrices with their corresponding dimensions $(q_1 \leq n,\ q_2 \leq n)$, they are almost everywhere continuous and bounded correspondingly on $[t_1, t_2]$ and $[t_3, t_4]$. $\alpha_1-(q_1\times 1), \alpha_2-(q_2\times 1)$ are vector-columns, elements of which are real numbers.

In general, there are possible cases when, on the time intervals $[t_1, t_2]$ and $[t_3, t_4]$ integral constraints (6.3) and (6.4) are not satisfied by all elements of state vector $x(t)$, but only by some elements of its components. In such cases, the corresponding elements of the matrices $F_1(t)$ and $F_2(t)$ are considered to be zero. It is believed that the system (6.1), with integral constraints (6.3) and (6.4), is completely controllable on the time interval $[t_0, T]$.

Let us have a look at the following problem.

Problem: It is required to establish the conditions for the existence of a program control $u = u(t),\ t \in [t_0, T]$ and program motion $x = x(t)$, which transfers the system (6.1) from the initial state $x(t_0)$ to final state $x(T)$ while satisfying the intermediate conditions (6.3) and (6.4), and to construct these functions.

6.3 Solution to the Problem

To solve the problem, we write the solution of (6.1) that starts from the initial state (6.2) in the form

$$x(t) = X[t, t_0]x(t_0) + \int_{t_0}^{t} X[t, \tau]B(\tau)u(\tau)d\tau + \int_{t_0}^{t} X[t, \tau]f(\tau)d\tau, \tag{6.5}$$

and by substituting expression (6.5) into the integral condition (6.3), we obtain

$$\begin{aligned}\int_{t_1}^{t_2} F_1(t)x(t)dt &= \left(\int_{t_1}^{t_2} F_1(t)X[t, t_0,]dt\right)x_0 + \int_{t_1}^{t_2}\left(\int_{t_0}^{t} F_1(t)X[t, \tau]B(\tau)u(\tau)d\tau\right)dt \\ &+ \int_{t_1}^{t_2}\left(\int_{t_0}^{t} F_1(t)X[t, \tau]f(\tau)d\tau\right)dt = \alpha_1,\end{aligned} \tag{6.6}$$

or

$$\begin{aligned}\int_{t_1}^{t_2}\left(\int_{t_0}^{t} F_1(t)X[t, \tau]B(\tau)u(\tau)d\tau\right)dt &= \alpha_1 - \left(\int_{t_1}^{t_2} F_1(t)X[t, t_0,]dt\right)x_0 \\ &- \int_{t_1}^{t_2}\left(\int_{t_0}^{t} F_1(t)X[t, \tau]f(\tau)d\tau\right)dt.\end{aligned} \tag{6.7}$$

By substituting expression (6.5) into the integral condition (6.4), we obtain

$$\begin{aligned}\int_{t_3}^{t_4} F_2(t)x(t)dt &= \left(\int_{t_3}^{t_4} F_2(t)X[t, t_0,]dt\right)x_0 + \int_{t_3}^{t_4}\left(\int_{t_0}^{t} F_2(t)X[t, \tau]B(\tau)u(\tau)d\tau\right)dt \\ &+ \int_{t_3}^{t_4}\left(\int_{t_0}^{t} F_2(t)X[t, \tau]f(\tau)d\tau\right)dt = \alpha_2,\end{aligned} \tag{6.8}$$

or

$$\int_{t_3}^{t_4}\left(\int_{t_0}^{t} F_2(t)X[t,\tau]B(\tau)u(\tau)d\tau\right)dt = \alpha_2 - \left(\int_{t_3}^{t_4} F_2(t)X[t,t_0,]dt\right)x_0 - \int_{t_3}^{t_4}\left(\int_{t_0}^{t} F_2(t)X[t,\tau]f(\tau)d\tau\right)dt. \tag{6.9}$$

The following lemma can be formulated.

Lemma 6.1. *Let the matrix function $F(\xi)$ be almost everywhere continuous and bounded on $[a,b]$ matrix functions $\Phi[\xi,\tau]$, $B(\tau)$ and vector-column $u(\tau)$ be almost everywhere continuous and bounded on $[t_0,T]$. Then for any a and b, $0 \le t_0 < a < b < T$ the following equality holds*

$$\int_{a}^{b}\left(\int_{t_0}^{\xi} F(\xi)\Phi\left[\xi,t_0\right]B(\tau)u(\tau)d\tau\right)d\xi = \int_{t_0}^{a} P(a)\Phi\left[t_0,\tau\right]B(\tau)u(\tau) + \int_{a}^{b} P(\tau)\Phi\left[t_0,\tau\right]B(\tau)u(\tau)d\tau, \tag{6.10}$$

where

$$P(\tau) = \int_{\tau}^{b} F(\xi)\Phi\left[\xi,t_0\right]d\xi, \quad P(a) = \int_{a}^{b} F(\xi)\Phi\left[\xi,t_0\right]d\xi. \tag{6.11}$$

The matrix $\Phi\left[\xi,\tau\right]$ possesses the composition property $\Phi\left[\xi,\tau\right] = \Phi\left[\xi,t_0\right]\Phi\left[t_0,\tau\right]$.

Proof. Consider the second-order integral of the left-hand side of the relation (6.10)

$$I = \int_{a}^{b}\left(\int_{t_0}^{\xi} F(\xi)\Phi\left[\xi,t_0\right]B(\tau)u(\tau)d\tau\right)d\xi.$$

According to Fubini's theorem [21] (concerning the interchange of the order of integration), interchanging the order of the iterated integral the value of it does not change. It is evident that the conditions of Fubini's theorem are satisfied.

Region of integration (Fig. 6.1) of integral I is indeed a union of rectangle $\{(\tau,\xi) : t_0 \le \tau \le a,\ a \le \xi \le b\}$ and a right triangle $\{(\tau,\xi) : a \le \tau \le \xi \le b\}$.

- in a rectangle: $\xi = a,\ \xi = b,\ \tau = t_0,\ \tau = a.$
- in a right triangle: $\xi = b,\ \xi = \tau,\ \tau = a,\ \tau = b.$

Second-order integral I can be decomposed into a sum of integrals over the rectangular and triangular regions:

Fig. 6.1 Region of integration.

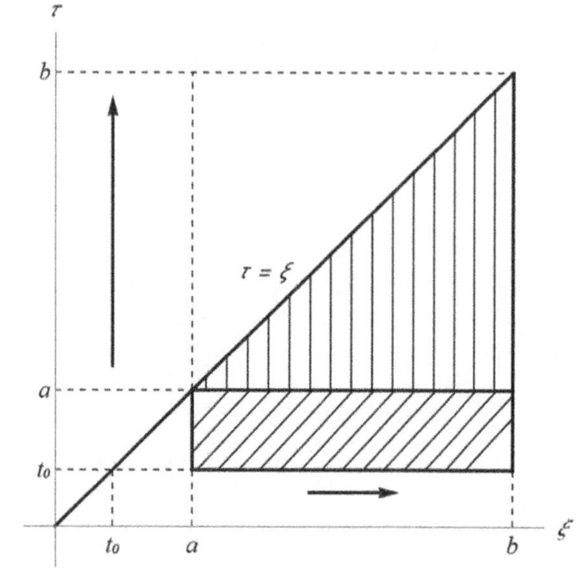

$$I = \int_a^b \left(\int_{t_0}^{a} F(\xi)\Phi[\xi,\tau]B(\tau)u(\tau)d\tau \right) d\xi + \int_a^b \left(\int_a^{\xi} F(\xi)\Phi[\xi,\tau]B(\tau)u(\tau)d\tau \right) d\xi.$$

Interchanging the order of integration, we obtain the following relation:

$$I = \int_{t_0}^{a} \left(\int_a^b F(\xi)\Phi[\xi,\tau]B(\tau)u(\tau)d\xi \right) d\tau + \int_a^b \left(\int_{\tau}^{b} F(\xi)\Phi[\xi,\tau]B(\tau)u(\tau)d\xi \right) d\tau.$$

Taking into account the composition property of the matrix

$$\Phi\left[\xi,\tau\right] = \Phi\left[\xi,t_0\right]\Phi\left[t_0,\tau\right],$$

we obtain

$$I = \int_{t_0}^{a} \left(\int_a^b F(\xi)\Phi\left[\xi,t_0\right]\Phi\left[t_0,\tau\right]B(\tau)u(\tau)d\xi \right) d\tau$$
$$+ \int_a^b \left(\int_{\tau}^{b} F(\xi)\Phi\left[\xi,t_0\right]\Phi\left[t_0,\tau\right]B(\tau)u(\tau)d\xi \right) d\tau.$$

since $\Phi\left[t_0,\tau\right]B(\tau)u(\tau)$ does not depend on ξ, it can be factored out of the inner integral in both terms

$$I = \int_{t_0}^{a} \left(\int_{a}^{b} F(\xi)\Phi\left[\xi, t_0\right] d\xi \right) \Phi\left[t_0, \tau\right] B(\tau)u(\tau)d\tau$$
$$+ \int_{a}^{b} \left(\int_{\tau}^{b} F(\xi)\Phi\left[\xi, t_0\right] d\xi \right) \Phi\left[t_0, \tau\right] B(\tau)u(\tau)d\tau.$$

Considering the notation (6.11), i.e.

$$P(\tau) = \int_{\tau}^{b} F(\xi)\Phi\left[\xi, t_0\right] d\xi, \quad P(a) = \int_{a}^{b} F(\xi)\Phi\left[\xi, t_0\right] d\xi,$$

we obtain

$$I = \int_{t_0}^{a} P(a)\Phi\left[t_0, \tau\right] B(\tau)u(\tau) + \int_{a}^{b} P(\tau)\Phi\left[t_0, \tau\right] B(\tau)u(\tau)d\tau.$$

The lemma is proved.

Let us apply the lemma to double integrals (6.7) and (6.9). For double integral (6.7) $a = t_1, \ b = t_2$, then according to (6.10), we have

$$\int_{t_0}^{t_1} P_1(t_1)X[t_0, \tau]B(\tau)u(\tau)d\tau + \int_{t_1}^{t_2} P_1(\tau)X[t_0, \tau]B(\tau)u(\tau)d\tau$$
$$= \alpha_1 - P_1(t_1)x_0 - \int_{t_1}^{t_2} \left(\int_{t_0}^{t} F_1(t)X[t, \tau]f(\tau)d\tau \right) dt, \tag{6.12}$$

where

$$P_1(t_1) \equiv \int_{t_1}^{t_2} F_1(\tau)X[\tau, t_0]d\tau, \quad P_1(t) \equiv \int_{t}^{t_2} F_1(\tau)X[\tau, t_0]d\tau. \tag{6.13}$$

For double integral (6.7) $a = t_3, \ b = t_4$, then according to (6.11), we have

$$\int_{t_0}^{t_3} P_2(t_3)X[t_0, \tau]B(\tau)u(\tau)d\tau + \int_{t_3}^{t_4} P_2(\tau)X[t_0, \tau]B(\tau)u(\tau)d\tau$$
$$= \alpha_1 - P_2(t_3)x_0 - \int_{t_3}^{t_4} \left(\int_{t_0}^{t} F_2(t)X[t, \tau]f(\tau)d\tau \right) dt, \tag{6.14}$$

where

$$P_2(t_3) \equiv \int_{t_3}^{t_4} F_2(\tau)X[\tau,t_0]d\tau, \quad P_2(t) \equiv \int_{t}^{t_4} F_2(\tau)X[\tau,t_0]d\tau. \tag{6.15}$$

We introduce the following notation:

$$H_1^{(1)}(t) = \begin{cases} P_1(t_1)X[t_0,t]B(t), & t \in [t_0,t_1], \\ 0, & t \in (t_1,T], \end{cases}$$
$$H_1^{(2)}(t) = \begin{cases} 0, & t \in [t_0,t_1), \\ P_1(t)X[t_0,t]B(t), & t \in [t_1,t_2], \\ 0, & t \in (t_2,T], \end{cases} \tag{6.16}$$

$$H_2^{(1)}(t) = \begin{cases} P_2(t_3)X[t_0,t]B(t), & t \in [t_0,t_3], \\ 0, & t \in (t_3,T], \end{cases}$$
$$H_2^{(2)}(t) = \begin{cases} 0, & t \in [t_0,t_3), \\ P_2(t)X[t_0,t]B(t), & t \in [t_3,t_4], \\ 0, & t \in (t_4,T]. \end{cases} \tag{6.17}$$

From notations (6.16) and (6.17), (6.12) and (6.14) can be represented in the following forms

$$\int_{t_0}^{T} H_1^{(1)}(t)u(t)dt + \int_{t_0}^{T} H_1^{(2)}(t)u(t)dt = \alpha_1 - P_1(t_1)x_0 - \int_{t_1}^{t_2}\left(\int_{t_0}^{t} F_1(t)X[t,\tau]f(\tau)d\tau\right)dt,$$

$$\int_{t_0}^{T} H_2^{(1)}(t)u(t)dt + \int_{t_0}^{T} H_2^{(2)}(t)u(t)dt = \alpha_2 - P_2(t_3)x_0 - \int_{t_3}^{t_4}\left(\int_{t_0}^{t} F_2(t)X[t,\tau]f(\tau)d\tau\right)dt$$

or

$$\int_{t_0}^{T} H_1(t)u(t)dt = \alpha_1 - P_1(t_1)x_0 - \int_{t_1}^{t_2}\left(\int_{t_0}^{t} F_1(t)X[t,\tau]f(\tau)d\tau\right)dt, \tag{6.18}$$

$$\int_{t_0}^{T} H_2(t)u(t)dt = \alpha_2 - P_2(t_3)x_0 - \int_{t_3}^{t_4}\left(\int_{t_0}^{t} F_2(t)X[t,\tau]f(\tau)d\tau\right)dt, \tag{6.19}$$

where

$$H_1(t) = H_1^{(1)}(t) + H_1^{(2)}(t) = \begin{cases} P_1(t_1)X[t_0,t]B(t), & t \in [t_0,t_1], \\ P_1(t)X[t_0,t]B(t), & t \in [t_1,t_2], \\ 0, & t \in [t_2,T], \end{cases} \tag{6.20}$$

$$H_2(t) = H_2^{(1)}(t) + H_2^{(2)}(t) = \begin{cases} P_2(t_3)X[t_0,t]B(t), \ t \in [t_0,t_3], \\ P_2(t)X[t_0,t]B(t), \ \ t \in [t_3,t_4], \\ 0, \qquad\qquad\qquad\quad t \in [t_4,T]. \end{cases} \tag{6.21}$$

As a result, instead of the integral conditions (6.12) and (6.13), we obtain the conditions (6.18) and (6.19).

From formula (6.5), for the final moment time $t = T$, we obtain the following integral condition

$$x(T) = X[T,t_0]x(t_0) + \int_{t_0}^{T} X[T,\tau]B(\tau)u(\tau)d\tau + \int_{t_0}^{T} X[T,\tau]f(\tau)d\tau. \tag{6.22}$$

In expressions (6.22), transferring all constant terms to the left-hand side, we obtain

$$\int_{t_0}^{T} X[T,\tau]B(\tau)u(\tau)d\tau = x_T - X[T,t_0]x_0 - \int_{t_0}^{T} X[T,\tau]f(\tau)d\tau. \tag{6.23}$$

Consequently, we obtain the expressions (6.20), (6.21) and (6.23). This means, that the sought control influence $u(t),\ t \in [t_0,T]$ must simultaneously satisfy the conditions (6.20)), (6.21) and (6.23).

Now, let us denote the following block matrices $H(t)$ and C,

$$H(t) = \begin{pmatrix} H_1(t) \\ H_2(t) \\ X[T,t]B(t) \end{pmatrix}, \quad C = \begin{pmatrix} \alpha_1 - P_1(t_1)x_0 - \int_{t_1}^{t_2}\left(\int_{t_0}^{t} F_1(t)X[t,\tau]f(\tau)\,d\tau\right)dt \\ \alpha_2 - P_2(t_3)x_0 - \int_{t_3}^{t_4}\left(\int_{t_0}^{t} F_2(t)X[t,\tau]f(\tau)\,d\tau\right)dt \\ x_T - X[T,t_0]x_0 - \int_{t_0}^{T} X[T,\tau]f(\tau)\,d\tau \end{pmatrix}, \tag{6.24}$$

where matrix $H(t)$ and vector C have the following dimensions $((q_1+q_2+n)\times r)$ and $((q_1+q_2+n)\times 1)$ correspondingly.

Taking into account the notations (6.20) and (6.21),

$$H(t) = \begin{cases} H^{(1)}(t), \ t \in [t_0,t_1], \\ H^{(2)}(t), \ t \in [t_1,t_2], \\ H^{(3)}(t), \ t \in [t_2,t_3], \\ H^{(4)}(t), \ t \in [t_3,t_4], \\ H^{(5)}(t), \ t \in [t_4,T], \end{cases}$$

where

$$H^{(1)}(t) = \begin{pmatrix} P_1(t_1)X[t_0,t]B(t) \\ P_2(t_3)X[t_0,t]B(t) \\ X[T,t]B(t) \end{pmatrix}, \quad H^{(2)}(t) = \begin{pmatrix} P_1(t)X[t_0,t]B(t) \\ P_2(t_3)X[t_0,t]B(t) \\ X[T,t]B(t) \end{pmatrix},$$

$$H^{(3)}(t) = \begin{pmatrix} 0_{q_1\times r} \\ P_2(t_3)X[t_0,t]B(t) \\ X[T,t]B(t) \end{pmatrix},$$

$$H^{(4)}(t) = \begin{pmatrix} 0_{q_1\times r} \\ P_2(t)X[t_0,t]B(t) \\ X[T,t]B(t) \end{pmatrix}, \quad H^{(5)}(t) = \begin{pmatrix} 0_{q_1\times r} \\ 0_{q_2\times r} \\ X[T,t]B(t) \end{pmatrix}.$$

According to the notation (6.24), integral conditions (6.20), (6.21) and (6.23), will be written in the subsequent form

$$\int_{t_0}^{T} H(t)u(t)dt = C. \tag{6.25}$$

Then, by denoting the matrix

$$Q = \int_{t_0}^{\mathrm{T}} H(t)H^{\mathrm{T}}(t)dt \tag{6.26}$$

of dimension $((q_1+q_2+n)) \times ((q_1+q_2+n))$ and considering, that $\det Q \neq 0$, control influence, which satisfies to (6.25), can be represented in the form

$$u(t) = H^{\mathrm{T}}(t)Q^{-1}C + v(t), \tag{6.27}$$

where letter "T" in the subscript indicates the operation of transposition, vector $v(t)$ is orthogonal to the block matrix $H(t)$, that is

$$\int_{t_0}^{T} H(t)v(t)dt = 0.$$

Following [18, 22], the subsequent assertions on the complete controllability of the system (6.1) with integral constraints (6.3) and (6.4), as well as on the existence of a control solution, are true.

The conditions for complete controllability of system (6.1) with integral conditions (6.3) and (6.4) can be formulated as the following theorem.

Theorem 6.1 (Controllability). *For system* (6.1) *with integral conditions* (6.3) *and* (6.4), *which is completely controllable on the interval* $[t_0, T]$, *it is necessary and sufficient that the column vectors of matrix* $H(t)$ (6.25) *be linearly independent on this interval.*

The condition of existence of a solution to the control problem for system (6.1) with integral conditions (6.3) and (6.4) can be formulated as the following theorem.

Theorem 6.2 (Control Influence). *A necessary and sufficient condition for the existence of a program control and its associated solution of system* (6.1), *satisfying conditions* (6.2)-(6.4), *is that matrix* Q (6.26) *be nonsingular (i.e.* $\det Q \neq 0$ *) or the ranks of the matrix* Q *and its augmented matrix* $\{Q, C\}$ *be equal.*

Note that, according to theorem 6.2, in the case of, $\det Q \neq 0$, control function $u(t)$, $t \in [t_0, T]$, solving problem (6.1)-(6.4) is not unique and can be represented in the form (6.27).

Taking into account the notation (6.24), consequently (6.20) and (6.21)) as well, control influence is represented in the form

$$u(t) = \begin{cases} B^{\mathrm{T}}(t)\left((P_1(t_1)X[t_0,t])^{\mathrm{T}}, (P_2(t_3)X[t_0,t])^{\mathrm{T}}, X^{\mathrm{T}}[T,t]\right) Q^{-1}C + v(t), \\ t \in [t_0, t_1], \\ B^{\mathrm{T}}(t)\left((P_1(t)X[t_0,t])^{\mathrm{T}}, (P_2(t_3)X[t_0,t])^{\mathrm{T}}, X^{\mathrm{T}}[T,t]\right) Q^{-1}C + v(t), \\ t \in [t_1, t_2], \\ B^{\mathrm{T}}(t)\left(0_{q_1 \times n}, (P_2(t_3)X[t_0,t])^{\mathrm{T}}, X^{\mathrm{T}}[T,t]\right) Q^{-1}C + v(t), \\ t \in [t_2, t_3], \\ B^{\mathrm{T}}(t)\left(0_{q_1 \times n}, (P_2(t)X[t_0,t])^{\mathrm{T}}, X^{\mathrm{T}}[T,t]\right) Q^{-1}C + v(t), \\ t \in [t_3, t_4], \\ B^{\mathrm{T}}(t)\left(0_{q_1 \times n}, 0_{q_2 \times n}, X^{\mathrm{T}}[T,t]\right) Q^{-1}C + v(t), \\ t \in [t_4, T]. \end{cases} \tag{6.28}$$

Note that the obtained control influence $u(t)$, $t \in [t_0, T]$ is a continuous function. It can be indeed be verified by direct computation that at the times $t = t_k$, $(k = 1, 2, 3, 4)$, the values of the function $u(t)$ are identical when approached from the left and right intervals. For example,

- for $t \in [t_0, t_1]$

$$u(t) = B^{\mathrm{T}}(t)\left((P_1(t_1)X[t_0,t])^{\mathrm{T}}, (P_2(t_3)X[t_0,t])^{\mathrm{T}}, X^{\mathrm{T}}[T,t]\right) Q^{-1}C + v(t),$$

when $t = t_1$

$$u(t_1) = B^{\mathrm{T}}(t_1)\left((P_1(t_1)X[t_0,t_1])^{\mathrm{T}}, (P_2(t_3)X[t_0,t_1])^{\mathrm{T}}, X^{\mathrm{T}}[T,t_1]\right) Q^{-1}C + v(t_1),$$

- for $t \in [t_1, t_2]$

$$u(t) = B^{\mathrm{T}}(t)\left((P_1(t)X[t_0,t])^{\mathrm{T}}, (P_2(t_3)X[t_0,t])^{\mathrm{T}}, X^{\mathrm{T}}[T,t]\right) Q^{-1}C + v(t),$$

when $t = t_1$

$$u(t_1) = B^{\mathrm{T}}(t_1)\left((P_1(t_1)X[t_0,t_1])^{\mathrm{T}}, (P_2(t_3)X[t_0,t_1])^{\mathrm{T}}, X^{\mathrm{T}}[T,t_1]\right) Q^{-1}C + v(t_1).$$

Similarly, it can be demonstrated that the function's value at the terminal point of one interval matches the value at the initial point of the subsequent interval.

Therefore, for the constructed control functions $u(t),\ t \in [t_0, T]$, the motion laws of system (6.1) take the following forms

- for $t \in [t_0, t_1]$

$$\begin{aligned} x(t) &= X[t,t_0]x_0 + \int_{t_0}^{t} X[t,\tau]B(\tau) \\ &\times \left[B^{\mathrm{T}}(\tau)\left((P_1(t_1)X[t_0,\tau])^{\mathrm{T}}, (P_2(t_3)X[t_0,\tau])^{\mathrm{T}}, X^{\mathrm{T}}[T,\tau]\right) Q^{-1}C \right. \\ &\left. + \ v(\tau)\right] d\tau, \end{aligned}$$

- for $t \in [t_1, t_2]$

$$\begin{aligned} x(t) &= X[t,t_0]x_0 + \int_{t_0}^{t_1} X[t,\tau]B(\tau) \\ &\times \left[B^{\mathrm{T}}(\tau)\left((P_1(t_1)X[t_0,\tau])^{\mathrm{T}}, (P_2(t_3)X[t_0,\tau])^{\mathrm{T}}, X^{\mathrm{T}}[T,\tau]\right) Q^{-1}C + v(\tau)\right] d\tau \\ &+ \int_{t_1}^{t} X[t,\tau]B(\tau)[B^{\mathrm{T}}(\tau)[(P_1(\tau)X[t_0,\tau])^{\mathrm{T}}, (P_2(t_3)X[t_0,\tau])^{\mathrm{T}}, X^{\mathrm{T}}[T,\tau]]Q^{-1}C \\ &+ v(\tau)]d\tau, \end{aligned}$$

- for $t \in [t_2, t_3]$

$$
\begin{aligned}
x(t) &= X[t,t_0]x_0 + \int_{t_0}^{t_1} X[t,\tau]B(\tau)\\
&\times \left[B^{\mathrm{T}}(\tau)\left((P_1(t_1)X[t_0,\tau])^{\mathrm{T}}, (P_2(t_3)X[t_0,\tau])^{\mathrm{T}}, X^{\mathrm{T}}[T,\tau]\right)Q^{-1}C + v(\tau)\right] d\tau\\
&+\int_{t_1}^{t_2} X[t,\tau]B(\tau)[B^{\mathrm{T}}(\tau)\left[(P_1(t_1)X[t_0,\tau])^{\mathrm{T}}, (P_2(t_3)X[t_0,\tau])^{\mathrm{T}}, X^{\mathrm{T}}[T,\tau]\right]Q^{-1}C\\
&+v(\tau)]d\tau + \int_{t_2}^{t} X[t,\tau]B(\tau)[B^{\mathrm{T}}(\tau)\left[(0_{q_1\times n}, (P_2(t_3)X[t_0,\tau])^{\mathrm{T}}, X^{\mathrm{T}}[T,\tau]\right]Q^{-1}C\\
&+v(\tau)]d\tau,
\end{aligned}
$$

- for $t \in [t_3, t_4]$

$$
\begin{aligned}
x(t) &= X[t,t_0]x_0 + \int_{t_0}^{t_1} X[t,\tau]B(\tau)\\
&\times \left[B^{\mathrm{T}}(\tau)\left[(P_1(t_1)X[t_0,\tau])^{\mathrm{T}}, (P_2(t_3)X[t_0,\tau])^{\mathrm{T}}, X^{\mathrm{T}}[T,\tau]\right]Q^{-1}C + v(\tau)\right] d\tau\\
&+\int_{t_1}^{t_2} X[t,\tau]B(\tau)[B^{\mathrm{T}}(\tau)\left[(P_1(t_1)X[t_0,\tau])^{\mathrm{T}}, (P_2(t_3)X[t_0,\tau])^{\mathrm{T}}, X^{\mathrm{T}}[T,\tau]\right]Q^{-1}C\\
&+v(\tau)]d\tau\\
&+\int_{t_2}^{t_3} X[t,\tau]B(\tau)[B^{\mathrm{T}}(\tau)\left[(0_{q_1\times n}, (P_2(t_3)X[t_0,\tau])^{\mathrm{T}}, X^{\mathrm{T}}[T,\tau]\right]Q^{-1}C + v(\tau)]d\tau\\
&+\int_{t_3}^{t} X[t,\tau]B(\tau)[B^{\mathrm{T}}(\tau)\left[(0_{q_1\times n}, (P_2(\tau)X[t_0,\tau])^{\mathrm{T}}, X^{\mathrm{T}}[T,\tau]\right]Q^{-1}C + v(\tau)]d\tau,
\end{aligned}
$$

- for $t \in [t_4, T]$

$$
\begin{aligned}
x(t) &= X[t,t_0]x_0 + \int_{t_0}^{t_1} X[t,\tau]B(\tau)\\
&\times \left[B^{\mathrm{T}}(\tau)\left((P_1(t_1)X[t_0,\tau])^{\mathrm{T}}, (P_2(t_3)X[t_0,\tau])^{\mathrm{T}}, X^{\mathrm{T}}[T,\tau]\right)Q^{-1}C + v(\tau)\right] d\tau\\
&+\int_{t_1}^{t_2} X[t,\tau]B(\tau)[B^{\mathrm{T}}(\tau)\left[(P_1(t_1)X[t_0,\tau])^{\mathrm{T}}, (P_2(t_3)X[t_0,\tau])^{\mathrm{T}}, X^{\mathrm{T}}[T,\tau]\right]Q^{-1}C\\
&+v(\tau)]d\tau
\end{aligned}
$$

$$+\int_{t_2}^{t_3} X[t,\tau]B(\tau)[B^{\mathrm{T}}(\tau)\left((0_{q_1\times n},(P_2(t_3)X[t_0,\tau])^{\mathrm{T}},X^{\mathrm{T}}[T,\tau]\right)Q^{-1}C+v(\tau)]d\tau$$

$$+\int_{t_3}^{t_4} X[t,\tau]B(\tau)[B^{\mathrm{T}}(\tau)\left((0_{q_1\times n},(P_2(\tau)X[t_0,\tau])^{\mathrm{T}},X^{\mathrm{T}}[T,\tau]\right)Q^{-1}C+v(\tau)]d\tau$$

$$+\int_{t_4}^{t} X[t,\tau]B(\tau)[B^{\mathrm{T}}(\tau)\left((0_{q_1\times n},0_{q_2\times n},X^{\mathrm{T}}[T,\tau]\right)Q^{-1}C+v(\tau)]d\tau.$$

The obtained phase motions are free of discontinuities, as the function values at the terminal point of each interval match those at the initial point of the subsequent interval. A case in point, one can verify that at the end of the first interval and the beginning of the second interval, the function $x(t)$ assumes the values

- for $t\in[t_0,t_1]$, when $t=t_1$

$$x(t_1)=X[t_1,t_0]x_0+\int_{t_0}^{t_1} X[t_1,\tau]B(\tau)$$

$$\times\left[B^{\mathrm{T}}(\tau)\left((P_1(t_1)X[t_0,\tau])^{\mathrm{T}},(P_2(t_3)X[t_0,\tau])^{\mathrm{T}},X^{\mathrm{T}}[T,\tau]\right)Q^{-1}C+v(\tau)\right]d\tau,$$

- for $t\in[t_1,t_2]$, when $t=t_1$

$$x(t_1)=X[t_1,t_0]x_0+\int_{t_0}^{t_1} X[t_1,\tau]B(\tau)$$

$$\times\left[B^{\mathrm{T}}(\tau)\left((P_1(t_1)X[t_0,\tau])^{\mathrm{T}},(P_2(t_3)X[t_0,\tau])^{\mathrm{T}},X^{\mathrm{T}}[T,\tau]\right)Q^{-1}C+v(\tau)\right]d\tau$$

$$+\int_{t_1}^{t_1} X[t_1,\tau]B(\tau)[B^{\mathrm{T}}(\tau)\left((P_1(\tau)X[t_0,\tau])^{\mathrm{T}},(P_2(t_3)X[t_0,\tau])^{\mathrm{T}},X^{\mathrm{T}}[T,\tau]\right)Q^{-1}C$$

$$+v(\tau)]d\tau,$$

where second term on right-hand side is zero; therefor, we obtain

$$x(t_1)=X[t_1,t_0]x_0+\int_{t_0}^{t_1} X[t_1,\tau]B(\tau)$$

$$\times\left[B^{\mathrm{T}}(\tau)\left((P_1(t_1)X[t_0,\tau])^{\mathrm{T}},(P_2(t_3)X[t_0,\tau])^{\mathrm{T}},X^{\mathrm{T}}[T,\tau]\right)Q^{-1}C+v(\tau)\right]d\tau.$$

Clearly, the values of $x(t_1)$ are identical. Thus, the control influence $u(t),\ t\in[t_0,T]$ and the phase trajectory $x(t),\ t\in[t_0,T]$ of the system are continuous.

6.4 Problem of Controlling a Material Point Moving in a Vertical Plane Under the Influence of a Reactive Force and Gravity with Integral Conditions

Consider the control problem of a point M, whose mass is m, moving in a vertical plane (ξ, η) under gravity [18]. The control influence is assumed to be the reactive force $\overline{f}$, applied to the point, generated by the ejection of particles of elementary mass $|dm_1|$. Thus, the point's mass $m = m(t)$ is variable, and its motion can be represented by a vector equation

$$m\frac{d\overline{v}}{dt} = \overline{P} + \overline{f}. \tag{6.29}$$

Here $m = m(t) = m_0 + m_1(t)$, where $m_0 = const$ - is the unchangeable part of the mass, $m_1(t) \geq 0$-is the reactive mass of a point; $f = (\overline{s} - \overline{v})\frac{dm_1}{dt}$; $\overline{v}$ - is the absolute velocity of point M; $\overline{s}$ is the vector of velocity of the particle dm_1 at time $t + dt$ following its separation, so that $\overline{a} = \overline{s} - \overline{v}$ is the relative speed of the separated particle; $\overline{P} = m\overline{g}$ - gravitational force. We will consider the reactive force as a control influence.

By projecting equation (6.29) onto the horizontal and vertical axes, the resulting equations of motion are

$$\begin{cases} m(t)\ddot{\xi} = \dot{m}a_\xi(t), \\ m(t)\ddot{\eta} = \dot{m}a_\eta(t) - m(t)g, \end{cases} \tag{6.30}$$

where a_ξ and a_η - are the projections of vector on axes ξ and η. Supposing that the absolute value of the vector $\overline{a}$ is specified and equal to σ, the system of projected equations can be expressed in normal form:

$$\dot{x_1} = x_2, \quad \dot{x_2} = u_1, \quad \dot{x_3} = x_4, \quad \dot{x_4} = u_2 - g, \tag{6.31}$$

where $x_1 = \xi$, $x_2 = \dot{\xi}$, $x_3 = \eta$, $x_4 = \dot{\eta}$, $u_1 = \sigma \cos\alpha_\xi \frac{\dot{m}}{m}$, $u_2 = \sigma \cos\alpha_\eta \frac{\dot{m}}{m}$; α_ξ and α_η - are angles, composed by the vector $\overline{a}$ along the axes ξ and η, such that $u_1^2 + u_2^2 = \sigma^2\left(\frac{\dot{m}}{m}\right)$.

Let the time instants $t_0 = 0$, $T = 30$ be given, along with the following initial and final states of the system:

$$x(t_0) = x(0) = (0, 25, 0, 40)^{\mathrm{T}}, \quad x(T) = x(30) = (1000, 30, 500, 50)^{\mathrm{T}}. \tag{6.32}$$

Let the following intermediate time instants also be given $t_1 = 3$, $t_2 = 7$, $t_3 = 10$, $t_4 = 15$ similarly the integral conditions of type (6.3) and (6.4) have the following representation

$$\int_{t_1}^{t_2} x_2(t)dt = \alpha_1,$$

$$\int_{t_3}^{t_4} [x_2(t) + x_4(t)]\, dt = \alpha_2$$

or

$$\int_{t_1}^{t_2} F_1(t)x(t)dt = \alpha_1, \tag{6.33}$$

$$\int_{t_3}^{t_4} F_2(t)x(t)dt = \alpha_2, \tag{6.34}$$

where matrices $F_1(t)$ and $F_2(t)$ are given as follows:

$$F_1(t) = (0\ \ 1\ \ 0\ \ 0),\quad F_2(t) = (0\ \ 1\ \ 0\ \ 1),\quad \alpha_1 = 55,\quad \alpha_2 = 68.$$

The fundamental matrix of the homogeneous part of the system (6.31) takes the form

$$X[t, t_0] = \begin{pmatrix} 1 & t - t_0 & 0 & 0 \\ 0 & 1 & 0 & 0 \\ 0 & 0 & 1 & t - t_0 \\ 0 & 0 & 0 & 1 \end{pmatrix}.$$

It is required to find the control influence $u(t) = (u_1(t), u_2(t))^{\mathrm{T}}$ under which the system (6.31) transfers from the initial state $x(t_0) = x_0$ to final state $x(T) = x_T$ while satisfying the integral conditions (6.33) and (6.34).

According to the formulas (6.20), (6.21) and (6.24), matrix $H(t)$ is represented in the form

$$H(t) = \begin{cases} H^{(1)}(t),\ t \in [0, 3], \\ H^{(2)}(t),\ t \in [3, 7], \\ H^{(3)}(t),\ t \in [7, 10], \\ H^{(4)}(t),\ t \in [10, 15], \\ H^{(5)}(t),\ t \in [15, 30], \end{cases}$$

where

$$H^{(1)}(t) = \begin{pmatrix} 4 & 0 \\ 5 & 5 \\ -t & 0 \\ 1 & 0 \\ 0 & -t \\ 0 & 1 \end{pmatrix},\quad H^{(2)}(t) = \begin{pmatrix} 7-t & 0 \\ 5 & 5 \\ -t & 0 \\ 1 & 0 \\ 0 & -t \\ 0 & 1 \end{pmatrix},\quad H^{(3)}(t) = \begin{pmatrix} 0 & 0 \\ 5 & 5 \\ -t & 0 \\ 1 & 0 \\ 0 & -t \\ 0 & 1 \end{pmatrix},$$

$$H^{(4)}(t) = \begin{pmatrix} 0 & 0 \\ 15-t & 15-t \\ -t & 0 \\ 1 & 0 \\ 0 & -t \\ 0 & 1 \end{pmatrix}, \quad H^{(5)}(t) = \begin{pmatrix} 0 & 0 \\ 0 & 0 \\ -t & 0 \\ 1 & 0 \\ 0 & -t \\ 0 & 1 \end{pmatrix}.$$

Then matrix Q (according (6.26)) and vector C (according to (6.24)) obtain the following values

$$Q = \begin{pmatrix} \frac{208}{3} & 100 & -\frac{158}{3} & 20 & 0 & 0 \\ 100 & \frac{1750}{3} & -\frac{2375}{6} & \frac{125}{2} & -\frac{2375}{6} & \frac{125}{2} \\ -\frac{158}{3} & -\frac{2375}{6} & 9000 & -450 & 0 & 0 \\ 20 & \frac{125}{2} & -450 & 30 & 0 & 0 \\ 0 & -\frac{2375}{6} & 0 & 0 & 9000 & -450 \\ 0 & \frac{125}{2} & 0 & 0 & -450 & 30 \end{pmatrix}, \quad C = \begin{pmatrix} -45 \\ 356.25 \\ 100 \\ 5 \\ -5414.5 \\ 304.3 \end{pmatrix}.$$

Since matrix Q is nonsingular ($\det Q = 8.14726 \times 10^{12} \neq 0$), then utilizing the obtained result, control functions $u_1(t)$ and $u_2(t)$ will be in the form

- for $t \in [t_0, t_1]$

$$u_1(t) = -1.3248 - 0.8439t, \qquad u_2(t) = 5.9154 - 0.121t;$$

- for $t \in [t_1, t_2]$

$$u_1(t) = -86215 - 1.5883t, \qquad u_2(t) = 5.9154 - 0.121t;$$

- for $t \in [t_2, t_3]$

$$u_1(t) = 8.4041 - 0.8439t, \qquad u_2(t) = 5.9154 - 0.121t;$$

- for $t \in [t_3, t_4]$

$$u_1(t) = -12.3145 + 1.2279t, \qquad u_2(t) = -14.8032 - 1.9508t;$$

- for $t \in [t_4, T]$

$$u_1(t) = -18.7635 - 0.8439t, \qquad u_2(t) = 16.2178 - 0.121t.$$

The plots of the control functions $u_1(t)$ and $u_2(t)$ over the interval $t \in [t_0, T]$, are shown in Figs. 6.2 and 6.3.

Fig. 6.2 Plot of $u_1 = u_1(t)$

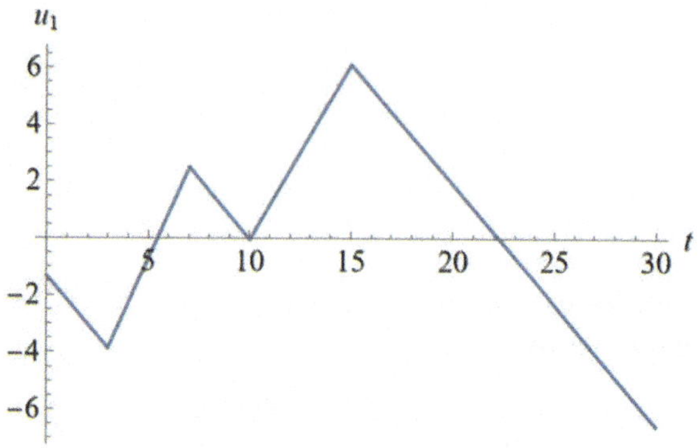

Fig. 6.3 Plot of $u_2 = u_2(t)$

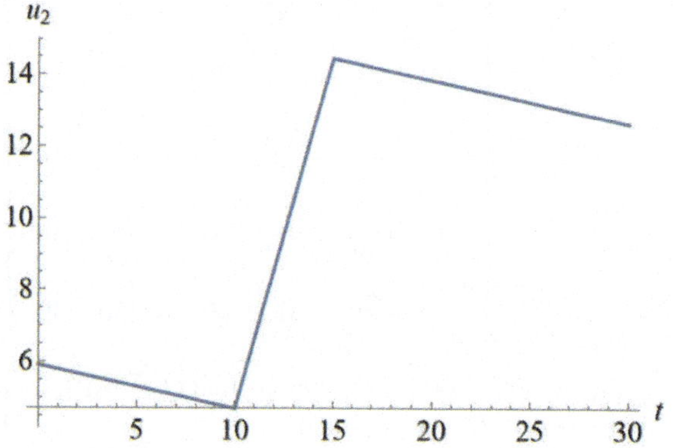

For the constructed control functions $u_1(t)$ and $u_2(t)$, the state coordinates of the system's motion are represented in the form

- for $t \in [t_0, t_1]$

$$x_1(t) = 25t - 0.6624t^2 - 0.1406t^3, \qquad x_2(t) = 25 - 1.3248t - 0.4219t^2,$$

$$x_3(t) = 40t - 1.9472t^2 - 0.0201t^3, \qquad x_4(t) = 40 - 3.8945t - 0.0605t^2,$$

- for $t \in [t_1, t_2]$

$$x_1(t) = -10.9451 + 35.9451t - 4.3107t^2 + 0.2647t^3,$$

$$x_2(t) = 35.9451 - 8.6215t + 0.7941t^2,$$

$$x_3(t) = 40t - 1.9472t^2 - 0.0201t^3, \qquad x_4(t) = 40 - 3.8945t - 0.0605t^2,$$

- for $t \in [t_2, t_3]$

$$x_1(t) = -128.098 - 23.6449t + 4.202t^2 - 0.1406t^3,$$

$$x_2(t) = -23.645 + 8.4041t - 0.422t^2,$$

$x_3(t) = 40t - 2t^2 - 0.0201t^3,\qquad x_4(t) = 40 - 3.8945t - 0.0605t^2,$

- for $t \in [t_3, t_4]$

$x_1(t) = -217.21 - 79.948t - 6.157t^2 + 0.204t^3,\quad x_2(t) = 79.94 - 12.314t - 0.61t^2,$

$x_3(t) = -345.31 + 143.59t - 12.3t^2 + 0.32t^3,\quad x_4(t) = 143.59 - 24.61t - 0.97t^2,$

- for $t \in [t_4, T]$

$x_1(t) = 948.2 - 153.1t + 9.38t^2 - 0.14t^3,\qquad x_2(t) = -153.1 + 18.76t - 0.42t^2,$

$x_3(t) = -820.11 - 89.49t - 3.23t^2 - 0.02t^3,\qquad x_4(t) = -89.4 + 6.46t - 0.06t^2.$

Note, that by substituting the obtained functions of the system motion in integral conditions (6.33) and (6.34) we can verify that these conditions are satisfied, i.e.

$$\int_3^7 x_2(t)dt = 55, \qquad \int_{10}^{15} [x_2(t) + x_4(t)]\, dt = 68.$$

The plots of functions $x_1(t),\ x_2(t),\ x_3(t),\ x_4(t)$ are shown in Figs. 6.4 - 6.7.

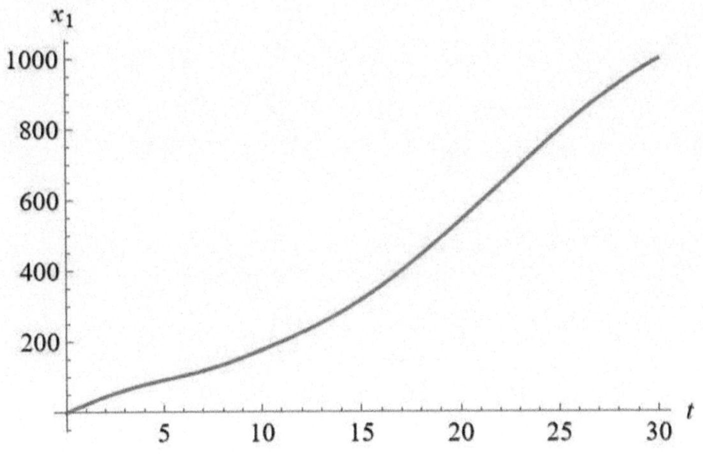

Fig. 6.4: Plot of $x_1 = x_1(t)$.

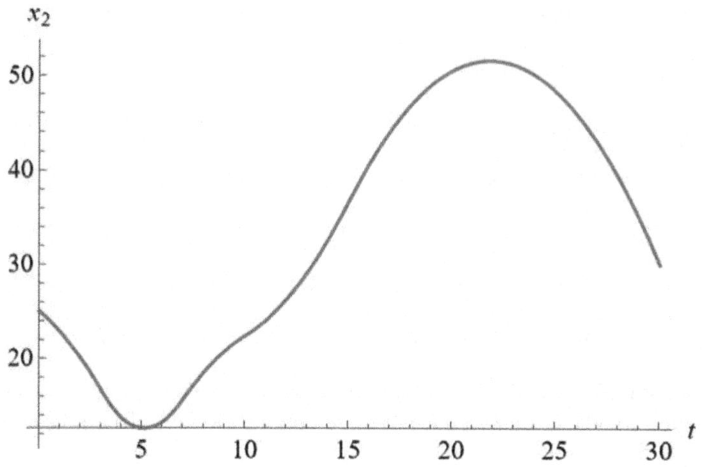

Fig. 6.5: Plot of $x_2 = x_2(t)$.

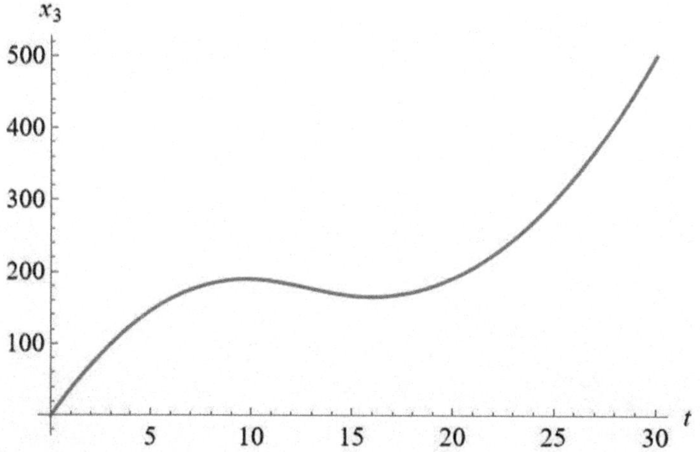

Fig. 6.6: Plot of $x_3 = x_3(t)$.

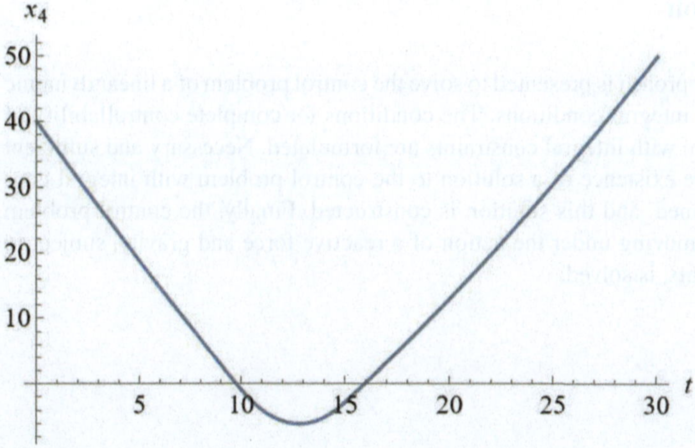

Fig. 6.7: Plot of $x_4 = x_4(t)$.

The trajectory of motion (i.e, the dependence on $x_1(t)\,x_3(t)$) is represented in Fig. 6.8.

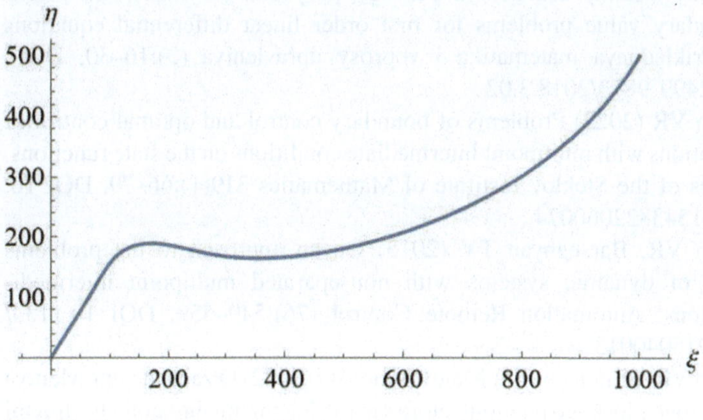

Fig. 6.8: The trajectory of an object's motion $\eta = \eta(\xi)$.

6.5 Conclusion

A constructive approach is presented to solve the control problem of a linear dynamic system with two integral conditions. The conditions for complete controllability of a dynamic system with integral constraints are formulated. Necessary and sufficient conditions for the existence of a solution to the control problem with integral constraints are obtained, and this solution is constructed. Finally, the control problem of a point mass moving under the action of a reactive force and gravity, subject to integral constraints, is solved.

References

[1] Abdullaev VM (2012) Resheniye differentsial'nykh uravneniy s nerazdelennymi mnogotochechnymi i integral'nymi usloviyami (in Russ., Solution of differential equations with unseparated multipoint and integral conditions). Sibirskiy zhurnal industrial'noy matematiki 15(3):3–15

[2] Avalishvili G, Avalishvili M, Gordeziani D (2011) On integral nonlocal boundary value problems for some partial differential equations. Bulletin of the Georgian National Academy of Sciences 5(1):31–37

[3] Baiburin MM (2018) O mnogotochechnykh zadachakh dlya lineynoy differentsial'noy sistemy uravneniya pervogo poryadka (in Russ., On multipoint boundary value problems for first order linear differential equations system). Prikladnaya matematika i voprosy upravleniya (3):16–30, DOI 10.15593/2499-9873/2018.3.02

[4] Barseghyan VR (2022) Problems of boundary control and optimal control of string vibrations with multipoint intermediate conditions on the state functions. Proceedings of the Steklov Institute of Mathematics 319(1):66–79, DOI 10.1134/S0081543822060074

[5] Barseghyan VR, Barseghyan TV (2015) On an approach to the problems of control of dynamic systems with nonseparated multipoint intermediate conditions. Automation Remote Control (76):549–559, DOI 10.1134/S0005117915040013

[6] Barseghyan VR, Simonyan TA, Matevosyan AG (2022) O zadache upravleniya dvukhzvennogo ploskogo manipulyatora s nerazdelennymi mnogotochechnymi promezhutochnymi usloviyami (in Russ., On the control problem of a two-link flat manipulator with nonseparate multipoint intermediate conditions). Vestnik Permskogo universiteta Matematika Mekhanika Informatika 2(57):53–60, DOI 10.17072/1993-0550-2022-2-53-60

[7] Barseghyan VR, Simonyan TA, Matevosyan AG (2024) Ob odnoy zadache upravleniya kvadrokopterom s zadannymi promezhutochnymi znacheniyami raznykh chastey koordinat (in Russ., On one problem of quadcopter control with given intermediate values of different parts of coordinates). Vest-

nik rossiyskikh universitetov Matematika 29(145):29–42, DOI 10.20310/2686-9667-2024-29-145-29-42
[8] Barseghyan V, Solodusha S (2022) Control of string vibrations by displacement of one end with the other end fixed, given the deflection form at an intermediate moment of time. Axioms 11(4):157 DOI 10.3390/axioms11040157
[9] Barseghyan VR, Solodusha SV (2021) On one boundary control problem of string vibrations with given velocity of points at an intermediate moment of time. Journal of Physics: Conference Series 1847(1):012016, DOI 10.1088/1742-6596/1847/1/012016
[10] Barseghyan V, Solodusha S (2021) On one problem in optimal boundary control for string vibrations with a given velocity of points at an intermediate moment of time. In: 2021 International Russian Automation Conference (RusAutoCon), pp 343–349, DOI 10.1109/RusAutoCon52004.2021.9537514
[11] Barseghyan V, Solodusha S (2022) On the optimal control problem for vibrations of the rod/string consisting of two non-homogeneous sections with the condition at an intermediate time. Mathematics 10(23):4444, DOI 10.3390/math10234444
[12] Barseghyan V, Solodusha S (2020) Zadacha granichnogo upravleniya kolebaniyami struny smeshcheniyem levogo kontsa pri zakreplennom pravom kontse s zadannymi znacheniyami funktsii progiba v promezhutochnyye momenty vremeni (in Russ., The problem of boundary control of string vibrations by displacement of the left end when the right end is fixed with the given values of the deflection function at intermediate times). Vestnik rossiyskikh universitetov Matematika 25(130):131–146, DOI 10.20310/2686-9667-2020-25-130-131-146
[13] Bažant Z, Jirásek M (2002) Nonlocal integral formulation of plasticity and damage: Survey of progress. Journal of Engineering Mechanics 128(11):1119–1149, DOI 10.1061/(ASCE)0733-9399(2002)128:11(1119)
[14] Bogatov A, Gilev A, Pul'kina L (2022) Zadacha s nelokal'nym usloviyem dlya uravneniya chetvertogo poryadka s kratnymi kharakteristikami (in Russ., A problem with a non-local condition for a fourth-order equation with multiple characteristics). Vestnik rossiyskikh universitetov Matematika 27(139):214–230, DOI 10.20310/2686-9667-2022-27-139-214-230
[15] Il'in V, Moiseev E (2000) Uniqueness of the solution of a mixed problem for the wave equation with nonlocal boundary conditions. Differential Equations 36(5):728–733, DOI 10.1007/BF02754231
[16] Korzyuk VI, Kozlovskaya IS (2011) Dvukhtochechnaya granichnaya zadacha dlya uravneniya kolebaniya struny s zadannoy skorost'yu v nekotoryy moment vremeni. II (in Russ., Two-point boundary problem for string oscillation equation with given velocity in arbitrary point of time. II). Izvestiya Nacional'noi Akademii Nauk Belarusi Trudy Instituta matematiki 19(1):62–70
[17] Korzyuk VI, Stolyarchuk II (2016) Klassicheskoye resheniye smeshannoy zadachi dlya volnovogo uravneniya s integral'nym usloviye (in Russ., Classical solution to the mixed problem for the wave equation with the integral condition). Doklady Natsional'noi akademii nauk Belarusi 60(6):22–27

[18] Krasovskii NN (1968) Teoriya upravleniya dvizheniyem. Lineynyye sistemy (in Russ., Theory of Control of Motion. Linear Systems). Nauka, Moscow
[19] Lions J (1988) Contrôlabilité exacte perturbations et stabilisation des systèmes distribués, vol 1: Contrôlabilité exacte. Masson, Paris, Milan & Barcelone
[20] Pul'kina LS (2012) Boundary-value Problems for a Hyperbolic Equation with Nonlocal Conditions of the I and II kind. Russian Mathematics 56(4):62–69, DOI 10.3103/S1066369X12040081
[21] Zorich VA (2016) Mathematical Analysis II, Universitext (UTX), vol 220. Springer, Berlin, Heidelberg, DOI 10.1007/978-3-662-48993-2
[22] Zubov VI (1975) Lektsii po teorii upravleniya (in Russ., Lectures on Control Theory). Nauka, Moscow

Chapter 7
Coordinate-Time Dependence of a Controlled Unmanned Aerial Vehicle and a Ground Object

Vanya R. Barseghyan, Manuk H. Mkrtchyan, and Aram G. Matevosyan

Abstract This paper examines the determination of the coordinate-time relationship between unmanned aerial vehicles (UAVs) and a ground target in an inertial coordinate system using a camera attached to the UAV. Based on data obtained from the UAV's video camera and the UAV's motion law, a mathematical model is proposed that describes the coordinate-time relationship between the UAV and the ground target. Using this mathematical model of the coordinate-time relationship between the UAV and target, in particular, it is possible to develop algorithms for autonomous UAV control, providing solutions to various problems.

Key words: Unmanned aerial vehicles, Ground target, Mathematical model of a video camera, Video image, Coordinate-time dependence of the UAV and target, Autonomous control mode of the UAV

7.1 Introduction

Currently, the use of unmanned aerial vehicles (UAVs) for various purposes, including military ones, is becoming increasingly widespread, which is highly relevant [7, 10, 12]. The diversity of applications requires the use of different types and

---

Vanya R. Barseghyan (corresponding author)
Institute of Mechanics, National Academy of Sciences of the Republic of Armenia, 24b Marshal Baghramyan Ave, Yerevan, 0019, & Mechanics, Yerevan State University, 1 Alec Manukyan, Yerevan, 0025, Republic of Armenia, e-mail: barseghyan@sci.am

Manuk H. Mkrtchyan
Institute of Mechanics, National Academy of Sciences of the Republic of Armenia, 24b Marshal Baghramyan Ave, Yerevan, 0019, Armenia, e-mail: mkmanuk@yandex.ru

Aram G. Matevosyan
Yerevan State University, 1 Alec Manukyan, Yerevan, 0025, Republic of Armenia, e-mail: matevosaram@gmail.com, amatevosyan@ysu.am

H. Altenbach and A. V. Sahakyan, *Contact Mechanics and Relevant Problems*,
Advanced Structured Materials 253, https://doi.org/10.1007/978-3-032-20869-9_7

different purposes of UAVs. Therefore, the practical application of controlled UAVs requires the use of various control approaches. In [1] approaches are applied involving Ant System algorithm, specifically algorithms simulating the actions of a human operator controlling an aircraft, based on the application of an approach strategy implemented using simple rules. In [8], the problem of controlling an UAV while pursuing a target moving along a curved trajectory in the pitch plane is solved. With the demand for UAVs for various purposes rapidly growing, the study of their control problems is undoubtedly a relevant area of research [1–6, 8, 9, 11]. In particular, autonomous control of UAVs has important practical implications, so research into such problems is in demand. For autonomous control and guidance of UAVs, the position and orientation of the vehicle in space during its flight play a major role [9]. During flight, the UAV can operate with varying degrees of autonomy. To implement autonomous flight, during the design process, appropriate software is installed in the UAV to perform various tasks in autonomous control mode, i.e. without human intervention. Tasks in autonomous UAV control—particularly those related to tracking, reconnaissance, pursuit and approaching—often include determining the location of ground objects (targets). Therefore, the development of UAV positioning systems that are independent of satellite navigation data is a relevant and important area of research. A promising solution to this problem is to equip UAVs with video cameras (one or more) and use algorithms and methods of machine vision [2, 11]. It is clear that increasing the efficiency of advanced UAV systems is directly related to improving control methods and their application during the approach of moving and stationary targets.

This paper examines a UAV equipped with control elements: engine thrust and aerodynamic control surfaces. The UAV, as a dynamic control object, is a dynamic system. It is assumed that the UAV has a positioning system and is equipped with a video camera, i.e. a video camera with a certain field of view is attached to the UAV using a gimbal. It is also assumed that the law of motion of the UAV is known and in the reference frame at some distance from the UAV on the ground there is an object called the target, which is in the field of view of the video camera. In this article, based on the data obtained from a video camera attached to a UAV, taking into account the law of motion, a mathematical model is proposed that describes the coordinate-time relationship between the UAV and a moving target. The importance of this coordinate-time relationship lies in the fact that, using this mathematical relationship between the UAV and the target, in particular, it is possible to develop algorithms for autonomous UAV control, providing solutions to problems for various purposes. An example with a numerical experiment is provided as an appendix. Results are presented in tables for various test values.

7.2 Statement of the Problem and Description of the Coordinate System

A UAV equipped with controls is considered: engine thrust and aerodynamic control surfaces (elevator and rudder deflection angles). As a dynamic control object, the UAV is a complex dynamic system. To control the UAV's motion, it is necessary to influence the forces and moments acting on the aircraft [3–7, 9]. It is assumed that the UAV has a positioning system and is equipped with a video camera, i.e., a video camera with a certain field of view is attached to the UAV using a gimbal (Fig. 7.1).

It is also assumed that the UAV's motion law is known and that an object, called the target, is located on the ground at a certain distance from the UAV in the reference frame. The target is within the field of view of the video camera. We will neglect the distance of the video camera from the UAV's center of gravity and the target's size, so the target is considered a point. We will denote the target's position at time t through $p^I_{\text{obj}}(t)$. This point corresponds to the line of sight - a straight line connecting the target with the UAV's center of gravity. It is assumed that at some point in time $t = t_0$ the coordinates of the UAV are known.

Let us consider the following problem:

> It is required, using data from a video camera attached to a UAV, to find a mathematical model that describes the coordinate-time dependence of the UAV and the so-called (moving) target.

To describe the movement of the UAV and determine the coordinate-time relationship between the UAV and the target coordinates, the following coordinate systems are used:

- $X_I Y_I Z_I$ – fixed (inertial) coordinate system,
- $X_v Y_v Z_v$ – moving coordinate system,
- $X_b Y_b Z_b$ – coordinate system relative to the UAV body.
- $X_g Y_g Z_g$ – coordinate system associated with the gimbal,
- $X_c Y_c Z_c$ – coordinate system associated with the visualization device (video camera), and
- $X_{\text{im}} Y_{\text{im}}$– coordinate system associated with the camera's image plane.

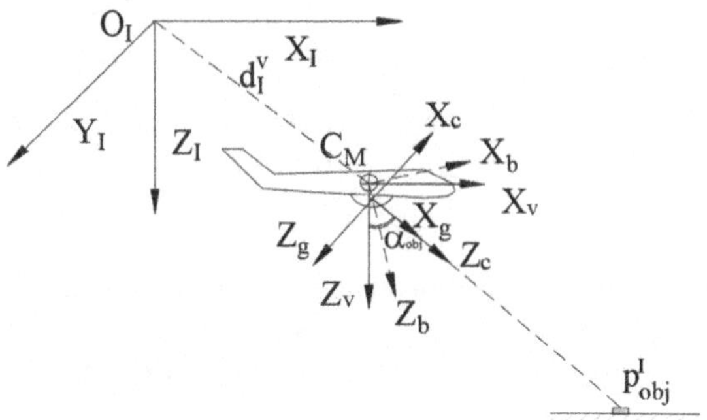

Fig. 7.1 Diagram of an aircraft-type UAV and coordinate system.

Note that these coordinate systems are necessary to describe the transition from one coordinate system to another using the corresponding transition matrices. So, if we have V^i vector v in the i- th system and must be represented in the j- th system, then we use the following transition formula

$$V^j = V^i T_i^j, T_i^j = \begin{bmatrix} R_i^j & -d_i^j \\ 0 & 1 \end{bmatrix},$$

where T_i^j is the transformation matrix from the ith coordinate system to the jth system, and R_i^j, d_i^j respectively are the vectors of the rotation and displacement matrices when moving from the ith system to the jth.

Below are the transition matrices used for transition from one coordinate system to another:

- T_I^v is the transition matrix from an inertial system $(X_I Y_I Z_I)$ to the UAV system $(X_v Y_v Z_v)$

$$T_I^v = \begin{bmatrix} I & -d_I^v \\ 0 & 1 \end{bmatrix}, \tag{7.1}$$

where

$$d_I^v = \begin{bmatrix} x_{UAV} \\ y_{UAV} \\ -h_{UAV} \end{bmatrix}, I = \begin{bmatrix} 1 & 0 & 0 \\ 0 & 1 & 0 \\ 0 & 0 & 1 \end{bmatrix}.$$

- T_v^b is the transition matrix from the UAV system $(X_v Y_v Z_v)$ to the UAV body system $(X_b Y_b Z_b)$

$$T_v^b = \begin{bmatrix} R_v^b & 0 \\ 0 & 1 \end{bmatrix}, \tag{7.2}$$

where

$$R_v^b = \begin{bmatrix} C_\theta C_\psi & C_\theta S_\psi & -S_\theta \\ S_\phi S_\theta C_\psi - C_\phi S_\psi & S_\phi S_\theta S_\psi + C_\phi C_\psi & S_\phi C_\theta \\ C_\phi S_\theta C_\psi + S_\phi S_\psi & C_\phi S_\theta S_\psi - S_\phi C_\psi & S_\phi C_\theta \end{bmatrix}.$$

- T_b^g is the transition matrix from the UAV body system $(X_b Y_b Z_b)$ to the gimbal system $(X_g Y_g Z_g)$

$$T_b^g = \begin{bmatrix} R_b^g & -d_b^g \\ 0 & 1 \end{bmatrix}, \tag{7.3}$$

where

$$R_b^g = R_{y,\alpha_{el}} R_{z,\alpha_{az}} = \begin{bmatrix} C_{el} & 0 & S_{el} \\ 0 & 1 & 0 \\ -S_{el} & 0 & C_{el} \end{bmatrix} \begin{bmatrix} C_{az} & S_{az} & 0 \\ -S_{az} & C_{az} & 0 \\ 0 & 0 & 1 \end{bmatrix} = \begin{bmatrix} C_{el} C_{az} & C_{el} S_{az} & S_{el} \\ -S_{el} & C_{az} & 0 \\ -S_{el} C_{az} & -S_{el} S_{az} & C_{el} \end{bmatrix}.$$

- T_g^c is the transition matrix from the system $(X_g Y_g Z_g)$ to the camera system $(X_c Y_c Z_c)$

$$T_g^c = \begin{bmatrix} R_g^c & -d_g^c \\ 0 & 1 \end{bmatrix}, \tag{7.4}$$

where

$$R_g^c = \begin{bmatrix} 0 & 0 & -1 \\ 0 & 1 & 0 \\ 1 & 0 & 0 \end{bmatrix}.$$

Let us note that $d_b^g = d_g^c = 0$ since it is assumed that the distances from the initial points of the coordinate systems $X_g Y_g Z_g$ and $X_c Y_c Z_c$ to the center of gravity of the UAV are negligibly small.

7.3 Mathematical Model of a Video Camera

A video camera model representing an ideal camera is considered. It is used to describe the relationship between the coordinates of points in an inertial frame and the coordinates of points in a two-dimensional image received from the camera. It is assumed that the model does not account for image distortions caused by the camera lens, as well as the error in imaging out-of-focus objects.

Now let us consider the mathematical model of such a video camera and the transition from the pixel coordinates of the target image obtained by the camera to the coordinate system $X_c Y_c Z_c$ (Fig. 7.2). Let us give some explanations of how a video camera obtains an image. We will denote the ground target observed by the camera as p_{obj}^c - target point in the coordinate system associated with the camera ($X_c Y_c Z_c$), where is the point O (projection center) is the origin of this coordinate system ($X_c Y_c Z_c$). From the point O at a distance f the image plane (the image received by the camera) is located where the projected image is q target point. The ($X_{im} Y_{im} Z_{im}$) coordinate system is fixed in this plane and coordinate distances are measured in meters, where $Z_{im} = Z_c - f$. The ($X_{ip} Y_{ip}$) coordinate system is fixed to

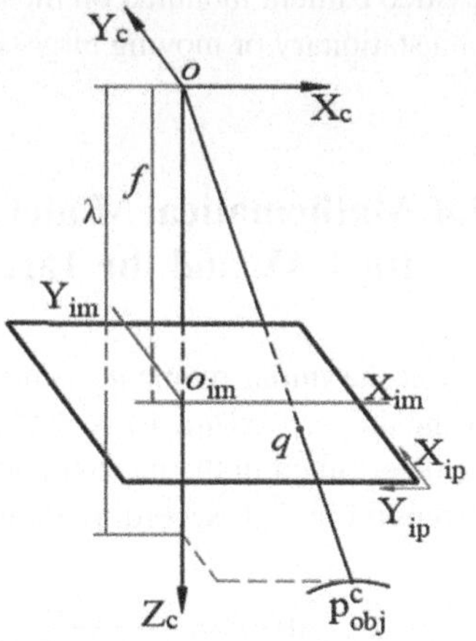

Fig. 7.2 Diagram of the mathematical model of a video camera and coordinate system.

the lower right corner of the image, which is measured in pixels. Optical axis OZ_c - a ray originating at a point O and perpendicular to the image plane.

Converting pixels to meters in the $(X_cY_cZ_c)$ image coordinate system according to the method of Trucco and Verri [2] it is represented as follows

$$x_{im} = (-y_{ip} + 0_y)S_y, \quad y_{im} = (-x_{ip} + 0_x)S_x.$$

Here, S_x, S_y denotes the width and height of one element of the photosensitive matrix.

From the properties of similar triangles $OO_{im}q$ and $OP_zP^c_{obj}$ (Fig. 7.2), we have the following relationships $x_{im}/f = p_x/p_z, y_{im}/f = p_y/p_z$, where (x_{ip}, y_{ip}) - coordinates in pixels, (x_{im}, y_{im}) - coordinates in meters, parameters $(0_x, 0_y)$ - distance from the center of the image to the coordinates (x_{ip}, y_{ip}) in pixels, and (S_x, S_y) - conversion factors from pixels to meters. We also denote by $\lambda = p_z$ projection of the distance from the camera to the target onto the Z_c-axis.

In the problem under consideration, it is necessary to determine the position of the target p^I_{obj} relative to the inertial system $(X_IY_IZ_I)$. If the value λ is known, then using the known transformation matrices $T^v_I, T^b_v, T^g_b, T^c_g$ from one system to another given in formulas (7.1)-(7.4), the position of the target p^I_{obj} is determined by the formula

$$p^I_{obj} = \left[CT^c_gT^g_bT^b_vT^v_I\right]^{-1} \Lambda q, \tag{7.5}$$

where

$$\Lambda = \begin{bmatrix} \lambda I & 0 \\ 0 & 1 \end{bmatrix}, \ \Lambda q = Cp^c_{obj}, \ C = \begin{bmatrix} 0 & f_x & 0_x & 0 \\ -f_y & 0 & 0_y & 0 \\ 0 & 0 & 1 & 0 \\ 0 & 0 & 0 & 1 \end{bmatrix}, \ f_x = \frac{f}{S_x}, \ f_y = \frac{f}{S_y}. \tag{7.6}$$

Here, C - calibration matrix, and the parameters f_x, f_y - focal lengths. Thus, using a video camera mounted on the body of a UAV, it is possible to find the coordinates of a stationary or moving target located on the ground.

7.4 Mathematical Model of the Coordinate-Time Dependence of the UAV and the Target

Let at the initial moment of time $t = t_0$ the coordinates of the UAV in the vertical plane are $x_0 = x(t_0)$, $y_0 = y(t_0)$, $z_0 = z(t_0)$. Then the target's position $p^I_{obj}(t_0)$ can be determined in the inertial coordinate system. From the geometric considerations given in Fig. 7.2, according to the work, and formula (7.5), we will have

$$\lambda(t_0) = \frac{z^I_{cc}(t_0)}{\cos \alpha_{obj}(t_0)}, \quad p^I_{obj}(t_0) = \left[CT^c_gT^g_bT^b_vT^v_I\right]^{-1} \Lambda q(t_0), \tag{7.7}$$

where $\lambda(t_0)$ - distance from camera to target, $\mathrm{p}_{cc}^{I} = \left(x_{cc}^{I}\ y_{cc}^{I}\ z_{cc}^{I}\ 1\right)^{\mathrm{T}}$ – camera position, $\alpha_{\mathrm{obj}}(t_0)$ – angle formed by the axes Z_I and $\mathrm{Z_c}$, $q(t_0)=\ (\mathrm{x_{ip}}(t_0),\ \mathrm{y_{ip}}(t_0),\ 1,\ 1)^{\mathrm{T}}$, and $\mathrm{x_{ip}}(t_0), \mathrm{y_{ip}}(t_0)$ – the coordinates of the target point in the camera image in pixels. Note that the explicit form of the matrix C given in (7.6). From here the letter "T" in the superscript means the transposition operation.

Let at the moment of time $t = t_j$, according to the law of motion of the UAV, starting the movement from the initial state (at $t = t_0$) the coordinates of the acquired position in the inertial system are $x(t_j)$, $y(t_j)$ and $z(t_j)$. Then the distance $\lambda(t_j)$ from the UAV to the target, according to (7.7), will be

$$\lambda(t_j) = \frac{z_{cc}^{I}(t_j)}{\cos\alpha_{\mathrm{obj}}(t_j)}. \tag{7.8}$$

Therefore, the coordinates of the target relative to the inertial coordinate system, according to formulas (7.7) and (7.8), are represented in the form

$$\mathrm{p}_{\mathrm{obj}}^{I}(t_j) = (CT_{\mathrm{all}})^{-1}\begin{bmatrix}\dfrac{z_{cc}^{I}(t_j)}{\cos\alpha_{\mathrm{obj}}(t_j)}I & 0\\ 0 & 1\end{bmatrix}q(t_j), \tag{7.9}$$

where $T_{\mathrm{all}} = T_g^{\mathrm{c}}T_b^{g}T_v^{b}T_I^{v}$.

Considering that

$$p_{cc}^{I}(t_j) = T_{\mathrm{all}}^{-1}B,\quad T_{all}^{-1} = p_{cc}^{I}(t_j)B^{+}, \tag{7.10}$$

where $B = p_{cc}^{\mathrm{c}} = \left(0\,0\,0\,1\right)^{\mathrm{T}}$, p_{cc}^{c} optical center of the camera, and B^{+} is the Moore-Penrose pseudoinverse of vector B. And $B^{+} = B^{\mathrm{T}}/||B||^2$, therefore, (7.9) is represented as

$$\mathrm{p}_{\mathrm{obj}}^{I}(t_j) = p_{cc}^{I}(t_j)C^{-1}B^{+}\begin{bmatrix}\dfrac{z_{cc}^{I}(t_j)}{\cos\alpha_{\mathrm{obj}}(t_j)}I & 0\\ 0 & 1\end{bmatrix}q(t_j). \tag{7.11}$$

In this case $B^{+} = \left(0\,0\,0\,1\right)$.

Now we represent formula (7.9) (or (7.11)) in the form

$$\mathrm{p}_{\mathrm{obj}}^{I}(t_j) = T_{\mathrm{all}}^{-1}C^{-1}\begin{bmatrix}\dfrac{z_{cc}^{I}(t_j)}{\cos\alpha_{\mathrm{obj}}(t_j)}I & 0\\ 0 & 1\end{bmatrix}q(t_j)$$

or according to (7.10) writing in the form

$$\mathrm{p}_{\mathrm{obj}}^{I}(t_j) = p_{cc}^{I}(t_j)B^{+}C^{-1}\begin{bmatrix}\dfrac{z_{cc}^{I}(t_j)}{\cos\alpha_{\mathrm{obj}}(t_j)}I & 0\\ 0 & 1\end{bmatrix}q(t_j).$$

Further taking into account that $p_{cc}^{I}(t_j) = T_{\text{all}}^{-1}B$, we will receive

$$\mathrm{p}_{\text{obj}}^{I}(t_j) = p_{cc}^{I}(t_j)B^{+}C^{-1}\Lambda q(t_j). \tag{7.12}$$

In the last expression, each of the factors has the following dimensions: $\mathrm{p}_{\text{cc}}(t_j)$ – (4×1), B^{+} – (1×4), C^{-1} – (4×4), Λ – (4×4), $q(t_j)$ – (4×1). Therefore, the product $B^{+}C^{-1}\Lambda q(t_j)$ is a scalar. If we denote this scalar by S_{c}, then the expression for $\mathrm{p}_{\text{obj}}^{I}(t_j)$ can be written in the following form:

$$\mathrm{p}_{\text{obj}}^{I}(t_j) = p_{cc}^{I}(t_j)S_{\text{c}}. \tag{7.13}$$

This shows that $\mathrm{p}_{\text{obj}}^{I}(t_j)$ - scaled version $p_{cc}^{I}(t_j)$.

It should be noted that, in general, using the mathematical relationship between two controlled dynamic objects, various problems of tracking or following one object by another are studied. Here, too, with an explicit mathematical expression for the coordinate-time relationship between a UAV and a ground target in the form of formula (7.12) (or (7.13)), in particular, it is possible to develop algorithms for autonomous UAV control to solve various problems.

7.5 Numerical Example

Using some test values for the UAV and video camera parameters, in the spatial case, some results of numerical calculations are presented. The following numerical values were used:

$$\begin{aligned}
&\phi = \pi/4,\ \psi = \pi/4,\ \theta = \pi/4,\ \alpha_{el} = \pi/6,\ \alpha_{az} = \pi/6,\\
&x_{uav} = 20,\ y_{uav} = 10,\ h_{uav} = 20,\\
&x_{bg} = 0.1,\ y_{bg} = 0.1,\ h_{bg} = 0,\\
&x_{gc} = 0,\ y_{gc} = 0,\ h_{gc} = 0,\\
&0_x = 960,\ 0_y = 540,\ f = 0.1,\ S_x = 0.004,\ S_y = 0.004,\\
&\alpha_{\text{obj}} = \pi/3,\ z_p = 500,
\end{aligned}$$

where z_p- is the height of the video camera point above the ground at the initial moment of time. Let the parameters of the image point on the video camera image, in the first case at the moment of time $t_0 = 0$ are

$$x_{\text{ip}} = 120,\ y_{\text{ip}} = 90,$$

and in the second case at the moment of time $t_1 = 5$, are

$$x_{\text{ip}} = 150,\ y_{\text{ip}} = 50.$$

These values are measured in pixels, and to convert to meters, the following formulas are used

$$x_{\text{im}} = \left(0_y - y_{\text{ip}}\right) S_y,\ y_{\text{im}} = \left(x_{\text{ip}} - 0_x\right) S_x,\ f_x = \frac{f}{S_x},\ f_y = \frac{f}{S_y}.$$

According to the formula (7.8) and (7.9), for moments in time $t_0 = 0$ and $t_1 = 5$, the following values were obtained

- $\lambda(t_0) = 424.366$, $\lambda(t_1) = 346.494$, i.e. the distance from the UAV (camera) to the target has decreased,
- for the position of the UAV (camera) we have

$$p^I_{cc}(t_0) = \left(80.036\ 80.116\ 300.072\ 1 \right)^{\text{T}},$$
$$p^I_{cc}(t_1) = \left(180.036\ 80.116\ 300.072\ 1 \right)^{\text{T}},$$

- for the target position we have

$$\text{p}^I_{\text{obj}}(t_0) = \left(332.628\ -103.881\ 621.842\ 1 \right)^{\text{T}},$$
$$\text{p}^I_{\text{obj}}(t_1) = \left(432.628\ -103.881\ 621.842\ 1 \right)^{\text{T}}.$$

Thus, a mathematical model is proposed that describes the coordinate-time dependence of the UAV (video camera) and a moving target, according to which, using the law of motion of the UAV, the position of the target can be determined.

7.6 Movement in a Plane

Considering the case when the movement of the UAV and the target are in the vertical plane, i.e., the movement of the UAV and the target occurs in the XZplane. The z_p parameter is changing - the height of the UAV. Let us assume that the target and the UAV are moving along the horizontal X-axis with constant speeds $V^x_{\text{obj}} = 5$ and $V^x_{uav} = 12$, respectively. After a period of time $\Delta t = 5$ the coordinates of the target are determined at a given moment in time $t = t_1$.

The numerical values of the parameters used in the calculation are as follows

$$\varphi = 0,\ \psi = 0,\ \theta = \pi/4,\ \alpha_{el} = \pi/4,\ \alpha_{az} = 0,$$
$$x_{uav} = 20,\ y_{uav} = 10,\ h_{uav} = 20,$$
$$x_{bg} = 0.1,\ y_{bg} = 0.1,\ h_{bg} = 0,$$
$$x_{gc} = 0,\ y_{gc} = 0,\ h_{gc} = 0.$$

The parameters of the video camera used have the following values

$$0_x = 960,\ 0_y = 540,\ f = 0.683,\ S_x = 0.004,\ S_y = 0.004.$$

The values of the target distance from the UAV were obtained at $z_p = 500$, $z_p = 700$, $z_p = 900$. Then, at that moment in time $t = t_0$ the distance $\lambda(t_0)$ to the target will be

$$\lambda(t_0) = 581.425, \qquad \lambda(t_0) = 823.686, \qquad \lambda(t_0) = 1065.94$$

and at a point in time $t = t_1$ the distance $\lambda(t_1)$ to the target will be

$$\lambda(t_1) = 562.014, \qquad \lambda(t_1) = 800.183, \qquad \lambda(t_1) = 1046.52.$$

Therefore, the coordinates of the UAV at the moment in time $t = t_0$ it will be

$$p^I_{cc}(t_0) = \begin{pmatrix} 20.036 & 0 & 20.072 & 1 \end{pmatrix}^{\mathrm{T}},$$

and at a point in time $t = t_1$ it will be

$$p^I_{cc}(t_1) = \begin{pmatrix} 80.091 & 0 & 20.072 & 1 \end{pmatrix}^{\mathrm{T}}.$$

Depending on the value z_p, it is possible to estimate the error in the target coordinates obtained by calculation. The calculation results are presented in Table 7.1. Error assessment $\varepsilon(x)$ is calculated taking into account that over a period of time Δt the target movement will be $V^x_{\mathrm{obj}}\Delta t = 25$, thus

$$\varepsilon(x) = \left| x^I_{\mathrm{obj}}(t_1) - x^I_{\mathrm{obj}}(t_0) \right|, \qquad \Delta x = V^x_{\mathrm{obj}}\Delta t - \varepsilon(x).$$

The last column shows the error as a percentage.

Let us consider an example for a stationary target. Let the UAV be at an altitude of $z_p = 500$, $z_p = 700$, $z_p = 900$ from the ground. Let us determine the coordinates of the target at moments in time $t = t_0$ and $t = t_1$. If we select the required parameters as in the previous example, then the target coordinates for the first case will be the same. The calculation results are presented in Table 7.2.

Table 7.1: Calculation results for the error in the target coordinates.

z_p	$p^I_{\mathrm{obj}}(t_0)$		$p^I_{\mathrm{obj}}(t_1)$		$x^I_{\mathrm{obj}}(t_1) - x^I_{\mathrm{obj}}(t_0)$	$\varepsilon(x)$	
	$x^I_{\mathrm{obj}}(t_0)$	$z^I_{\mathrm{obj}}(t_0)$	$x^I_{\mathrm{obj}}(t_1)$	$z^I_{\mathrm{obj}}(t_1)$		Δx	%
500	348.2	500.006	372.48	500.006	24.28	0.72	2.88
700	484.908	700.009	509.865	700.009	24.95	0.05	0.2
900	621.618	900.006	646.541	900.006	24.92	0.07	0.32

Table 7.2: Calculation results for the error in the target coordinates.

z_p	$p^I_{\mathrm{obj}}(t_0)$		$p^I_{\mathrm{obj}}(t_1)$		$\varepsilon(x)$	
	$x^I_{\mathrm{obj}}(t_0)$	$z^I_{\mathrm{obj}}(t_0)$	$x^I_{\mathrm{obj}}(t_1)$	$z^I_{\mathrm{obj}}(t_1)$	Δx	%
500	348.2	500.006	348.13	500.006	0.07	0.24
700	484.908	700.009	484.94	700.009	0.032	0.15
900	621.618	900.006	621.804	900.006	0.186	1.15

7.7 Conclusion

This paper proposes a method for determining the coordinate-time relationship between controlled UAVs equipped with a video camera and a ground target in an inertial coordinate system. Using a mathematical model of the coordinate-time relationship between UAV and target motion, in particular, it is possible to develop algorithms for autonomous UAV control, providing solutions for various tasks. The proposed approach may be of interest for developing control algorithms for various types of UAVs when solving problems such as pursuing and approaching moving and stationary targets, landing, and so on.

Acknowledgements The work was carried out with the support of the Committee on Higher Education and Sciences of the Republic of Armenia, within the framework of research project No. 23-2DP-1B001.

References

[1] Abramov NS, Khachumov MV (2013) Modelirovaniye provodki po marshrutu bespilotnogo letatel'nogo apparata kak zadachi presledovaniya tseli (in Russ., Air route simulation of unmanned aerial vehicles as the pursuit-evasion problem). Aviakosmicheskoye Priborostroyeniye (9):9–22

[2] Barber DB, Redding JD, McLain TW, Beard RW, Taylor CN (2006) Vision-based target geo-location using a fixed-wing miniature air vehicle. Journal of Intelligent & Robotic Systems 47(4):361–382, DOI 10.1007/s10846-006-9088-7

[3] Barseghyan VR, Matevosyan AG (2023) Ob odnoy zadache upravleniya bespilotnym letatel'nym apparatom samoletnogo tipa s zadannymi promezhutochnymi znacheniyami raznykh chastey koordinat (in Russ., On one problem of controlling an airplane-type unmanned aerial vehicle with given intermediate values of different parts of coordinates). Differentsial'nye Uravneniya i Protsessy Upravleniya (2):86–96, DOI 10.21638/11701/spbu35.2023.205

[4] Barseghyan VR, Mkrtchyan MH, Matevosyan AG (2024) Ob odnoi zadache upravleniya BPLA po linii vizirovaniya (in Russ., On one problem of uav control along the line of sight. In: Antonik VG, Arguchintsev AV, Falaleyev MV (eds) Dynamic Systems and Computer Science: Theory and Applications (DYSC 2024). Proceedings of the 6th International Conference, September 16–20, 2024, Irkutsk State University, Irkutsk, pp 199–202, DOI 10.26516/978-5-9624-2309-8.2024.1-224

[5] Barseghyan VR, Simonyan TA, Matevosyan AG (2024) Ob odnoy zadache optimal'nogo upravleniya kvadrokopterom s zadannym promezhutochnym znacheniyem chasti koordinat fazovogo vektora (in Russ., On one problem of optimal control of a quadcopter with a given intermediate value of part of

the coordinates of the phase vector. Differentsial'nye Uravneniya i Protsessy Upravleniya (2):59–72, DOI 10.21638/11701/spbu35.2024.205

[6] Barseghyan VR, Simonyan TA, Matevosyan AG (2024) Ob odnoy zadache upravleniya kvadrokopteroms zadannymi promezhutochnymi znacheniyamiraznykh chastey koordinat (in Russ., Ob odnoy zadache upravleniya kvadrokopteroms zadannymi promezhutochnymi znacheniyamiraznykh chastey koordinaton one problem of quad-copter control with given intermediate values of different parts of coordinates. Vestnik rossiyskikh universitetov Matematika 29(145):29–42, DOI 10.20310/2686-9667-2024-29-145-29-42

[7] Epov MI, Zlygostev IN (2012) Primenenie bespilotnykh letatel'nykh apparatov v aerogeofizicheskoi razvedke (in Russ., The use of unmanned air vehicles for aero geophysical survey). InterExpo GeoSiberia 2(3):22–27, https://cyberleninka.ru/article/n/primenenie-bespilotnyh-letatelnyh-apparatov-v-aerogeofizicheskoy-razvedke

[8] Khachumov MV (2014) Zadacha avtomaticheskogo upravleniya letatel'nym apparatom v protsesse presledovaniya tseli (in Russ., The task of control of an unmanned aerial vehicle in the process of pursuit of the target). Programmnyye Sistemy: Teoriya i Prilozheniya 5(5(23)):67–77, http://psta.psiras.ru/read/psta2014_5_67-77.pdf

[9] Latonov VV, Tikhomirov VV (2018) Line-of-sight guidance control using video images. Moscow University Mechanics Bulletin 73(1):11–17, DOI 10.3103/S002713301801003X

[10] Lemeshonok TY, Sizova AA, Baranov NE, Sannikov VA (2020) Matematicheskie modeli dinamiki dvizheniya letatel'nykh apparatov (in Russ., Mathematical Models of Aircraft Motion Dynamics). Baltic State Technical University, St. Petersburg

[11] Stepanov DN, Tishchenko IP (2011) Zadacha modelirovaniya poleta bespilotnogo letatel'nogo apparata na osnove sistemy tekhnicheskogo zreniya (in Russ., The problem of modeling the flight unmanned aerial vehicle based on vision systems). Programmnye Sistemy: Teoriya i Prilozheniya (4(8)):33–43, http://psta.psiras.ru/read/psta2011_4_33-43.pdf

[12] Telukhin SV, Matveev VV (2008) Bespilotnyy letatel'nyi apparat kak obyekt upravleniya (in Russ.,Unmanned aerial vehicles as a control object. Mechatronika, Avtomatizaciya, Upravlenie (10):7–10, https://cyberleninka.ru/article/n/algoritmy-avtomaticheskogo-upravleniya-bokovym-dvizheniem-udarnogo-bespilotnogo-letatelnogo-apparata-samoletnogo-tipa-na-vozdushnom

Chapter 8
Low-Frequency Propagating and Evanescent Waves in thin Elastic Laminates Composed of High-Contrast Materials

Illia Chernomorets, Julius Kaplunov and Danila Prikazchikov

Abstract High-contrast elastic structures find numerous applications in modern engineering, including but not limited to photovoltaic panels, laminated glass, lightweight solutions, e.g., aerogel-based components. The main peculiarities of structural response of thin strongly inhomogeneous laminates involve slowly varying propagating and evanescent low-frequency modes. The focus of the present contribution is on the relation between these modes investigated by a multi-parametric asymptotic procedure. The key problem parameters are relative thickness, stiffness, and density of the layers, as well as the traditional geometric parameter corresponding to small thickness. As an illustration, anti-plane harmonic motion of sandwich elastic plate is analysed. Explicit estimates for the parametric ranges supporting various limiting behaviours are investigated. Comparisons of asymptotic and exact results are demonstrated.

Key words: High contrast, Asymptotic, Elastic, Propagating, Evanescent

8.1 Introduction

Mechanics of layered structures have been actively studied since long ago, see, e.g., [1, 15, 16, 22] and references therein. Within the last decade, an important focus was on high-contrast laminates, motivated by modern engineering applications in photovoltaic panels, laminated glass, as well as advanced sandwich-type panels, see [2, 3, 6–8] to name a few.

Two of the remarkable phenomena often arising in high-contrast laminates include the well-known degeneration of the static Saint-Venant's boundary layers, transforming to slowly decaying solutions, e.g., see [4, 5, 11, 24], as well as a more

Illia Chernomorets · Julius Kaplunov · Danila Prikazchikov (corresponding author)
School of Computer Science and Mathematics, Keele, Staffordshire, ST5 5BG, UK, e-mail: i.chernomorets@keele.ac.uk, j.kaplunov@keele.ac.uk, d.prikazchikov@keele.ac.uk

H. Altenbach and A. V. Sahakyan, *Contact Mechanics and Relevant Problems*,
Advanced Structured Materials 253, https://doi.org/10.1007/978-3-032-20869-9_8

recently investigated effect of the lowest non-zero cut-off frequencies being small, see [12]. The latter is also typical of almost rigid body motions of a three-component rod [13]. The existence of the second low-frequency shear mode in a three layered plate gave rise to the two-mode long-wave low-frequency asymptotic theories for asymmetric three-layered strip, treated within the framework of antiplane elasticity in [14, 19]. Moreover, the presence of contrast inspired generalisation of the Saint-Venant's principle for laminates, leading to the formulation of asymptotically consistent boundary conditions, see [14, 18].

For a long time, these two phenomena, i.e., the degenerate boundary layer and the extra low-frequency mode, were usually studied separately, apart from an initial insight in [14]. A recent consideration in [21] establishes a link between them by studying a transition by means of low-frequency long-wave evanescent modes, predicting both the static Saint-Venant limit and the lowest cut-off frequency. The cited paper deals with the setup corresponding to dispersion of plane waves in a three-layered plate with a core aerogel layer and aluminum skin layers.

The main goal of the current contribution is to make the relation between the propagating and evanescent modes more clear by treating a simpler antiplane problem for a three-layered strongly inhomogeneous plate. In this case, we restrict ourselves to analysis of antisymmetric waves propagating along a geometrically symmetric structure, in order to concentrate on the lowest shear cut-off. As a result, it is demonstrated that two phenomena of interest do not always occur simultaneously. In particular, we discover the parameter range supporting an extra low-frequency mode, which, however, does not originate from a degenerate slow varying boundary layer. This scenario corresponds to non-uniform asymptotic behaviour of the related dispersion curve, valid only over a smallest cut-off vicinity.

The paper is organised as follows. Following the introduction, in Section 2 the problem is formulated and the dispersion relation is presented. Then, in Section 3 low-frequency multiparametric analysis (for independent ratios of stiffness, thickness, and density) is developed.

8.2 Low-frequency dispersion of antiplane shear waves in three-layered elastic laminate

Consider antiplane shear motion of a three-layered symmetric elastic laminate, with the inner layer and two outer layers located within the domains

$$\{-\infty < x < \infty,\ |y| \le h_1\}$$

and

$$\{-\infty < x < \infty,\ h_1 \le |y| \le h_1 + h_2\},$$

respectively, see Fig. 8.1. Let us denote the out-of plane displacement component by $u = u(x, y)$. Note that in view of the symmetry, the problem may be considered for

Fig. 8.1 Infinite three-layered symmetric laminate

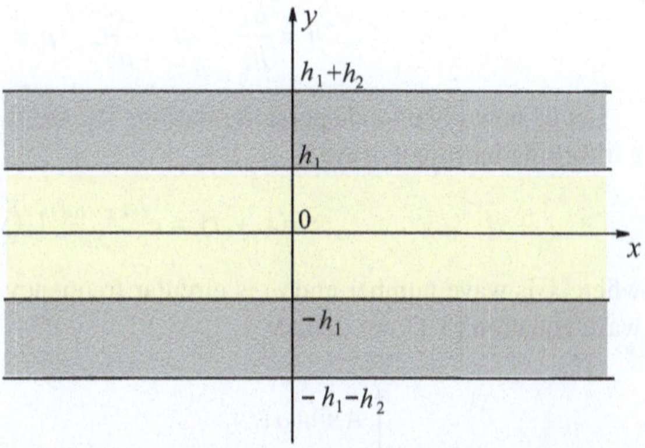

the upper half of the structure only ($y \geq 0$). In this case, concentrate on the problem, for which the function above is odd in y, i.e $u(x, -y) = -u(x, y)$.

The conventional wave equation governing antiplane motion can be written as

$$\frac{\partial^2 u}{\partial x^2} + \frac{\partial^2 u}{\partial y^2} - \frac{1}{c^2}\frac{\partial^2 u}{\partial t^2} = 0, \tag{8.1}$$

where

$$c = \begin{cases} c_1 = \sqrt{\dfrac{\mu_1}{\rho_1}}, & |y| \leq h_1, \\ c_2 = \sqrt{\dfrac{\mu_2}{\rho_2}}, & h_1 \leq |y| \leq h_1 + h_2 \end{cases}$$

is a piecewise shear wave speed, with μ_i and ρ_i ($i = 1, 2$) denoting the shear moduli and the volume mass densities of the skin ($i = 2$) and core ($i = 1$) layers, respectively.

The outer face $y = h_1 + h_2$ is assumed to be traction-free, i.e.,

$$\left.\frac{\partial u}{\partial y}\right|_{y=h_1+h_2} = 0 \tag{8.2}$$

with continuity of the displacement and shear stress along the interface $y = h_1$ dictating

$$u\big|_{y=h_1-0} = u\big|_{y=h_1+0} \tag{8.3}$$

and

$$\mu_1 \left.\frac{\partial u}{\partial y}\right|_{y=h_1-0} = \mu_2 \left.\frac{\partial u}{\partial y}\right|_{y=h_1+0}. \tag{8.4}$$

In addition, we impose decay or radiation conditions at $x \to \infty$ associated with evanescent and propagating modes, respectively.

In spite of apparent simplicity, the formulated problem involves three dimensionless parameters given by

$$h = \frac{h_1}{h_2}, \quad \mu = \frac{\mu_1}{\mu_2}, \quad \rho = \frac{\rho_1}{\rho_2}. \tag{8.5}$$

Let us now obtain a dispersion relation. The solution is sought for in the form of a travelling harmonic wave

$$u(x,y,t) = e^{i(kx-\omega t)}U(y), \tag{8.6}$$

where k is wave number and ω is circular frequency. On substituting (8.6) into the wave equation (8.1), we deduce

$$U(y) = \begin{cases} A\sin(\alpha_1\zeta), & 0 \le \zeta \le \dfrac{h}{h+1}, \\ B\sin(\alpha_2\zeta) + C\cos(\alpha_2\zeta), & \dfrac{h}{h+1} \le \zeta \le 1, \end{cases} \tag{8.7}$$

where A, B and C are arbitrary constants, whereas

$$\zeta = \frac{y}{h_1 + h_2} \tag{8.8}$$

is the dimensionless transverse coordinate ($0 \le \zeta \le 1$), and quantities α_i defined by

$$\alpha_i = \sqrt{(1+h)^2 h^{2(i-2)}\Omega_i^2 - K^2}, \qquad i = 1,2, \tag{8.9}$$

with

$$K = k(h_1 + h_2), \quad \Omega_i = \frac{\omega h_i}{c_i}, \qquad i = 1,2 \tag{8.10}$$

denoting dimensionless wave number and frequency, respectively.

Next, inserting (8.7) into the boundary and interfacial conditions (8.2) - (8.4), we arrive at a homogeneous set of algebraic equations in the constants A, B and C. Equating its determinant to zero, we get the following dispersion relation.

$$\frac{\alpha_2}{\alpha_1}\tan\left(\frac{\alpha_1 h}{1+h}\right)\tan\left(\frac{\alpha_2}{1+h}\right) = \mu. \tag{8.11}$$

Below, we restrict ourselves to long-wave low-frequency approximations of the dispersion relation (8.11), when

$$|K| \ll 1 \quad \text{and} \quad \Omega_i \ll 1. \tag{8.12}$$

This regime may be referred to as a global one, see [13], since it does not support sinusoidal displacement variations across each of the layers due to the limitations on both dimensionless frequencies Ω_1 and Ω_2. The absolute value of K accounts for both the propagating and evanescent modes, with K being imaginary for the latter.

8.3 Multiparametric Analysis

First, consider briefly the cut-off frequencies that emerge from the dispersion relation (8.11) at $K = 0$. As follows from the previous considerations [12, 13] for multi-span elastic rods, the global low-frequency regime occurs, provided that the problem parameters satisfy the problem parameters.

$$\mu \ll h \ll \frac{1}{\rho}. \tag{8.13}$$

In this case, the lowest non-zero cut-off frequency

$$\Omega_2^2 = \frac{\mu}{\rho h^2}\Omega_1^2 \approx \frac{\mu}{h}. \tag{8.14}$$

Another limiting case at $\Omega_i = 0$ corresponds to the recently examined static high-contrast problem, see [9]. For the latter, the degenerated boundary layer is supported by a small root $K = K_0$ of the secular equation (cf. Eq. (9) in [9], see also the earlier consideration in [5]). At leading order, it is

$$K_0 \approx \sqrt{\frac{\mu}{h}}\,(h+1) \ll 1, \tag{8.15}$$

provided that

$$\mu \ll \frac{h}{(h+1)^2}. \tag{8.16}$$

It is worth mentioning that for the small and large values of the parameter h, when

$$\mu \ll h \ll 1, \tag{8.17}$$

or

$$\mu \ll \frac{1}{h} \ll 1, \tag{8.18}$$

respectively, the one-parametric condition in [5] $\mu \ll 1$, valid at $h \sim 1$, has to be substituted by a stronger one. It might be expected that multi-parametric analyses of a more elaborated plane strain problem, see [10], may lead to non-trivial conclusions.

Here we extend all these findings to the whole parametric range (8.12). Comparison between the strong inequalities (8.13) and (8.16) on the contrast parameters μ and h indicates that over the interval

$$\frac{1}{h} \ll \mu \ll h, \qquad (h \gg 1) \tag{8.19}$$

a propagating low-frequency mode may exist along with a strongly localised (non-degenerate) static boundary layer, supported by the canonical Saint-Venant's principle. On the contrary, for $h \lesssim 1$, i.e., for no contrast in thickness or for thick skin and thin core layers, the aforementioned mode always occurs along with the degenerate (slowly decaying) boundary layer.

Further insight is passed on the derivation of a shortened form of the dispersion relation (8.11) by taking the leading order Taylor expansions of involved functions over the parameter range (8.13). Thus, we have

$$\alpha_2^2 = \mu \frac{(1+h)^2}{h}, \tag{8.20}$$

or

$$\Omega_2^2 - \frac{K^2}{(1+h)^2} = \frac{\mu}{h}. \tag{8.21}$$

Then, we readily deduce from the underlying assumptions

$$\frac{\alpha_1 h}{1+h} \ll 1 \quad \text{and} \quad \frac{\alpha_2}{1+h} \ll 1 \tag{8.22}$$

the sought for constraint on the wave number. It is given by

$$K \ll \min\left\{1, \sqrt{\frac{\mu}{\rho}}\right\} \tag{8.23}$$

at

$$\Omega_2^2 > \frac{\mu}{h},$$

or

$$|K| \ll 1 \quad \text{and} \quad |K| \leq K_0, \tag{8.24}$$

at

$$\Omega_2^2 < \frac{\mu}{h}.$$

Next, using (8.23) and (8.24) in the shortened dispersion relation (8.21), we have for the vibration frequency

$$\Omega_2^2 - \frac{\mu}{h} \ll \frac{1}{(1+h)^2} \min\left\{1, \frac{\mu}{\rho}\right\} \tag{8.25}$$

at

$$\Omega_2^2 > \frac{\mu}{h},$$

and

$$\frac{\mu}{h} - \Omega_2^2 \ll \frac{1}{(1+h)^2}, \tag{8.26}$$

at

$$\Omega_2^2 < \frac{\mu}{h}.$$

In the formulae above, (8.23), (8.25) and (8.24), (8.26) correspond to propagating and evanescent modes, respectively. The last formula (8.26), under the condition (8.16), can be simplified to

$$0 \leq \Omega_2 \leq \sqrt{\frac{\mu}{h}}. \tag{8.27}$$

We also remark that the dispersion relation (8.21) supports a quasi-static regime, at

$$\Omega_2 \ll \sqrt{\frac{\mu}{h}}. \tag{8.28}$$

Let us now present a numerical illustration for the approximation (8.21). Figure 8.2 shows the exact and approximate curves corresponding to (8.11) and (8.21), shown by solid and dashed lines, respectively, covering both real and imaginary domains of the wavenumber K. From this Figure we can also observe the values of the decay scale K_0 at the static limit $\Omega \to 0$, and the cut-off frequency Ω_{sh}, being the lowest positive solution of the dispersion relation in the long-wave limit $K \to 0$.

The geometrical and material parameters in Fig. 8.2 below are the same as in [21], namely,

$$h_1 = 0.01\,\text{m}, \qquad h_2 = 0.002\,\text{m},$$
$$\rho_1 = 100\,\text{kg/m}^3,\ \rho_2 = 2750\,\text{kg/m}^3,$$
$$\mu_1 = 93.75\,\text{kPa}, \ \ \mu_2 = 26.93\,\text{GPa},$$

corresponding to an Al-aerogel-Al sandwich structure with a strongly pronounced high contrast in material properties. In this case we obtain the exact and approximate values for the slow decay scale K_0 as 0.00501 and 0.00502, respectively. In a similar

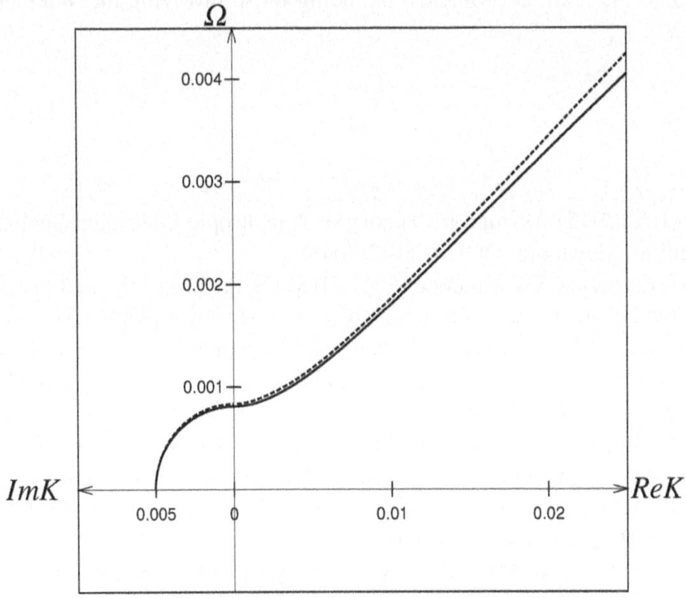

Fig. 8.2: The approximation (8.21) and the exact solution of the dispersion relation (8.11) plotted by dashed and continuous lines, respectively..

manner, the exact and approximate values for the cut-off frequency Ω_{sh} are given by 0.000812 and 0.000837, respectively.

Figure 8.2 clearly shows that for the analysed parameter setup the derived relation (8.21) nicely approximates both propagating and evanescent waves, including accurate estimates for the limiting values of K_0 and Ω_{sh}.

8.4 Concluding Remarks

The results of multiparametric analysis of long-wave low-frequency propagating and evanescent modes in a three-layered, strongly inhomogeneous laminate have been reported. The main focus was on the transition from zero to the lowest cut-off frequency.

The polynomial approximation (8.21) of the dispersion relation was obtained. The ranges of the problem parameters corresponding to limiting static and near-cut-off behaviours have been evaluated. It was demonstrated that (8.21) could be either uniformly valid for both propagating and evanescent modes at the long-wave low-frequency limit, or justified, only locally near the cut-off or zero frequencies. A case study of Al-aerogel-Al sandwich plate was further developed, illustrating numerically the antiplane setup, for another high-contrast case study see [17].

Possible extension of the developed approach may include analysis of local low-frequency regimes [13], as well as tackling multi-layered thin laminates [18, 23] and more elaborated 2D and 3D problems, including those involving the Winkler foundation, see [20].

References

[1] Aghalovyan LA (2015) Asymptotic Theory of Anisotropic Plates and Shells. World Scientific, Singapore, DOI 10.1142/9048

[2] Altenbach H, Eremeyev VA, Naumenko K (2015) On the use of the first order shear deformation plate theory for the analysis of three-layer plates with thin soft core layer. ZAMM - Journal of Applied Mathematics and Mechanics / Zeitschrift für Angewandte Mathematik und Mechanik 95(10):1004-1011, DOI 10.1002/zamm.201500069

[3] Eisenträger J, Naumenko K, Altenbach H, Köppe H (2015) Application of the first-order shear deformation theory to the analysis of laminated glasses and photovoltaic panels. International Journal of Mechanical Sciences 96-97:163-171, DOI 10.1016/j.ijmecsci.2015.03.012

[4] Avila-Pozos O, Movchan AB (2003) Slow decay of end effects in layered structures with an imperfect interface. Journal of Engineering Mathematics 45(2):15-168, DOI 10.1023/A:1022125917959

[5] Baxter SC, Horgan CO (1995) End effects for anti-plane shear deformations of sandwich structures. Journal of Elasticity 40(2):123-164, DOI 10.1007/BF00042458

[6] Boutin C, Viverge K (2016) Generalized plate model for highly contrasted laminates. European Journal of Mechanics A/Solids 55:149-166, DOI 10.1016/j.euromechsol.2015.08.008

[7] Boutin C, Viverge K, Hans S (2021) Dynamics of contrasted stratified elastic and viscoelastic plates-application to laminated glass. Composites Part B: Engineering 212:108551, DOI 10.1016/j.compositesb.2020.108551

[8] Ivanov IV (2006) Analysis, modelling, and optimization of laminated glasses as plane beam. International Journal of Solids and Structures 43(22-23):6887-6907, DOI 10.1016/j.ijsolstr.2006.02.014

[9] Chernomorets I, Kaplunov J, Prikazchikov D (2024) Two-parametric analysis of a semi-infinite three-layered high-contrast elastic strip under antiplane shear deformation. In: Altenbach H, Bogdanov V, Grigorenko AY, Kushnir RM, Nazarenko VM, Eremeyev VA (eds) Selected Problems of Solid Mechanics and Solving Methods, Advanced Structured Materials, Vol 204, pp 99-109. Springer, Cham, DOI 10.1007/978-3-031-54063-9_7

[10] Choi I, Horgan C (1978) Saint-Venant end effects for plane deformation of sandwich strips. International Journal of Solids and Structures 14(3):187-195, DOI 10.1016/0020-7683(78)90024-0

[11] Horgan C (1998) Saint-Venant end effects for sandwich structures. In: Fourth International Conference on Sanwich Construction, vol 1, pp 191–200, EMAS Publishing, UK

[12] Kaplunov J, Prikazchikov DA, Prikazchikova LA (2017) Dispersion of elastic waves in a strongly inhomogeneous three-layered plate. International Journal of Solids and Structures 113-114:169-179, DOI 10.1016/j.ijsolstr.2017.01.042

[13] Kaplunov J, Prikazchikov DA, Sergushova O (2016) Multi-parametric analysis of the lowest natural frequencies of strongly inhomogeneous elastic rods. Journal of Sound and Vibration 366:264–276, DOI 10.1016/j.jsv.2015.12.008

[14] Kaplunov J, Prikazchikova L, Alkinidri M (2021) Antiplane shear of an asymmetric sandwich plate. Continuum Mechanics and Thermodynamics 33(4):1247-1262, DOI 10.1007/s00161-021-00969-6

[15] Mikhasev GI, Altenbach H (2019) Thin-Walled Laminated Structures. Vibrations, and Their Suppression. Advanced Structured Materials, vol 106. Cham:, Springer, DOI 10.1007/978-3-030-12761-9

[16] Liew K, Pan Z, Zhang L (2019) An overview of layerwise theories for composite laminates and structures: development, numerical implementation and application. Composite Structures 216:240-259, DOI 10.1016/j.compstruct.2019.02.074

[17] Prikazchikov D, Kaplunov J, Prikazchikova L, Rogerson GA (2025) Multiparametric dynamic modelling of lightweight elastic structures. In: Aston P (ed) More UK Success Stories in Industrial Mathematics, Mathematics in Industry, Vol 42, pp 71-77. Springer, Cham, DOI 10.1007/978-3-031-48683-8_10

[18] Prikazchikova L (2022) Decay conditions for antiplane shear of a high-contrast multi-layered semi-infinite elastic strip. Symmetry, 14(8):1697, DOI 10.3390/sym14081697

[19] Prikazchikova L, Ece Aydın Y, Erbaş B, Kaplunov J (2020) Asymptotic analysis of an anti-plane dynamic problem for a three-layered strongly inhomogeneous laminate. Mathematics and Mechanics of Solids 25(1):3-16, DOI 10.1177/1081286518790804

[20] Prikazchikova L, Nolde E, Miszuris W, Kaplunov J (2024) On the refined boundary condition at the edge of a thin elastic strip supported by a Winkler-type foundation under antiplane shear deformation. International Journal of Engineering Science 205:104152. DOI10.1016/j.ijengsci.2024.104152

[21] Prikazchikova L, Rege A, Kaplunov J, Prikazchikov D (2024) Low-frequency propagating and evanescent waves in strongly inhomogeneous sandwich plates. Zeitschrift für angewandte Mathematik und Physik 75(6):211, DOI 10.1007/s00033-024-02347-z

[22] Reddy JN (2003) Mechanics of Laminated Composite Plates and Shells: Theory and Analysis. 2nd ed. CRC Press, Boca Raton, DOI 10.1201/b12409

[23] Kaplunov J, Prikazchikov DA, Prikazchikova L, Sergushova O (2019) The lowest vibration spectra of multi-component structures with contrast material properties. Journal of Sound and Vibration 445:132-147, DOI 10.1016/j.jsv.2019.01.013

[24] Wijeyewickrema AC, Horgan CO, Dundurs J (1996) Further analysis of end effects for plane deformations of sandwich strips. International Journal of Solids and Structures 33(29):4327-4336, DOI 10.1016/0020-7683(95)00232-4

Chapter 9
Contact Problem for a Piecewice-Homogeneous Plane with an Interphase Slit Under Static Friction

Lilit L. Dashtoyan and Lusine V. Hakobyan

Abstract The plane-deformed state of a composite elastic plane with an interphase crack is considered on one of the bank of which, with adhesion, an absolutely rigid stamp is pressed that does not reach the end points of the crack. Using discontinuous solutions of the Lame equations for a composite plane with an interphase crack, the solution of the problem is reduced to solving a singular integral equation. The solution of the latter is constructed by the numerical-analytical method of mechanical quadratures. A numerical experiment has been carried out and regularities of changes in contact stresses and the Cherepanov-Rice integral have been studied depending on the physical-mechanical and geometrical parameters of the problem.

Key words: Mixed boundary value problem, Crack, Stamp, Interphase defect

9.1 Introduction

Identifying the features of changes in the stress state near stress concentrators of various types located simultaneously in homogeneous and piecewise homogeneous massive bodies and the patterns of their mutual influence has always remains one of the most important fields of contact and mixed boundary value problems of the theory of elasticity and fracture mechanics. A huge number of works have been published in this area, the fundamental results of which are presented in [1, 2, 16–18]. In [3–5, 7, 12, 13], exact solutions are constructed for a number of contact and mixed boundary value problems for a homogeneous, piecewise homogeneous plane and space with interphase cracks. In [7, 10, 14] closed solutions for a number of contact problems for an elastic homogeneous and orthotropic plane with finite cracks are

---

Lilit L. Dashtoyan (cooresponding author) · Lusine V. Hakobyan
Institute of Mechanics of the National Academy of Sciences of Armenia, 24/2 M.Baghramyan, Yerevan 0019, Republic of Armenia, e-mail: lilit.dashtoyan@sci.am, lusine.vahram.hakobyan@gmail.com

H. Altenbach and A. V. Sahakyan, *Contact Mechanics and Relevant Problems*,
Advanced Structured Materials 253, https://doi.org/10.1007/978-3-032-20869-9_9

discussed, on one or two edges of which absolutely rigid stamps act simultaneously, not reaching the end points of the crack. Also, in [6, 9] the effective solutions to problems for a piecewise homogeneous plane and space with an interphase crack are constructed, into one of the edges of which a rigid stamp with a flat base with adhesion is pressed, not reaching the end points of the crack. The plane stress state of a piecewise homogeneous elastic plane with an interphase inclusion partially detached from the matrix, considered under static friction [11] was studied in [8].

Here, within the framework of the contact model taking into account static friction, a plane stress state of a piecewise homogeneous elastic plane with an interphase finite crack is considered, onto one of the edges of which an absolutely rigid stamp is pressed, not reaching the end points of the crack.

9.2 Statement of the Problem and Derivation of the Governing Equation

Let us consider the plane stress state of a piecewise homogeneous elastic plane, formed by two heterogeneous half-planes with Lamé coefficients $\lambda_1,\ \mu_1$ and $\lambda_2,\ \mu_2$, respectively, and related to the Cartesian coordinate system Oxy, the direction of the axis Ox of which coincides with the line of connection of heterogeneous materials. It is assumed that a piecewise homogeneous plane on the line of connection of heterogeneous materials $y = 0$ on the interval $(-a, a)$ is weakened by the finite slit and is deformed by an absolutely rigid stamp with a flat base, which is pressed into the lower bank of the slit in $(-b, b,)$ $(0 < b < a)$ by a concentrated normal load P_0 applied at the central point of the stamp. It is also assumed that under the stamp, in addition to the normal stress flows, there are also shear static stresses, which are directly proportional to the normal contact pressure, the proportionality coefficient of which depends on the coordinates of the contacting points, and the rest of the edges of the slit are free from stresses (Fig. 9.1).

The problem is to construct an effective solution, to study the patterns of changes in contact stresses under the stamp, crack opening and the Cherepanov-Rice J-integral

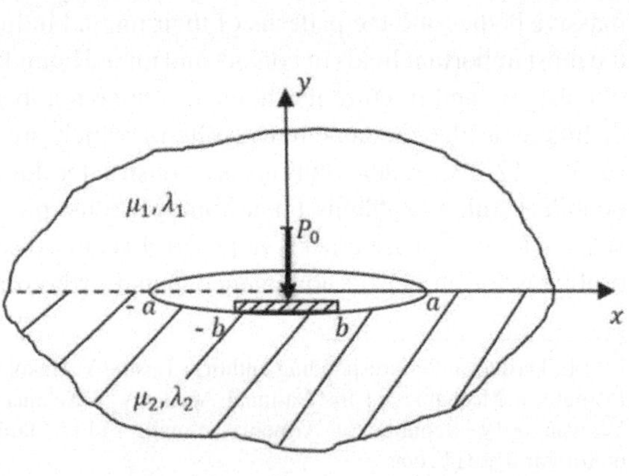

Fig. 9.1 Piecewise-homogeneous plane with an interphase slit.

depending on the physical, mechanical and geometric parameters. Using indices 1 and 2 for the components of stresses and displacements of points of the upper and lower half-planes, the problem mathematically is formulated as the following boundary value problem

$$\begin{cases} \sigma_y^{(1)}(x,0) - i\tau_{xy}^{(1)}(x,0) = \sigma_y^{(2)}(x,0) - i\tau_{xy}^{(2)}(x,0), \\ u_1(x,0) + iv_1(x,0) = u_2(x,0) + iv_2(x,0), \end{cases} \quad (|x| > a), \tag{9.1a}$$

$$\begin{cases} \sigma_y^{(1)}(x,0) = \tau_{xy}^{(1)}(x,0) = 0, & (-a < x < a), \\ \sigma_y^{(2)}(x,0) = \tau_{xy}^{(2)}(x,0) = 0, & (-a < x < -b \ \ \& \ b < x < a), \\ v_2(x,0) = \delta, & (-b < x < b), \\ \tau_{xy}^{(2)}(x,0) = f\dfrac{x}{b}P(x), & (-b < x < b). \end{cases} \tag{9.1b}$$

Here $\sigma_y^{(j)}(x,y)$, $\tau_{xy}^{(j)}(x,y)$ $(j = 1,2)$ are the components of the stresses acting in the corresponding half-planes, $u_j(x,y)$ and $v_j(x,y)$ $(j = 1,2)$ are the components of the displacements of the points of these half-planes,

$$P(x) = -\sigma_y^{(2)}(x,0)$$

is the contact pressure, f is the coefficient of friction, and δ is the rigid displacement of the stamp.

To construct a solution to the boundary value problem (9.1), we will use the influence function for a piecewise homogeneous plane with an interphase crack, constructed on the basis of discontinuous solutions of the plane elasticity theory for a piecewise homogeneous plane with a slit [6], which satisfy conditions (9.1a). Using these solutions, the complex combination of displacements of the points of the lower bank of the slit is written as follows

$$\frac{d}{dx}\left[u_2(x,0) + iv_2(x,0)\right]$$
$$= A_1\left\{\chi(x) - \frac{q_*}{\pi i}\int_{-b}^{b}\frac{\chi(s)\,ds}{s-x} + \int_{-b}^{b}K(s,x)\,\chi(s)\,ds - f(x)\right\} \tag{9.2}$$
$$(-b < x < b).$$

Here we use the following notation

$$\chi(x) = -\left[\sigma_y^{(2)}(x,0) - i\tau_{xy}^{(2)}(x,0)\right] \quad (-b < x < b),$$

$$K(s,x) = \frac{2\vartheta_2^{(2)}\mu_1^2}{\pi i æ_1\vartheta_1^{(2)}l_3}\,\frac{[\omega(x) - \omega(s)]}{\omega(s)(s-x)}, \quad f(x) = \frac{4P_0 l_2\mu_1^2\mu_2^2\cosh(\pi\gamma)\,\omega(x)}{\pi\, æ_1 æ_2\vartheta_1^{(2)}\vartheta_2^{(2)}l_3^2},$$

$$A_1 = \frac{\vartheta_1^{(2)}}{2\left[\left(\vartheta_2^{(2)}\right)^2 - \left(\vartheta_1^{(2)}\right)^2\right]}, \quad q_* = \frac{\vartheta_2^{(2)}}{\vartheta_1^{(2)}}; \quad \vartheta_1^{(j)} = \frac{\mu_j}{2+\alpha_j},$$

$$\vartheta_2^{(j)} = \frac{(1+\alpha_j)\,\mu_j}{2+\alpha_j}, \quad \alpha_j = \frac{1}{1-2\nu_j}, \quad (j=1,2),$$

$$\omega(x) = (x+a)^{-1/2+i\gamma}(a-x)^{-1/2-i\gamma}, \quad \gamma = \frac{1}{2\pi}\ln|g| = \frac{1}{2\pi}\ln\frac{\mu_1 + æ_1\mu_2}{\mu_2 + æ_2\mu_1},$$

$$l_2 = \vartheta_1^{(1)}\vartheta_2^{(2)} + \vartheta_2^{(1)}\vartheta_1^{(2)}, \qquad l_3 = 2\frac{\vartheta_1^{(2)} l_2 - \vartheta_2^{(2)} l_0}{\vartheta_2^{(2)}},$$

$$æ_j = \frac{\lambda_j + 3\mu_j}{\lambda_j + \mu_j} = 3 - 4\nu_j \quad (j=1,2),$$

ν_j $(j = 1,2)$ are Poisson's ratios of the corresponding half-planes

We separate the imaginary part in representation (9.2) and satisfy the third condition (9.1b), initially differentiating it with respect to x. Finally, taking into account the last equation (9.1b), to determine the normal contact pressure, the following singular integral equation with variable coefficients is obtained $(-b < x < b)$

$$f\frac{x}{b}P(x) + \frac{q_*}{\pi}\int_{-b}^{b}\frac{P(s)\,ds}{s-x} + \mathrm{Im}\int_{-b}^{b}K(s,x)(1+ifs/b)P(s)\,ds = \mathrm{Im}\,f(x). \quad (9.3)$$

This equation should be considered under the condition of stamp equilibrium

$$\int_{-b}^{b}P(s)\,ds = P_0. \quad (9.4)$$

The solution of the problem is reduced to the solution of the governing singular integral equation (9.3) under condition (9.4). The solution of the governing singular integral equation is built by the numerical-analytical method of mechanical quadratures [19].

Using a change of variables $x = b\eta, \;\; s = b\xi$ and notation

$$\varphi(\eta) = bP(b\eta)/P_0, \;\; f_*(\eta) = \mathrm{Im}\left[bf(b\eta)/P_0\right],$$
$$K_*(\eta,\xi) = \mathrm{Im}\left[b(1+if\xi)K_*(b\eta,b\xi)\right],$$

Eq. (9.3) can be rewritten in the following form $(-1 < \eta < 1)$

$$f\eta\varphi(\eta) + \frac{q_*}{\pi}\int_{-1}^{1}\frac{\varphi(\xi)\,d\xi}{\xi-\eta} + \int_{-1}^{1}K_*(\xi,\eta)\,\varphi(\xi)\,ds = f_*(\eta). \quad (9.5)$$

The condition (9.4) the form is taken

$$\int_{-1}^{1} \varphi(\xi)\, d\xi = 1. \tag{9.6}$$

The study of the principal part of Eq. (9.5) shows that the unknown function $\varphi(\eta)$ at the end points of the integration interval has an exponential singularity [11] and, therefore, it can be represented in the following form:

$$\varphi(\eta) = \frac{\varphi_*(\eta)}{\left(1-\eta^2\right)^{1/2-\beta}}, \tag{9.7}$$

where

$$\beta = \frac{1}{2\pi} \arctan \frac{f}{q_*}$$

and $\varphi_*(\eta)$ is continuous bounded function on the interval $[-1, 1]$. Substituting the values of the functions $\varphi(\eta)$ from (9.7) into (9.5) and (9.6), using the quadrature formulas given in [19], according to the usual procedure, the system of algebraic equations with respect to the values of the function $\varphi_*(\eta)$ at the collocation points $\xi_k \ (k = 1 - N)$ is obtained.

Usually, after solving the resulting system, having the values of the functions $\varphi_*(\eta)$ at the collocation points the unknown function $\varphi(\eta)$ is restored and all the necessary mechanical characteristics of the problem are determined using the Lagrange formula. In the stated problem, the main mechanical characteristics are the contact stresses acting under the stamp, the Cherepanov-Rice J-integral at the end points of the crack [15], and also the crack opening. To determine the Cherepanov-Rice J-integral, we will use the formulas for the destructive stresses acting outside the crack and the complex intensity factor of the destructive stresses obtained in [6]. In the given formulas, replacing the dimensionless complex combination of contact stresses $\psi(\eta)$ with $(1 + f\xi)\,\varphi(\eta)$ we find

$$\begin{gathered} \frac{\sigma_y^{(1)}(x,0) - i\tau_{xy}^{(1)}(x,0)}{P_0} = -\frac{\operatorname{sign} x\, \omega_1(x)}{\pi\Delta} \\ \times \left[il_2 - \frac{b_* l_3}{l_* \cosh(\pi\gamma)} \int_{-1}^{1} \frac{(1+f\xi)\,\varphi(\xi)\, d\xi}{\omega_*(\xi)\,(\xi - x/b)} \right] + \Phi_*(x) \quad (|x| > a), \end{gathered} \tag{9.8}$$

$$\begin{aligned} K_I(\pm a) - iK_{II}(\pm a) &= \lim_{x\to(-1)^j(a+0)} \frac{\omega_1^{-1}(x)\left[\sigma_y^{(1)}(x,0) - i\tau_{xy}^{(1)}(x,0)\right]}{P_0} \\ &= \mp\frac{1}{\pi\Delta}\left[il_2 - \frac{b_* l_3}{l_* \cosh(\pi\gamma)} \int_{-1}^{1} \frac{(1+if\xi)\,\varphi(\xi)\, d\xi}{\omega_*(\xi)\,(\xi \mp l_*)} \right], \end{aligned} \tag{9.9}$$

$$J(\pm a) = \tilde{\mu} K(\pm a)\, \bar{K}(\pm a), \quad \tilde{\mu} = \frac{P_0}{2a}\left(\frac{1-\nu_1^2}{E_1} + \frac{1-\nu_2^2}{E_2} \right), \tag{9.10}$$

where

$$\Phi_*(x) = \frac{iP_0(l_2 + a_*l_3 - ib_*\tanh(\pi\gamma))}{\pi b\Delta}\int_{-1}^{1}\frac{(1+f\xi)\varphi(\xi)\,d\xi}{(\xi - x/b)},$$

$$\omega_1(x) = |a+x|^{-\frac{1}{2}+i\gamma}|a-x|^{-\frac{1}{2}-i\gamma};\quad \omega_*(\xi) = (\xi + l_*)^{-\frac{1}{2}+i\gamma}(l_* - \xi)^{-\frac{1}{2}-i\gamma},$$

$$b_* = \frac{æ_2\left(\vartheta_2^{(2)}\right)^2}{2\mu_2^2};\quad a_* = \frac{æ_2\vartheta_1^{(2)}\vartheta_2^{(2)}}{2\mu_2^2};\ d = \frac{l_2\cosh(\pi\gamma)}{\pi l_3}P_0;\quad l_* = a/b > 1.$$

Let us also write a formula for determining the crack opening. For this, we will use the formula for the complex combination of displacement dislocations through the complex combination of reduced displacement dislocations, obtained in [6]. In this case the formula is represented in the following form

$$\frac{d}{dx}\{[u_1(x,0) + iv_1(x,0)] - [u_2(x,0) + iv_2(x,0)]\} = w'_*(x) = \frac{w'(x)}{\vartheta_2^{(2)}}$$

$$= \frac{d\omega(x)}{\vartheta_2^{(2)}} - \frac{A}{\left(l_1^2 - l_3^2\right)\vartheta_2^{(2)}}\begin{cases}\dfrac{1}{q_*}\chi(x) + \dfrac{\omega(x)}{\pi i}\displaystyle\int_{-b}^{b}\frac{\chi(s)\,ds}{\omega(s)(s-x)} & (|x| \le b)\\[2ex] \dfrac{\omega(x)}{\pi i}\displaystyle\int_{-b}^{b}\frac{\chi(s)\,ds}{\omega(s)(s-x)} & (b < |x| < a),\end{cases} \tag{9.11}$$

where

$$A = l_0l_3 - l_1l_2 = -2\Delta\left(\left(\vartheta_2^{(1)}\right)^2 - \left(\vartheta_1^{(1)}\right)^2\right),$$

$$l_1^2 - l_3^2 = -4\left(\left(\vartheta_2^{(2)}\right)^2 - \left(\vartheta_1^{(2)}\right)^2\right)\left(\left(\vartheta_2^{(1)}\right)^2 - \left(\vartheta_1^{(1)}\right)^2\right)\Delta/\left(\vartheta_2^{(2)}\right)^2.$$

Then, (9.11) on the interval $(-1, 1)$ will be written in the following form

$$w'_*(bx) = \frac{d\omega_*(x)}{b\,\vartheta_2^{(2)}}$$

$$-\frac{(1-\nu_2)P_0^*}{\mu}\begin{cases}\dfrac{1+ifx}{q_*}\varphi(x) + \dfrac{\omega_*(x)}{\pi i}\displaystyle\int_{-1}^{1}\frac{(1+ifs)\varphi(s)\,ds}{\omega_*(s)(s-x)} & (|x| < 1),\\[2ex] \dfrac{\omega_*(x)}{\pi i}\displaystyle\int_{-1}^{1}\frac{(1+ifs)\varphi(s)\,ds}{\omega_*(s)(s-x)} & (1 < |x| < l_*)\end{cases}$$

with

$$P_0^* = P_0/b\mu_1, \;\; \mu = \mu_2/\mu_1, \;\; \frac{d}{b\,\vartheta_2^{(2)}} = \frac{\mu_1 l_2 P_0^*}{2\left(\vartheta_1^{(2)} l_2 - \vartheta_2^{(2)} l_0\right)}.$$

Therefore, the dimensionless crack opening is determined by the formula

$$v_*(x) = \frac{v(bx)}{a} = \frac{1}{a}\,\mathrm{Im}\int_{-l_*}^{x} w'(bx)\,dx \quad (-l_* < x < l_*)\,. \tag{9.12}$$

9.3 Numerical Calculations

Numerical calculations were performed and patterns of change in the reduced contact pressure $\varphi(\eta)$ and shear stresses $T(\eta) = f\eta\varphi(\eta)$ acting under the stamp, and dimensionless Cherepanov-Rice integrals $J(\pm a)$ at the end points of cracks were identified depending on the static friction coefficient, ratio $\mu = \mu_2/\mu_1$, and parameter $l_* = a/b$ in the case of fixed values of Poisson's ratios ν_j $(j = 1, 2)$. It is assumed that $P_0/a\mu_1 = 1$.

The results of the numerical calculations are presented as plots in Fig. 9.2 - 9.5. Figures 9.2 and 9.3 show plots of the normal contact pressure and shear contact stresses, respectively, for different values of the coefficient of static friction f, in the case when $\nu_1 = 0.1$, $\nu_2 = 0.4$, $\mu = 3$ and $l_* = 1.3$. Figures 9.4 and 9.5 display the plots for Cherepanov-Rice J-integral at the end-points of crack depending on parameters l_* and μ for values of friction coefficients $f = 0.1, 0.3, 0.6$ in case of $\nu_1 = 0.2$, $\nu_2 = 0.3$, $\mu = 3$ and $\nu_1 = 0.2$, $\nu_2 = 0.3$, $l_* = 2$. The coefficient of static friction also has little effect on the value of the Cherepanov-Rice J-integral, which decreases both with the removal of the stamp from the end points of the crack and with an increase in the rigidity of the lower half-plane.

Figures 9.6 and 9.7 show plots of dimensionless Cherepanov-Rice J-integrals at the end points of a crack, respectively, depending on the parameter l_* for different values μ and depending on the parameter μ for different values l_* in the case

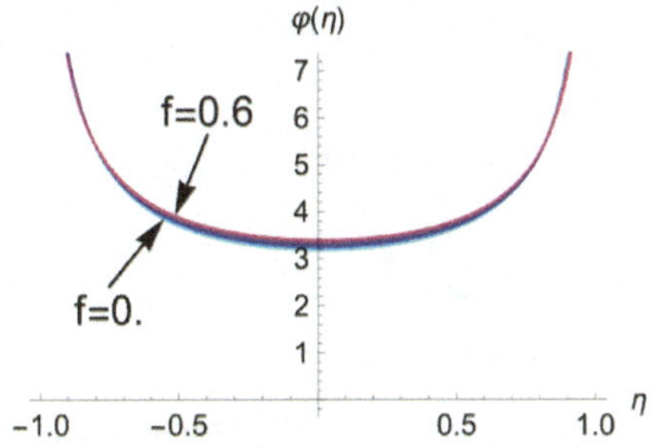

Fig. 9.2: Normal contact pressure.

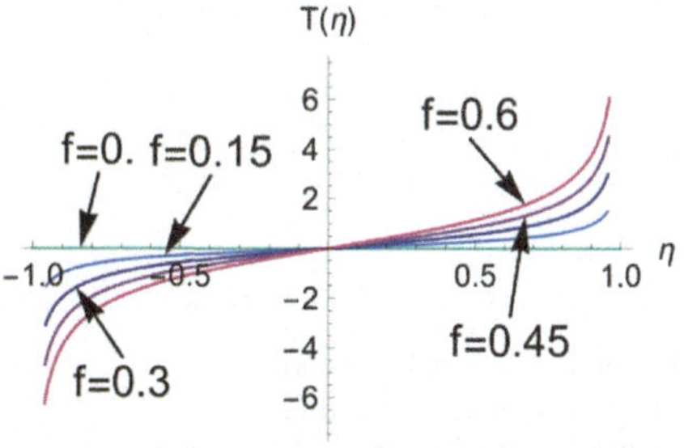

Fig. 9.3: Shear contact stresses.

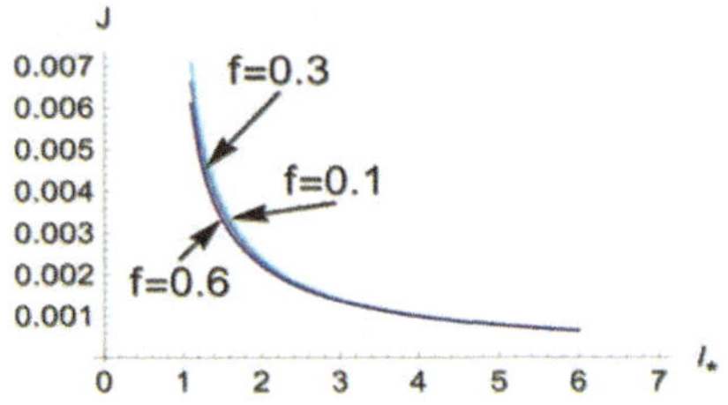

Fig. 9.4: J-integrals depending on parameter l_* for $\mu = 3$.

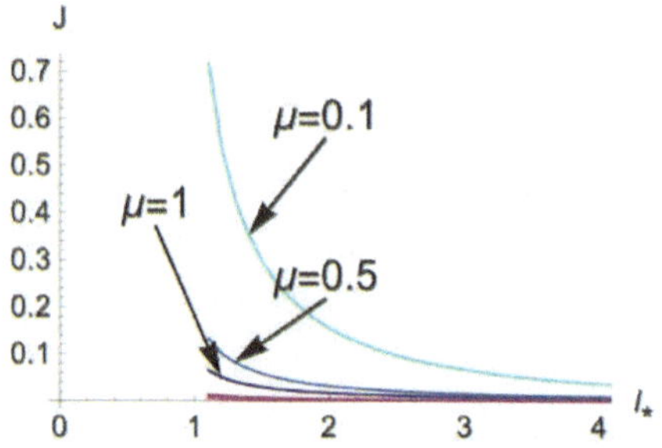

Fig. 9.5: J-integrals depending on parameter μ for $l_* = 2$.

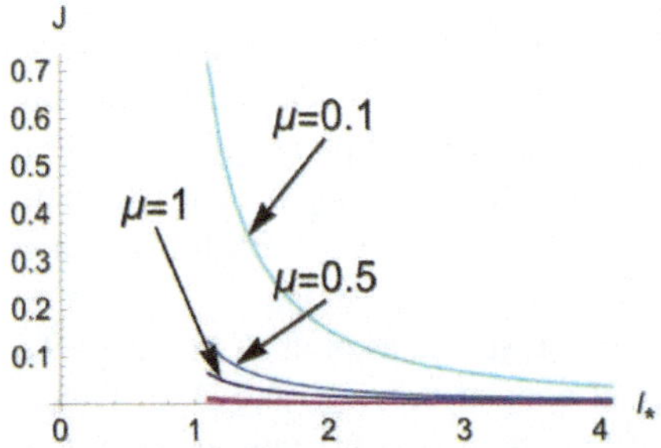

Fig. 9.6: J-integrals depending on parameter l_* for different values of μ in case $\nu_1 = 0.2$, $\nu_2 = 0.3$ and $f = 0.3$.

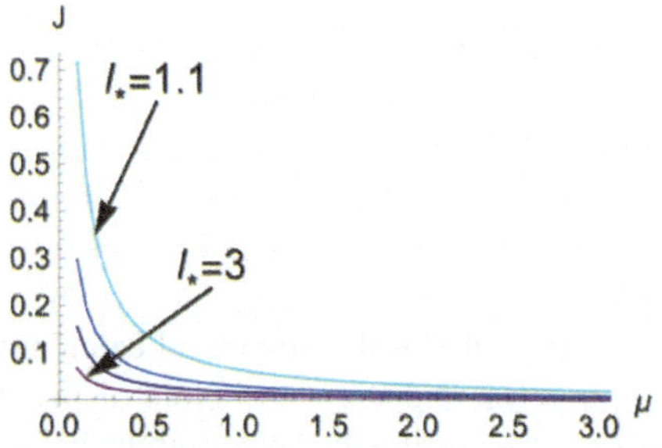

Fig. 9.7: J-integrals depending on parameter μ for different values of μ in case $\nu_1 = 0.2$, $\nu_2 = 0.3$ and $f = 0.3$.

$\nu_1 = 0.2$, $\nu_2 = 0.3$ and $f = 0.3$. As in [6], the plots show that as the stamp moves away from the end points of the crack, the Cherepanov-Rice J-integrals decrease. Moreover, the softer the lower half-plane relative to the upper half-plane, the greater the value of the J-integral.

Figure 9.8 displays the plots of the dimensionless Cherepanov-Rice J-integrals at the end points of the crack, respectively, depending on the parameter l_* for different values ν_2 in the case $\nu_1 = 0.2$ and $f = 0.3$. It is evident that with an increase in the

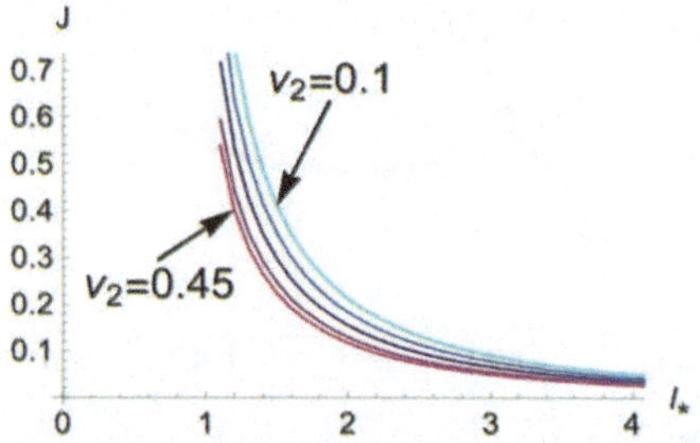

Fig. 9.8 J-integrals depending on parameter l_* for different values of ν_2 in case $\nu_1 = 0.2$ and $f = 0.3$

Poisson's ratio of the lower half-plane of the Cherepanov-Rice J-integrals at the end points of the crack decreases.

9.4 Conclusion

Using the methods of singular integral equations and mechanical quadratures, the problem of indentation an absolutely rigid stamp, not reaching the crack endpoints, into one of the edges of an interphase crack in a composite elastic plane, considering under static friction is solved. The patterns of change in contact stresses and the Cherepanov-Rice J-integral at the crack endpoints were studied. It has been shown that the closer the stamp endpoints are to the crack endpoints, the greater the probability of crack propagation.

References

[1] Berezhnitski LT, Panasyuk VV, Stashyuk NG (1983) Vzaimodeystviye zhestkikh lineynykh vklyucheniy i treshchin v deformiruyemom tele (in Russ., Interaction of Hard Linear Inclusions and Cracks in Deformed Solid). Naukova Dumka, Kiev

[2] Galin LA (1980) Kontaktnyye zadachi teorii uprugosti i vyazkouprugosti (in Russ., Contact Problems in the Theory of Elasticity and Viscoelasticity). Nauka, Moscow

[3] Hakobyan VN (2007) Napryazheniya vozle absolyutno zhostkogo monetoobraznogo vklyucheniya v kusochno-odnorodnom prostranstve (in Russ., The stresses near the absolutely rigid coin-type inclusion in piecewise homogeneous space). In: Proceedings of International Conference dedicated to the 95th Anniversary of Academician Nagush Kh. Arutyunyan "Topical Problems of Continuum Mechanics", pp 45 - 51

[4] Hakobyan VN (1995) Ob odnoy smeshannoy zadache dlya sostavnoy ploskosti, oslablennoy treshchinoy (in Russ., On one mixed problem for composite plate, weakened by a crack). Mechanics. Proceedings of NAS RA 48(4):57-65

[5] Hakobyan VN (2022) Stress Concentrators in Continuous Deformable Bodies. Advanced Structured Materials, Vol 181. Springer, Cham, DOI 10.1007/978-3-031-16023-3

[6] Hakobyan VN, Amirjanyan HA, Hakobyan LV (2022) Kontaktnaya zadacha dlya sostavnoy ploskosti s razrezom (in Russ., Contact problem for a compound plane with a slit). Proceedings of NAS RA. Mechanics 75(1-2)17 - 26, DOI 10.54503/0002-3051-2022.75.1-2-17

[7] Hakobyan VN, Dashtoyan LL (2013) Contact problem for an orthotropic plane with a slit. Mechanics of Composite Materials 49(5):507 - 518, DOI 10.1007/s11029-013-9367-x

[8] Hakobyan V, Dashtoyan L, Amirjanyan H (2024) Contact problem for a piecewise homogeneous plane with a finite crack under the static friction. In: Altenbach H, Öchsner A (eds) State of the Art and Future Trends in Materials Modelling 2, Advanced Structured Materials, Vol 200, pp 243 - 253. Springer, Cham, DOI 10.1007/978-3-031-72900-3_12

[9] Hakobyan VN, Grigoryan AH, Amirjanyan HA (2023) On an axisymmetric contact problem for a piecewise homogeneous space with disc-shaped crack. In: Altenbach H, Hakobyan V, Mkhitaryan S, Sahakyan A (eds) Solid Mechanics, Theory of Elasticity and Creep, Advanced Structured Materials, Vol 185, pp 133 -147, DOI 10.1007/978-3-031-18564-9

[10] Hakobyan VN, Hakobyan LV (2020) Kontaktnaya zadacha dlya odnorodnoy ploskosti s treshchinoy (in Russ., Contact problem for a homogeneous plane with crack). Procedings of NAS RA. Mechanics 73(4)3 - 12, DOI 10.33018/73.4.1

[11] Hakobyan VN, Hakobyan LV (2023) Ob odnoy modeli treniya primenitel'no k kontaktnym zadacham teorii uprugosti (in Russ., On a model of friction for contact problems of the theory of elasticity). Proceedings of NAS RA. Mechanics 76(2)20 - 31, DOI 10.54503/0002-3051-2023.76.2-20

[12] Hakobyan VN, Mirzoyan ST, Dashtoyan LL (2015) Osesimmetrichnaya smeshannaya zadacha dlya sostavnogo prostranstva s monetoobraznoy treshchinoy (in Russ., Axis-symmetric mixed boundary value problem for composite space with coin-shaped crack). Vestnik MGTU imeni NE Baumana, Seriya Yestestvennye nauki (3):31-46

[13] Ilyina I, Silvestrov V (2005) The problem of thin rigid interphase inclusion, detached from media along one side. Proceedings Russian Academy of Sciences. Mechanics of Solids 40(3):123 - 136

[14] Mkhitaryan SM (2019) On the stress-strain state of an elastic infinite plate with a crack expanding by means of smooth thin inclusion indentation. Proceedings of NAS RA. Mechanics 72(4)38 - 64, DOI 10.33018/72.4.4

[15] Murakami Y (2003) Handbook of Stress Intensity Factors, Elsevier

[16] Muskhelishvili NI (1977) Some Basic Problems of the Mathematical Theory of Elasticity. Springer, Dordrecht, DOI 10.1007/978-94-017-3034-1

[17] Panasyuk VV, Savruk MP, Datsyshin AP (1976) Raspredeleniye napryazheniy okolo treshchin v plastinakh i obolochkakh (in Russ., Stress Distribution Around Cracks in Plates and Shells). Naukova Dumka, Kiev,

[18] Popov GY (1982) Kontsentratsiya uprugikh napryazheniy vozle shtampov, razrezov, tonkikh vklyucheniy i podkrepleniy (in Russ., Concentration of Elastic Stresses Near the Punches, Slits, Thin Inclusions and Stiffened Constructions). Nauka, Moscow

[19] Sahakyan AV, Amirjanyan HA (2018) Method of mechanical quadratures for solving singular integral equations of various types. IOP Conference Series: Journal of Physics 991:012070, DOI 10.1088/1742-6596/991/1/012070

Chapter 10
Manufacturing Technology and Characteristics of Three-Layer Magnetoelectric Gyrators Based on an Amorphous Alloy and Piezoceramics

Tigran A. Galichyan, Dmitry A. Filippov, Stanislav O. Sapel'nikov, and Sergey B. Sapozhkov

Abstract This work examines the fabrication technology of a multilayer magnetoelectric (ME) gyrator based on the amorphous alloy AMAG 212N and lead zirconate titanate PZT-19. The study considers the specific features of the fabrication process using a bonding method, as well as the structural design of the device. Frequency and load characteristics of gyrators made from different layers of the amorphous alloy are investigated. It is shown that increasing the number of amorphous alloy layers improves the magnetoelectric performance of the gyrator, although this also complicates the fabrication process. The optimal number of amorphous alloy layers is determined, taking into account both the manufacturing technology and the device's performance characteristics.

Key words: Gyrator, Composite structure, Magnetostriction, Piezoelectricity, Magnetoelectric effect

10.1 Introduction

In the high-tech environment, there is a clear trend toward miniaturization: mobile phones and laptops are becoming thinner, and various devices are becoming "smart", such as glasses, watches, and other gadgets. A notable manifestation of this trend is the gradual transition to wireless data transmission and charging. In recent years, "smart devices" have increasingly incorporated artificial intelligence functionality,

Tigran A. Galichyan (corresponding author)
Institute of Mechanics, National Academy of Sciences of the Republic of Armenia, 24b Marshal Baghramyan Ave, Yerevan, 0019, Republic of Armenia, e-mail: Galichyantigran@gmail.com

Dmitry A. Filippov · Stanislav O. Sapel'nikov · Sergey B. Sapozhkov
Yaroslav-the-Wise Novgorod State University, 41 Bolshaya St., Petersburg Str., Veliky Novgorod, Russian Federation, e-mail: Dmitry.Filippov@novsu.ru, Stanislav.Sapelnikov@novsu.ru, ssb@novsu.ru

H. Altenbach and A. V. Sahakyan, *Contact Mechanics and Relevant Problems*,
Advanced Structured Materials 253, https://doi.org/10.1007/978-3-032-20869-9_10

significantly expanding their range of applications. The hardware of these devices is also evolving: creating lightweight and thin equipment requires very compact components. Against this background, the demand for device miniaturization has grown.

Creating gyrators based on the magnetoelectric (ME) effect is particularly promising, as it allows significant simplification of fabrication technology while reducing device dimensions. The concept of a gyrator was first introduced in 1948 by the Danish scientist Bernard Tellegen in his work "The Gyrator: A New Network Element", where he described the gyrator as a new electrical circuit element [9]. An ideal gyrator performs the function of converting current to voltage and vice versa, or equivalently, converting inductive impedance into capacitive impedance. Gyrators have proven useful in the design of duplex filters, high-power wireless transmitters, and other applications [1, 2, 8].

Traditionally, gyrators were implemented using circuit based solutions with operational amplifiers. With the discovery of the ME effect, it became possible to simplify gyrator implementation. A classic ME gyrator consists of an inductive coil with a core made of an ME material. A pioneering study [3] demonstrated that gyrator properties are inherent in a system combining an inductive coil with an ME core material.

For such applications, it is advisable to use layered magnetostrictive–piezoelectric composite structures, in which the magnitude of the ME conversion effect is several orders of magnitude higher than in single crystals. Unlike single crystals, the ME effect in composites occurs in the presence of a bias magnetic field, the magnitude of which depends on the magnetostrictive layer material and can vary from tens to hundreds of Oersteds (Oe). A schematic representation of the ME gyrator structure is shown in Fig. 10.1.

The current-to-voltage (I/V conversion) occurs as follows: an alternating current I_{in} flowing through the inductance coil generates an alternating magnetic field H. Due to the magnetostrictive effect in the magnetic component of the ME sample, mechanical strains are induced, which are transmitted across the interface to the piezoelectric phase, resulting in the generation of an electric voltage V_{out}. The I/V conversion coefficient K_{IV} is defined as the ratio of the output voltage V_{out} to the input current I_{in} QUOTEIin, i.e.

$$K_{\text{IV}} = \frac{V_{\text{out}}}{I_{\text{in}}}.$$

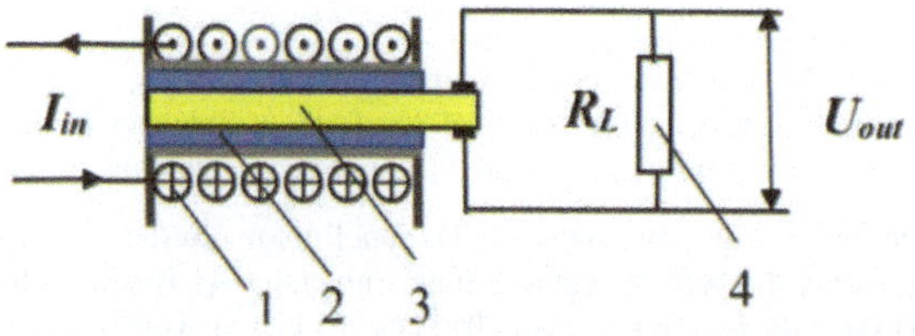

Fig. 10.1 Schematic diagram of the gyrator: 1 – inductance coil, 2 – magnetic layer (Metglas), 3 – piezoelectric layer (PZT), 4 – load resistance.

This coefficient depends on both the frequency of the alternating input signal and the parameters of the ME sample, as well as the characteristics of the inductance coil and the load resistance. Its behaviour is typically described using frequency and load characteristics. In the present work, the dependence of these characteristics on the thickness of the magnetostrictive layers is investigated.

10.2 Materials and Methods

This study examines the fabrication technology of an ME gyrator based on a symmetric magnetostrictive–piezoelectric structure. When fabricating the ME core of the gyrator, it is preferable to select magnetostrictive materials with a low bias magnetic field. From this perspective, the most suitable materials are FeNiBSiC-based amorphous alloys, such as Metglas (USA) and AMAG (Russia). For these materials, the optimal bias field H_{bias}^{opt} corresponding to the maximum ME response depends on the thickness of the magnetostrictive layer 15-30 Oe [7], whereas for permendur it is equal to H_{bias}^{opt} = 125 Oe [6].

Accordingly, for the fabrication of the gyrator, the amorphous nanocrystalline alloy AMAG 212N ("Mstator", Borovichi, Novgorod Region, Russia) was selected as the magnetostrictive material. The piezoelectric layer was made from lead zirconate titanate-based ceramics PZT-19 ("Piezopribor", Rostov-on-Don, Russia). The choice of an amorphous alloy for the magnetostrictive component was dictated by its high magnetostriction at low bias fields, on the order of 10-20 Oe. The AMAG 212N material is supplied as a ribbon with a width of 20 mm and a thickness of 20 μm. The piezoelectric material was fabricated as a rectangular plate with dimensions 20 × 10 × 0.3 mm, coated with a silver layer of thickness 1-3 μm. As seen in Figs. 10.2a and 10.2b, the materials exhibit very low surface roughness and have different material natures, which limits the available methods for bonding. An

(a) Surface of AMAG 212N alloy. (b) Surface of PZT-19 ceramic.

Fig. 10.2: High magnification images of the sample surfaces (50× zoomed in).

adhesive bonding technique was used to fabricate the magnetostrictive–piezoelectric structure, in which the layers were joined together using an adhesive.

To create the AMAG layer, it was necessary to prepare rectangular plates with dimensions 20 × 5 mm. These plates were fabricated by cutting. Figure 10.3 shows photographs of the edges of the ribbon prepared using different cutting tools. The images clearly show the burrs left by tools with different blade sharpening and surface finishing. In Fig. 10.3a, the edge cut with Stanley 2-14-563 scissors (China), which have a serrated blade and a ground inner surface, is shown. Figure 10.3b shows the edge cut with Deli 4 scissors (China), which have a straight blade and a coarsely ground inner surface. Figure 10.3c shows the edge cut with Kraftool Alligator scissors (China), which have a straight blade and a polished inner surface. As seen from the images, the best results were obtained using the Kraftool Alligator scissors with a straight, polished blade.

The next stage involved selecting an adhesive for bonding the PZT-19 and AMAG plates. The adhesive must be rigid after curing to ensure efficient vibration transmission between the layers of the gyrator structure. The adhesives considered included cyanoacrylate adhesives (Loctite 415, China; Loctite 435, Ireland; Loctite 496, China; and MTE, Russia), an epoxy resin (Superplast EP-SM-PRO, Russia), and BF-2 glue (Russia). Adhesion between layers was tested using a tape peel method. All selected adhesives showed poor adhesion due to the low roughness, lack of porosity and the characteristics of the surfaces being bonded.

As an alternative, a bonded frame approach for the gyrator structure was proposed. Figure 10.4 schematically shows a cross-section of the three-layer structure. This design allows for a minimal adhesive layer between the layers 1-3 μ m while preserving the physical integrity of the gyrator structure.

For the experimental part, a gyrator was fabricated, consisting of an inductive coil with a core. The core had the form of a three layer magnetostrictive–piezoelectric structure (Fig. 10.4), to which AMAG 212N plates were bonded on both sides. A symmetric structure was chosen because placing an asymmetric magnetostrictive–piezoelectric structure in a magnetic field can induce both longitudinal and bending vibrations. These vibration modes interact destructively, reducing the overall effi-

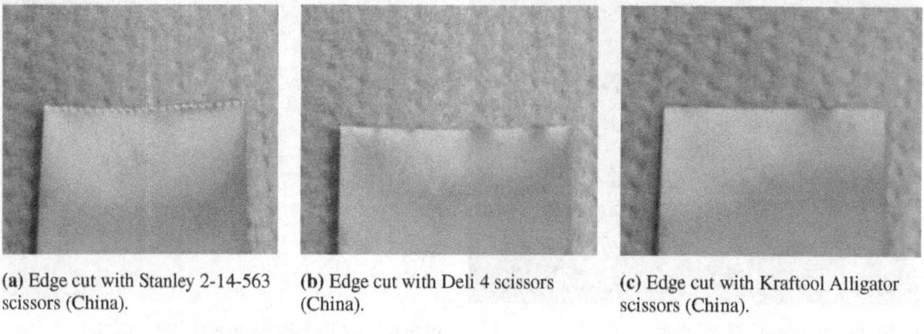

(a) Edge cut with Stanley 2-14-563 scissors (China).

(b) Edge cut with Deli 4 scissors (China).

(c) Edge cut with Kraftool Alligator scissors (China).

Fig. 10.3: Edge of AMAG ribbon cut with scissors of different blade sharpening types.

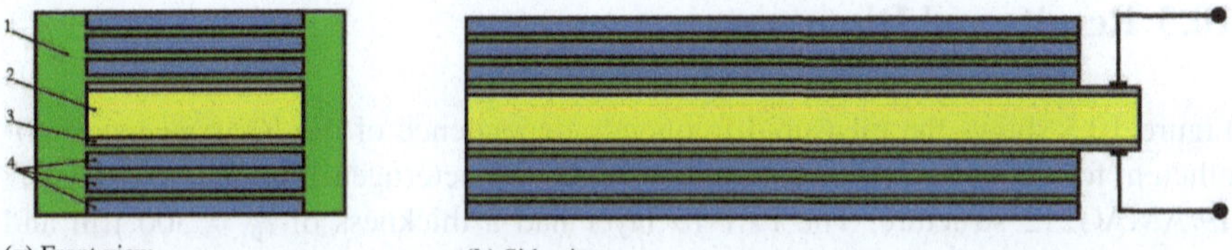

(a) Front view. (b) Side view.

Fig. 10.4: Cross-section of the three-layer structure: 1 – adhesive frame, 2 – PZT-19 layer, 3 – silver layer, 4 – AMAG 212N layer.

ciency of magnetoelectric conversion [5]. In a symmetric structure, only longitudinal vibrations occur, leading to more efficient ME interaction.

The gyrator coil was wound on a frame with dimensions, 20 × 10 × 5 mm which contained a rectangular through hole of size 6 × 1.5 mm for housing the sample. The coil was made from copper wire with a diameter of $d = 0.15$ mm and contained 150 turns. The core consisted of a PZT-19 ceramic plate with electrodes obtained by firing silver paste, and two plates of the amorphous nanocrystalline alloy AMAG 212N. The magnetostrictive layers were fabricated by bonding 20 μm thick foils of the amorphous alloy. For the first sample, the core contained three layers of foil, and for the second sample, five layers. Thus, the total thickness of the magnetostrictive layer was $t_1^m = 120$ μm for the first sample and $t_2^m = 200$ μm for the second.

The entire assembly was placed inside a solenoid with a diameter of $D = 40$ mm and a length of, $L = 100$ mm containing $N = 500$ turns of copper wire with a diameter of 0.3mm. The bias magnetic field was created by applying a DC voltage to the solenoid from a UT 3010E power supply. The alternating magnetic field was generated by applying a voltage to the gyrator coil from an AKIP-3410/4 signal generator. A resistance box (R-33) was used to provide the load, and the voltage across the load was measured with an AKIP-2101 voltmeter and simultaneously monitored with an AKIP-4115/6A oscilloscope.

At the experimental researches first, the dependence of the output voltage on the bias field was measured at an alternating magnetic field frequency of 10 kHz and a fixed current through the coil. Then, with the bias field value corresponding to the signal maximum, the frequency and load characteristics of the gyrator were measured. During the frequency and load characteristic measurements, the alternating current through the gyrator coil was in the range of 10 – 15 mA, corresponding to an alternating magnetic field intensity of approximately 1 Oe. The non-resonant voltage across the load resistance was several tens of millivolts. The magnitude of the I/V conversion coefficient was calculated as the ratio of the voltage across the load to the input current in the gyrator coil.

10.3 Results and Discussion

Figure 10.5 shows the measured frequency dependence of the I/V conversion coefficient for the gyrator, whose core consisted of a heterogeneous AMAG212/PZT-19/AMAG212 structure. The PZT-19 layer had a thickness of $t_p = 300$ μm and the AMAG 212N layers had a thickness of $t_{1m} = 80$ μm for the first sample and $t_{2m} = 160$ μm for the second sample. As seen in Fig. 10.5, increasing the thickness of the magnetostrictive layer from 80 μm to 160 μm results in an increase of the low frequency value of the I/V conversion coefficient from 0.8 V/A to 1.8 V/A at a frequency of 20 kHz and a change in its resonant value from 60 V/A to 116 V/A accompanied by an increase in the resonant frequency from 79.6 kHz to 80.5 kHz.

The change in the resonant frequency is due to variations in the average structural density and the effective Young's modulus of the system. In the first approximation, according to [4], the resonant frequency is given by

$$f_{res} \approx \frac{1}{2L_{eff}}\sqrt{\frac{Y_{eff}}{\rho_{eff}}}, \qquad (10.1)$$

where

$$L_{eff} = (L_p t_p + L_m t_m)/(t_p + t_m),$$
$$Y_{eff} = (Y_p t_p + Y_m t_m)/(t_p + t_m),$$
$$\rho_{eff} = (\rho_p t_p + \rho_m t_m)/(t_p + t_m)$$

are the effective length, Young's modulus, and density of the magnetostrictive–piezoelectric structure, respectively, and $L_p, L_m, Y_p, Y_m, \rho_p, \rho_m, t_p, t_m$ are the length, Young's modulus, density, and thickness of the magnetostrictive and piezoelectric layers, respectively.

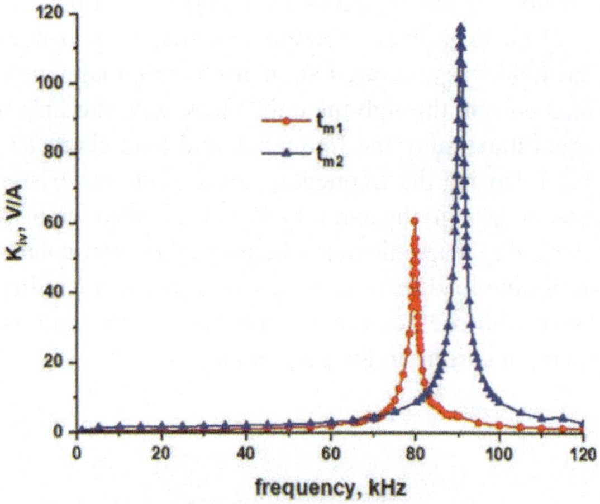

Fig. 10.5 Frequency dependence of the I/V conversion coefficient

The observed increase in the resonant frequency with increasing magnet layer thickness (t_m) has a twofold mechanism. The first mechanism is geometric: due to the shorter length of the magnetostrictive layer relative to the piezoelectric layer, increasing t_m reduces the effective length L_{eff} of the system, directly raising the frequency f_{res}. The second mechanism is related to differences in elastic properties: the Young's modulus of AMAG ($Y_m = 110$ GPa) significantly exceeds that of PZT-19 ($Y_p = 67$ GPa) while their densities are similar. As a result, with increasing t_m, the effective modulus Y_{eff} grows faster than the effective density ρ_{eff}, which according to Eq. (10.2) further increases the resonant frequency.

Reference [4] shows that the low-frequency value of the I/V conversion coefficient is determined by the expression

$$|K_{IV}| = \frac{Y^p d_{31}^p q_{11}^m}{\varepsilon_{33}} \times \frac{Y^m t^m}{\overline{Y} t} t^p n \times \frac{(R_L/X_C)}{\sqrt{1 + ((R_L/X_C))^2}}, \quad (10.2)$$

where d_{31}^p, q_{11}^m are the piezoelectric and piezomagnetic coefficients, $\varepsilon_{33} = \varepsilon_{33}^p \varepsilon_0$, ε_{33}^p is the relative permittivity of the piezoelectric, $\varepsilon_0 = 8.85 \times 10^{-12}$ F/m is the vacuum permittivity,

$$X_C = \frac{1}{\omega C}$$

is the capacitive resistance,

$$C = \frac{\varepsilon_{33} W L}{t^p}$$

is the capacitance of the piezoelectric plate, W is the plate width, $\omega = 2\pi f$ is the angular frequency, and n is the number of turns per unit length of the inductance coil.

As follows from Eq. (10.1), the magnitude of the I/V conversion coefficient increases with the thickness of the magnetostrictive layer, which is consistent with the experimental data. Figure 10.6 shows the load characteristic of the gyrator based on the AMAG212/PZT-19/AMAG212 structure, with the AMAG212 layer thickness of 60 μm on each side. As seen in Fig. 10.6, there is good agreement between theory and experiment. At low load resistances, the magnitude of K_{IV} increases linearly with the load resistance; then the dependence becomes nonlinear, and at high load resistances, the ratio

$$\frac{R_L}{X_C} \gg 1$$

reaches saturation.

One of the important characteristics of the gyrator is the output power

$$P_{out} = \frac{V_{out}^2}{R_L}.$$

We get the output voltage expression from the relation $V_{out} = K_{IV} I_{in}$, where I_{in} is an input current. Using the Eq. (10.2) we can obtain the output power expression in the

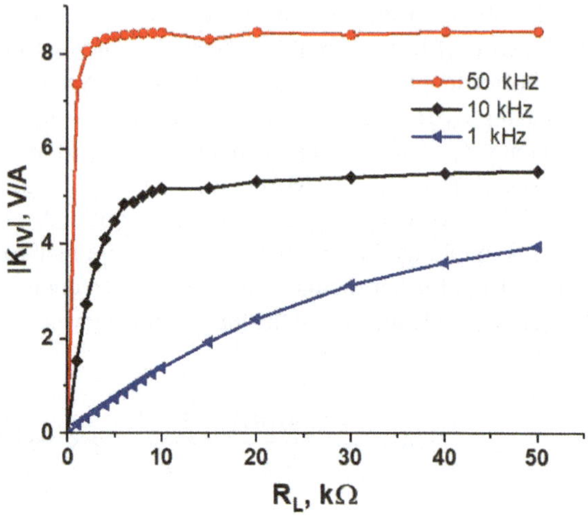

Fig. 10.6 Load characteristic of the gyrator based on the AMAG212/PZT-19/AMAG212 structure at different frequencies.

following form

$$P_{\text{out}} = \left(\frac{Y^{\text{p}} d_{31}^{\text{p}} q_{11}^{m}}{\varepsilon_{33}} \frac{Y^{\text{m}} t^{\text{m}}}{\overline{Y} t} t^{\text{p}} I_{\text{in}} n \right) \frac{R_{\text{L}}}{R_{\text{L}}^2 + X_{\text{C}}^2}. \tag{10.3}$$

It follows, from Eq. (10.3) that the output power has maximum, when load resistance equals the capacitive resistance, i.e. $R_{\text{L}} = X_{\text{C}}$. The piezoelectric plate capacitance has the value near $C \cong 5$ nF, the resonance frequency is near the value $f_{\text{res}} \cong 80$ kHz. This gives the value of the capacitive reactance equal to the value $X_{\text{C}} \cong 380$ kHz.

The dependence of the output power on the load resistance at the resonant frequency, for different magnetostrictive layer thicknesses and the alternating input current, $I_{\text{in}} = 4$ mA, is shown in Fig. 10.7.

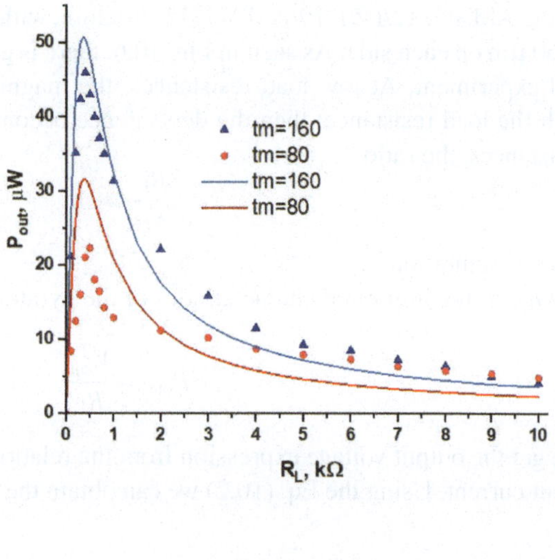

Fig. 10.7 The output power dependencies from load resistance at the resonance frequency.

10.4 Conclusion

The use of an adhesive frame for bonding the gyrator structure allows the fabrication of a robust layered construction from different components with weak interlayer adhesion, while achieving high frequency and load characteristics. nThe I/V conversion coefficient of the gyrator depends both on the frequency of the alternating input signal and on the parameters of the magnetostrictive–piezoelectric heterostructure, as well as on the load resistance. With increasing thickness of the magnetostrictive layer, the output voltage and output power initially increase sharply, then the growth slows, and their values tend toward saturation when the thickness of the magnetostrictive layer is much greater than that of the piezoelectric layer. At the frequency corresponding to the electromechanical resonance, a peak increase in the I/V conversion coefficient of the gyrator is observed.

References

[1] Bosco S, Haupt F, DiVincenzo DP (2017) Self-impedance-matched hall-effect gyrators and circulators. Physical Review Applied 7(2):024030, DOI 10.1103/PhysRevApplied.7.024030

[2] Cervera A, Evzelman M, Peretz MM, Ben-Yaakov S (2015) A high-efficiency resonant switched capacitor converter with continuous conversion ratio. IEEE Transactions on Power Electronics 30(3), 1373–1382, DOI 10.1109/TPEL.2014.2317758

[3] Dong S, Zhai J, Li JF, Viehland D, Bichurin MI (2006) Magnetoelectric gyration effect in $Tb_{1-x}Dy_xFe_{2-y}$/Pb(Zr,Ti)O_3 laminated composites at the electromechanical resonance. Applied Physics Letters 89(24):252904, DOI 10.1063/1.2404977

[4] Filippov DA, Galkina TA, Manicheva IN, Zhang J (2023) A magnetoelectric gyrator based on three-layer symmetric structures: theory and experiment. Technical Physics 68(9):1253-1258

[5] Filippov D, Liu Y, Zhou P, Ge B, Liu J, Zhang J, Zhang T, Srinivasan G. (2021) Low-frequency magnetoelectric effects in magnetostrictive–piezoelectric bilayers: longitudinal and bending Ddeformations. Journal of Composites Science 5(11):287, DOI 10.3390/jcs5110287

[6] Laletin VM, Paddubnaya NN, Filippov DA (2020) Field dependence of magnetoelectric effect in the electromechanical resonance region on the Permendur -Quartz - Permendur structures. IOP Conference Series: Materials Science and Engineering 939:012040, DOI 10.1088/1757-899X/939/1/012040

[7] Saveliev D, Fetisov L, Chashin D, Fetisov Y, Khon A, Shamonin M (2021) Effects of ferromagnetic-material thickness on magnetoelectric voltage transformation in a multiferroic heterostructure. Smart Materials and Structures 30(6):067002, DOI 10.1088/1361-665X/abf6c0

[8] Singer S (1987) Gyrators application in power processing circuits. IEEE Transactions on Industrial Electronics 34(3):313-318, DOI 10.1109/TIE.1987.350962
[9] Tellegen BDH (1948) The gyrator, a new electric network element. Philips Research Reports 3:81-101

Chapter 11
On Bending Vibrations and Stability of Circular Beams of Variable Radius According to a Refined Theory

Gnun Z. Gevorgyan and Artavazd Z. Darbinyan

Abstract The problem of bending vibrations and stability of transversely isotropic beams with variable circular cross-sections is considered. Equations of refined plate theory are used to determine both the bending vibration frequencies and the axial force values at which the beam loses stability. The equations are solved using the numerical collocation method for various values of the geometric and physical parameters characterizing the variable-radius beam and the boundary conditions for elastic end restraint. A numerical analysis of the problem is performed.

Key words: Bending vibrations, Stability, Variable cross-section, Refined plate theory

11.1 Introduction

The problem of bending vibrations and stability of transversely isotropic beams of variable circular cross-section is considered, taking into account transverse effects. In [1, 2, 6, 9, 11], various refined theories of plates and shells based on hypotheses, asymptotic theories, iterative theories, etc. are presented. In [7, 10], wave problems for prestressed plates are considered. Here the problem is considered using the methodology proposed in [8, 9].

Gnun Z. Gevorgyan (corresponding author) · Artavazd Z. Darbinyan
Institute of Mechanics, National Academy of Sciences of the Republic of Armenia, 24b Marshal Baghramyan Ave, Yerevan, 0019, Republic of Armenia, e-mail: gnungev@gmail.com, darbinyan_1954@mail.ru

H. Altenbach and A. V. Sahakyan, *Contact Mechanics and Relevant Problems*,
Advanced Structured Materials 253, https://doi.org/10.1007/978-3-032-20869-9_11

11.2 Problem Statement

Consider a transversely isotropic beam of variable radius $r(x)$ and length a. The coordinate axis Ox is aligned with the beam's midline, and the axes Oy and Oz are in the plane of the cross-section, which is the plane of isotropy. The beam is prestressed by two equal, oppositely directed forces P acting at the beam's ends along the axis Ox. Assuming that bending occurs in the plane Oxz, we neglect the displacement v and stresses σ_y, τ_{yz}, and assume all other quantities to be independent of y.

Using the equations obtained in [2, 10], the equations of bending vibrations in the presence of an axial force P can be represented as

$$\frac{\partial M_x}{\partial x} - N_x = -\frac{\pi r^4 \rho}{4}\left(\frac{\partial^3 w}{\partial x \partial t^2} - a_{55}\frac{\partial^2 \varphi_1}{\partial t^2}\right), \tag{11.1}$$

where

$$N_x = \frac{\pi r^2}{4}\left(3\varphi_1 - B_{11} r\left(\frac{\partial^2 w}{\partial x^2} - a_{55}\frac{\partial \varphi_1}{\partial x}\right)\frac{\partial r}{\partial x}\right),$$
$$M_x = -\frac{B_{11}\pi r^4}{4}\left(\frac{\partial^2 w}{\partial x^2} - a_{55}\frac{\partial \varphi_1}{\partial x}\right), \tag{11.2}$$
$$u_x = z\left(\frac{\partial w}{\partial x} - a_{55}\varphi_1\right).$$

Here a_{55} and B_{11} are the elastic constants of the material, $\varphi_1(x,t)$ is a function characterizing the distribution of shear stresses , w is the deflection of the beam axis points, t is time, ρ is the density of the material.

We consider problems of stability or oscillations, assuming that w and φ_1 have a factor $e^{i\omega t}$, and use nondimensionalizing notations

$$x = a\xi, \quad r = r_0 R, \quad \zeta = r_0 z, \; s = r_0/a, \; N_x = N B_{11} r_0^2 e^{i\omega t},$$
$$M_x = M B_{11} r_0^3 e^{i\omega t}, \quad u_x = r_0 u e^{i\omega t}, \quad \varphi_1 = \varphi B_{11} e^{i\omega t}, \tag{11.3}$$
$$P = -T B_{11} r_0^2, \; w = r_0 f e^{i\omega t}, \; \Omega^2 = \frac{\pi r_0^2 \rho \omega^2}{4 B_{11}}, \; \chi = B_{11} a_{55},$$

where r_0 is a coefficient having the dimension of length, and $r_0/a \ll 1$.

After some transformations we obtain a system of two equations for dimensionless $f(\xi)$ and $\varphi(\xi)$

$$\left(-s^2 T R - s^4 R^4 R''\right) f'' + \Omega^2 s^2 R^4 R' f' - 4\Omega^2 R^3 f - 3sR^3 + \chi s^3 R^4 R'' \varphi'$$
$$+ \left(9sR^2 r' - \chi \Omega^2 s r^4 r'\right)\varphi = 0,$$
$$- s^4 R^3 R' f''' + \left(-s^2 T - 3s^4 R^2 R'^2 - s^4 R^3 R''\right) f'' - 4\Omega^2 R^2 f \tag{11.4}$$
$$+ \chi s^3 R^3 R' \varphi'' + \left(3sR^2 + 3\chi s^3 R^2 R'^2 + \chi s^3 R^3 R''\right)\varphi' + 6sRR'\varphi = 0,$$

where, for brevity, the argument ξ at $\varphi(\xi)$ and $R(\xi)$ is omitted. Boundary conditions must be added to Eqs. (11.4). We will use the boundary conditions of elastic embedment, which in dimensionless notation we will represent as

$$l_1 u'_{\varsigma} \pm (1 - l_1) M = 0 \tag{11.5}$$

or

$$l_1 s f' \pm (1 - l_1) M = 0, \tag{11.6}$$

$$l_2 f + (1 - l_2) N = 0, \tag{11.7}$$

where $l_1,\ l_2 \in [0, 1]$. At $\chi = 0$ conditions (11.5) and (11.6) coincide, which corresponds to the classical formulation of the problem.

From these conditions, the most commonly used boundary conditions can be obtained for various $l_1,\ l_2$

a) $l_1 = 0,\ l_2 = 0\ (M = 0,\ N = 0)$ free end,
b) $l_1 = 0,\ l_2 = 1\ (M = 0,\ f = 0)$ hinged support,
c) $l_1 = 1,\ l_2 = 0\ (sf' = 0,$ or $u'_{\varsigma} = 0,\ N = 0)$ symmetry conditions, which can also be called sliding embedding, and
d) $l_1 = 1,\ l_2 = 1\ (sf' = 0,$ or $u'_{\varsigma} = 0,\ f = 0)$ hard seal.

Dimensionless displacement, shear force and moment have the form

$$\begin{aligned} u &= \varsigma\,(sf' - \chi\varphi)\,, \\ N &= \frac{R^2}{4}\left[3\varphi - R^2 s^3 \left(sf'' - \chi\varphi'\right) R'\right], \\ M &= \frac{R^4}{4}\,(sf'' - \chi\varphi')\,. \end{aligned} \tag{11.8}$$

11.3 Numerical Solution and Analysis

In the case of constant thickness, ignoring transverse effects ($\chi = 0$), Eqs. (11.4) can be reduced to a single equation and the solution obtained analytically. In the general case, the equation is solved numerically using the collocation method. To solve Eqs. (11.4) using the collocation method, we represent the functions and as

$$\begin{aligned} f &= a_0 + a_1\xi + a_2\xi^2 + \sum_{i=1}^{n-1} a_{i+2}\xi^{i+2}(1-\xi)^3, \\ \varphi &= b_0 + b_1\xi + \sum_{i=1}^{n} b_{i+1}\xi^{i+1}(1-\xi)^2. \end{aligned} \tag{11.9}$$

By satisfying the boundary conditions (11.5), (11.6) or (11.5), (11.7), the constants $a_0,\ a_1,\ a_2,\ b_0,\ b_1$ can be expressed in terms of one of them, for example a_0. The sums in the representations for f and φ satisfy zero boundary conditions at the endpoints. Next, by choosing the zeros

$$\xi_k = \frac{1}{2}\left[1 + \cos\frac{\pi(2k-1)}{2n}\right], \quad k = \overline{1,n}$$

of the shifted Chebyshev polynomials of the first kind as collocation points and writing equations (11.4) at these points, we obtain a homogeneous system of $2n$ equations for a_0, a_i $(i = \overline{3, n+1})$, and b_i $(i = \overline{2, n+1})$.

Assuming $\Omega = 0$, in one case, $T = T_0$, in the other case, and equating the determinant of this system to zero, in the first case we obtain the critical values of the force T at which the beam loses stability, and in the second case, the frequencies Ω of bending vibrations in the presence of an axial force $T = T_0$. At $T = 0$, we obtain the values of free vibrations of the beam.

Once these values are determined, we can determine the type of functions f and φ corresponding to these values. To do this, divide the resulting equations by one of the coefficients a_m or b_m, move the corresponding column to the right side, and discard one of the equations. By solving these equations, we can determine the functions f and φ with a constant factor accuracy.

Let us consider beams of variable radius, for which the dimensionless radius $R(\xi)$ changes according to the laws (Fig. 11.1)

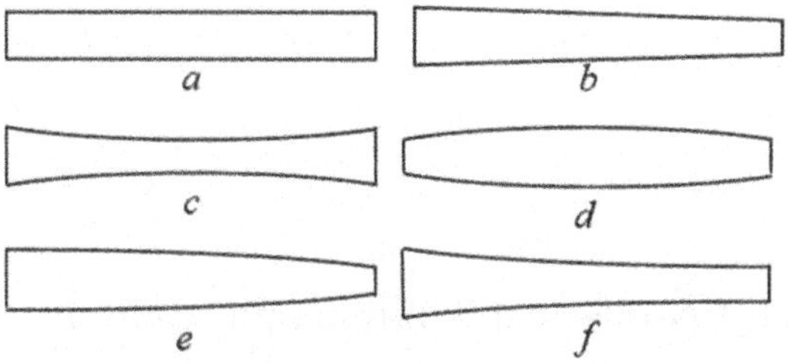

Fig. 11.1 Types of longitudinal sections of beams of variable thickness.

$$\begin{aligned}
&a: R(\xi)=1,\\
&b: R(\xi)=\sqrt{3}\frac{(1-k)\xi+k}{\sqrt{k^2+k+1}}, \qquad k>1,\\
&c: R(\xi)=\sqrt{15}\frac{4(k-1)(\xi-0.5)^2+1}{\sqrt{3k^2+4k+8}},\\
&d: R(\xi)=\sqrt{15}\frac{4(1-k)(\xi-0.5)^2+k}{\sqrt{8k^2+4k+3}},\\
&e: R(\xi)=\sqrt{15}\frac{(1-k)\xi^2+k}{\sqrt{8k^2+4k+3}},\\
&f: R(\xi)=\sqrt{15}\frac{4(k-1)(\xi-1)^2+1}{\sqrt{3k^2+4k+8}}.
\end{aligned} \tag{11.10}$$

They all have the same volume and, with variations in thickness, the thick part is k times larger than the thin one. In the calculations it is assumed that $s = 0.05$, $\chi = 0,\ 3,\ 30,\ \ k = 2$.

Table 11.1 shows the dimensionless values of the first two fundamental frequencies $\Omega_1^1,\ \Omega_1^2$ and the value of the frequency Ω_2^1 due to taking into account transverse shears and the first value of the critical compressive force T_{cr} at which loss of stability occurs. For each value of the critical force or frequency, the type of functions f and φ is determined, which makes it possible to determine the form of loss of stability or the nature of oscillations. Frequencies Ω_i^1 are mainly related to transverse vibrations, which are accompanied by small-amplitude shear vibrations. Frequencies Ω_i^2 are related to shear vibrations, which are accompanied by small-amplitude transverse vibrations. The amplitude values of the dimensionless displacement f at Ω_i^1 are two orders of magnitude greater than the values of the function φ characterizing the rotation, and at Ω_i^2, the amplitude values f and φ are of the same order. In the classical formulation, there are no Ω_i^2.

Table 11.2 shows the dimensionless values of the first frequency Ω_1^1 at $k = 2$ for one compressive $T = 0.5T_{cr}^1$ and three tensile forces. Let us consider the case where both ends of the beam are rigidly clamped. We consider two clamping options (11.5), (11.7) or (11.6), (11.7). In the first case (11.5), (11.7) the tangent $f'_{\xi} = 0$ at the end points, while in the second (11.6), (11.7) $u'_{\varsigma} = 0$ it is at both ends. The case where both ends are hinged is given in [4]. Note that, according to (11.3), positive values T correspond to a compressive load.

Figure 11.2 shows graphs Ω_1^1 for all beams except for the beam of constant thickness as a function of $k \in [1, 4]$ and $\chi = 0$ at the boundary conditions (11.6), (11.7). The graphs show that Ω_1^1 as a function of $k \in [1, 4]$ increases in case c, insignificantly in case f, and decreases in all other cases, and most rapidly Ω_1^1 in case d under the boundary conditions (11.6), (11.7). Figure 11.3 shows graphs Ω_1^1 for beams a, c, d depending on the $l_1 \in [0, 1]$ characterizing elasticity of the fixation ($l_1 = 0$ -hinge, $l_1 = 1$ rigid fixation) under boundary conditions (11.6), (11.7). Under the boundary conditions hinge-hinge ($l_1 = 0,\ l_2 = 1$), the greatest value Ω_1^1 is in case d), and the least value is in case c) [4], and under the boundary conditions rigid

Table 11.1: Dimensionless values of frequencies and critical forces for k = 2 and boundary conditions (11.5), (11.7) or (11.6), (11.7).

		Ω_1^1		Ω_1^2		Ω_2^1		T_{cr}	
		(11.6)	(11.5)	(11.6)	(11.5)	(11.6)	(11.5)	(11.6)	(11.5)
a 	$\chi = 0$	0.0278	0.0278	0.076	0.076	-	-	0.0987	0.0987
	$\chi = 3$	0.0274	0.0263	0.0722	0.067	1.014	1.014	0.0898	0.089
	$\chi = 30$	0.0244	0.0186	0.0523	0.0399	0.107	0.0894	0.0496	0.0496
b 	$\chi = 0$	0.0268	0.0268	0.073	0.073	-	-	0.0723	0.0723
	$\chi = 3$	0.0264	0.0254	0.0691	0.0655	1.014	0.9269	0.065	0.065
	$\chi = 30$	0.0237	0.0186	0.051	0.0394	0.105	0.0884	0.0527	0.0415
c 	$\chi = 0$	0.0395	0.0395	0.0882	0.0882	-	-	0.088	0.088
	$\chi = 3$	0.0383	0.0376	0.0823	0.079	1.177	0.2379	0.077	0.077
	$\chi = 30$	0.0297	0.0271	0.0820	0.0465	0.2347	0.237	0.026	0.0249
d 	$\chi = 0$	0.0196	0.0196	0.0614	0.0614	-	-	0.0727	0.0727
	$\chi = 3$	0.0191	0.0186	0.059	0.0552	0.897	0.242	0.068	0.066
	$\chi = 30$	0.0182	0.0132	0.0500	0.0336	0.128	0.127	0.047	0.044
e 	$\chi = 0$	0.0247	0.0247	0.0704	0.0704	-	-	0.070	0.070
	$\chi = 3$	0.0245	0.0235	0.0674	0.0631	0.976	0.189	0.064	0.064
	$\chi = 30$	0.0231	0.0184	0.0394	0.039	0.1056	0.0884	0.056	0.0372
f 	$\chi = 0$	0.0295	0.0295	0.076	0.076	-	-	0.0721	0.0721
	$\chi = 3$	0.0289	0.0280	0.076	0.068	0.1901	0.33	0.065	0.065
	$\chi = 30$	0.0247	0.0204	0.0519	0.041	0.105	0.088	0.034	0.0353

fixation-rigid fixation, ($l_1 = 1,\ l_2 = 1$), vice versa. Under $l_1 = 0.12$ for both options $\Omega_1^1 = 0.0165$, and for a beam of constant thickness $\Omega_1^1 = 0.0168$.

Figure 11.4 shows graphs of functions f for three values of dimensionless frequency Ω_1^1 - Ω_3^1 -for a linear change in thickness at $k = 2$, $\chi = 0$. Under the rigid-fixation boundary conditions $k = 2$, $\chi = 0$, the highest values $\Omega_1^1 = 0.0395$ of the first frequency are taken in case c, when the smallest radius is at the beam's midpoint. The lowest values $\Omega_1^1 = 0,0196$ of the first frequency are taken in case d, when the largest radius is at the beam's midpoint. The critical force $T_{cr}^1 = 0.0987$ takes its highest value for a beam of constant thickness, and its lowest value for $T_{cr}^1 = 0.0702$ beam e.

In the problem of free vibrations, the greatest difference Ω_1^1 between the refined and classical formulations occurs at for $k = 2$ beam c. The value Ω_1^1 of calculated by the refined theory at $\chi = 30$ is less than the value calculated by the classical theory ($\chi = 0$) by 24.8% under boundary conditions (11.5), (11.7), and by 31.4% under

Table 11.2: Dimensionless values of the first frequency at $k = 2$ for 5 values of T under boundary conditions (11.6), (11.7).

		a	b	c	d	e	f
	$T = 0$	0.0278	0.0268	0.0395	0.0196	0.0247	0.0295
	$T = 0.5T^1_{cr}$	0.0210	0.0210	0.0300	0.0150	0.0190	0.0230
$\chi = 0$	$T = -0.5T^1_{cr}$	0.0328	0.0328	0.0476	0.0240	0.0300	0.0350
	$T = -T^1_{cr}$	0.0388	0.0368	0.0542	0.0276	0.0334	0.0403
	$T = -2T^1_{cr}$	0.0471	0.0441	0.0647	0.0337	0.0408	0.0481
	$T = 0$	0.0274	0.0264	0.0383	0.0191	0.0245	0.0295
	$T = 0.5T^1_{cr}$	0.0200	0.0210	0.0290	0.0160	0.0190	0.0230
$\chi = 3$	$T = -0.5T^1_{cr}$	0.0334	0.0318	0.0458	0.0243	0.0295	0.0350
	$T = -T^1_{cr}$	0.0393	0.0361	0.0519	0.0280	0.0336	0.0394
	$T = -2T^1_{cr}$	0.0348	0.0433	0.0618	0.0342	0.0405	0.0398
	$T = 0$	0.0244	0.0237	0.0297	0.0182	0.0231	0.0247
	$T = 0.5T^1_{cr}$	0.019	0.0150	0.026	0.0190	0.0130	0.0200
$\chi = 30$	$T = -0.5T^1_{cr}$	0.034	0.0302	0.0337	0.0308	0.0287	0.0293
	$T = -T^1_{cr}$	0.042	0.0352	0.0371	0.0355	0.0349	0.0332
	$T = -2T^1_{cr}$	0.056	0.0436	0.0429	0.0435	0.0434	0.0396

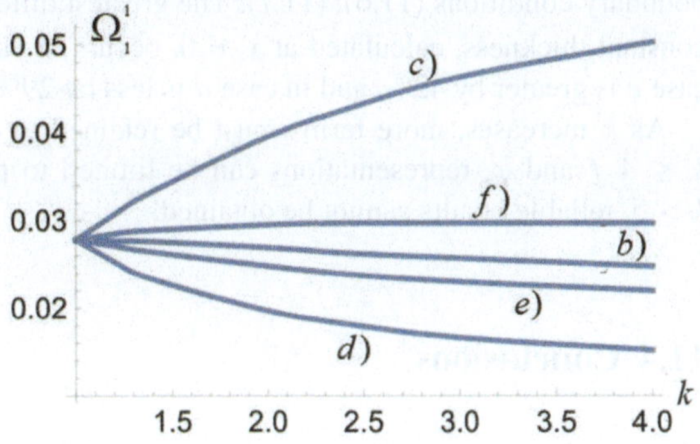

Fig. 11.2 Dependence of Ω_1^1 on parameter k.

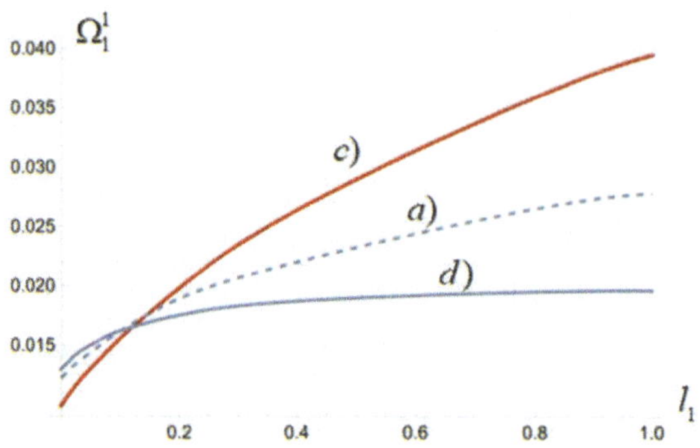

Fig. 11.3 Dependence of Ω_1^1 on parameter l_1.

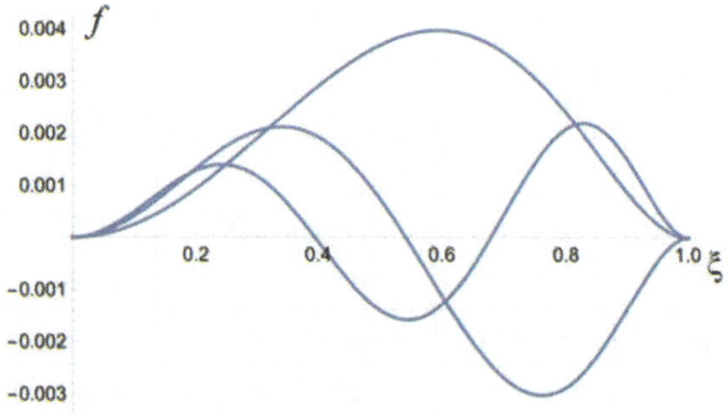

Fig. 11.4 The first three modes of vibration of a beam with a linearly changing thickness.

boundary conditions (11.6), (11.7). The greatest difference Ω_1^1 relative to a plate of constant thickness, calculated at $\chi = 0$, occurs in cases c and d. The value Ω_1^1 in case c is greater by 42%, and in case d is less by 29%.

As k increases, more terms must be retained to obtain acceptable results. For $k \leq 4$ f and φ, representations can be limited to polynomials of order 7-9. For $k > 5$, reliable results cannot be obtained.

11.4 Conclusions

The stability and free vibrations of beams with variable circular cross-sections for different longitudinal beam cross-sections under axial forces were investigated. The effect of accounting for transverse shear on the magnitude of the critical force leading to loss of stability and on the natural frequencies of vibrations was identified. The following conclusions can be made:

1. Accounting for the influence of rotational inertia from transverse shears leads to the emergence of new, secondary vibration modes, which are predominantly shear and have frequencies of the order of the 6th to 8th frequency of the fundamental vibration mode.

2. Accounting for thickness variability in the considered variation variants leads to a change in the fundamental frequency of up to 42%.
3. The collocation method used demonstrated fairly good convergence; increasing the number of collocation points beyond 6-7 has virtually no effect on the fundamental frequency.

References

[1] Aghalovyan LA (2015) Asymptotic Theory of Anisotropic Plates and Shells. World Scientific DOI 10.1142/9048

[2] Ambartsumyan SA (1991) Theory Of Anisotropic Plates: Strength, Stability, and Vibrations. CRC Press

[3] Birger AI, Panovko YG (eds) (1964) Prochnost', ustoychivost', kolebaniya. Spravochnik (in Russ., Strength, stability, vibrations. Handbook), Vol 3. Mashinostroenie, Moscow

[4] Gevorgyan GZ, Darbinyan AZ (2025) On flexural oscillations of pre-stressed circular beams of variable radius taking into account transverse effects. In: Altenbach, H (ed) Current Developments in Solid Mechanics and Their Applications, Advanced Structured Materials, Vol 223, pp 223-232. Springer, Cham, DOI 10.1007/978-3-031-90022-8_15

[5] Gevorgyan GZ, Kirakosyan RM (2002) Svobodnyye poperechnyye kolebaniya pryamougol'nykh ortotropnykh plastin peremennoy tolshchiny pri uchete poperechnykh sdvigov i inertsii vrashcheniya (in Russ., Free transverse vibrations of rectangular orthotropic plates of variable thickness taking into account transverse shears and rotational inertia). In: Problems of Mechanics of Thin Deformable Bodies, devoted to the 80th anniversary of Academician of NAS RA S.A. Ambartsumyan, pp 137-146. Yerevan

[6] Gol'denveizer AL (2014) Theory of Elastic Shells. Pergamon, Oxford

[7] Kaplunov JD, Nolde EV, Rogerson GA (2003) A low-frequency model for dynamic motion in pre-stressed incompressible elastic structures. Proceedings of the Royal Society of London. Series A: Mathematical, Physical and Engineering Sciences 456(2003):2589-2610, DOI 10.1098/rspa.2000.0627

[8] Kirakosyan RM (2014) Non-classical problem of bending of an orthotropic beam with an elastically clamped support. Reports of the NAS RA 114(2):101-107

[9] Kirakosyan RM (2002) Osnovnyye uravneniya ortotropnykh plastin peremennoy tolshchiny pri uchete poperechnykh sdvigov i obzhatiya (in Russ., Fundamental equations of orthotropic plates of variable thickness taking into account transverse shears and compression). Proceedings of NAS RA. Mechanics 55(4):12-20

[10] Pichugin AV, Rogerson GA (2002) An asymptotic membrane-like theory for long-wave motion in a pre-stressed elastic plate. Proceedings of the Royal

Society of London. Series A: Mathematical, Physical and Engineering Sciences 458(2022):1447-1468, DOI 10.1098/rspa.2001.0932

[11] Vasiliev VV, Lurie SA (1990) K probleme postroyeniya neklassicheskikh teoriy plastin (in Russ., On the problem of constructing a non-classical theory of plates). Izvestiya AN MTT (2):158-167

Chapter 12
Dynamic Problems of the Thin Elastic Composed Panel with Rigid-Clamped Edge Generator

Gurgen R. Ghulghazaryan and Lusine G. Ghulghazaryan

Abstract The free boundary and interfacial vibrations of the composite cylindrical panel with the rigidly clamped edge generator consisting of two finite orthotropic thin cylindrical panels with different elastic properties and having full contact with the generators are studied. Using the system of equations of the related classical theory of orthotropic cylindrical shells dispersion equations and asymptotic formulas are derived to obtain eigenfrequencies of possible interfacial and boundary vibrations of such composite cylindrical panels. An algorithm is presented to separate possible interfacial and boundary vibrations. Asymptotic links are established between the dispersion equations of the problem at hand and analogous problems for a composed rectangular plate. Examples of cylindrical panels with different widths of constituents are given, along with approximate values of dimensionless characteristics of eigenfrequencies.

Key words: Composed cylindrical panel, Boundary and interfacial vibrations, Asymptotic formulae

12.1 Introduction

Investigation of free vibrations of composed cylindrical panels plays an important role in the studies of dynamics of deformable solid bodies. Such studies are required

Gurgen R. Ghulghazaryan (corresponding author)
Khachatur Abovian Armenian State Pedagogical University, 17 Tigran Mets Ave, Yerevan, 0010, Republic of Armenia, e-mail: ghulghazaryangurgen08@aspu.am

Lusine G. Ghulghazaryan
Institute of Mechanics, National Academy of Sciences of the Republic of Armenia, 24b Marshal Baghramyan Ave, Yerevan, 0019, & Khachatur Abovian Armenian State Pedagogical University, 17 Tigran Mets Ave, Yerevan, 0010, Republic of Armenia,
e-mail: ghulghazaryanlusine08@aspu.am

H. Altenbach and A. V. Sahakyan, *Contact Mechanics and Relevant Problems*,
Advanced Structured Materials 253, https://doi.org/10.1007/978-3-032-20869-9_12

by the needs of the theory itself and by the practical needs of different branches of engineering industry, construction, instrument engineering, seismic survey, etc. [13]. In many cases, the objects of investigations are finite thin-walled composed cylindrical panels. For such panels, much attention is attracted to the investigation of free vibrations localized near the edges of the panels, i.e. edge vibrations, and vibrations localized near the interface of material properties, i.e. interfacial vibrations.

It is known that, at the free edge of an orthotropic plate, independently of each other, planar and bending vibrations can occur [10, 13, 15]. When the plate is bent, these types of vibration are coupled, giving two new types of vibration localized at the free edge: predominantly tangential and predominantly bending vibrations. Transformation of one type of vibration into the other occurs at the free edge of a thin cylindrical elastic panel. For this transformation of vibrations, a complex distribution picture of the frequencies of natural vibrations occurs depending on the geometrical and mechanical parameters of the finite and infinite cylindrical panel [8–10, 15]. With an increasing number of free edges in a cylindrical panel, this picture becomes increasingly complex [4–7]. Therefore, investigations into the edge resonance of composed plates and composed cylindrical panels with free edges are the most difficult problems in the theory of vibrations of plates and shells [15]. These difficulties are overcome by using a combination of analytical and asymptotic theories, as well as numerical methods.

The investigation of free interfacial vibrations was initiated by Zilbergleit et al. [17] and Gertman et al. [2], where Stoneley wave analogues were investigated [14]. In [17] transverse vibrations running along the contact line of two semi-infinite plates and concentrated close to it were studied. In [2] the plane interfacial vibrations near the interface of two bonded semi-strips with different elastic properties are investigated. Later, Kaplunov et al., using the special asymptotic method, studied the free interfacial vibrations of composed circular cylindrical shells as well as the shells of revolution [15].

In the present article, free interfacial and edge vibrations of a composed cylindrical panel with rigid-clamped edge generator, composed of finite orthotropic cylindrical panels with different elastic properties and with longitudinal section of material properties, are studied. On the line of material interface the complete contact conditions are set. Such elements are important components of modern structures and constructions; therefore, the question of free vibrations of these elements is vital enough and demands attention.

The dispersion equations are obtained to determine the appropriate frequencies of interfacial and edge vibrations of a composed cylindrical panel with a rigid-clamped edge generator. An asymptotic link between the dispersion equations of the problem considered and the analogous problem for a plate with a rigid-clamped side, composed of orthotropic plates with different elastic properties, is established. The derived dispersion equations and related asymptotic formulae can be used to control the spectrum of frequencies of the stated problem by varying the geometry of the panel and mechanical properties of materials. In particular, one can control the spectrum by shifting either the origin of the spectrum or the points of condensation from the undesirable resonance region.

12.2 Statement of the Problem and Basic Equations

Let the generatrices of the cylindrical panel be orthogonal to the ends of the panel. On the middle surface of the panel , curvilinear coordinates (α, β) are introduced, where $\alpha(0 \le \alpha \le l)$ and $\beta(-s^{(2)} \le \beta \le s^{(1)})$ are the orientated length of the generatrice and the length of the arc of the directing circle, respectively. l is the length of the cylindrical panel (Fig. 12.1). $\beta = 0$ corresponds to the interface of separation of material properties. Figure 12.2 shows the composite rectangular plate corresponding to the panel in Fig. 12.1. All values corresponding to the right panel $(0 \le \beta \le s^{(1)})$ in (Fig. 12.1) are marked with a superscript (1). Similarly, for the left panel $(-s^{(2)} \le \beta \le 0)$ the superscript (2) is used.

As the initial equations describing vibrations of the cylindrical panels on the left and right sides we will use the equations corresponding to the classical theory of orthotropic cylindrical shells written in curvilinear coordinates α and β chosen (Fig. 12.1) [1]

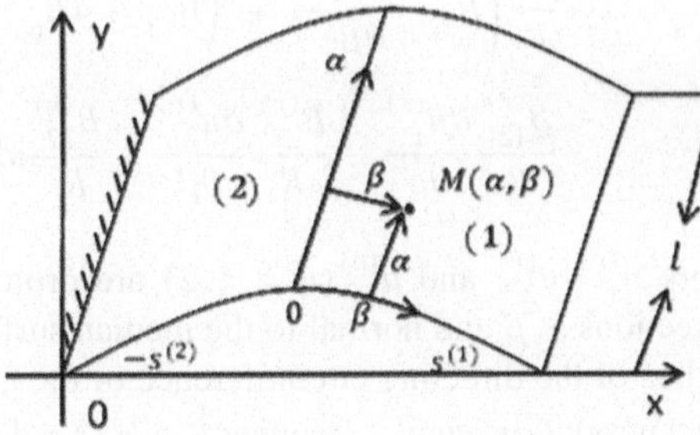

Fig. 12.1 Composite cylindrical panel.

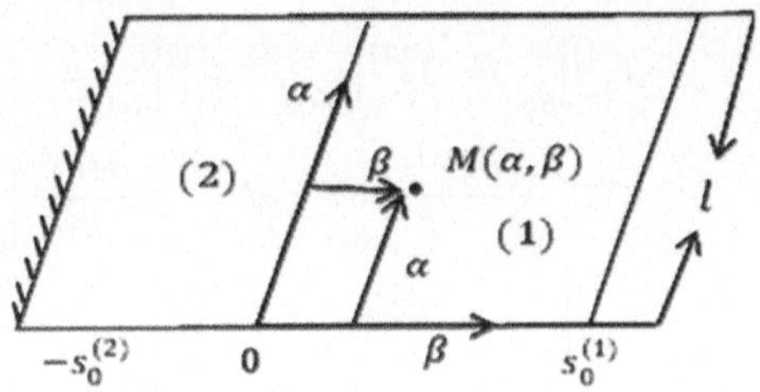

Fig. 12.2 Corresponding composite rectangular plate.

$$
\begin{aligned}
&-B_{11}^{(r)}\frac{\partial^2 u_1^{(r)}}{\partial \alpha^2}-B_{66}^{(r)}\frac{\partial^2 u_1^{(r)}}{\partial\beta^2}-\left(B_{12}^{(r)}+B_{66}^{(r)}\right)\frac{\partial^2 u_2^{(r)}}{\partial\alpha\partial\beta}\\
&+\frac{B_{12}^{(r)}}{R}\frac{\partial u_3^{(r)}}{\partial\alpha}=\rho^{(r)}\omega^2 u_1^{(r)},\\
&-\left(B_{12}^{(r)}+B_{66}^{(r)}\right)\frac{\partial^2 u_1^{(r)}}{\partial\alpha\partial\beta}-B_{66}^{(r)}\frac{\partial^2 u_2^{(r)}}{\partial \alpha^2}-B_{22}^{(r)}\frac{\partial^2 u_2^{(r)}}{\partial\beta^2}+\frac{B_{22}^{(r)}}{R}\frac{\partial u_3^{(r)}}{\partial\beta}\\
&-\frac{\mu^4}{R^2}\left(4B_{66}^{(r)}\frac{\partial^2 u_2^{(r)}}{\partial \alpha^2}+B_{22}^{(r)}\frac{\partial^2 u_2^{(r)}}{\partial\beta^2}\right)\\
&-\frac{\mu^4}{R}\left(B_{22}^{(r)}\frac{\partial^3 u_3^{(r)}}{\partial\beta^3}+\left(B_{12}^{(r)}+4B_{66}^{(r)}\right)\frac{\partial^3 u_3^{(r)}}{\partial\alpha^2\partial\beta}\right)=\rho^{(r)}\omega^2 u_2^{(r)},\\
&\mu^4\left(B_{11}^{(r)}\frac{\partial^4 u_3^{(r)}}{\partial\alpha^4}+2\left(B_{12}^{(r)}+2B_{66}^{(r)}\right)\frac{\partial^4 u_3^{(r)}}{\partial\alpha^2\partial\beta^2}+B_{22}^{(r)}\frac{\partial^4 u_3^{(r)}}{\partial\beta^4}\right)\\
&+\frac{\mu^4}{R}\left(B_{22}^{(r)}\frac{\partial^3 u_2^{(r)}}{\partial\beta^3}+\left(B_{12}^{(r)}+4B_{66}^{(r)}\right)\frac{\partial^3 u_2^{(r)}}{\partial\alpha^2\partial\beta}\right)\\
&-\frac{B_{12}^{(r)}}{R}\frac{\partial u_1^{(r)}}{\partial\alpha}-\frac{B_{22}^{(r)}}{R}\frac{\partial u_2^{(r)}}{\partial\beta}+\frac{B_{22}^{(r)}}{R^2}u_3^{(r)}=\rho^{(r)}\omega^2 u_3^{(r)},r=1,2.
\end{aligned}
\tag{12.1}
$$

Here $u_1^{(r)}, u_2^{(r)}$ and $u_3^{(r)} (r = 1,2)$ are projections of the displacement vector in directions α, β and normal to the median surface of the panel, respectively. R is the radius of the directing circumference of the median surface; $\mu^4 = h^2/12$ (h is shell thickness); ω is angular frequency, $\rho^{(r)} (r = 1,2)$ is the density of material; $B_{ij}^{(r)} (r = 1,2)$ are the coefficients of elasticity. The boundary conditions for cylindrical panel has the form [1]

$$
\begin{aligned}
&T_2^{(1)}\Big|_{\beta=0}=T_2^{(2)}\Big|_{\beta=0},\quad S_{21}^{(1)}\Big|_{\beta=0}=S_{21}^{(2)}\Big|_{\beta=0},\quad M_2^{(1)}\Big|_{\beta=0}=M_2^{(2)}\Big|_{\beta=0},\\
&N_2^{(1)}+\frac{\partial H^{(1)}}{\partial\alpha}\Big|_{\beta=0}=N_2^{(2)}+\frac{\partial H^{(2)}}{\partial\alpha}\Big|_{\beta=0},\\
&u_1^{(1)}\Big|_{\beta=0}=u_1^{(2)}\Big|_{\beta=0},\quad u_2^{(1)}\Big|_{\beta=0}=u_2^{(2)}\Big|_{\beta=0},\\
&u_3^{(1)}\Big|_{\beta=0}=u_3^{(2)}\Big|_{\beta=0},\quad \frac{\partial u_3^{(1)}}{\partial\beta}\Big|_{\beta=0}=\frac{\partial u_3^{(2)}}{\partial\beta}\Big|_{\beta=0}
\end{aligned}
\tag{12.2}
$$

$$
T_2^{(1)}\Big|_{\beta=s^{(1)}}=0,\ S_{21}^{(1)}\Big|_{\beta=s^{(1)}}=0,\ M_2^{(1)}\Big|_{\beta=s^{(1)}}=0,\ N_2^{(1)}+\frac{\partial H^{(1)}}{\partial\alpha}\Big|_{\beta=s^{(1)}}=0,
\tag{12.3}
$$

$$u_1^{(2)}\Big|_{\beta=-s^{(2)}} = 0,\ u_2^{(2)}\Big|_{\beta=-s^{(2)}} = 0,\ u_3^{(2)}\Big|_{\beta=-s^{(2)}} = 0,\ \frac{\partial u_3^{(2)}}{\partial \beta}\Big|_{\beta=-s^{(2)}} = 0, \tag{12.4}$$

$$T_1^{(r)}\Big|_{\alpha=0,l} = S_{12}^{(r)} + \frac{H^{(r)}}{R}\Big|_{\alpha=0,l} = N_1^{(r)} + \frac{\partial H^{(r)}}{\partial \beta}\Big|_{\alpha=0,l} = 0,\ M_1^{(r)}\Big|_{\alpha=0,l} = 0, \tag{12.5}$$

$$T_1^{(r)} = hB_{11}^{(r)}\left[\frac{\partial u_1^{(r)}}{\partial \alpha} + \frac{B_{12}^{(r)}}{B_{11}^{(r)}}\left(\frac{\partial u_2^{(r)}}{\partial \beta} - \frac{u_3^{(r)}}{R}\right)\right],$$

$$T_2^{(r)} = hB_{22}^{(r)}\left[\frac{B_{12}^{(r)}}{B_{22}^{(r)}}\frac{\partial u_1^{(r)}}{\partial \alpha} + \frac{\partial u_2^{(r)}}{\partial \beta} - \frac{u_3^{(r)}}{R}\right],$$

$$M_1^{(r)} = \frac{h^3}{12}B_{11}^{(r)}\left[\frac{\partial^2 u_3^{(r)}}{\partial \alpha^2} + \frac{B_{12}^{(r)}}{B_{11}^{(r)}}\left(\frac{\partial^2 u_3^{(r)}}{\partial \beta^2} + \frac{1}{R}\frac{\partial u_2^{(r)}}{\partial \beta}\right)\right],$$

$$M_2^{(r)} = \frac{h^3}{12}B_{22}^{(r)}\left[\frac{B_{12}^{(r)}}{B_{22}^{(r)}}\frac{\partial^2 u_3^{(r)}}{\partial \alpha^2} + \frac{\partial^2 u_3^{(r)}}{\partial \beta^2} + \frac{1}{R}\frac{\partial u_2^{(r)}}{\partial \beta}\right],$$

$$S_{12}^{(r)} + \frac{H^{(r)}}{R} = hB_{66}^{(r)}\left[\frac{\partial u_1^{(r)}}{\partial \beta} + \frac{\partial u_2^{(r)}}{\partial \alpha} + \frac{h^2}{3R}\left(\frac{\partial^2 u_3^{(r)}}{\partial \alpha \partial \beta} + \frac{1}{R}\frac{\partial u_2^{(r)}}{\partial \alpha}\right)\right],$$

$$S_{21}^{(r)} = hB_{66}^{(r)}\left(\frac{\partial u_1^{(r)}}{\partial \beta} + \frac{\partial u_2^{(r)}}{\partial \alpha}\right), \tag{12.6}$$

$$N_1^{(r)} + \frac{\partial H^{(r)}}{\partial \beta} = \frac{h^3}{12}B_{11}^{(r)}\left[\frac{\partial^3 u_3^{(r)}}{\partial \alpha^3} + \frac{B_{12}^{(r)} + 4B_{66}^{(r)}}{B_{11}^{(r)}}\left(\frac{\partial^3 u_3^{(r)}}{\partial \beta^2 \partial \alpha} + \frac{1}{R}\frac{\partial^2 u_2^{(r)}}{\partial \alpha \partial \beta}\right)\right],$$

$$N_2^{(r)} + \frac{\partial H^{(r)}}{\partial \alpha} = \frac{h^3}{12}B_{22}^{(r)}\left[\frac{\partial^3 u_3^{(r)}}{\partial \beta^3} + \frac{B_{12}^{(r)} + 4B_{66}^{(r)}}{B_{22}^{(r)}}\frac{\partial^3 u_3^{(r)}}{\partial \alpha^2 \partial \beta} + \right.$$

$$\left.\frac{1}{R}\frac{\partial^2 u_2^{(r)}}{\partial \beta^2} + \frac{4B_{66}^{(r)}}{B_{22}^{(r)}}\frac{1}{R}\frac{\partial^2 u_2^{(r)}}{\partial \alpha^2}\right].$$

Relations (12.2) are complete contact conditions at $\beta = 0$. Relations (12.3) and (12.5) are the conditions of free edges at $\beta = s^{(1)}$ and $\alpha = 0, l$, respectively. Relations (12.4) are the conditions of rigid-clamped edge at $\beta = -s^{(2)}$ (Fig. 12.1. It can be proved that problem (12.1)-(12.5) is self–conjugate and has a non–negative discrete spectrum with a limit point at $+\infty$.

Problems (12.1) - (12.5) do not allow for the separation of variables. Therefore, based on the nonnegative definiteness of the corresponding operator of problem (12.1) - (12.5), to find the natural frequencies and the corresponding natural forms, a generalized Kantorovich – Vlasov method of reduction to ordinary differential equations can be applied [11, 12, 16].

As the basic functions, eigenfunctions of the problem

$$w^{VIII} = \theta^8 w, \quad w|_{\alpha=0,l} = w'\Big|_{\alpha=0,l} = w''\Big|_{\alpha=0,l} = w'''\Big|_{\alpha=0,l} = 0, \quad 0 \le \alpha \le l \tag{12.7}$$

are used. Problem (12.7) is self-conjugate and positive definite. To the eigenvalues θ_m^8, $m = \overline{1,\infty}$ of problem (12.7) correspond the eigenfunctions

$$w_m(\theta_m\alpha) = \frac{\Delta_1}{\Delta} x_1(\theta_m\alpha) + \frac{\Delta_2}{\Delta} x_2(\theta_m\alpha) + \frac{\Delta_3}{\Delta} x_3(\theta_m\alpha) + x_4(\theta_m\alpha),$$
$$0 \le \alpha \le l, \quad m = \overline{1,\infty},$$
$$x_1(\theta_m\alpha) = \cosh(\theta_m\alpha) - \cosh\frac{\theta_m\alpha}{\sqrt{2}}\cos\frac{\theta_m\alpha}{\sqrt{2}} - \sinh\frac{\theta_m\alpha}{\sqrt{2}}\sin\frac{\theta_m\alpha}{\sqrt{2}},$$
$$x_2(\theta_m\alpha) = \sinh(\theta_m\alpha) - \sqrt{2}\cosh\frac{\theta_m\alpha}{\sqrt{2}}\sin\frac{\theta_m\alpha}{\sqrt{2}},$$
$$x_3(\theta_m\alpha) = \sin(\theta_m\alpha) - \sqrt{2}\sinh\frac{\theta_m\alpha}{\sqrt{2}}\cos\frac{\theta_m\alpha}{\sqrt{2}},$$
$$x_4(\theta_m\alpha) = \cos(\theta_m\alpha) - \cosh\frac{\theta_m\alpha}{\sqrt{2}}\cos\frac{\theta_m\alpha}{\sqrt{2}} + \sinh\frac{\theta_m\alpha}{\sqrt{2}}\sin\frac{\theta_m\alpha}{\sqrt{2}},$$
$$\Delta = \begin{vmatrix} x_1(\theta_m\alpha) & x_2(\theta_m\alpha) & x_3(\theta_m\alpha) \\ x_1'(\theta_m\alpha) & x_2'(\theta_m\alpha) & x_3'(\theta_m\alpha) \\ x_1''(\theta_m\alpha) & x_2''(\theta_m\alpha) & x_3''(\theta_m\alpha) \end{vmatrix}, \tag{12.8}$$
$$\Delta_1 = -\begin{vmatrix} x_4(\theta_m\alpha) & x_2(\theta_m\alpha) & x_3(\theta_m\alpha) \\ x_4'(\theta_m\alpha) & x_2'(\theta_m\alpha) & x_3'(\theta_m\alpha) \\ x_4''(\theta_m\alpha) & x_2''(\theta_m\alpha) & x_3''(\theta_m\alpha) \end{vmatrix},$$
$$\Delta_2 = -\begin{vmatrix} x_1(\theta_m\alpha) & x_4(\theta_m\alpha) & x_3(\theta_m\alpha) \\ x_1'(\theta_m\alpha) & x_4'(\theta_m\alpha) & x_3'(\theta_m\alpha) \\ x_1''(\theta_m\alpha) & x_4''(\theta_m\alpha) & x_3''(\theta_m\alpha) \end{vmatrix},$$
$$\Delta_3 = -\begin{vmatrix} x_1(\theta_m\alpha) & x_2(\theta_m\alpha) & x_4(\theta_m\alpha) \\ x_1'(\theta_m\alpha) & x_2'(\theta_m\alpha) & x_4'(\theta_m\alpha) \\ x_1''(\theta_m\alpha) & x_2''(\theta_m\alpha) & x_4''(\theta_m\alpha) \end{vmatrix}$$

and form with its first and second derivations orthogonal basis in a Hilbert space $L_2[0,l]$ [12]. Where θ_m, $m = \overline{1,\infty}$, are the positive zeros of the Wronskian of functions (12.8) at the point $\alpha = l$. Let us introduce the designations

$$\beta_m' = \frac{\int_0^l \left(w_m'(\theta_m\alpha)\right)^2 d\alpha}{\int_0^l \left(w_m(\theta_m\alpha)\right)^2 d\alpha}, \qquad \beta_m'' = \frac{\int_0^l \left(w_m''(\theta_m\alpha)\right)^2 d\alpha}{\int_0^l \left(w_m'(\theta_m\alpha)\right)^2 d\alpha}. \tag{12.9}$$

In formulas (12.8) and (12.9), derivatives are taken with respect to $\theta_m\alpha$ and $\beta_m' \to 1$, $\beta_m'' \to 1$ at $m \to +\infty$.

12.3 Derivation and an Analysis of Characteristic Equations

In the first, second, and third equations of system (12.1), the angular frequency ω is formally replaced by ω_1, ω_2 and ω_3, respectively. The solution of system (12.1) is sought in the form

$$\left(u_1^{(r)}, u_2^{(r)}, u_3^{(r)}\right) = \left(u_m^{(r)} w_m^{'}(\theta_m \alpha), v_m^{(r)} w_m(\theta_m \alpha), w_m(\theta_m \alpha)\right)$$
$$\times \exp\left((-1)^r x^{(r)} \theta_m \beta + x^{(r)} \theta_m s^{(r)}\right), \; r = 1, 2; m = \overline{1, \infty}. \quad (12.10)$$

Here, $w_m(\theta_m \alpha), m = \overline{1, \infty}$, are determined from formula (12.8) and $u_m^{(r)}, v_m^{(r)}$ and $x^{(r)}$ are indeterminate constants. In this case, conditions (12.5) are obeyed automatically. Let us insert Eq. (12.10) into Eq. (12.1). The equations found are scalarly multiplied by the vector-function

$$\left(w_m^{'}(\theta_m \alpha), w_m(\theta_m \alpha), w_m(\theta_m \alpha), w_m^{'}(\theta_m \alpha), w_m(\theta_m \alpha), w_m(\theta_m \alpha)\right) \quad (12.11)$$

and then integrated between the limits from 0 to l. From the first two pairs of equations, we obtain the following

$$\left(c_m^{(r)} + \varepsilon_m^2 a^2 g_m^{(r)} d_m^{(r)}\right) u_m^{(r)} =$$
$$\varepsilon_m \left\{ a_m^{(r)} + a^2 \frac{B_{22}^{(r)}\left(B_{12}^{(r)} + B_{66}^{(r)}\right)}{B_{11}^{(r)} B_{66}^{(r)}} \left(x^{(r)}\right)^2 l_m^{(r)} + \varepsilon_m^2 a^2 \frac{B_{22}^{(r)} B_{12}^{(r)}}{B_{11}^{(r)} B_{66}^{(r)}} d_m^{(r)} \right\}, \quad (12.12)$$

$$\left(c_m^{(r)} + \varepsilon_m^2 a^2 g_m^{(r)} d_m^{(r)}\right) v_m^{(r)} = (-1)^r \varepsilon_m x^{(r)} \left\{ b_m^{(r)} - a^2 g_m^{(r)} l_m^{(r)} \right\}, \; r = 1, 2. \quad (12.13)$$

However, from the third equation, taking into account relations (12.12) and (12.13), we obtain the characteristic equations

$$R_{mm}^{(r)} c_m^{(r)} + \varepsilon_m^2 \left\{ c_m^{(r)} - b_m^{(r)} \left(x^{(r)}\right)^2 + \frac{B_{12}^{(r)}}{B_{22}^{(r)}} \beta_m^{'} a_m^{(r)} \right.$$
$$+ a^2 \left(R_{mm}^{(r)} g_m^{(r)} d_m^{(r)} + 2 b_m^{(r)} l_m^{(r)} \left(x^{(r)}\right)^2 \right) + \varepsilon_m^2 a^2 d_m^{(r)} \left(b_m^{(r)} - \frac{B_{12}^{(r)}}{B_{22}^{(r)}} \beta_m^{'} \right)$$
$$\left. - a^4 \left(x^{(r)}\right)^2 g_m^{(r)} \left(l_m^{(r)}\right)^2 \right\} = 0, \; r = 1, 2; m = \overline{1, \infty}, \quad (12.14)$$

$$a_m^{(r)} = -\left(\frac{B_{12}^{(r)}}{B_{11}^{(r)}} \left(x^{(r)}\right)^2 + \frac{B_{12}^{(r)}}{B_{11}^{(r)}} \beta_m^{'} - \frac{B_{12}^{(r)}}{B_{11}^{(r)}} \left(\eta_{2m}^{(r)}\right)^2 \right),$$

$$b_m^{(r)} = \frac{B_{22}^{(r)}}{B_{11}^{(r)}}\left(x^{(r)}\right)^2 + \frac{B_{22}^{(r)}}{B_{11}^{(r)}}\left(\eta_{1m}^{(r)}\right)^2 - B_1^{(r)}, a^2 = \mu^4\theta_m^2,$$

$$c_m^{(r)} = \frac{B_{22}^{(r)}}{B_{11}^{(r)}}\left(x^{(r)}\right)^4 - B_2^{(r)}\left(x^{(r)}\right)^2$$

$$+\left(\frac{B_{22}^{(r)}}{B_{11}^{(r)}}\left(\eta_{1m}^{(r)}\right)^2 + \frac{B_{66}^{(r)}}{B_{11}^{(r)}}\left(\eta_{2m}^{(r)}\right)^2\right)\left(x^{(r)}\right)^2$$

$$+\left(\beta_m' - \left(\eta_{2m}^{(r)}\right)^2\right)\left(\beta_m'' - \frac{B_{66}^{(r)}}{B_{11}^{(r)}}\left(\eta_{1m}^{(r)}\right)^2\right),$$

$$d_m^{(r)} = \left(x^{(r)}\right)^2 - \frac{4\,B_{66}^{(r)}}{B_{11}^{(r)}}\beta_m',\; g_m^{(r)} = \frac{B_{22}^{(r)}}{B_{11}^{(r)}}\left(x^{(r)}\right)^2 + \frac{B_{22}^{(r)}}{B_{11}^{(r)}}\left(\eta_{1m}^{(r)}\right)^2$$

$$-\frac{B_{22}^{(r)}}{B_{66}^{(r)}}\beta_m'',\; l_m^{(r)} = \left(x^{(r)}\right)^2 - \frac{B_{12}^{(r)} + 4B_{66}^{(r)}}{B_{22}^{(r)}}\beta_m', \tag{12.15}$$

$$R_{mm}^{(r)} = a^2\left(\left(x^{(r)}\right)^4 - \frac{2\left(B_{12}^{(r)} + 2B_{66}^{(r)}\right)}{B_{22}^{(r)}}\beta_m'\left(x^{(r)}\right)^2 + \frac{B_{11}^{(r)}}{B_{22}^{(r)}}\beta_m'\beta_m''\right.$$

$$\left.-\frac{B_{66}^{(r)}}{B_{22}^{(r)}}\left(\eta_{3m}^{(r)}\right)^2\right),\; \left(\eta_{im}^{(r)}\right)^2 = \frac{\rho^{(r)}\omega_i}{B_{66}^{(r)}\theta_m^2},\; i = 1,2,3,$$

$$\varepsilon_m = \frac{1}{R\theta_m}, \qquad B_1^{(r)} = \frac{B_{11}^{(r)}B_{11}^{(r)}\beta_m'' - \left(B_{12}^{(r)}\right)^2\beta_m' - B_{12}^{(r)}B_{66}^{(r)}\beta_m'}{B_{11}^{(r)}B_{66}^{(r)}},$$

$$B_2^{(r)} = \frac{B_{11}^{(r)}B_{11}^{(r)}\beta_m'' - \left(B_{12}^{(r)}\right)^2\beta_m' - 2B_{12}^{(r)}B_{66}^{(r)}\beta_m'}{B_{11}^{(r)}B_{66}^{(r)}}.$$

Let $x^{(r)}$ ($j = 1,2,3,4$), be pairwise different roots of the Eq. (12.14) with nonnegative real parts and $x_{j+4}^{(r)} = -x_j^{(r)}$, $j = 1,2,3,4$. Let $\left(u_{1j}^{(r)}, u_{2j}^{(r)}, u_{3j}^{(r)}\right)$, be nontrivial solutions of type (12.10) of system (12.1) at $x^{(r)} = x_j^{(r)}$, $j = 1, 2, ..., 8$, respectively. The solution of problem (12.1)-(12.5) is sought in the form

$$u_i^{(r)} = \sum_{j=1}^{8} u_{ij}^{(r)} w_j^{(r)}\,,\; i = 1,2,3; r = 1,2. \tag{12.16}$$

Let as insert Eq. (12.16) into the boundary conditions (12.2) - (12.4). As a result, we obtain the system of equations

$$\sum_{j=1}^{8} \frac{M_{ij}^{(1)} exp\left(x_j^{(1)}\theta_m s^{(1)}\right) w_j^{(1)}}{c_{mj}^{(1)} + \varepsilon_m^2 a^2 g_{mj}^{(1)} d_{mj}^{(1)}} - \sum_{j=1}^{8} \frac{c\, M_{ij}^{(2)} \exp\left(x_j^{(2)}\theta_m s^{(2)}\right) w_j^{(2)}}{c_{mj}^{(2)} + \varepsilon_m^2 a^2 g_{mj}^{(2)} d_{mj}^{(2)}} = 0,$$
$$i = \overline{1,4},$$
$$\sum_{j=1}^{8} \frac{M_{ij}^{(1)} \exp\left(x_j^{(1)}\theta_m s^{(1)}\right) w_j^{(1)}}{c_{mj}^{(1)} + \varepsilon_m^2 a^2 g_{mj}^{(1)} d_{mj}^{(1)}} - \sum_{j=1}^{8} \frac{M_{ij}^{(2)} \exp\left(x_j^{(2)}\theta_m s^{(2)}\right) w_j^{(2)}}{c_{mj}^{(2)} + \varepsilon_m^2 a^2 g_{mj}^{(2)} d_{mj}^{(2)}} = 0,$$
$$i = \overline{5,8}, \tag{12.17}$$
$$\sum_{j=1}^{8} \frac{M_{ij}^{(1)} w_j^{(1)}}{c_{mj}^{(1)} + \varepsilon_m^2 a^2 g_{mj}^{(1)} d_{mj}^{(1)}} = 0, i = \overline{9,12}, \sum_{j=1}^{8} \frac{M_{ij}^{(2)} w_j^{(2)}}{c_{mj}^{(2)} + \varepsilon_m^2 a^2 g_{mj}^{(2)} d_{mj}^{(2)}} = 0,$$
$$i = \overline{13,16},$$

$$M_{1j}^{(r)} = \left(x_j^{(r)}\right)^2 b_{mj}^{(r)} - \frac{B_{12}^{(r)}}{B_{22}^{(r)}}\beta_m' a_{mj}^{(r)} - c_{mj}^{(r)} - a^2 l_{mj}^{(r)} b_{mj}^{(r)} \left(x_j^{(r)}\right)^2$$
$$-\varepsilon_m^2 a^2 d_{mj}^{(r)} \left(b_{mj}^{(r)} - \frac{B_{12}^{(r)}}{B_{22}^{(r)}}\beta_m'\right),\ M_{2j}^{(r)} = \frac{B_{66}^{(r)}}{B_{11}^{(r)}} x_j^{(r)} \left\{a_{mj}^{(r)} + b_{mj}^{(r)}\right. \tag{12.18}$$
$$\left. + a^2 l_{mj}^{(r)} \left[\frac{B_{22}^{(r)} B_{12}^{(r)}}{B_{11}^{(r)} B_{66}^{(r)}} \left(x_j^{(r)}\right)^2 + \frac{B_{22}^{(r)}}{B_{66}^{(r)}}\beta_m'' - \frac{B_{22}^{(r)}}{B_{11}^{(r)}} \left(\eta_{1m}^{(r)}\right)^2\right] + \varepsilon_m^2 a^2 \frac{B_{22}^{(r)} B_{12}^{(r)}}{B_{11}^{(r)} B_{66}^{(r)}} d_{mj}^{(r)}\right\},$$

$$M_{3j}^{(r)} = \left(\left(x_j^{(r)}\right)^2 - \frac{B_{12}^{(r)}}{B_{22}^{(r)}}\beta_m'\right) c_{mj}^{(r)} + \varepsilon_m^2 \left[b_{mj}^{(r)} \left(x_j^{(r)}\right)^2\right.$$
$$\left. + 4a^2 \frac{B_{66}^{(r)} B_{12}^{(r)}}{\left(B_{22}^{(r)}\right)^2} g_{mj}^{(r)} \left(\beta_m'\right)^2\right], \tag{12.19}$$

$$M_{4j}^{(r)} = \varkappa_j^{(r)} \left(l_{mj}^{(r)} c_{mj}^{(r)} + \varepsilon_m^2 b_{mj}^{(r)} d_{mj}^{(r)}\right), \qquad c = \frac{B_{22}^{(2)}}{B_{22}^{(1)}},$$

$$M_{5j}^{(r)} = a_{mj}^{(r)} + a^2 \beta_m' \frac{B_{22}^{(r)} (B_{12}^{(r)} + B_{66}^{(r)})}{B_{11}^{(r)} B_{66}^{(r)}} l_{mj}^{(r)} \left(x_j^{(r)}\right)^2 + \varepsilon_m^2 a^2 \frac{B_{22}^{(r)} B_{12}^{(r)}}{B_{11}^{(r)} B_{66}^{(r)}} d_{mj}^{(r)},$$
$$M_{6j}^{(r)} = x_j^{(r)} \left(b_{mj}^{(r)} - a^2 g_{mj}^{(r)} l_{mj}^{(r)}\right), \tag{12.20}$$
$$M_{7j}^{(r)} = c_{mj}^{(r)} + \varepsilon_m^2 a^2 g_{mj}^{(r)} d_{mj}^{(r)},$$
$$M_{8j}^{(r)} = x_j^{(r)} (c_{mj}^{(r)} + \varepsilon_m^2 a^2 g_{mj}^{(r)} d_{mj}^{(r)}),$$
$$M_{9j}^{(1)} = M_{1j}^{(1)},\ M_{10j}^{(1)} = M_{2j}^{(1)},\ M_{11j}^{(1)} = M_{3j}^{(1)},\ M_{12j}^{(1)} = M_{4j}^{(1)},\ j = \overline{1,8},$$
$$M_{13j}^{(2)} = M_{5j}^{(2)},\ M_{14j}^{(2)} = M_{6j}^{(2)},\ M_{15j}^{(2)} = M_{7j}^{(2)},\ M_{16j}^{(2)} = M_{8j}^{(2)},\ j = \overline{1,8}$$

The superscript j means that the corresponding function is taken at $x^{(r)} = x_j^{(r)}$. In order that system (12.17) had a nontrivial solution, it is necessary and sufficient that

$$\Delta = \exp\left(-\sum_{r=1}^{2}\sum_{j=1}^{4} z_j^{(r)}\right) Det \left\|T_{ij}\right\|_{i,j=1}^{4} = 0\,,\ m = \overline{1,\infty}, \tag{12.21}$$

$$\begin{aligned}
&T_{11} = \left\|M_{ij}^{(1)}\right\|_{i,j=1}^{4}, \quad T_{12} = \left\|(-1)^{i-1} M_{ij}^{(1)} \exp\left(z_j^{(1)}\right)\right\|_{i,j=1}^{4},\\
&T_{13} = c\left\|M_{ij}^{(2)}\right\|_{i,j=1}^{4}, \quad T_{14} = c\left\|(-1)^{i} M_{ij}^{(2)} \exp\left(z_j^{(2)}\right)\right\|_{i,j=1}^{4},\\
&T_{21} = \left\|M_{ij}^{(1)}\right\|_{i=5,j=1}^{8,4}, \quad T_{22} = \left\|(-1)^{i-1} M_{ij}^{(1)} \exp\left(z_j^{(1)}\right)\right\|_{i=5,j=1}^{8,4},\\
&T_{23} = \left\|M_{ij}^{(2)}\right\|_{i=5,j=1}^{8,4}, \quad T_{24} = \left\|(-1)^{i} M_{ij}^{(2)} \exp\left(z_j^{(2)}\right)\right\|_{i=5,j=1}^{8,4},\\
&T_{31} = T_{12}, \quad T_{32} = T_{11}, \quad T_{33} = 0, \quad T_{34} = 0, \quad T_{41} = 0,\\
&T_{42} = 0, \quad T_{43} = T_{24}, \quad T_{44} = T_{23}, \qquad z_j^{(r)} = -\theta_m\, x_j^{(r)} s^{(r)}.
\end{aligned} \tag{12.22}$$

Performing elementary operations with columns of determinant (12.21), we obtain

$$\Delta = \exp\left(-\sum_{r=1}^{2}\sum_{j=1}^{4} z_j^{(r)}\right)\left(K^{(1)}\right)^2 \left(K^{(2)}\right)^2 Det \left\|t_{ij}\right\|_{i,j=1}^{4} = 0,$$

$$m = \overline{1,\infty}, \tag{12.23}$$

$$\begin{aligned}
K^{(r)} = &\left(x_1^{(r)} - x_2^{(r)}\right)\left(x_1^{(r)} - x_3^{(r)}\right)\left(x_1^{(r)} - x_4^{(r)}\right)\left(x_2^{(r)} - x_3^{(r)}\right)\\
&\times\left(x_2^{(r)} - x_4^{(r)}\right)\left(x_3^{(r)} - x_4^{(r)}\right), \quad r = 1, 2.
\end{aligned} \tag{12.24}$$

$$\begin{aligned}
&t_{11} = \left\|m_{ij}^{(1)}\right\|_{i,j=1}^{4}, t_{12} = \left\|m_{ij}^{(1)}\right\|_{i=1,j=5}^{4,8}, t_{13} = c\left\|m_{ij}^{(2)}\right\|_{ij=1}^{4}, t_{14} = c\left\|m_{ij}^{(2)}\right\|_{i=1,j=5}^{4,8},\\
&t_{21} = \left\|m_{ij}^{(1)}\right\|_{i=5,j=1}^{8,4}, t_{22} = \left\|m_{ij}^{(1)}\right\|_{i,j=5}^{8}, t_{23} = \left\|m_{ij}^{(2)}\right\|_{i=5,j=1}^{8,4}, t_{24} = \left\|m_{ij}^{(2)}\right\|_{ij=5}^{8},\\
&t_{31} = t_{12}, \ t_{32} = t_{11}, \ t_{33} = 0, \ t_{34} = 0; \ T_{41} = 0, \ t_{42} = 0, \ t_{43} = t_{24}, \ t_{44} = t_{23}.
\end{aligned}$$

Expressions for $m_{ij}^{(r)}, r = 1, 2$ are not presented.

It follows from (12.23) that Eqs. (12.21) are equivalent to the equations

$$\nabla = \text{Det} \left\|t_{ij}\right\|_{i,j=1}^{4} = 0\,,\ m = \overline{1,\infty}. \tag{12.25}$$

Taking into account the possible relations between $\eta_{1m}^{(r)}$, $\eta_{2m}^{(r)}$ and $\eta_{3m}^{(r)}$, we conclude that Eqs. (12.25) determine the frequencies of the interface and the boundary types of the corresponding vibrations. At $\eta_{1m}^{(r)}=\eta_{2m}^{(r)}=\eta_{3m}^{(r)}=\eta_m^{(r)}$ the Eqs. (12.14) are the characteristic equations of system (12.1), and Eqs. (12.25) at $\theta=\theta_m, m\in$N the dispersion equations of problem (12.1) - (12.5).

12.4 About Planar Vibrations of Composed Rectangular Plate with Rigid-Clamped Edge

Free interfacial and edge vibrations of rectangular plate composed of finite thin elastic orthotropic rectangular plates with different elastic coefficients are considered. Let us introduce rectilinear oriented orthogonal coordinates (α,β) on the median plane, where $0\le\alpha\le l,\ -s_0^{(2)}\le\beta\le s_0^{(1)}$. The line $\beta=0$ corresponds to the interface of material properties. All values related to the right plate $0\le\beta\le s_0^{(1)}$ are indicated by the superscript (1), to the left plate $-s_0^{(2)}\le\beta\le 0$ by the index (2). As initial equations, the equations of small planar vibrations of the left and right plates are used, which correspond to the classical theory of orthotropic plates [1]

$$-B_{11}^{(r)}\frac{\partial^2 u_1^{(r)}}{\partial\alpha^2}-B_{66}^{(r)}\frac{\partial^2 u_1^{(r)}}{\partial\beta^2}-\left(B_{12}^{(r)}+B_{66}^{(r)}\right)\frac{\partial^2 u_2^{(r)}}{\partial\alpha\partial\beta}=\rho^{(r)}\omega^2 u_1^{(r)}, \tag{12.26}$$
$$-\left(B_{12}^{(r)}+B_{66}^{(r)}\right)\frac{\partial^2 u_1^{(r)}}{\partial\alpha\partial\beta}-B_{66}^{(r)}\frac{\partial^2 u_2^{(r)}}{\partial\alpha^2}-B_{22}^{(r)}\frac{\partial^2 u_2^{(r)}}{\partial\beta^2}=\rho^{(r)}\omega^2 u_2^{(r)},\ r=1,2,$$

where $\alpha(0\le\alpha\le l)$ and $\beta(-s_0^{(2)}\le\beta\le s_0^{(1)})$ are the orthogonal rectilinear coordinates of a point in the midplane; $u_1^{(r)}, u_2^{(r)}\,(r=1,2)$ are the projections of the displacement vector in the directions α and β, respectively; $B_{ik}^{(r)},\ i,k=1,2,6\ (r=1,2)$, are coefficients of elasticity; ω - is the natural frequency; $\rho^{(r)}(r=1,2)$–are the density of the materials. The boundary conditions are

$$T_2^{(1)}\Big|_{\beta=0}=T_2^{(2)}\Big|_{\beta=0},\quad S_{21}^{(1)}\Big|_{\beta=0}=S_{21}^{(2)}\Big|_{\beta=0},$$
$$u_1^{(1)}\Big|_{\beta=0}=u_1^{(2)}\Big|_{\beta=0},\quad u_2^{(1)}\Big|_{\beta=0}=u_2^{(2)}\Big|_{\beta=0}, \tag{12.27}$$

$$T_2^{(1)}\Big|_{\beta=s_0^{(1)}}=0,\quad S_{21}^{(1)}\Big|_{\beta=s_0^{(1)}}=0, \tag{12.28}$$

$$u_1^{(2)}\Big|_{\beta=-s_0^{(2)}}=0,\quad u_2^{(2)}\Big|_{\beta=-s_0^{(2)}}=0, \tag{12.29}$$

$$T_1^{(r)}\Big|_{\alpha=0,l}=0,\quad S_{12}^{(r)}\Big|_{\alpha=0,l}=0,\ r=1,2. \tag{12.30}$$

$$T_1^{(r)}=hB_{11}^{(r)}\left[\frac{\partial u_1^{(r)}}{\partial\alpha}+\frac{B_{12}^{(r)}}{B_{11}^{(r)}}\frac{\partial u_2^{(r)}}{\partial\beta}\right],$$

$$T_2^{(r)} = hB_{22}^{(r)} \left[\frac{B_{12}^{(r)}}{B_{22}^{(r)}} \frac{\partial u_1^{(r)}}{\partial \alpha} + \frac{\partial u_2^{(r)}}{\partial \beta} \right], \tag{12.31}$$

$$S_{21}^{(r)} = S_{12}^{(r)} = hB_{66}^{(r)} \left[\frac{\partial u_1^{(r)}}{\partial \beta} + \frac{\partial u_2^{(r)}}{\partial \alpha} \right], \ r = 1,2,$$

where relations (12.27) are complete contacts conditions at $\beta = 0$. Relations (12.28) and (12.30) are the conditions of free edges at $\beta = s_0^{(1)}$ and $\alpha = 0, l$, respectively (Fig. 12.2). In analogous way as in Sect. 12.3 we can determine the characteristic equations of problem (12.26) - (12.30)

$$c_m^{(r)} = \frac{B_{22}^{(r)}}{B_{11}^{(r)}} \left(y^{(r)}\right)^4 - B_2^{(r)} \left(y^{(r)}\right)^2 + \frac{B_{22}^{(r)} + B_{66}^{(r)}}{B_{11}^{(r)}} \left(\eta_m^{(r)}\right)^2 \left(y^{(r)}\right)^2 +$$
$$\left(\beta_m' - \left(\eta_m^{(r)}\right)^2\right) \left(\frac{B_{22}^{(r)}}{B_{11}^{(r)}} \beta_m'' - \frac{B_{66}^{(r)}}{B_{11}^{(r)}} \left(\eta_m^{(r)}\right)^2 \right) = 0 \ , r = 1,2; \ m = \overline{1, \infty}. \tag{12.32}$$

Let $y_1^{(r)}, y_2^{(r)} (r = 1,2)$ be various roots of Eqs. (12.32) with nonnegative real parts and $y_{j+2}^{(r)} = -y_j^{(r)} (j = 1,2)$. The dispersion equation has the forms

$$\mathrm{Det} \left\| e_{ij} \right\|_{i,j=1}^{8} = 0, \ m = \overline{1, \infty}.$$

$$e_{11} = \frac{B_{66}^{(1)}}{B_{22}^{(1)}} \left(\left(y_j^{(1)}\right)^2 + \frac{B_{12}^{(1)}}{B_{22}^{(1)}} \left(\beta_m' - \left(\eta_m^{(1)}\right)^2 \right) \right),$$
$$e_{12} = \frac{B_{66}^{(1)}}{B_{22}^{(1)}} \left(y_1^{(1)} + y_2^{(1)}\right), \ e_{13} = e_{11} \exp\left(z_1^{(1)}\right),$$
$$e_{14} = e_{12} \exp\left(z_2^{(1)}\right) + e_{11} \left[z_1^{(1)} z_2^{(1)}\right],$$
$$e_{15} = -c \frac{B_{66}^{(2)}}{B_{22}^{(2)}} \left(\left(y_j^{(2)}\right)^2 + \frac{B_{12}^{(2)}}{B_{22}^{(2)}} \left(\beta_m' - \left(\eta_m^{(2)}\right)^2 \right) \right),$$
$$e_{16} = -c \frac{B_{66}^{(2)}}{B_{22}^{(2)}} \left(y_1^{(2)} + y_2^{(2)}\right), \ e_{17} = c e_{15} \exp\left(z_1^{(2)}\right), \tag{12.33}$$
$$e_{18} = e_{16} \exp\left(z_2^{(2)}\right) + e_{15} \left[z_1^{(2)} z_2^{(2)}\right],$$
$$e_{21} = \frac{B_{66}^{(1)}}{B_{22}^{(1)}} y_1^{(1)} \left(\left(y_1^{(1)}\right)^2 + \frac{B_{12}^{(1)}}{B_{22}^{(1)}} \beta_m' + \frac{B_{66}^{(1)}}{B_{22}^{(1)}} \left(\eta_m^{(1)}\right)^2 \right),$$

$$e_{22} = \frac{B_{66}^{(1)}}{B_{22}^{(1)}}\left(y_1^{(1)}y_2^{(1)} + B^{(1)} - \left(\eta_m^{(1)}\right)^2\right),$$

$$e_{23} = -e_{21}\exp\left(z_1^{(1)}\right), e_{24} = -e_{22}\exp\left(z_2^{(1)}\right) - e_{21}\left[z_1^{(1)}z_2^{(1)}\right],$$

$$e_{25} = \frac{B_{66}^{(2)}}{B_{22}^{(2)}}y_1^{(2)}\left(\left(y_1^{(2)}\right)^2 + \frac{B_{12}^{(2)}}{B_{22}^{(2)}}\beta_m' + \frac{B_{66}^{(2)}}{B_{22}^{(2)}}\left(\eta_m^{(2)}\right)^2\right),$$

$$e_{26}^{(1)} = \frac{B_{66}^{(2)}}{B_{22}^{(2)}}\left(y_1^{(2)}y_2^{(2)} + B^{(2)} - \left(\eta_m^{(2)}\right)^2\right), \tag{12.34}$$

$$e_{27} = -e_{25}\exp\left(z_1^{(2)}\right), e_{28} = -e_{26}\exp\left(z_2^{(2)}\right) - e_{25}\left[z_1^{(2)}z_2^{(2)}\right],$$

$$e_{31} = \left(y_1^1\right)^2 - \frac{B_{66}^{(1)}}{B_{22}^{(1)}}\left(\beta_m' - \left(\eta_m^{(1)}\right)^2\right), e_{32} = y_1^{(1)} + y_2^{(1)}, e_{33} = e_{31}\exp\left(z_1^{(1)}\right),$$

$$e_{34} = e_{32}\exp\left(z_2^{(1)}\right) + e_{31}\left[z_1^{(1)}z_2^{(1)}\right], e_{35} = -\left(\left(y_1^{(2)}\right)^2 - \frac{B_{66}^{(2)}}{B_{22}^{(2)}}\left(\beta_m' - \left(\eta_m^{(2)}\right)^2\right)\right),$$

$$e_{36} = -\left(y_1^{(2)} + y_2^{(2)}\right), e_{37} = e_{35}\exp\left(z_1^{(2)}\right), e_{38} = e_{36}\exp\left(z_2^{(2)}\right) + e_{35}\left[z_1^{(2)}z_2^{(2)}\right].$$

$$e_{41} = y_1^{(1)}\frac{B_{12}^{(1)} + B_{66}^{(1)}}{B_{22}^{(1)}}, e_{42} = \frac{B_{12}^{(1)} + B_{66}^{(1)}}{B_{22}^{(1)}}, e_{43} = -e_{41}\exp\left(z_1^{(1)}\right),$$

$$e_{44} = -e_{42}\exp\left(z_2^{(1)}\right) - e_{41}\left[z_1^{(1)}z_2^{(1)}\right], e_{45} = y_1^{(2)}\frac{B_{12}^{(2)} + B_{66}^{(2)}}{B_{22}^{(2)}}, e_{46} = \frac{B_{12}^{(2)} + B_{66}^{(2)}}{B_{22}^{(2)}},$$

$$e_{47} = -e_{45}\exp\left(z_1^{(2)}\right), e_{48} = -e_{46}\exp\left(z_2^{(2)}\right) - e_{45}\left[z_1^{(2)}z_2^{(2)}\right],$$

$$e_{51} = e_{13}, e_{52} = e_{14}, e_{53} = e_{11}, e_{54} = e_{12}, e_{5j} = 0, j = \overline{5,8},$$

$$e_{61} = e_{23}, e_{62} = e_{24}, e_{63} = e_{21}, e_{64} = e_{22}, e_{6j} = 0, j = \overline{5,8},$$

$$e_{7j} = 0, j = \overline{1,4}, e_{75} = e_{37}, e_{76} = e_{38}, e_{77} = e_{35}, e_{78} = e_{36},$$

$$e_{8j} = 0, j = \overline{1,4}, e_{85} = e_{47}, e_{86} = e_{48}, e_{87} = e_{45}, e_{88} = e_{46},$$

$$z_j^{(r)} = -y_j^{(r)}\theta_m s_0^{(r)}, \left[z_1^{(r)}z_2^{(r)}\right] = \frac{-\theta_m s_0^{(r)}\left(\exp\left(z_2^{(r)}\right) - \exp\left(z_1^{(r)}\right)\right)}{\left(z_2^{(r)} - z_1^{(r)}\right)},$$

$$j = 1,2; r = 1,2; c = \frac{B_{22}^{(2)}}{B_{22}^{(1)}}.$$

For $\theta_m s_0^{(1)} \to \infty$ and for $\theta_m s_0^{(2)} \to \infty$ the set of Eqs. (12.34) have the form

$$\operatorname{Det}\left\|e_{ij}\right\|_{i,j=1}^{8} = \left(\frac{B_{66}^{(1)}}{B_{22}^{(1)}}\right)^{2} \frac{B_{12}^{(1)}+B_{66}^{(1)}}{B_{22}^{(1)}} \frac{B_{12}^{(2)}+B_{66}^{(2)}}{B_{22}^{(2)}} Q^{(2)}\left(\eta_m^{(2)}\right) K_2^{(1)}\left(\eta_m^{(1)}\right)$$

$$\times L\left(\eta_m^{(1)},\eta_m^{(2)}\right) + \sum_{r=1}^{2}\sum_{j=1}^{2} \exp\left(z_j^{(r)}\right) = 0,\ m=\overline{1,\infty}. \qquad (12.35)$$

$$L\left(\eta_m^{(1)},\eta_m^{(2)}\right) = \frac{B_{12}^{(1)}+B_{66}^{(1)}}{B_{22}^{(1)}} \frac{B_{12}^{(2)}+B_{66}^{(2)}}{B_{22}^{(2)}} \left\{\left(\frac{B_{66}^{(1)}}{B_{22}^{(1)}}\right)^{2} K_2^{(1)}\left(\eta_m^{(1)}\right) Q^{(2)}\left(\eta_m^{(2)}\right)\right.$$

$$\left. + \left(\frac{B_{22}^{(2)}}{B_{11}^{(2)}}\right)^{2}\left(\frac{B_{66}^{(2)}}{B_{22}^{(2)}}\right)^{2} K_2^{(2)}\left(\eta_m^{(2)}\right) Q^{(1)}\left(\eta_m^{(1)}\right)\right\} + \frac{B_{22}^{(2)}}{B_{11}^{(2)}}\left[\left(l_{11}^{(1)}l_{22}^{(2)}+l_{21}^{(1)}l_{12}^{(2)}\right)\right.$$

$$\times\left(l_{32}^{(1)}l_{41}^{(2)}+l_{31}^{(1)}l_{42}^{(2)}\right)+\left(l_{11}^{(2)}l_{21}^{(1)}+l_{21}^{(2)}l_{12}^{(1)}\right)\left(l_{32}^{(2)}l_{41}^{(1)}+l_{31}^{(2)}l_{42}^{(1)}\right)$$

$$\left.-\left(l_{11}^{(1)}l_{21}^{(2)}+l_{11}^{(2)}l_{21}^{(1)}\right)\left(l_{31}^{(1)}l_{42}^{(2)}+l_{31}^{(2)}l_{42}^{(1)}\right)-\left(l_{12}^{(1)}l_{22}^{(2)}+l_{12}^{(2)}l_{22}^{(1)}\right)\left(l_{31}^{(1)}l_{41}^{(2)}+l_{31}^{(2)}l_{41}^{(1)}\right)\right],$$

$$K_2^{(r)}\left(\eta_m^{(r)}\right) = \left(\beta_m' - \left(\eta_m^{(r)}\right)^2\right)\left(\frac{B_{11}^{(r)}B_{22}^{(r)}\beta_m'' - \left(B_{12}^{(r)}\right)^2\beta_m'}{B_{11}^{(r)}B_{66}^{(r)}} - \left(\eta_m^{(r)}\right)^2\right)$$

$$-\left(\eta_m^{(r)}\right)^2 y_1^{(r)}y_2^{(r)},\quad Q^{(2)}\left(\eta_m^{(r)}\right) = y_1^{(r)}y_2^{(r)} + \frac{B_{66}^{(r)}}{B_{22}^{(r)}}\left(\beta_m' - \left(\eta_m^{(r)}\right)^2\right), \qquad (12.36)$$

$$l_{11}^{(r)} = \frac{B_{66}^{(r)}}{B_{22}^{(r)}}\left(\left(y_1^{(r)}\right)^2 + \frac{B_{12}^{(r)}}{B_{22}^{(r)}}\left(\beta_m' - \left(\eta_m^{(r)}\right)^2\right)\right),\quad l_{12}^{(r)} = \frac{B_{66}^{(r)}}{B_{22}^{(r)}}\left(y_1^{(r)}+y_2^{(r)}\right),$$

$$l_{21}^{(r)} = \frac{B_{66}^{(r)}}{B_{22}^{(r)}}y_1^{(r)}\left(\left(y_1^{(r)}\right)^2 + \frac{B_{12}^{(r)}}{B_{22}^{(r)}}\beta_m' + \frac{B_{66}^{(r)}}{B_{22}^{(r)}}\left(\eta_m^{(r)}\right)^2\right),$$

$$l_{22}^{(r)} = \frac{B_{66}^{(r)}}{B_{22}^{(r)}}\left(y_1^{(r)}y_2^{(r)} + B_1^{(r)} - \left(\eta_m^{(r)}\right)^2\right),$$

$$l_{31}^{(r)} = \left(y_1^{(r)}\right)^2 - \frac{B_{66}^{(r)}}{B_{22}^{(r)}}\left(\beta_m' - \left(\eta_m^{(r)}\right)^2\right),\quad l_{32}^{(r)} = y_1^{(r)} + y_2^{(r)},$$

$$l_{41}^{(r)} = y_1^{(r)}\frac{B_{12}^{(r)}+B_{66}^{(r)}}{B_{22}^{(r)}},\quad l_{42}^{(r)} = \frac{B_{12}^{(r)}+B_{66}^{(r)}}{B_{22}^{(r)}},\qquad r=1,2.$$

It follows from (12.35) that for $\theta_m s_0^{(1)} \to \infty$ and for $\theta_m s_0^{(2)} \to \infty$ the set of equations from (12.34) is split into sets of equations

$$K_2^{(1)}\left(\eta_m^{(1)}\right) = 0,\ m=\overline{1,\infty},\qquad L\left(\eta_m^{(1)},\eta_m^{(2)}\right) = 0,\ m=\overline{1,\infty}. \qquad (12.37)$$

The first and second sets of equations from (12.37) are analogs of the Rayleigh type equation, which determine the frequencies of the edge oscillations of the generator $\beta = s_0^{(1)}$, the second set of equations from (12.37) is an analogue of the Stoneley's dispersion equations, which determines the frequencies of the interface planar vibrations at the interface line of material properties $\beta = 0$.

12.5 On Bending Vibrations of Composed Rectangular Plate with Rigid-Clamped Edge

The question of the existence of free bending interface and edge vibrations of a rectangular plate composed of thin elastic orthotropic rectangular plates with different elastic properties is investigated. We introduce rectilinear orthogonal coordinates (α, β) in the middle plane of a rectangular plate, where $0 \le \alpha \le l$, $-s_0^{(2)} \le \beta \le s_0^{(1)}$. The straight line $\beta = 0$ corresponds to the interface of the material properties. All values related to the right plate $0 \le \beta \le s_0^{(1)}$ are indicated by the superscript (1), and to the left plate $-s_0^{(2)} \le \beta \le 0$ by the index (2). As initial equations, the equations of small bending vibrations of the left and right plates are used, which correspond to the classical theory of orthotropic plates [1]

$$\mu^4 \left(B_{11}^{(r)} \frac{\partial^4 u_3^{(r)}}{\partial \alpha^4} + 2\left(B_{12}^{(r)} + 2B_{66}^{(r)}\right) \frac{\partial^4 u_3^{(r)}}{\partial \alpha^2 \partial \beta^2} + B_{22}^{(r)} \frac{\partial^4 u_3^{(r)}}{\partial \beta^4} \right) = \rho^{(r)} \omega^2 u_3^{(r)},$$

$$r = 1, 2. \tag{12.38}$$

Here $u_3^{(r)}$ $(r = 1, 2)$ are the normal components of the displacement vector of the right and left plates; $B_{ij}^{(r)}$ $(r = 1, 2)$ - elasticity coefficients; ω - angular frequency; $\rho^{(r)} (r = 1, 2)$ - the density of materials; $\mu^4 = h^2/12$ (h- shell thickness). The following boundary conditions are considered

$$M_2^{(1)}\Big|_{\beta=0} = M_2^{(2)}\Big|_{\beta=0}, \quad N_2^{(1)} + \frac{\partial H^{(1)}}{\partial \alpha}\Big|_{\beta=0} = N_2^{(2)} + \frac{\partial H^{(2)}}{\partial \alpha}\Big|_{\beta=0}, \tag{12.39}$$

$$u_3^{(1)}\Big|_{\beta=0} = u_3^{(2)}\Big|_{\beta=0}, \quad \frac{\partial u_3^{(1)}}{\partial \beta}\Big|_{\beta=0} = \frac{\partial u_3^{(2)}}{\partial \beta}\Big|_{\beta=0}, \tag{12.40}$$

$$M_2^{(1)}\Big|_{\beta=s_0^{(1)}} = 0, \quad N_2^{(1)} + \frac{\partial H^{(1)}}{\partial \alpha}\Big|_{\beta=s_0^{(1)}} = 0, \tag{12.41}$$

$$u_3^{(2)}\Big|_{\beta=-s_0^{(2)}} = 0, \quad \frac{\partial u_3^{(2)}}{\partial \beta}\Big|_{\beta=-s_0^{(2)}} = 0, \tag{12.42}$$

$$M_1^{(1)}\Big|_{\alpha=0,l}=0,\quad N_1^{(1)}+\frac{\partial H^{(1)}}{\partial\beta}\Big|_{\alpha=0,l}=0. \tag{12.43}$$

$$M_1^{(r)}=\frac{h^3}{12}B_{11}^{(r)}\left[\frac{\partial^2u_3^{(r)}}{\partial\alpha^2}+\frac{B_{12}^{(r)}}{B_{11}^{(r)}}\frac{\partial^2u_3^{(r)}}{\partial\beta^2}\right], \tag{12.44}$$

$$M_2^{(r)}=\frac{h^3}{12}B_{22}^{(r)}\left[\frac{B_{12}^{(r)}}{B_{22}^{(r)}}\frac{\partial^2u_3^{(r)}}{\partial\alpha^2}+\frac{\partial^2u_3^{(r)}}{\partial\beta^2}\right], \tag{12.45}$$

$$N_1^{(r)}+\frac{\partial H^{(r)}}{\partial\beta}=\frac{h^3}{12}B_{11}^{(r)}\left[\frac{\partial^3u_3^{(r)}}{\partial\alpha^3}+\frac{B_{12}^{(r)}+4B_{66}^{(r)}}{B_{11}^{(r)}}\frac{\partial^3u_3^{(r)}}{\partial\beta^2\partial\alpha}\right], \tag{12.46}$$

$$N_2^{(r)}+\frac{\partial H^{(r)}}{\partial\alpha}=\frac{h^3}{12}B_{22}^{(r)}\left[\frac{\partial^3u_3^{(r)}}{\partial\beta^3}+\frac{B_{12}^{(r)}+4B_{66}^{(r)}}{B_{22}^{(r)}}\frac{\partial^3u_3^{(r)}}{\partial\alpha^2\partial\beta}\right]. \tag{12.47}$$

Where relations (12.40) are complete contacts conditions at $\beta=0$. Relations (12.41) and (12.43) are the conditions of free edges at $\beta=s_0^{(1)}$ and $\alpha=0,l$, where l is the length of the plate, respectively (Fig. 12.2)

$$R_{mm}^{(r)}=a^2\left(\left(y^{(r)}\right)^4-\frac{2\left(B_{12}^{(r)}+2B_{66}^{(r)}\right)}{B_{22}^{(r)}}\beta_m'\left(y^{(r)}\right)^2+\frac{B_{11}^{(r)}}{B_{22}^{(r)}}\beta_m'\beta_m''\right.$$

$$\left.-\frac{B_{66}^{(r)}}{B_{22}^{(r)}}\left(\eta_m^{(r)}\right)^2\right)=0,\quad r=1,2,\quad m=\overline{1,\infty}, \tag{12.48}$$

where a^2, $\eta_m^{(r)}$, β_m', β_m'' are defined in (12.15) and (12.9).

Let $y_3^{(r)}$, $y_4^{(r)}$ be the different roots of Eqn. (12.48) with positive real parts and $y_{j+2}^{(r)}=-y_j^{(r)}$ $(j=3,4)$

$$\mathrm{Det}\left\|b_{ij}\right\|_{i,j=1}^{8}=0,\ m=\overline{1,\infty}. \tag{12.49}$$

$$b_{11}=\left(y_3^{(1)}\right)^2-\frac{B_{12}^{(1)}}{B_{22}^{(1)}}\beta_m',\ b_{12}=y_3^{(1)}+y_4^{(1)},\ b_{13}=b_{11}\exp\left(z_3^{(1)}\right),$$

$$b_{14}=b_{12}\exp\left(z_4^{(1)}\right)+b_{11}\left[z_3^{(1)}z_4^{(1)}\right],\ b_{15}=-c\left(\left(y_3^{(2)}\right)^2-\frac{B_{12}^{(2)}}{B_{22}^{(2)}}\beta_m'\right),$$

$$b_{16}=-c\left(y_1^{(2)}+y_2^{(2)}\right),\ b_{17}=b_{15}\exp\left(z_3^{(2)}\right),$$

$$b_{18}=b_{16}\exp\left(z_4^{(2)}\right)+b_{15}\left[z_3^{(2)}z_4^{(2)}\right],$$

$$b_{21}=y_3^{(1)}\left(\left(y_3^{(1)}\right)^2-\frac{B_{12}^{(1)}+4B_{66}^{(1)}}{B_{22}^{(1)}}\beta_m'\right),b_{22}=y_3^{(1)}y_4^{(1)}+\frac{B_{12}^{(1)}}{B_{22}^{(1)}}\beta_m',$$

$$b_{23} = -b_{21}\exp\left(z_3^{(1)}\right),\ b_{24} = -b_{22}\exp\left(z_4^{(1)}\right) - b_{21}\left[z_3^{(1)}z_4^{(1)}\right],$$

$$b_{25} = cy_3^{(2)}\left(\left(y_3^{(2)}\right)^2 - \frac{B_{12}^{(2)} + 4B_{66}^{(2)}}{B_{22}^{(2)}}\beta_m'\right),\ b_{26} = c\left(y_3^{(2)}y_4^{(2)} + \frac{B_{12}^{(2)}}{B_{22}^{(2)}}\beta_m'\right),$$

$$b_{27} = -b_{25}\exp\left(z_3^{(2)}\right),\ b_{31} = 1, b_{32} = 0,\ b_{28} = -b_{26}\exp\left(z_4^{(2)}\right) - b_{25}\left[z_3^{(2)}z_4^{(2)}\right],$$

$$b_{33} = \exp\left(z_3^{(1)}\right),\ b_{34} = \left[z_3^{(1)}z_4^{(1)}\right],\ b_{35} = -1,\ b_{36} = 0,\ b_{37} = -\exp\left(z_3^{(2)}\right),$$

$$b_{38} = -\left[z_3^{(3)}z_4^{(4)}\right],\ b_{41} = y_3^{(1)}, b_{42} = 1,\ b_{43} = -y_3^{(1)}\exp\left(z_3^{(1)}\right), \tag{12.50}$$

$$b_{44} = -\exp\left(z_4^{(1)}\right) - y_3^{(1)}\left[z_3^{(1)}z_4^{(1)}\right]\ b_{45} = y_3^{(2)}, b_{46} = 1,$$

$$b_{47} = -y_3^{(2)}\exp\left(z_3^{(2)}\right),\ b_{48} = -\exp\left(z_4^{(2)}\right) - y_3^{(2)}\left[z_3^{(2)}z_4^{(2)}\right],$$

$$b_{51} = b_{13},\ b_{52} = b_{14},\ b_{53} = b_{11},\ b_{54} = b_{12},\quad b_{5j} = 0,\ j = \overline{5,8},$$

$$b_{61} = b_{23},\ b_{62} = b_{24},\ b_{63} = b_{21},\ b_{64} = b_{22},\quad b_{6j} = 0,\ j = \overline{5,8},$$

$$b_{7j} = 0,\ j = \overline{1,4},\ \ b_{75} = b_{37},\ b_{76} = b_{38},\ b_{77} = b_{35},\ b_{78} = b_{36},$$

$$b_{8j} = 0,\ j = \overline{1,4},\ \ b_{85} = b_{47},\ b_{86} = b_{48},\ b_{87} = b_{45},\ b_{88} = b_{46},$$

$$z_j^{(r)} = -y_j^{(r)}\theta_m s_0^{(r)},\qquad \left[z_3^{(r)}z_4^{(r)}\right] = \frac{-\theta_m s_0^{(r)}\left(\exp\left(z_4^{(r)}\right) - \exp\left(z_3^{(r)}\right)\right)}{\left(z_4^{(r)} - z_3^{(r)}\right)},$$

$$j = 3, 4; r = 1, 2\,.$$

For $\theta_m s_0^{(1)} \to \infty$ and for $\theta_m s_0^{(2)} \to \infty$ the set of Eqs. (12.49) have the form $m = \overline{1,\infty}$

$$\mathrm{Det}\left\|b_{ij}\right\|_{i,j=1}^{8} = G\left(\eta_m^{(1)}, \eta_m^{(2)}\right)K_1^{(1)}\left(\eta_m^{(1)}\right) + \sum_{r=1}^{2}\sum_{j=3}^{4}\exp\left(z_j^{(r)}\right) = 0, \tag{12.51}$$

$$G\left(\eta_m^{(1)}, \eta_m^{(2)}\right) = -K_1^{(1)}\left(\eta_m^{(1)}\right) - c^2K_1^{(2)}\left(\eta_m^{(2)}\right) + c\left[\left(b_{11}^{(1)}b_{22}^{(2)} + b_{22}^{(1)}b_{11}^{(2)}\right)\right.$$
$$\left. + \left(b_{21}^{(1)}b_{12}^{(2)} + b_{12}^{(1)}b_{21}^{(2)}\right) - \left(y_3^{(1)} + y_3^{(2)}\right)\left(b_{12}^{(1)}b_{22}^{(2)} + b_{22}^{(1)}b_{12}^{(2)}\right)\right],$$

$$K_1^{(r)}\left(\eta_m^{(r)}\right) = \left(y_3^{(r)}\right)^2\left(y_4^{(r)}\right)^2 + \frac{4B_{66}^{(r)}}{B_{22}^{(r)}}\beta_m' y_3^{(r)}y_4^{(r)} - \left(\frac{B_{12}^{(r)}}{B_{22}^{(r)}}\right)^2\left(\beta_m'\right)^2,$$

$$m = \overline{1,\infty}. \tag{12.52}$$

$$b_{11}^{(r)} = \left(y_3^{(r)}\right)^2 - \frac{B_{12}^{(r)}}{B_{22}^{(r)}}\beta_m',\ b_{22}^{(r)} = y_3^{(r)}y_4^{(r)} + \frac{B_{12}^{(r)}}{B_{22}^{(r)}}\beta_m',\ r = 1, 2,$$

$$b_{12}^{(r)} = y_3^{(r)} + y_4^{(r)},\ b_{21}^{(r)} = y_3^{(r)}\left(\left(y_3^{(r)}\right)^2 - \frac{B_{12}^{(r)} + 4B_{66}^{(r)}}{B_{22}^{(r)}}\beta_m'\right).$$

It follows from (12.51) that, for $\theta_m s_0^{(1)} \to \infty$ and for $\theta_m s_0^{(2)} \to \infty$ Eqs. (12.51) splits into sets of equations

$$G\left(\eta_m^{(1)}, \eta_m^{(2)}\right) = 0, \quad m = \overline{1, \infty}\,; \; K_1^{(1)}\left(\eta_m^{(1)}\right) = 0, \quad m = \overline{1, \infty}\,. \tag{12.53}$$

12.6 Asymptotics of the Dispersion Equation (12.25) for $\varepsilon_m \to 0$

Using the formula in Sect. 12.2, we assume that $\eta_{1m}^{(r)} = \eta_{2m}^{(r)} = \eta_{3m}^{(r)} = \eta_m^{(r)}$ $(r = 1, 2)$. Then, when $\varepsilon_m \to 0$ Eq. (12.14) is transformed into a set of equations

$$c_m^{(r)} = \frac{B_{22}^{(r)}}{B_{11}^{(r)}}\left(x^{(r)}\right)^4 - B_2^{(r)}\left(x^{(r)}\right)^2 + \frac{B_{22}^{(r)} + B_{66}^{(r)}}{B_{11}^{(r)}}\left(\eta_m^{(r)}\right)^2\left(x^{(r)}\right)^2$$
$$+\left(\beta_m' - \left(\eta_m^{(r)}\right)^2\right)\left(\beta_m'' - \frac{B_{66}^{(r)}}{B_{11}^{(r)}}\left(\eta_m^{(r)}\right)^2\right) = 0, \quad m = \overline{1, \infty}, \tag{12.54}$$

$$R_{mm}^{(r)} = a^2\left(\left(x^{(r)}\right)^4 - \frac{2\left(B_{12}^{(r)} + 2B_{66}^{(r)}\right)}{B_{22}^{(r)}}\beta_m'\left(x^{(r)}\right)^2 + \frac{B_{11}^{(r)}}{B_{22}^{(r)}}\beta_m'\beta_m''\right) -$$
$$\frac{B_{66}^{(r)}}{B_{22}^{(r)}}\left(\eta_m^{(r)}\right)^2 = 0, \; m = \overline{1, \infty}, \; r = 1, 2. \tag{12.55}$$

which are sets of characteristic equations for the equations of planar and bending vibrations of the rectangular components of the left and right with two opposite free edges, respectively [4, 5]. The roots of Eqs. (12.54) and (12.55) with positive real parts are denoted by $y_1^{(r)}, y_2^{(r)}$ and $y_3^{(r)}, y_4^{(r)}$, respectively.

In the same way as in [3], it is proved that under the condition

$$\varepsilon_m \ll 1, y_i^{(r)} \neq y_j^{(r)}, \; i \neq j, \tag{12.56}$$

the roots $\left(x^{(r)}\right)^2$ of equation (12.14) can be represented as

$$\left(x_j^{(r)}\right)^2 = \left(y_j^{(r)}\right)^2 + \alpha_{jm}^{(r)}\varepsilon_m^2 + \beta_{jm}^{(r)}\varepsilon_m^4 + \ldots, \quad j = \overline{1, 4}. \tag{12.57}$$

Under conditions (12.56), taking into account relations (12.18) and (12.57) and the fact that

$$M_{3j}^{(r)} = M_{4j}^{(r)} = M_{7j}^{(r)} = M_{8j}^{(r)} = O\left(\varepsilon_m^2\right), \; j = 1, 2; r = 1, 2. \tag{12.58}$$

The set of equations (12.25) is reduced to the form

$$\text{Det}\,\|t_{ij}\|_{i,j=1}^{4} = \left(N^{(1)}\left(\eta_m^{(1)}\right)N^{(2)}\left(\eta_m^{(2)}\right)\right)^2\left(K_3^{(1)}\left(\eta_m^{(1)}\right)K_3^{(2)}\left(\eta_m^{(2)}\right)\right)^2,$$

$$\text{Det}\,\|e_{ij}\|_{i,j=1}^{8}\,\text{Det}\,\|b_{ij}\|_{i,j=1}^{8} + O\left(\varepsilon_m^2\right) = 0,\ m = \overline{1,\infty}, \tag{12.59}$$

$$N^{(r)}\left(\eta_m^{(r)}\right) = \left(y_1^{(r)} + y_3^{(r)}\right)\left(y_1^{(r)} + y_4^{(r)}\right)\left(y_2^{(r)} + y_3^{(r)}\right)\left(y_2^{(r)} + y_4^{(r)}\right),$$

$$K_3^{(r)}\left(\eta_m^{(r)}\right) = \left\{\left(\beta_m - \left(\eta_m^{(r)}\right)^2\right)\left(\beta_m'' - \frac{B_{66}^{(r)}}{B_{11}^{(r)}}\left(\eta_m^{(r)}\right)^2\right)\left(p^{(r)}\right)^2\right.$$

$$\left.\times\left[\frac{B_{11}^{(r)}}{B_{22}^{(r)}}B_2^{(r)} - \frac{B_{22}^{(r)} + B_{66}^{(r)}}{B_{22}^{(r)}}\left(\eta_m^{(r)}\right)^2\right]p^{(r)}q^{(r)} + \left(q^{(r)}\right)^2\right\}\left(\frac{B_{22}^{(r)}}{B_{12}^{(r)} + B_{66}^{(r)}}\right)^2\left(\frac{1}{\beta_m'}\right)^2,$$

$$p^{(r)} = \frac{B_{22}^{(r)}}{B_{11}^{(r)}} + a^2\left(\frac{B_{66}^{(r)}}{B_{11}^{(r)}}\left(\eta_m^{(r)}\right)^2 + \frac{\left(B_{12}^{(r)}\right)^2 + 3B_{12}^{(r)}B_{66}^{(r)} + 4\left(B_{66}^{(r)}\right)^2}{B_{11}^{(r)}B_{66}^{(r)}}\beta_m'\right), \tag{12.60}$$

$$q^{(r)} = \frac{B_{22}^{(r)}}{B_{11}^{(r)}}\left(\eta_m^{(r)}\right)^2 - B_1^{(r)} + a^2\left(\frac{B_{66}^{(r)}}{B_{11}^{(r)}}\left(\eta_m^{(r)}\right)^2 - \beta_m''\right)\left(\left(\eta_m^{(r)}\right)^2 + \frac{B_{12}^{(r)} + 3B_{66}^{(r)}}{B_{66}^{(r)}}\beta_m'\right),$$

and determinants $\text{Det}\,\|e_{ij}\|_{i,j=1}^{8}$, $\text{Det}\,\|b_{ij}\|_{i,j=1}^{8}$ are defined in (12.34) and (12.49), respectively.

It follows from Eq. (12.59) that, when $\varepsilon_m \to 0$ the set of equations (12.25) splits into sets of equations

$$Det\,\|e_{ij}\|_{i,j=1}^{8} = 0, m = \overline{1,\infty},\ Det\,\|b_{ij}\|_{i,j=1}^{8} = 0, m = \overline{1,\infty},$$
$$K_3^{(1)}\left(\eta_m^{(1)}\right) = 0, m = \overline{1,\infty}\ K_3^{(2)}\left(\eta_m^{(2)}\right) = 0, m = \overline{1,\infty}\,. \tag{12.61}$$

The first and second sets of equations from (12.61) are, respectively, the dispersion equations of planar and bending interface and edge vibrations of a composite plate. The roots of the third and fourth set of equations correspond to planar vibrations of the components of the cylindrical panel. They appear as a result of using the equation of the corresponding classical theory of orthotropic cylindrical shells.

For $\varepsilon_m \to 0$, $\theta_m s^{(1)} \to \infty$ and $\theta_m s^{(2)} \to \infty$ Eqs. (12.25) takes the form

$$Det\,\|t_{ij}\|_{i,j=1}^{4} = -\left(\frac{B_{66}^{(1)}}{B_{22}^{(1)}}\right)^2\frac{B_{12}^{(1)} + B_{66}^{(1)}}{B_{22}^{(1)}}\frac{B_{12}^{(2)} + B_{66}^{(2)}}{B_{22}^{(2)}}\,\cdot$$

$$\left(N^{(1)}\left(\eta_m^{(1)}\right)N^{(2)}\left(\eta_m^{(2)}\right)\right)^2\left(K_3^{(1)}\left(\eta_m^{(1)}\right)K_3^{(2)}\left(\eta_m^{(2)}\right)\right)^2Q^{(2)}\left(\eta_m^{(2)}\right)\cdot$$

$$K_1^{(1)}\left(\eta_m^{(1)}\right)K_2^{(1)}\left(\eta_m^{(1)}\right)L\left(\eta_m^{(1)},\eta_m^{(2)}\right)G\left(\eta_m^{(1)},\eta_m^{(2)}\right) + O\left(\varepsilon_m^2\right) +$$

$$\sum_{r=1}^{2}\sum_{j=1}^{4}O\left(\exp\left(z_j^{(r)}\right)\right) = 0,\ m = \overline{1,\infty}. \tag{12.62}$$

From (12.62) it follows that for $\varepsilon_m \to 0$, $\theta_m s^{(1)} \to \infty$ and $\theta_m s^{(2)} \to \infty$ the set of Eqs. (12.25) splits into the totality of the equations

$$L\left(\eta_m^{(1)}, \eta_m^{(2)}\right) = 0,\ m = \overline{1,\infty}\ ;\ G\left(\eta_m^{(1)}, \eta_m^{(2)}\right) = 0,\ m = \overline{1,\infty},$$
$$K_1^{(1)}\left(\eta_m^{(1)}\right) = 0,\ m = \overline{1,\infty},\ K_2^{(1)}\left(\eta_m^{(1)}\right) = 0, m = \overline{1,\infty},$$
$$K_3^{(r)}\left(\eta_m^{(r)}\right) = 0,\ m = \overline{1,\infty}, r = 1,2\,. \qquad (12.63)$$

The first and second sets of equations from (12.63) are, respectively, the dispersion equations of the planar and bending interface vibrations for a sufficiently wide composite plate with free edges. The third and fourth sets of equations from (12.63) are, respectively, analogs to the Konenkov and Rayleigh equations for vibrations of a plate made of material (1) localized at the free edge.

12.7 Numerical Investigation

Table 12.1 shows some of the roots of Eqs. (12.37) and (12.53) and dispersion equations (12.34) and (12.49) of planar and bending vibrations for a rectangular composite plate, with a rigid-clamped edge generator, made of boroplastic and special paper with mechanical parameters [4, 6]

- Boroplastic

$$\rho^{(1)} = 2\ 10^3\text{kg/m}^3, E_1^{(1)} = 2.646\ 10^{11}\text{N/m}^2, E_2^{(1)} = 1.323\ 10^{10}\text{N/m}^2,$$
$$G^{(1)} = 9.604\ 10^9\text{N/m}^2,\ \nu_1^{(1)} = 0.2\,,\ \nu_2^{(1)} = 0.01\,, \qquad (12.64)$$

- Paper

$$\rho^{(2)} = 0.16\text{kg/m}^3, E_1^{(2)} = 2.95281\ 10^9\text{N/m}^2, E_2^{(2)} = 2.2106\ 10^9\text{N/m}^2,$$
$$G^{(2)} = 9.77076\ 10^8\text{N/m}^2,\ \nu_1^{(2)} = \nu_2^{(2)} E_1^{(2)}/E_2^{(2)},\ \nu_2^{(2)} = 0.23 \qquad (12.65)$$

and geometric parameters $l = 4, h = \frac{1}{50}$, $s_0^{(1)} = 15, s_0^{(2)} = 5$. The roots of the dispersion equations (12.37) and (12.53) of the planar and bending edge and the vibrations of the interface of a rectangular plate are given. Note that the connection between $\eta_m^{(1)}$ and $\eta_m^{(2)}$ has the form

$$\eta_m^{(2)} = \frac{\rho^{(2)}}{\rho^{(1)}} \frac{B_{66}^{(1)}}{B_{66}^{(2)}} \eta_m^{(1)}\,. \qquad (12.66)$$

Tables 12.2 and 12.3 show some dimensionless characteristics $\eta_m^{(1)}$ of the natural frequencies of interface and edge vibrations for a composite cylindrical panel, made

Table 12.1: Characteristics of natural frequencies for planar and bending vibrations of a rectangular plate.

N	θ_m	$K_2^{(1)}\left(\eta_m^{(1)}\right)=0,$	$L\left(\eta_m^{(1)},\eta_m^{(2)}\right)=0$	Det $\|e_{ij}\|_{i,j=1}^{8}=0$	
1	1.95473			0.91197 $e^{(1)}$	0.92512 ine
2	2.74891	0.98307	0.98398	0.98172 $e^{(1)}$	0.98312 ine
3	3.52957	1.00099	1.00225	0.99965 $e^{(1)}$	1.00100 ine
4	4.27693	0.97897	0.98012	0.97897 $e^{(1)}$	0.99790 ine
11	11.6577	0.98580	0.98696	0.98436 $e^{(1)}$	0.98580 ine
15	16.0962	0.98580	0.98696	0.98436 $e^{(1)}$	0.98580 ine
16	17.1935	0.98580	0.98696	0.98436 $e^{(1)}$	0.98580 ine
N	θ_m	$K_1^{(1)}\left(\eta_m^{(1)}\right)=0,$	$G\left(\eta_m^{(1)},\eta_m^{(2)}\right)=0$	Det $\|b_{ij}\|_{i,j=1}^{8}=0$	
1	1.95473	0.04872	0.04874	0.04889 $b^{(1)}$	0.04930 inb
2	2.74891	0.08576	0.08580	0.08580 $b^{(1)}$	0.08679 inb
3	3.52957	0.10921	1.0927	0.10930 $b^{(1)}$	0.11140 inb
4	4.27693	0.12789	0.12795	0.12790 $b^{(1)}$	0.13065 inb
11	11.6577	0.35363	0.35382	0.35363 $b^{(1)}$	0.35462 inb
15	16.0962	0.48828	0.48853	0.48828 $b^{(1)}$	0.49469 inb
16	17.1935	0.52156	0.52183	0.52156 $b^{(1)}$	0.52839 inb

of boroplastic and special paper, with mechanical parameters (12.64) and (12.65) and geometric parameters:

$$R=45,\ l=4,\ h=\frac{1}{50},\ s^{(1)}=15.2927,\ s^{(2)}=5.01035.$$

In Tables 12.1-12.3, after the characteristics of the natural frequencies, the type of interface vibrations is indicated: ine is predominantly planar, inb is predominantly bending; edge vibrations are: e$^{(1)}$- predominantly planar types, $b^{(1)}$ - predominantly bending types, $n^{(r)}$, $r=1,2$ - new types of vibrations corresponding to materials (1) and (2), respectively. Note that the new types of vibrations are predominantly planar types. The latest manifests itself as a result of using the basic equations corresponding to the classical theory of orthotropic cylindrical shells. In Table 12.2 the case $\eta_{1m}^{(r)}=\eta_{2m}^{(r)}=\eta_m^{(r)}, \eta_{3m}^{(r)}=0$ corresponds to problem (12.1) - (12.5), in which there is no normal component of the inertia force, i.e. we have a predominantly planar type of interface and edge vibrations. Similarly, the case $\eta_{1m}^{(r)}=\eta_{2m}^{(r)}=0, \eta_{3m}^{(r)}=\eta_m^{(r)}$ corresponds predominantly bending type. In Table 12.3, the case $\eta_{1m}^{(r)}=\eta_{2m}^{(r)}=\eta_{3m}^{(r)}=\eta_{fm}^{(r)}$ corresponds to the problem (12.1) - (12.5). For $\varepsilon_m\to 0$ the interface and edge vibrations of problem (12.1) - (12.5), are splitting on quasi-transverse and quasi tangential vibrations. Meanwhile, the frequencies of this problem tend to the frequencies of a similar problem for a composite plate. In vibrations of a predominantly tangential type ($\eta_{1m}^{(r)}=\eta_{2m}^{(r)}=\eta_m^{(r)},\ \eta_{3m}^{(r)}=0$), in addition to planar interface and edge

Table 12.2: Characteristics of the natural frequencies for predominantly planar and predominantly bending vibrations of cylindrical panel.

N	θ_m	$K_3^{(1)}(\eta_m^{(1)}) = 0$, $K_3^{(2)}(\eta_m^{(2)}) = 0$	$\eta_{1m}^{(r)} = \eta_{2m}^{(r)} = \eta_m^{(r)}$, $\eta_{3m}^{(r)} = 0, r = 1, 2$	$\eta_{3m}^{(r)} = \eta_m^{(r)}, r = 1, 2$	$\eta_{1m}^{(r)} = 0, \eta_{2m}^{(r)} = 0$
1	1.95391	4.25538	0.91110e$^{(1)}$	0.92492 ine	0.04987 $b^{(1)}$
		41.3822	4.25538$n^{(1)}$	41.4240$n^{(2)}$	0.05037 inb
2	2.74776	4.94711	0.98383e$^{(1)}$	0.99608 ine	0.08621 $b^{(1)}$
		44.7479	4.89137$n^{(1)}$	44.6619$n^{(2)}$	0.08682 inb
3	3.52810	4.81097	1.00216e$^{(1)}$	1.01580 ine	0.10944 $b^{(1)}$
		44.0542	4.76037$n^{(1)}$	43.7886$n^{(2)}$	0.10995 inb
4	4.27542	4.75564	0.99289e$^{(1)}$	0.99684 ine	0.12803 $b^{(1)}$
		43.8201	4.75343$n^{(1)}$	43.7954$n^{(2)}$	0.12895 inb
11	11.6577	4.78850	0.98656e$^{(1)}$	0.99965 ine	0.35381 $b^{(1)}$
		44.0115	4.70067$n^{(1)}$	44.0190$n^{(2)}$	0.35500 inb
15	16.1102	4.78555	0.98614e$^{(1)}$	0.999959 ine	0.49626 $b^{(1)}$
		44.0396	4.70876$n^{(1)}$	44.0292$n^{(2)}$	0.50340 inb
16	17.2065	4.78465	0.98597e$^{(1)}$	0.99956 ine	0.52160 $b^{(1)}$
		44.0479	4.70338$n^{(1)}$	44.0331$n^{(2)}$	0.52409 inb

Table 12.3: Characteristics of the natural frequencies of a cylindrical panel.

N	θ_m	$\eta_{1m}^{(r)} = \eta_{2m}^{(r)} = \eta_{3m}^{(r)} = \eta_m^{(r)}, r = 1, 2$					
1	1.95391	0.91206e$^{(1)}$	0.92683 ine	0.04987$b^{(1)}$	0.08621 inb	4.25906$n^{(1)}$	41.4248$n^{(2)}$
2	2.74776	0.98430e$^{(1)}$	0.99654 ine	0.04987 $b^{(1)}$	0.08682 inb	4.94292$n^{(1)}$	44.6413$n^{(2)}$
3	3.52810	1.00405e$^{(1)}$	1.01913 ine	0.10944 $b^{(1)}$	0.10995 inb	4.77267$n^{(1)}$	43.7883$n^{(2)}$
4	4.27542	0.98246e$^{(1)}$	0.99621 ine	0.12803 $b^{(1)}$	0.12895 inb	4.75646$n^{(1)}$	43.7927$n^{(2)}$
11	11.6577	0.98805e$^{(1)}$	0.99943 ine	0.35381 $b^{(1)}$	0.35500 inb	4.70684$n^{(1)}$	44.0235$n^{(2)}$
15	16.1102	0.98615e$^{(1)}$	0.99959 ine	0.49626 $b^{(1)}$	0.50340 inb	4.70706$n^{(1)}$	44.0321$n^{(2)}$
16	17.2065	0.98597e$^{(1)}$	0.99957 ine	0.52163$b^{(1)}$	0.52409 inb	4.70662$n^{(1)}$	44.0317$n^{(2)}$

vibrations of the Stoneley and Rayleigh type, new vibrations can also appear due to the longitudinal and torsional components of the inertia force.

12.8 Conclusion

Using the system of equations of dynamic equilibrium of orthotropic cylindrical shells of the corresponding classical theory, dispersion equations are obtained to determine the eigenfrequencies of possible interface and edge vibrations of the composite cylindrical panel with rigid-clamped edge generator. The frequencies of intrinsic interface and edge vibrations of a composite cylindrical panel composed of two orthotropic thin elastic cylindrical panels with different elastic coefficients and having full contact along the generators are determined by the set of equations (12.25). The frequencies of natural interface and edge vibrations for the composite rectangular plate with rigid-clamped edge, are determined by the sets of equations (12.34) and (12.49).

The existence of interface and boundary vibrations depends on the radius of the circle, the length and width of the components of the cylindrical panels and the elastic coefficients. The obtained asymptotic formulas and numerical analysis show that for large θ_m or small curvature of the composed panel, all the characteristics of the intrinsic interface and edge vibrations of the cylindrical panel tend to the characteristics of the interface and edge vibrations of the composed rectangular plate, respectively.

The first frequencies of natural vibrations depend on the selected basis functions satisfying the same boundary conditions, and also, when $\theta_m \rightarrow \infty$ the vibration frequencies at the free generators become independent of the basic functions and the boundary conditions on the ends [4]-[6].

The numerical results show that the asymptotic formulas (12.59) and (12.62) of the dispersion equation (12.25) are a good guide to find the eigenfrequencies of problem (12.1) - (12.5).

References

[1] Ambartsumyan SA (1964) Theory of Anisotropic Shells, NASA Technical Translation TT F-118. National Aeronautics and Space Administration, Washington, D.C.

[2] Gertman IP, Lisitskiy ON (1988) Reflection and transmission of acoustic waves at the interface of separation of two elastic semi-strips. Journal of Applied Mathematics and Mechanics 52(6):816-820, DOI 10.1016/0021-8928(88)90024-X

[3] Gulgazaryan GR (2004) Vibrations of semi-infinite, orthotropic, cylindrical shells of open profile. International Applied Mechanics 40(2):199-212, DOI 10.1023/B:INAM.0000028599.17681.19

[4] Ghulghazaryan GR, Ghulghazaryan LG (2023) Free localized vibrations of a thin elastic composite panel. In: Altenbach H, Prikazchikov D, Nobili A (eds.) Mechanics of High-Contrast Elastic Solids. Advanced Structured Materials, Vol 187, Springer, Cham, pp 91-118, DOI 10.1067/978-3-031-24141-3_7

[5] Ghulghazaryan GR, Ghulghazaryan LG (2023) Free vibrations of thin elastic orthotropic cylindrical panel with rigid-clamped edge generator. ASPU after Kh. Abovyan. Scientific Bulletin 2(45):46-72, DOI 10.24234/scientific.v2i45.93

[6] Ghulghazaryan GR, Ghulghazaryan LG (2025) Svobodnyye kolebaniya tonkoy uprugoy ortotropnoy tsilindricheskoy paneli s sharnirno zakreplennoy granichnoy obrazuyushchey (in Russ., Free vibrations of a thin elastic orthotropic cylindrical panel with a hinged boundary generatrix). Prikladnaya Matematika i Mekhanika 89(1):59-78, DOI 10.31857/S0032823525010053

[7] Ghulghazaryan GR, Ghulghazaryan LG, Kudish II (2019) Free vibrations of a thin elastic orthotropic cylindrical panel with free edges. Mechanics of Composite Materials 55(5):557-574, DOI 10.1007/s11029-019-09834-9

[8] Gulgazaryan GR, Gulgazaryan RG, Khachanyan AA (2013) Vibrations of an orthotropic cylindrical panel with various boundary conditions. International Applied Mechanics 49(5):534-554, DOI 10.1007/s10778-013-0588-4

[9] Ghulghazaryan GR, Ghulghazaryan RG, Srapionyan DL (2013) Localized vibrations of a thin-walled structure consisted of orthotropic elastic non-closed cylindrical shells with free and rigid-clamped edge generators. ZAMM - Journal of Applied Mathematics and Mechanics / Zeitschrift für Angewandte Mathematik und Mechanik 93(4):269-283, DOI 10.1002/zamm.201200024

[10] Gol'denveizer AL, Lidskii VB, Tovstik PE (1979) Svobodnye kolebaniya tonkikh uprugikh obolochek (in. Russ., Free Vibrations of thin Elastic Shells), Nauka, Moscow

[11] Kantorovich LV (1933) Odin pryamoi metod priblizhennogo resheniya zadachi o minimume dvoinogo integrala (in Russ., A direct method for approximate solution of a problem on the minimum of a double integral). Izvestiya AN SSSR, Seriya VII Otdelenie Matematicheskikh i Estestvennykh Nauk (5):647-652

[12] Mikhlin SG (1964) Variational Methods in Mathematical Physics, International Series of Monographs in Pure and Applied Mathematics, Vol 50. Pergamon Press, Oxford et al.

[13] Norris AN (1994) Flexural edge waves. Journal of Sound and Vibration 171(4):571-573, DOI 10.1006/jsvi.1994.1141

[14] Stoneley R (1924) The elastic waves at the interface of two solids. Proceedings of the Royal Society of London. Series A, Containing Papers of a Mathematical and Physical Character 106(738):416-429, DOI 10.1098/rspa.1924.0079

[15] Vilde MV, Kaplunov YD, Kossovich LY (2010) Krayevyye i interfeysnyye rezonansnyye yavleniya v uprugikh telakh (in Russ., Boundary and Interfase Resonant Phenomena in Elastic Bodies). Fizmatlit, Moscow

[16] Vlasov VZ (1932) Novyy prakticheskiy metod rascheta skladchatykh pokrytiy i obolochek Novyy prakticheskiy metod rascheta skladchatykh pokrytiy i obolochek(in Russ., A new practical method for calculating folfed coverings and shells). Stroitel'naya Promyshlennost' (11):33-38 and (12):21-26

[17] Zilbergleit AS, Suslova EB (1983) Kontaktnye volny izgiba v ton'kikh plastinakh (in Russ., Contact waves of bend in thin plates). Akusticheskii Zhurnal 29(2):186-191

Chapter 13
Investigation of Shear Strengths and Shear Creep of Swellings ClAY Soils with and without Basalt Fiber Reinforcement

Sarkis G. Hairoyan, David H. Grigoryan, and Yeghish S. Manukyan

Abstract The article presents the results of an investigation into the shear strength and shear creep behavior of swelling clays, both unreinforced and reinforced with basalt fibers. Shear strength was determined using the standard procedure with M-5 torsional shear apparatus on solid cylindrical specimens. Experimental studies established that basalt fiber reinforcement—at volumetric contents of 1% for the reddish clays and 2% for the bentonitic clays—leads to an increase in shear strength, primarily due to an increase in cohesion, while the internal friction angle remains practically unchanged. It was further determined that the shear creep of the soils, both without reinforcement and with basalt fiber reinforcement, is nearly identical at equal levels of relative shear stress.

Key words: Shear strength, Adhesion, Clay swelling, Reinforcement with basalt fibers, Internal friction, Creep

13.1 Introduction

The determination of shear strength and shear creep of highly dispersed swelling clay soils, both unreinforced and reinforced with basalt fibers, constitutes one of the relevant problems in soil mechanics [3, 7]. The objective of the present study is to investigate the effect of basalt fiber reinforcement on the shear strength and shear creep behavior of swelling bentonitic and reddish clay soils.

Sarkis G. Hairoyan (corresponding author)
Yerevan State University, Yerevan, Republic of Armenia, e-mail: hairoyan@ysu.am

David H. Grigoryan · Yeghish S. Manukyan
Institute of Mechanics, National Academy of Sciences of the Republic of Armenia, 24b Marshal Baghramyan Ave, Yerevan, 0019, Republic of Armenia, e-mail: davit-grigoryan@yandex.com, exishe.manukyan@gmail.com

H. Altenbach and A. V. Sahakyan, *Contact Mechanics and Relevant Problems*,
Advanced Structured Materials 253, https://doi.org/10.1007/978-3-032-20869-9_13

13.2 Materials and Methods

To investigate the effect of basalt fiber reinforcement on the shear strength and shear creep characteristics of swelling soils, bentonitic (montmorillonitic) clays from the Sarigyugh deposit (Republic of Armenia) and reddish swelling clays of Sarmatian age with disturbed structure from the vicinity of Yerevan were used as test materials [1, 5].

To determine swelling pressure, swelling deformation, shear strength, and shear creep of the reddish and bentonitic soils, disturbed-structure specimens were tested both without reinforcement and with basalt fiber reinforcement. To obtain homogeneous specimens, basalt fibers were introduced into the clay soils in a fluid consistency state, mixed thoroughly, and distributed uniformly throughout the mass. For bentonitic clays, the basalt fiber content was 2%, while in the reddish clay the fiber content was 1%. Short basalt fibers were evenly distributed within the soil matrix, and their orientation was random.

Specimens prepared in a fluid consistency state were compacted in oedometer and shear devices under a compaction pressure of $\sigma = 0.4$MPa. After stabilization of compaction deformation, the specimens were fully unloaded and left without load until rebound deformation had stabilized. After stabilization of the rebound deformation, one portion of the bentonitic and reddish clay specimens was tested for shear strength and shear creep without basalt fiber reinforcement, and the other portion was tested with basalt fiber reinforcement.

The physical and mechanical properties of the basalt fibers are presented in Table 13.1. Table 13.2 shows the mean values of the hydro-physical properties of soils after preliminary compaction.

In addition to determining the hydro-physical properties of the soils, the swelling pressure and swelling deformation were also evaluated. The swelling pressure (σ_{sw}) was determined by the compensation method, while the swelling deformation (δ_{sw}) was measured using an oedometer apparatus [6]. The results of the swelling pressure and relative swelling deformation tests for the reddish and bentonitic clays—both unreinforced and reinforced with basalt fibers - are presented in Table 13.3. The table provides the mean values of swelling deformation and swelling pressure for the reddish and bentonitic clays determined in the unreinforced state and with basalt fiber reinforcement.

When reinforcing reddish and bentonitic clays with basalt fibers, the high tensile strength of the fibers enables the three-dimensional reinforcing framework to resist tensile and compressive stresses that develop during hydration swelling of the soils.

Table 13.1: Properties of the basalt fibers.

Tensile strength MPa	Fiber diameter µm	Fiber length mm	Modulus of elasticity GPa	Density g/cm^2
3500	13-17	10,0-15,0	75,0	2,60

Table 13.2: properties of soils: a - reddish clays without basalt fiber reinforcement, b - reddish clays with basalt fiber reinforcement, c - bentonite clays without basalt fiber reinforcement and d - bentonite clays with basalt fiber reinforcement.

Types of soils	Density (g/cm^3) of			Porosity factor, e	Humidity, W	Plasticity index			Consis, I_L
	Soil, ρ	Mineral particles, ρ_s	Dry density, ρ_d			W_L	W_P	I_P	
a)	1,82	2,69	1,48	0,817	0,232	0,510	0,250	0,260	<0
b)	1,84	2,69	1,49	0,805	0,233	0,510	0,250	0,260	<0
c)	1,77	2,74	1,25	1,192	0,413	1,705	0,404	1,301	0,007
d)	1,79	2,74	1,27	1,157	0,413	1,705	0,404	1,301	0,007

Table 13.3: Results of the swelling pressure and relative swelling deformation tests.

Soil type	Swelling pressure, σ_{sw}, MPa	Relative swelling deformations, δ_{sw}
Reddish clays without basalt fiber reinforcement	0,1525	0,18
Reddish clays with basalt fiber reinforcement	0,025	0,0035
Bentonite clays without basalt fiber reinforcement	0,475	0,325
Bentonite clays with basalt fiber reinforcement	0,110	0,009

13.3 Results of Experimental Investigations of Shear Strength of Reddish and Bentonitic Clays, both Unreinforced and Reinforced with Basalt Fibers

To investigate the effect of reinforcement on the shear strength and shear creep of reddish and bentonitic clays, shear tests were conducted using the M-5 torsional shear apparatus [3, 4, 7]. Experiments repeated twice. This testing method provides the most reliable simulation of the stress-strain state of the soil; therefore, the shear strength and shear creep data obtained are more accurate compared to direct shear tests. The results of shear testing of reddish and bentonitic clays, using both unreinforced and basalt fiber-reinforced specimens, are shown in Figs. 13.1 and 13.2. For unreinforced reddish clays, the shear strength is determined by the following formula

$$\tau_f = 0,325\sigma + 0,0471. \tag{13.1}$$

For reddish clays with basalt fiber reinforcement, the shear strength is determined by the following formula

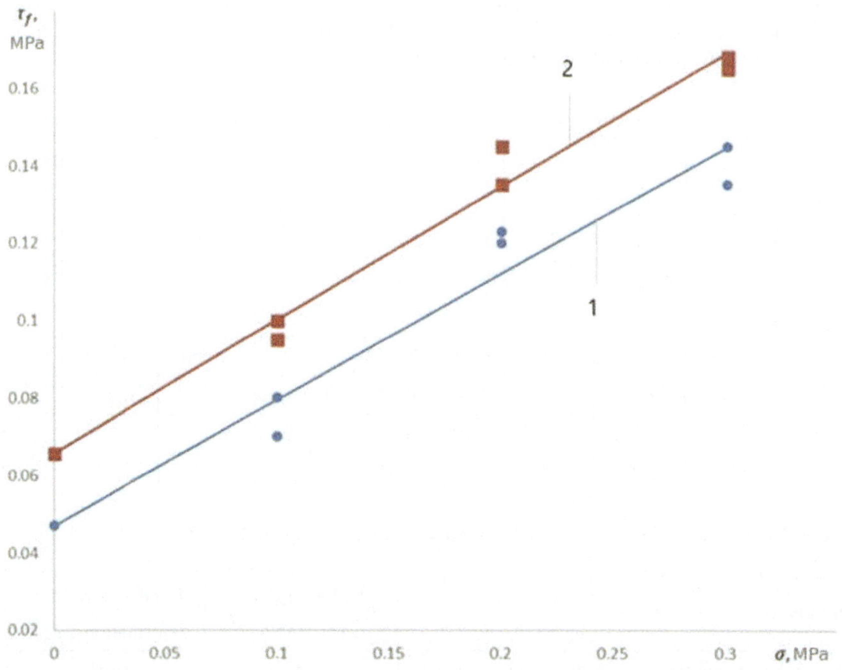

Fig. 13.1: Shear strength diagrams: 1 - unreinforced reddish clays, 2 - basalt fiber reinforced reddish clays.

$$\tau_f = 0,344\sigma + 0,0658. \tag{13.2}$$

For bentonite clays without reinforcement, the shear strength is determined by the following formula

$$\tau_f = 0,313\sigma + 0,0608. \tag{13.3}$$

For basalt fiber reinforced reddish clays, the shear strength is determined by the following formula

$$\tau_f = 0,323\sigma + 0,080. \tag{13.4}$$

Reinforcement of reddish and bentonitic clays results in increased shear strength, primarily due to enhanced cohesion, while the internal friction coefficient remains virtually unchanged. basalt fibers bind clay particles together within the soil, forming a three-dimensional reinforcing network. It follows from this that the basalt fibers within the soil bind the clay particles, forming a three-dimensional soil network. Under the application of shear stresses, this three-dimensional network restrains the displacement of the particles.

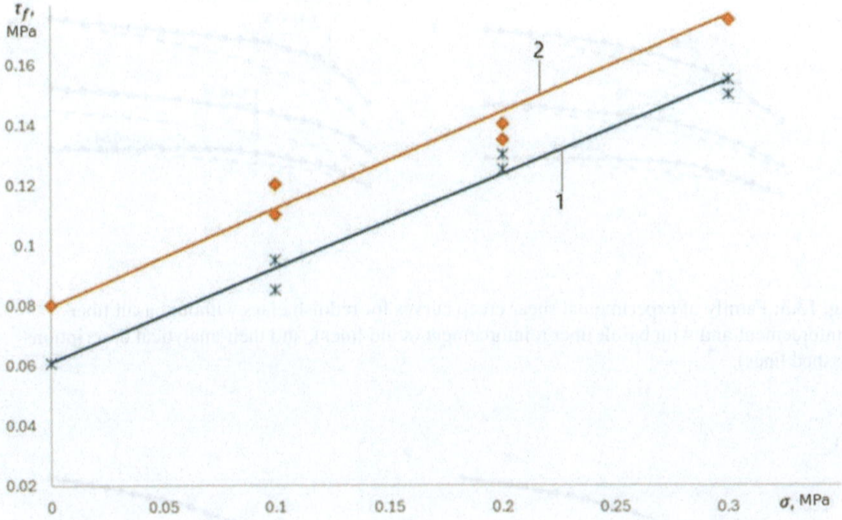

Fig. 13.2: Shear strength diagrams: 1 - bentonite clays without reinforcement, 2 - with basalt fiber reinforcement.

13.4 Results of Experimental Investigations on the Shear Creep of Reddish and Bentonite Clays with and without Basalt Fiber Reinforcement

Shear creep tests on reddish and bentonitic clays, both unreinforced and reinforced with basalt fibers, were conducted using the M-5 shear apparatus[3]. Prior to testing, specimens in a fluid-consistency state were compacted in the shear devices under a preliminary compacting pressure of $\sigma = 0.4$ MPa. After stabilization of compaction deformation, the specimens were fully unloaded and allowed to stand without load until stabilization of the rebound deformation. Once rebound deformation had stabilized, the reddish and bentonitic clay specimens - both without and with basalt fiber reinforcement - were recompacted under a compaction pressure of $\sigma = 0.3$ MPa.

For the reddish clays, the shear creep tests were performed at three different levels of relative shear stress: $\tau/\tau_{st} = 0,58;\ 0,70;\ 0,82$. For the bentonitic clays, the relative shear stress levels were $\tau/\tau_{st} = 0,54; 0,70;\ 0,830$. The shear creep tests for the reddish and bentonitic clays, both unreinforced and reinforced with basalt fibers, were conducted simultaneously at nearly identical relative shear stress levels.

The results of the shear creep tests for the reddish and bentonitic clays, with and without basalt fiber reinforcement, are presented in Figs. 13.3 and 13.4. The shear creep deformation at constant values of τ/τ_{st} is determined according to the following relationship [3]

$$\gamma\left[\left(t, f\left(\tau/\tau_{st}\right)\right)\right] = \omega\left(t\right) f\left(\tau/\tau_{st}\right), \tag{13.5}$$

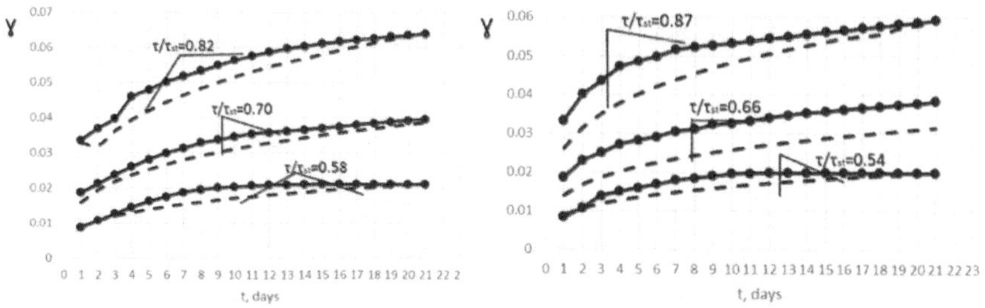

Fig. 13.3: Family of experimental shear creep curves for reddish clays without basalt fiber reinforcement and with basalt fiber reinforcement (solid lines), and their analytical descriptions (dashed lines).

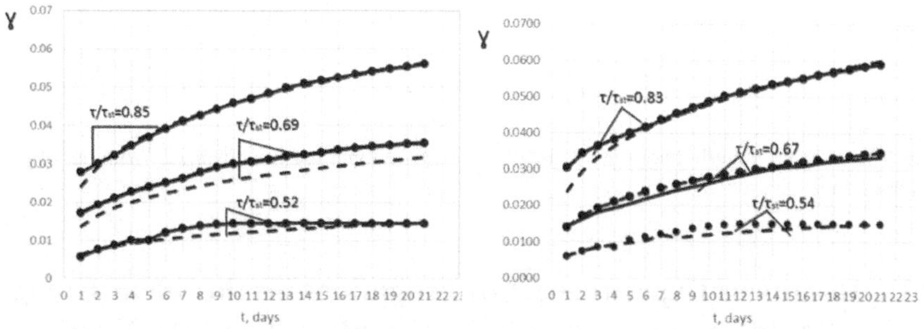

Fig. 13.4: Family of experimental shear creep curves for bentonitic clays without basalt fiber reinforcement and with basalt fiber reinforcement (solid lines), and their analytical descriptions (dashed lines) according to expressions (13.14) and (13.15).

where $\omega(t)$ is the shear creep measure (creep deformation), $f(\tau/\tau_{st})$ is the function of the relative shear stress, at $\tau/\tau_{st} = 1$, τ is the applied shear stress, τ_{st} is the standard shear strength, and τ/τ_{st} is the relative shear stress level.

To express the creep measure, the following power-law relationship was used

$$\omega(t) = At^n \tag{13.6}$$

and for the function of the shear stress level

$$f(\tau/\tau_{st}) = (\tau/\tau_{st})^m, \tag{13.7}$$

where A, n and m are experimentally determined parameters. It should be emphasized that the shear creep measure is a fictitious quantity. It is used only for calculating the shear creep deformation when $\tau/\tau_{st} < 1$.

For the reddish clays without basalt fiber reinforcement, the following expression for the shear creep measure was obtained

$$\omega\,(t, \tau/\tau_{st} = 0,58\,) = 0,0086\,t^{0,29484} \tag{13.8}$$

and for the shear stress level function at $t = 20$ days

$$\gamma\,[(t = 20\text{ day},\ f\,(\tau/\tau_{st}\,))] = 0,11975(\tau/\tau_{st}\,)^{3,20442}, \tag{13.9}$$

$$f\,(\tau/\tau_{st}\,) = 0,11975(\tau/\tau_{st}\,)^{3,20442}/0,0209, \tag{13.10}$$

$$f\,(\tau/\tau_{st}\,) = 5,730(\tau/\tau_{st}\,)^{3,20442}, \tag{13.11}$$

which satisfies the condition

$$f\,[(\tau/\tau_{st}\,) = 0,58] = 1.$$

Thus, for the shear creep deformation of reddish clays without basalt fiber reinforcement, the following relationship was obtained

$$\gamma\,[(t, f\,(\tau/\tau_{st}\,))] = 0,0086t^{0,29484} \cdot 5,730(\tau/\tau_{st}\,)^{3,20442}. \tag{13.12}$$

For the reddish clays reinforced with basalt fibers, an analogous analysis yielded

$$\gamma\,[(t, f\,(\tau/\tau_{st}\,))] = 0,0088t^{0,2673} \cdot 4,0606(\tau/\tau_{st}\,)^{2,275}. \tag{13.13}$$

For the shear creep deformation of bentonitic clays without and with basalt fiber reinforcement, the following relationships were obtained, respectively:

1. For bentonitic clay without reinforcement

$$\gamma\,[(t, f\,(\tau/\tau_{st}\,))] = 0,0060t^{0,30192} \cdot 7,229(\tau/\tau_{st}\,)^{3,20191}, \tag{13.14}$$

2. For bentonitic clay reinforced with basalt fibers

$$\gamma\,[(t, f\,(\tau/\tau_{st}\,))] = 0,0062t^{0,280275} \cdot 6,04895(\tau/\tau_{st}\,)^{2,754233}. \tag{13.15}$$

A comparison of the shear creep deformation values for the reddish and bentonitic clays, both unreinforced and reinforced with basalt fibers, shows that at equal levels of relative shear stress $(\tau/\tau_{st}\,)$ the creep deformations coincide (see Figs. 13.3 and 13.4).

13.5 Conclusions

Based on the experimental investigations, the following conclusions can be drawn:

1. Basalt fiber reinforcement at a volume content of 1% for the reddish clays and 2% for the bentonitic clays leads to an increase in shear resistance and a decrease in both the relative swelling deformation and swelling pressure. Due to its high

tensile strength, the basalt fiber and the resulting reinforcing framework inhibit the swelling of the soils.

2. Reinforcement of the reddish and bentonitic clays increases their shear strength, primarily due to an increase in cohesion, while the internal friction angle does not change significantly. The shear strength equation for clays reinforced with basalt fibers is given by expression (13.5).
3. At stress levels not exceeding the standard shear strength, the shear deformation of the soils occurs in the stages of decelerating and steady-state creep, without transitioning into progressive flow.
4. The shear creep of soils with and without basalt fiber reinforcement is practically identical at equal relative shear stress levels.

References

[1] Avakyan GS (1980) Betonity (in Russ., Bentonites). Science, Moscow
[2] Hayroyan SG (2005) Impact Shrinkable Fractuning of Shear Strength in Clay. Madrid, 2005, 6-14G-005.
[3] Meschyan SR (1995) Experimental Rheology of Clayey Soil, A.A. Balkema Publishers, Rotterdam
[4] Meschyan SR, Hayroyan SG (1987) Vliyaniye nachal'nogo sostoyaniya na soprotivleniye sdvigu nabukhayushchego bentonitovogo grunta (in Russ., Effect of the initial state on the shear resistance of swelling bentonite soil). Reports of the Academy of Sciences of the Armenian SSR LXXXIV:119-123
[5] Sargsyan O.A. Regional'naya geotektonika Armenii (in Russ., Regional Geotectonics of Armenia). Yerevan State University, Yerevan
[6] Sorochan EA (1989) Stroitel'stvo sooruzheniy na nabukhayushchikh gruntakh (in Russ., Construction of Structures on Swelling Soils). Stroyizdat, Moscow
[7] Trofimov VT, Korolev VA, Voznesensky EA, Golodkovskaya GA, Vasilchuk YK, Ziangirov RS (2005) Gruntovedeniye (in Russ., Soil Science. Moscow State University Publisher, Nauka

Chapter 14
Forced Shear Oscillations of a Piecewise-Homogeneous Half-Space with a Periodic System of Interphase Thin Inclusions Partially Detached from the Matrix

Vahram N. Hakobyan, Harutyun A. Amirjanyan, and Sona V. Sahakyan

Abstract The paper considers stationary shear oscillations of a piecewise homogeneous half-space obtained by connecting a homogeneous layer and a halfspace with a periodic system of interphase strip thin absolutely rigid inclusions, one of the long sides of which is torn off from the matrix, forming a crack.

It is assumed that the half-space is deformed under the action of periodically time-varying concentrated loads acting on the inclusions. The governing system of the second-kind singular integral equations is written for two specific combinations of contact stress amplitudes acting under the inclusion and the dislocation displacements of the crack edge points.

Simple formulas for determining the intensity factors at end points of crack and crack openings were obtained. The solution to the governing equations was constructed using the numerical-analytical method of mechanical quadratures. Numerical calculations were performed, and the patterns of change in important mechanical characteristics of the problem were studied depending on the physical, mechanical, and geometric characteristics of the problem.

Key words: Dynamic mixed boundary value problem, Interphase inclusion, Crack, Vibrations

14.1 Introduction

The study of forced stationary vibrations of homogeneous and piecewise homogeneous deformable massive bodies containing defects such as cracks and absolutely rigid inclusions is highly relevant from both theoretical and practical points of view.

Vahram N. Hakobyan (corresponding author) · Harutyun A. Amirjanyan · Sona V. Sahakyan
Institute of Mechanics, National Academy of Sciences of the Republic of Armenia, 24b Marshal Baghramyan Ave, Yerevan, 0019, Republic of Armenia,
e-mail: vhakobyan@sci.am, amirjanyan@gmail.com, sonasahakyan1999@gmail.com

H. Altenbach and A. V. Sahakyan, *Contact Mechanics and Relevant Problems*,
Advanced Structured Materials 253, https://doi.org/10.1007/978-3-032-20869-9_14

The effective methods for solving dynamic problems for layered media with various interphase defects have been developed and proposed, and a number of topical and interesting problems in this direction have been posed and solved in [1–4] and [5]. On the other hand, few works are devoted to the study of the stress state of homogeneous and piecewise homogeneous deformable massive bodies simultaneously containing different types of defects and the identification of patterns of their mutual influence depending on the physical, mechanical and geometric parameters of the problems. Similar tasks are often encountered in seismology, earthquake-resistant construction, seismic exploration and flaw detection.

A series of anti-plane and plane problems on forced vibrations of a composite half-space and a composite plane with interphase cracks and completely or partially coupled absolutely rigid inclusions are considered in [6] and [7–10]. The pattern of mutual influence of these stress concentrators are revealed. Problems of shear forced oscillations of a composite half-space with a periodic system of interphase inclusions partially detached from the matrix are considered.

14.2 Problem Statement and Governing Equaiton

Let us consider the shear vibrations of a piecewise homogeneous half-space, formed by layer of thickness h and a half-space made of different materials with shear moduli μ_1 and μ_2, respectively. The half-space contains a periodic system of interphase rigid, thin strip inclusions with a period of $2l$. One of the long sides of inclusions has detached from the matrix, forming main cracks.

It is assumed that in the Cartesian coordinate system $Oxyz$, the Ox axis is directed along the line of junction of the layer and the half-space. On the plane of the junction $y = 0$ the cracks occupy the region

$$\left\{ y = 0;\ -\infty < z < \infty, \qquad x \in L = \bigcup_{n=-\infty}^{\infty} (2nl - a < x < a + 2nl) \right\}.$$

The upper banks of cracks are free from loads, and an absolutely rigid thin inclusions are fused on the lower banks of cracks. It is also assumed that the composite half-space is deformed by the intensity of periodically changing shear concentrated loads $T_0 e^{i\omega t}$ applied to the inclusions (Fig. 14.1)

It is necessary to determine the pattern of changes in contact stresses under inclusions, crack opening and the intensity factor of fracture stresses at the end points of the crack depending on the physical, mechanical and geometric characteristics, as well as frequency of forced vibrations. The problem in the Oxy plane is equivalent to the following boundary-value problem:

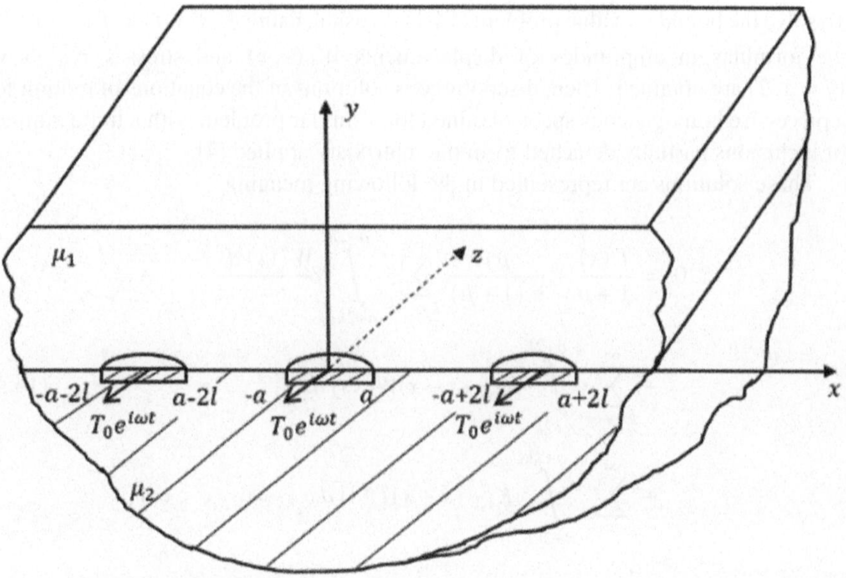

Fig. 14.1: Geometry and loading of the cracked area.

$$
\begin{aligned}
&W_1(x,0,t) = W_2(x,0,t) && (x \notin L),\\
&\tau_{yz}^{(1)}(x,0,t) = \tau_{yz}^{(2)}(x,0,t) && (x \notin L,\\
&\tau_{yz}^{(1)}(x,h,t) = 0 && (-\infty < x < \infty),\\
&\tau_{yz}^{(1)}(x,\ +0,t) = 0 && (x \in L),\\
&W_2(x,-0,t) = \text{const.}\ \mathrm{e}^{i\omega t} && (x \in L).
\end{aligned}
\tag{14.1}
$$

Here $W_j(x,y,t)$ $(j = 1,2)$ are, respectively, the displacements of the points of the layer and half-spaces in the direction of the Oz axis, satisfying, each in the domain, the equation of motion

$$
\frac{\partial^2 W_j}{\partial x^2} + \frac{\partial^2 W_j}{\partial y^2} = \frac{1}{\left(c_2^{(j)}\right)^2}\frac{\partial^2 W_j}{\partial t^2},
\tag{14.2}
$$

where $c_2^{(j)}$ $(j = 1,2)$ are the propagation speeds of shear waves in the layer and half-space, respectively, and $\tau_{yz}^{(j)}(x,y,t)$ $(j = 1,2)$ are the shear stresses acting in the layer and half-space related with the displacements by known formulas

$$
\tau_{yz}^{(j)}(x,y,t) = \mu_j \frac{\partial W_j(x,y,t)}{\partial y} \quad (j = 1,2).
\tag{14.3}
$$

To solve the boundary value problem (14.1), as usual, using $f(x,y,t) = f(x,y)\,\mathrm{e}^{i\omega t}$ the formulas in amplitudes of displacements $W_j(x,y)$ and stresses $\tau_{yz}^{(j)}(x,y)$ $(j = 1,2)$ are obtained. Then, discontinuous solutions of the equations of motion for a piecewise homogeneous space obtained for a similar problem with a finite number of inclusions partially detached from the matrix are applied [7].

These solutions are represented in the following meaning

$$
\begin{aligned}
\tau_{yz}^{(1)}(x,0) &= \frac{T(x)}{1+\mu} + \frac{\mu_2}{\pi(1+\mu)} \sum_{k=-\infty}^{\infty} \int_{-a+2kl}^{a+2kl} \frac{W'(s)\,ds}{s-x} \\
&+ \sum_{k=-\infty}^{\infty} \int_{-a+2k}^{a+2k} K_{1,1}(s-x) W'(s)\,ds \\
&+ \sum_{k=-\infty}^{\infty} \int_{-a+2kl}^{a+2kl} K_{1,2}(s-x) T(s)\,ds \quad (-\infty < x < \infty),
\end{aligned} \tag{14.4}
$$

$$
\begin{aligned}
W'_2(x,0) &= -\frac{W'(x)}{1+\mu} - \frac{1}{\pi\mu_1(1+\mu)} \sum_{k=-\infty}^{\infty} \int_{-a+2kl}^{a+2kl} \frac{T(s)\,ds}{s-x} \\
&+ \sum_{k=-\infty}^{\infty} \int_{-a+2kl}^{a+2kl} K_{2,1}(s-x) W'(s)\,ds \\
&+ \sum_{k=-\infty}^{\infty} \int_{-a+2kl}^{a+2kl} K_{2,2}(s-x) T(s)\,ds \quad (-\infty < x < \infty)
\end{aligned}
$$

The following notations are introduced here

$$
\begin{aligned}
& W_1(x,0) - W_2(x,0) = W(x) \quad (x \in L), \\
& \tau_{yz}^{(1)}(x,0) - \tau_{yz}^{(2)}(x,0) = T(x) \quad (x \in L), \\
& K_{ij}(t) = \frac{1}{\pi} \int_{-\infty}^{\infty} \frac{Q_{ij}(s)}{\Delta(s)} \mathrm{e}^{ist}\,ds \quad (i,j = 1,2), \\
& Q_{11}(s) = \frac{\mu_2 \left[2(1+\mu)\chi_1(s)\chi_2(s)\sinh(\chi_1 h) - |s|\,\Delta(s)\right]}{2i(1+\mu)s}, \\
& Q_{12}(s) = Q_{21}(s) = \frac{\mu\left[2(1+\mu)\chi_2(s)\cosh(\chi_1 h) - \Delta(s)\right]}{2(1+\mu)}, \\
& Q_{22}(s) = -\frac{\left[2(1+\mu)s\cosh(\chi_1 h) - \operatorname{sign}(s)\,\Delta(s)\right]}{2i\mu_1(1+\mu)}, \\
& \Delta(s) = 2\left[\chi_1(s)\sinh(\chi_1 h) + \mu\chi_2(s)\cosh(\chi_1 h)\right], \quad \mu = \mu_2/\mu_1.
\end{aligned}
$$

In this case, those branches of the multi-valued function

$$\chi_j(s) = \sqrt{s^2 - k_j^2}, \qquad \left(k_j = \omega / c_2^j,\ j = 1, 2\right)$$

are chosen that behave at infinity like $|s|$ $(j = 1, 2)$ [11].

Note that the given solutions satisfy the first three conditions of the boundary value problem (14.1) written in the amplitudes. Using representations (14.4), we satisfy the last two conditions of (14.1) by differentiating the last of them with respect to x. As a result, to determine the contact stresses $T(x)$ under inclusions and the dislocation displacements $W'(x)$ of the crack edge points, the following system of singular integral equations is obtained:

$$\begin{gathered}
\frac{T(x)}{1+\mu} + \frac{\mu_2}{\pi(1+\mu)} \sum_{k=-\infty}^{\infty} \int_{-a+2kl}^{a+2kl} \frac{W'(s)\,ds}{s-x} \\
+ \sum_{k=-\infty}^{k} \int_{-a+2k}^{a+2k} K_{1,1}(s-x) W'(s)\,ds + \sum_{k=-\infty}^{\infty} \int_{-a+2k}^{a+2k} K_{1,2}(s-x) T(s)\,ds = 0, \\
-\frac{W'(x)}{1+\mu} - \frac{1}{\pi \mu_1 (1+\mu)} \sum_{k=-\infty}^{\infty} \int_{-a+2kl}^{a+2kl} \frac{T(s)\,ds}{s-x} \\
+ \sum_{k=-\infty}^{\infty} \int_{-a+2kl}^{a+2kl} K_{2,1}(s-x) W'(s)\,ds + \sum_{k=-\infty}^{\infty} \int_{-a+2kl}^{a+2kl} K_{2,2}(s-x) T(s)\,ds = 0, \\
(-a + 2ml < x < a + 2ml;\ \ m \in R).
\end{gathered} \tag{14.5}$$

We will consider system (14.1) with equations of motion of inclusions without taking into account their mass and continuity of displacements at the end points of the crack:

$$\int_{-a+2kl}^{a+2kl} T(x)\,dx = T_0; \quad \int_{-a+2kl}^{a+2kl} W'(x)\,dx = 0 \quad (k \in R). \tag{14.6}$$

Using the periodicity of the unknown function, by changing the variables $s = \xi + 2kl$, $x = \eta + 2ml$, we write the system (14.5) and conditions (14.6) on the interval $(-a, a)$. Then, changing the places of the integrals in the representations of the function $K_{ij}(\xi + 2kl, \eta + 2ml)$ with summation signs and taking into account the known relations [12]

$$\sum_{k=-\infty}^{\infty} \frac{1}{\xi - \eta + 2kl} = \frac{\pi}{2l} \cot \frac{\pi(\xi - \eta)}{2l}, \quad \sum_{k=-\infty}^{\infty} e^{2ikls} = \frac{\pi}{l} \sum_{k=-\infty}^{\infty} \delta\left(s + \frac{\pi k}{l}\right),$$

where $\delta(\xi)$ is the known Dirac delta function, we rewrite the system (14.5) and conditions (14.6) in the form

$$\begin{gathered}
\frac{T(\eta)}{1+\mu}+\frac{\mu_2}{2l(1+\mu)}\int_{-a}^{a}\cot\frac{\pi(\xi-\eta)}{2l}W'(\xi)\,d\xi \\
+\int_{-a}^{a}K_{1,1}^*(\xi-\eta)W'(\xi)\,d\xi+\int_{-a}^{a}K_{1,2}^*(\xi-\eta)T(\xi)\,d\xi=0, \\
-\frac{W'(\eta)}{1+\mu}-\frac{1}{2l\mu_1(1+\mu)}\int_{-a}^{a}\cot\frac{\pi(\xi-\eta)}{2l}T(\xi)\,d\xi \\
+\int_{-a}^{a}K_{2,1}^*(\xi-\eta)W'(\xi)\,d\xi+\int_{-a}^{a}K_{2,2}^*(\xi-\eta)T(\xi)\,d\xi=0, \\
(-a<x<a),
\end{gathered}\tag{14.7}$$

$$\int_{-a}^{a}T(x)\,dx=T_0;\quad\int_{-a}^{a}W'(x)\,dx=0.\tag{14.8}$$

Here

$$K_{ij}^*(t)=\frac{1}{l}\begin{cases}\dfrac{Q_{ij}(0)}{\Delta(0)}+\displaystyle\sum_{k=1}^{\infty}\frac{2Q_{ij}(\pi k/l)\cos\left(\frac{\pi k}{l}t\right)}{\Delta(\pi k/l)} & (i\neq j=1,2),\\[2ex] i\displaystyle\sum_{k=1}^{\infty}\frac{2Q_{ij}(\pi k/l)\sin\left(\frac{\pi k}{l}t\right)}{\Delta(\pi k/l)} & (i=j=1,2).\end{cases}$$

To solve the system of integral equations (14.7) under the condition (14.8), we multiply the first equation by $(-1)^j\sqrt{\mu}/\mu_2$ $(j=1,2)$ and sum it with the second. Next, introducing new dimensionless unknown functions using the formulas

$$\varphi_j(\eta)=W'(\eta)+(-1)^{j+1}\sqrt{\mu}T(\eta)/\mu_2\quad(j=1,2)\tag{14.9}$$

and representing the singular part of the obtained equations in the form of the Cauchy kernel, the following system of singular integral equations is obtained

$$\begin{gathered}
\varphi_j(\eta)+\frac{(-1)^{j+1}\sqrt{\mu}}{\pi}\int_{-a}^{a}\frac{\varphi_j(\xi)\,d\xi}{\xi-\eta}+\sum_{i=1}^{2}\int_{-a}^{a}R_{j,i}(\xi-\eta)\varphi_i(\xi)\,d\xi=0, \\
(-a<\eta<a;\ j=1,2).
\end{gathered}\tag{14.10}$$

In this case, conditions (14.8) will be written as

$$\int_{-a}^{a}\varphi_i(\xi)\,d\xi=(-1)^{i+1}\sqrt{\mu}T_0/\mu_2\ \ (i=1,2).\tag{14.11}$$

The following notations are introduced here:

$$R_{11}(t) = -\frac{(1+\mu)}{2}\left[-\frac{\sqrt{\mu}}{\mu_2}K^*_{1,1}(t) + \frac{\mu_2}{\sqrt{\mu}}K^*_{2,2}(t)\right] + \frac{\sqrt{\mu}}{2l}\left[\cot\frac{\pi t}{2l} - \frac{2l}{\pi t}\right],$$

$$R_{12}(t) = -\frac{(1+\mu)}{2}\left[-\frac{\sqrt{\mu}}{\mu_2}K^*_{1,1}(t) + K^*_{2,1}(t) + K^*_{1,2}(t) - \frac{\mu_2}{\sqrt{\mu}}K^*_{2,2}(t)\right],$$

$$R_{21}(t) = -\frac{(1+\mu)}{2}\left[\frac{\sqrt{\mu}}{\mu_2}K^*_{1,1}(t) + K^*_{2,1}(t) + K^*_{1,2}(t) + \frac{\mu_2}{\sqrt{\mu}}K^*_{2,2}(t)\right],$$

$$R_{22}(t) = -\frac{(1+\mu)}{2}\left[\frac{\sqrt{\mu}}{\mu_2}K^*_{1,1}(t) - \frac{\mu_2}{\sqrt{\mu}}K^*_{2,2}(t)\right] - \frac{\sqrt{\mu}}{2l}\left[\cot\frac{\pi t}{2l} - \frac{2l}{\pi t}\right].$$

Thus, the solution to the problem is reduced to solving the system of integral equations (14.10) under conditions (14.11). We will construct solutions to the system using numerical-analytical method of mechanical quadratures. For this purpose, using a change of variables, we write system (14.10) on the interval $(-1, 1)$

$$\psi_j(x) + \frac{(-1)^{j+1}\sqrt{\mu}}{\pi}\int_{-1}^{1}\frac{\psi_j(s)\,ds}{s-x} + \sum_{i=1}^{2}\int_{-1}^{1}R^*_{j,i}(s-x)\psi_i(s)\,ds = 0, \tag{14.12}$$

$$(-1 < x < 1;\quad j = 1, 2).$$

In this case, conditions (14.11) will take the form

$$\int_{-1}^{1}\psi_j(\xi)\,d\xi = (-1)^{i+1}T_0^* \quad (j = 1, 2). \tag{14.13}$$

Here

$$\psi_j(x) = \varphi_j(ax),\; R^*_{j,i}(x) = aR_{j,i}(ax),\; T_0^* = \sqrt{\mu}T_0/a\mu_2.$$

As in [7], it is not difficult to verify that the unknown functions at the points ± 1 have a power singularity and they can be represented in the following forms [13]

$$\psi_1(t) = \frac{\psi_1^*(t)}{(1+t)^{\gamma}(1-t)^{1-\gamma}},\quad \psi_2(t) = \frac{\psi_2^*(t)}{(1+t)^{1-\gamma}(1-t)^{\gamma}}, \tag{14.14}$$

where

$$\gamma = \frac{1}{2\pi}\arctan\frac{2\sqrt{\mu}}{1-\mu}\;(\mu < 1),\quad \gamma = \frac{1}{2} - \frac{1}{2\pi}\arctan\frac{2\sqrt{\mu}}{\mu-1}\;(\mu > 1).$$

$\psi_j^*(x)$ be continuous functions, bounded up to the ends of the interval $[-1, 1]$. Substituting the values of the functions $\psi_j(x)$ $(j = 1-2)$ into system (14.12) and conditions (14.13), using the relationships given in [13], according to the standard procedure, the following system of $2n$ algebraic equations with respect to the values $\psi_j^*(\xi_i)$ $\left(j = 1-2;\ i = \overline{1,n}\right)$ is obtained.

After determining the functions $\psi_j^*(\xi_i)$ using the Lagrange interpolation polynomial, the functions $\psi_j(x)$ $(j = 1,2)$ are restored and all important characteristics for the stress-strain state in a piecewise homogeneous half-space are obtained. In this case, the reduced fracture stresses are written in the form

$$\frac{\tau_{yz}^{(j)}(at,0)}{\mu_2} = \frac{1}{2\pi(1+\mu)} \int_{-1}^{1} \frac{[\psi_1(\xi) + \psi_2(\xi)]\, d\xi}{\xi - t} + F_j(t) \quad (|t| > 1,\ j = 1,2), \tag{14.15}$$

where the functions $F_j(t)$ $(j = 1,2)$ are bounded at both endpoints of the interval $(-1,1)$.

Substituting into (14.15) the value of the functions $\psi_1(t)$ and $\psi_2(t)$ from (14.14) and taking into account the relation [14]

$$\int_a^b \frac{(x-a)^{\alpha-1}(b-x)^{-\alpha}}{x-y} dx = \frac{\pi}{(b-y)\sin\pi\alpha} \left|\frac{a-y}{b-y}\right|^{\alpha-1}$$

$$(0 < \operatorname{Re}\alpha < 1,\quad y < a < b;\quad a < b < y).$$

The dimensionless stresses $\tau_*^{(j)}(x)$ are represented in the form

$$\tau_*^{(j)}(x) = \frac{\tau_{yz}^{(j)}(t,0)}{\mu_2} = -\frac{\operatorname{sign}(t)}{2(1+\mu)\sin\pi\gamma}$$
$$\times \left\{ \frac{\psi_1^*(\pm 1)}{|1+x|^{\gamma}|1-x|^{1-\gamma}} + \frac{\psi_2^*(\pm 1)}{|1+x|^{1-\gamma}|1-x|^{\gamma}} \right\} + F_j^*(t),$$

where $F_j^*(t)$ $(j = 1,2)$ are the bounded functions at both ends of the crack, the values of which are not given here due to their cumbersomeness.

Therefore, just as $\gamma < 1/2$ the formulas for determining the dimensionless intensity factors of fracture stresses fracture at the end points of a crack and crack opening in the case under consideration take the following forms

$$K_{III}^*(a) = \sqrt{2\pi} \lim_{t\to 1+0} (t-1)^{1-\gamma} \frac{\tau_{yz}^{(j)}(at,0)}{\mu_2} = -\frac{2^{-\gamma-1/2}\sqrt{\pi}\psi_1^*(1)}{(1+\mu)\sin\pi\gamma},$$

$$K_{III}^*(-a) = \sqrt{2\pi} \lim_{t\to -1-0} (t+1)^{1-\gamma} \frac{\tau_{yz}^{(j)}(at,0)}{\mu_2} = \frac{2^{-\gamma-1/2}\sqrt{\pi}\psi_2^*(-1)}{(1+\mu)\sin\pi\gamma},$$

$$K_{III}(\pm a, t) = K_{III}^*(\pm a)\, e^{i\omega t} = \left|K_{III}^*(\pm a)\right| e^{i\left(\omega t + \arg K_{III}^*(\pm a)\right)},$$

$$V(t) = w(at)/a = \frac{1}{2} \int_{-1}^{t} [\psi_1(\xi) + \psi_2(\xi)]\, d\xi.$$

To determine the dimensionless contact stresses acting under the inclusion the following formula is obtained:

$$\tau(t) = \frac{\tau_{yz}^{(2)}(at,0)}{\mu_2} = \frac{T(at)}{\mu_2} = \frac{\psi_1(t) - \psi_2(t)}{2\sqrt{\mu}} \qquad (-1 < t < 1).$$

14.3 Some Numerical Calculations

Numerical calculations were carried out and the patterns of change in reduced contact stresses, intensity factors of fracture stresses and crack opening were studied depending on the ratio $L = l/a$ in the case $/h = 1$; $c_2^{(1)}/c_2^{(2)} = 2$, $\omega/c_2^{(1)} = 0.5$ $T_0^* = 0.1$, $\mu = 1.5$. The results of the calculations are presented in Table 14.1 and Figs. 14.2 and 14.3. Table 14.1 displays the values of the amplitude modules of the dimensionless intensity factors of fracture stresses and their arguments at the end points of the crack. It becomes clear that with an increase of reduced period L, the modulus of the intensity factor of fracture stresses oscillates and decreases, tending to a certain limit corresponding to the value of the intensity factor of fracture stresses in the case of one inclusion partially detached from the matrix, and their arguments change sinuously. Figures 14.2 and 14.3 show plots of the absolute value of crack opening amplitude and dimensionless amplitudes of contact stresses under inclusion in the case when $L = 1.1, 2, 6, 11$. It is evident that with an increase of reduced period L, the absolute value of the crack opening amplitude decreases while

Table 14.1: Dependence of intensity factors of fracture stresses on distance between stamps.

L	1.1	1.5	2	5	10	50
$\lvert K_{III}^*(\pm 1)\rvert$	0.12112	0.03209	0.02392	0.02593	0.01958	0.00794
$\arg K_{III}^*(\pm 1)$	0.31386	0.54159	0.54553	0.08378	0.17853	1.72722

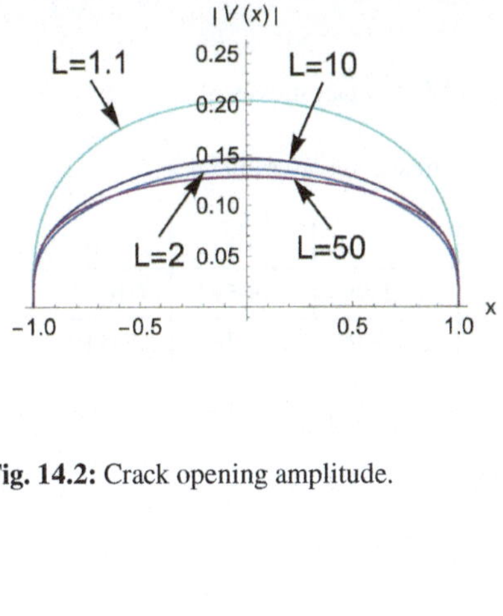

Fig. 14.2: Crack opening amplitude.

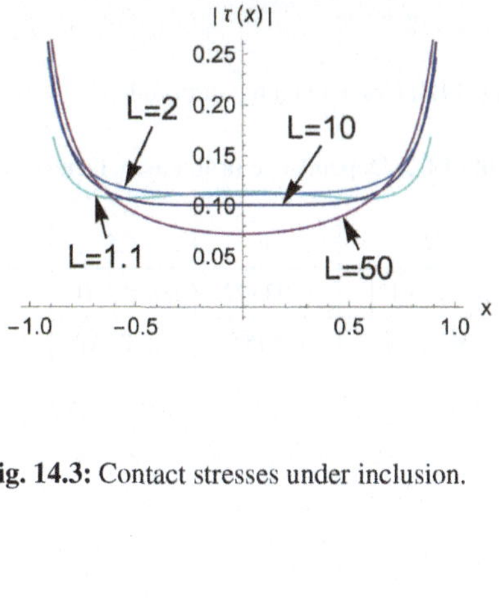

Fig. 14.3: Contact stresses under inclusion.

oscillating, tending to a certain limit corresponding to the absolute value of the crack opening amplitude in the case where there is only one interphase inclusion partially detached from the matrix. In this case, contact stresses also decrease in the central part of the contact zone, while they increase at the edges of the contact zone.

Figures 14.4 and 14.5 display plots of the absolute values of the crack opening amplitude and the dimensionless amplitudes of contact stresses under inclusion for different values of the reduced layer height $H = h/a$ in the case when $L = 2$, $c_2^{(1)}/c_2^{(2)} = 2/3$, $\omega/c_2^{(1)} = 0.5$ $T_0^* = 0.1$, $\mu = 2$. Figure 14.4 shows that the absolute value of the crack opening amplitude first increases with increasing layer height and then decreases, tending to the value corresponding to the absolute value of the crack opening amplitude in the case where the layer becomes a half-space. It is evident from Fig. 14.5 that with an increase in the layer height, the absolute value of the contact stresses in the middle part of the contact zone initially decreases and then increases, tending to the value corresponding to the absolute value of the amplitude of the contact stresses in the case where the layer becomes a half-space.

The values of the amplitude moduli of the intensity factors of fracture stresses at the end points of the crack for different values of the reduced layer height are represented in Table 14.2. It is clear that with an increase in the parameter H, which can be interpreted as an increase in the layer height h with a constant crack length $2a$, the absolute values of the amplitudes of the intensity factors of fracture stresses change sinuously, and their arguments, starting from the value $H = 2$, increase.

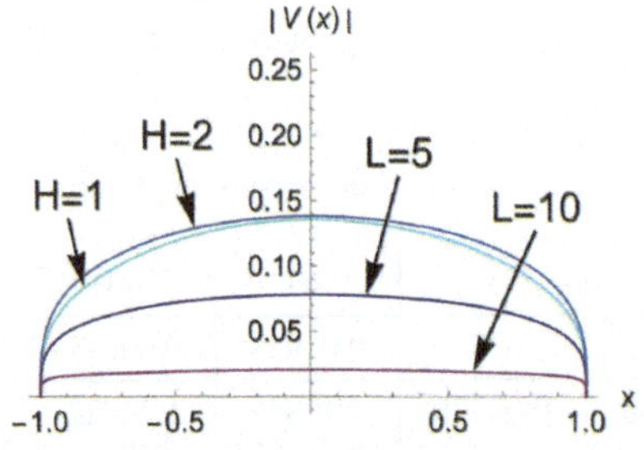

Fig. 14.4: Crack opening amplitude.

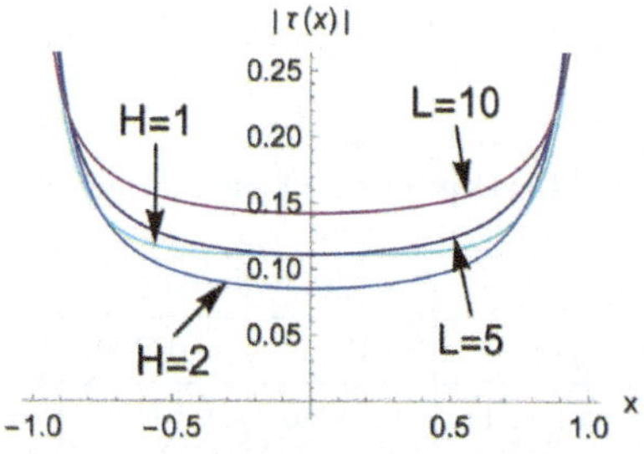

Fig. 14.5: Contact stresses.

Table 14.2: Dependence of intensity factors of fracture stresses on layer thickness.

H	1	1.5	2	5	10	50
$\lvert K_{III}^*(\pm 1)\rvert$	0.02392	0.01560	0.02079	0.02399	0.06572	0.01562
$\arg K_{III}^*(\pm 1)$	0.5455	0.78131	0.51455	2.62836	2.93281	2.33105

14.4 Conclusion

The forced vibrations of a piecewise homogeneous half-space of heterogeneous layers and a half-space with a periodic system of absolutely rigid thin interphase inclusions partially detached from the matrix and subjected to time-harmonic concentrated loads were studied. Using numerical-analytical method of mechanical quadratures, the features of the patterns of change in the absolute values of the modules of contact stresses under inclusions, the coefficients of fracture stresses at the end points of cracks formed during the detachment of inclusions and the opening of cracks depending on the values of the reduced period of inclusions and the reduced layer height were identified. It is shown that with an increase in the reduced period of inclusions, both the values of the factors of fracture stresses and the absolute values of contact stresses and crack opening tend to the values of these quantities in the case where there is only one inclusion.

Acknowledgements The study was carried out with the financial support of the Higher Education and Science Committee RA of MoESCS within the framework of the scientific project 25RG-2C221.

References

[1] Babeshko VA (1989) K probleme dinamicheskogo razrusheniya treshchinovatykh sloistykh tel (in Russ., Problem of dynamic fracture of cracked layered bodies). Doklady Akademii Nauk SSSR 207(2):324-327

[2] Babeshko VA (2000) Sredy s neodnorodnostyami (sluchay sovokupnostey vklyucheniy i treshchin) (in Russ., Media with Heterogeneities (a set of inclusions and heterogeneities)). Izvestiya Rossiyskoy Akademii Nauk. Mechanika Tverdogo Tela (3):5-9

[3] Babeshko VA, Pavlova AV, Ratner SV, Williams RT (2002) K resheniyu zadachi o vibratsiiuprugogo tela, soderzhashchego sistemu vnutrennikh polostey (in Russ., Solving the vibration problem for an elastic body containing internal cavities). Doklady Akademii Nauk SSSR 382(5):625-628

[4] Pryakhina OD, Smirnova AV (2004) An effective method of solving dynamic problems for layered media with discontinuous boundary conditions. Journal of Applied Mathematics and Mechanics 68(3):447-453, DOI 10.1016/S0021-8928(04)00059-0

[5] Abrahamyan BL (1998) Prostranstvennyye zadachi teorii uprugosti (in Russ., Spatial Problems of the Theory of Elasticity). NAS RA

[6] Hakobyan VN (2022) Stress Concentrators in Continuous Deformable Bodies, Advanced Structured Materials, Vol 181. Springer, Cham, DOI 10.1007/978-3-031-16023-3

[7] Hakobyan VN, Amirjanyan HA, Dashtoyan LL (2020) Forced shift vibrations of stamp on the border of composite half-space with interfacial partially de-

tached thin inclusions from the matrix. Journal of Physics: Conference Series 1474:012016, DOI 10.1088/1742-6596/1474/1/012016

[8] Hakobyan VN, Sahakyan AV, Sargsyan AH (2005) The plane deformation state of elastic plane with finite rigid inclusion under harmonic loading. Proceedings of the Twelfth International Conference on Composites or Nano Engineering. ICCE-12 August 1-6, Spain

[9] Hakobyan VN, Sargsyan AO (2006) On One Dynamic Mixed Problem for a Composite Space with a Crack under Antiplane Deformation. In: Selected Problems of the Theory of Elasticity, Plasticity and Creep, ded. 75th Anniversary of Academician M.A. Zadoyan. Yerevan, Gitutyun, pp 50-56

[10] Amirjanyan H, Hakobyan L (2018) Antiplane dynamic contact problem for compound half-space with interphase crack. Proceedings of IX International Conference 01-06 October Goris, pp 44-48

[11] Noble B (1958) Methods Based on the Wiener-Hopf Technique for the Solution of Partial Differential Equations. Pergamon Press, New York

[12] Vladimirov VS (1976) Obobshchennyye funktsii v matematicheskoy fizike (in Russ., Generalized Functions in Mathematical Physics). Nauka, Moscow

[13] Sahakyan AV, Amirjanyan HA (2018) Method of mechanical quadratures for solving singular integral equations of various types. IOP Conf. Series: Journal of Physics: Conf. Series 991:012070, DOI 10.1088/1742-6596/991/1/012070

[14] Prudnikov AP, Brychkov YA, Marichev OI (1981) Integraly i ryady (in Russ., Integrals and Series). Nauka, Moscow

Chapter 15
Vibrations of a Transversally Isotropic Plate on a Coupled Thermoelasticity Problem

Samvel H. Jilavyan, Tatevik G. Vasilyan, Karen S. Sargsyan, and Tigran A. Galichyan

Abstract Thermoelastic transverse vibrations of a transversally isotropic rectangular plate are examined with allowance for transverse shear deformations. In the formulation of the vibration problem, thermoelastic dissipation of mechanical energy is taken into account; that is, the equations and boundary conditions of the coupled thermoelasticity problem are adopted on the basis of a refined theory of anisotropic plates. Heat exchange occurs between the surfaces of the simply supported plate and the external environment, and transverse thermoelastic oscillations of the plate under a moving linear heat source are considered. The characteristic features of the vibrational process under this formulation of the thermoelasticity problem are identified.

Key words: Vibrations, Coupled thermoelasticity, Energy dissipation, Plate, Refined plate theory, Transverse shear deformation

15.1 Introduction

The magnitudes of thermal stresses, and consequently the resulting strains, in an elastic body are of fundamental importance for ensuring the reliability of individual structural elements and the structural system as a whole. Numerous engineering and technological problems have been described that investigate the stress–strain state of structural elements and the fluctuations arising in these elements due to nonuniform

Samvel H. Jilavyan (corresponding author)
Faculty of Mathematics and Mechanics, Yerevan State University, Alex Manoogian 1, 0025, Yerevan, Republic of Armenia, e-mail: samjilavyan@ysu.am

Tatevik G. Vasilyan · Karen S. Sargsyan · Tigran A. Galichyan
Institute of Mechanics, National Academy of Sciences of the Republic of Armenia, 24b Marshal Baghramyan Ave, Yerevan, 0019, Republic of Armenia, e-mail: tat.vasilyan@gmail.com, sarkaren1999@gmail.com, galichyantigran@gmail.com

H. Altenbach and A. V. Sahakyan, *Contact Mechanics and Relevant Problems*,
Advanced Structured Materials 253, https://doi.org/10.1007/978-3-032-20869-9_15

heating in the presence of large thermal fluxes and fields. It is necessary to perform physical and mathematical modelling of the physicomechanical and technological problems encountered in practice when elastic structures operate under high thermal and mechanical loads.

Thermoelastic deformations are most often non-stationary and dynamic in nature, arising from the interaction of thermal and mechanical strains within the material. From a thermodynamic standpoint, in certain formulations of thermoelasticity problems it is also important to consider temperature variations in the elastic body due to deformation. This means that, in studying heat conduction in elastic media and examining non-stationary thermoelastic processes (among others), one should account for thermoelastic dissipation of mechanical energy [3, 7, 9]. Research in the field of coupled dynamic thermoelasticity problems has been developing intensively. The results of these studies are presented in monographs and research papers. The equations of the coupled theory of thermoelasticity are applied to mechanics problems in which thermoelastic dissipation of mechanical energy is of interest or essential [2, 6, 7, 9]. General representations of solutions to coupled thermoelasticity problems have been developed, along with methods for solving the differential equations of thermoelasticity; fundamental energy and variational theorems have been obtained. Theorems on the existence and uniqueness of solutions to basic boundary value problems for the differential equations of coupled thermoelasticity have also been proven.

The characteristics of thermoelastic deformations depend on numerous factors: material properties, geometry, rate and magnitude of temperature change, boundary and initial conditions, and external forces and loads. Thermoelastic vibrational phenomena are especially relevant in engineering applications where structures are subjected to significant temperature gradients. Ensuring the strength and integrity of structural elements in the form of thin-walled plates is of crucial importance, as this is an essential requirement for the reliable operation of the entire structure under thermal loading.

15.2 Statement of the Problem

Thermoelastic vibrations of an elastic plate made of a homogeneous, transversally isotropic material is considered. A rectangular plate of thickness $2h$ is described in a rectangular coordinate system x, y, z, with the planes of isotropy parallel to the xOy coordinate plane coinciding with the mid-surface of the plate. The plate is subjected to a temperature field, and thermoelastic dissipation of mechanical energy is taken into account by considering the coupled thermoelasticity problem.

The formulation is based on the assumptions of a refined plate theory free from the principal hypothesis of classical plate theory, namely, the hypothesis of inextensible normals [1]. Therefore, transverse shear deformations are included. Classical theory bending-vibration models of anisotropic plates are not always suitable for many applied problems. Certain anisotropic plates combine high tensile compressive stiffness with relatively low shear resistance. It is assumed that the shear stresses, or the corresponding shear strains, vary through the plate thickness according to a prescribed quadratic law. To determine the transverse displacement function $w(x, y, \tau)$ of the plate vibrations, we derive the governing differential equation [1, 4]

$$D\Delta\Delta w + \left(1 - h_0 h^2 \Delta\right) \frac{\partial^2 w}{\partial \tau^2} + D\alpha \frac{1+\nu}{h} \Delta T = 0, \qquad (15.1)$$

where

$$D = \frac{Eh^2}{3\rho\left(1-\nu^2\right)}, \qquad h_0 = \frac{2E}{5G'\left(1-\nu^2\right)}, \qquad \Delta = \frac{\partial^2}{\partial x^2} + \frac{\partial^2}{\partial y^2}.$$

E, α – Young's modulus and the coefficient of linear thermal expansion for directions in the plane of isotropy; G' – the shear modulus for planes normal to the plane of isotropy; ν - the Poisson's ratio characterizing lateral contraction in the plane of isotropy under tension in the same plane; τ - the time parameter and ρ – the material density. It is assumed that temperature changes do not lead to significant variations in the physicomechanical properties of the material.

$$T(x, y, \tau) = \frac{3}{2h^2} \int_{-h}^{h} tz dz$$

is the integral characteristic of the temperature field $t(x, y, z, \tau)$. Note that if one sets $h_0 = 0$, the resulting equation for transverse vibrations reduces to the classical thin plate theory, i.e. the model that neglects transverse shear.

The vibrations of the elastic plate are governed by the interaction of the strain field and the temperature field. Thermal strains and stresses in the elastic body, as well as heat conduction, arise due to boundary conditions, heat exchange with the environment, heat input from thermal sources, and the evolution of the temperature field caused by deformation. The actual process of thermoelastic deformation of a body driven by a temperature gradient is, strictly speaking, nonequilibrium and irreversible [2]. Rapid heating or cooling gives rise to dynamic thermal stresses. Along with Eq. (15.1), the differential equation of heat conduction is also considered.

Strong anisotropy in planes perpendicular to the midsurface of the plate introduces substantial modifications to the classical temperature-field theory of thin plates, which, being based on the hypothesis of inextensible normals is insensitive to the transverse mechanical properties of the material. Determining the temperature field in thin walled structural elements such as plates is carried out by reducing the three dimensional heat conduction problem to a two dimensional one while accounting for thermoelastic dissipation of mechanical energy. Significant difficulties arise when one must satisfy the thermal boundary conditions on the plate surfaces. To address this, the operator (symbolic) method is applied in conjunction with averaging of the temperature across the thickness of the plate [4, 9].

As is well known, in the theory of elastic plates, the components of the stress tensor are replaced by integral quantities (resultants, moments) and plate problems are reduced to integrating differential equations of finite order on the mid-surface. In the coupled thermoelasticity problem for thin plates, the unknown functions are likewise introduced as integral characteristics of the temperature field. After taking the appropriate limiting transition, a two dimensional second order differential equation is obtained for the integral temperature characteristic [4]

$$\begin{aligned} h^2\frac{\lambda}{\lambda_z}\Delta T - h^2\frac{1+\varepsilon_0\chi}{a_z}\frac{\partial T}{\partial\tau} - \mu T \\ = \frac{h^3\varepsilon_0}{\alpha\, a_z}\frac{1-2\nu}{1+\nu}\frac{\partial}{\partial\tau}\left(\frac{6\rho}{5G'}\frac{\partial^2 w}{\partial\tau^2} - \Delta w\right) - \mu\frac{Bi}{1+Bi}t_2 - \frac{h^2}{\lambda_z}q. \end{aligned} \tag{15.2}$$

The motion of an elastic body is governed by the interaction between the deformation field and the temperature field. This interaction has a complex mathematical description and is formulated on the basis of a coupled system of the equations of motion and the heat-conduction equation, since the latter contains deformation-dependent terms that characterize the dissipation of mechanical energy. A distinctive feature of these equations is that, like the equations of the full three-dimensional thermoelasticity problem, they incorporate thermoelastic dissipation of mechanical energy. As follows from Eqs. (15.1) and (15.2), the inclusion of transverse shear deformations introduces substantial modifications into the classical coupled thermoelasticity theory for anisotropic thin-walled structural elements such as plates. Thermoelasticity problems are also important for analysing temperature variations in an elastic body resulting from deformation. The heat-balance equations make it possible to account for heat transfer from the surfaces of the thin-walled plate. Heat transfer, like heat conduction, affects the equalization of temperature oscillations within the plate and, consequently, influences the thermoelastic processes occurring during deformation. λ, λ_z – are the thermal-conductivity coefficients along the coordinate axes x, y, z, respectively,

$$\chi = \frac{\nu_1}{\nu_2}\frac{1-2\nu}{1+\nu}\frac{((1-\nu)\alpha_1 + 2\nu_2\alpha)^2}{\alpha^2(1-\nu-2\nu_1\nu_2)}.$$

ε_0, χ characterize the thermoelastic dissipation of the mechanical energy in an anisotropic material. In the case of an isotropic plate, $\chi = 1$, and if one additionally assumes $\varepsilon_0 = 0$, then equation (15.2) reduces to the ordinary non-stationary heat conduction equation without thermoelastic energy dissipation [1, 5, 8].

$$Bi = h\frac{\lambda_c}{\lambda_z}$$

characterizes heat exchange with the external environment (Biot criterion),

$$\mu = 15\frac{1+Bi}{6+Bi},$$

λ_c – are the heat-transfer coefficients at the surfaces $z = \pm h$, $2t_2 = \left(t_c^+ - t_c^-\right)$, $t_c^\pm$ – the temperatures of the ambient medium in contact with the surfaces $z = \pm h$, q – the average value of the intensity of a heat source linearly distributed across the thickness of the plate. ν_1, ν_2 are the contraction coefficient in the plane of isotropy under tension in the perpendicular direction, and contraction coefficient in the direction normal to the plane of isotropy under tension in the plane of isotropy; α_1 is the linear thermal expansion coefficient in the direction perpendicular to the plane of isotropy.

For an isotropic or transversally isotropic material

$$\varepsilon_0 = \frac{c_\sigma - c_\varepsilon}{3c_\varepsilon}\frac{1+\nu}{1-\nu}, \quad c_\sigma > c_\varepsilon,$$

c_σ, c_ε are the specific heat capacities, h^2/a_z – is the characteristic heating time of the plate, and

$$a_z = \frac{\lambda_z}{c_\varepsilon}$$

is the thermal diffusivity is the fundamental parameter of unsteady heat conduction (describing the relation between the material's ability to conduct heat and its ability to store heat). Physically, it indicates how quickly a substance equalizes temperature under thermally nonequilibrium conditions [3, 4, 9]. In thermoelastic vibrations whose period is comparable to the heat-propagation time across the plate thickness, an irreversible process occurs in which a significant portion of the mechanical energy is converted into heat. In this regime, energy dissipation reaches its maximum, in agreement with experimental observations. The mechanical energy accumulated in the oscillating body (in this case, the plate) is dissipated over time, gradually decreasing from its initial value to a minimum possible level. Ultimately, this leads to the establishment of an equilibrium state. The rate of change of the accumulated energy in the system represents the amount of mechanical energy dissipation per unit time.

Equations (15.1) and (15.2) form a system of differential equations describing transverse thermoelastic vibrations of plates with allowance for transverse shear and thermoelastic dissipation of mechanical energy. Under the refined plate theory, where a quadratic law is adopted for the distribution of transverse shear deformations through the thickness [1], the displacements, calculated strains, and stresses in the plate vary through the thickness according to cubic laws. The adoption of a cubic temperature profile through the thickness is fully consistent with the refined theory [1, 4]. In the classical theory of thermoelastic bending of thin plates, a linear temperature distribution through the thickness may be assumed, which corresponds to the hypothesis of inextensible normals. It should be noted that the heat-conduction equation (15.2) is obtained by imposing heat-exchange boundary conditions on the plate surfaces $z = \pm h$

$$h\frac{\partial t}{\partial z} \pm Bi\left(t - t_c^\pm\right) = 0. \tag{15.3}$$

The equation of transverse vibrations (15.1) is obtained by assuming the shear stresses to have the form [1]

$$
\begin{aligned}
2\sigma_{xz} &= \left(z^2 - h^2\right) \varphi(x, y, \tau), \\
2\sigma_{yz} &= \left(z^2 - h^2\right) \psi(x, y, \tau).
\end{aligned}
\tag{15.4}
$$

Taking into account the relations that connect the shear strains with the tangential displacements, we obtain

$$
\begin{aligned}
u_x &= u - z\frac{\partial w}{\partial x} - \frac{z}{6G'}\left(z^2 - 3h^2\right) \varphi(x, y, \tau), \\
u_y &= v - z\frac{\partial w}{\partial y} - \frac{z}{6G'}\left(z^2 - 3h^2\right) \psi(x, y, \tau).
\end{aligned}
\tag{15.5}
$$

$u(x, y, \tau), v(x, y, t)$ - the tangential displacements of the points of the mid-surface. The expressions for the integral quantities (the internal transverse forces and bending moments) are determined by the formulas [1, 4]

$$
\begin{aligned}
N_x &= \frac{2}{3}h^3\varphi(x, y, t), \\
N_y &= \frac{2}{3}h^3\psi(x, y, t), \\
M_x &= -\frac{2}{3}h^3\frac{E}{1-\nu^2}\left(\frac{\partial^2 w}{\partial x^2} + \nu\frac{\partial^2 w}{\partial y^2}\right) \\
&\quad + \frac{4h^5E}{15G'\left(1-\nu^2\right)}\left(\frac{\partial \varphi}{\partial x} + \nu\frac{\partial \psi}{\partial y}\right) - \frac{2}{3}h^2\frac{E}{1-\nu}\alpha T, \\
M_y &= -\frac{2}{3}h^3\frac{E}{1-\nu^2}\left(\frac{\partial^2 w}{\partial y^2} + \nu\frac{\partial^2 w}{\partial x^2}\right) \\
&\quad + \frac{4h^5E}{15G'\left(1-\nu^2\right)}\left(\frac{\partial \psi}{\partial y} + \nu\frac{\partial \varphi}{\partial x}\right) - \frac{2}{3}h^2\frac{E}{1-\nu}\alpha T, \\
M_{xy} &= -\frac{2}{3}h^3\frac{E}{1+\nu}\frac{\partial^2 w}{\partial x\partial y} + \frac{2h^5E}{15G'(1+\nu)}\left(\frac{\partial \varphi}{\partial y} + \frac{\partial \psi}{\partial x}\right).
\end{aligned}
\tag{15.6}
$$

If necessary, the sought functions for the distribution of shear stresses (transverse forces) $\varphi(x, y, \tau)$, $\psi(x, y, \tau)$ can be determined from the equations [4]

$$
\begin{aligned}
\frac{\partial M_x}{\partial x} + \frac{\partial M_{xy}}{\partial y} &= N_x, \\
\frac{\partial M_y}{\partial y} + \frac{\partial M_{xy}}{\partial x} &= N_y.
\end{aligned}
$$

In the interconnected thermoelasticity problem for transversally isotropic thin plates, the parameters [4]

$$
\frac{\lambda}{\lambda_z}\frac{h}{L}, \quad \frac{h^2(1+\varepsilon_0\chi)}{a_z\tau_a}
$$

characterize the thinness of the plate from the standpoint of heat-conduction theory and indicate the error introduced by replacing the three dimensional equations with two dimensional ones. L is the characteristic dimension (width) of the plate, and τ_a is the characteristic time for a significant rise in temperature. Together with the criteria of the refined plate-bending theory that account for transverse shear deformations,

$$\left(\frac{\lambda}{\lambda_z}\frac{h}{L}\right)^2 \ll 1, \qquad \left(\frac{h^2(1+\varepsilon_0\chi)}{a_z\tau_a}\right)^2 \ll 1$$

they represent the criteria for the thinness of a transversally isotropic plate for the given coupled thermoelasticity problem.

15.3 Moving Heat Source

Let a rectangular plate, simply supported along its entire boundary $x = 0, x = a$, $y = 0, y = b$, be subjected to heat exchange with the environment and a heat source of intensity q. On these same surfaces, a constant temperature is maintained

$$\begin{aligned} &w = 0, \quad M_x = 0, \quad \psi = 0, \quad T = 0, \quad \text{where } x = 0, x = a, \\ &w = 0, \quad M_y = 0, \quad \varphi = 0, \quad T = 0, \quad \text{where } y = 0, y = b. \end{aligned} \tag{15.7}$$

Along with the boundary conditions, the initial conditions must also be taken into account

$$w = 0, \quad \frac{\partial w}{\partial \tau} = 0, \quad T = 0, \quad \text{where } \tau = 0. \tag{15.8}$$

Heat conduction (heat transfer) in solids occurs significantly more slowly than the propagation of sound or mechanical vibrations. From a technical standpoint, for example, laser heating is widely used due to its distinctive thermal effects. This and other similar heating devices are modelled as a concentrated heat source applied over a small area and moving along the surface of the heated body. Such an approach allows modelling a heat exchanger with a heat source as a heat conduction system with a moving source. The heat source does not perform mechanical work but alters the temperature field, generating heat fluxes. A concentrated heat source is represented by a Dirac delta function.

At time $\tau = 0$, a linearly distributed heat source (distributed through the thickness of the plate) begins to move along the plate from the point $(0, r_0)$, parallel to side Oa with a constant velocity V. In Eq. (15.2) it is assumed that

$$\frac{h^2}{\lambda_z}q = q_0 S(\tau)\delta(x - V\tau)\delta(y - r_0), \quad q_0 = \text{const}, \tag{15.9}$$

$S(\tau) = 1$, where $\tau \in [0, \tau_0]$, $S(\tau) = 0$ where $\tau > \tau_0$, τ_0 - the time for the heat source to traverse the plate. Assuming in equation (15.2) that $t_c^+ = t_c^-$, and under the boundary conditions (15.7), the solutions of the system of equations (15.1) and

(15.2) are expressed in the form

$$w(x,y,\tau)=\sum_{m=1}^{\infty}\sum_{n=1}^{\infty}w_{mn}(\tau)\,\Phi_{mn}(x,y),$$

$$T(x,y,\tau)=\sum_{m=1}^{\infty}\sum_{n=1}^{\infty}T_{mn}(\tau)\,\Phi_{mn}(x,y),\tag{15.10}$$

$$\Phi_{mn}(x,y)=\sin\frac{\pi mx}{a}\sin\frac{\pi ny}{b},\quad m,n=1,2,...$$

From Eq. (15.1), we find that

$$\Theta_{mn}(\vartheta)=\lambda_{mn}W_{mn}(\vartheta)+\frac{1+h_0\lambda_{mn}}{\lambda_{mn}}\frac{d^2W_{mn}}{d\vartheta^2},\tag{15.11}$$

where the following notations are used

$$\Theta_{mn}(\tau)=\alpha(1+\nu)T_{mn}(\tau),\quad W_{mn}(\tau)=\frac{w_{mn}(\tau)}{h},$$

$$\lambda_{mn}=\pi^2h^2\left(\frac{m^2}{a^2}+\frac{n^2}{b^2}\right),\quad \vartheta=\omega_0\tau,\quad \omega_0^2=\frac{E}{3\rho\left(1-\nu^2\right)h^2}.$$

From Eq. (15.2), we obtain third order differential equations for the unknown functions $W_{mn}(\vartheta)$, $m,n=1,2,...$

$$\left(\frac{d^2}{d\vartheta^2}+\lambda_{mn}^2H_{mn}\right)\left(\frac{dW_{mn}}{d\vartheta}+A_{mn}W_{mn}\right)+\varepsilon_0B_{mn}\frac{dW_{mn}}{d\vartheta}=Q_{mn}S(\vartheta)\sin\left(\pi m\frac{\vartheta}{v_*}\right).\tag{15.12}$$

Here it is assumed that

$$\begin{aligned}H_{mn}&=\frac{1}{1+h_0\lambda_{mn}},\\B_{mn}&=\frac{\lambda_{mn}^2H_{mn}^2(1-2\nu)}{1+\varepsilon_0\chi+\varepsilon_0(1-2\nu)(1-H_{mn})},\\A_{mn}&=\frac{1}{\upsilon_0}\frac{\lambda_{mn}\lambda_1+\mu}{1+\varepsilon_0\chi+\varepsilon_0(1-2\nu)(1-H_{mn})},\ \lambda_1=\frac{\lambda}{\lambda_z},\ r=\frac{r_0}{b},\\Q_{mn}&=\frac{4}{\upsilon_0}\frac{q_0H_{mn}\lambda_{mn}\sin(\pi nr)}{ab\left(1+\varepsilon_0\chi+\varepsilon_0(1-2\nu)(1-H_{mn})\right)},\end{aligned}\tag{15.13}$$

$S(\vartheta)=1$, where $\vartheta\in[0,\upsilon_*]$, $S(\vartheta)=0$, where

$$\upsilon_0=\frac{h^2\omega_0}{a_z}$$

characterizes the ratio of the plate heating time to the period of elastic vibrations, $\upsilon_*=\tau_0\omega_0$ characterizes the ratio of the elastic wave propagation speed to the speed of the moving heat source. Obviously,

$$\frac{\tau_0 a_z}{h^2} = \frac{v_*}{v_0}$$

is the ratio of the time for the heat source to traverse the plate to the characteristic heating time of the plate.

For the dimensionless coefficient characterizing the thermoelastic dissipation of mechanical energy, we have

$$\varepsilon_0 = \frac{E t_0 \alpha^2}{c_\varepsilon (1-2\nu)} \frac{1+\nu}{1-\nu} > 0,$$

t_0 is the absolute temperature of the natural state [4, 9]. If in the relations (15.13) and in Eqs. (15.12) we set $\varepsilon_0 = 0$, we obtain the uncoupled problem of thermoelastic vibrations (without accounting for energy dissipation). When $H_{mn} = 1$ ($h_0 = 0$), the problem corresponds to the thermoelastic vibrations of a plate without consideration of transverse shear deformations.

The solutions of Eqs. (15.12) are expressed in the form

$$W_{mn}(\vartheta) = \theta_{mn} \exp(\Omega_{mn}\vartheta), \quad \theta_{mn} = \text{const}, \quad m, n = 1, 2, \ldots \tag{15.14}$$

We obtain the characteristic equation of the considered thermoelastic process.

$$\left(\Omega_{mn}^2 + \lambda_{mn}^2 H_{mn}\right)(\Omega_{mn} + A_{mn}) + \varepsilon_0 B_{mn}\Omega_{mn} = 0, \; m, n = 1, 2, \ldots \tag{15.15}$$

The study of thermoelastic processes is based on the principles, considerations, and hypotheses of thermodynamics. Thermoelastic deformation is a nonuniform, irreversible process, and its analysis must rely on the laws of thermodynamics of irreversible processes. From thermodynamic considerations, $\varepsilon_0 \ll 1$, and since $0 < B_{mn} < 1$ then $\varepsilon_0 B_{mn} \ll 1$. From these considerations, it can be concluded that equation (15.15) always has a complex solution with a negative real part, i.e., the vibration process is inherently damped. When $\tau \in [0, \tau_0]$, satisfying conditions (15.8), we obtain

$$W_{mn}(\vartheta) = W_{mn}^0(\vartheta) = \frac{1}{P_{mn}} \int_0^\vartheta Q_{mn} G_{mn}(\vartheta - s) \sin\frac{\pi m \vartheta}{V_*} ds, \tag{15.16}$$

$$\begin{aligned} G_{mn}(s) &= \beta_{mn}\exp(-\gamma_{mn}s) \\ &- \exp(-\alpha_{mn}s)\left((\alpha_{mn} - \gamma_{mn})\sin(\beta_{mn}s) + \beta_{mn}\cos(\beta_{mn}s)\right). \end{aligned} \tag{15.17}$$

Here, $-\gamma_{mn}$, $-\alpha_{mn} \pm i\beta_{mn}$ are the roots of the characteristic equation (15.15), and moreover

$$\gamma_{mn} > 0, \alpha_{mn} > 0, \beta_{mn} > 0, P_{mn} = \beta_{mn}\left((\alpha_{mn} - \gamma_{mn})^2 + \beta_{mn}^2\right), m, n = 1, 2, \ldots$$

It should be noted that the thermodynamic justification for the roots of the characteristic equation (15.15) leads to the following system of conditions for these quantities

$$\begin{aligned}
&\gamma_{mn} + 2\alpha_{mn} = A_{mn},\\
&2\alpha_{mn}\gamma_{mn} + \alpha_{mn}^2 + \beta_{mn}^2 = \lambda_{mn}^2 H_{mn} + \varepsilon_0 B_{mn},\\
&\gamma_{mn}\left(\alpha_{mn}^2 + \beta_{mn}^2\right) = \lambda_{mn}^2 H_{mn} A_{mn}.
\end{aligned} \tag{15.18}$$

When $\tau > \tau_0$, i.e., after the heat source leaves the plate, the values of the functions $W_{mn}(\vartheta)$ take the form

$$\begin{aligned}
W_{mn}(\vartheta) = W_{mn}^1(\vartheta) &= C_{mn}\exp\left(-\gamma_{mn}\vartheta\right)\\
&+ \exp\left(-\alpha_{mn}\vartheta\right)\left(M_{mn}\sin\left(\beta_{mn}\vartheta\right) + N_{mn}\cos\left(\beta_{mn}\vartheta\right)\right).
\end{aligned} \tag{15.19}$$

A damped oscillatory process takes place; the constants C_{mn}, M_{mn}, are determined from the continuity conditions for the functions

$$W_{mn}(\vartheta), \quad \frac{dW_{mn}}{d\vartheta}, \quad \Theta_{mn}(\vartheta),$$

where $\tau = \tau_0$ and

$$\begin{aligned}
&W_{mn}^0(\vartheta) = W_{mn}^1(\vartheta),\\
&\frac{dW_{mn}^0}{d\vartheta} = \frac{dW_{mn}^1}{d\vartheta},\\
&\frac{d^2W_{mn}^0}{d\vartheta^2} = \frac{d^2W_{mn}^1}{d\vartheta^2},
\end{aligned} \tag{15.20}$$

where $\vartheta = \upsilon_*$. It may be noted once again that the attenuation is due to the thermoelectric dissipation of mechanical energy.

15.4 Conclusion

By examining the principal characteristic equations of the thermoelastic process (15.12) and (15.15), we find that the relative damping coefficient is small for both very large and very small values of the parameter ν_0. When $\nu_0 \to 0$ (low vibration frequencies and short heating times of the plate), the thermoelastic process proceeds almost as if thermoelastic dissipation of mechanical energy were negligible. It should be noted that, for finite heat transfer coefficients, the maximum heat losses through the body's surface are observed experimentally when vibrations occur at sufficiently low frequencies. When $\nu_0 \to \infty$, the heat generated by deformation does not have time to propagate across the plate thickness, and the thermoelastic process takes place under conditions close to adiabatic; the vibrations then occur with almost no damping.

We note that, under other thermomechanical conditions, thermal moments relax faster than mechanical forces, which leads to transverse vibrations decaying more rapidly than longitudinal vibrations. In thermoelastic dynamics, accounting for mechanical energy dissipation qualitatively changes the nature of the solution to the vibration problem: free vibrations are suppressed, and amplitude growth is limited (even in the case of resonances). For certain values of the coefficient ν_0 the damping coefficient reaches its maximum. It is known that plate vibrations decay more rapidly for large values of the thermoelastic dissipation coefficient ε_0. Energy dissipation leads to a slight increase in the natural frequency of the elastic plate. With an increase in the ratio $\nu_*/\nu_0 = a_z\tau_0/h^2$ (which characterizes the relation between the heat-propagation time across the plate and the plate-heating time), the maximum deflection increases. For large values of the coefficient ν_*, as expected, the dynamic thermoelastic deflections differ only slightly from the quasi-static thermal ones.

The coupled theory of thermoelasticity, describing the interaction between deformation and temperature, can also be used to determine damping parameters of free vibrations in metallic plates. Under dynamic loading and high-temperature fields, when analyzing structural elements such as thin anisotropic plates together with refined theories of the strain-stress state, it is necessary to consider, if required, the mechanics of coupled fields. The regularities of deformation of solids are investigated, and the features arising from the coupling of processes, as well as the effect of transverse shear deformations, are identified. Accounting for shear deformations becomes significant as the coefficient h_0 (characterizing the material's resistance to transverse shear) increases; the damping coefficient decreases and differs substantially from the case $h_0 = 0$. As is known, the influence of transverse shear is very substantial for the vibration frequencies of a plate: they decrease markedly with increasing h_0. For small values of ν_*, inclusion of transverse shear deformations strongly affects the vibration amplitude. In this case, the maximum value of the amplitude function is reached after the heat source leaves the plate ($\tau > \tau_0$). From a thermodynamic standpoint, thermoelastic vibrations in structural components can cause additional energy losses due to thermoelastic dissipation of mechanical energy, which increases energy consumption and reduces the efficiency of the process.

References

[1] Ambartsumyan SA (1970) Theory of Anisotropic Plates. Technomic Publishing Company

[2] Green AE, Lindsay KA (1972) Thermoelasticity. Journal of Elasticity 2(1):1-7, DOI 10.1007/BF00045689

[3] de Groot SR, Mazur P (1984) Non-Equilibrium Thermodynamics (Dover Books on Physics). Dover Publications Inc

[4] Jilavyan SH (1986) K voprosu o termouprugikh kolebaniyakh transtropnykh plastin (in Russ., On the problem of thermoelastic vibrations of transversally isotropic plates). Mechanics, Yerevan State University (5):230–241.

[5] Motovilolets IA, Kozlov VI (1987) Termouprugost'. Mekhanika svyazannykh poley v elementakh konsktruktsiy (in Russ., Thermoelasticity. Mechanics of Coupled Fields in Structural Elements). Naukova Dumka, Kiev
[6] Mountfort M (1966) Transformations in problems of coupled thermoelasticity. Transactions of the ASME, Applied Mechanics 33(1):196-198, DOI 10.1115/1.3624980
[7] Nowacki W (1975) Dynamic Problems of Thermoelasticity. Noordhoff, Leyden
[8] Podstrigach YS, Koliano YM (1972) Neustanovivshiyesya temperaturnyye polya i napryazheniya v tonkikh plastinkakh (in Russ., Non-Steady Temperature Fields and Stresses in Thin Plates). Naukova Dumka, Kiev
[9] Podstrigach YS, Shvets RN (1978) Termouprugost' tonkikh obolochek (in Russ., Thermoelasticity of Thin Shells). Naukova Dumka, Kiev

Chapter 16
Comprehensive Approach to Designing Thin-Walled Hollow Shafts Made of Woven Fiberglass-Plastic with Enhanced Torsional Load Capacity and Operational Reliability

Koryun A. Karapetyan

Abstract Based on a comparison of test data from simple torsion of symmetrically and asymmetrically reinforced tubular elements made of woven fiberglass-plastic, as well as torsion combined with internal hydrostatic pressure in symmetrically reinforced tubes, a comprehensive design approach is proposed for transmission hollow shafts made of such material. These shafts exhibit increased torsional load capacity and operational reliability. It is shown that the development of such approaches is best carried out through a comparative analysis of specific features, resistance to failure, and deformation behavior of the studied object, which serves as a prototype of the final product, under conditions close to actual operation.

Key words: Thin-walled laminated fiberglass tubes, Plain-weave fiberglass fabric with primary crossover, Torsion, Internal hydrostatic pressure, Torsional load capacity, Operational reliability

16.1 Introduction

One of the main structural elements of modern mechanisms and machines, as well as their individual components, are thin-walled rotational elements made of reinforced plastics (RP) with layered or fibrous structure. Practical experience confirms the feasibility, and in some cases also the economic advantage, of using thin-walled tubular elements and cylindrical rotating shells of various geometries made of layered RP in the manufacture of automotive cardan shafts [9], hollow shafts of high-speed rotors with different profiles [4], high-speed centrifugal gas generators, rotary machine motors, and others [2]. These structural components, produced in most

Koryun A. Karapetyan
Institute of Mechanics, National Academy of Sciences of the Republic of Armenia, 24b Marshal Baghramyan Ave, Yerevan, 0019, Republic of Armenia, e-mail: koryun.a.karapetyan@gmail.com

H. Altenbach and A. V. Sahakyan, *Contact Mechanics and Relevant Problems*, Advanced Structured Materials 253, https://doi.org/10.1007/978-3-032-20869-9_16

cases by winding veneer or prepreg based on woven fabric, are typically operated under simple torsion or torsion combined with other force factors. This means that the material of such tubular elements is subjected to pure shear in service, and quite often to shear combined with heterogeneous stresses (a complex stress–strain state).

It is also known that weak shear resistance, both interlayer and in the reinforcement plane, as well as pronounced shear creep-arising from the structural nature of the material–are among the main specific disadvantages of layered and fibrous RP [1]. As a result of these shortcomings, tangential stress and strain components arising in one of these planes may become the primary causes of premature structural failure. The situation may become even more unfavorable when shear stresses act in combination with heterogeneous stresses [3]. Under experimental conditions, a stress–strain state equivalent to the mentioned complex loading can be reproduced by applying to the specimen one of two basic load combinations: tension with torsion or compression with torsion [7].

In the present work, based on comparative analysis of data obtained from direct experiments, the feasibility of using a comprehensive approach to the optimal design of hollow shafts made of fiberglass-fabric-reinforced plastics–with increased torsional load capacity and stable performance characteristics over their required service life–is justified.

16.2 Experimental Results

The results discussed here were obtained from short-term tests performed under simple torsion of symmetrically and asymmetrically reinforced thin-walled tubular fiberglass specimens, as well as under torsion of symmetrically reinforced tubes preloaded with internal hydrostatic pressure at various levels (combined loading of tubes). It is known that one of the methods for increasing the strength and deformation characteristics of reinforced plastics (RPs) is to improve the degree to which the strength of the reinforcing material component is realized. This realization is associated with the development and application of technological processes for converting the semi-finished product (prepreg, i.e., a resin-impregnated reinforcing component) into a final part, ensuring that the reinforcing component operates in the structural material at its maximum potential.

Both the strength and elastic properties of RPs are related to the uniform tension of the reinforcing fibers and to the uniform compaction of the material across all cross-sections of the product during processing. For structural elements such as thin-walled tubes, shells, and other products of complex geometry made of RPs, the most effective processing method that satisfies the above requirements is the method known in the literature as "Self-pressurization" [8].

The technological process for manufacturing thin-walled tubes from RPs using this method is described below. The prepreg is wound onto a mandrel with preliminary tension; the mandrel material has a thermal coefficient of linear expansion (TCLE) α_1. Once the required winding thickness is achieved, a foil layer of a certain thickness with the same TCLE α_1 is wound over the prepreg as a damping layer. Then, four layers of a metallic tape (outer mold) are wound on top, the material of which has a TCLE α_2, smaller than α_1. The entire assembly is then placed into a thermal chamber for curing the binder, where pressing of the prepreg simultaneously occurs due to the difference in thermal expansion between the inner mandrel and the outer mold. During heating, not only the inner mandrel expands, but also the damping foil layer, allowing the prepreg fibers to straighten and thus become uniformly tensioned. However, due to the restricted expansion of the damping layer—limited by the steel tape wound over it — the prepreg is pressed. After the binder-curing process is completed, the mold is disassembled, and the finished tube, whose TCLE is lower than that of the mandrel, easily slides off the mandrel after cooling.

Typically, the inner mandrel is made of duralumin ($\alpha_1 \approx 23.4\, 10^{-6}K^{-1}$), aluminum foil is used as the damping element ($\alpha_1 \approx 27.5\, 10^{-6}K^{-1}$), and low-carbon steel serves as the material of the outer tape ($\alpha_2 \approx 11.5 \ldots 14.1\, 10^{-6}K^{-1}$). It should also be noted that the specific pressure during prepreg winding onto the mandrel is approximately 0.18 MPa, which, as practice shows, ensures the production of tubes with the highest strength properties. To carry out the full research program according to the above technology, the required number of tubular specimens was manufactured from laminated fiberglass, with an inner diameter of 38 mm, wall thickness of 2.25 mm, and length of 285 mm.

As the reinforcing component, plain-weave fiberglass fabric with primary overlap of grade T23 (TU 6-11-231-76), thickness 0.27 mm and density 36:20 (ratio of warp to weft fibers per 1 cm^2 of fabric), was used. The tensile breaking load of the fabric (strip 25×100 mm) in the warp direction is 2.00–2.35 kN, and in the weft direction is 1.10–1.40 kN. A modified epoxy-phenolic resin (ED-20+BF-4) was used as the matrix (binder). The reinforcement coefficient of the fiberglass plastic is $\mu = 0.45$ ($\mu_{\text{warp}} = 0.29$, $\mu_{\text{weft}} = 0.16$).

The fiberglass tubular specimens were produced so that the directions of the tube axis and the main yarns of the woven reinforcement fabric were parallel (symmetrically reinforced relative to the tube axis; initial reinforcement angle $\varphi = 0°$). Fiberglass tubes with an initial distortion of reinforcement symmetry were also produced, in which the angle between these directions was $\varphi = 6 - 8°$(asymmetrically reinforced relative to the tube axis).

It should be noted that misalignment of reinforcement symmetry within the mentioned range was detected when manufacturing laminated fiberglass tubular specimens in laboratory conditions, even though they were intended to be symmetrically reinforced relative to the axis [6]. However, in a portion of the finished tubes (more than 8% of the total), a deviation from reinforcement symmetry was observed, with the maximum magnitude reaching $\pm(6 - 8°)$.

Before performing the short-term torsion tests and the torsion tests under constant internal hydrostatic pressure at levels 0.4, 0.6, and 0.8 $\sigma^P_{\theta\theta}$, the circumferential failure stress under uniaxial internal hydrostatic pressure ($\sigma^P_{\theta\theta} = 458.2$ MPa) and the torsional shear failure stress ($\tau^P_{\theta z} = 36,2$ MPa) were determined. During short-term tests, the applied torque was increased stepwise in increments corresponding to $0.07 - 0.08$ of the failure torque under simple torsion. The duration of each test was 6 – 10 minutes. A statistically sufficient number of specimens was used in each test series.

16.3 Discussion of Results

It should be mentioned that some data from works [10] and [5] were used in addressing the problem covered in this study, and these contributed significantly to formulating the comprehensive approach. The development of this approach is based on comparative analysis of specific features, failure resistance, and deformation behavior of the test object under conditions close to operational.

According to the results of the short-term experiments, increasing the level of internal hydrostatic pressure pre-applied to symmetrically reinforced fiberglass tubular specimens leads to an initial increase and subsequent decrease in torsional failure resistance. The maximum increase-approximately 40% compared with the value obtained under simple torsion-was observed in tubes preloaded with internal pressure equal to $0{,}4\sigma_{\theta\theta}$.

During the further discussion in this work, the fiberglass tubular specimens exposed to a constant internal hydrostatic pressure equal to $\sigma_{\theta\theta} = 0,4\sigma^{p}_{\theta\theta}$ = const, will be referred to as mechanically reinforced tubes. The above-mentioned effect, identified through direct measurement, indicates that a significant increase in torsional load capacity can be achieved in symmetrically reinforced tubular elements (with initial angle $\varphi = 0°$) made of plain-weave fiberglass fabric with primary overlap by ensuring a constant internal pressure equal to 0.4 of the burst pressure $(\sigma^{p}_{\theta\theta})$.

Application of this approach can support the design of transmission hollow shafts made of woven fiberglass plastic with increased torsional load capacity. The data presented in Fig. 16.1 also support the feasibility of this proposal.

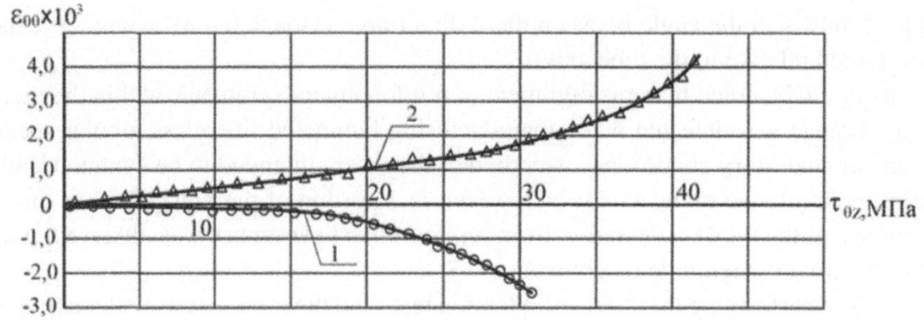

Fig. 16.1: Curves describing circumferential deformation of symmetrically reinforced fiberglass tubes subjected to simple torsion (curve 1), and of identical tubes subjected to torsion combined with internal hydrostatic pressure corresponding to $\sigma_{\theta\theta} = 0{,}4\sigma_{\theta\theta}$ = const (curve 2).

According to these data, during simple torsion, as the torque ($\tau_{\theta z}$) is increased stepwise over a period of 6-10 minutes, a monotonic decrease-with a variably changing rate–of circumferential deformation is observed in symmetrically reinforced fiberglass tubes (Fig. 16.1, curve 1), that is, the tubes undergo circumferential contraction.

In contrast, in the case of complex loading of symmetrically reinforced thin-walled fiberglass tubular specimens (under constant internal hydrostatic pressure corresponding to $\sigma_{\theta\theta} = 0,4\sigma^{P}_{\theta\theta}$) stepwise increases of the applied torque lead to an increase in circumferential deformation (Fig.1, curve 2), meaning that the tubes expand. Such a deformation state of tubular elements intended for torque transmission may be considered more favorable compared to the contraction observed in identical tubes under simple torsion, since in this expanded state the tubes may exhibit greater compliance with respect to torsion.

At the same time, to substantiate the proposal developed in this work, we turn our attention to the axial deformation behavior during torsion for the previously mentioned symmetrically reinforced thin-walled fiberglass tubular elements strengthened by the mechanical method, and for similar tubes with reinforcement-symmetry deviation up to ±(6-8°), subjected to simple torsion (Fig. 16.2).

As can be seen from the data in Fig. 16.2 (see curve 1), during the stepwise increase of the applied torque, an initial elongation followed by subsequent shortening is observed in the mechanically strengthened, symmetrically reinforced thin-walled tubes (initial reinforcement angle $\varphi = 0°$). This means that at a certain level of applied torque (in this case corresponding to $\tau_{\theta z} = 10$ MPa), the direction of axial deformation of the fiberglass tubes changes. This indicates that, under cyclically pulsating torsional loading, the repeated reversal of axial deformation may lead to the formation of an alternating pulsating stress–strain state within the material. Repeated occurrence of such a stress–strain state may pose a particular danger to the organic matrix, which is the most vulnerable component ensuring material

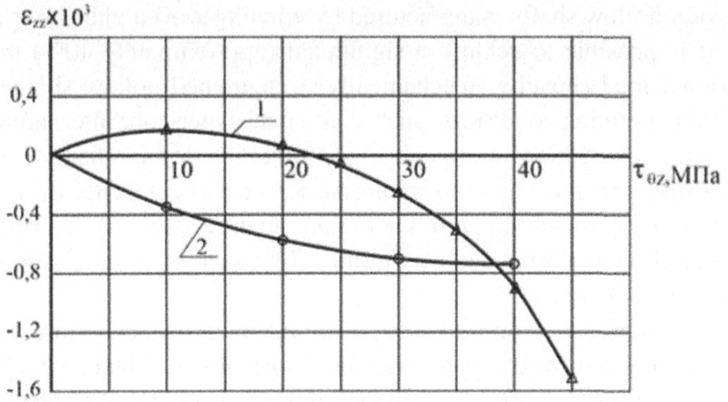

Fig. 16.2: Curves describing axial deformation during torsion for symmetrically reinforced fiberglass tubes under internal hydrostatic pressure $\sigma_{\theta\theta} = 0{,}4\sigma_{\theta\theta} = \text{const}$ (curve 1), and for similar tubes reinforced asymmetrically relative to the axis under simple torsion (curve 2).

integrity. To substantiate the comprehensive approach proposed in this work-aimed at mitigating, to some extent, the above-mentioned phenomenon that negatively affects the operational reliability and performance of structural transmission hollow shafts made of woven fiberglass plastic, which serve as prototypes of thin-walled tubes of this material-let us consider the results of the experimental study described below.

In these experiments, the axial deformation resistance and failure behavior of thin-walled woven-fiberglass tubes reinforced symmetrically and asymmetrically (initial reinforcement angle $\varphi = 6 - 8°$) were investigated under simple torsion, with the focus on increasing the initial reinforcement angle. The test specimens were thin-walled tubes made of fiberglass-reinforced plastics with the same dimensions and structure as the tubes described above.

The measurements showed that the torsional failure resistance of the two types of fiberglass thin-walled tubes differed only slightly. According to the behavior of the curve describing axial deformation in asymmetrically reinforced fiberglass tubular specimens, a monotonic shortening of the tubes is observed as the torque ($\tau_{\theta z}$) increases, with an initially high deformation rate gradually diminishing (Fig. 16.2, curve 2). From comparison of the data in Fig. 16.2, we note that the maximum elongation of 0.2×10^{-3} observed in the mechanically strengthened, symmetrically reinforced thin-walled fiberglass tubes occurs at a torque level corresponding to $\tau_{\theta z} = 10MPa$ (curve 1), whereas in similar unstrengthened asymmetrically reinforced tubes, shortening is recorded at the same torque level, reaching a magnitude of up to 0.38×10^{-3}(curve 2). This comparison allows us to conclude that during torsion, the occurrence of axial elongation in mechanically strengthened fiberglass tubular elements with an initial reinforcement angle $\varphi = 6 - 8°$is unlikely.

16.4 Conclusion

Thus, by applying internal hydrostatic pressure of a certain level to symmetrically reinforced transmission hollow shafts manufactured by winding woven plain-weave fiberglass prepreg, it is possible to achieve a significant (approximately 40%) increase in their torsional load capacity (mechanically strengthened hollow shafts). However, under certain operating conditions, such shafts may experience alternating axial deformation cycles (elongation–shortening), leading to the development of an alternating pulsating stress–strain state in the material. Repeated occurrence of this phenomenon may pose a significant threat to the organic matrix, which is the most vulnerable component of the material, and may ultimately lead to premature failure of the transmission hollow shafts.

The comprehensive approach proposed here, justified by direct experimental measurements, allows mitigation of this phenomenon that negatively affects reliable operation of transmission hollow shafts. According to this approach, the entire technological process of manufacturing woven-fiberglass hollow shafts with increased torsional load capacity and stable operational characteristics should be performed as follows:

- The winding process of the woven-fabric-based prepreg onto the mandrel should be carried out so that the angle between the directions of the main yarns of the reinforcing fabric and the shaft axis is 6 – 8°.
- After completion of the manufacturing process, the finished hollow shafts should be subjected to mechanical strengthening. Applying such a comprehensive approach eliminates the alternating pulsating stress-strain state in the material during operation and therefore increases the reliability of the shafts.

References

[1] Baltin YuP, Tamuzh VP (975) Cyclic strength of fiberglass under shear loading in the reinforcement plane. Polymer Mechanics (11(5):994-996, DOI 10.1007/BF00859663
[2] Daneshjou K, Talebitooti M (2014) Free vibration analysis of rotating stiffened composite cylindrical shells by using the layerwise-differential quadrature (LW-DQ) method. Mechanics of Composite Materials 50(1):21-38, DOI 10.1007/s11029-014-9390-6
[3] Gnuni VV (2000) Proyektirovaniye szhatykh osevoy siloy tsilindricheskikh obolochek iz kompozitsionnykh materialov (in Russ., Design of axially compressed cylindrical shells made of composite materials). Yerevan: Gitutyun, Yerevan
[4] Grāpis O, Tamužs V, Olson N, Anderson J, Vilks U (1995) Application of CFRP as a rotor shaft material. Mechanics of Composite Materials 31(2):163-173, DOI 10.1007/BF00616285
[5] Karapetyan KA 2005) Vliyaniye nachal'noy razoriyentatsii armirovaniya na mekhanicheskoye povedeniye sloistykh stekloplastikov pri staticheskikh nagruzheniyakh (in Russ., Influence of initial reinforcement misorientation on the mechanical behavior of laminated fiberglass under static loading). Reports of the National Academy of Sciences of Armenia, 105(3):249-255
[6] Karapetyan KA (2001) O prochnosti i deformativnykh svoystvakh stekloplastikovykh trub pri povtorno-staticheskikh nagruzheniyakh v zavisimosti ot otkloneniy oriyentatsii armirovaniya (in Russ., Strength and deformation properties of fiberglass tubes under repeated static loading depending on deviations in reinforcement orientation). Proceedings of the National Academy of Sciences of Armenia. Mechanics 54(2):70-79
[7] Limonov VA, Razin AF, Mikelson MY (1992) Strength and fatigue limit of fabric base composites under combined static shear and cyclic compressive stresses. Mechanics of Composite Materials 28(3):229-236, DOI 10.1007/BF00604914
[8] Martirosyan MM (1971) Polucheniye pressovannykh tonkostennykh trub iz stekloplastikov (in Russ., Production of pressed thin-walled fiberglass tubes. Industry of Armenia (10):56-57

[9] Nepershin RI, Klimenov VV (1987) Optimum computer design of a composite cardan shaft according to the criteria of cost and weight. Mechanics of Composite Materials 22(4):490-494, DOI 10.1007/BF00692262

[10] Shermelor TD (1977) Teoriya uprugosti mikroneodnorodnykh sred (in Russ., Theory of Elasticity of Micro-Inhomogeneous Media). Nauka, Moscow

Chapter 17
Integral Equations in the 3D Aerodynamics of Thin Rotating Blade

Eugene A. Kazakov and Alexei A. Bondarchuk

Abstract The paper studies the aerodynamics of a rotating blade without incident flow (a helicopter "hover“ mode). On the simplest model of a thin blade rectangular in plan, by comparison with a direct numerical calculation in the ANSYS CFX software, the accuracy of known two-dimensional and three-dimensional theories is estimated, which turns out to be very rough even for the main physical parameter - thrust. In this regard, two new semi-analytical theories are proposed, which significantly improve the accuracy of calculations. Both of them are reduced to two-dimensional integral equations similar to the classical equation from the linear theory of a thin wing.

Key words: Two-dimensional integral equations, Aerodynamics, Rotating blade, Semi-analytical model

17.1 Introduction

In propeller aerodynamics, experimental data on aerodynamic characteristics has been accumulated over many years of research. In order to adequately evaluate the experimental data, various engineering and semi-engineering theories have been proposed [4, 6]. Alternatively, in recent decades, direct numerical methods such as the Finite Element Method, Finite Difference Method, Finite Volume Method and others have been developed [10, 13, 18], which provide higher accuracy compared to experimental data than theoretical models. The main disadvantage of the numerical methods is that they require long calculations (weeks or months) even on modern supercomputers, which reduces their efficiency. Hence, a lot of effort was expended to find suitable models, which would permit explicit assessment of aerodynamic

Eugene A. Kazakov (corresponding author) · Alexei A. Bondarchuk
Vorovich Institute of Mathematics, Mechanics and Computer Science, Southern Federal University, Milchakova Street 8a, Rostov-on-Don, 344090, Russian Federation, e-mail: `Eugene.A.KazakovV@yandex.ru`, `melchior@list.ru`

H. Altenbach and A. V. Sahakyan, *Contact Mechanics and Relevant Problems*,
Advanced Structured Materials 253, https://doi.org/10.1007/978-3-032-20869-9_17

parameters in real time. Unfortunately, nowadays there is no analytical theory with acceptable accuracy even for thrust force.

The current situation seems to be strange, since for finite-span wings there is an elegant semi-analytical theory of small perturbations (a linear theory), the so called "discrete vortex method" (DV method) [1, 8], or the "panel method" [7]. They are reduced to two-dimensional integral equations and provide a good accuracy within one hour on a personal computer, even for very dense grids over the wing surface [1, 7, 8, 14].

The first works in propeller aerodynamics belong to the classics [5, 11, 16], who described the qualitative structure of the vortex field. Unfortunately, the general qualitative approach provide only a rough approximation to main characteristics, such as thrust, overturning moment, power, etc. The next section shows that 2d and even rough 3d analytical models give a serious error compared with a direct numerical solution.

In the present work, we propose a semi-analytical approach that allows us to significantly reduce the error of theoretical results. In order to test the accuracy of various theories, we limit ourselves to the case of rotation of a single absolutely thin blade in the form of a flat rectangular plate installed at a fixed angle α to the plane of rotation, without an incident flow (helicopter "hover" mode), see Fig. 17.1, which is not effective in the aerodynamic sense, but the simplest for comparison with a numerical simulation by the ANSYS CFX.

17.2 Review of Existing Methods in the Theory of Rotating Blade

Let the thin blade rotate around the horizontal axis x_1 with a constant angular velocity ω and an installation angle α, so that at the current moment of time its line of symmetry along the span coincides with the axis x_2. It is meant that any three-dimensional point in the space has its natural Cartesian coordinates: $x = (x_1, x_2, x_3)$, $y = (y_1, y_2, y_3)$, etc. Obviously, if the plate is not twisted, then the installation angle α is constant both along the chord and along the span. In the linear theory α is small, the plate is slightly inclined to the vertical plane x_2x_3, and the no-through boundary

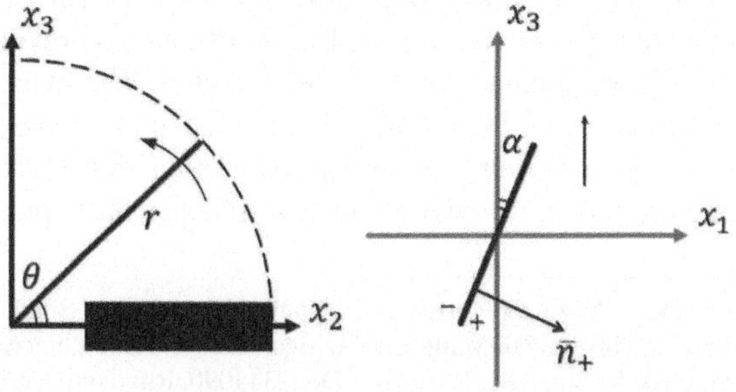

Fig. 17.1 Rotation of a thin rectangular blade in a stationary, unbounded volume of a non-viscous incompressible fluid

condition can be satisfied not on the blade itself, but transferred to the vertical plane x_2x_3 – the projection of the blade on this plane. Let the blade along the span be $l \le x_2 \le L$, and along the chord $-b \le x_3 \le b$. Thus, blade's span is $L - l$, and the chord length is $2b$.

The simplest 2d model is that in each normal section orthogonal to the axis x_2, a flow around the rotating blade is considered as a plane (2d) problem. Hence, the flow does not depend on the stream in other normal sections, and the local incident velocity is $V_0 = \omega x_2$, if the chord length is considerably smaller than blade's length along the span. Since in the 2d problem the lift coefficient is $c_L = 2\pi\alpha$ [2, 9], where α is blade's installation angle (in the 2d theory it is equal to the angle of attack), then the total thrust force is equal to

$$T = c_L \int_l^L \frac{\rho V_0^2 (2b)\, dx_2}{2} = (2\pi\alpha)(\rho\omega^2 b)\int_l^L x_2^2\, dx_2 = \frac{2}{3}\pi\rho b\,\omega^2\alpha(L^3 - l^3), \quad (17.1)$$

where ρ is mass density of the fluid. The comparison, for a test blade, between this theory and the direct numerical calculation in the ANSYS CFX, is presented in the 2nd row of Table 17.1. This theory is the worst.

The classical theory of propeller by Joukowsky [11] is not applicable to the helicopter "hover" mode. It does not describe heavily loaded propellers and without incident flow gives a zero value for axial efficiency.

Then, let us pass from the rough 2d model to a 3d model, with a classical linear "small perturbations" theory of a thin finite-span wing [1, 2, 8, 14]. In this theory the problem is reduced to a two-dimensional integral equation (with $(x_2, x_3) \in S$)

$$\frac{1}{4\pi}\iint_S \left[\frac{y_3 - x_3}{\sqrt{(x_3 - y_3)^2 + (x_2 - y_2)^2}} + 1\right]\frac{g(y_2, y_3)\, dy_2 dy_3}{(x_2 - y_2)^2} = -\rho V_0^2\alpha\,, \quad (17.2)$$

Table 17.1: Comparison of different models for the thrust force T (in Newtons), calculated for a test blade.

α, degrees	3	5	8	10	14	17
ANSYS CFX,	0.139	0.243	0.429	0.558	0.816	0.994
2D model, Eq. (17.1),	0.280	0.467	0.746	0.933	1.306	1.586
relative error	101 %	92 %	74 %	67 %	60 %	60 %
3D model, Sect. 17.2,	0.170	0.283	0.453	0.566	0.792	0.962
relative error	22 %	16 %	6 %	1 %	3 %	3 %
3D model, Sect. 17.4,	0.163	0.271	0.434	0.543	0.760	0.922
relative error	17 %	12 %	1 %	3 %	7 %	7 %
3D model, Sect. 17.5,	0.107	0.219	0.404	0.532	0.791	0.982
relative error	23 %	10 %	6 %	5 %	3 %	1 %

where S is a 2d projection of the thin wing onto plane x_2x_3, V_0 is the incoming velocity along the x_3-axis, α is a small installation angle, and function $g = (p_- - p_+)$, $(x_2, x_3) \in S$ is the pressure jump between the lower and upper faces of the wing (a local lift). Usually, the discretization in Eq. (17.2) is performed by the DV method [1, 8]. By applying this equation to a rotating blade, the incoming velocity becomes different for each section along the span, in the right-hand side of (17.2), being proportional to the distance from the rotation axis: $V_0 = \omega x_2$. The thrust T is a 2d integral over S, of function $g(y_2, y_3)$ found in this way.

The thrust by this theory is presented in the 3rd row of Table 17.1. Obviously, the accuracy of even rough 3d theory is noticeably better than in the 2d theory, which operates with the profile, not with the spatial surface of the blade. However, it is still rough, and some improved models are proposed in the next Sections.

17.3 A 3D Model of the Thin Rotating Blade

This approach is based on the linear theory of "small perturbations" for non-viscous fluid. This means that if the attack angle is zero and initially the fluid mass was at rest, then this remains at rest even with the rotating blade. After that, if $\alpha \neq 0$, $\alpha \ll 1$, then for every non-trivial physical quantity we neglect all higher-order powers, keeping only its linear (i.e. first-power) term. Let the working surface of the blade be $x_1 = f(x_2, x_3)$, see Fig. 17.1. Let the blade twist and angle of attack $\alpha = \partial f/\partial x_3$ be small: $|\partial f/\partial x_2| \ll 1$, $|\partial f/\partial x_3| \ll 1$. Under such assumptions, the angle between the normal vector to the blade surface and the x_1 axis is also small. This permits to transfer the no-through boundary condition from blade's surface to its projection S in the plane of rotation x_2x_3. It is assumed that there is no vortex everywhere except the vortex sheet, which slides off away from the trailing edge. The problem is stationary in the rotating coordinate system.

The kinematic condition implies for the velocity potential ($\overline{v} = \text{grad}\varphi$):

$$\varphi(x_1, r, \omega t + \theta, t) = \varphi(x_1, r, \theta, 0)\,, \Longrightarrow \omega\frac{\partial\varphi}{\partial\theta} + \frac{\partial\varphi}{\partial t} = 0\,, \sim \frac{\partial\varphi}{\partial t} = -\omega\frac{\partial\varphi}{\partial\theta}\,, \tag{17.3}$$

where r, θ, x_1 are coordinates of the natural cylindrical coordinate system.

The linearization means that perturbations arise additionally to the background of the main flow. If the latter has the following pressure and velocity: p_0, $\overline{V}_0$ in the incoming flow – far from the rotating blade, then full characteristics at any point are: $p = p_0 + p'$, $\overline{v} = \overline{V}_0 + \overline{v}'$, where quantities with prime are related to the perturbed flow. The linear theory means that the quantities with the prime are significantly smaller than in the main flow (with subscript "zero"), then higher powers of perturbed quantities may be neglected. This is valid if $V_0 \neq 0$, i.e. with an incoming velocity, like in the case of wind turbine aerodynamics [14]. Strictly speaking, this is not correct in the helicopter "hover" regime, when $V_0 = 0$. In the meantime, it is accepted that linear theory can be applied also in this case, i.e. only first power of the velocity may be kept in all equations. In this case, velocity's perturbation $|\overline{v}'|$ is small, say if compared with $\sqrt{p_0/\rho} \approx 318\, m/s$.

In the "hover" regime at hand, with trivial incident velocity: $V_0 = 0$, the perturbed and full velocities are the same physical fields: $\overline{v}' = \overline{v}$. The velocity is thus a small vector, hence the prime may be omitted.

The Cauchy-Lagrange integral, after the linearization, leads to the following relations:

$$\frac{p}{\rho} + \frac{v^2}{2} + \frac{\partial \varphi}{\partial t} = \frac{p_0}{\rho}, \sim \frac{p}{\rho} + \frac{\partial \varphi}{\partial t} = \frac{p_0}{\rho}, \implies \frac{p'}{\rho} = -\frac{\partial \varphi}{\partial t}, \sim p' = \rho\omega \frac{\partial \varphi}{\partial \theta}. \quad (17.4)$$

The no-through boundary condition implies: $(\overline{v} \cdot \overline{n}) = (\overline{v}_b \cdot \overline{n})$, where the subscript "$b$" is taken from the word "blade". The components of the unit normal are known

$$\overline{n} = \frac{\{1, -\partial f/\partial x_2, -\partial f/\partial x_3\}}{\left[1 + (\partial f/\partial x_2)^2 + (\partial f/\partial x_3)^2\right]^{1/2}} \approx \left\{1, -\frac{\partial f}{\partial x_2}, -\frac{\partial f}{\partial x_3} = -\alpha\right\}, \quad (17.5)$$

where the linearization over small partial derivatives of function f has been applied.

To be more specific, we also assume that the blade is narrow: $|x_3|/x_2 \ll 1$, so that the product of quantity x_3 multiplied by any other small quantity should be neglected too, if considered over blade's surface. Now, since $\overline{v}_b = \{0, -\omega x_3, \omega x_2\}$, the no-through boundary condition implies, after linearization ($v_1 = \partial\varphi/\partial x_1$):

$$v_1 - v_2 \frac{\partial f}{\partial x_2} - v_3 \alpha = \omega x_3 \frac{\partial f}{\partial x_2} + \omega x_2(-\alpha), \Rightarrow v_1 = -\omega x_2 \alpha\,, \ (x_2, x_3) \in S\,, \quad (17.6)$$

since the second, the third, and the fourth terms in the first equality in (17.6) are asymptotically small.

It follows from Eq. (17.4) that, since for incompressible fluid potential φ satisfies the Laplace equation: $\Delta\varphi = 0$, then the aerodynamic pressure satisfies the same equation too: $\Delta p' = 0$. In this case, the basic integral representation of the classical potential theory [2, 3, 15]

$$p'(x) = \iint_S \left[p'(y) \frac{\partial G(y,x)}{\partial n_y} - \frac{\partial p'(y)}{\partial n_y} G(y,x) \right] ds_y, \ G(y,x) = \frac{1}{4\pi|y-x|} \quad (17.7)$$

gives the pressure at arbitrary point x in the medium in terms of an integral over the boundary surface. Here $G(y,x)$ is the Green's function. Note that (17.7) is written for perturbed, not full, pressure function, to avoid a contribution of infinitely distant surface, since the perturbed pressure vanishes in the far-zone. Here S is a closed surface placed around the blade at infinitely small distance, $\overline{n}_y$ is the outer unit normal to S.

Let us introduce the natural cylindrical coordinate system, so that

$$x = (x_1, x_2, x_3) = (r, \theta, x_1), \qquad y = (y_1, y_2, y_3) = (\mu, \psi, y_1)\,. \quad (17.8)$$

Then, by using Eq. (17.4), representation (17.7) can be rewritten as follows:

$$\rho\omega\frac{\partial\varphi(x)}{\partial\theta} = \iint_S \left[p'(y)\,\frac{\partial G(y,x)}{\partial n_y} - \rho\omega\frac{\partial v_n(y)}{\partial\psi}\,G(y,x)\right]ds_y\,, \tag{17.9}$$

where S is still a double-sided closed surface surrounding the blade. On the positive side of this surface, $v_n(y) = v_1(y)$ for $y_1 = +0$, and on the negative one: $v_n(y) = -v_1(y)$ for $y_1 = -0$. Since boundary condition (3.4) predetermines the same value of v_1 on the both faces of the blade, the contribution of the second term of the integrand in (17.9), over two faces of the closed surface S, cancel each other out, as in summation of two opposite quantities. In the same way, quantity $\partial G(y,x)/\partial n_y$ in the first term of the integrand (17.9) has opposite signs over the opposite faces of surface S, with $y_1 = +0$ and $y_1 = -0$. As a result, this generates the "jump" of the pressure when crossing blade's surface, i.e. the difference of the aerodynamic pressure for $y_1 = -0$ and $y_1 = +0$. If one designates this difference as a new unknown function g: $p'(y_1 = -0, y_2, y_3) - p'(y_1 = +0, y_2, y_3) = g(y_2, y_3)$, which is again a local thrust like in Eq. (17.2), then representation (17.9) can be rewritten as follows:

$$\rho\omega\frac{\partial\varphi(x)}{\partial\theta} = -\iint_{S_+} g(y_2, y_3)\,\frac{\partial G(y,x)}{\partial y_1}\,dy_2dy_3\,, \tag{17.10}$$

where S_+ is the upper face of the blade ($y_1 = +0$). Further, for the sake of brevity, subscript "+" is omitted.

Obviously, Eq. (17.10) can be rewritten in the equivalent form

$$\rho\omega\frac{\partial\varphi(x)}{\partial\theta} =$$

$$-\frac{1}{4\pi}\iint_S g(y_2, y_3)\frac{\partial}{\partial y_1}\left\{\frac{1}{[(y_1-x_1)^2+(y_2-x_2)^2+(y_3-x_3)^2]^{1/2}}\right\}_{y_1=0} dy_2dy_3 \tag{17.11}$$

$$= -\frac{x_1}{4\pi}\iint_S \frac{g(y_2, y_3)\,dy_2dy_3}{[x_1^2 + (y_2 - x_2)^2 + (y_3 - x_3)^2]^{3/2}}\,.$$

Now, the derivative applied to Eq. (17.11) with respect to x_1 implies

$$\rho\omega\left.\frac{\partial^2\varphi}{\partial\theta\partial x_1}\right|_{x_1=0} = -\frac{1}{4\pi}\iint_S \frac{g(y_2, y_3)\,dy_2dy_3}{[(y_2 - x_2)^2 + (y_3 - x_3)^2]^{3/2}}\,. \tag{17.12}$$

The last step, to reduce the problem to a two-dimensional integral equation is to integrate Eq. (17.12) with respect to θ, since the quantity $v_1 = \partial\varphi/\partial x_1$ is involved in the boundary condition on blade's surface (17.6). However, direct integration is not simple. That is why we first consider an approximate approach in the next Section.

17.4 A Simplified 3D Theory on the Basis of Representation (17.12)

An approach may be founded on the following approximate relation: $d\theta = x_2 dx_3/(x_2^2 + x_3^2)$, which heuristically follows from a view to Fig. 17.1. A more detailed analysis shows that this relation is only approximate. In the meantime, in frames of such an approach, one obtains from (17.12)

$$\rho\omega \frac{\partial^2 \varphi}{\partial x_1 \partial x_3}\bigg|_{x_1=0} = -\frac{x_2}{4\pi(x_2^2+x_3^2)} \iint\limits_S \frac{g(y_2,y_3)\,dy_2 dy_3}{[(y_2-x_2)^2+(y_3-x_3)^2]^{3/2}}. \tag{17.13}$$

Now, by integrating (17.13) along variable x_3 upstream, from the current value to infinity, one obtains

$$\rho\omega v_1|_{x_1=0} = \rho\omega \frac{\partial \varphi}{\partial x_1}\bigg|_{x_1=0}$$

$$= \frac{x_2}{4\pi} \int\limits_{x_3}^{\infty} \frac{dz}{z^2+x_2^2} \iint\limits_S \frac{g(y_2,y_3)\,dy_2 dy_3}{[(y_2-x_2)^2+(y_3-z)^2]^{3/2}}, \quad (v_1|_{x_3=+\infty}=0). \tag{17.14}$$

Then, by boundary condition (17.6), the problem is reduced to the two-dimensional integral equation (where $(x_2, x_3) \in S$)

$$\frac{1}{4\pi} \iint\limits_S I(x_2,x_3,y_2,y_3) g(y_2,y_3) dy_2 dy_3 = -\rho\omega^2\alpha,$$

$$I = \int\limits_{x_3}^{\infty} \frac{dz}{(z^2+x_2^2)[(z-y_3)^2+(y_2-x_2)^2]^{3/2}}. \tag{17.15}$$

Integral I in (17.15) can be calculated applying integration by parts

$$u = \frac{1}{z^2+x_2^2}, \quad du = -\frac{2z dz}{(z^2+x_2^2)^2}, \quad dv = \frac{dz}{[(z-y_3)^2+(y_2-x_2)^2]^{3/2}},$$

$$v = \frac{z-y_3}{(x_2-y_2)^2\sqrt{(z-y_3)^2+(y_2-x_2)^2}}, \tag{17.16}$$

therefore,

$$I = -\frac{x_3 - y_3}{(x_3^2 + x_2^2)(x_2 - y_2)^2\sqrt{(x_3 - y_3)^2 + (y_2 - x_2)^2}} + \frac{2}{(x_2 - y_2)^2}\int\limits_{x_3}^{\infty}\frac{z(z - y_3)\,dz}{(z^2 + x_2^2)^2\sqrt{(z - y_3)^2 + (y_2 - x_2)^2}}. \tag{17.17}$$

Let us designate the last integral in (17.17) as I_1. Since variables x_3 and y_3 are small on the blade, let us approximately set in Eq. (17.17): $x_3 \approx 0$, $y_3 \approx 0$, coming to the following equality

$$I_1 \approx \int\limits_0^{\infty}\frac{z^2 dz}{(z^2+x_2^2)^2\sqrt{z^2+(y_2-x_2)^2}} = \int\limits_0^{\infty}\frac{dz}{(z^2+x_2^2)\sqrt{z^2+(y_2-x_2)^2}} - x_2^2\int\limits_0^{\infty}\frac{dz}{(z^2+x_2^2)^2\sqrt{z^2+(y_2-x_2)^2}}. \tag{17.18}$$

The both integrals in (17.18) are tabulated [12]

$$I_2 = \int\limits_0^{\infty}\frac{dz}{(z^2+x_2^2)\sqrt{z^2+(y_2-x_2)^2}} = \frac{1}{x_2\sqrt{|x_2^2-(x_2-y_2)^2|}}\begin{cases}\ln\dfrac{\sqrt{x_2^2-(x_2-y_2)^2}}{|x_2-y_2|}, & |x_2-y_2| < x_2, \\ \operatorname{atan}\dfrac{\sqrt{(x_2-y_2)^2-x_2^2}}{x_2}, & |x_2-y_2| \ge x_2\end{cases} \tag{17.19}$$

and

$$I_1 = \frac{1}{2\left[(x_2 - y_2)^2 - x_2^2\right]} \times \left[(x_2 - y_2)^2 I_2 - 1\right]. \tag{17.20}$$

As a result, the two-dimensional integral equation (17.15) is reduced to the following form:

$$\iint\limits_S K(x_2, x_3, y_2, y_3)\, g(y_2, y_3)\, dy_2 dy_3 = -\rho\,\omega^2\alpha, \qquad (x_2, x_3) \in S, \tag{17.21}$$

where

$$K = \frac{y_3 - x_3}{(x_3^2 + x_2^2)(x_2 - y_2)^2\sqrt{(x_3 - y_3)^2 + (x_2 - y_2)^2}} + \frac{1}{x_2^2(x_2 - y_2)^2} + \frac{I_2}{(x_2 - y_2)^2 - x_2^2} - \frac{1}{x_2^2\left[(x_2 - y_2)^2 - x_2^2\right]}. \tag{17.22}$$

Here two first hyper-singular terms (as $y_2 \to x_2$) in Eq. (17.22) are analogues to the kernel in the classical 3d wing theory, Eq. (17.2). The third and the fourth regular terms in (17.22) are responsible for blade's rotation.

The qualitative similarity of kernels in Eqs. (17.22) and (17.2) permits direct treatment of equation (17.21) by the DV method or by the Panel method, by analogy to Eq. (17.2) [1, 7, 8, 14]. The results of such a treatment for a test blade, whose geometry is described below, are presented in the 4th row of Table 17.1.

17.5 An Advanced 3D Theory on the Basis of Representation (17.12)

Relation $d\theta = x_2 dx_3/(x_2^2 + x_3^2)$, which is the key point for the method of Sect. 17.4, is approximate. Exact integration relatively variable θ in Eq. (17.12) can be performed if the derivative with respect to θ is rewritten in Cartesian coordinates:

$$\left.\frac{\partial^2\varphi}{\partial\theta\partial x_1}\right|_{x_1=0} = \left.\frac{\partial v_1}{\partial\theta}\right|_{x_1=0} = r\left.\frac{\partial v_1}{\partial\tau}\right|_{x_1=0} = r\left(\nabla v_1 \cdot \overline{\tau}\right)\big|_{x_1=0} = r\left[\frac{\partial v_1}{\partial x_2}\left(\frac{-x_3}{r}\right) + \frac{\partial v_1}{\partial x_3}\left(\frac{x_2}{r}\right)\right]\Bigg|_{x_1=0}, \tag{17.23}$$

since $\overline{\tau} = \{0, -x_3/r, x_2/r\}$ is the unit vector tangential to the radius. Then Eq. (17.12) takes the form:

$$\rho\omega\left.\left(x_2\frac{\partial v_1}{\partial x_3} - x_3\frac{\partial v_1}{\partial x_2}\right)\right|_{x_1=0} = -\frac{1}{4\pi}\iint_S \frac{g(y_2, y_3)\, dy_2 dy_3}{[(y_2 - x_2)^2 + (y_3 - x_3)^2]^{3/2}}. \tag{17.24}$$

Let us apply to (17.24) the integration relatively variable x_3 upstream, from the current value to infinity, like in the previous Sect. 17.4. Then, for the right-hand side of (17.24) one obtains:

$$\int_{x_3}^{\infty}\left(-\frac{dz}{4\pi}\right)\iint_S \frac{g(y_2, y_3)\, dy_2 dy_3}{[(y_2 - x_2)^2 + (y_3 - z)^2]^{3/2}} = -\frac{1}{4\pi}\iint_S\left[\frac{1}{(y_2 - x_2)^2} + \frac{y_3 - x_3}{(y_2 - x_2)^2\sqrt{(y_2 - x_2)^2 + (y_3 - x_3)^2}}\right] g(y_2, y_3) dy_2 dy_3 \tag{17.25}$$

where a tabulated integral over variable z has been used [12]. Integrating the first term of the left-hand side in (17.24) gives

$$\int_{x_3}^{\infty} x_2 \frac{\partial v_1(0, x_2, z)}{\partial z} dz = x_2 \left[v_1(0, x_2, \infty) - v_1(0, x_2, x_3) \right]$$
$$= x_2 \left[0 - (-\omega x_2 \alpha) \right] = \omega x_2^2 \alpha \,, \tag{17.26}$$

where boundary condition (17.6) has been used, and the fact that the velocity decays upstream.

The integration of the second term in (17.24) is divided to the two intervals:

(1) $x_3 < z < b$ (on the blade), and
(2) $b < z < \infty$ (on the fluid).

On the first interval the value of $\partial v_1 / \partial x_2$ is known from the boundary condition (17.6)

$$-\int_{x_3}^{b} z \left. \frac{\partial v_1}{\partial x_2} \right|_{x1=0} dz = \omega \alpha \int_{x_3}^{b} z dz = \omega \alpha \, \frac{b^2 - x_3^2}{2} \,. \tag{17.27}$$

Outside the blade, from b to ∞ the value of v_1 is unknown. Physically, this is the velocity of the "suction stream" generated by the blade itself in its rotation, and there is no *a priori* information about the structure of this suction stream over the fluid, in the plane of blade's rotation. Therefore, we accept the assumption that $v_1 = 0$ outside the round domain covered by the rotating blade, and inside this ring (where $\ell \leq r \leq L$ the distribution follows the law predetermined by the boundary condition (17.6). With the evident observation that in Eq. (17.6), which is valid on the blade, the quantities x_2 and r are approximately the same: $r = \sqrt{x_2^2 + x_3^2} \approx x_2$, we accept in the region (2) the quantity v_1 to be simply proportional to the polar radius r. The behavior of v_1 over polar angle is more complex, and one should take into account the qualitative properties of the flaw. Namely, one can notice that the aerodynamic field is more intensive near the trailing edge of the blade than near the leading edge, since the vortex sheet slides off away just from the trailing edge [17]. Thus, we accept the hypothesis that the behavior of v_1 inside the ring $l < r < L$, round from the trailing to the leading edge decreases monotonically as a certain function, say exponentially:

$$v_1|_{x_1=0} = -\omega r \alpha \, e^{k(2\pi - \theta)} \,, \qquad (k < 0) \,. \tag{17.28}$$

In this case, $\theta = 0$ is related to velocity near the leading edge, and $\theta = 2\pi$ – near the trailing edge. Besides, since by the Kutta-Joukowsky law [8,13,14,18], the solution should be regular near the trailing edge, the hypothesis (17.28) automatically provides continuity of v_1 when approaching from the fluid to the trailing edge, as Eq. (17.28) with $\theta = 2\pi$ confirms the boundary condition (17.6) on the blade surface. A concrete choice of the monotonically decreasing function of θ in (17.28) has only a slight influence on the precision of this model.

Let us calculate the integral along the second interval $b < z < \infty$, analogous to (17.27)), with v_1 taken in the form (17.28)

$$-\frac{\partial}{\partial x_2}\int_b^\infty x_3 v_1 dx_3 = \omega\alpha\frac{\partial}{\partial x_2}\int_b^{\sqrt{L^2-x_2^2}} x_3 r e^{k(2\pi-\theta)}dx_3$$

$$= \omega\alpha\frac{\partial}{\partial x_2}\int_b^{\sqrt{L^2-x_2^2}} x_3\sqrt{x_2^2+x_3^2}\,\mathrm{e}^{k[2\pi-\arctan(x_3/x_2)]}dx_3$$

$$= \omega\alpha\int_b^{\sqrt{L^2-x_2^2}} x_3\frac{\partial}{\partial x_2}\left\{\sqrt{x_2^2+x_3^2}\,\mathrm{e}^{k[2\pi-\arctan(x_3/x_2)]}\right\}dx_3 \tag{17.29}$$

$$+\omega\alpha\frac{-x_2}{\sqrt{L^2-x_2^2}}\sqrt{L^2-x_2^2}\,L\,\mathrm{e}^{k[2\pi-\arctan(\sqrt{L^2-x_2^2}/x_2)]}$$

$$= \omega\alpha\, h(x_2,k) - \omega L x_2\alpha\,\mathrm{e}^{k[2\pi-\arctan(\sqrt{L^2-x_2^2}/x_2)]},$$

where function $h(x_2, k)$ in Eq. (5.7a) is calculated numerically:

$$h(x_2,k) = \int_b^{\sqrt{L^2-x_2^2}} \frac{x_2x_3 + x_3^2 k}{\sqrt{x_2^2+x_3^2}}\,\mathrm{e}^{k[2\pi-\arctan(x_3/x_2)]}dx_3\,. \tag{17.30}$$

Collecting all formulas together, one comes from (17.24), (17.25) to the following integral equation:

$$-\iint_S\left[\frac{1}{(y_2-x_2)^2}+\frac{y_3-x_3}{(y_2-x_2)^2\sqrt{(y_2-x_2)^2+(y_3-x_3)^2}}\right]g_1(y_2,y_3)dy_2dy_3$$

$$= x_2^2+\frac{b^2-x_3^2}{2} - Lx_2\,\mathrm{e}^{k[2\pi-\arctan(\sqrt{L^2-x_2^2}/x_2)]} + h(x_2,k)\,, \tag{17.31}$$

$$g_1(y_2,y_3) = \frac{g(y_2,y_3)}{4\pi\omega\alpha}\,.$$

Now, the only remaining unknown parameter is parameter k in Eq. (17.28). In order to determine this, we use the impulse balance for a half-space upstream, i.e. for the half-space $x_1 > 0$. The geometry of the chosen closed domain, for the impulse balance theorem is shown in Fig. 17.2, where the green line on the right, top and bottom, shows the outer infinite boundary of the half-space.

The factors changing the impulse for a half-space are

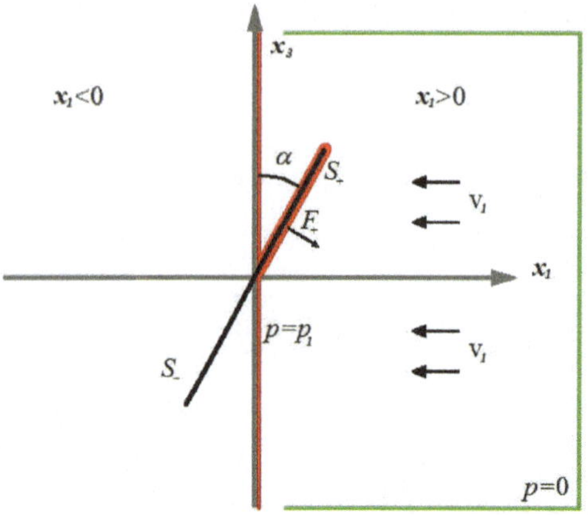

Fig. 17.2: Geometry of a closed region with red and green boundary lines, to which the impulse balance theorem is applied.

1. the pressure on that part of the blade, which is immersed in this half-space;
2. the pressure on the boundary $x_1 = 0$;
3. the transfer of the impulse by the flow, which, according to the assumption above, is not equal to zero only in the disk swept by the blade.

To define the pressure in the fluid, let us start from Eq. (17.11) written in a more general form:

$$p' = \rho\omega\frac{\partial\varphi}{\partial\theta} = \frac{1}{4\pi}\iint\limits_S \frac{(y_1 - x_1)g(y_2, y_3)\,dy_2dy_3}{[(y_1 - x_1)^2 + (y_2 - x_2)^2 + (y_3 - x_3)^2]^{3/2}}\,. \tag{17.32}$$

This formula is used, to calculate pressure $p' = p'_1$ in the plane $x_1 = 0$. Since the blade is inclined at a small angle α to the plane, then under the integral in the linear approximation we have $y_1 = \alpha y_3$, which leads to the following relationship

$$p'_1 = \frac{\alpha}{4\pi}\iint\limits_S \frac{y_3\, g(y_2, y_3)\,dy_2dy_3}{[\alpha^2 y_3^2 + (y_2 - x_2)^2 + (y_3 - x_3)^2]^{3/2}}\,. \tag{17.33}$$

Then the total force acting to the half-space from its boundary is as follows

$$
\begin{aligned}
F_1 &= \iint\limits_{-\infty}^{\infty} p'_1 dx_2 dx_3 \\
&= \frac{\alpha}{4\pi} \iint\limits_{S} y_3 g(y_2, y_3) dy_2 dy_3 \iint\limits_{-\infty}^{\infty} \frac{dx_2 dx_3}{[\alpha^2 y_3^2 + (x_2 - y_2)^2 + (x_3 - y_3)^2]^{3/2}} \\
&= \frac{\alpha}{4\pi} \iint\limits_{S} y_3 g(y_2, y_3) dy_2 dy_3 \iint\limits_{-\infty}^{\infty} \frac{dx_2 dx_3}{(\alpha^2 y_3^2 + x_2^2 + x_3^2)^{3/2}} \\
&= \frac{1}{2} \iint\limits_{S} \operatorname{sign}(y_3) g(y_2, y_3) dy_2 dy_3 \\
&= \frac{1}{2} \iint\limits_{S_+} g(y_2, y_3) dy_2 dy_3 - \frac{1}{2} \iint\limits_{S_-} g(y_2, y_3) dy_2 dy_3, \qquad (\alpha > 0),
\end{aligned}
\tag{17.34}
$$

where the integral

$$
\begin{aligned}
\iint\limits_{-\infty}^{\infty} \frac{dx_2 dx_3}{(x_2^2 + x_3^2 + b^2)^{3/2}} &= 4 \iint\limits_{0}^{\infty} \frac{dx_2 dx_3}{(x_2^2 + x_3^2 + b^2)^{3/2}} \\
&= 4 \int\limits_{0}^{\infty} \frac{dx_3}{x_3^2 + b^2} = 4 \frac{\pi}{2|b|} = \frac{2\pi}{|b|}
\end{aligned}
\tag{17.35}
$$

can easily be calculated by successive one-fold integration [12]. In Eq. (17.34) S_+ is a part of the blade for which $y_3 > 0$ (this part is immersed into the half-space at hand); and S_- is a part for which $y_3 < 0$.

The total force acting to the half-space from the part of the blade S_+, immersed in this half-space, is

$$
F_+ = - \iint\limits_{S_+} g(y_2, y_3) dy_2 dy_3 \,. \tag{17.36}
$$

Obviously, the following relation takes place

$$
\begin{aligned}
F_1 + F_+ &= -\tfrac{1}{2} \iint\limits_{S_+} g(y_2, y_3) dy_2 dy_3 \\
&- \frac{1}{2} \iint\limits_{S_-} g(y_2, y_3) dy_2 dy_3 = -\frac{1}{2} \iint\limits_{S} g(y_2, y_3) dy_2 dy_3 .
\end{aligned}
\tag{17.37}
$$

The impulse flow, by taking into account Eq. (17.28), can be written as follows

$$
\begin{aligned}
P &= \rho \iint v_1^2 r dr d\theta = \rho\omega^2\alpha^2 \int_l^L r^3 \int_0^{2\pi} e^{2k(2\pi-\theta)} d\theta dr \\
&= \frac{\rho\omega^2\alpha^2}{8k}(L^4 - l^4)(e^{4\pi k} - 1).
\end{aligned}
\tag{17.38}
$$

For the impulse balance to be valid, the following equality should be satisfied

$$
F_1 + F_+ + P = 0\,, \tag{17.39}
$$

which, by taking into account Eqs. (17.37) and (17.38), is equivalent to

$$
\begin{aligned}
\iint_S g(y_2, y_3)dy_2dy_3 &= \frac{\rho\omega^2\alpha^2}{4k}(L^4 - l^4)(e^{4\pi k} - 1), \\
\iint_S g_1(y_2, y_3)dy_2dy_3 &= \frac{\rho\omega\alpha}{16\pi k}(L^4 - l^4)(e^{4\pi k} - 1).
\end{aligned}
\tag{17.40}
$$

Therefore, this method implies that, for any value of parameter k, the two-dimensional integral equation (17.31) can be solved, like the classical lifting-surface Eq. (17.2), by the standard DV method [1, 7, 8, 14]. The true value of k must be chosen to ensure that equality (17.40) is satisfied. In Table 17.2 for the test blade described below, we demonstrate the values of parameter k for various values of the installation angle. The mesh on blade's rectangular surface, in application of the DV method, is $N_2 = 135$ (along x_2 axis), $N_3 = 35$ (along x_3 axis). The results by this model are presented in the last row of Table 17.1.

The physical parameters of the test blade correspond to those of a mini quadcopter DJI Mavic Air (China): the angular velocity $f = 200$ Hz, hence $\omega = 2\pi f = 1257$ rad/sec, the span is from $l = 0.5$ cm to $L = 6.5$ cm, the chord $2b = 1$ cm, the mass density of the air $\rho = 1.177$ kg/m^3. The dynamic viscosity of the medium $\mu = 1.831 \cdot 10^{-7}$ kg/(m $\cdot$ sec) used in the ANSYS CFX is specially taken to be two orders of magnitude smaller than the real viscosity of air, to provide a closer comparison with theoretical models, which operate with a non-viscous medium. The temperature is 25 degrees Celsius. The thin plate rectangular in plan, in the ANSYS CFX digital model, is taken plane (untwisted) with a ceratin angle of installation, constant along both span and chord directions, which varies in different regimes of calculations from 3 to 17 degrees. The values of the thrust force obtained by the ANSYS CFX computations are used as test values, to evaluate precision of the proposed models presented in Table 17.1.

Table 17.2: The values of parameter k for the test blade.

α^0	3	5	8	10	14	17
k	-0.1940	-0.2805	-0.3978	-0.4736	-0.6263	-0.7438

17.6 Conclusions

1. The precision of the simple 2d theory is too poor. It is very surprising that this is still widely used in the engineering practice.
2. To improve the accuracy of the theoretical prediction of blade's thrust force, we propose two new 3d models, Sects. 17.4 and 17.5. The both these models are allied with the lifting surface theory in the 3d uniform flow, being in their essence some linear theories of small perturbations.
3. None of the theories considered provides satisfactory accuracy for extremely small angles $\alpha < 5^0$, when compared with the direct numerical solution by the ANSYS CFX program. It seems that qualitative aerodynamic properties in the small-angle regime are absolutely different from the case of moderately small angles $\alpha \geq 5^0$, hence some other physical ideas should be applied as an adequate model in this range.
4. Solution by the method of Section 5 seems to be the best among all others, for moderately small angles $\alpha \geq 5^0$. It is easily seen that this model is the only, which provides an acceptable and a monotonically decreasing relative error with increasing α, out of the range of extremely small angles.
5. A possible explanation, why the theory of Sect. 17.5, which demonstrates the best accuracy among all others, is so rough for extremely small α, may be connected with the fact that this theory implicitly evaluates the influence of suction, which is considerable for larger angles, being insignificant for $\alpha < 5^0$.

References

[1] Belotserkovsky CM, Lifanov IK (1992) Method of Discrete Vortices. CRC Press, Boca Raton, Florida

[2] Bisplinghoff RL, Ashley H, Halfman RL (1955) Aeroelasticity. Addison-Wesley, New York

[3] Colton D, Kress R (1983) Integral Equation Methods in Scattering Theory. John Wiley, New York

[4] Johnson W (1980) Helicopter Theory. Princeton University Press, New York

[5] Glauert H (1935) Airplane Propellers, In: Durand WF (ed.) Aerodynamic Theory: A General Review of Progress Under a Grant of the Guggenheim Fund for the Promotion of Aeronautics, **4(50)**, 169–360, Springer, New York, DOI 10.1007/978-3-642-91487-4_3

[6] Hansen MOI (2007) Aerodynamics of Wind Turbines. Taylor & Francis: Boca Raton, Florida

[7] Katz J, Plotkin A (1991) Low-Speed Aerodynamics. From Wing Theory to Panel Methods. McGraw-Hill, New York

[8] Lifanov IK, Poltavskii LN, Vainikko GM (2004) Hypersingular Integral Equations and Their Applications. CRC Press, Boca Raton, Florida

[9] Loitsyanskii LG (1973) Mechanics of Liquids and Gases. Pergamon Press, London

[10] Maukalled F., Mangani L, Darwish M (2016) The Finite Volume Method in Computational Fluid Dynamics, Springer, New York

[11] Okulov VL, Sorensen JN, Wood DH (2015) The rotor theories by Professor Joukowsky: Vortex theories. Progress in Aerospace Sciences 73:19–46, DOI 10.1016/j.paerosci.2014.10.002

[12] Prudnikov AP, Brychkov YA, Marichev OI (1986) Integrals and Series, Vol.1. Gordon and Breach, New York

[13] Roache PJ (1972) Computational Fluid Dynamics. Hermosa Publications, Albuquerque

[14] Sumbatyan MA, Bondarchuk AA, Mescheryakov KI (2018) An efficient numerical algorithm in the classical 3d theory of thin lifting surface in a flow of non-viscous incompressible fluid. Mechanics Research Communications 89:18–22, DOI 10.1016/j.mechrescom.2018.03.001

[15] Sumbatyan MA, Scalia A (2005) Equations of Mathematical Diffraction Theory. CRC Press, Boca Raton, Florida

[16] Wald QR (2006) The aerodynamics of propellers. Progress in Aerospace Sciences 42(2):85–128, DOI 10.1016/j.paerosci.2006.04.001

[17] Zampoli V, Sumbatyan MA (2019) Flow-induced vortex field generated by a thin oscillating plate in an aeroacoustics framework. Zeitschrift für Angewandte Mathematik und Physik (ZAMP) 70, article 33, DOI 10.1007/s00033-019-1081-7

[18] Zienkiewicz OC, Taylor RL (2000) The Finite Element Method, Vol.3: Fluid Dynamics (5th ed.). Butterworth-Heinemann, Oxford

Chapter 18
The Problem of the Interaction of an Axisymmetric Insert and a Pipe with an Internal Heterogeneous Coating Placed in a Rigid Mandrel

Kirill E. Kazakov

Abstract The paper describes the mechanical and mathematical formulation of the problem of the interaction of an axisymmetric rigid insert and a pipe made of viscoelastic aging material, the inner surface of which is covered with an inhomogeneous layer. It is assumed that the pipe is placed in a rigid mandrel. For this problem, an algorithm is specified for constructing an analytical solution in series based on the use of basic functions based on complex functions included in the mathematical model, representing the solution in a special form, and using the eigenfunctions of the Fredholm operator. The type of analytical solution is given. This approach makes it possible to isolate the function associated with the heterogeneity of the coating in the solution by a separate multiplier, which allows calculations to be performed with high accuracy even in the case when the heterogeneity of the coating is described by a rapidly changing function, since it is sufficient to limit the calculations to a relatively small number of terms of the series.

Key words: Contact problem, Aging material, Pipe, Coating, Complex profiles, System of integral equations

18.1 Introduction

The use of multilayer pipes in industry and machinery is due to the fact that such pipes allow taking into account many technological requirements. Some layers provide the necessary structural rigidity, while others, for example, act as electrical insulators, while others protect against aggressive environments, etc. However, the application of such layers can affect the stress-strain state in areas where such multi-

---

Kirill E. Kazakov
Ishlinsky Institute for Problems in Mechanics of the Russian Academy of Sciences, Vernadskogo Ave 101 Bldg 1, Moscow, 119526 & Bauman Moscow State Technical University, 2nd Baumanskaya str., 5, p. 1, Moscow, Russian Federation, e-mail: kazakov@ipmnet.ru, kazakov-ke@bmstu.ru

H. Altenbach and A. V. Sahakyan, *Contact Mechanics and Relevant Problems*,
Advanced Structured Materials 253, https://doi.org/10.1007/978-3-032-20869-9_18

layer pipes are attached to each other or to other structures using various elements: couplings, inserts, nipples, etc. This effect can have both a positive and a negative effect, therefore, taking into account additional layers is important when calculating structures for strength.

Previously, various options for the composition and fastening of pipes were considered. Cases have been considered where the outer coating is heterogeneous and/or has a variable profile, and the pipe is enclosed by a coupling, the inner surface of which can also be complex (see, for example, [4]). The [5] take into account the presence of a rigid shaft inside the pipe. The articles [6, 7] describe the construction of solutions for cases where the coating is internal, and the rigid element is placed inside such a pipe.

This article discusses the case when a rigid insert is placed in a rigid mandrel tube with an internal coating. The purpose of the work is to predict changes in contact stresses over time. As a method of solving the problem under consideration, it is proposed to use a special approach that allows us to construct an analytical solution to the problem in rows, taking into account both the wear of the inner layer and its heterogeneity.

18.2 Physical Formulation of the Problem

An extended axisymmetric cylindrical tube consists of two layers: a viscoelastic, aging, homogeneous outer layer and a thin inner layer, the elastic properties of which depend on the axial coordinate. The layers are interconnected with each other, that is, there is perfect contact between the layers. The described inner-coated pipe is placed in a rigid mandrel, the inner diameter of which coincides with the outer diameter of the pipe, as a result of which there is no tension and gaps. It is assumed that there is no friction between the mandrel and the pipe and smooth contact is carried out. At some point in time, a rigid axisymmetric insert is placed inside such a pipe so that full contact occurs between the inner layer and the insert, since the outer diameter of the insert is larger than the inner diameter of the coating (the length of the contact line coincides with the length of the insert). It is considered that smooth contact is carried out at the border of the insert and the coating. As a result of this interaction, the pipe is deformed. The Fig. 18.1 schematically shows the described interaction. Over time, the stresses and deformations that have arisen in the pipe change, since one of the pipe layers is made of viscoelastic aging material. Assuming that the thickness of the coating is significantly less than all other geometric linear parameters, and its stiffness is less than that of the main pipe layer, it is necessary to find the evolution of contact pressures in the area of interaction between the insert and the coating.

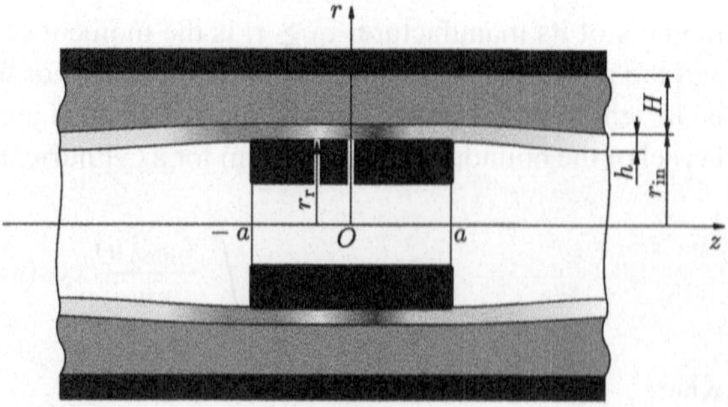

Fig. 18.1 Contact interaction of a two-layer pipe and a rigid insert

18.3 Mathematical Formulation of the Problem

To construct a mathematical model, one can use the results of solving boundary value problems on the deformation of thick homogeneous

$$u_{\text{thick}}(z,t) = \frac{2(1-\nu_t^2)}{\pi} \int_{-a}^{a} k_{\text{ims}}\left(\frac{z-\zeta}{H}\right) \frac{q(\zeta,t)}{E_t} d\zeta \tag{18.1}$$

and thin inhomogeneous

$$u_{\text{thin}}(z,t) = \frac{q(z,t)h}{R_c(z)} \tag{18.2}$$

pipes placed in a rigid mandrel under the action of internal distributed pressure $q(z,t)$ (see, for example, [3, 8, 9]). Their sum and generalization to the case of a viscoelastic aging material of the thick layer allow us to construct a solution of boundary value problem for a two-layer pipe placed in a rigid mandrel under the action of an internally distributed load

$$u_{\text{2layer}}(z,t) = \frac{q(z,t)h}{R_c(z)} + \frac{2(1-\nu_t^2)}{\pi}\left[\int_{-a}^{a} k_{\text{ims}}\left(\frac{z-\zeta}{H}\right) \frac{q(\zeta,t)}{E_t(t-\tau_t)} d\zeta \right.$$

$$\left. - \int_{\tau_0}^{t} K_t(t-\tau_t, \tau-\tau_t) \int_{-a}^{a} k_{\text{ims}}\left(\frac{z-\zeta}{H}\right) \frac{q(\zeta,\tau)}{E_t(\tau-\tau_t)} d\zeta d\tau \right]. \tag{18.3}$$

In Eqs. (18.1)–(18.3), $q(z,t)$ is function describing the contact stresses in the interaction zone,

$$R_c(z) = \frac{E_c(z)[1-\nu_c(z)]}{1-\nu_c(z)-2\nu_c^2(z)}$$

is contact rigidity, $\nu_c(z)$, $E_c(z)$ are the Poisson's ratio and Young's modulus of the inhomogeneous coating, h is the thickness of this coating in the undeformed state, ν_t, $E_t(t)$ are the Poisson's ratio and Young's modulus of the main viscoelastic aging pipe layer; $K_t(t,\tau)$ is the creep kernel of this layer (see, for example, [1–3]), τ_t is the moment of its manufacture, $\tau_0 \geq \tau_t$ is the moment of interaction between the insert and the pipe with the coating, H is the thickness of the viscoelastic aging layer, $2a$ is the length of the insert, coinciding with contact area length, $k_{\text{ims}}(s)$ is the known kernel of the boundary value problem for a cylindrical pipe placed in a rigid mandrel

$$k_{\text{ims}}(s) = \int_0^\infty \frac{L_{\text{ims}}(u)}{u} \cos(su)du,$$

where

$$L_{\text{ims}}(u) = -\frac{uD_1^2(u)}{S_{\text{ims}}(u)}, \quad S_{\text{ims}}(u) = f_{\text{in}}(u) - f(k_{\text{in}}, u)D_1^2(u), \quad k_{\text{in}} = \frac{r_{\text{in}}}{H},$$

$$f_{\text{in}}(u) = k_{\text{in}} u^2 B_1^2(u) - \frac{1}{k_{\text{in}}}, \quad f(k,u) = \frac{2(1-\nu_t)}{k} + u^2 k, \quad kout = k_{\text{in}} + 1,$$

$$B_1(u) = I_0(k_{\text{in}}u)K_1(koutu) + I_1(koutu)K_0(k_{\text{in}}u),$$

$$D_1(u) = I_1(k_{\text{in}}u)K_1(koutu) - I_1(koutu)K_1(k_{\text{in}}u).$$

In these formulas, r_{in} is the inner radius of the viscoelastic layer in the undeformed state; $I_0(u)$, $I_1(u)$, $K_0(u)$, and $K_1(u)$ are modified Bessel functions of the first and second kind.

On the other hand, the displacement of the inner layer is known: it is equal to the difference between the outer radius of the insert and the inner radius of the layer

$$u_{2\text{layer}}(z,t) = r_{\text{r}} + h - r_{\text{in}}, \tag{18.4}$$

where r_{r} is the outer radius of the rigid insert.

A mathematical model of the contact problem considered in this article can be obtained by equating the right-hand sides of equations (18.3) and (18.4):

$$\frac{q(z,t)h}{R_{\text{c}}(z)} + \frac{2(1-\nu_t^2)}{\pi} \Bigg[\int_{-a}^{a} k_{\text{ims}}\left(\frac{z-\zeta}{H}\right) \frac{q(\zeta,t)}{E_t(t-\tau_t)} d\zeta$$
$$- \int_{\tau_0}^{t} K_t(t-\tau_t, \tau-\tau_t) \int_{-a}^{a} k_{\text{ims}}\left(\frac{z-\zeta}{H}\right) \frac{q(\zeta,\tau)}{E_t(\tau-\tau_t)} d\zeta d\tau \Bigg]$$
$$= r_{\text{r}} + h - r_{\text{in}}, \quad z \in [-a,a], \quad t \geq \tau_0. \tag{18.5}$$

The resulting mathematical model is valid if coating is softer than thick layer, it's thickness is much smaller than all other linear dimensions, and there are no plastic deformations.

It is necessary to find the function $q(z,t)$ describing the contact pressure in the area of interaction between the insert and the coating from the obtained integral equation (18.5). Note that this integral equation contains operators of various types: with variable integration limits and with constant integration limits. This equation also includes a function that describes the coating inhomogeneity, which can be rapidly changing.

18.4 Mathematical Model in a Dimensionless Form

By making a change of variables in Eq. (18.5) according to the formulas

$$z^* = \frac{z}{a}, \quad \zeta^* = \frac{\zeta}{a}, \quad t^* = \frac{t}{\tau_0}, \quad \tau_t^* = \frac{\tau_t}{\tau_0}, \quad c(t^*) = \frac{E_t(t-\tau_t)}{E_0},$$

$$m(z^*) = \frac{h}{2a}\frac{E_0}{R_c(z)(1-\nu_t^2)}, \quad g^* = \frac{r_r + h - r_{in}}{a},$$

$$q^*(z^*,t^*) = \frac{2(1-\nu_t^2)q(z,t)}{E_t(t-\tau_t)}, \quad \mathbf{V}_t f(t^*) = \int_1^{t^*} K_t^*(t^*,\tau^*)f(\tau^*)d\tau^*,$$

$$\mathbf{F}f(z^*) = \int_{-1}^{1} k^*(z^*,\zeta^*)f(\zeta^*)d\zeta^*, \quad \mathbf{I}f(z^*,t^*) = f(z^*,t^*),$$

$$K_t^*(t^*,\tau^*) = K_t(t-\tau_t,\tau-\tau_t)\tau_0, \quad k^*(z^*,\zeta^*) = \frac{1}{\pi}k_{ims}\left(\frac{z-\zeta}{H}\right),$$

we obtain a mixed integral equation in dimensionless form

$$c(t^*)m(z^*)q^*(z^*,t^*) + (\mathbf{I}-\mathbf{V}_t)\mathbf{F}q^*(z^*,t^*) = g^*, \quad z^* \in [-1,1], \quad t^* \geq 1. \quad (18.6)$$

The constant E_0 can be any value, for example, $E_0 = E_c(0)$.

Thus, it is necessary to find the function $q^*(z^*,t^*)$ from the obtained integral equation (18.6). Note that the new dimensionless function $m(z^*)$ is related to the inhomogeneity of the coating and can be rapidly changing. Moreover, the resulting operator equation contains both the Fredholm operator and the Volterra operator, which also complicates the algorithm in finding its solution.

18.5 Algorithm of Analytical Solution

To solve Eq. (18.6), a special approach must be applied, since it is a mixed integral equation and contains a multiplier $m(z^*)$ associated with the elastic properties of the coating, which can be described by complex rapidly changing function. The approach is based on the following sequence of actions.

1. By making the substitution

$$q^*(z^*,t^*) = \frac{Q(z^*,t^*)}{\sqrt{m(z^*)}}$$

in (18.6), we can obtain a new operator equation with the Fredholm operator $\tilde{\mathbf{F}}$ (with kernel $\tilde{k}(z^*,\zeta^*)$), which has the same properties as the operator $\mathbf{F}$ (symmetric, positive definite, self-adjoint)

$$c(t^*)Q(z^*,t^*)+(\mathbf{I}-\mathbf{V}_t)\tilde{\mathbf{F}}Q(z^*,t^*)=\frac{g^*}{\sqrt{m(z^*)}}, \quad z^* \in [-1,1], \quad t^* \geq 1. \quad (18.7)$$

2. The solution of Eq. (18.7) at each moment of time t should be sought in Hilbert space $L_2[-1,1]$ with a basis $\{w_0(z^*), w_1(z^*), w_2(z^*), \ldots\}$ based on the orthogonalization of the system of linearly independent functions

$$\left\{\frac{1}{\sqrt{m(z^*)}}, \frac{z^*}{\sqrt{m(z^*)}}, \frac{(z^*)^2}{\sqrt{m(z^*)}}, \ldots\right\}.$$

3. Based on the basis functions $\{w_0(z^*), w_1(z^*), w_2(z^*), \ldots\}$, it is necessary to construct a system of eigenfunctions $\{\varphi_0(z^*), \varphi_1(z^*), \varphi_2(z^*), \ldots\}$ of the operator $\tilde{\mathbf{F}}$, which will be the system of basis functions used to represent the solution.
4. Finally, by equating the corresponding coefficients for the eigenfunctions in the left and right sides of the resulting equation, we obtain the following formulas for calculating the desired contact pressures $q^*(z^*,t^*)$.

18.6 Analytical Solution

Without going into the details of the analytical calculations, we will give only the final formulas ($z^* \in [-1,1], t^* \geq 1$)

$$q^*(z^*,t^*) = \frac{1}{\sqrt{m(z^*)}} \sum_{k=0}^{\infty} f_k^*(t^*) \sum_{m=0}^{\infty} \psi_{km} w_m(z^*) = \frac{1}{m(z^*)} \sum_{k=0}^{\infty} f_k^*(t^*) \sum_{m=0}^{\infty} \psi_{km} p_m(z^*), \quad (18.8)$$

where

$$f_k^*(t^*) = \sqrt{\int_{-1}^{1} \frac{d\zeta^*}{m(\zeta^*)}} \psi_{k0} g^* \left\{ \frac{1}{c^*(t^*)+\gamma_k} + \int_1^{t^*} \frac{R_k(t^*,\tau^*)d\tau^*}{c^*(\tau^*)+\gamma_k} \right\},$$

$$p_k(z^*) = w_k(z^*)\sqrt{m(z^*)}, \quad k = 0, 1, 2, \ldots,$$

where the system $\{p_0(z^*), p_1(z^*), p_2(z^*), \ldots\}$ is a system of polynomials, the expansion coefficients γ_k and ψ_{km} are the eigenvalues and components of the eigenvectors of $\tilde{\mathbf{F}}$, respectively, which can be found from the spectral problem

$$\sum_{l=0}^{\infty} K_{il}\psi_{kl} = \gamma_k \psi_{ki}, \quad i = 0, 1, 2, \ldots,$$

where K_{il} are the expansion coefficients of the kernel $\tilde{k}(z^*, \zeta^*)$ in a double series in the system of basis functions $\{w_0(z^*), w_1(z^*), w_2(z^*), \ldots\}$, i.e.,

$$\tilde{k}(z^*, \zeta^*) = \sum_{i,l=0}^{\infty} K_{il} w_i(z^*) w_l(\zeta^*), \qquad K_{il} = \int_{-1}^{1}\int_{-1}^{1} \tilde{k}(z^*, \zeta^*) w_i(z^*) w_l(\zeta^*) dz^* d\zeta^*.$$

Using the inverse change of variables, Eq. (18.8) can be reduced to dimensional form, thereby obtaining a formula for finding the contact pressures ($z \in [-a, a], t \geq \tau_0$)

$$q(z,t) = \frac{a}{h} \frac{E_c(z)[1 - \nu_c(z)]}{1 - \nu_c(z) - 2\nu_c^2(z)} \frac{E_t(t - \tau_t)}{E_0} \sum_{k=0}^{\infty} f_k^* \left(\frac{t}{\tau_0}\right) \sum_{m=0}^{\infty} \psi_{km} p_m \left(\frac{z}{a}\right). \tag{18.9}$$

18.7 Numerical Example

Let us take the following dimensional parameters for numerical calculation:
$r_{in} = 0.02$ m, $H = 0.015$ m, $h = 0.001$ m, $a = 0.01$ m, $r_r = 0.02$ m,
$E_t(t) \equiv 100$ GPa, $\nu_t = 0.3$, $R_c(z) = (20 + \sin(2\pi z/a))$ GPa.
We assume that the properties of viscoelasticity and aging correspond to the properties of concrete (with $\tau_t = 0$), and the interaction start time between the insert and the pipe is $\tau_0 = 15$ days. The dimensionless contact pressures at the initial (*1*), intermediate (*2*), and sufficiently large time (*3*) points are shown in Fig. 18.2. It can be seen that contact stresses become less over time, but the influence of coating heterogeneity is everywhere present: contact pressure has local maxima near places where the rigidity is maximum, and local minima near places where the rigidity is minimum.

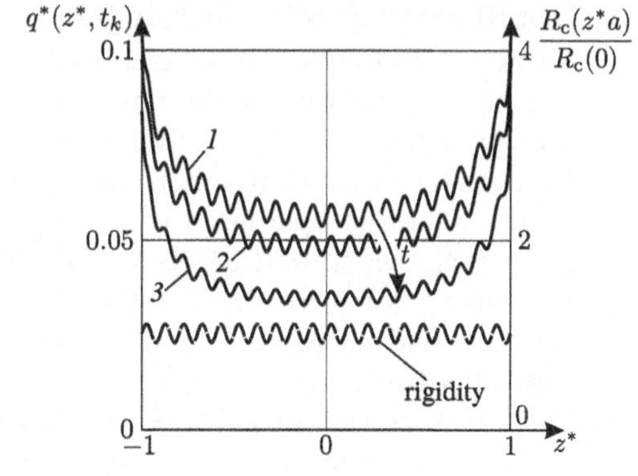

Fig. 18.2 Contact pressures initial(*1*), intermediate (*2*), and sufficiently large time (*3*) and rigidity.

18.8 Conclusions

Thus, the multiplier associated with the functions describing the elastic properties of the coating is explicitly isolated in the solution (18.9). This allows achieving high accuracy in numerical calculations even while retaining a relatively small number of terms. This representation of the solution makes it possible to perform efficient calculations when the physical and mechanical properties of the mating parts are described by complex rapidly changing functions. Solutions constructed in the form of a series in bases that do not take into account the complex functions included in the equations do not yield an acceptable result.

Acknowledgements The study was carried out on the topic of the state assignment (state registration number 124012500437-9).

References

[1] Arutyunyan NK (1966) Some Problems in the Theory of Creep, International Series of Monographs in Civil Engineering, Vol 1. Pergamon Press, Oxford

[2] Arutyunyan NK, Kolmanovskii VB (1983) Teoriya polzuchesti neodnorodnykh telTeoriya polzuchesti neodnorodnykh tel (in Russ., Creep Theory of Inhomogeneous Solids). Nauka, Moscow

[3] Arutyunyan NK, Manzhirov AV (1999) Kontaktnyye zadachi teorii polzuchestiKontaktnyye zadachi teorii polzuchesti (in Russ., Contact Problems in the Theory of Creep). Izdatel'stvo Instituta Mekhaniki NAN Armenii, Erevan

[4] Kazakov KE (2020) On the interaction of rigid bush and pipe with nonuniform coating. AIP Conference Proceedings 2293:210012, DOI 10.1063/5.0026560

[5] Kazakov KE (2024) Interaction of a circular bush and a layered pipe fitted on a shaft. Proceedings of the XII All Russian Scientific Conference on Current Issues of Continuum Mechanics and Celestial Mechanics, CICMCM 2023, Springer Proceedings in Physics, Vol. 1067. Springer Singapore, Singapore, pp 116-121, DOI 10.1007/978-981-97-1872-6_17

[6] Kazakov KE, Kurdina SP (2021) Interaction of viscoelastic tube with inner nonuniform coating of variable thickness and rigid cylindrical insert. Key Engineering Materials 894:67-71, DOI 10.4028/www.scientific.net/kem.894.67

[7] Kazakov KE, Kurdina SP (2021) Wear problem for tube with thin nonuniform inner coating and rigid insert with complex profile. E3S Web of Conference 243:02008, DOI 10.1051/e3sconf/202124302008

[8] Manzhirov AV, Chernysh VA (1988) On the interaction of a rigid reinforcing sleeve and inhomogeneous aging high-pressure pipes. Mechanics of Solids 23(6):104–110

[9] Chernysh VA (1999) Kontaktnyye zadachi teorii polzuchesti dlya stareyushchikh tsilindricheskikh tel (in Russ., Contact Problems of Creep Theory for Aging Cylindrical Bodies). PhD thesis. IPMech RAS, Moscow

Chapter 19
Healing Microdefects and Improving the Mechanical Behavior of Additively Manufactured Steel with High-Energy Electropulsing Treatment

Konstantin K. Kukudzhanov

Abstract Low 3D print quality leads to both deterioration and significant dispersion of the mechanical properties of materials produced using additive manufacturing (AM) methods. By applying a pulsed high-energy electropulsing treatment (HeEPT) to 316L stainless steel printed samples, produced using selective laser melting (SLM) technology with low print quality, a significant increase in the ultimate strain was achieved with only a slight decrease in the ultimate tensile strength. A new approach to determine the preferred route of HeEPT is proposed to improve the plastic behavior of samples and to heal internal microdefects in additive materials. Applying the preferred route of HeEPT to additively manufactured 316L steel samples made the mechanical properties more predictable and less variable, as evidenced by a significant reduction in the standard deviations of the ultimate tensile strength and ultimate strain. This macro-behavior is a result of healing microdefects and a considerable decrease in porosity after HeEPT.

Key words: Additive materials, Selective laser melting, Healing, High-energy electropulsing treatment, High-density current pulses, Defects, Microcracks, Micropores

19.1 Introduction

Additive manufacturing (AM) is aimed at the creation of three-dimensional objects of any complexity based on digital models. One of the most common AM methods is selective laser melting (SLM). The mechanical properties of SLM-manufactured parts differ from those of conventionally produced parts due to rapid cooling and crystallization, which causes residual strain and high local residual stresses that can reach several hundred MPa.

Konstantin K. Kukudzhanov
Ishlinsky Institute for Problems in Mechanics of the Russian Academy of Sciences, Vernadskogo Ave 101 Bldg 1, Moscow, 119526, Russian Federation, e-mail: kkukudzhanov@yandex.ru

H. Altenbach and A. V. Sahakyan, *Contact Mechanics and Relevant Problems*,
Advanced Structured Materials 253, https://doi.org/10.1007/978-3-032-20869-9_19

The main distinctions are as follows:

1. increased porosity, which reduces plastic and fatigue characteristics due to the propagation of macrocracks, especially from irregularly shaped pores and cracks, as well as from grouped pores and cracks;
2. a fine microstructure of the material, which increases strength and hardness but reduces the ductility of steels.

The industrial application of SLM technology, as well as other AM methods, does not allow defect-free metal parts to be produced. Incorrect and/or inaccurate management of process parameters leads to increased porosity and the formation of many kinds of internal defect (such as cracks and pores of incomplete fusion and melting, pores which contain some unmelted and unconsolidated powder, gas pores, etc.). These defects reduce the mechanical properties of the products and increase the dispersion of the properties, making additive materials of a lower quality than conventionally produced ones, thus limiting their application in production.

To improve the characteristics of SLM- manufactured parts, thermomechanical methods (high-temperature annealing, hot plastic deformation, hot isostatic pressing) and electropulsing treatment (EPT) methods are used. Thermomechanical methods increase plasticity by relieving residual stress and reducing porosity [3], but also reduce tensile strength and fatigue properties. Thermal degradation of the microstructure also occurs [4]. The porosity healing mechanism within thermomechanical processing methods implies the thermal diffusion of atoms at high temperatures. These methods require significant energy and time, making them less effective. By healing internal microdefects in additive material, here and further we mean the restoration of the material's continuity by reducing the volume of internal pores and cracks, as well as smoothing the shapes of these defects, and reducing the concentration of mechanical stresses at their sharp tips.

Electropulsing treatment (EPT) methods for steel involve inducing high-density current pulses in a conductive material by applying an intense external electromagnetic field. For conventionally produced steel products, a wide range of EPT routes is used: pulse durations from 10^{-5} s to 10^{-2} s and current densities from 10^{1} A/mm^{2} to 10^{3} A/mm^{2}. This results in a specific energy dissipation in the material in the range of $10^{5}-10^{9}$ J/m^{3}. Here and in the following, the EPT will be considered a high-energy EPT (HeEPT) in the case of specific energy dissipation per single pulse in steels between 10^{7}-10^{9} J/m^{3} within relatively short pulse durations of $10^{-5}-10^{-4}$ s [11]. The other means of EPT with pulse durations $> 10^{-4}$ s and specific energy dissipation per single pulse $< 10^{7}$ J/m^{3} will be called low-energy EPT (LeEPT). This distinction is based on the significant differences in the core physical processes that occur in additive materials and so in the mechanisms to improve their mechanical properties while HeEPT and LeEPT [11].

It should be noted that the topic of improving the mechanical properties of additive materials with EPT is insufficiently studied. A small number of works have been devoted to the study of the impact of EPT on the mechanical properties of additive materials [5, 6, 14, 16, 17]. In these studies, the authors consider numerous additive materials (AlSi10Mg, Ti-6Al-4V, T11, 316L); however, all of them rely on LeEPT as the base treatment method. [18] and [2] should also be mentioned here, since they observed phenomena similar to those studied in the present work. However, the authors of these studies did not pay much attention to the changes in plastic and strength properties after EPT. In [18], after electromechanical processing (simultaneous electropulsing and pressing force), the authors observed smoothing of the defect boundaries in the additively produced Ti-6Al-4V alloy sample. In [2], it was possible to partially heal long macrocracks by welding their edges during HeEPT with successively increasing current pulses (75 pulses in total) in two samples made of TiC/316L composite material. The cumulative heating caused by multiple current pulses resulted in strong thermal contraction of the crack edges and facilitated welding of the edges and crack healing. The mechanical properties of the samples after HeEPT were not studied in [2]. In the above-mentioned papers, qualitative changes in the fine microstructure caused by cumulative high-temperature heating of samples of additive materials after LeEPT were observed. The effect of LeEPT in that case was similar to that of annealing [15]

The authors of [18] and [2] observe and confirm a significant increase in temperature and note the similarity of the effect of LeEPT with thermomechanical treatment. Both in terms of changes in microstructure and in terms of improved mechanical properties [7]. However, significantly different evolutionary changes will occur in additive materials (as well as in conventionally produced materials) during HeEPT than during thermomechanical processing or LeEPT.

Within HeEPT there is a different defect healing mechanism that leads to improvement of mechanical characteristics of the materials rather than with thermo- or electromechanical treatment (LeEPT + compressive stress). A strong concentration of the electromagnetic field occurs in the sharp tips of internal microcracks and micropores, causing intense release of Joule heat that occurs in a short period of time. Compressive thermal stresses arise with simultaneous softening and melting of the metal in these areas then.

With all of the above, this leads to changing the shape of the internal microcracks and micropores, decreasing their volume, and the edges of the defects are welded together by molten metal. The numerical simulation results in [9, 10, 13] confirm this particular mechanism for healing internal defects in a conductive material during HeEPT. In these studies, the finite element electrothermoelastoplastic model has shown that during HeEPT of a metal with microdefects, the volume of the microdefects in it decreases. Therefore, the porosity of the material reduces up to complete healing of some micropores and microcracks by welding their edges with molten metal.

A numerical study of macrocrack healing in 316L steel during HeEPT using another finite element model [1] also indirectly confirms the mechanism described above. In addition, edge welding and porosity healing were observed in macrocracks healing experiments in additive materials during HeEPT [18], as well as in conventionally produced materials [8, 12]. In contrast, during thermomechanical processing (annealing, hot isostatic pressing, and hot plastic deformation) and LeEPT thermal energy is released over longer time intervals (compared to HeEPT) and healing of microcracks and micropores occurs through thermal atom diffusion at high heating temperatures. So, there is no intensive targeted melting at the tips of microdefects, and welding of their edges does not occur during LeEPT. Thus, studies of the effect of HeEPT on additive materials aimed at improving their mechanical properties have not been conducted yet. The ability to heal internal microdefects during HeEPT has also not been studied.

Regardless of HeEPT it should also be noted that an extremely wide range of pulsing parameters (current densities, pulse shapes, and durations) was used in the mentioned LeEPT studies [7]. In [2, 4–6, 14, 16–18] the authors neither explained nor justified the parameters for the EPT mode. So, the chosen EPT mode was carried out purely empirically or for any other reason not mentioned.

This paper attempts to fill these gaps. The study develops and uses a new approach to determine the HeEPT route that aims at healing internal defects in metallic materials to improve their mechanical properties in the end. The study is carried out on AISI 316L stainless steel, manufactured using the SLM method. This grade of steel is popular in industry because of its excellent physical and mechanical properties, high oxidation resistance, and good weldability. It is widely used in SLM printing, therefore. The steel samples used for the experiments were of poor print quality and contained large defects, such as lack of fusion with linear dimensions up to 250 μm. The presence of these defects is preferable to demonstrate the healing abilities of HeEPT.

The objectives of the present study are:

1. Experimentally confirm healing of micropores and microcracks (in a wide range of linear dimensions ranging from 5 to 250 μm) in additive steel 316L, under the preferred route of HeEPT (previously there have been no such confirmations);
2. Investigate changes in porosity, microstructure of defects, and their influence on the macroscopic mechanical properties of additively manufactured steel 316L.

The study of fine microstructure was optional as it affected porosity healing and mechanical properties.

19.2 Materials and Methods

For the purposes of the study, the thin neck central area samples with dimensions 120×6×3 mm were produced from AISI 316L stainless steel powder (see Fig. 19.1a). The samples were manufactured using the SLM method on an SLM Kings 3D Printer M280HD (China).

In the experiment, 10 randomly chosen samples were equally divided into two groups. The first group (UNT - untreated) was tested for tensile strength without preliminary HeEPT. The second group (TR - treated) underwent HeEPT and then proceeded to tensile testing. All samples were stretched up to rupture on a GOST MIM.2-100 test machine (Russian Federation) at a strain rate of 0.15 mm/min (initial distance between clamps $l_0 = 35$ mm).

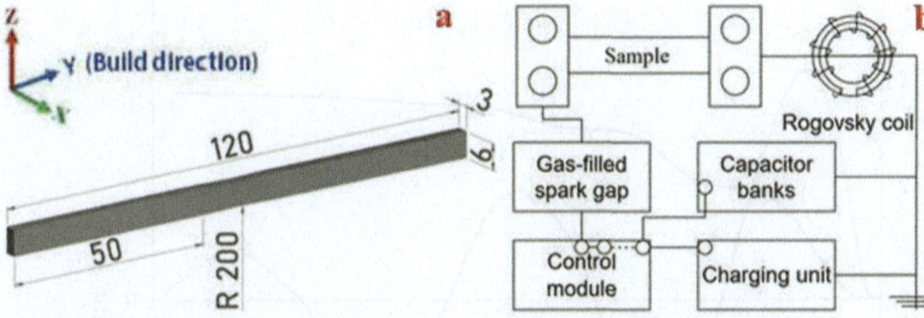

Fig. 19.1: Printed 316L samples (a) and schematic diagram of experimental setup (b).

Two additional untreated samples and two treated samples were used to produce the sections. Pieces 10 × 6 × 3mm were cut from each one. Sections for microstructural studies were made along the wide edge of the samples (parallel to the YZ plane in Fig. 19.1a). The defects were examined in three randomly selected planes perpendicular to the X-axis. Porosity was measured on the total area of 9.9 mm^2 of the microsections. The microstructure before and after HeEPT was examined using a Coxem EM-30AX Plus scanning electron microscope (SEM) (China) and a Zeiss Axio Imager.D1m optical microscope (Germany).

The TR-samples, rigidly fixed between the copper clamp electrodes (38 mm between the electrodes), were treated with HeEPT using a GIT-20-40 pulse generator (Russia) in a specially designed laboratory setup, schematically shown in Fig. 19.1b. The change in current in the circuit over time was measured using a calibrated Rogowski coil connected to an oscilloscope. During HeEPT, current pulses were induced in the samples, having the shape shown in Fig. 19.2. For GIT-20-40 pulse generator $f = \omega/(2\pi) = 14.76$ kH) and $\zeta = 0.15$, $\tau_0 = 162$ μs. In the experiments carried out, the amplitude of the current density $j_{\max}$ varied, the duration τ_0 of all pulses was the same. There has been a 120 s interval between pulses while the samples cooled to T_0, so there has been no cumulative heating during HeEPT.

The HeEPT route was selected according to the approach proposed in [11]. The approach implies only one control parameter which is the specific energy of the current pulse q instead of selecting several such parameters as current density, duration $t = \tau_0$ and impulse form as was proposed in previous studies [2]. The specific energy of the current pulse is calculated with Eq. (19.1)

$$q = \int_0^{\tau_0} \frac{j_\infty^2(t)}{\sigma^E}\, dt, \tag{19.1}$$

where σ^E is the electrical conductivity coefficient of the material.

A wide variety of sizes and shapes of the microdefects observed makes the non-empirical determination of the parameters of the HeEPT for their healing and

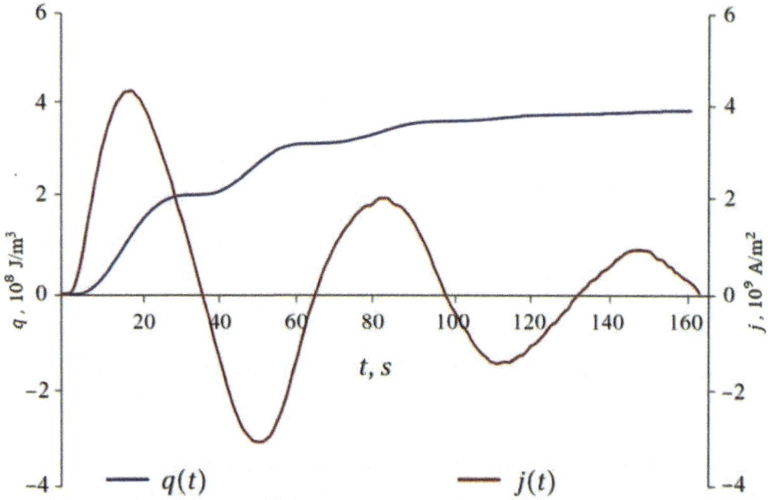

Fig. 19.2: The experimentally measured dependence of the current density and the calculated dependence of the specific dissipated energy of current pulse in the sample on time.

improvement of the plastic macrocharacteristics of the material complicated. Qualitative analysis of internal defects shows that the vast majority of them (except for gas pores) typically have a non-smooth shape with sharp tips and protrusions (Fig. 19.3).

One of the features of the approach that aims to heal defects, presented in [12], is that it requires knowledge of the specific shape and relative geometric dimensions of the defects to heal. For the purposes of the current study, all the defects in samples are assumed to be of an "elliptical" cylinder shape with a generatrix along the Z axis (Fig. 19.1a). The cylinders are based on ellipses with the major semi-axis a and the minor semi-axis b. All the ellipses are oriented identically with their minor semi-axis b parallel to the Y-axis of the sample, so as the current direction. The radius of curvature of the defect is defined $R = b^2/a$.

To determine the HeEPT route, we use Eqs. (31) and (32) of [11]. As follows from these equations the relation between the specific energy of the current pulse q and the temperature of the metal T at the apex of the elliptical defect is specified with

$$q = \frac{R}{a}\left(1 - \sqrt{\frac{R}{a}}\right)^2 \frac{\pi a}{2W} \cot\left(\frac{\pi}{2}\frac{a}{W}\right) \rho c(T - T_0), \tag{19.2}$$

where a is the major semi-axis of the defect, $2W$ is the average distance between the defects, $R = b^2/a$ is the radius of curvature at the tip, ρ is the density and c the specific heat of the material, T_0 is the initial temperature at the tip. Equation (19.2) is self-similar to b/a and a/W.

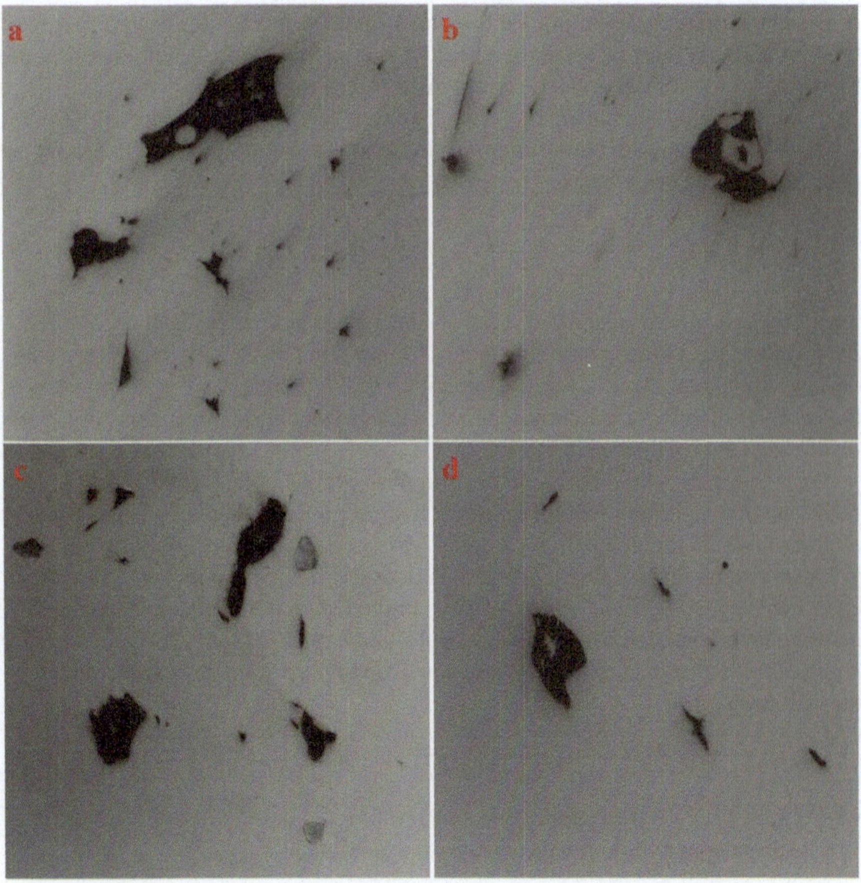

Fig. 19.3: Various types of defects in the sample material observed: lack of fusion and gas pores (a, b, c, d), microcracks (a, c, d). Optical microscope image with 200× magnification.

Assuming that the defects are of elongated shape, so the aspect ratio of defects $b/a < 1$, thus $b^2/a^2 = R/a \ll 1$ we will neglect the terms $o(R/a)$ further in (19.2). We will also assume that the average distance between the defects $2W$ significantly exceeds their maximum size a, so that $a/W \ll 1$, thus neglecting the terms $o(a/W)$ in (19.2). This assumption is consistent with the geometry of the observed microdefects and the distances between them in sections of samples.

In the introduction, it was reported that microcracks or micropores healing by welding the edges occurs at its tips, where the current pulse causes simultaneous melting of the metal and compression. It is a step-by-step process when the crack or defect tip moves toward the center along the major semi-axis with each pulse.

Let the HeEPT route consist of N pulses numbered $n = 0, 1, 2...N$. In [12] it was proposed to maintain constant internal energy at the tip of the defect for each of the $n = 0, 1, 2..., N$ pulses, equal to the energy of the first pulse ($n = 0$), to ensure consistent melting and welding. At the same time, the specific dissipated energy of the first pulse q_0 must be set so that healing and melting will certainly occur at the tips of the defects. Such an HeEPT route will be called preferable.

The measure of internal energy at the tip of the defect is the temperature $(T - T_0)$ in Eq. (19.2). Thus, with the assumptions made above, it follows from (19.2) that the preferred route is described by

$$\frac{q_{n+1}}{q_n} = \frac{R_{n+1}\,a_n}{R_n\,a_{n+1}}, \quad n = 0, 1, 2..., N, \tag{19.3}$$

where q_{n+1}, q_n is the specific energy of any subsequent current pulses, R_{n+1}, R_n, a_{n+1}, a_n are the corresponding geometric dimensions of the defect, which change during subsequent electric pulses. Equation (19.3) rationalizes the relationship of the energies of subsequent current pulses with the geometric dimensions of defects that change during the healing process.

We assume that the defects of the elliptical cylinders heal uniformly: for each pulse, an identical small section Δa of the main semi-axis a at the tip is welded, and after n pulses its initial length a_0 will decrease to $a_n = a_0 - n\Delta a$ with $\Delta a/a_0 \ll 1$. We also assume that the minor semiaxis b does not change significantly throughout electropulsing, so $b_n = b_0$, where b_0 is the initial length of the minor semiaxis. This assumption is consistent with the numerical simulation results in [1]. Therefore, after n pulses the radius of curvature R_n at the new tip of the healed defect is defined with the following

$$R_n = \frac{b_n^2}{a_n} = \frac{R_0}{1 - n\,(\Delta a/a_0)}, \quad n = 0, 1, \ldots, N, \tag{19.4}$$

where $R_0 = b_0^2/a_0$ is the initial radius of curvature.

Electropulsing to be conducted satisfy the condition

$$n\frac{\Delta a}{a_0} \ll 1, \quad n = 0, 1, \ldots, N.$$

Therefore, we will neglect the terms $o(n(\Delta a/a_0))$ in further calculations. Then with all the assumptions made above, the preferred HeEPT route that causes melting and healing at the tip of the defect to occur within each of the N subsequent pulses is defined by

$$q_n = q_0\left(1 + 2n\frac{\Delta a}{a_0}\right) \tag{19.5}$$

accordingly, from (19.3) and (19.4), where q_0 is the energy dissipated by the initial pulse. Equation (19.5) describes the function of pulse energies versus n for the healing of defects in the preferred route.

The following considerations are used to specify (19.5) for the purposes of the current study. Refer to Fig. 19.3. The samples contain microdefects with various initial aspect ratios b_0/a_0, however, we assume that they can all be approximated by ellipses. As follows from (19.2) for each microdefect with it's b_0/a_0 a different q_0 is required to initiate melting and healing. In addition, for various b_0/a_0 the parameter $\Delta a/a_0$ will vary as melting and healing discern between different defect shapes.

However, for simplicity here, we assume that in the sample material there are only two characteristic types of defect sizes:

a) elongated, where $0.07 < b_0/a_0 < 0.11$ with average 0.09,
b) convex, where $0.27 < b_0/a_0 < 0.33$ with average 0.3.

To heal these types of defects, we use the HeEPT route with two linear branches of type (19.5):

1) $q_n = q_0'(1 + 2n\Delta a'/a_0)$,
 $n = 0, 1, \ldots, 4, \quad q_0' = 0.14 \cdot 10^8 J/m^3, \quad \Delta a'/a_0 = 7.5 \cdot 10^{-3}$,
2) $q_n = q_0''(1 + 2(n-5)\Delta a''/a_0)$,
 $n = 5, 6, \ldots, 9, \quad q_0'' = 1.51 \cdot 10^8 J/m^3, \quad \Delta a''/a_0 = 4 \cdot 10^{-2}$.

The function of the only parameter q_n of the preferred route of our experiments is presented in Fig. 19.4, and the numerical values in Table 19.1.

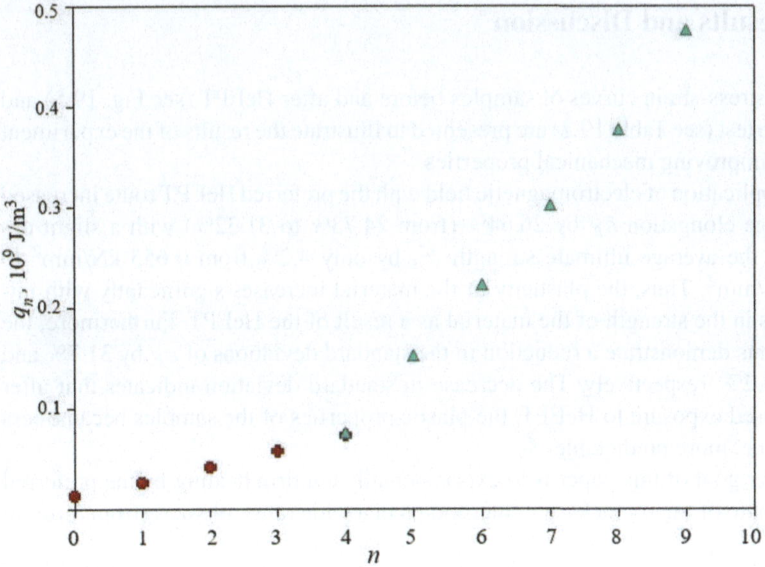

Fig. 19.4: Function of specific dissipated energy of a pulse q_n from the pulse serial number n in the preferred HeEPT route..

Table 19.1: Function of specific dissipated energy of a pulse q_n from the pulse serial number n in the preferred HeEPT route.

Pulse number, n	q_n, 10^8 J/m^3
0	0.14
1	0.27
2	0.42
3	0.58
4	0.74
5	1.51
6	2.21
7	3.01
8	3.76
9	4.75
Total:	17.39

19.3 Results and Discussion

Nominal stress-strain curves of samples before and after HeEPT (see Fig. 19.5) and the tensile test (see Table 19.2) are presented to illustrate the results of the experiment aimed at improving mechanical properties.

The application of electromagnetic field with the preferred HeEPT route increased the average elongation $\bar{\varepsilon}_F$ by 26.64% (from 24.73% to 31.32%) with a slight decrease in the average ultimate strength $\bar{\sigma}_B$ by only 4.2% from 0.655 kN/mm^2 to 0.627 kN/mm^2. Thus, the plasticity of the material increases significantly with minor losses in the strength of the material as a result of the HeEPT. Furthermore, the experiments demonstrate a reduction in the standard deviations of ε_F by 31.7% and σ_B by 45.2%, respectively. The decrease in standard deviation indicates that after the proposed exposure to HeEPT, the plastic properties of the samples became less variable, i.e. more predictable.

Another goal of this paper is to experimentally confirm healing by the preferred HeEPT route of microcracks and micropores in a wide range of sizes (from 5 μm to

Table 19.2: Tensile test numerical results.

Group	$\bar{\sigma}_B$, kN/mm^2	$\bar{\varepsilon}_F$, %	$Sd\ \ \sigma_B$, kN/mm^2	$Sd\ \ \varepsilon_F$, %	$\sum q_n$, 10^9 J/m^3
Untreated	0,655	24,73%	0,041	7,78%	-
Treated	0,627	31,32%	0,028	4,26%	1,73

where $\bar{\sigma}_B$ - average ultimate strength, $\bar{\varepsilon}_F$ - average ultimate strain before fracture (elongation), $Sd\ \ \bar{\sigma}_B$ - standard deviation of the ultimate strength, $Sd\ \ \varepsilon_F$ - standard deviation of the ultimate strain (elongation), $\sum q_n$ - total specific dissipated energy deployed

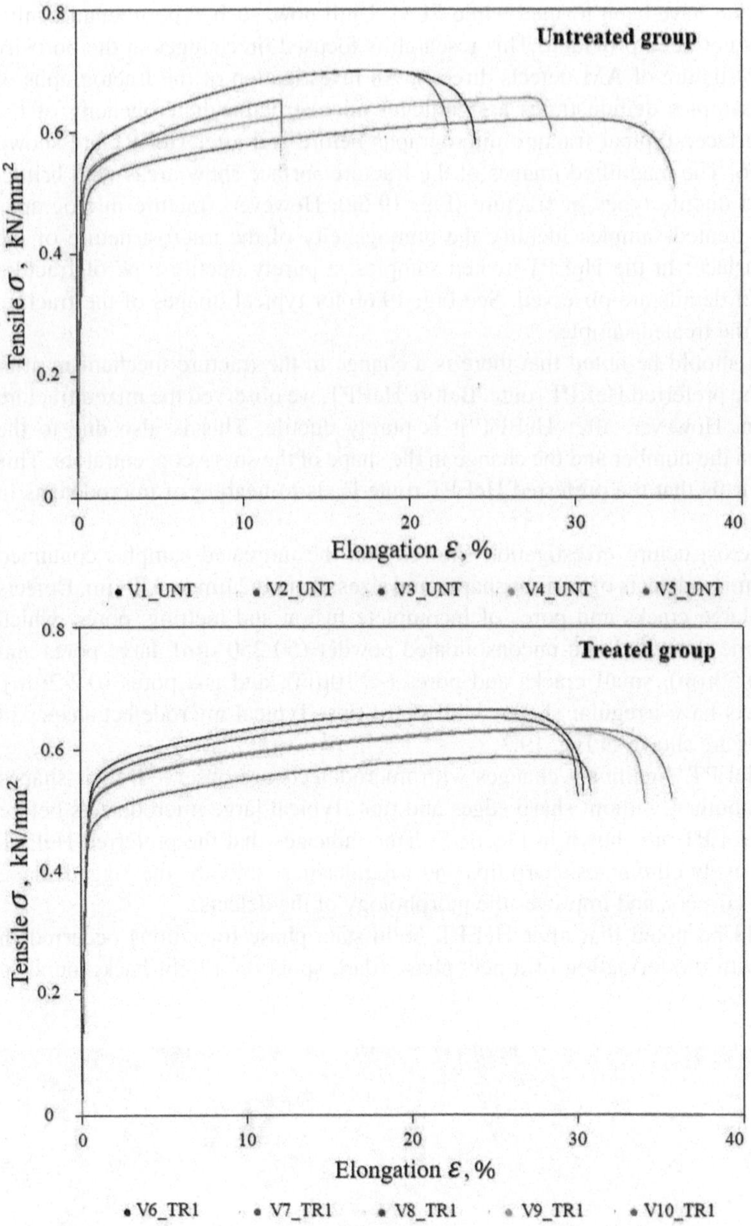

Fig. 19.5: Nominal stress-strain curves of untreated and HEPT-treated samples.

250 μm), that have been formed while SLM. Until now, such experimental confirmation has not been provided. This research is focused on changes in the porosity and microstructure of AM defects directly. An investigation of the fractographs of untreated samples demonstrates a significant microstructure heterogeneity of the fracture surface. Typical fracture micrographs before and after HeEPT are shown in Fig. 19.6. The magnified images of the fracture surface show areas with brittle, mixed, and ductile types of fracture (Fig. 19.6a). However, fracture micrographs of HeEPT-treated samples identify the homogeneity of the microstructure of the fracture surface. In the HeEPT-treated samples, a purely ductile type of fracture micro-relief details are observed. See Fig. 19.6b for typical images of the fracture surface of the treated sample.

Thus, it should be noted that there is a change in the fracture mechanism after applying the preferred HeEPT route. Before HeEPT, we observed the mixed fracture mechanism. However, after HeEPT it is purely ductile. This is also due to the reduction in the number and the change in the shape of the stress concentrators. This result confirms that the preferred HeEPT route leads to healing of microdefects in the material.

The microstructure investigation showed that the untreated samples contained numerous microdefects of various shapes and sizes: from 0.2 μm to 250 μm. Defects observed: large cracks and pores of incomplete fusion and melting, pores, which contain some unmelted and unconsolidated powder (50-250 μm), large pores and cracks (10-50μm), small cracks and pores (< 10μm), and gas pores (0.2-2μm). Most defects have irregular shapes with sharp tips. Typical microdefect shapes in the samples are shown in Fig. 19.3.

After HeEPT, significant changes with microdefects are observed: their shapes become smoother, without sharp edges and tips. Typical large microdefects before and after HeEPT are shown in Fig. 19.7. This indicates that the preferred HeEPT route effectively eliminates sharp tips and irregularities, smooths the jagged shape with sharp corners, and improves the morphology of the defects.

It should be noted that after HeEPT, solid-state phase transitions occurred in the steel with the formation of a new phase (dark spots on a light background in

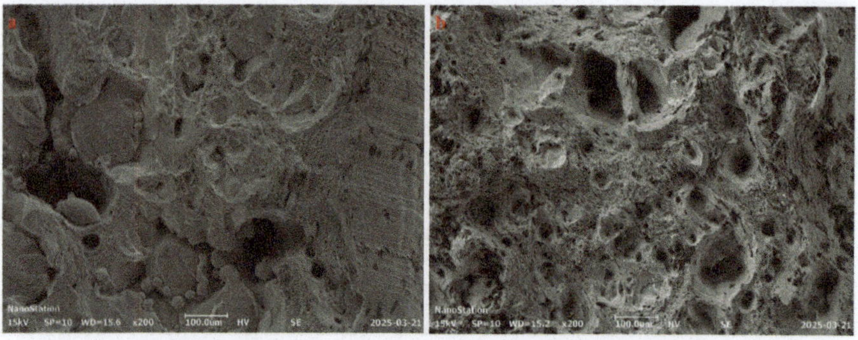

Fig. 19.6: Microstructure of fracture surfaces before (a) and after HeEPT (b).

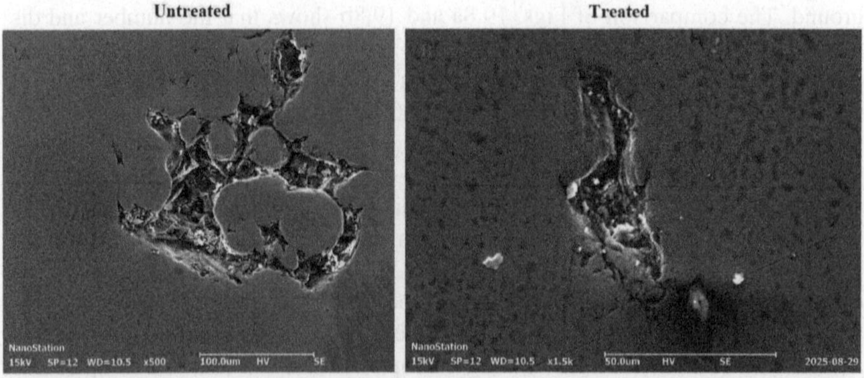

Fig. 19.7: Typical internal microdefects manufacturing microdefects in samples (a) before HeEPT and (b) after HeEPT.

Fig. 19.7b). However, in the vicinity of the defects of the untreated samples, no new phase was observed (light area around the defect in Fig. 19.7a). In our opinion, the presence of a new phase indicates the melting that occurred in the vicinity, thus healing of the microdefects according to the mechanism described in the introduction. The changes in phase composition also apparently influenced the modification of the treated steel type of fracture.

For porosity analysis, refer to Fig. 19.8a showing the microstructure of the untreated sample section and Fig. 19.8b showing the microstructure of the sample section after HeEPT. Microdefects can be identified as white areas on a black back-

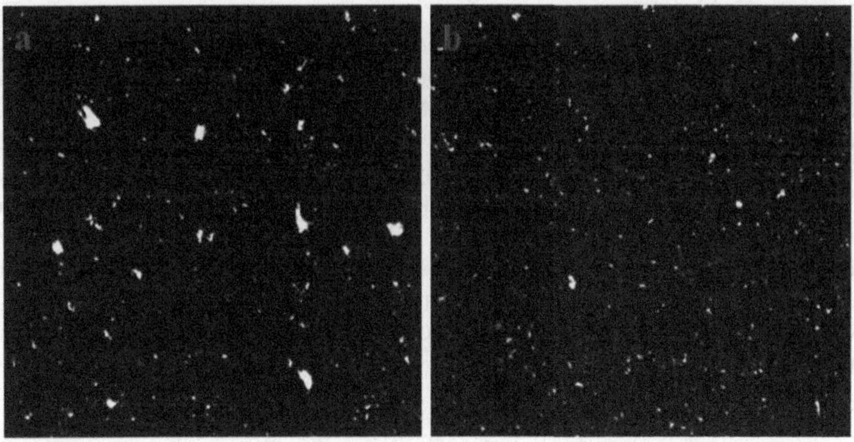

Fig. 19.8: Panoramic images of a typical section taken with an optical microscope (a) before HeEPT, (b) after HeEPT. The horizontal and vertical dimensions of the image (black field) are 1825μm ×1825μm.

ground. The comparison of Figs. 19.8a and 19.8b shows that the number and distribution of microdefects differ essentially. The untreated sample is more porous and contains larger microdefects (Fig. 19.8a), while after HeEPT, the number and size of the defects are significantly smaller (Fig. 19.8b). It is visually evident that the preferred HeEPT route effectively reduces porosity and the number of large microdefects.

The open source software Porosity Calculator Matlab was used for a quantitative analysis. The porosity was calculated from photographs of the surfaces of thin sections, similar to those shown in Fig. 19.8. The area of 9.9 mm^2 of thin sections of samples before and after HeEPT (in three randomly selected sample sections) was analyzed to calculate the porosity. The calculations indicated that the porosity healed from 0.93% within the untreated samples to 0.63% within the samples after HeEPT, that is, there was a decrease in porosity by 32%.

With all of the above mentioned, the research conducted confirms that in 316L SLM-manufactured steel healing of microdefects occurs after applying the preferred HeEPT route with the parameters defined by the proposed model. The mechanism of this healing is as follows: while HeEPT, a strong concentration of the electromagnetic field occurs in the sharp tips of internal microcracks and micropores, causing an intense release of Joule heat that occurs in a short period of time. The compressive thermal stresses arise with the simultaneous softening and melting of the metal in these areas. Together, this results in welding and melting of the edges at the tips of the internal microcracks and micropores. The defects are smoothed and their volume is decreasing.

19.4 Conclusions

The preferred route proposed for high-energy electro pulse treatment (HeEPT) is capable of healing microcracks and micropores with linear dimensions of 5 - 250μm in SLM-manufactured 316L stainless steel. As a result of internal microdefect healing, there was a significant improvement in the mechanical characteristics of the samples: increased plasticity (by 26.64%) with only a slight decrease in strength (by 4.2%). The studies carried out show that HeEPT allows the mechanical characteristics even of low-quality printed additive materials to be improved, increasing predictability and reducing the variability of its macroscopic behavior. We strongly believe that since HeEPT has great potential to heal internal microdefects and improve the mechanical properties of additively manufactured parts, more research is required in this field.

References

[1] Cai Q, Zhou M, Bagherpour E, Hosseini S, Mendis C, Chang I, Assadi H (2023) New insight into crack-healing mechanism via electropulsing treatment. Metallurgical and Materials Transactions A 54(7):2960–2974, DOI 10.1007/s11661-023-07073-1

[2] Cai Q, Rodriguez PR, Santos SC, Castro G, Mendis CL, Chang IT, Assadi H (2024) Crack healing via electropulsing treatment applied to additive-manufactured tic/316L stainless steel composites. Materials Letters 365:136410, DOI 10.1016/j.matlet.2024.136410

[3] Chadha K, Tian Y, Spray JG, Aranas Jr C (2020) Effect of annealing heat treatment on the microstructural evolution and mechanical properties of hot isostatic pressed 316L stainless steel fabricated by laser powder bed fusion. Metals 10(6):753, DOI 10.3390/met10060753

[4] Dryepondt S, Nandwana P, Fernandez-Zelaia P, List III F (2021) Microstructure and high temperature tensile properties of 316L fabricated by laser powder-bed fusion. Additive Manufacturing 37:101723, DOI 10.1016/j.addma.2020.101723

[5] Fan S, He B, Liu M (2022) Effect of pulse current density on microstructure of Ti-6Al-4V alloy by laser powder bed fusion. Metals 12(8):1327, DOI 10.3390/met12081327

[6] Guo H, Liu P, Qin X, Song Y, Qian D, Xie L, Wang L, Zhang LC, Hua L (2021) Electroshock treatment dependent microstructural evolution and mechanical properties of near-β titanium alloy manufactured by directed energy deposition. Materials & Design 212:110286, DOI 10.1016/j.matdes.2021.110286

[7] Kukudzhanov KK (2025) A review of studies on improving the plastic properties of additive materials under strong pulsed current. Mechanics of Solids 60(4):2353–2369, DOI 10.1134/S0025654425600114

[8] Kukudzhanov KV (2024) Healing of discontinuities in metal by the strong electromagnetic field treatment. Journal of Physics: Conference Series, IOP Publishing 2817:012032, DOI 10.1088/1742-6596/2817/1/012032

[9] Kukudzhanov KV (2018) Modeling the healing of damage of metal by high-energy pulsed electromagnetic field. Letters on Materials 8(1):27–32, DOI 10.22226/2410-3535-2018-1-27-32

[10] Kukudzhanov KV (2015) Modelirovaniye vozdeystviya vysokoenergeticheskogo impul'snogo elektromagnitnogo polya na mikrotreshchiny v polikristallicheskom metalle (in Russ., Modeling the treatment of high-energy pulsed electromagnetic field of the micro-cracks in a polycrystalline metal). PNRPU Mechanics Bulletin (4):138–158, DOI 10.15593/perm.mech/2015.4.09

[11] Kukudzhanov KV, Chentsov AV (2024) The energy of short current pulses required for healing cracks in conductive materials. Mechanics of Solids 59(8):3929–3948, DOI 10.1134/S0025654424607031

[12] Kukudzhanov KV, Korznikova EA, Chentsov AV, Dmitriev SV (2025) A novel approach to determining the parameters of electric pulse healing of fatigue cracks. Materials Letters 404:139565, DOI 10.1016/j.matlet.2025.139565

[13] Kukudzhanov KV, Levitin AL (2016) Protsessy transformatsii i vzaimodeystviya mikrotreshchin v metalle pod vozdeistviem vysokoenergeticheskogo impul'snogo elektromagnitnogo polya (in Russ., Transformation and interaction of microcracks in metal under high-energy pulsed electromagnetic field). PNRPU Mechanics Bulletin (2):89–110, DOI 10.15593/perm.mech/2016.2.07
[14] Liu C, Yin F, Xie L, Qian D, Song Y, Wu W, Wang L, Zhang LC, Hua L (2022) Evolution of grain boundary and texture in tc11 titanium alloy under electroshock treatment. Journal of Alloys and Compounds 904:163969, DOI 10.1016/j.jallcom.2022.163969
[15] Maskery I, Aboulkhair N, Corfield M, Tuck C, Clare A, Leach RK, Wildman RD, Ashcroft I, Hague RJ (2016) Quantification and characterisation of porosity in selectively laser melted Al–Si10–Mg using X-ray computed tomography. Materials Characterization 111:193–204, DOI 10.1016/j.matchar.2015.12.001
[16] Noell PJ, Rodelas JM, Ghanbari ZN, Laursen CM (2020) Microstructural modification of additively manufactured metals by electropulsing. Additive Manufacturing 33:101128, DOI 10.1016/j.addma.2020.101128
[17] Xue L, Liao C, Wu M, Li Q, Hu Z, Yang Y, Liu J (2024) Improvement of mechanical properties and corrosion resistance of slm-alsi10mg alloy by an eco-friendly electric pulse treatment. Journal of Cleaner Production 439:140864, DOI 10.1016/j.jclepro.2024.140864
[18] Yan X, Xu X, Wu C, Zhao Y, Li D, Zhou Y, Wu Z, Wei L (2023) A novel electropulsing treatment to improve the surface strength and repair the pore of additively manufactured Ti-6Al-4V alloy. Surface and Coatings Technology 458:129364, DOI 10.1016/j.surfcoat.2023.129364

Chapter 20
On the Determination of the Route of High-Energy Electropulsing Treatment for Healing Cracks in Metals and Alloys

Konstantin V. Kukudzhanov and Alexander V. Chentsov

Abstract Incorrect selection of parameters of the electropulsing treatment (EPT) route often leads to unhealed or poorly healed cracks. Therefore, solving the problem of determining the route parameters of the EPT is critical for successful restoration of the material integrity. Using the results of successful EPT welding of long fatigue crack edges in a thin stainless steel plate, this paper develops a new approach to determining the preferred route of the healing by EPT. An analytical model is proposed that allows limiting the parameter range to high-energy pulses. An energy analysis based on this model shows that a single control parameter, the specific energy of the current pulse, is sufficient for welding of crack edges. This eliminates the need for the empirical selection of several parameters (current density, shape, duration), as it was done previously. The proposed model is used to obtain the certain dependence of the specific energy of the current pulse on its ordinal number. The dependence ensures a constant change in the internal energy at the crack tip created with each pulse. This allows achieving the preferred route of EPT. The applicability of the proposed approach was tested in experiments on fatigue crack healing. The developed 13-step route with 132 pulses demonstrated significant effectiveness in healing a complex crack in stainless steel. Its use ensured a relatively high quality welded joint between the crack faces and significantly improved the mechanical properties of the healed sample.

Key words: Crack healing, Electropulsing treatment

Konstantin V. Kukudzhanov
Ishlinsky Institute for Problems in Mechanics of the Russian Academy of Sciences, Vernadskogo Ave 101 Bldg 1, Moscow, 119526, Russian Federation, e-mail: kconstantin@ipmnet.ru

Alexander V. Chentsov
Moscow State University of Civil Engineering, Moscow, Russian Federation, e-mail: chentsov@ipmnet.ru

H. Altenbach and A. V. Sahakyan, *Contact Mechanics and Relevant Problems*, Advanced Structured Materials 253, https://doi.org/10.1007/978-3-032-20869-9_20

20.1 Introduction

To extend the service life of metal components and reduce maintenance costs, it is common to use the approaches slowing down the initiation and propagation of fatigue cracks and leading to cracks repairing (restoring integrity). Both groups of approaches can be called crack healing approaches. The former include shot peenling, pulsed laser peenling, drilling holes, blow holes or craters formation. The creation of compressive residual stresses in the surface layers by mechanical shot peening or pulsed laser peening of surfaces, as well as the creation of chemical bonds that bond the edges of surface cracks by applying protective coatings (protective chemical coating), strengthen the surface of metal components, which prevents the initiation and propagation of fatigue cracks in materials [9, 23, 26]. The creation of holes near the crack tips by pulsed electromagnetic action (blow holes or craters) or drilling (drilling holes) reduces the concentration of mechanical stresses at the crack tips, which prevents their further propagation/growth and can be considered as one of the methods for healing fatigue cracks [5–7, 25, 28].

Approaches that restore integrity in places of cracks include thermal (high-temperature annealing or heat treatment) and thermo-mechanical (thermal-mechanical coupling, hot plastic deformation, hot isostatic pressing) treatments, as well as electropulsing treatment (EPT). The most common (known) approaches to eliminating cracks in metals and alloys are thermal and thermomechanical treatments, which can form mechanical bonds in the form of metallic bridges between crack edges through atomic diffusion [22, 30]. However, these methods have significant drawbacks: they degrade mechanical properties due to thermal degradation of the metal microstructure and require significant labor and energy costs. Furthermore, high-temperature anneling only eliminates nanometer-sized defects.

EPT is a promising crack healing approach that concentrates electromagnetic energy on defects, providing high energy efficiency. In recent decades, this approach has been actively developed to eliminate both fatigue macrocracks and internal microdefects in conductive materials [1–4, 8, 9, 12, 15, 18, 19, 29, 32–39]. In [2, 3, 18], it is shown that EPT healing of macrocracks can be accomplished by welding their edges. This requires determining the appropriate parameters of current pulses: amplitude, shape, duration, number and duty cycle, as well as their change over time (hereinafter referred to as the "route" of EPT). The preferred route of EPT should ensure the maximum length/volume of crack healing, while preventing crater formation, material evaporation, and thermal degradation of its microstructure. Incorrect selection of parameters can leave the crack unhealed, so the choice of route is critical to the success of EPT. In [1–4, 8, 9, 12, 15, 18, 19, 29, 32–39], the problem of route selection was solved empirically, by trial and error. Their authors did not provide a theoretical justification for the choice of a particular route. This paper attempts to fill this gap and consider a new approach to determining the preferred EPT route. It will be shown that in order to heal cracks by welding their edges, it is necessary to expose the samples to pulses of a high-energy electromagnetic field (hereinafter referred to as HEMF).

We will name the electromagnetic field to be HEMF in the case of dissipation of specific EM energy in the material in one pulse of $q \geq 10^7$ J/m^3 at relatively short durations of pulse $\tau_0 \sim 10^{-4}$s. Electropulsing treatment of samples with HEMF pulses will be further referred to as high-energy electropulsing treatment (HeEPT). It is easy to show that the EPT of stainless steels used in [2, 3, 18, 35, 38, 39] is high-energy.

An analytical model [13] is proposed that allows limiting the parameter range to high-energy pulses. The energy analysis, based on this model, shows that a single control parameter - the specific energy of the current pulse - is sufficient for crack tip welding. This eliminates the need for empirically selecting multiple parameters (current density, shape, and duration), as was done previously. The model yields a certain dependence of specific energy of the current pulse on its serial number, ensuring a constant change in the internal energy at the crack tip with each pulse. This enables the preferred route of HeEPT to be achieved. With the exception of [3, 8, 9, 18], EPT studies have been focused on healing short fatigue cracks in thick plates and strips. In [3, 17, 18] the challenges of healing long macrocracks in thin plates by HeEPT (for which the crack length to plate thickness ratio is large) were addressed.

The applicability of the proposed approach was tested in experiments on welding the edges of long fatigue cracks in thin stainless steel plates by HeEPT [17]. The developed 13-step, 132-pulse preferred route demonstrated significant effectiveness. Its application ensured a sufficiently high quality of the welded crack edge joint and significantly improved the mechanical properties of the healed sample.

20.2 The Model

Let us consider the effect of a strong pulsed electromagnetic field on an ellipsoidal crack with a boundary Γ in a conducting body (Fig. 20.1), the linear dimensions of which (or the distances between cracks in which) L are much greater than the maximum linear size of the crack a, i.e., we will assume that the body is infinitely large (or the cracks are not interacting with each other). We will assume that at

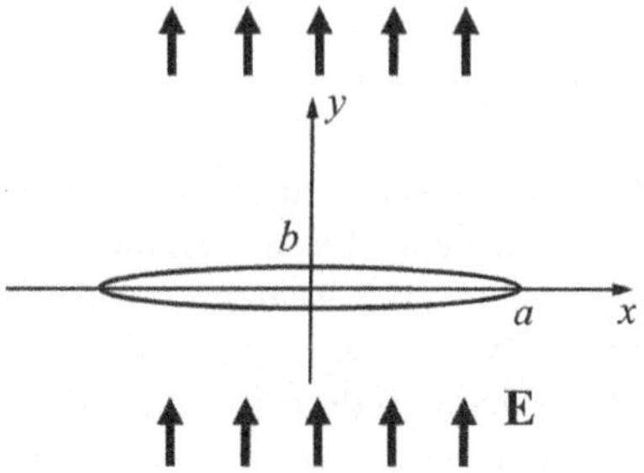

Fig. 20.1 Ellipsoidal crack in the medium.

the outer boundary of the body there is a given electric field intensity vector, with characteristic time of variation τ_0 being much greater than L/v_c, where v_c is the speed of light. Since the primary interest for the study is the temperature distribution in the vicinity of the crack tip, the electromagnetic field under the considered effect can be considered quasi-stationary. Magnetic effects are neglected. Due to the fact that the thermomechanical state in the immediate vicinity of the crack tip is most significant for the analysis of crack behavior, and also since we will subsequently apply the considered analysis to cracks in plates and strips, the spatial problem will be replaced by a planar one (we will assume the size of the ellipsoid along the z axis to be infinite).

The electric field far from the crack is assumed to be uniform, and the electric field intensity vector is applied perpendicular to the major semiaxis of the ellipse (Fig. 20.1). Therefore, below we consider the plane problem of the action of a short electric field (current) pulse at infinity on an elliptical crack. As a result, from Maxwell's equations and Ohm's law we obtain [20, 31]

$$\mathbf{E} = -\nabla\varphi = \frac{\partial\varphi}{\partial\mathbf{x}}, \quad \operatorname{div}\mathbf{j} = 0, \tag{20.1}$$

$$\mathbf{j} = \sigma^{\mathrm{E}}(T)\mathbf{E} = -\sigma^{\mathrm{E}}(T)\frac{\partial\varphi}{\partial\mathbf{x}}, \tag{20.2}$$

where $\mathbf{j}$ is the current density vector, $\mathbf{E}\,(\mathbf{x})$ is the electric field intensity vector, φ is the electric potential, $\sigma^{E}\,(T)$ is the material's electrical conductivity scalar, T is the temperature, and ∇ is the gradient operator. For simplicity, we will assume that the material is electrically isotropic and its electrical conductivity is independent of temperature. Assuming that the crack does not pass the current and that far from it the field $\mathbf{E}\,(\mathbf{x})$ is uniform and parallel to the y axis

$$E_x = 0, \quad E_y = E_\infty(t) = \frac{j_\infty(t)}{\sigma^E},$$

we obtain an equation and boundary conditions for the electric potential

$$\Delta\varphi = 0, \tag{20.3}$$

$$\left.\frac{\partial\varphi}{\partial n}\right|_\Gamma = 0, \quad \left.\frac{\partial\varphi}{\partial y}\right|_{y=\pm\infty} = -E_\infty, \tag{20.4}$$

$$\Gamma: \frac{x^2}{a^2} + \frac{y^2}{b^2} = 1, \quad R = \frac{b^2}{a}, \quad c^2 = a^2 - b^2, a \neq b, \ a > 0, \ b \geq 0, \ b < a,$$

where R is the curvature radius at the crack tip. The dependencies of the change in the electric field intensity at infinity $E_\infty(t)$ on time t, realized in the experiments, are shown in Fig. 20.2, where $j_\infty(t)$ is the modulus of the current density vector induced in the body at its outer boundary.

Note that exactly the same equation describes the irrotational (potential) flow of an ideal fluid around an elliptical cylinder. Then, using the results of [10] and considering, that the circulation of the electric field intensity vector along any closed

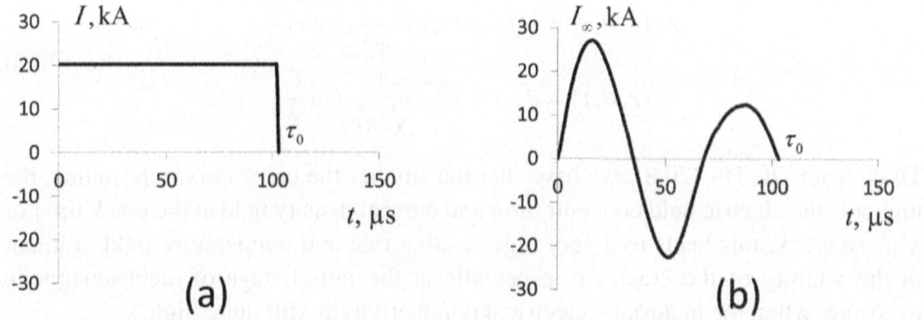

Fig. 20.2: Pulse shape: (a) - rectangular, (b) - damped sinusoid.

path is zero, we obtain for the electric potential at $x \geq 0, y \geq 0$

$$\varphi = \frac{E_\infty(t)}{a-b}\left(by - \frac{a}{\sqrt{2}}\sqrt{\sqrt{\left((x-c)^2+y^2\right)\left((x+c)^2+y^2\right)} - \left(x^2-y^2-c^2\right)}\right). \tag{20.5}$$

From (20.1), (20.2) and (20.5) we obtain for the components of the current density vector

$$j_x = -\sigma^{\mathrm{E}}(T)\frac{\partial\varphi}{\partial x}, \quad j_y = -\sigma^{\mathrm{E}}(T)\frac{\partial\varphi}{\partial y}, \tag{20.6}$$

$$j_x(x,y,t) = -\frac{a\sigma^E E_\infty(t)\,x\left(-c^2+x^2+y^2-g(x,y)\right)}{\sqrt{2}(a-b)\,g(x,y)\sqrt{c^2-x^2+y^2+g(x,y)}},$$

$$j_y(x,y,t) = \frac{\sigma^E E_\infty(t)}{(a-b)}\left(b - \frac{ay\left(c^2+x^2+y^2+g(x,y)\right)}{\sqrt{2}g(x,y)\sqrt{c^2-x^2+y^2+g(x,y)}}\right), \tag{20.7}$$

$$g(x,y) = \sqrt{\left((c-x)^2+y^2\right)\left((c+x)^2+y^2\right)}.$$

Below, we will consider cracks that have an elongated shape in the x-axis direction, such that

$$\frac{b^2}{a^2} = \frac{R}{a} \ll 1$$

(see Fig. 20.1). As the radius of curvature at the crack tip tends to zero $R \to 0$ $(b \to 0)$, the magnitudes of the electric field intensity vectors and, accordingly, the current density at the crack tip tend to infinity (as $|\mathbf{E}| \to \infty$, $|\mathbf{j}| \to \infty$). Indeed, let us introduce the polar coordinates $x = a + r\cos\theta$, $y = r\sin\theta$ and set in (20.7) $b = 0$ (the crack being a mathematical cut), then for the components of the current density vector at $r \to 0$ in the vicinity of the tip $(a, 0)$ we obtain at $0 \leq \theta \leq \pi/2$

$$
\begin{aligned}
j_x(r,\theta,t) &= -j_\infty(t)\sqrt{\pi a}\frac{1}{\sqrt{2\pi r}}\cos\frac{\theta}{2},\\
j_y(r,\theta,t) &= j_\infty(t)\sqrt{\pi a}\frac{1}{\sqrt{2\pi r}}\sin\frac{\theta}{2}.
\end{aligned} \tag{20.8}
$$

Thus, from (20.7) - (20.8), we have that the smaller the crack curvature radius, the higher is the electric field concentration and current density field at the crack tip. For a sharp crack, this leads to a very high heating rate and temperature field gradient in the vicinity of the crack tip, especially at the initial stage of electromagnetic exposure, when the material's electrical conductivity is still quite high.

From (20.1), (20.6) and (20.7), it is easy to see that the electric field intensity and current density at the point $x = 0,\ y = b$ are equal to zero. The magnitudes of these fields in the region of the crack center are close to zero.

To obtain the temperature field, we use the law of conservation of energy [27]:

$$
\int_V \rho \dot{U} dV = \int_S q dS + \int_V r^E dV, \tag{20.9}
$$

where V is an arbitrary volume bounded by a piecewise smooth surface S; $\rho = \rho(T)$ is the density of the material, $\dot{U}$ is the material derivative of the internal energy, q is the heat flux per unit area along the outward normal $\mathbf{n}$ to the surface S, $r^E = r^E(x, y, t)$ is the heat released due to the flow of electric current in a unit of volume per unit time (the specific power of volumetric heat sources). If we accept the Fourier law of thermal conductivity [27] to be valid for describing the thermal processes under consideration

$$
\mathbf{q} = -K\nabla T, \tag{20.10}
$$

where K is the thermal conductivity coefficient, then, applying the Gauss-Ostrogradsky theorem to Eq. (20.9), using (20.10), we obtain the thermal conductivity equation for determining T

$$
\frac{\partial T}{\partial t} = \zeta \Delta T + r^{\mathrm{E}}, \quad \zeta = \frac{K}{\rho c}, \tag{20.11}
$$

where $c = c(T)$ is the specific heat capacity. This is the very equation used in [2, 7]. Let us introduce the dimensionless variables

$$
\bar{T} = \frac{T}{T_0}, \quad \bar{t} = \frac{t}{\tau_0}, \quad \bar{x} = \frac{x}{R}, \quad \bar{y} = \frac{y}{R},
$$

then (20.6) will have the form

$$
\begin{aligned}
\frac{\partial \bar{T}}{\partial \bar{t}} &= \frac{\zeta \tau_0}{R^2}\Delta \bar{T} + \frac{\tau_0}{T_0} r^{\mathrm{E}},\\
\frac{\zeta \tau_0}{R^2} &= \frac{K\tau_0}{\rho c R^2} = \frac{1}{\mathrm{Pe}}.
\end{aligned} \tag{20.12}
$$

Pe is the dimensionless Péclet number.

Further, in this work we will consider model (artificial) cracks in zinc and copper plates with curvature radius at the tip being 150 μm. For the problem of a crack in zinc, we have

$$K = 111\ \mathrm{W/(mK)},\ \rho = 7140\ \mathrm{kg/m^3},\ c = 394\ \mathrm{J/(kgK)},\ R = 150\,\mu\mathrm{m},$$
$$\tau_0 = 100\ \mu\mathrm{s},\ \frac{1}{Pe} \approx 0.17 \ll 1.$$

For the problem of a crack in copper, we have

$$K = 389.6\ \mathrm{W/(mK)},\ \rho = 8933\ \mathrm{kg/m^3},\ c = 385\ \mathrm{J/(kgK)},\ R = 150\,\mu\mathrm{m},$$
$$\tau_0 = 100\ \mu\mathrm{s},\ \frac{1}{Pe} \approx 0.5 < 1.$$

In addition, the time of electromagnetic action τ_0 is quite small and does not exceed $\tau_0 = 10^{-4}$ s, therefore, thermal conductivity during the current pulse can be neglected and the heating processes can be considered adiabatic.

The thermal energy per unit volume of a body per unit time, released due to the dissipation of electromagnetic energy during the current in a conductor, is determined by the Joule-Lenz law. The kinetic energy of the directed motion of conduction electrons, moving under the influence of an electric field, is converted into the internal energy U of the chaotic motion of phonons (and electrons) in the crystal lattice U. The measure of the internal energy U is the temperature T.

Thus, for the temperature field, the evolution equation is valid

$$\rho c \dot{T} = r(x, y, t),$$
$$r^{\mathrm{E}}(x, y, t) = \begin{cases} r^{\mathrm{E}}(x, y) & t \le \tau_0 \\ 0 & t > \tau_0 \end{cases}, \qquad r^{\mathrm{E}}(x, y) = \mathbf{j} \cdot \mathbf{E} = \sigma^{\mathrm{E}} \nabla\varphi \cdot \nabla\varphi. \tag{20.13}$$

For simplicity, let us consider the rectangular current pulse shown in Fig. 20.2(a). All subsequent discussions and results will be independent of the pulse shape, since the heating process during the current pulse is adiabatic.

From (20.5) and (20.13), we obtain the temperature distribution for $x \ge 0$, $y \ge 0$, $t \le \tau_0$

$$T(x, y, t) = T_0 + \frac{\sigma^{\mathrm{E}} E_\infty^2 t}{\rho c\,(a-b)^2} \left[b^2 + \frac{a^2\left(x^2+y^2\right)}{\sqrt{c^4 + 2c^2\left(-x^2+y^2\right) + \left(x^2+y^2\right)^2}} - \frac{\sqrt{2}aby\left(c^2 + x^2 + y^2 + \sqrt{\left(c^2-x^2\right)^2 + 2\left(c^2+x^2\right)y^2 + y^4}\right)}{\sqrt{c^4+2c^2\left(-x^2+y^2\right)+\left(x^2+y^2\right)^2}\sqrt{c^2-x^2+y^2+\sqrt{\left(c^2-x^2\right)^2+2\left(c^2+x^2\right)y^2+y^4}}} \right]. \tag{20.14}$$

When integrating, we assumed that the initial temperature was $T(x, y, 0) = T_0$. It is clear that for $x \ge 0$ the temperature distribution is symmetrical about the x-axis: $T(x, y, t) = T(x, -y, t)$. Then, for the temperature at the crack tip, we have

$$T(a,0,t) = T_0 + \frac{\sigma^{\mathrm{E}} E_\infty^2 t}{\rho c \left(1 - \sqrt{p}\right)^2} \left(p + \frac{1}{p}\right), \tag{20.15}$$

where

$$p = \frac{b^2}{a^2} = \frac{R}{a},$$

R is the radius of curvature at the crack tip. Note that for cracks in the experiments below

$$p = \frac{b^2}{a^2} \ll 1.$$

From (20.15) it is evident that at $p \to 0$ the solution for the temperature at the crack tip has a singularity of $\sim 1/p$. Or, in other words, as the radius of curvature at the crack tip $R \to 0$, the temperature at the crack tip $T \to \infty$.

From (20.14), a strong concentration of the temperature field around the crack tip is evident. While the heating at point $x = 0, y = b$ is zero, the heating near the crack center is close to zero. Since the main heating occurs in a small region at the crack tip, the change in the average temperature is approximately equal to the change in temperature in the region of current uniformity at some distance from the crack $T_* - T_0 \approx \sigma^{\mathrm{E}} E_\infty^2 t\, (\rho c)^{-1}$.

Let us define the specific electromagnetic energy dissipated in the sample material when a current pulse passes through it in time interval from 0 to t as follows:

$$q(t) = \int\limits_0^t \frac{j_\infty^2(t)}{\sigma^E} dt. \tag{20.16}$$

Knowing this dependence, Eq. (20.15) for the temperature at the crack tip can be written in the form

$$T(a,0,t) = T_0 + \frac{q(t)}{\rho c \left(1 - \sqrt{p}\right)^2} \left(p + \frac{1}{p}\right). \tag{20.17}$$

In a similar form, Eq. (20.14) can be written for the temperature in the entire region by making the substitution $\sigma^{\mathrm{E}} E_\infty^2 t = q(t)$.

At $t = \tau_0$ the expression (20.16) will define the electromagnetic energy dissipated in a unit volume of the material during the entire current pulse. For brevity, we will henceforth refer to this specific energy as the "specific energy of the current pulse". This terminology is introduced because crack healing experiments involve recording oscillogram of the current in the samples. From the oscillogram, it is easy to calculate the dependence $j_\infty(t)$, and from it, calculate the energy of the current pulse

$$q = q(\tau_0) = \int\limits_0^{\tau_0} \frac{j_\infty^2(t)}{\sigma^E} dt. \tag{20.18}$$

The quantity $q = q(\tau_0)$ acts as a generalized time, knowing which we can calculate the temperature at the crack tip (20.17) or in the entire region at a given moment.

Let us determine the duration of the current pulse $\tau_0 = t_m$, during which the material at the crack tip will melt. To do this, we multiply the right and left sides of equation (15) by dt/t and, integrate it over time t from 0 to t_m, where t_m is the moment in time at which the material will change to a molten state, and obtain

$$\int_{T_0}^{T_m} \rho c dT = \frac{1}{p\left(1-\sqrt{p}\right)^2} \int_0^{t_m} \frac{j_\infty^2}{\sigma^E} dt, \tag{20.19}$$

where T_m is the melting temperature of the material. We have neglected terms of order p ($p \ll 1$) in the numerator of the right-hand side of Eq. (20.19). Since the process is adiabatic, considering any other pulse shape does not change the form of Eq. (20.19), which, in the case of a non-rectangular pulse, must take into account the time dependence $j_\infty = j_\infty(t)$.

The integral on the left-hand side of (20.19) represents the amount of heat Q_m that must be imparted to a unit volume of the material at the crack tip for it to melt. The integral on the right-hand side of (20.19) represents the electromagnetic energy q_m that must be dissipated in a unit volume of material at the crack tip over a period of time from 0 to t_m. So, q_m is the energy of current pulse at which the material at the crack tip begins to melt. Thus, we have

$$Q_m = \alpha q_m, \tag{20.20}$$

$$\alpha = \frac{1}{p\left(1-\sqrt{p}\right)^2} = \frac{a}{R\left(1-\sqrt{R/a}\right)^2}, \tag{20.21}$$

$$Q_m = \int_{T_0}^{T_m} \rho c dT, \quad q_m = \int_0^{t_m} \frac{j_\infty^2(t)}{\sigma^E} dt, \tag{20.22}$$

where the dimensionless coefficient α will be called the electromagnetic energy concentration factor at the crack tip. Equation (20.20) can be written as:

$$q_m = \frac{1}{\alpha} Q_m = \frac{1}{\alpha} \rho c \left(T_m - T_0\right). \tag{20.23}$$

Equation (20.23) determines the minimum specific electromagnetic energy q_m required to be dissipated in the material (during the single current pulse) for the material to melt at the crack tip. For $p = \frac{b^2}{a^2} \ll 1$, we have $\alpha \gg 1$, that means a small energy of current pulse is required to melt the material at the tip of a crack elongated along x. Melting at the tip triggers the crack healing process by welding its edges; without melting, it is impossible to weld the crack edges. Therefore, the specific energy of the current pulse q_m can be chosen as the minimum threshold energy for the considered range of possible current pulse energies.

This equation was derived under the assumption that the density, heat capacity, and electrical conductivity of the material are independent of temperature. This is generally not the case in the considered range of crack tip temperature variation. Specifically, the amount of heat required to heat most metals by 1°C increases with increasing temperature, while density decreases with increasing temperature. However, overall, the amount of heat required for melting increases. As a first approximation, these temperature dependencies of material's physical properties can be taken into account in (20.23) by assuming the average values of these quantities over the interval from T_0 to T_m.

Similarly to (20.23), one can determine the minimum specific electromagnetic energy q_{ev} that must be dissipated in the material during a single current pulse of duration t_{ev}, in order for the material at the crack tip to begin to evaporate. Evaporation at the tip will obviously mean that healing the crack by welding its edges is impossible. The specific energy of the current pulse q_{ev} can be chosen as the maximum (threshold) energy for the considered range of possible specific energies of current pulses

$$q_{ev} = \frac{1}{\alpha} Q_{ev} = \frac{1}{\alpha} \rho c \left(T_m - T_0\right) + \frac{1}{\alpha} \rho c \left(T_{ev} - T_m\right), \tag{20.24}$$

$$Q_{ev} = \int\limits_{T_0}^{T_{ev}} \rho c dT, \qquad q_{ev} = \int\limits_{0}^{t_{ev}} \frac{j_\infty^2 (t)}{\sigma^E} dt. \tag{20.25}$$

The temperature dependencies of c and ρ can be approximately taken into account by assuming in (24) the average values of these quantities on the intervals from T_0 to T_m and from T_m to T_{ev}, respectively.

For a crack that does not interact with other cracks or free boundaries of the body, the coefficient α is found to depend only on the geometric dimensions of the crack. More generally, this coefficient should depend not only on the defect geometry but also on the distances to free surfaces or to other defects, the presence of which increases the current concentration at the crack tip. In particular, the presence of free boundaries can be taken into account in the following simple manner.

The solution of the antiplane problem of an edge crack (mathematical cut) in an infinitely thick strip of width W for mode III is known and can be found, for example, in [7, 24]. Applying the electromechanical analogy, for the current density field near the tip of such a crack at $r \to 0$, $0 \le \theta \le \pi/2$ we have:

$$\begin{aligned} j_x &= -j_\infty (t) \sqrt{\pi a} \sqrt{\frac{2W}{\pi a} \tan\left(\frac{\pi}{2} \frac{a}{W}\right)} \frac{1}{\sqrt{2\pi r}} \cos \frac{\theta}{2}, \\ j_y &= j_\infty (t) \sqrt{\pi a} \sqrt{\frac{2W}{\pi a} \tan\left(\frac{\pi}{2} \frac{a}{W}\right)} \frac{1}{\sqrt{2\pi r}} \sin \frac{\theta}{2}. \end{aligned} \tag{20.26}$$

From a comparison of (20.8) and (20.26), it is evident that the presence of free boundaries leads to an increase in the current density in the vicinity of the tip by $\sqrt{\frac{2W}{\pi a}\tan\left(\frac{\pi}{2}\frac{a}{W}\right)}$. Therefore, taking into account (20.20) - (20.26), the following estimate can be obtained for the electromagnetic energy concentration factor α' at the tip of an edge crack in the strip:

$$\alpha' = \alpha\frac{2W}{\pi a}\tan\left(\frac{\pi}{2}\frac{a}{W}\right). \tag{20.27}$$

Taking into account (20.27), the Eqs. (20.23) and (20.24) can be written as

$$q_m = \frac{R}{a}\left(1-\sqrt{\frac{R}{a}}\right)^2\frac{\pi a}{2W}\cot\left(\frac{\pi}{2}\frac{a}{W}\right)\rho c\,(T_m - T_0)\,, \tag{20.28}$$

$$q_{ev} = q_m + \frac{R}{a}\left(1-\sqrt{\frac{R}{a}}\right)^2\frac{\pi a}{2W}\cot\left(\frac{\pi}{2}\frac{a}{W}\right)\rho c\,(T_{ev} - T_m)\,. \tag{20.29}$$

We will use the Eqs. (20.28) and (20.29) below to calculate the values of q_m and q_{ev} for zinc and copper plates with edge model cracks.

If internal microcracks in the continuum are arranged in an orderly fashion with crack centers at the nodes of a rectangular or hexagonal lattice and a period (the distance between nodes equal to $2W$), then formulas (20.28) and (20.29) can also be considered approximately valid for them. If $a/W \ll 1$, microcracks do not interact with each other, and (20.28) and (20.29) transform into (20.23) and (20.24).

Thus, in order for a crack in a material to be healed by welding its edges, we must act on the material with a high-energy EMF pulse, such that the energy q is dissipated in a unit volume of the material lies in the range:

$$\begin{aligned} &q_m \le q \le q_{ev}, \\ &q_{ev} = q_m + \Delta q_{ev}, \\ &\Delta q_{ev} = \frac{R}{a}\left(1-\sqrt{\frac{R}{a}}\right)^2\frac{\pi a}{2W}\cot\left(\frac{\pi}{2}\frac{a}{W}\right)\rho c\,(T_m - T_0)\,. \end{aligned} \tag{20.30}$$

Inequality (20.30) defines the minimum and maximum threshold values of the specific electromagnetic energy that must be dissipated in the sample during the current pulse for the crack to be healed by welding. However, this condition is only a necessary one, but not sufficient, for such healing. Even if the specific energy of the current pulse q satisfies inequality (20.30), this does not guarantee successful welding of the crack faces. A certain value energy of current pulse from the permissible range (20.30) must be selected for the crack edges to contact along a certain section of its length and form a strong weld along this part. This energy must be sufficient to melt the material not only at the crack tip, but also along this entire section of length. Simultaneously, compressive stresses must act in this molten part of the crack, ensuring the reliable contact between the crack edges. For this to be possible, the released Joule heat must be concentrated in a small region near the tip, while the surrounding material remains significantly cooler or cold. To achieve this, the influence of heat transfer by conduction must be minimized. This is possible if the pulse duration is sufficiently short and energy of current pulse is sufficiently high. Thus, we arrive to the requirement for high-energy EPT route for healing by welding.

To neglect thermal conductivity and consider the heating process adiabatic, the pulse duration must not exceed several hundred microseconds. Experiments below will demonstrate that, for pulse durations $\tau_0 \sim 10^{-4}$ s corresponding to those used for model (artificial) cracks in copper, zinc, and aluminum alloys, the thermal conductivity can be neglected.

In EPT, to heal a fatigue crack in a plate or strip along all or part of its length, it is necessary to apply several current pulses to the sample (as a rule, more than 10). Typically, these pulses have the same maximum current density, shape, and duration [1, 4, 8, 9, 12, 15, 19, 29, 32–39]. However, it follows from Eqs. (20.19) and (20.28) and (20.29) that the current concentration at the crack tip is determined by the geometric dimensions of the crack $b/a$and the sample a/W. During the EPT welding, both b/a and a/W change. Therefore, as the crack heals, it is necessary to adjust the energy of a pulse. During EPT, we will apply current pulses with varying pulse energy to the sample with a crack. We will change this energy during EPT welding so that the change in the internal energy at the crack tip with each pulse is constant and equal to the change in the internal energy with the first pulse. This condition is a requirement for maintaining a constant welding temperature in the moving (due to healing) crack tip. Let us examine how accurately this condition is met during successful welding of edges of a fatigue crack.

In [2], it was proposed to perform healing in two steps. In each of these two steps, current pulses of the same shape but different densities and durations were passed through the sample. The current in the sample was induced by completely discharging the capacitor bank. In each step, the maximum and minimum current densities in the pulses were selected empirically (not by the results of numerical simulations).

In the first step of the exposure, 20 pulses in the form of a damped sinusoid with the maximum current density $2.0\,\mathrm{kA/mm^2}$ and a total duration of approximately 500 μs were induced in a steel strip with an edge fatigue crack of length $a_1 = 3350\,\mu\mathrm{m}$ (including the notch length of 2000 μm). As a result, part of the crack was welded with molten metal. However, further healing of the crack did not occur, it was necessary to increase the maximum current density and the pulse duration. By the beginning of the second step, 52% of the absolute length of the fatigue crack (1350 μm) was healed and the total length of the edge crack decreased to (including the notch length of 2000 μm). In the second step, after exposure to 30 pulses with the maximum current density $2.15\,\mathrm{kA/mm^2}$, the remaining part of the crack (up to the notch) was welded. Due to the lack of experimental data in [2], we assume that the radii of curvature in the first and second steps were unchanged $R_1 = R_2 = 10\,\mu\mathrm{m}$.

From (20.28) and (20.29) it follows that the energy of the current pulse at the second step q_2 is related to the energy of the pulse q_1 at the first stage inversely proportional to the ratio of the corresponding the electromagnetic energy concentration factors

$$q_2/q_1 = \alpha_1'/\alpha_2'. \tag{20.31}$$

Since for fatigue cracks from work [2] we have $R/a \ll 1$, $a/W \ll 1$, then from (20.31) we obtain

$$q_2/q_1 = R_2\, a_1/R_1\, a_2. \quad (20.32)$$

Based on the oscillogram presented in [2], using (20.18), we can calculate that, according to the experiment, $q_2/q_1 = 1.32$. On the other hand, in accordance with (20.31), $q_2/q_1 \approx R_2\, a_1/R_1\, a_2 = 1.273$. The difference in the energies of current pulse obtained experimentally and according to the considered model, assuming a constant thermal state at the crack tip, is 4.7%.

Thus, by (20.28) and (20.29) we can determine of the energy of current pulse in the first step, then using (20.32), we can calculate the required pulse energy in the second step. If necessary, using (20.18), we can reconstruct the maximum current density and pulse duration in the second step. After that, using this algorithm, we can determine the energy of current pulse in the third step, etc. Further, this algorithm will be explained using a specific example of healing long fatigue cracks in thin stainless steel plates.

We will call the EPT route preferable if condition (20.31) or (20.32) is satisfied for two consecutive pulses of this route. However, we believe that when determining the EPT steps in an experiment, for optimal healing of fatigue macrocracks, it is more convenient to use not the current density, pulse duration, and shape, but a generalized parameter such as the specific energy of the current pulse.

From (20.28) it follows that as the macrocrack heals (its length decreases or the radius of curvature at the apex or the distance from the apex to the free boundary increases), the specific energy of the current pulse must increase for the healing process to continue successfully. This is why a 2-step route was selected in [2] for healing a short macrocrack in a strip. However, increasing the pulse energy is especially important for healing long macrocracks in thin samples. Therefore, a 6-step method was selected for this purpose in [18].

Equations (20.19) and (20.28) represent an attempt to propose a relationship according to which the energy of current pulse should be increased for successful crack healing by HeEPT. This relationship can be used to adjust the pulse parameters as the crack length decreases and radius of curvature changes. It follows from this relationship that the best route is not a two-step route [2] or a six-step route [3, 18], but a multi-step route, in which we repeatedly, ideally for each subsequent pulse, increase the pulse energy in accordance with equation (20.31) or (20.32). Below in this paper, we will demonstrate how the proposed approach to determining the preferred route is implemented in experiments.

20.3 Experiments

To validate the model and estimates obtained above, the following experiments with model and fatigue cracks were conducted. Polycrystalline zinc grade Z0 (99.975% Zn) (the closest American equivalent is Zn Z15001) and copper grade M1 (99.90%

Cu) (the closest equivalent is Cu C11000) were chosen as the study materials. Their chemical composition complies with GOST 3640-94 and GOST 859-2014. Samples were cut using the electric spark method from hot-rolled metal sheets along the rolling direction. The geometric dimensions of the copper samples are shown in Fig. 20.3. The dimensions of the zinc samples differ from those of the copper samples only in their thickness, which for the zinc plate was 1.0 (±0.05) mm. The properties of Zinc and Copper are presented in Table 20.1.

A model edge crack (notch) of 4.5×0.3 mm was cut in the plates using the electric spark method at a distance of 40 mm from one end. The crack had parallel edges and a rounded tip with a radius of curvature at the tip $R = 150\ (\pm 5)$ μm (Fig. 20.4). To study the temperature field at the tip of the macrocrack at pulsed electromagnetic field, the current pulses were applied to the samples which were rigidly clamped in copper electrode clamps. This caused a certain amount of electromagnetic energy from the electric current to dissipate in the sample. The distance between the electrodes was 38 mm. The crack was located in the middle, between the electrodes (Fig. 20.5).

The samples were exposed to current pulses using a GIT-20-40 pulsed electromagnetic field generator on a specially developed laboratory setup, schematically shown in Fig. 20.5a. The generator control unit adjusts the parameters of the charging unit in the charging mode and connects it to the capacitor bank. In the discharge mode, as shown in the diagram, the capacitor bank is connected to the sample through a

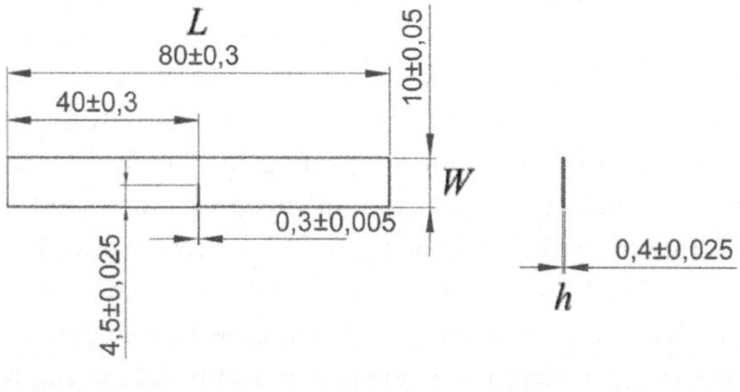

Fig. 20.3 Drawing of a sample made of copper Cu M1.

Table 20.1: Properties of Zinc and Copper.

Zinc Zn	ρ_{sol}, kg/m^3	ρ_{liq}, kg/m^3	c_{sol}, J/kgK	c_{liq}, J/kgK
$T_0 = 20°C$	7140	-	394	
$T_m = 419°C$	7140	6600	394	502
$T_{ev} = 906°C$	-	6600	-	502
Copper Cu				
$T_0 = 20°C$	8933	-	385	-
$T_m = 1084,6°C$	8361	7990	525,2	513,9
$T_{ev} = 2560°C$	-	7960	-	513,9

Fig. 20.4 View of a zinc sample in the region of a long model crack. Coordinate system, XY - in the plane of the plate.

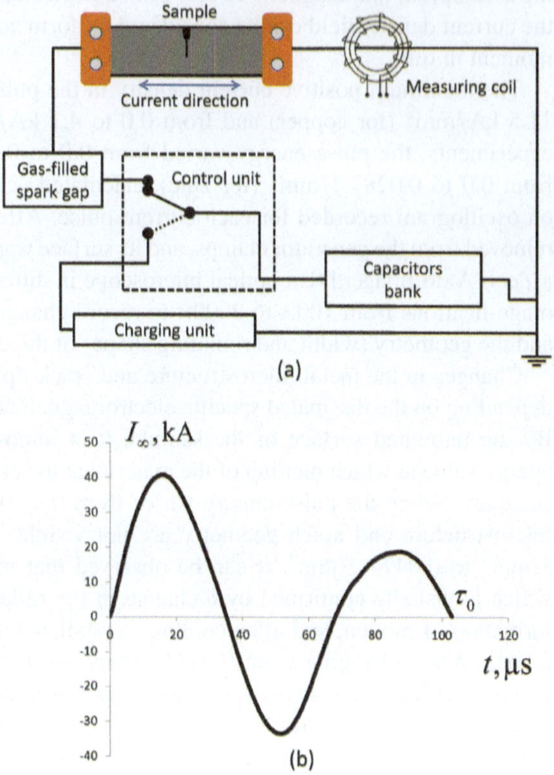

Fig. 20.5 Laboratory setup circuit (a), typical shape of current pulse in samples (b).

gas spark gap. As soon as the ignition pulse is sent from the control unit to the spark gap, the capacitor bank begins to discharge through the clamps-electrodes through the sample until the circuit is interrupted by the spark gap. The setup induced a current pulse in the sample with the shape shown in Fig. 20.5b. In the experiments, the amplitude of the current density in the pulses was varied, while the duration remained approximately constant. The direction of the current density vector on the electrodes was perpendicular to the crack edges (faces), i.e., the current was directed along the y-axis. The current density is approximated by a damped sine wave from the time $t \in [0, \tau_0]$. The eigen frequency was $f = \omega/(2\pi) \sim 13.5\,\text{kHz} - 14.76\,\text{kHz}$, damping coefficient $\delta = 0.15$, duration $\tau_0 = 102 - 103\ \mu\text{s}$.

The specific electromagnetic energy dissipated in the sample during one such pulse (energy of current pulse) was calculated based on equation (18) using the current oscillogram in the samples obtained experimentally. The temperature dependence of the electrical conductivity of zinc $1.69\ 10^7\ 1/(\text{Ohm} \cdot \text{m})$ and copper $5.9\ 10^7\ 1/(\text{Ohm} \cdot \text{m})$ was not taken into account in the calculations, since the temperature change $T_* - T_o$ for zinc and copper was 20°K and 10°K, respectively.

The skin depth at this frequency in copper and zinc is ~ 0.6 mm and 1.2 mm, respectively, which is significantly greater than the sample thicknesses. Therefore, the assumption that magnetic effects can be neglected in our analysis is correct, and the current density field can be considered uniform across the plate thickness at any moment in time.

The maximum positive current density in the pulses varied widely, from 0.0 to $11.5\ \text{kA/mm}^2$ (for copper) and from 0.0 to $4.1\ \text{kA/mm}^2$ (for zinc). Thus, in the experiments, the pulse energy varied from 0.0 to $0.0681\ \text{J/mm}^3$(for copper) and from 0.0 to $0.0287\ \text{J/mm}^3$ (for zinc), calculated according to Eq. (20.18), based on oscillogram recorded for each current pulse. After each pulse, the sample was removed from the generator clamps, and its surface was examined on both sides using a Zeiss Axio Imager.D1m optical microscope in differential interference contrast at magnifications from 100× to 2000× to record changes in the metal microstructure and the geometry (width and rounding shape) of the crack in the crack tip region.

Changes in the metal microstructure and crack tip geometry in a copper sample depending on the dissipated specific electromagnetic energy are shown in Fig. 20.6. By the untreated surface of the sample, it is impossible to determine the pulse energy value at which melting of the material at the crack tip begins with acceptable accuracy. When the pulse energy varies from 0 to $0.0156\ \text{J/mm}^3$, changes in the microstructure and notch geometry are not visible. When changing from 0.0156 J/mm^3 to $0.0474\ \text{J/mm}^3$, it can be observed that melting begins in the tip area, which is visually confirmed by a change in the reflectivity of the surface and the formation of molten, and upon cooling, crystallized, copper grain boundaries (Fig. 20.6c). When changing from $0.0474\ \text{J/mm}^3$ to $0.0552\ \text{J/mm}^3$, it is evident that intense melting occurs in the apex region, leading to changes in the cutout geometry, with cavities forming (Fig. 20.6d). This may indicate that the temperature at the tip is approaching the evaporation temperature of the metal. Increasing the specific energy of the current pulse to $0.0681\ \text{J/mm}^3$ causes a microexplosion and the formation of a crater at the crack tip (Fig. 20.6e-f). Melted metal is ejected from the tip in a radial direction, as evidenced by traces of crystallized melt on the plate surface (the direction is shown by the thick red arrow in Fig. 20.6e-f).

Observing the copper microstructure as a function of the measured specific energy of the current pulse will give us the record how this microstructure changes over time. In other words, the specific energy of the current pulse acts as a generalized time. Knowing this energy and using (20.18), we can determine the moment in time for an arbitrary current pulse shape and energy.

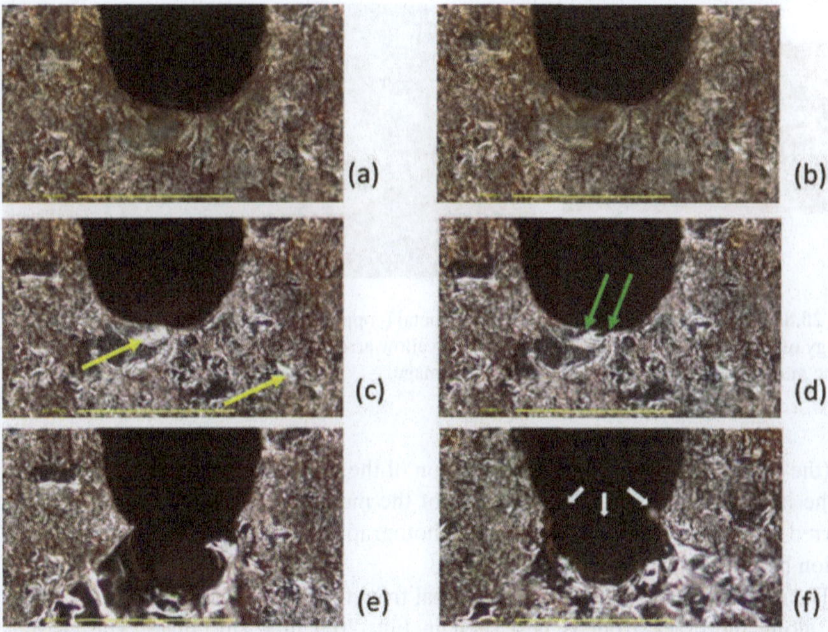

Fig. 20.6: Changes in the copper microstructure on the sample surface depending on the pulse energy: (a) 0.0 J/mm^3, (b) 0.0156 J/mm^3, (c) 0.0474 J/mm^3 (yellow arrows - molten grains), (d) 0.0552 J/mm^3 (green arrows - cavities on the cut surface), (e) 0.0681 J/mm^3 (crater, front side of the sample), (f) 0.0681 J/mm^3 (crater, back surface of the sample). The direction of the Lorentz force is shown by white arrows. The straight yellow line is the 300 μm scale bar.

At specific energy of the current pulse of 0.0620 J/mm^3, a molten metal ejection is observed at the crack tip, both inside the crack along the crack's edges and along its front (Fig. 20.7b). The direction of this ejection is shown by the bold yellow arrow. Based on the energy value, we can see that this ejection precedes crater formation. Figure 20.8 shows an enlarged photograph of the molten metal ejection at the crack

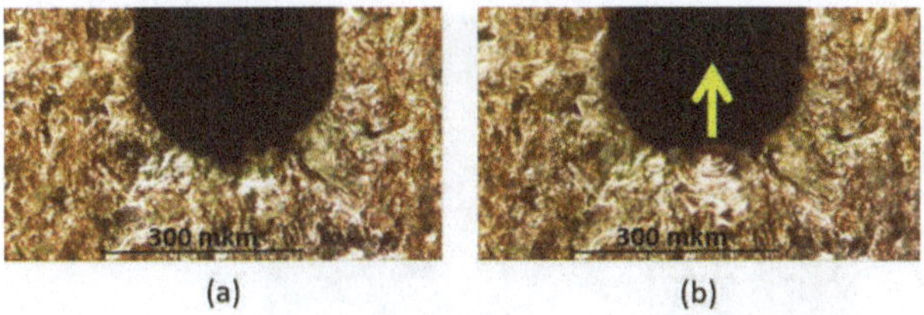

Fig. 20.7: Ejection of molten metal (copper) along the crack apex. (a) initial apex boundary (at a pulse energy of 0.0 J/mm^3); (b) ejection (displacement) of the melt (specific energy of the current pulse of 0.0620 J/mm^3); the yellow arrow shows the direction of the melt ejection. The straight black line is the 300 μm scale bar.

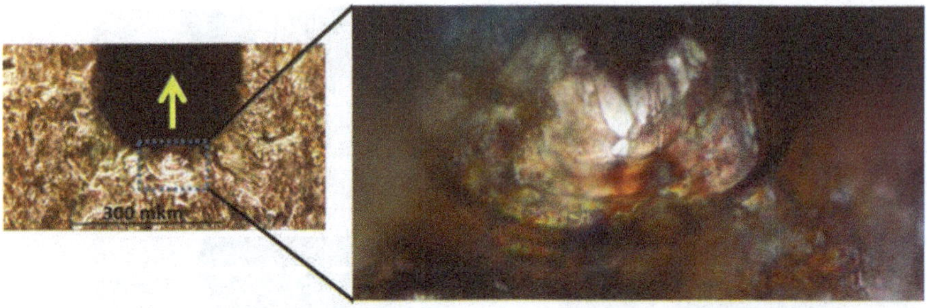

Fig. 20.8: An enlarged photograph of a molten metal (copper) ejection at the crack tip (specific energy of the current pulse 0.0620 J/mm^3). The yellow arrow (on the right) indicates the direction of the molten metal ejection in the plane of the image.

tip (the yellow arrow indicates the direction of the ejection - along the crack edges). In the image on the right, the boundary of the melting region (zone) separates the blurred and unblurred portions of the photograph. The oval shape of the molten region boundary is clearly visible.

The ejection (expulsion) of molten metal from the crack tip along the crack edges and along the crack front is best seen in Fig. 20.9 in a zinc plate. The crystallized "stream" of molten zinc is clearly visible in both the enlarged and unenlarged photographs.

Crater formation, as reported above in the introduction, has been observed by many researchers and is not new. Crater formation is a complex, essentially three-dimensional process; mathematical modeling of this process has not been done, and the physical mechanism is not fully understood. The overwhelming majority of researchers believe that the crater formation process occurs as follows: under the action of the Lorentz force, molten material is ejected radially from the crack tip. The Lorentz force occurs during the current pulse acting through the sample and is directed along the radius of a circle inscribed in the crack tip. Meanwhile, the ejection/expulsion of melt along the crack front was observed only in Gallo's work during video recording of crater formation in aluminum plates. Thus, during the

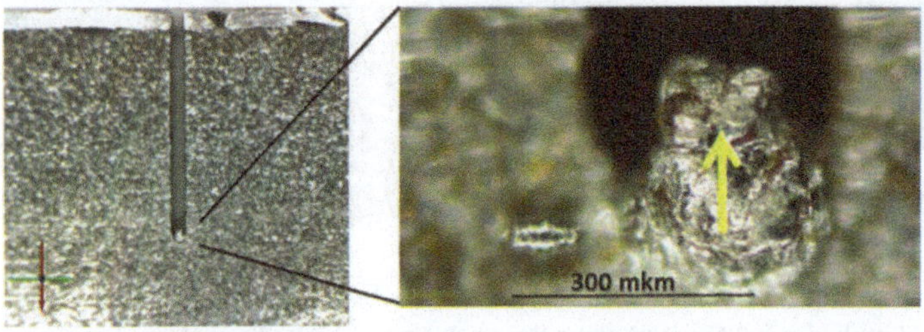

Fig. 20.9: Ejection of metal (zinc) along the crack front at the tip. The yellow arrow shows the direction of melt ejection (specific energy of the current pulse 0.0282 J/mm^3) .

analysis of video images [18], it was noted that molten metal from the interior of the sample is first expelled along the crack front and then ejected radially.

In [9, 26], again based on video data, it was noted that during crater formation, molten metal "fountained" from the crack tip on both sides of the plate, i.e., microexplosion products were observed moving in a direction perpendicular to the plane of the plate. Therefore, the crater formation itself was called a "microexplosion" in [26]. Results of modeling the healing of microcracks in the material [11, 14] demonstrated the possibility of molten metal being expelled from the defect tip along the crack edges, forming a "stream".

Furthermore, results of three-dimensional numerical modeling of the effect of a current pulse on a long crack in a zinc plate [16] show that the current density, heating, and thermal expansion within the plate (near its midplane) are higher than near its surface (the numerical modeling was done in the absence of heat exchange with the environment). The temperature at this point reaches the evaporation temperature and evaporation of the metal begins near the middle plane of the plate.

The Lorentz force at the crack tip is directed radially, i.e., opposite to the direction of ejection, which occurs primarily along the crack edges into its center (Fig. 20.8b). Presumably, the cause of this expulsion is that the ejection begins with the evaporation of metal at the crack tip, specifically, in a small region on the midplane of the plate due to the rapid heating of the material at this location during the pulse. During this localized evaporation, the surrounding molten metal is expelled from the tip both into the crack (inwards, along its edges) and at an angle to its surface. However, this assumption requires detailed verification, including three-dimensional numerical modeling and additional experiments.

In addition to melt ejection from the crack tip, we observed minor overall displacements (10-15 μm) of points on the rounded surface toward the center of the inscribed circle of radius R, i.e., in the direction opposite to the radial direction and into the crack. This indicates that residual deformations (stresses) exist in the material near the tip after temperature relaxation.

The experiments allowed us to accurately determine the boundary of the metal melting zone at the crack tip. This region is the oval area at the crack tip, clearly visible in Figs. 20.7 - 20.10. In our opinion, this is the region in which the metal

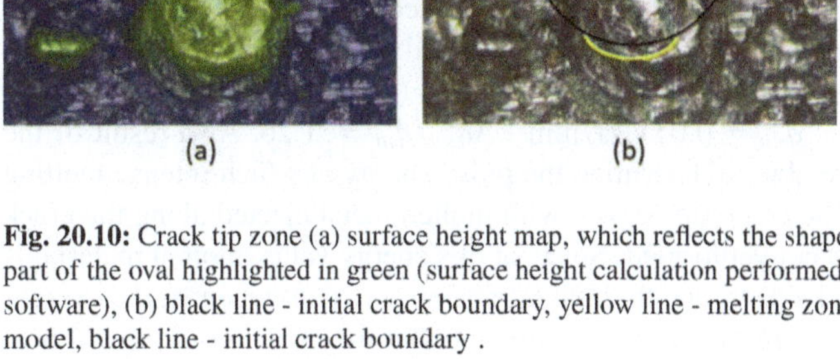

Fig. 20.10: Crack tip zone (a) surface height map, which reflects the shape of the melting region - part of the oval highlighted in green (surface height calculation performed in PICOLAY software), (b) black line - initial crack boundary, yellow line - melting zone calculated by the model, black line - initial crack boundary .

transitioned from a solid to a liquid state at the end of the pulse.

Determining the boundaries of the melting zone allows us to compare the obtained analytical solution for temperature with the experimental results. Furthermore, this allows us to evaluate the validity of the obtained constraints (30) on the pulse energy, obtained analytically based on the model assumptions.

Figure 20.10b shows the melting region boundary calculated according to (20.14) - (20.17) using the experimentally obtained dependence $j(t)$. The boundary of this region is an isoline of the temperature field T_m = 419°C. The maximum linear dimension of this region along the x-axis is 20 μm, and along the y-axis, 58 μm. However, the analogous dimensions of the melting region observed experimentally are along the x-axis- 80 μm and along the y-axis - 100 μm. Thus, the model underestimates the melting region dimensions compared to the experimental data.

In our opinion, this underestimation of the dimensions and, consequently, temperature occurred because the model does not explicitly take into account the temperature dependence of the electrical conductivity, heat capacity, and density of the metal. Meanwhile, the changes in these physical quantities for zinc are quite large and must be taken into account in the model. For example, the electrical conductivity of zinc decreases by a factor of 5.76 in the range between the solid (20°C) and liquid (420°C) states [11, 14]. This decrease in electrical conductivity leads to the displacement of current from the heated regions of the metal to colder ones. That is, the maximum current density shifts from the tip in the direction of the greatest temperature gradient, namely, from the crack tip along the x-axis, and the current density field itself becomes smoother. This in turn leads to smoothing of the temperature field in the vicinity of the tip and a disproportionate increase in the size of the melting zone. A more significant increase in size will occur along the x-axis. Our calculations using a finite-element model that takes into account changes in all physical properties of the metal with temperature showed a significant increase in the size of this region, primarily along the x-axis [11, 14]. Calculations conducted in Gallo's work, using a different finite-element model that takes into account only the dependence of electrical conductivity on temperature, lead to the melting region becoming elongated along the x-axis.

Furthermore, it appears that the actual geometry of a model crack with parallel edges differs significantly from an ellipse, which also increases the electric field concentration and current density in the vicinity of the crack tip and, accordingly, increases the temperature recorded experimentally as an increase in the melting region size and proportional to the minimum threshold pulse energy (20.28).

Let us now evaluate the fulfillment of theoretical constraints (20.30) on the pulse energy based on the melt expulsion/ejection observed experimentally. To do this, we calculate the q_m, q_{ev}values for zinc using formulas (20.28) and (20.29), obtaining q_m = 0.0252 J/mm^3, q_{ev} = 0.0571 J/mm^3, $\Delta q_{ev}/q_m$ = 1.26. As a result of the experiments, we were able to determine the pulse energy at which intense melting of the material near the crack tip begins, with molten metal ejected along the crack front, but a crater has not yet formed. Since, at this energy value, molten material is ejected along the crack edges, and thermal expansion and, consequently, the stresses compressing the edges are maximum, we assume that the model crack would heal best at this energy. Therefore, this specific pulse energy value corresponds to the first pulse of the preferred healing route. It is also understood that the model crack cannot be healed by welding its edges due to the excessively large distance between them. Thus, the healing of model cracks is somewhat speculative. The pulse energy for the model crack in a zinc sample, calculated according to (20.18), based on experimental data, is $q_{pr} = 0.0282\ \mathrm{J/mm^3}$. Thus, the optimal value of the pulse energy obtained in the experiment, necessary for healing a crack in a zinc sample, satisfies relation (20.30), namely, $q_m < q_{pr} < q_{ev}$. Similar values of q_m, q_{ev} for a crack in copper are, $q_m = 0.0950\ \mathrm{J/mm^3}$, $q_{ev} = 0.2341\ \mathrm{J/mm^3}$, $\Delta q_{ev}/q_m = 1,46$.

The optimal pulse energy for a model crack in a copper sample was $q_{pr} = 0.0620\ \mathrm{J/mm^3}$ (Fig. 7). This means that this energy is lower than q_m, and inequality (20.30) is not satisfied for copper. Therefore, the experimentally obtained optimal pulse energy required to heal a crack in a copper sample does not satisfy the relation $q_m < q_{pr} < q_{ev}$. As noted above, the analytical model underestimates the crack tip temperature and the melting region size, and consequently overestimates the minimum energy required to heal a model crack in copper compared to experiment.

In our opinion, this overestimation of the pulse energy occurs for the model crack in copper plates because the proposed model does not account for the three-dimensional effect of free boundaries (in the plane with the normal) on the electric field concentration at the crack tip. Meanwhile, the free surfaces of the plates lead to an additional concentration of the electric field and current at the tip of the model crack in copper. Moreover, this concentration increases with the thinness of the plate. We believe that, therefore, (20.30) is satisfied for the zinc plate. Conversely, it is not satisfied for the copper plate, which is 2.5 times thinner, and its thickness is comparable to the distance between the crack faces.

Finally, the imperfections of the actual geometry of the crack tip in copper, the radius of which varies depending on the direction of the angle θ, and the walls (surfaces along the z-axis) have numerous microroughnesses and microcracks resulting from the process of electric spark cutting of the metal. These microcracks and microroughnesses are clearly visible in copper in Fig. 20.7a and act as additional concentrators of the electric field at the tip, leading to more rapid heating at the tip. This contributes to premature melting at the tip and to the discrepancy between the data obtained by the model and the experiment.

Let us consider the healing of a model crack (notch) in a $30 \times 10 \times 0,8$ mm plate made of Al6061-T6 aluminum alloy. For this, we will use the experimental data obtained in [7]. The total initial length of the edge model crack in the plate was $a = 4000\ \mu\mathrm{m}$. The radius of curvature was $R = 200\ \mu\mathrm{m}$. Table 20.2 presents the properties of the Al6061-T6 aluminum alloy, taken in accordance with the data of [7], which we used in our calculations. Note that in this case $1/\mathrm{Pe} \approx 0.17 \ll 1$. Using (20.28) and (20.29), we obtain for this crack in the aluminum alloy plate $q_m = 0,0356\ \mathrm{J/mm^3}$, $q_{ev} = 0,082\,\mathrm{J/mm^3}$.

Here, as for model cracks in zinc and copper, we consider healing to occur when molten metal is ejected along the crack front onto its edges, but a crater has not yet formed. According to [7], this moment immediately precedes crater formation. Therefore, the value of the specific electromagnetic energy dissipated

Table 20.2: Properties of the Al6061-T6 aluminum alloy.

Alloy Al6061-T6	ρ_{sol}, kg/m^3	ρ_{liq}, kg/m^3	c_{sol}, J/kgK	c_{liq}, J/kgK
$T_0 = 20°C$	2700	-	899	-
$T_m = 582°C$	2700	2700	899	801
$T_{ev} = 2519°C$	-	2700	-	801

in the sample material at this moment determines the optimal healing route. This moment can be determined using the data presented in [7] (Fig. 20.11), according to which the first crater formation in the plate occurs at a time of 80 μmm. The crater radius is approximately 20-25 μm, and the maximum current density in the pulse is 4.7 kA/mm^2.

Based on the current oscillogram and the data on the frequency and attenuation coefficient of the pulse in the plate with a crack, given in [7], it is easy to conclude that the first period is 227 μm, and the total pulse duration is approximately 1000 μm. Note that the skin depth at this frequency (about 4.4 kHz) in Al6061-T6 is of the order of the sample thickness. Using (20.16), one can calculate the specific electromagnetic energy dissipated in the alloy over a time interval from 0 to 80 μm at a maximum pulse current density of 4.7 kA/mm^2. This energy is equal to the pulse energy at which healing occurs. The calculations used the electrical conductivity of the Al6061-T6 alloy $0,25\ 10^8\ \Omega m^{-1}$ [7]. Thus, the pulse energy for a model crack in an aluminum alloy sample, calculated on the basis of experimental data [7], is $q_{pr} = 0,0438\ \text{J/mm}^3$. It follows that the experimentally obtained optimal pulse energy required to heal the model crack in the Al6061-T6 sample satisfies relation (20.30), namely, $q_m < q_{pr} < q_{ev}$.

Similarly to the crack in zinc, using (20.14) - (20.17), it is easy to calculate the melting region boundary at a time of 80 μm at $q_{pr} = 0,0438\ \text{J/mm}^3$. The boundary

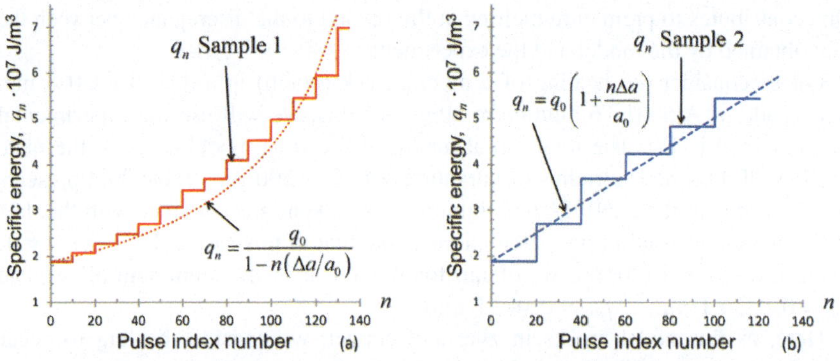

Fig. 20.11: Routes for samples 1 (a) and 2 (b).

of this region is an isoline of the temperature field $T = T_m = 582°C$. The maximum linear size of this region along the x-axis is 15 μm, and along the y-axis - 51 μm. However, according to experimental data [7], the crater radius and, accordingly, the melting region size along the x-axis are approximately 20-25 μm. Thus, we are once again convinced that the model underestimates the melting region size and the temperature at the crack tip compared to the experimental data, but quantitatively correctly reproduces the experimental data. The reasons for this underestimation were analyzed above.

Next, we will consider determining the preferred route of HeEPT for fatigue cracks. The preferred route should ensure a long crack healing length while preventing thermal degradation of the microstructure, crater formation, and metal evaporation. Incorrect selection of the route can result in no healing or poor-quality crack healing, which is critical for EPT. Therefore, determining such an HeEPT route is a key task. Due to the complexity of this task, theoretical approaches to selecting the HeEPT route have not been developed.

Samples 1-3, of $120 \times 18.2 \times 0.8$ mm (length $L\times$ width $W\times$ thickness h), were cut from an industrial strip of AISI201 stainless steel in the as-received condition. Long macrocracks were formed in the plates using fatigue loading according to the method [21], propagating from an edge notch $0.4 \times 2.0 \times 0.8$ mm in size. Our approach uses a single control parameter - the specific energy of the current pulse q, instead of selecting several ones (current densities, pulse durations, and pulse shapes), as this was done in previous studies. The specific energy of the current pulse is calculated using formula (20.18).

As noted above, during EPT, crack healing by welding the edges occurs stepwise in the vicinity of the crack tip. With each subsequent pulse, the new crack tip moves along the crack front from the initial tip to the notch in the plate. Suppose that a single pulse welds an identical small crack length Δa at the new tip, and after n pulses, its initial length a_0 decreases to $a_n = a_0 - n \cdot \Delta a$. We will assume that with each n-th pulse, the radius of curvature at the new crack tip R_n changes insignificantly after healing, i.e. $R_n = R_0$. Then, from equation (32), it follows that the specific energy of each n-th pulse q_n should increase in accordance with the relationship

$$q_n = \frac{q_0}{1 - n\,(\Delta a/a_0)}, \quad n = 1, 2, \ldots, N, \tag{20.33}$$

where $q_0 = 1.89\ 10^7\ \mathrm{J/mm^3}$ is the initial pulse energy, and N is the total number of pulses in the route.

For Sample 1, the EPT route was selected in accordance with Eq. (20.33). In the experiments for Sample 1, we have $N = 132$, and the duty cycle were assumed to be $1.145\ 10^{-5}$ and dependence (20.33) was approximated by a piecewise constant function (Fig. 20.11a) consisting of 14 steps (13 steps of 10 and 1 step of 2 = 132) pulses. For Sample 2, the following route was selected. From (20.33), a linear dependence follows for $q_n = q_0\,(1 + n\,(\Delta a/a_0))$. In the experiments we had $q_0 = 1.89\ 10^7\ \mathrm{J/m^3}$, and the duty cycle assumed to be $1.145\ 10^{-5}$ and the linear dependence was approximated by a piecewise constant function (Fig. 20.11b) consisting of 6 steps. Sample 3 was not subjected to EPT and was used to evaluate the healing quality.

After HeEPT, the samples were cut to a new width (W') so that the initial crack tips were located 3.0 mm from the new plate edges. The new width was calculated in mm. The cut samples were stretched to rupture on an Instron testing machine at a constant strain rate of 1 mm/min (initial clamp spacing was l_0 = 60 mm). The HeEPT treatment routes for samples 1 and 2 allowed for healing of an average of 1.36 mm and 0.82 mm of crack length, respectively (Fig. 20.12). The fractograms (Fig. 20.12) show zones of notch (zone 1), fatigue growth (zone 2), healing (zone 3), and ductile fracture (zone 4). The microstructure of zone 3 differs from zones 2 and 4, making the boundaries between them distinct.

Neck formation is observed in zones 3 and 4 of all samples (Fig. 20.12). Neck formation in zones 4 of samples 1 and 2 (Fig. 20.12) confirms the welding of the crack faces during HeEPT and subsequent ductile fracture of the welded material. The fractures in the healing zone are non-uniform, with various local areas differing in microrelief and microorientation of pits, which correspond to different mechanisms of sample destruction in this zone under conditions of a complex stress-strain state. Sample 2 healed worse than sample 1. In the fractogram, between the two necks in the healing and ductile fracture zones, there is a region with very little crack adhesion. In this region, there are only small localized areas (bridges) of crack adhesion, visible as equiaxed dimples in the fractogram. Meanwhile, in the well-healed zone, where the neck formed in sample 2, equiaxed and elongated dimples are visible everywhere. The crater formed by the ejection of molten metal is observed.

Figure 20.13 shows the dependencies of tensile load F on elongation $\Delta l = l - l_0$ for samples 1–3. The specific work of deformation and fracture of the sample in zones 3 and 4 is calculated as $Q = A/V$, where

$$A = \int_0^{\Delta l_m} F dl$$

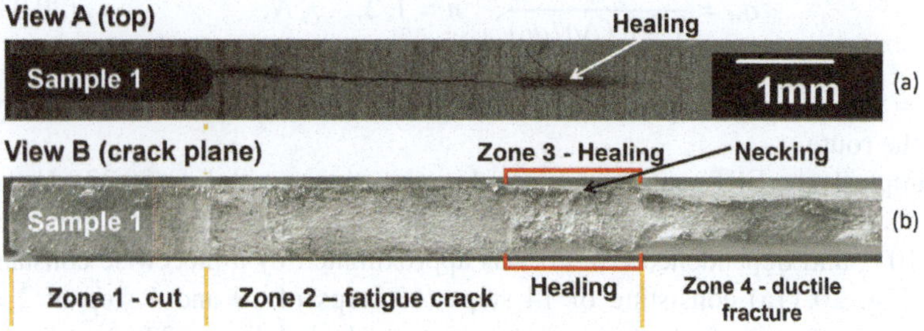

Fig. 20.12: (a) Outer surface of sample 1 with a healed crack after EPT, view through an optical microscope; (b) fractogram of sample 1 in SEM in SE mode.

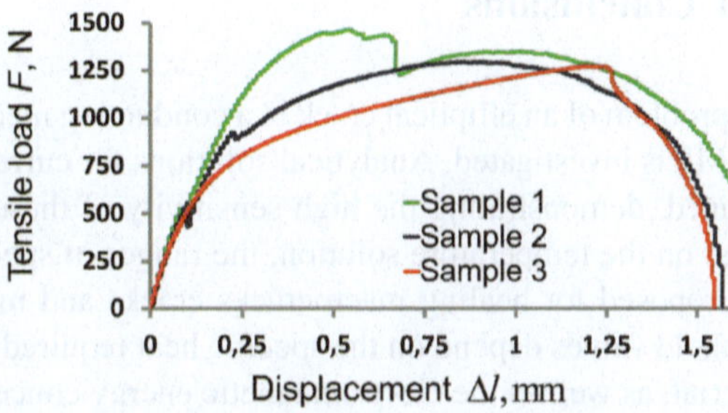

Fig. 20.13 Tension diagram of samples cut to a new width.

is the external work spent on deformation and fracture of the sample, and $V = (W' - a_N)\,h$ is the volume of the sample of unit length. The calculation results for A and Q are presented in Table 20.3.

Samples 1 and 2 required 27.1% and 8.9% more energy to fracture, respectively, compared to sample 3. The specific work required to fracture healed samples 1 and 2 was only 12.58% and 14.50% less than that of the unhealed sample. This energy estimate indicates that the quality of the EPT welded crack face joint is comparable to the quality of the strip from which the samples were made, i.e., quite high. Furthermore, ESD healing increased the ultimate load for samples 1 and 2 by 13.8% and 1.4%, respectively, compared to sample 3. Healing qualitatively changed the shape of the curves (Fig. 13), leading to a significant increase in the average load that the healed samples could withstand. For samples 1 and 2, it was 18.8% and 7.1% higher than for the unhealed sample.

Effective welding of crack edges requires the use of high-energy pulses, such that a single current pulse dissipates specific electromagnetic energy in the material within a range 10^7 J/m^3 $\leq q \leq 10^9$ J/m^3 under the single pulse durations $\tau_0 \sim 10^{-4}$ s. This is confirmed by the specific energies and durations of current pulse used in experiments conducted on steel, zinc, copper, and aluminum alloy.

Table 20.3: Parameters of cracks and mechanical characteristics of Samples.

Description	Sample 1 (132 pulses)	Sample 2 (120 pulses)	Sample 3 (untreated)
A – work of force F, J (compared to Sample 3)	1.883 +27.1 %	1.614 +8.9 %	1.482 0
Q, J/m^3 (compared to Sample 3)	5.40×10^8 -12.6 %	5.28×10^8 -14.5 %	6.18×10^8 0

20.4 Conclusions

The problem of an elliptical crack in a conducting medium exposed to short pulses of HeEMF is investigated. Analytical solutions for current density and temperature are obtained, demonstrating the high sensitivity of these quantities to crack geometry. Based on the temperature solution, the ranges of specific energies of current pulses are proposed for healing macrocracks cracks and microdefects by welding. These threshold values depend on the specific heat required for melting and vaporizing the material, as well as the electromagnetic energy concentration factor at the crack tip, which is determined by the crack geometry.

Efficient welding of crack faces requires the use of high-energy pulses, such that a single current pulse dissipates specific electromagnetic energy in the material within a range 10^7 J/m^3 $\leq q \leq 10^9$ J/m^3 under the single pulse durations $\tau_0 \sim 10^{-4}$ s. Furthermore, it is sufficient to adjust only one control parameter - the specific pulse energy - rather than several, as in previous studies.

The results obtained using the model are in a good enough quantitative agreement with the experimental data for the model cracks. The experimentally observed motion (injection) of the molten material along the edges into the model cracks is confirmed by numerical modeling, conducted previously by us, and by the experiments done by other scientists. Our experiments allowed to accurately determine the boundary of the metal melting zone at the tip of the model cracks.

Comparison of analytical estimates with experimental data confirms the validity of the assumptions made, including the one on adiabaticity of the process. However, the model underestimates the size of the melting zone and changes the temperature distribution in the vicinity of the crack tip. The reasons for this are: neglection of changes in electrical conductivity with temperature and elliptic simplification of the crack geometry of the model cracks.

The Fourier's law used to describe the energy transfer reduces the temperature and prevents reaching the melting point at the model crack tip at experimentally used energies of the current pulses. Modeling HeEPT using this law yields only qualitative results.

For preferred healing by welding, the ratio of energies of current pulses in two successive steps must be inversely proportional to the ratio of the electromagnetic energy concentration factors at the crack tips for these steps. This will ensure a constant welding thermal regime from one step to next step.

As the crack length decreases or the tip curvature increases, the pulse energy must increase, making the multi-step HeEPT route preferable. This provides the effective healing of macrocracks by gradually increasing the energy of the current pulses.

The applicability of the proposed approach was confirmed in experiments on welding the edges of long fatigue cracks in thin stainless steel plates. The developed 13-step preferred route with 132 pulses demonstrated significant effectiveness. Its use ensured a sufficiently high quality of the welded crack edge joint and significantly improved the mechanical properties of the healed sample.

References

[1] Bryzgalov VA, Morkina AYU, Abdullina DU, Semenov AS, Tatarinov PS, Yakushev IA, Bebikhov YV, Kukudzhanov KV, Korznikova YA, Dmitriyev SV (2024) Obzor issledovaniy po zalechivaniyu makrotreshchin v metallakh pod deystviyem impul'snogo toka vysokoy plotnosti (in Russ., High-density pulsed current: a review of its role in healing macrocracks in metals). Materials Technologies Design 6(2(17)):38–58, DOI 10.54708/26587572_2024_621738

[2] Cai Q, Zhou M, Bagherpour E, Hosseini S, Mendis C, Chang I, Assadi H (2023) New insight into crack-healing mechanism via electropulsing treatment. Metallurgical and Materials Transactions A 54(7):2960–2974, DOI 10.1007/s11661-023-07073-1

[3] Cai Q, Rey Rodriguez P, Carracelas Santos S, Castro G, Mendis CL, Chang ITH, Assadi H (2024) Crack healing via electropulsing treatment applied to additive-manufactured TiC/316L stainless steel composites. Materials Letters 365:136410, DOI 10.1016/j.matlet.2024.136410

[4] Dmitriev SV, Morkina AY, Tarov DV, Khalikova GR, Abdullina DU, Tatarinov PS, Tatarinov VP, Semenov AS, Naimark OB, Khokhlov AV, Stolyarov VV (2024) Effect of repetitive high-density current pulses on plastic deformation of copper wires under stepwise loading. Spectrum of Mechanical Engineering and Operational Research 1(1):27–43, DOI 10.31181/smeor1120243

[5] Finkel VM, Golovin YI, Sletkov AA (1977) Razrushenie vershiny treshhiny silovym elektromagnitnym polem (in Russ., Fracture of crack tip by high-energy electromagnetic field). Doklady AN SSSR 237(2):325-327

[6] Finkel VM, Pashevitsky MM, Golovin YI, Ivanov VM, Kiperman VA (1985) Elektrotermicheskoe razrushenie metalla jelektromagnitnym polem (in Russ., Electrothermal destruction of metal by electromagnetic field). Fizika i khimiya obrabotki materialov (2):16–23

[7] Gallo F, Satapathy S, Ravi-Chandar K (2010) Melting and crack growth in electrical conductors subjected to short-duration current pulses. International Journal of Fracture 167(2):183–193, DOI 10.1007/s10704-010-9543-0

[8] Hosoi A, Nagahama T, Ju Y (2012) Fatigue crack healing by a controlled high density electric current field. Materials Science and Engineering: A 533:38–42, DOI 10.1016/j.msea.2011.11.024

[9] Hosoi A, Kishi T, Ju Y (2013) Healing of fatigue crack by high-density electropulsing in austenitic stainless steel treated with the surface-activated precoating. Materials 6(9):4213–4225, DOI 10.3390/ma6094213

[10] Kochin NE, Kibel' IA, Roze NV (1964) Theoretical Hydromechanics. Interscience Publisher, New York

[11] Kukudzhanov KV (2015) Modelirovaniye vozdeystviya vysokoenergeticheskogo impul'snogo elektromagnitnogo polya na mikrotreshchiny v polikristallicheskom metale (in Russ., Modeling the treatment of high-energy pulsed electromagnetic field of the micro-cracks in a polycrystalline metal). PNRPU Mechanics Bulletin (4):138–158, DOI 10.15593/perm.mech/2015.4.09

[12] Kukudzhanov KV, Chentsov AV (2020) O zalechivanii poristosti v metalle pod deystviyem vysokoenergeticheskogo elektromagnitnogo polya (in Russ., About healing of porosity in metal under high energy electromagnetic field). Bulletin of the Chuvash State Pedagogical University named after IY Yakovlev, Series: Limit State Mechanics (2(44)):116–131, DOI 10.37972/chgpu.2020.44.2.012
[13] Kukudzhanov KV, Chentsov AV (2024) The energy of short current pulses required for healing cracks in conductive materials. Mechanics of Solids 59(8):3929–3948, DOI 10.1134/s0025654424607031
[14] Kukudzhanov KV, Levitin AL (2015) Modeling the healing of microcracks in metal stimulated by a pulsed high-energy electromagnetic field. Part I. Nanomechanics Science and Technology: An International Journal 6(3):233–249, DOI 10.1615/nanomechanicsscitechnolintj.v6.i3.60
[15] Kukudzhanov KV, Levitin AL (2019) Modelling of some mechanism of metal electroplasticity under pulsed high-energy electromagnetic field action. Journal of Physics: Conference Series 1205:012032, DOI 10.1088/1742-6596/1205/1/012032
[16] Kukudzhanov KV, Levitin AL, Ugurchiev UK (2021) Healing of cracks in plates by strong electromagnetic field. Journal of Samara State Technical University, Ser Physical and Mathematical Sciences 25(1):193–202, DOI 10.14498/vsgtu1831, URL http://dx.doi.org/10.14498/vsgtu1831
[17] Kukudzhanov KV, Korznikova EA, Chentsov AV, Dmitriev SV (2026) A novel approach to determining the parameters of electric pulse healing of fatigue cracks. Materials Letters 404:139565, DOI 10.1016/j.matlet.2025.139565
[18] Kukudzhanov KV, Khalikova GR, Korznikova EA, Chentsov AV, Dmitriev SV (2024) Healing of long fatigue cracks in steel plates by high-density current pulses. Mechanics of Solids 59(5):3223–3234, DOI 10.1134/s0025654424605949
[19] Kumar A, Paul SK (2020) Healing of fatigue crack in steel with the application of pulsed electric current. Materialia 14:100906, DOI 10.1016/j.mtla.2020.100906
[20] Landau LD, Lifshitz EM, Sykes JB, Bell JS, Dill EH (1961) Electrodynamics of continuous media. Physics Today 14(10):48–50, DOI 10.1063/1.3057154
[21] Murugova OV, Nagimov MI, Khalikova GR, Dmitriev SV (2024) Zalechivanie dlinnykh treshchin v rzherzhaveyushchei stali 12Ch15G9ND s pomoshch'yu elektroimpul'snoi obrobotki (in Russ., Healing of long cracks in stainless steel 12Ch15G9ND using electric pulse treatment). Fundamentalnye Problemy Sovremennogo Materialovedenia 21(3):294–303, DOI 10.25712/astu.1811-1416.2024.03.003
[22] Paladugu SRM, Sreekanth PSR, Sahu SK, Naresh, K, Karthick SA, Venkateshwaran N, Ramoni M, Mensah RA, Das O, Shanmugam RA (2022) A comprehensive review of self-healing polymer, metal, and ceramic matrix composites and their modeling aspects for aerospace applications. Materials 15(23):8521, DOI 10.3390/ma15238521
[23] Que Z, Volpe L, Toivonen A, Burke MG, Scenini F, Ehrnstén U (2021) Effects of surface treatments on environmentally-assisted cracking susceptibility of

Alloy 182 in BWR environment. Corrosion Science 188:109555, DOI 10.1016/j.corsci.2021.109555
[24] Rabotnov YN (1988) Mekhanika deformiruemogo tverdogo tela (in Russ., Mechanics of Deformable Solids). Nauka, Moscow
[25] Razavi N, Ayatollahi MR, Sommitsch C, Moser C (2017) Retardation of fatigue crack growth in high strength steel s690 using a modified stop-hole technique. Engineering Fracture Mechanics 169:226–237, DOI 10.1016/j.engfracmech.2016.11.013
[26] Sano Y, Obata M, Kubo T, Mukai N, Yoda M, Masaki K, Ochi Ya (2006) Retardation of crack initiation and growth in austenitic stainless steels by laser peening without protective coating. Materials Science and Engineering: A 417(1–2):334–340, DOI 10.1016/j.msea.2005.11.017
[27] Sedov LI (1971) A Course in Continuum Mechanics, Vol I. Wolters-Noordhoff Publishing
[28] Sharma D, Reddy BS, Kumar P (2018) Fracture of pre-cracked metallic conductors under combined electric current and mechanical loading. International Journal of Fracture 212(2):167–182, DOI 10.1007/s10704-018-0298-3
[29] Sheng Y, Hua Y, Wang X, Zhao X, Chen L, Zhou H, Wang J, Berndt CC, Li W (2018) Application of high-density electropulsing to improve the performance of metallic materials: Mechanisms, microstructure and properties. Materials 11(2):185, DOI 10.3390/ma11020185
[30] Singh S, Narayanan JA, Dehgahi S, Qureshi AJ, Palani IA, Paul CP, Prashanth KG (2024) Revealing the impact of Hot Isostatic Pressing temperature on the microstructure and mechanical characteristics of Selective Laser Melted CuAlNiMn shape memory alloy. Materials Letter 365:136452, DOI 10.1016/j.matlet.2024.136452
[31] Sivukhin DV (2014) Termodinamika i molekulyarnaya fizika (in Russ., Thermodynamics and Molecular Physics). Fizmatlit, Moscow
[32] Xu W, Yang C, Yu H, Jin X, Yang G, Shan D, Guo B (2021) Combination of eddy current and heat treatment for crack healing and mechanical-property improvement in magnesium alloy tube. Journal of Magnesium and Alloys 9(5):1768–1781, DOI 10.1016/j.jma.2020.08.022
[33] Yang C, Xu W, Guo B, Shan D, Zhang J (2016) Healing of fatigue crack in 1045 steel by using eddy current treatment. Materials 9(8):641, DOI 10.3390/ma9080641
[34] Yin Y, Chen H, Morita Y, Toku Y, Ju Y (2018) Effect of electropulsing treatment on the fatigue crack growth behavior of copper. Materials 11(11):2168, DOI 10.3390/ma11112168
[35] Yizhou Z, You Z, Guanhu H, Benlian Z (2001) The healing of quenched crack in 1045 steel under electropulsing. Journal of Materials Research 16(1):17–19, DOI 10.1557/jmr.2001.0005
[36] Yu T, Deng D, Wang G, Zhang H (2016) Crack healing in sus304 stainless steel by electropulsing treatment. Journal of Cleaner Production 113:989–994, DOI 10.1016/j.jclepro.2015.12.060

[37] Zheng X, Shi YN, Lu K (2013) Electro-healing cracks in nickel. Materials Science and Engineering: A 561:52–59, DOI 10.1016/j.msea.2012.10.080
[38] Zhou Y, Qin R, Xiao S, He G, Zhou B (2000) Reversing effect of electropulsing on damage of 1045 steel. Journal of Materials Research 15(5):1056–1061, DOI 10.1557/jmr.2000.0152
[39] Zhou Y, Guo J, Gao M, He G (2004) Crack healing in a steel by using electropulsing technique. Materials Letters 58(11):1732–1736, DOI 10.1016/j.matlet.2003.10.049

Chapter 21
The Problem of Precise Positioning of the Center of Mass of the Executive Link of a Two-Link Elastic Manipulator

Svetlana P. Kurdina and Hrant A. Gevorgyan

Abstract The development of computational methods and computational algorithms for the dynamic analysis of elastic mechanisms is due to the increasing needs of the theory of control of elastic dynamic systems, involving either correction of motion or refinement of driving factors, taking into account the elasticity of the components of these systems. This category of control tasks includes the so-called tasks for the accuracy of the positioning of actuators, namely: in such tasks, it is necessary to clarify the laws of motion, taking into account undesirable elastic oscillations, which must be minimized or smoothed by making appropriate adjustments to the drive forces. The proposed article considers the problem of precise positioning of the center of mass of an elastic link of a two-link manipulator.

Key words: Dynamic analysis, Control theory, Newton–Euler method, Two-link elastic manipulator, Center of mass, Positioning accuracy

21.1 Introduction

The formalization and generalization of methods for dynamic analysis of multibody systems without inverting their mass matrices in the case of elastically deformable components of these systems has become a significant milestone in the development of this discipline. In a number of scientific publications, such computational strategies were later called the generalized Newton–Euler method [1–3, 7, 8].

Svetlana P. Kurdina (corresponding author)
Bauman Moscow State Technical University, 2nd Baumanskaya str., 5, p. 1, Moscow, Russian Federation, e-mail: kurdinasp@bmstu.ru

Hrant A. Gevorgyan
Institute of Mechanics of the National Academy of Sciences of the Republic of Armenia, Baghramyan Ave. 24B, Yerevan, Armenia, e-mail: hrgevorkian@mail.ru

H. Altenbach and A. V. Sahakyan, *Contact Mechanics and Relevant Problems*,
Advanced Structured Materials 253, https://doi.org/10.1007/978-3-032-20869-9_21

The special importance of high-speed computational strategies lies in their application to solving problems of controlling mechanisms containing links with distributed elasticity in their structure. This circumstance makes it possible in many cases to make available a computational process that cannot be carried out in real time due to the use of computational strategies involving the procedure for inverting the mass matrix of the systems under study, especially when it comes to multibody dynamic systems. The numerical solution of one of these control tasks for tracking the manipulator's executive body in advance of a given trajectory, taking into account the elasticity of one of the component links, was carried out by the author and presented by him in stages in publications [4, 5].

The modeling of elastic mechanisms in a slightly different formulation of the problems of control theory includes the case of accurate positioning of a dynamic system, taking into account the distributed elasticity of the constituent links. Here it is proposed to consider the movement of a two-link manipulator with an elastic actuating link [9], the center of gravity of which describes a trajectory with respect to which strict requirements are imposed on elastic deviations from a strictly horizontal position.

21.2 Dynamic Model of Elastic Manipulators

A manipulator of an arbitrary kinematic chain is considered (Fig. 21.1), consisting of n linearly deformable links of various geometric configurations $C_1, C_2, \ldots, C_n$ interconnected by n rotational or translational connecting nodes (kinematic pairs) identified by the logical variable [2, 3] σ_j ($j = 1, 2, \ldots, n$):

$$\sigma_j = \begin{cases} 0 & \text{if kinematic pair } j \text{ is rotational,} \\ 1 & \text{if kinematic pair } j \text{ is translational.} \end{cases}$$

It is known from the deformable body theory that small linear and angular deformations of an elastic link C_j are determined by displacements of a certain point $E_{0j} \in C_j$ from an undeformed S_{0j} to an other point E_j in deformed configuration S_j (Fig. 21.1) and are calculated in a rigidly connected with link C_j mobile reference

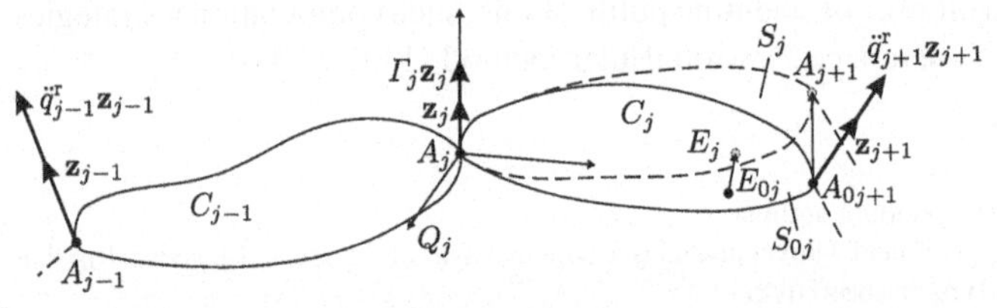

Fig. 21.1: Kinematic chain with elastically deformable link.

frame Q_j based on the following expansions [2, 3]

$$
{}^j\mathbf{u}_j(E_j,t) = \sum_{i=1}^{N_j} {}^j\mathbf{\Phi}_{\mathrm{d}j,i}(E_{0j})q^{\mathrm{e}}_{j,i}(t),
$$
$$
{}^j\varphi_j(E_j,t) = \sum_{i=1}^{N_j} {}^j\mathbf{\Phi}_{\mathrm{r}j,i}(E_{0j})q^{\mathrm{e}}_{j,i}(t), \quad j = 1,2,\ldots,n.
$$

Here $\mathbf{\Phi}_{\mathrm{d}j,i}$ and $\mathbf{\Phi}_{\mathrm{r}j,i}$ ($j = 1,2,\ldots,n$, $i = 1,2,\ldots,N_j$) are linear and angular Rayleigh–Ritz coordinate functions in mobile reference frame Q_j, N_j ($j = 1,2,\ldots,n$) is the number of interpolation vector polynomials of the elastic link, $q^{\mathrm{e}}_{j,i}(t)$ ($j = 1,2,\ldots,n$, $i = 1,2,\ldots,N_j$) are components of elastic variables. Rayleigh–Ritz coordinate functions are subject to vector equality [2, 3]

$$
{}^j\mathbf{\Phi}_{\mathrm{r}j,i}(E_{0j}) = \frac{1}{2}\mathrm{rot}\,{}^j\mathbf{\Phi}_{\mathrm{d}j,i}(E_{0j}), \quad i = 1,2,\ldots,N_j, \; j = 1,2,\ldots,n.
$$

They form a generalized matrix of shapes at point A_{0j+1} ($j = 1,2,\ldots,n$) (the point at which links j and $(j+1)$ interact in an undeformed configuration) [2, 3]

$$
{}^j\mathbf{\Phi}_j = \begin{bmatrix} {}^j\mathbf{\Phi}_{\mathrm{d}j,1}(A_{0j+1}) & {}^j\mathbf{\Phi}_{\mathrm{d}j,2}(A_{0j+1}) & \cdots & {}^j\mathbf{\Phi}_{\mathrm{d}j,N_j}(A_{0j+1}) \\ {}^j\mathbf{\Phi}_{\mathrm{r}j,1}(A_{0j+1}) & {}^j\mathbf{\Phi}_{\mathrm{r}j,2}(A_{0j+1}) & \cdots & {}^j\mathbf{\Phi}_{\mathrm{r}j,N_j}(A_{0j+1}) \end{bmatrix}.
$$

The dynamic model of elastic manipulators is expressed by a system of nonlinear algebraic-differential equations, which include: the system of nonlinear differential equations of dynamic equilibrium of elastic link C_j ($j = 1,2,\ldots,n$) (Fig. 21.2) [2, 3]

$$
\begin{bmatrix} {}^j\mathbf{J}_{\mathrm{rr}j} & {}^j\mathbf{J}_{\mathrm{re}j} \\ {}^j\mathbf{J}_{\mathrm{er}j} & {}^j\mathbf{J}_{\mathrm{ee}j} \end{bmatrix} \begin{bmatrix} {}^j\dot{\mathbf{V}}_j \\ \ddot{\mathbf{q}}^{\mathrm{e}}_j \end{bmatrix} + \begin{bmatrix} {}^j\mathbf{C}^{\mathrm{r}}_j \\ {}^j\mathbf{C}^{\mathrm{e}}_j \end{bmatrix} = \begin{bmatrix} {}^j\mathbf{F}_j - {}^{j+1}\mathbf{T}^T_j\,{}^{j+1}\mathbf{F}_{j+1} \\ -{}^j\mathbf{\Phi}^T_j\,{}^{j+1}\mathbf{R}^T_j\,{}^{j+1}\mathbf{F}_{j+1} \end{bmatrix},
$$

where ${}^j\mathbf{J}_{\mathrm{rr}j}$, ${}^j\mathbf{J}_{\mathrm{re}j}$, ${}^j\mathbf{J}_{\mathrm{er}j}$, and ${}^j\mathbf{J}_{\mathrm{ee}j}$ are known block inertia matrices of "rigid-rigid", "rigid-elastic", "elastic-rigid", and "elastic-elastic" aggregates, respectively; ${}^{j+1}\mathbf{R}_j$ and ${}^{j+1}\mathbf{T}_j$ are known matrices of dimension (6×6), which are commonly called six-dimensional rotation and transformation matrices, respectively; ${}^j\mathbf{C}^{\mathrm{r}}_j$ and ${}^j\mathbf{C}^{\mathrm{e}}_j$ are vectors containing centrifugal and gyroscopic terms of rigid and elastic aggregates, respectively, are also assumed to be known; ${}^j\dot{\mathbf{V}}_j$ is a vector of accelerations of dimension (6×1), which combines the accelerations of the link C_j and is expressed in terms of known three-dimensional vectors of linear ${}^j\dot{\mathbf{v}}_j$ and angular ${}^j\dot{\omega}_j$ accelerations

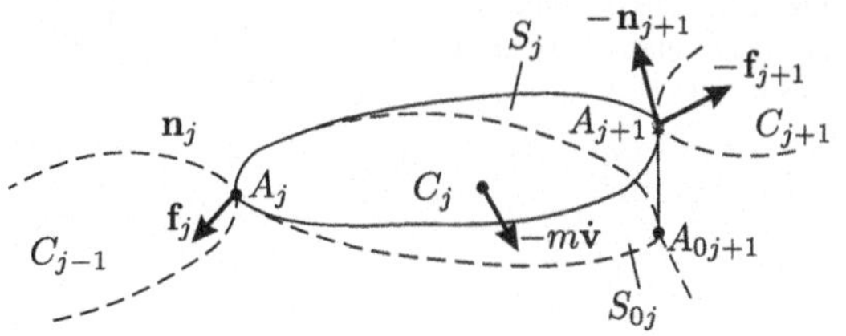

Fig. 21.2 The scheme of dynamic characteristics of an elastic link.

(Fig. 21.3)

$$ {}^j\dot{\mathbf{V}}_j = \begin{bmatrix} {}^j\dot{\mathbf{v}}_j \\ {}^j\dot{\omega}_j \end{bmatrix}, $$

${}^j\mathbf{F}_j$ is the vector of internal and driving forces given by the equation

$$ {}^j\mathbf{F}_j = \begin{bmatrix} {}^j\mathbf{f}_j \\ {}^j\mathbf{n}_j \end{bmatrix} + \tau_j(\Gamma_j, \ddot{\mathbf{q}}^{\mathrm{r}}_j)\, {}^j\mathbf{A}_j, $$

$$ {}^j\mathbf{A}_j = \begin{bmatrix} \sigma_j\, {}^j\mathbf{a}_j \\ \bar{\sigma}_j\, {}^j\mathbf{a}_j \end{bmatrix}, $$

in which τ_j is jth driving force function, ${}^j\mathbf{f}_j$ and ${}^j\mathbf{n}_j$ are jth vectors of internal force and moment, Γ_j is jth generalized force function, ${}^j\mathbf{a}_j$ is a three-dimensional unit vector directed along the axis of motion of the jth kinematic pair. It should also be noted that the vector $\ddot{\mathbf{q}}^{\mathrm{e}}_j$ includes accelerations of elastic variables of the deformable link C_j, and its dimension coincides with the number of "elastic" degrees of freedom of this link, i.e.

$$ \ddot{\mathbf{q}}^{\mathrm{e}}_j = \begin{bmatrix} \ddot{q}^{\mathrm{e}}_{j,1} \\ \ddot{q}^{\mathrm{e}}_{j,2} \\ \vdots \\ \ddot{q}^{\mathrm{e}}_{j,N_j} \end{bmatrix}. $$

A pair of vector algebraic equations are also performed that express the kinematic dependencies between the velocities (linear and angular) and accelerations (linear and angular) of the points A_j and A_{j-1} of the elastic link C_j (Fig. 21.3):

$$ {}^j\mathbf{V}_j = {}^j\mathbf{T}_{j-1}\, {}^{j-1}\mathbf{V}_{j-1} + {}^j\mathbf{T}_{j-1}\, {}^{j-1}\mathbf{\Phi}_{j-1}\mathbf{q}^{\mathrm{e}}_{j-1} + \dot{q}^{\mathrm{r}}_j\, {}^j\mathbf{A}_j, $$
$$ {}^j\dot{\mathbf{V}}_j = {}^j\mathbf{T}_{j-1}\, {}^{j-1}\dot{\mathbf{V}}_{j-1} + {}^j\mathbf{T}_{j-1}\, {}^{j-1}\mathbf{\Phi}_{j-1}\mathbf{q}^{\mathrm{e}}_{j-1} + \ddot{q}^{\mathrm{r}}_j\, {}^j\mathbf{A}_j + {}^j\mathbf{h}_j. $$

Here q^{r}_j are generalized coordinates and vector ${}^j\mathbf{h}_j$ includes nonlinear kinematic terms formed by "rigid-rigid" and "rigid-elastic" aggregates of the deformable link C_j explained in publication [3].

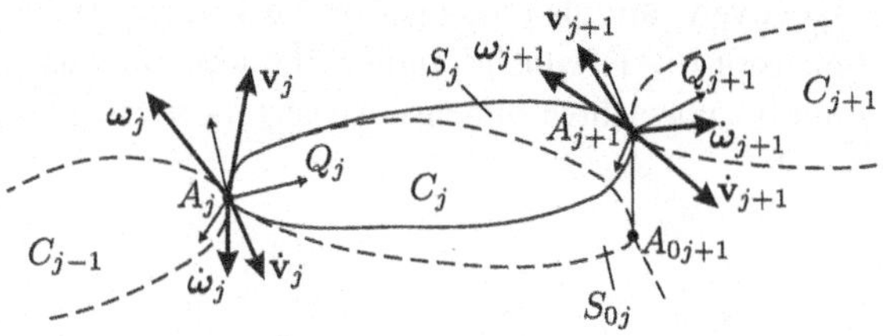

Fig. 21.3 The scheme of kinematic characteristics of an elastic link.

21.3 Formulation and Numerical Solution of a Mixed Problem of Elastic Manipulator Dynamics Using the Generalized Newton–Euler Method

The set of links $M = \{C_1, C_2, \ldots, C_n\}$ in the kinematic chain (Fig. 21.1) is divided into two subsets $M_1 = \{C'_1, C'_2, \ldots, C'_l\}$ and $M_2 = \{C''_1, C''_2, \ldots, C''_m,\}$ $(l + m = n)$ according to the following criteria: the link C_j belongs to the set M_1 if the driving force Γ_j is set in the kinematic pair j, and to the set M_2 if the generalized acceleration $\ddot{q}^{\mathrm{r}}_j$ is set in the kinematic pair j. Based on this principle, we will introduce the logical variable

$$\chi_j = \begin{cases} 0 & \text{if the driving force } \Gamma_j \text{ is set in the kinematic pair } j, \\ 1 & \text{if if the generalized acceleration } \ddot{q}^{\mathrm{r}}_j \text{ is set in the kinematic pair } j. \end{cases}$$

Then the mixed problem of elastic manipulator dynamics is formulated as follows. It is assumed that the laws of variation of the generalized forces Γ_i $(i = 1, 2, \ldots, l)$ and generalized accelerations $\ddot{q}^{\mathrm{r}}_k$ $(k = 1, 2, \ldots, m)$, belonging to subsets M_1 and M_2, respectively, are given. It is required to determine the laws of variation of the generalized forces Γ_k $(k = 1, 2, \ldots, m)$ and generalized accelerations $\ddot{q}^{\mathrm{r}}_i$ $(i = 1, 2, \ldots, l)$, belonging to subsets M_2 and M_1, respectively.

The algorithm for solving the mixed dynamic problem of elastic manipulators without inverting their mass matrices is compiled using elements of symbolic-iterative calculus, which has found successful application in the analysis of elastic manipulators [4]. It includes three recursive loops sequentially performed for each time point $t \in [0, T]$ [7].

The *first*, ascending, *recursive cycle* consists in carrying out the kinematic calculation of the manipulator based on the specified geometric and inertial characteristics of the constituent links, namely (j passes values from 1 to n):

$${}^j\mathbf{V}_j = {}^j\mathbf{T}_{j-1}\,{}^{j-1}\mathbf{V}_{j-1} + {}^j\mathbf{R}_{j-1}\,{}^{j-1}\mathbf{\Phi}_{j-1}\dot{\mathbf{q}}^{\mathrm{e}}_j + \dot{q}^{\mathrm{r}}_j\,{}^j\mathbf{A}_j$$

with initialisations

$${}^0\mathbf{V}_0 = \mathbf{0}_6, \quad \dot{\mathbf{q}}^{\mathrm{e}}_0 = \mathbf{0}_{N_0}.$$

The *second*, descending, *recursive cycle* provides for the definition of symbolic-iterative quantities (matrices, vectors, and scalars), the role of which is to ensure the proportionality of the computational process of the algorithm on the system dimension [7].

After initialization using formulas

$$\begin{gathered} {}^n\mathbf{J}^*_{\mathrm{rr}n} = {}^n\mathbf{J}_{\mathrm{rr}n}, \quad {}^n\mathbf{J}^*_{\mathrm{re}n} = {}^n\mathbf{J}_{\mathrm{re}n}, \quad {}^n\mathbf{J}^*_{\mathrm{er}n} = {}^n\mathbf{J}_{\mathrm{er}n}, \quad {}^n\mathbf{J}^*_{\mathrm{ee}n} = {}^n\mathbf{J}_{\mathrm{ee}n}, \\ {}^n\mathbf{C}^{\mathrm{r}*}_n = {}^n\mathbf{C}^{\mathrm{r}}_n + {}^n\mathbf{F}_{nt}, \quad {}^n\mathbf{C}^{\mathrm{e}*}_n = {}^n\mathbf{C}^{\mathrm{e}}_n + {}^n\mathbf{\Phi}^T_n\,{}^n\mathbf{F}_{nt}, \end{gathered}$$

where ${}^{n}\mathbf{F}_{nt}$ is generalized external force acting on the manipulator, the auxiliary inertial parameters ${}^{j}\mathbf{J}^{*}_{\mathrm{rr}j}$, ${}^{j}\mathbf{J}^{*}_{\mathrm{re}j}$, ${}^{j}\mathbf{J}^{*}_{\mathrm{er}j}$, ${}^{j}\mathbf{J}^{*}_{\mathrm{ee}j}$, ${}^{j}\mathbf{C}^{\mathrm{r}*}_{j}$, and ${}^{j}\mathbf{C}^{\mathrm{e}*}_{j}$ ($j = n-1, n-2, \dots, 1$) are determined in the next cycle (j passes values from n to 1):

1) determining parameters

$$\mathbf{K}_j = {}^{j}\mathbf{J}^{*}_{\mathrm{rr}j} - {}^{j}\mathbf{J}^{*}_{\mathrm{re}j}({}^{j}\mathbf{J}^{*}_{\mathrm{ee}j})^{-1}\,{}^{j}\mathbf{J}^{*}_{\mathrm{er}j},$$
$${}^{j}\mathbf{m}_j = -\,{}^{j}\mathbf{J}^{*}_{\mathrm{re}j}({}^{j}\mathbf{J}^{*}_{\mathrm{ee}j})^{-1}\,{}^{j}\mathbf{C}^{\mathrm{e}*}_{j} + {}^{j}\mathbf{C}^{\mathrm{r}*}_{j},$$

2) if $\chi_j = 0$ or $j \neq 1$ then calculate following parameters:

$$\mathbf{G}_j = \mathbf{K}_j\,{}^{j}\mathbf{T}_{j-1}, \quad \mathbf{H}_j = \mathbf{K}_j\,{}^{j}\mathbf{R}_{j-1}\,{}^{j-1}\mathbf{\Phi}_{j-1}, \quad {}^{j}\mathbf{p}_j = \mathbf{K}_j\,{}^{j}\mathbf{h}_j + {}^{j}\mathbf{m}_j,$$

3) if $\chi_j = 0$ then

$$Z_j = \frac{1}{{}^{j}\mathbf{A}^{T}_{j}\mathbf{K}_j\,{}^{j}\mathbf{A}_j},$$

4) if $j \neq 1$ then calculate auxiliary inertial parameters

$$\begin{aligned}
{}^{j-1}\mathbf{J}^{*}_{\mathrm{rr}j-1} &= {}^{j-1}\mathbf{J}_{\mathrm{rr}j-1} + {}^{j}\mathbf{T}^{T}_{j-1}\tilde{\mathbf{G}}_j,\\
{}^{j-1}\mathbf{J}^{*}_{\mathrm{re}j-1} &= {}^{j-1}\mathbf{J}_{\mathrm{re}j-1} + {}^{j}\mathbf{T}^{T}_{j-1}\tilde{\mathbf{H}}_j,\\
{}^{j-1}\mathbf{J}^{*}_{\mathrm{er}j-1} &= {}^{j-1}\mathbf{J}_{\mathrm{er}j-1},\\
{}^{j-1}\mathbf{J}^{*}_{\mathrm{ee}j-1} &= {}^{j-1}\mathbf{J}_{\mathrm{ee}j-1} + {}^{j-1}\mathbf{\Phi}^{T}_{j-1}\,{}^{j}\mathbf{R}^{T}_{j-1}\,{}^{j}\tilde{\mathbf{p}}_j,\\
{}^{j-1}\mathbf{C}^{\mathrm{r}*}_{j-1} &= {}^{j-1}\mathbf{C}^{\mathrm{r}}_{j-1} + {}^{j}\mathbf{T}^{T}_{j-1}\,{}^{j}\tilde{\mathbf{p}}_j,\\
{}^{j-1}\mathbf{C}^{\mathrm{e}*}_{j-1} &= {}^{j-1}\mathbf{C}^{\mathrm{e}}_{j-1} + {}^{j-1}\mathbf{\Phi}^{T}_{j-1}\,{}^{j}\mathbf{R}^{T}_{j-1}\,{}^{j}\tilde{\mathbf{p}}_j,
\end{aligned}$$

where

$$\tilde{\mathbf{G}}_j = \begin{cases} \mathbf{G}_j - \mathbf{U}_j\,{}^{j}\mathbf{T}_{j-1} & \text{for } \chi_j = 0,\\ \mathbf{G}_j & \text{for } \chi_j = 1,\end{cases}$$

$$\tilde{\mathbf{H}}_j = \begin{cases} \mathbf{H}_j - \mathbf{U}_j\,{}^{j}\mathbf{R}_{j-1}\,{}^{j-1}\mathbf{\Phi}_{j-1} & \text{for } \chi_j = 0,\\ \mathbf{H}_j & \text{for } \chi_j = 1,\end{cases}$$

$${}^{j}\tilde{\mathbf{p}}_j = \begin{cases} {}^{j}\mathbf{p}_j + \mathbf{K}_j\,{}^{j}\mathbf{A}_j(\tau_j - {}^{j}\mathbf{A}^{T}_{j}\,{}^{j}\mathbf{p}_j)Z_j & \text{for } \chi_j = 0,\\ {}^{j}\mathbf{p}_j + \mathbf{K}_j\,{}^{j}\mathbf{A}_j\ddot{q}^{\mathrm{r}}_{j} & \text{for } \chi_j = 1,\end{cases}$$

$$\mathbf{U}_j = Z_j\mathbf{K}_j\,{}^{j}\mathbf{A}_j\,{}^{j}\mathbf{A}^{T}_{j}\mathbf{K}_j.$$

The *third*, final, ascending *recursive cycle* [4] allows us to determine

- generalized accelerations $\ddot{q}^{\mathrm{r}}_{j}$ for subset M_1 (j passes values from 1 to n; j such as $\chi_j = 0$)

$$\ddot{q}^{\mathrm{r}}_{j} = Z_j(\tau_j - {}^{j}\mathbf{A}^{T}_{j}(\mathbf{G}_j\,{}^{j-1}\dot{\mathbf{V}}_{j-1} + \mathbf{H}_j\ddot{\mathbf{q}}^{\mathrm{e}}_{j-1} + {}^{j}\mathbf{p}_j));$$

- acceleration of links ${}^{j}\dot{\mathbf{V}}_{j}$, acceleration of variable parameters of elastic displacements $\ddot{\mathbf{q}}_{j}^{\mathrm{e}}$, and, if required, internal forces in pairs ${}^{j}\mathbf{F}_{j}$ for the set M (j passes values from 1 to n)

$$
\begin{aligned}
{}^{j}\dot{\mathbf{V}}_{j} &= {}^{j}\mathbf{T}_{j-1}\,{}^{j-1}\dot{\mathbf{V}}_{j-1} + {}^{j}\mathbf{R}_{j-1}\,{}^{j-1}\boldsymbol{\Phi}_{j-1}\ddot{\mathbf{q}}_{j-1}^{\mathrm{e}} + \ddot{q}_{j}^{\mathrm{r}}\,{}^{j}\mathbf{A}_{j} + {}^{j}\mathbf{h}_{j},\\
\ddot{\mathbf{q}}_{j}^{\mathrm{e}} &= -({}^{j}\mathbf{J}_{\mathrm{ee}j}^{*})^{-1}({}^{j}\mathbf{J}_{\mathrm{er}j}^{*}\,{}^{j}\dot{\mathbf{V}}_{j} + {}^{j}\mathbf{C}_{j}^{\mathrm{e}*}),\\
{}^{j}\mathbf{F}_{j} &= \mathbf{K}_{j}\,{}^{j}\dot{\mathbf{V}}_{j} + {}^{j}\mathbf{m}_{j};
\end{aligned}
$$

- generalized forces Γ_j for a subset M_2 (j passes values from 1 to n; j such as $\chi_j = 1$)

$$
\Gamma_j = {}^{j}\mathbf{A}_{j}^{T}(\mathbf{K}_{j}\,{}^{j}\dot{\mathbf{V}}_{j} + {}^{j}\mathbf{m}_{j}) + F_{\mathrm{d}j}\,\mathrm{sgn}\,\dot{q}_{j}^{\mathrm{r}} + F_{\mathrm{v}j}\dot{q}_{j}^{\mathrm{r}},
$$

  where $F_{\mathrm{d}j}$ and $F_{\mathrm{v}j}$ are coefficients of dry and viscous friction in pairs;

taking as initialisations

$$
{}^{0}\dot{\mathbf{V}}_{0} = \begin{bmatrix} {}^{0}\mathbf{g}^{T} \\ \mathbf{0}_{3}^{T} \end{bmatrix}, \quad \ddot{\mathbf{q}}_{0}^{\mathrm{e}} = \mathbf{0}_{N_0},
$$

where ${}^{0}\mathbf{g}$ is acceleration of gravity.

The described computational algorithm for the numerical solution of the mixed dynamic problem of elastic manipulators allows us to illustrate the functioning of the structural unit of the kinematic chain corresponding to the Newton–Euler method [1–3, 7, 8]. The algorithm for the numerical solution of the mixed dynamic problem of flexible manipulators based on the generalized Newton-Euler method assumes a schematic illustration of the transformation of kinematic and dynamic characteristics at the level of the structural unit of the kinematic chain. The figurative converter of velocities, accelerations, and forces formed in this way is commonly referred to in the scientific literature as the kinetostatic element [9]. A schematic diagram of such a converter is shown in Fig. 21.4.

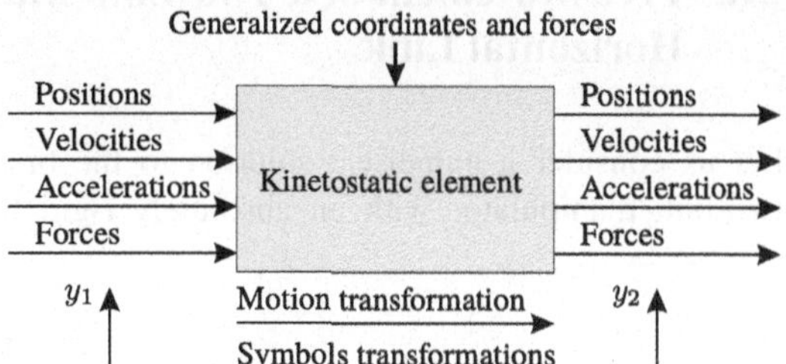

Fig. 21.4 Schematic diagram of kinetostatic element.

21.4 Method of Numerical Integration of Equations of Motion

A method of numerical integration of nonlinear equations of motion is based on the Newmark general scheme of numerical integration [9]

$$\begin{aligned}&\mathbf{x}_j(t+\Delta t) = \mathbf{x}_j(t) + \dot{\mathbf{x}}_j(t)\Delta t + [(0.5-\beta)\ddot{\mathbf{x}}_j(t) + \beta\ddot{\mathbf{x}}_j(t+\Delta t)]\Delta t^2,\\ &\dot{\mathbf{x}}_j(t+\Delta t) = \dot{\mathbf{x}}_j(t) + [(1-\gamma)\ddot{\mathbf{x}}_j(t) + \gamma\ddot{\mathbf{x}}_j(t+\Delta t)]\Delta t,\\ &t \in [0, T-\Delta t], \quad j = 1, 2, \ldots, n.\end{aligned}$$

where $\mathbf{x}_j = [q^{\mathrm{r}}_j, (\mathbf{q}^{\mathrm{e}}_j)^T]^T$; $\beta \in [0, 0.5]$ and $\gamma \in [0, 1]$ are constants. In the case when $\beta = \gamma = 0$, the system of equations (14) degenerates into an explicit Newmark integration scheme, the use of which often requires the choice of an extremely small time integration step Δt [9]. If $\beta \neq 0$ or $\gamma \neq 0$, then we get an implicit scheme for numerical integration of Newmark.

The solution of the system of linearized equations of motion of elastic manipulators is carried out using Newton–Raphson method. An exact calculation of the Jacobian (iteration matrix) in the calculation formula of the Newton–Raphson method, as is known, accelerates the convergence process, but involves a cumbersome calculation of tangential stiffness and damping matrices, especially with a large total number of degrees of freedom of the systems under consideration. At the same time, the assimilation of the Jacobian with the mass matrix of the manipulator, while justified for rigid systems, raises serious doubts if there are elastic links in the kinematic chain [9]. In this regard, a compromise approach to overcoming these computational difficulties is of particular practical interest, consisting in replacing the exact expression of the iteration matrix with its reduced equivalent [7] (Fig. 21.5).

21.5 Free Movement of a Two-Link Manipulator with an Elastic Horizontal Link

Let us consider a numerical solution to the first problem of the dynamics of a two-link manipulator with an absolutely rigid link *1* and an elastic link *2* that

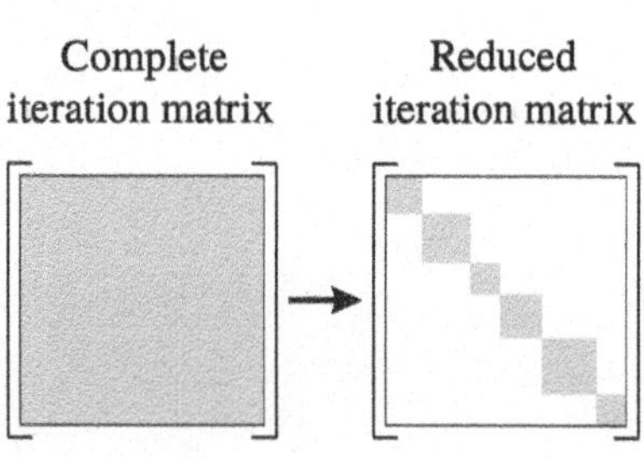

Fig. 21.5 Structural image of iteration matrices.

maintains a strictly horizontal position during movement [6] (Fig. 21.6). The links of the manipulator have equal lengths $l_1 = l_2 = 0.8\,\text{m}$ and have the same cross-sectional areas $S_1 = S_2 = 4 \times 10^{-4}\,\text{m}^{-2}$. The masses of the links are also equal $m_1 = m_2 = 2.512\,\text{kg}$. The inertia moment of the section of the elastic link is $I_{z2} = 1.333 \times 10^{-8}\,\text{m}^4$, and the Young's modulus is $E = 7 \times 10^{10}\,\text{N/m}^2$. Dynamic modeling of the flat two-link manipulator (Fig. 21.6) is supposed to be carried out taking into account the effect of gravity.

Described manipulator starts moving from its initial position with the initial conditions in generalized coordinates and velocities

$$q_1^{\text{r}}(0) = q_2^{\text{r}}(0) = 0, \quad \dot{q}_1^{\text{r}}(0) = 10\,\text{rad/s}, \quad \dot{q}_2^{\text{r}}(0) = -10\,\text{rad/s},$$

as well as in elastic displacements and velocities

$$q_1^{\text{e}}(0) = q_2^{\text{e}}(0) = 0, \quad \dot{q}_1^{\text{e}}(0) = \dot{q}_2^{\text{e}}(0) = 0.$$

In the absence of friction in the rotational pairs, the the inverse dynamic problem of elastic manipulators under consideration (Fig. 21.6) is formulated at zero generalized accelerations:

$$\ddot{q}_1^{\text{r}}(t) = \ddot{q}_2^{\text{r}}(t) = 0, \quad t \in [0, T],$$

where the final time T is fixed by the full rotation period of the link *1*.

The problem is modeled with a single coordinate function involved in the shape function, which is expressed by the polynomial of the fourth degree, i.e.

$$\Phi_{\text{d}2}(x) = k_1 x^4 + k_2 x^3 + k_3 x^2 + k_4 x + k_5. \tag{21.1}$$

The coordinate function (21.1) must meet the geometric and dynamic conditions

$$\begin{aligned} \Phi_{\text{d}2}(0) = \Phi'_{\text{d}2}(0) = 0, \\ \Phi''_{\text{d}2}(0) = \Phi'''_{\text{d}2}(0) = 0, \end{aligned} \tag{21.2}$$

and also comply with the normalization condition

$$\Phi_{\text{d}2}(l_2) = 1. \tag{21.3}$$

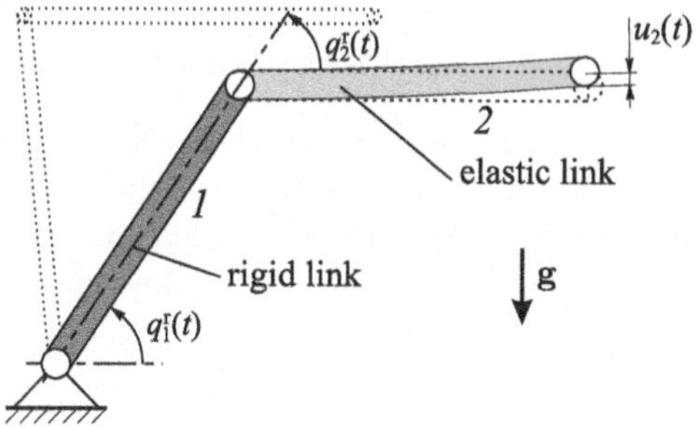

Fig. 21.6 Scheme of a two-link manipulator with an elastic actuating link.

The polynomial (21.1), taking into account the conditions (21.2) and (21.3), takes the form

$$\Phi_{d2}(x) = \frac{1}{3}\left(\frac{x}{l_2}\right)^4 - \frac{4}{3}\left(\frac{x}{l_2}\right)^3 + 2\left(\frac{x}{l_2}\right)^2$$

on the basis of which it is possible to obtain an expression for the rotational coordinate function

$$\Phi_{r2}(x) = \frac{4}{3l_2}\left(\frac{x}{l_2}\right)^3 - \frac{4}{l_2}\left(\frac{x}{l_2}\right)^2 + \frac{4}{l_2}\frac{x}{l_2}.$$

Figure 21.7 shows graphs of the bending displacement function (*a*) of the end point of the elastic link of the manipulator, obtained respectively in publications [7] (curve *1*) and [6] (curve *2*) and generalized force functions (*b*).

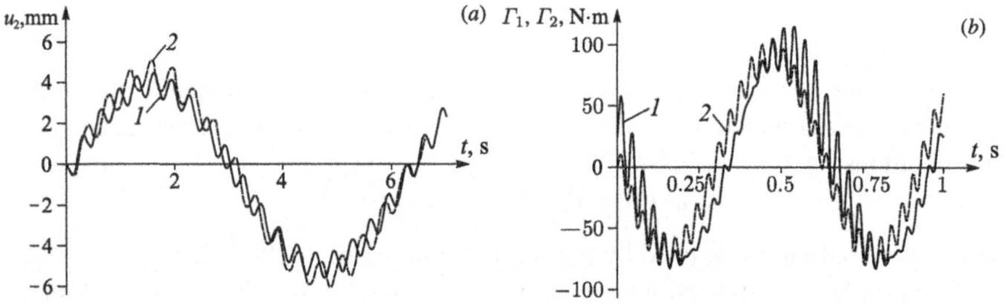

Fig. 21.7: Elastic displacements $u_2(t)$ from papers [4] (curve *1*) and [9] (curve *2*) (*a*) and generalized force functions $\Gamma_1(t)$ (curve *1*) and $\Gamma_2(t)$ (curve *2*) (*b*).

21.6 Smoothing Movement of a Two-Link Manipulator with an Elastically Horizontal Link

In order to smooth out undesirable elastic vibrations at the level of the center of mass of the second link, the following adjustments are introduced to the law of motion of the second link

$$q_2(t) = -q_1(t) - 2 \times 10^{-6} \sin(104.7t), \quad t \in [0, T].$$

Taking into account the introduced adjustments to the laws of motion of the elastic manipulator, we obtain a smoothed function of the elastic vibrations of the center of mass of the second link (Fig. 21.8*a*) and the corresponding functions of the driving torques (Fig. 21.8*b*) that carry out this smoothing motion.

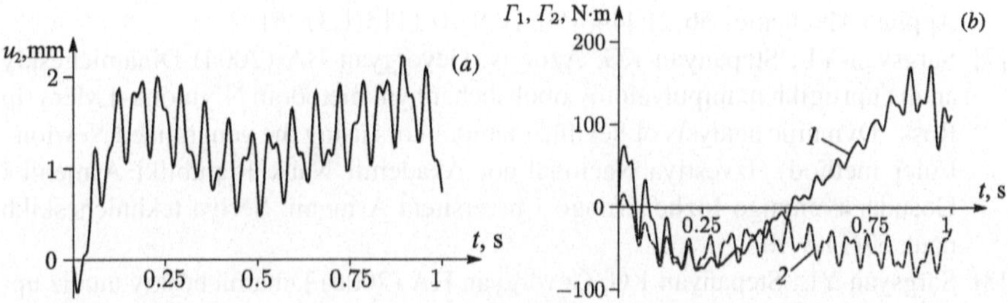

Fig. 21.8: Elastic displacement $u_2(t)$ (*a*) and generalized force functions $\Gamma_1(t)$ (curve *1*) and $\Gamma_2(t)$ (curve *2*) (*b*) in the smoothing process of the elastic vibrations of the second link.

21.7 Conclusion

The article discusses the possibility of smoothing out undesirable elastic vibrations of deformable manipulator links in precision manufacturing by introducing simple correction to the laws of motion of the dynamic system under study. The results of numerical implementation show that such adjustments can be successfully applied in engineering.

References

[1] Boyer F, Coiffet P (1996) Generalization of Newton–Euler model for flexible manipulators. Journal of Robotic Systems 13(1):11-24, DOI 10.1002/(SICI)1097-4563(199601)13:1<11::AID-ROB2>3.0.CO;2-Y

[2] Boyer F, Coiffet P (1996) Symbolic modeling of a flexible manipulator via assembling of its generalized Newton–Euler model. Mechanism and Machine Theory 31(1):45-56, DOI 10.1016/0094-114X(95)00038-Z

[3] Boyer F, Khalil W (1998) An efficient calculation of flexible manipulators inverse dynamics. The International Journal of Robotics Research 17(3):282-293, DOI 10.1177/027836499801700305

[4] Gevorgyan HA (2010) Prilozheniye obobshchennogo metoda N'yutona–Eylera k zadacham optimal'nogo upravleniya uprugikh mekhanizmov (in Russ., An application of the generalized Newton–Euler method in the optimal control problems of flexible mechanisms). Izvestiya Nacional'noi Akademii Nauk Respubliki Armenii i Gosudarstvennogo Inzhenernogo Universiteta Armenii. Seriya tekhnicheskikh nauk 63(2):133-138

[5] Gevorgyan HA (2021) Problem on optimizing the motion of an elastic tracking manipulator. Mechanics od Solids 56(2):242-249, DOI 10.3103/S0025654421020072

[6] Ider SK, Amirouche FML (1989) Nonlinear modeling of flexible multibody dynamics subjected to variable constraints. Transaction of ASME. Journal of

Applied Mechanics 56(2):444-450, DOI 10.1115/1.3176103

[7] Sargsyan YL, Stepanyan KG, Azuz N, Gevorgyan HA (2004) Dinamicheskiy analiz uprugikh manipulyatorov obobshchennym metodom N'yutona–Eylera (in Russ., Dynamic analysis of flexible manipulators using the generalized Newton–Euler method). Izvestiya Nacional'noi Akademii Nauk Republiki Armenii i Gosudarstvennogo Inzhenernogo Universiteta Armenii. Seriya tekhnicheskikh nauk 57(1):3-10

[8] Sargsyan YL, Stepanyan KG, Gevorgyan HA (2006) Dinamicheskiy analiz uprugikh drevovidnykh mekhanicheskikh sistem bez vneshnikh svyazey (in Russ., Dynamic analysis of flexible tree-like mechanical systems without external constraints). Izvestiya Nacional'noi Akademii Nauk Republiki Armenii i Gosudarstvennogo Inzhenernogo Universiteta Armenii. Seriya tekhnicheskikh nauk 59(1):3-9

[9] Verlinden O (1994) Simulation du comportement dynamique de systèmes multicorps flexibles comportant des membrures de forme complexe. These de doctorat de la Faculte Polytechnique de Mons

Chapter 22
Three-Dimensional Analogues of the Blasius–Chaplygin Formulas

Dmitrii Legatiuk and Heikki Orelma

Abstract The classical Blasius–Chaplygin formula provides an elegant method for calculating the lift force on a two-dimensional body in steady, irrotational flow. The key ingredient is the definition of a complex-valued potential function

$$f(z) = \varphi(x, y) + \mathbf{i}\psi(x, y),$$

which can then be integrated using Cauchy's theorem along any closed contour surrounding the body.

In this paper, we propose a three-dimensional extension of the classical Blasius–Chaplygin formula using quaternionic analysis. After presenting the basics of quaternionic analysis, we discuss how *monogenic functions*—the quaternionic analog of classical holomorphic functions—can be used to describe problems in fluid dynamics. Finally, we present the Blasius–Chaplygin formula in quaternionic form.

Key words: Monogenic functions, Quaternionic analysis, Fluid mechanics, Flow potential, Blasius-Chaplygin formula

22.1 Introduction

The velocity potential ϕ and stream function ψ are used to describe the components of the velocity field in the case of the classical two-dimensional ideal (irrotational and inviscid) gas dynamics through the following expressions [16, 17]:

Dmitrii Legatiuk
Chair of Mathematics, Universität Erfurt, Nordhäuser Str. 63, 99089 Erfurt, Germany, e-mail: dmitrii.legatiuk@uni-erfurt.de

Heikki Orelma (corresponding author)
Tampere Institute of Mathematics, Hepolamminkatu 51, 33720 Tampere, Finland, e-mail: heikki.orelma@proton.me

H. Altenbach and A. V. Sahakyan, *Contact Mechanics and Relevant Problems*, Advanced Structured Materials 253, https://doi.org/10.1007/978-3-032-20869-9_22

$$
\begin{aligned}
v_1 &= \frac{\partial \phi}{\partial x} = \frac{\partial \psi}{\partial y}, \\
v_2 &= \frac{\partial \phi}{\partial y} = -\frac{\partial \psi}{\partial x}.
\end{aligned}
\tag{22.1}
$$

Stream functions are geometrically related to the flow since streamlines are their level curves. Moreover, equations (22.1) form the *Cauchy-Riemann system* known from complex analysis, which consequently leads to defining the complex potential function $f(z) := \phi + \mathbf{i}\psi$. This complex function describes the flow's potential and stream function fields such that its derivative $f'(z)$ corresponds to the local flow velocity. Through this, complex analysis opens an exceptionally efficient and elegant mathematical tool for fluid mechanics, especially for the study of ideal two-dimensional flows, see again the classical references [16, 17].

A central problem in the ideal flow mechanics is a calculation of forces, which requires computing pressure forces arising from the distribution of flow velocity on the body's surface. In the two-dimensional case, the complex potential provides a method for this, where the force acting on the body is given by

$$
F_x - \mathbf{i}F_y = \frac{\mathbf{i}\rho}{2} \oint_C (f'(z))^2 \, dz.
$$

This formula is called the *Blasius-Chaplygin formula*. It is noted that from a mathematical point of view, holomorphy of the potential inside the region bounded by the path C implies that the flow exerts no forces on the body. Consequently, forces appear only when the holomorphy is lost at some point of the body. Complex analysis provides many tools to analyze such situations.

The complex integral form of the force formula known as the Blasius–Chaplygin formula was developed independently by two researchers in the early 20th century. Let us now provide a brief overview of these discoveries:

- *Paul Blasius* (1883–1970), a German physicist and student of Prandtl, published the formula in 1911 in his article "Grenzschichten in Flüssigkeiten mit kleiner Reibung" (Boundary layers in fluids with small viscosity) [1–3]. Although the article is mostly known for Blasius' boundary layer solution, he also used the complex flow theory for force calculations in a two-dimensional potential flow, presenting an integral formula to determine lift and drag on a body.
- *Sergei A. Chaplygin* (1869–1942), a Russian mathematician and mechanics researcher, published similar formulas almost simultaneously (in 1902 and more extensively in 1904) as part of his studies on flows over lifting surfaces. His work appeared in the publications of the Russian Academy of Sciences under titles such as "On gas streams with uniform motion" [8, 9]. Chaplygin applied complex functions to analyze the flow and calculated forces on bodies using a similar integral method.

Although P. Blasius and S.A. Chaplygin mainly studied forces, a formula for the moment (about the origin) has also been derived based on their works:

$$M = -\frac{\rho}{2} \operatorname{Re} \oint_C z\,(f'(z))^2 \, dz.$$

Thus, the force and moment formulas combine complex analysis and potential flow theory, forming an elegant method to calculate the force and moment acting on a body in a two-dimensional inviscid flow.

It is important to emphasize that the Blasius formulas are derived for a two-dimensional ideal flow where the body considered is essentially infinitely long in the third dimension. Consequently, the flow is homogeneous in the third dimension and unaffected along the length. Therefore, the formulas give the values of flow force and moment per cross-sectional length, meaning they are *force and moment densities*:

- The unit of force is therefore Newtons per meter [N/m], which means the force acting on a one-meter long portion of the body.
- The moment is similarly expressed as moment per unit length, i.e., in units of Newton-meters per meter [Nm/m], which typically simplifies to Newton-meters [Nm] in the two-dimensional case.

In practice, this means that the Blasius formulas apply to describing, for example, the lift and torque generated by the cross-section of an airplane wing, assuming the wing is long and the flow is homogeneous in two dimensions.

This article proposes a three-dimensional counterpart of the complex potential and related Blasius-Chaplygin formulas. This three-dimensional extension has become possible with the advanced in the field of *quaternion analysis*, which is a hypercomplex complex function theory in $\mathbb{R}^3$ and $\mathbb{R}^4$, see [12, 13] and references therein for an overview of a hypercomplex analysis. The core idea of quaternionic analysis is to work with the so-called *monogenic functions*, which generalize the concept of holomorphic functions to quaternions. In the context of boundary value problems in mechanics, two general strategies of using tools of quaternionic analysis can be distinguished: first strategy is to apply the so-called *quaternionic operator calculus*, which allows to construct explicit representation formulas for the solution of boundary value problems by using singular integral operators, such as Cauchy-Bitsadze operator and Teodorescu transform, see [14, 15] for examples of using this strategy for elasticity theories; the second strategy is to generalize the classical analytical solution approaches to the quaternionic case, see for example works [4, 5] for spatial analogues of the classical Kolosov-Muskhelishvili formulas. In this paper, we follow the second strategy.

While two-dimensional potential flows are well understood through complex analysis, the extension to three dimensions is far from trivial. Quaternion analysis offers promising tools for this extension, enabling novel analytical methods for ideal flow problems in 3D. First ideas towards developing quaternionic theory for three-dimensional flow and the construction of monogenic flow potentials have been presented in [18, 19]. This paper presents a continuation of these works. Moreover, to make the paper as self-contained and complete as possible, we recall also some results from [18, 19] to support the reader.

22.2 Preliminaries

In this section, we provide a brief introduction into quaternions and quaternionic analysis. We refer to [12, 13] and references therein for more detailed information, as well as for applications of quaternionic setting.

22.2.1 *Quaternions*

Quaternions are a generalization of complex numbers $\mathbb{C}$ to $\mathbb{R}^4$. The algebra of quaternions is defined by the basis elements $\{1, \mathbf{i}, \mathbf{j}, \mathbf{k}\}$, which satisfy the well-known multiplication rules

$$\mathbf{i}^2 = \mathbf{j}^2 = \mathbf{k}^2 = \mathbf{ijk} = -1.$$

The algebra of quaternions is associative, but not commutative, as illustrated by the following multiplication rules for the unit elements

$$\begin{aligned} \mathbf{ij} = \mathbf{k}, \quad \mathbf{jk} = \mathbf{i}, \quad \mathbf{ki} = \mathbf{j}, \\ \mathbf{ji} = -\mathbf{k}, \mathbf{kj} = -\mathbf{i}, \mathbf{ik} = -\mathbf{j}. \end{aligned}$$

The element 1 is the multiplicative identity in the algebra of quaternions.

A general quaternion $q \in \mathbb{H}$ is of the form

$$q = q_0 + q_1\mathbf{i} + q_2\mathbf{j} + q_3\mathbf{k},$$

where $q_0, q_1, q_2, q_3 \in \mathbb{R}$. Since a complex number $z \in \mathbb{C}$ is of the form $z = x + iy$, a quaternion can be viewed as a "complex number" with three imaginary units instead of one. Geometrically, quaternions can be identified with a four-dimensional vector space, that is, $\mathbb{H} \cong \mathbb{R}^4$, where the vector space isomorphism is defined as

$$q_0 + q_1\mathbf{i} + q_2\mathbf{j} + q_3\mathbf{k} \quad \mapsto \quad (q_0, q_1, q_2, q_3).$$

This identification holds only when quaternions are regarded as a vector space, since the space $\mathbb{R}^4$ does not have a naturally defined multiplication.

For convenience reasons, we will sometimes switch to the following denotation of the basis elements:

$$\mathbf{e}_0 := 1, \quad \mathbf{e}_1 := \mathbf{i}, \quad \mathbf{e}_2 := \mathbf{j}, \quad \mathbf{e}_3 := \mathbf{k}. \tag{22.2}$$

Further, a quaternion can be written in the form

$$q = q_0 + \mathbf{q},$$

where q_0 is the *scalar part* of the quaternion q and $\mathbf{q} = q_1\mathbf{e}_1 + q_2\mathbf{e}_2 + q_3\mathbf{e}_3$ is the *vector part*, which are denoted as $\mathrm{Sc}(q) := q_0$ and $\mathrm{Vec}(q) := \mathbf{q}$, respectively. The *conjugate* of a quaternion is defined by

$$\overline{q} := q_0 - \mathbf{q}.$$

Quaternion conjugation reverses the order of multiplication, that is,

$$\overline{pq} = \overline{q}\,\overline{p}, \qquad p, q \in \mathbb{H}.$$

The scalar and vector parts of a quaternion $q \in \mathbb{H}$ can be computed using the conjugate as

$$\mathrm{Sc}(q) = \frac{1}{2}(q + \overline{q}), \text{ and } \mathrm{Vec}(q) = \frac{1}{2}(q - \overline{q}).$$

Additionally, we have the following identity

$$\overline{q}q = q\overline{q} = q_0^2 + q_1^2 + q_2^2 + q_3^2.$$

This means that the Euclidean norm can be defined by setting

$$|q|^2 = \overline{q}q = q\overline{q}.$$

Consequently, the multiplicativity of the norm with respect to the product follows immediately:

$$|pq| = |p||q|.$$

The Euclidean inner product induces the Euclidean norm, and it can be expressed using the quaternionic product as

$$\langle q, p\rangle = \mathrm{Sc}(q\overline{p}) = \sum_{n=0}^{3} p_n q_n$$

for $p, q \in \mathbb{H}$. Moreover, the scalar part of the product satisfies the symmetry relations

$$\mathrm{Sc}(q\overline{p}) = \mathrm{Sc}(p\overline{q}) = \mathrm{Sc}(\overline{q}p) = \mathrm{Sc}(\overline{p}q).$$

22.2.2 *Quaternionic Analysis*

Although there are several alternative approaches to develop a quaternionic analysis, see again [12, 13], the underlying idea is always to construct a class of functions analogous to analytic functions in complex analysis. Keeping in mind applications to fluid mechanics, let us consider functions of the form

$$f : U \to \mathbb{H},$$

where $U \subset \mathbb{R}^3$. Accordingly, the functions are of the form

$$f(x, y, z) = f_0(x, y, z) + f_1(x, y, z)\mathbf{i} + f_2(x, y, z)\mathbf{j} + f_3(x, y, z)\mathbf{k},$$

where the component functions $f_0, f_1, f_2, f_3 \colon U \to \mathbb{R}$. The usual properties of quaternion-valued functions – such as differentiability, boundedness, integrability etc. – are defined in terms of the corresponding properties of the component functions. In this article, the points (x, y, z) are identified with the *reduced quaternions*

$$\mathrm{x} = x + y\mathbf{i} + z\mathbf{j}.$$

Next, we define a class of functions in quaternionic analysis, which generalize the concept of complex analytic functions to $\mathbb{R}^3$. First, we need to introduce the the *generalized Cauchy-Riemann operator*:

$$D := \frac{\partial}{\partial x} + \mathbf{i}\frac{\partial}{\partial y} + \mathbf{j}\frac{\partial}{\partial z}$$

and its conjugate

$$\overline{D} := \frac{\partial}{\partial x} - \mathbf{i}\frac{\partial}{\partial y} - \mathbf{j}\frac{\partial}{\partial z}.$$

These operators give rise to the definition:

Definition 22.1 (Monogenicity in a domain U**)** *Let* $U \subset \mathbb{R}^3$ *be an open set and* $f \colon U \to \mathbb{H}$ *a differentiable function. The function is called left-monogenic if* $Df = 0$ *in* U*, and correspondingly right-monogenic if* $fD = 0$ *in* U*.*

As it has been already mentioned, the monogenicity concept introduced in this definition corresponds to the complex analyticity. The following proposition explores this connection further by introduce system of partial differental equations analogous to the Cauchy-Riemann equations:

Proposition 22.2 *Let* $f \colon U \to \mathbb{H}$ *be a differentiable function. The following conditions are equivalent:*

a) $Df = 0$ *(*f *is monogenic),*
b) The components of the function $f = f_0 + f_1\mathbf{i} + f_2\mathbf{j} + f_3\mathbf{k}$ *are a solution of the Moisil-Theodorescu system*

$$\begin{cases} \dfrac{\partial f_0}{\partial x} - \dfrac{\partial f_1}{\partial y} - \dfrac{\partial f_2}{\partial z} = 0, \\[2ex] \dfrac{\partial f_1}{\partial x} + \dfrac{\partial f_0}{\partial y} + \dfrac{\partial f_3}{\partial z} = 0, \\[2ex] \dfrac{\partial f_2}{\partial x} + \dfrac{\partial f_0}{\partial z} - \dfrac{\partial f_3}{\partial y} = 0, \\[2ex] \dfrac{\partial f_3}{\partial x} + \dfrac{\partial f_2}{\partial y} - \dfrac{\partial f_1}{\partial z} = 0. \end{cases}$$

The principal difference between monogenic functions in quaternionic analysis and analytic functions in complex analysis is that, unlike the powers of complex numbers $(x+iy)^n$, the powers of the quaternionic variable $(x+\mathbf{i}y+\mathbf{j}z)^n$ are no longer monogenic. This can be seen, for example, by computing $D(x+\mathbf{i}y+\mathbf{j}z) = -1$. In quaternionic analysis, much attention has been devoted to constructing various bases of monogenic polynomials, see for example works and references therein [6, 10, 11].

The definition of monogenicity can also be associated with the conjugate Cauchy–Riemann operator. In this case, instead of monogenicity, one speaks of antimonogenicity.

Definition 22.3 (Antimonogenicity in a domain U) *Let* $U \subset \mathbb{R}^3$ *be an open set and* $f : U \to \mathbb{H}$ *a differentiable function. The function is called left-antimonogenic if* $\overline{D} f = 0$ *in* U*, and correspondingly right-antimonogenic if* $f\overline{D} = 0$ *in* U.

One of the most important property of the generalized Cauchy–Riemann operator and its conjugate is the factorization of the *Laplace operator*:

$$D\overline{D} = \overline{D}D = \Delta \text{ with } \Delta := \frac{\partial^2}{\partial x^2} + \frac{\partial^2}{\partial y^2} + \frac{\partial^2}{\partial z^2}.$$

Consequently, the component functions of every monogenic function are harmonic. Conversely, monogenic functions can be generated from harmonic functions. Let us now consider two methods to generate a monogenic function from a harmonic function. The first method follows directly from the factorization property:

Proposition 22.4 *Let* $u : \Omega \to \mathbb{R}$ *be a harmonic function in an open set* $\Omega \subset \mathbb{R}^3$. *Then* $\overline{D}u$ *is a monogenic function in the set* Ω.

The second method, which is more useful in fluid dynamics, is based on the formula presented in [7], where the case $n = 2$ corresponds to quaternionic analysis. To make the paper self-containing, we also provide a proof of this result, since the original source does not present a proof of the formula either. The proof uses the *Euler operator* defined as follows

$$E := x\frac{\partial}{\partial x} + y\frac{\partial}{\partial y} + z\frac{\partial}{\partial z},$$

which can be expressed using the generalized Cauchy–Riemann operator and inner product as $E = \langle \mathrm{x}, D\rangle$. Hence, we have the following proposition:

Proposition 22.5 *Let* $\Omega \subset \mathbb{R}^3$ *be an open set star-shaped with respect to the origin and let* $u : \Omega \to \mathbb{R}$ *be a harmonic function. Then*

$$f(\mathrm{x}) = u(\mathrm{x}) + Vec\left(\int_0^1 \overline{D}u(\mathrm{x}t)\mathrm{x}t\, dt\right)$$

is a monogenic function whose real part is the function u.

Proof. To make construction conciser, we use the notation (22.2) for the unit elements along with the notations $x_0 := x$, $x_1 := y$, and $x_2 := z$.

We start by writing the function f as follows

$$f(\mathrm{x}) = u(\mathrm{x}) + \int_0^1 \overline{D}u(\mathrm{x}t)\mathrm{x}t\,dt - \mathrm{Sc}\left(\int_0^1 \overline{D}u(\mathrm{x}t)\mathrm{x}t\,dt\right).$$

Let us now introduce the notation

$$I(\mathrm{x}) := \int_0^1 \overline{D}u(\mathrm{x}t)\mathrm{x}t\,dt,$$

implying that the function f can now be rewritten as

$$f(\mathrm{x}) = u(\mathrm{x}) + I(\mathrm{x}) - \mathrm{Sc}(I(\mathrm{x})).$$

By applying the fact that $D\overline{D}u = \Delta u = 0$ and the characterization of reduced quaternions[1], we obtain

$$\begin{aligned} DI(\mathrm{x}) &= \int_0^1 D\overline{D}u(\mathrm{x}t)\mathrm{x}t\,dt + \int_0^1 \sum_{j=0}^{2} \mathbf{e}_j \overline{D}u(\mathrm{x}t)\frac{\partial \mathrm{x}}{\partial x_j} t\,dt \\ &= \int_0^1 \sum_{j=0}^{2} \mathbf{e}_j \overline{D}u(\mathrm{x}t)\mathbf{e}_j t\,dt = -\int_0^1 Du(\mathrm{x}t)t\,dt. \end{aligned}$$

Additionally, we have the following relation

$$\mathrm{Sc}(\overline{D}u(\mathrm{x}t)\mathrm{x}) = \langle Du(\mathrm{x}t), \mathrm{x}\rangle = \sum_{j=0}^{2} t x_j \frac{\partial}{\partial(t x_j)} u(\mathrm{x}t) = Eu(\mathrm{x}t).$$

Thus,

$$Df(\mathrm{x}) = Du(\mathrm{x}) - \int_0^1 Du(\mathrm{x}t)t\,dt - \int_0^1 DEu(\mathrm{x}t)t\,dt.$$

A direct computation yields

$$DEu = Du + EDu.$$

Hence,

[1] A quaternion $p \in \mathbb{H}$ is a reduced quaternion (i.e., of the form $q = q_0\mathbf{e}_0 + q_1\mathbf{e}_1 + q_2\mathbf{e}_2$) if and only if

$$\sum_{j=0}^{2} \mathbf{e}_j q \mathbf{e}_j = -\overline{q}.$$

$$Df(\mathrm{x}) = Du(\mathrm{x}) - 2\int_0^1 Du(\mathrm{x}t)t\,dt - \int_0^1 EDu(\mathrm{x}t)t\,dt.$$

Now, $EDu(t\mathrm{x}) = t\frac{d}{dt}Du(\mathrm{x}t)$, so by integration by parts we get

$$\int_0^1 EDu(\mathrm{x}t)t\,dt = \int_0^1 t^2\frac{d}{dt}Du(\mathrm{x}t)\,dt = [t^2 Du(\mathrm{x}t)]_0^1 - 2\int_0^1 tDu(\mathrm{x}t)\,dt$$

where $[t^2 Du(\mathrm{x}t)]_0^1 = Du(\mathrm{x})$. Therefore,

$$Df(\mathrm{x}) = Du(\mathrm{x}) - 2\int_0^1 Du(\mathrm{x}t)t\,dt - Du(\mathrm{x}) + 2\int_0^1 tDu(\mathrm{x}t)\,dt = 0.$$

The next step is to recall basics of integration theory for monogenic functions, where the integration has to be computed now over surfaces, and not just curves as in the two-dimensional case. Assume that $\Omega \subset \mathbb{R}^3$ is a compact set. Suppose that the boundary surface $\partial\Omega$ of the set Ω is piecewise smooth, meaning that an outward unit normal can be defined almost everywhere on the surface. Such a set Ω will henceforth be called *regular*. We define the quaternion-valued 2-differential form

$$d\sigma = dy \wedge dz - \mathbf{i}dx \wedge dz + \mathbf{j}dx \wedge dy, \tag{22.3}$$

and assume that $n = n_1 + \mathbf{i}n_2 + \mathbf{j}n_3$ is a quaternionic representation of the outward unit normal on the surface $\partial\Omega$. It can be shown[2] that

$$d\sigma = n\,dS \tag{22.4}$$

when restricted to the surface $\partial\Omega$, where dS is the scalar surface element.

The significance of the above quaternion-valued surface element $d\sigma$ in quaternionic analysis is revealed in the following integral formula. The proof of the result is a "one-line proof" based on the classical Stokes' theorem.

Theorem 22.6 (Quaternionic Stokes' theorem) *Let Ω be a regular domain and let $f, g\colon \Omega \to \mathbb{H}$ be continuous functions such that $f, g\colon \Omega^\circ \to \mathbb{H}$ are differentiable, where Ω° denotes the interior points of Ω. Then*

$$\int_{\partial\Omega} g\,d\sigma f = \int_{\Omega} ((gD)f + g(Df))dV,$$

[2] The proof is straightforward if one assumes that the surface has a smooth parametrization on some chart $r(u,v) = (x(u,v), y(u,v), z(u,v))$. One then simply restricts $d\sigma$ to this chart and observes that $n(r(u,v)) = \pm\frac{r_u(u,v)\times r_v(u,v)}{|r_u(u,v)\times r_v(u,v)|}$. In this case, $dS = \pm|r_u(u,v)\times r_v(u,v)|\,du\wedge dv$. The plus-minus combinations above are chosen so that the unit normal vector points outward.

where $dV = dx \wedge dy \wedge dz$.

An important example of a function that is monogenic on both the left and the right in the integration theory is the *Cauchy kernel*, defined as follows

$$\mathcal{E}(\mathsf{x}) = \frac{1}{4\pi}\frac{\overline{\mathsf{x}}}{|\mathsf{x}|^3},$$

where $\mathsf{x} = x+y\mathbf{i}+z\mathbf{j}$. A direct computation shows that $D\mathcal{E}(\mathsf{x}) = \mathcal{E}(\mathsf{x})D = 0$ whenever $\mathsf{x} \neq 0$. By substituting the fundamental solution for g in the Stokes' formula, and assuming monogenicity of f, we immediately obtain the Cauchy formula:

Theorem 22.7 (Cauchy Integral Formula) *Let* Ω *be a regular domain and let* $f : \Omega \to \mathbb{H}$ *be continuous such that* $f : \Omega^\circ \to \mathbb{H}$ *is monogenic. Then*

$$f(\mathsf{x}) = \int_{\partial\Omega} \mathcal{E}(\mathsf{y}-\mathsf{x})d\sigma(\mathsf{y})f(\mathsf{y}).$$

In flow dynamics, however, we largely work with surface integrals. For this reason, the quaternionic Stokes' theorem is particularly important, as it connects monogenicity with surface integrals. Hence, we obtain the following result:

Proposition 22.8 *Let* Ω *be a regular domain and let* $f, g : \Omega \to \mathbb{H}$ *be continuous such that* $g : \Omega^\circ \to \mathbb{H}$ *is right-monogenic and* $f : \Omega^\circ \to \mathbb{H}$ *is left-monogenic. Then*

$$\int_{\partial\Omega} g d\sigma f = 0.$$

In order to make use of the monogenicity concept, for example in the form of a three-dimensional residue calculus, all surface integrals must be expressed, in one way or another, in terms of integrals of the type shown above.

22.3 Monogenic Flow Potential

In this section, we present the use of monogenic functions to introduce a flow potential. Let us consider a velocity vector field of an ideal flow $\vec{v} = (v_1, v_2, v_3)$ defined in an open set $U \subset \mathbb{R}^3$, and assume that $\phi : U \to \mathbb{R}$ is a *potential of the velocity vector field*, that is,

$$v_1 = \frac{\partial \phi}{\partial x}, \quad v_2 = \frac{\partial \phi}{\partial y}, \quad v_3 = \frac{\partial \phi}{\partial z}.$$

Further, we can consider the *stream function* is $\Psi = \psi_1\mathbf{i} + \psi_2\mathbf{j} + \psi_3\mathbf{k}$ such that

$$w = \phi + \Psi$$

is a monogenic function in U. The function w is called a *monogenic flow potential.* Because

$$v := v_1 + v_2\mathbf{i} + v_3\mathbf{j} = D\phi,$$

and $Dw = 0$, then

$$D\Psi = -v_1 - v_2\mathbf{i} - v_3\mathbf{j}.$$

From this it follows that

$$v = v_1 + v_2\mathbf{i} + v_3\mathbf{j} = \frac{1}{2}D\overline{w}. \tag{22.5}$$

Note that since the components of the monogenic flow potential are harmonic functions, in ideal flow, the velocity is a left anti-monogenic function

$$\overline{D}v = 0.$$

A monogenic flow potential is a function, which is constructed starting with a harmonic velocity potential ϕ.

Proposition 22.5 leads to the following result:

Proposition 22.9 *Let $U \subset \mathbb{R}^3$ be a star-shaped open set with respect to the origin. Then a monogenic flow potential exists in this domain.*

In domains more general than star-shaped domains, the construction of a monogenic flow potential is an open question. Moreover, the stream function, and thus the monogenic flow potential, is not unique, because it is always possible to add a monogenic vector-valued function to the stream function Ψ. More precisely, the transformation $\Psi \mapsto \Psi + \mathbf{H}$ is a gauge transformation that leaves the flow properties unchanged. In particular, the vector-valued monogenic function $\mathbf{H}$ does not depend on the velocity potential ϕ.

In general, it is difficult to geometrically interpret the stream function Ψ. Nonetheless, the following theorem presents a monogenic flow potential, which produces a stream functions, that can be easily interpreted geometrically:

Theorem 22.10 ([19]) *If the monogenic flow potential is of the form*

$$w(x, y, z) = \phi(x, y, z) + \psi_1(x, y)\mathbf{i} + \psi_2(x, z)\mathbf{j} + \psi_3(y, z)\mathbf{k},$$

satisfying the conditions

$$\frac{\partial \psi_1}{\partial y} = \frac{1}{2}v_1, \quad \frac{\partial \psi_2}{\partial z} = \frac{1}{2}v_1, \quad \frac{\partial \psi_3}{\partial z} = -\frac{1}{2}v_2,$$

$$\frac{\partial \psi_1}{\partial x} = -\frac{1}{2}v_2, \quad \frac{\partial \psi_2}{\partial x} = -\frac{1}{2}v_3, \quad \frac{\partial \psi_3}{\partial y} = \frac{1}{2}v_3,$$

then

$$\nabla\phi \cdot \nabla\psi_1 = \nabla\phi \cdot \nabla\psi_2 = \nabla\phi \cdot \nabla\psi_3 = 0$$

and

$$v_1 = \frac{\partial \psi_1}{\partial y} + \frac{\partial \psi_2}{\partial z}, \quad v_2 = -\frac{\partial \psi_1}{\partial x} - \frac{\partial \psi_3}{\partial z}, \quad v_3 = -\frac{\partial \psi_2}{\partial x} + \frac{\partial \psi_3}{\partial y}.$$

The idea of the above stream function is to search for two-dimensional stream functions on the coordinate planes. Accordingly, the streamlines lie on the intersection of the level surfaces defined by them. We refer to [19] for further details.

22.4 Blasius–Chaplygin Formula for the Force

Let U be a body that is assumed to form a regular domain, and let $p(x, y, z)$ be the pressure field on its surface ∂U. Then the total force caused by the pressure

$$F = F_1 + F_2\mathbf{i} + F_3\mathbf{j}$$

can be calculated using the formula [16]:

$$F = -\int_{\partial U} p \, d\sigma.$$

Let us recall the famous *Bernoulli equation* along each streamline, neglecting the effect of gravity, see e.g. [16]), which is given by

$$p + \frac{\rho}{2}|v|^2 = \text{const}.$$

Note that in the case of an ideal flow, the density ρ is constant. Substituting the Bernoulli equation into the expression for the force yields

$$F = \frac{\rho}{2} \int_{\partial U} |v|^2 \, d\sigma.$$

Above, the integral over the constant term vanishes by Theorem 22.6. Substituting the expression for the potential from (22.5), we obtain

$$F = \frac{\rho}{8} \int_{\partial U} |D\overline{w}|^2 \, d\sigma. \tag{22.6}$$

The above expression is a spatial analog of the *Blasius–Chaplygin force formula*, which allows the force acting on the body to be calculated using the monogenic flow potential. However, the previous integral is not of the form presented in the integration theory, since $|D\overline{w}|^2$ does not represent a monogenic function when w is monogenic. Nevertheless, by taking the conjugate of the force expression on both sides and extracting the scalar part, we obtain

$$F_1 = \frac{\rho}{8}\mathrm{Sc}\int_{\partial U} |D\overline{w}|^2\,\overline{d\sigma} = \frac{\rho}{8}\mathrm{Sc}\int_{\partial U} (w\overline{D})(D\overline{w})\overline{d\sigma}$$

and by multiplying both sides by $\mathbf{i}$, we obtain

$$F_2 = \frac{\rho}{8}\mathrm{Sc}\int_{\partial U} |D\overline{w}|^2\, d\sigma\mathbf{i} = \frac{\rho}{8}\mathrm{Sc}\int_{\partial U} (w\overline{D})(D\overline{w})\overline{d\sigma\mathbf{i}}.$$

Similarly to the above, we obtain

$$F_3 = \frac{\rho}{8}\mathrm{Sc}\int_{\partial U} (w\overline{D})(D\overline{w})\overline{d\sigma\mathbf{j}}.$$

Assume that w is of the same form as in Theorem 22.10. Define a quaternion-valued 1-form by setting

$$U = 2(-\psi_3 + \phi\mathbf{k})dx + 2(-\psi_2 - \phi\mathbf{j})dy + 2(\psi_1 + \phi\mathbf{i})dz.$$

By direct computation, we obtain

$$dU = D\overline{w}\overline{d\sigma}.$$

If the stream lines lie on the surface ∂U, then $d\psi_1 = d\psi_2 = d\psi_3 = 0$, from which it follows that $dU = d\overline{U}$. Restricting to the surface, we thus have

$$D\overline{w}\overline{d\sigma} = -d\sigma w\overline{D}.$$

Under these assumptions, the force integral can be expressed using integrals of the form in Proposition 22.8. Hence, we have:

Theorem 22.11 (Monogenic Blasius–Chaplygin).

$$F = -\frac{\rho}{8}\int_{\partial U} \Big(Sc\big((w\overline{D})d\sigma(w\overline{D})\big) + Sc\big((w\overline{D})d\sigma(w\overline{D})\mathbf{i}\big)\mathbf{i} + Sc\big((w\overline{D})d\sigma(w\overline{D})\mathbf{j}\big)\mathbf{j}\Big).$$

This result has an important consequence, namely:

Corollary 22.12 *If* $w\colon U \to \mathbb{H}$ *is monogenic, then* $F = 0$.

Proof. The result follows from Proposition 22.8, by noting that $w\overline{D}D = \Delta w = 0$ and $D(wD) = (Dw)D = 0$.

It is necessary to make the following remark:

Remark 22.13 *The idea in the proof of the previous theorem is to restrict the Cauchy–Riemann operator to the surface, using suitably chosen coordinates. This idea can be developed considerably further, and we refer to [20] for further details.*

2D Reduction of the Blasius–Chaplygin Force Formula

In this section, we illustrate how the classical two-dimensional Blasius–Chaplygin force formula can be obtained from the monogenic formula introduced in the previous section. Let us assume that the flow is invariant in the z-direction, that is, the velocity potential is $\phi = \phi(x, y)$. Then the monogenic flow potential is also invariant in the z-direction, and the Cauchy–Riemann operator reduces to

$$D_{2d} = \frac{\partial}{\partial x} + \mathbf{i}\frac{\partial}{\partial y}.$$

The monogenicity condition for the flow potential then reduces to

$$D_{2d}w = D_{2d}(\phi + \psi_1\mathbf{i}) + D_{2d}(\psi_2\mathbf{j} + \psi_3\mathbf{k}) = 0,$$

which holds if equations

$$D_{2d}(\phi + \psi_1\mathbf{i}) = \partial_x\phi - \partial_y\psi_1 + (\partial_y\phi + \partial_x\psi_1)\mathbf{i} = 0$$

and

$$D_{2d}(\psi_2\mathbf{j} + \psi_3\mathbf{k}) = (\partial_x\psi_2 - \partial_y\psi_3)\mathbf{j} + (\partial_x\psi_3 + \partial_y\psi_2)\mathbf{k} = 0$$

are satisfied.

As it can clearly seen from the equations presented above, the functions ψ_2 and ψ_3 are independent of the velocity potential ϕ. This is because, as mentioned previously, a vector-valued monogenic function can be added to the stream function, that is, $\Psi \mapsto \Psi + \mathbf{H}$, and this monogenic function is naturally independent of the velocity potential. Consequently, in the above, $\mathbf{H} = \psi_2\mathbf{j} + \psi_3\mathbf{k}$, and we may consider the potential

$$w = \phi + \psi_1\mathbf{i},$$

which is precisely the two-dimensional analytic flow potential that satisfies conditions (22.1).

Assume now that C is a piecewise smooth curve in the plane with positive orientation (counterclockwise), that is, if $n = (n_1, n_2)$ is the outer unit normal and

$$(n_1 + \mathbf{i}\,n_2)ds = -\mathbf{i}\,dz, \tag{22.7}$$

where $dz = dx + \mathbf{i}\,dy$. The curve C defines in the space $\mathbb{R}^3$ a cylinder $C \times [-h, h]$, where $h > 0$. Let U_h be a regular domain consisting of the aforementioned cylinder together with its top and bottom surfaces, that is,

$$U_h = C \times [-h, h] \cup U_{\text{top}} \cup U_{\text{bottom}}.$$

Next, we apply formula (22.6), which gives

$$F = \frac{\rho}{8} \int_{C\times[-h,h]} |D_{2d}\overline{w}|^2\, d\sigma + \frac{\rho}{8} \int_{U_{\text{top}}} |D_{2d}\overline{w}|^2\, d\sigma + \frac{\rho}{8} \int_{U_{\text{bottom}}} |D_{2d}\overline{w}|^2\, d\sigma.$$

Since the integrand $|D_{2d}\overline{w}|^2$ does not vary in the z-direction, and by using (22.4), we have $d\sigma = \mathbf{j}dS$ on the top surface and $d\sigma = -\mathbf{j}dS$ on the bottom surface. Therefore, the last two integrals cancel each other out, since the integration goes over the same region. Thus,

$$F = \frac{\rho}{8} \int_{C\times[-h,h]} |D_{2d}\overline{w}|^2\, d\sigma.$$

Let now the curve C be parametrized by arc length as $(x(s), y(s))$ for $s \in [0, L]$, and then the cylindrical surface is parametrized as

$$r(s,t) = (x(s), y(s), t),$$

with $t \in [-h, h]$. Then we have

$$dx = x'(s)\, ds, \qquad dy = y'(s)\, ds, \qquad dz = dt,$$

and the surface element (22.3) becomes

$$d\sigma|_{C\times[-h,h]} = (y'(s) - \mathbf{i}x'(s))\, ds \wedge dt.$$

Since the tangent vector of the curve is $(x'(s), y'(s))$, the outward normal vector is obtained by rotating this vector $\frac{\pi}{2}$ clockwise, that is, $(y'(s), -x'(s))$. In terms of complex numbers, rotation corresponds to multiplication by the number $-\mathbf{i}$, hence

$$n(s) = -\mathbf{i}(x'(s) + \mathbf{i}y'(s)) = y'(s) - \mathbf{i}x'(s).$$

This is a unit normal vector because the parametrization is by arc length. Using (22.7), we can rewrite the integral as

$$F = -\frac{\mathbf{i}\rho}{8} \int_{-h}^{h} \oint_C |D_{2d}\overline{w}|^2\, dz\, dt.$$

To obtain a two-dimensional formula for the force density, we define

$$F_{2d} := \frac{F}{2h} = -\frac{\mathbf{i}\rho}{16h} \int_{-h}^{h} dt \oint_C |D_{2d}\overline{w}|^2\, dz = -\frac{\mathbf{i}\rho}{8} \oint_C |D_{2d}\overline{w}|^2\, dz.$$

Let now $f(z) = \phi + \mathbf{i}\psi_1$, so that $f'(z) = v_1 - \mathbf{i}v_2$. Then

$$D_{2d}\overline{w} = 2\overline{f'(z)},$$

and thus

$$F_{2d} = -\frac{\mathbf{i}\rho}{2} \oint_C |f'(\mathrm{z})|^2 \, dz,$$

where $F_{2d} = F_x + \mathbf{i}F_y$. From this point onward, the derivation of the formula is a standard task in fluid mechanics textbooks. That is, by taking the complex conjugate we obtain

$$F_x - \mathbf{i}F_y = \frac{\mathbf{i}\rho}{2} \oint_C f'(\mathrm{z})\overline{f'(\mathrm{z})} \, d\bar{\mathrm{z}} = \frac{\mathbf{i}\rho}{2} \oint_C f'(\mathrm{z}) \, d\bar{f},$$

where $df = d\phi + \mathbf{i}d\psi_1$. Since $d\psi_1 = 0$ along the streamlines, we have $d\bar{f} = df = f'(\mathrm{z})dz$, hence

$$F_x - \mathbf{i}F_y = \frac{\mathbf{i}\rho}{2} \oint_C (f'(\mathrm{z}))^2 \, dz,$$

which is the holomorphic form of the Blasius–Chaplygin formula.

22.5 Quaternionic Blasius–Chaplygin Formula for the Moment

In this section, we introduce a quaternionic Blasius–Chaplygin formula for the moment. Assume that U is a solid body and that the pressure field $p(x, y, z)$ is known on its surface. Suppose that $\mathrm{x}_0 \in \mathbb{R}^3$ is an arbitrary point. We work with reduced quaternions and define the moment by

$$M := r \times F = \begin{vmatrix} 1 & \mathbf{i} & \mathbf{j} \\ r_1 & r_2 & r_3 \\ F_1 & F_2 & F_3 \end{vmatrix} = r_2F_3 - r_3F_2 + (r_3F_1 - r_1F_3)\mathbf{i} + (r_1F_2 - r_2F_1)\mathbf{j},$$

where $r = r_1 + r_2\mathbf{i} + r_3\mathbf{j}$ and $F = F_1 + F_2\mathbf{i} + F_3\mathbf{j}$.

Let $\mathrm{x} \in \partial U$ and set $r(\mathrm{x}) = \mathrm{x} - \mathrm{x}_0$, then the moment induced by the pressure field with respect to the point x_0 is given by

$$M_{\mathrm{x}_0} = -\int_{\partial U} p(\mathrm{x})(r(\mathrm{x}) \times n(\mathrm{x}))dS(\mathrm{x}),$$

where the cross product is interpreted in the sense of reduced quaternions, as in the definition of the moment presented above.

From the definition of the moment, we immediately obtain the following proposition:

Proposition 22.14 *If the pressure p is constant on the entire surface ∂U, then* $M_{\mathrm{x}_0} = 0$.

Proof. The moment can alternatively be computed as

$$M_{\mathrm{x}_0} = (\mathrm{x}_c - \mathrm{x}_0) \times F,$$

where x_c is the center of mass of the body, and F is the total force as computed in the previous result. When the pressure is constant, then $F = 0$.

Substituting Bernoulli's formula, we obtain

$$M_{\mathsf{x}_0} = \frac{\rho}{2} \int_{\partial U} |v(\mathsf{x})|^2 (r(\mathsf{x}) \times n(\mathsf{x})) dS(\mathsf{x}),$$

leading to the *Blasius–Chaplygin formula for the moment*

$$M_{\mathsf{x}_0} = \frac{\rho}{8} \int_{\partial U} |D\overline{w}(\mathsf{x})|^2 r(\mathsf{x}) \times d\sigma(\mathsf{x}).$$

The force formula was presented in Theorem 22.11 using monogenic functions. For the moment, the situation is more complicated, and the formula is left as an open problem in this article. However, by using the same proof technique as in Proposition 22.14, it can be shown that:

Proposition 22.15 *If $w : U \to \mathbb{H}$ is monogenic, then $M_{\mathsf{x}_0} = 0$.*

2D Reduction of the Blasius–Chaplygin Moment Formula

Finally, in this section, we show how to obtain the classical two-dimensional Blasius–Chaplygin moment formula from the monogenic formula introduced in the previous section. Assume $\mathsf{x}_0 = 0$, in which case

$$M = \frac{\rho}{8} \int_{\partial U} |D\overline{w}(\mathsf{x})|^2 (\mathsf{x} \times n(\mathsf{x})) dS(\mathsf{x}).$$

Again, suppose the flow is invariant in the z-direction, and write the integral over the previously defined surface U_h, which leads to

$$M = \frac{\rho}{8} \int_{-h}^{h} \int_{0}^{L} |D_{2d}\overline{w}|^2 (\mathsf{x}(s,t) \times n(s))\, ds\, dt,$$

where

$$\mathsf{x}(s,t) = x(s) + y(s)\mathbf{i} + t\mathbf{j}$$

and

$$n(s) = y'(s) - \mathbf{i}x'(s),$$

implying that

$$\mathsf{x}(s,t) \times n(s) = (x'(s) + y'(s)\mathbf{i})t - (x(s)x'(s) + y(s)y'(s))\mathbf{j}.$$

Substituting this expression into the definition of M and performing the outer integration yields

$$M = -\mathbf{j}\frac{\rho h}{4}\int_0^L |D_{2d}\overline{w}|^2 (x(s)x'(s) + y(s)y'(s))\, ds.$$

Let $D_{2d}\overline{w} = 2\overline{f'(\mathrm{z})}$, so we obtain

$$M = -\mathbf{j}\rho h \int_0^L |f'(x(s), y(s))|^2 (x(s)x'(s) + y(s)y'(s))\, ds.$$

Note that $dx = x'(s)ds$ and $dy = y'(s)ds$, meaning that $\mathrm{Re}(\mathrm{z}d\overline{\mathrm{z}}) = (x(s)x'(s) + y(s)y'(s))ds$. Consequently, the moment density becomes

$$M_{2d} = \frac{M}{2h} = -\frac{\rho}{2}\mathrm{Re}\left(\oint_C \mathrm{z}|f'(\mathrm{z})|^2\, d\overline{\mathrm{z}}\right)\mathbf{j}.$$

Proceeding analogously as with the force formula, we finally obtain

$$M_{2d} = -\frac{\rho}{2}\mathrm{Re}\left(\oint_C \mathrm{z}(f'(\mathrm{z}))^2\, d\mathrm{z}\right)\mathbf{j},$$

which is the two-dimensional Blasius–Chaplygin formula for the moment around the **j**-axis.

Finally, it should be noted that, unlike in two dimensions, the three-dimensional Blasius–Chaplygin formulas yield actual forces and moments instead of force and moment distributions. However, the assumption of two-dimensionality is always an idealization of the physical problem.

22.6 Conclusion

The main result in this article is the proof of a Blasius–Chaplygin type force formula in three dimensions when the flow potential is quaternion-valued. This result is presented in Theorem 22.11. It demonstrates that quaternion analysis can be used in the study of 3D ideal flow in a manner analogous to complex analysis, and in fact, the complex 2D idealization is always obtained as a special case. The results presented in this article open up possibilities for the investigation of three-dimensional fluid dynamics problems, see for example [17].

From the perspective of the mathematical tools employed, quaternion algebra is used instead of complex numbers, with the most significant difference being the non-commutativity of multiplication. Moreover, instead of contour integrals, the integrals are surface integrals in the space $\mathbb{R}^3$, whose effective handling requires the use of differential forms. The aforementioned mathematics demands a somewhat different mindset from the practitioner, although all concepts and tools from complex analysis have their counterparts in quaternion analysis.

References

[1] Blasius P (1908) Grenzschichten in Flüssigkeiten mit kleiner Reibung. Zeitschrift für Mathematik und Physik 56(1):1–37
[2] Blasius P (1911) Grenzschichten in Flüssigkeiten mit kleiner Reibung. Mitteilungen über Forschungsarbeiten auf dem Gebiete des Ingenieurwesens 131:1–41
[3] Blasius P (1950) Boundary layers in fluids with little friction. National Advisory Committee for Aeronautics, Technical Memorandum 1256
[4] Bock S, Gürlebeck K (2009) On a spatial generalization of the Kolosov–Muskhelishvili formulae, Mathematical Methods in the Applied Sciences, 32(2):223-240, DOI 10.1002/mma.1033
[5] Bock S, Gürlebeck K, Legatiuk D, Nguyen HM (2015) ψ-Hyperholomorphic functions and a Kolosov–Muskhelishvili formula. Mathematical Methods in the Applied Sciences 38(18):5114-5123, DOI 10.1002/mma.3431
[6] Bock S (2014) On monogenic series expansions with applications to linear elasticity. Advances in Applied Clifford Algebras 24(4):931-943, DOI 10.1007/s00006-014-0490-0
[7] Brackx F, Delanghe R, Sommen F (1982) Clifford analysis, Research Notes in Mathematics, Vol 76, Pitman, London
[8] Chaplygin SA (1902) O gazovykh struiakh pri ravnomernom dvizhenii (in Russ., On gas streams with uniform motion), Scientific Memoirs by the Professors of the Imperial Moscow University, 21
[9] Chaplygin SA (1949) O gazovykh struiakh (in Russ., On gas streams), Gosudarstvennoe izdatel'stvo tekhniko-teoreticheskoi literatury, Moscow-Leningrad
[10] Cruz C, Falcão MI, Malonek HR (2014) Monogenic pseudo-complex power functions and their applications. Mathematical Methods in the Applied Sciences 37(12):1717-1872, DOI 10.1002/mma.2931
[11] Grigor'ev Y (2018) Quaternionic functions and their applications in a viscous fluid flow. Complex Analysis and Operator Theory 12(2):491-508, DOI 10.1007/s11785-017-0715-z
[12] Gürlebeck K, Habetha K, Sprößig W (2008) Application of Holomorphic Functions in Two and Higher Dimensions. Birkhäuser, Basel, DOI 10.1007/978-3-0348-0964-1
[13] Gürlebeck K, Habetha K, Sprößig W (2008) Holomorphic Functions in the Plane and n-dimensional Space. Birkhäuser, Basel, DOI 10.1007/978-3-7643-8272-8

[14] Gürlebeck K, Legatiuk D (2019) Quaternionic operator calculus for boundary value problems of micropolar elasticity. In: Bernstein, S. (ed) Topics in Clifford Analysis. Trends in Mathematics. Birkhäuser, Cham, DOI 10.1007/978-3-030-23854-4_10

[15] Gürlebeck K, Legatiuk D, Webber K (2022) Operator calculus approach to comparison of elasticity models for modelling of masonry structures, Mathematics, 10(10):1670, DOI 10.3390/math10101670

[16] Landau LD, Lifshitz EM (1987) Fluid Mechanics, 2nd ed., Pergamon Press, Oxford

[17] Lavrentev MA, Shabat BV (1977) Problemy gidrodinamiki i ikh matematicheskiye modeli (in Russ., Hydrodynamics problems and their mathematical models). Izdatel'stvo Nauka, Moscow

[18] Orelma H (2025) Modeling the third dimension: Quaternion techniques in mechanics and flow dynamics, In: Proceedings IX International Conference Topical Problems of Continuum Mechanics, Tsaghkadzor, Armenia, Sep. 2025.

[19] Orelma H (2025) Three dimensional stream functions and monogenic flow potentials, In: Altenbach, H. (ed) Current Developments in Solid Mechanics and Their Applications, Advanced Structured Materials, Vol 223, Springer, Cham, DOI 10.1007/978-3-031-90022-8_32

[20] Orelma H, Sommen F (2011) A differential form approach to Dirac operators on surfaces. In: Sabadini I, Sommen F (eds.) Hypercomplex Analysis and Applications, Trends in Mathematics, Springer, Basel, pp. 213-232, DOI 10.1007/978-3-0346-0246-4_15

Chapter 23
Necessary Conditions for Optimality in Variable Structure Problems

Irina S. Maksimova

Abstract In this paper, the necessary conditions for optimality are obtained for a problem with a variable structure. The change of phase spaces is due to the presence of several controlled objects in the problem with a sequential mode of operation in time. Problems of this kind have both physical and economic applications.

Key words: Optimal control problem, Phase space change, Problem with a variable structure

23.1 Problem Statement

There is an $m-1$ phase space $X_1 = R^{m_1}, \ldots, X_{m-1} = R^{m_{m-1}}$, of the variables $x_i = (x_{i1}, \ldots, x_{im_k}), k = \overline{1, m-1}$, respectively, having different dimensions. The motion of the controlled object is described by an $(m-1)$-system of differential equations, each of which is defined on its own time interval and in its own phase space, i.e.

$$\dot{x}_i(t) = f_i(t, x_i(t), u_i(t)), \tag{23.1}$$

$$x_i(t) \in X_i, \quad t \in [t_i, t_{i+1}], i = \overline{1, m-1},$$

provided that $u_i(t) \in U_i(t)$. The functions $f_i(t, x_i(t), u_i(t)), i = \overline{1, m-1}$ are continuously differentiable with respect to (x_i, u_i) for t, measurable with respect to t for any fixed (x_i, u_i) together with their partial derivatives with respect to (x_i, u_i).

The class of admissible controls consists of a set of all possible measurable, essentially bounded functions $u_i(\cdot)$, each defined on its own finite interval and satisfying the conditions $u_i(t) \in U_i(t)$ for almost all t. Here, $U_i(t)$ are given multivalued mappings that assign to each t nonempty subsets $U_i(t) \subseteq R^{m_i}, i = \overline{1, m-1}$.

Irina S. Maksimova
Peoples' Friendship University of Russia named after Patrice Lumumba, Mikluho-Maklaya st., 6, Moscow, 117198, Russian Federation, e-mail: irismax@yandex.ru

H. Altenbach and A. V. Sahakyan, *Contact Mechanics and Relevant Problems*, Advanced Structured Materials 253, https://doi.org/10.1007/978-3-032-20869-9_23

The following definition is needed for the subsequent discussions:

Definition 23.1. A measurable function v is called a measurable selector of a multi-valued mapping U if $v(t) \in U(t)$ for almost all t [1, 2].

Multi-valued mappings U_i are assumed to be measurable, i.e., there exist sequences of measurable selectors $\{v^i_j\}(i = \overline{1, m-1})$ such that the sets $\{v^i_j\}(i = \overline{1, m-1}, j = 1, 2, \ldots)$ are dense in $U_i(t)$, respectively, for almost all t. Also, the multivalued mappings U_i are assumed to be semi-continuous from above on the left at the points t_i and semi-continuous from above on the right at the points t_{i+1}, and they have measurable selectors $\{v^i_j\}$, continuous from the left at the points t_i and from the right at the points t_{i+1}, for which the sequences $\{v^i_j(t_s)\}$ are dense everywhere in the sets $U_i(t_s)(s = \overline{1, m})$. The solutions of the systems (23.1) at $t \in [t_i, t_{i+1}](i = \overline{1, m-1})$ are called absolutely continuous functions satisfying almost everywhere on $[t_i, t_{i+1}]$, respectively, the systems (23.1).

In each space $X_i(i = \overline{1, m-2})$ mappings

$$q_i : X_i \to X_{i+1}, i = \overline{1, m-2}$$

are given, by means of which, at given moments of time t_i, a sequential transition from one space to another is carried out (each of these mappings q_i is assumed to be smooth). The functional is given by

$$K_0(x_1, \cdots, x_m), \tag{23.2}$$

where $x_1 = x(t_1), x_m = x(t_m)$,

$$x_i = x(t_i - 0), x(t_i + 0) = q_{i-1}(x(t_i - 0)), \qquad i = \overline{2, m-1}.$$

The time points $t_1, \cdots, t_m$ are considered to be given. The function

$$K_0(x_1, \cdots, x_m)$$

is assumed to be continuously differentiable with respect to $x_1, \cdots, x_m$.

The motion of the object is carried out according to the following scheme. At time t_1 in space X_1, the object moves according to the solutions of the first of the systems (23.1) until time t_2, then it transitions to space X_2 under the action of the mapping $q_1 : X_1 \to X_2$, and the object's further motion is carried out according to the solutions of the second system from (23.1) with the initial condition $x_2(t_2) = q_1(x_1(t_2))$ and so on until the last space X_{m-1}. In space X_{m-1}, the object moves according to the solutions $(m-1)$th system from (23.1) with the boundary condition $x(t_{m-1} + 0) = q_{m-2}(x(t_{m-1} - 0))$ until the final time moment t_m. It is assumed that at times t_1 and t_m, no restrictions are imposed on the initial and final positions, i.e., the ends of the trajectory are free.

The optimization problem consists in finding, among all admissible controls $u_i(i = \overline{1, m-1})$ that move the object from the initial position to the final position, those for which the functional

$$K_0(x_1, \cdots, x_m)$$

would take the smallest possible value. Similar problems have been studied previously, for example, in [3, 4]. The problem of controllability with a similar motion pattern is considered by the author in [5–7]. Let us denote by $p = (x_1, x_2, \ldots, x_m), p^* = (x_1^*, x_2^*, \ldots, x_m^*)$, where (x^*, p^*, u^*)– is the optimal process in the given problem with free ends. Hereinafter, we omit the notation $*$ for brevity.

23.2 Main Results

In this paper, we obtain the necessary conditions for optimality for the given problem, which can be formulated in the form of the following theorem.

Theorem 23.1. *Let (x^*, p^*, u^*) be the optimal process in the given problem, thus it satisfies the following conditions: there exist absolutely continuous m_i-dimensional vector functions $\psi_i (i = \overline{1, m-1})$ such that*

(i) on each interval $[t_i, t_{i+1}]$ the function ψ_i satisfies the conjugate system

$$\dot{\psi}_i(t) = -\frac{\partial f_i(t, x_i(t), u_i(t))}{\partial x_i} \psi_i(t), \tag{23.3}$$

(ii) the transversality conditions are satisfied at the boundary points t_1 and t_m

$$\psi_1(t_1) = \frac{\partial K_0(p)}{\partial x_1},\ \psi_{m-1}(t_m) = -\frac{\partial K_0(p)}{\partial x_m} \tag{23.4}$$

and conditions of transversality at the "joint" points

$$\psi_{i-1}(t_i) = -\frac{\partial K_0(p)}{\partial x_i} + (\frac{\partial q_{i-1}}{\partial x_{i-1}})^* \psi_i(t_i), i = \overline{2, m-1}, \tag{23.5}$$

(iii) for almost all t, the conditions for the maximum of the Pontryagin function are satisfied

$$H_i(t, u_i^*(t)) = \sup_{u_i \in U_i(t)} H_i(t, u_i),\ i = \overline{1, m-1}, \tag{23.6}$$

where

$$H_i(t, x_i, u_i, \psi_i) = \langle f_i(t, x_i, u_i), \psi_i \rangle,\ i = \overline{1, m-1}.$$

Proof. Let us obtain the transversality conditions at times $t_1, \ldots, t_m$ for the given problem. We will prove that there exist absolutely continuous m_i-dimensional vector functions $\psi_i (i = \overline{1, m-1})$ such that

(i) on each interval $[t_i, t_{i+1}]$ the function ψ_i satisfies the conjugate system (23.3), denoting

$$\frac{\partial f_i(t, x_i(t), u_i(t))}{\partial x_i}$$

by $A_i^*(t)$, we obtain system (23.3) in the form

$$\dot{\psi}_i(t) = -A_i^*(t)\psi_i(t); \tag{23.7}$$

(ii) the transversality conditions (23.4) are satisfied at the boundary points t_1 and t_m and the transversality conditions are satisfied at the "junction" points (23.5).

Let us consistently derive the conditions of transversality from left to right, examining time intervals one by one. First, let us examine the interval $[t_1, t_2]$. The right-hand side of the equation

$$\dot{\psi}_1(t) = -A_1^*(t)\psi_1(t) \tag{23.8}$$

linear by the variable ψ_1. Therefore, by virtue of the existence theorem for linear differential equations, Eq. (23.8) has a solution ψ_1 on the interval $[t_1, t_2]$ with the initial condition (23.4), i.e.

$$\psi_1(t_1) = \frac{\partial K_0(p)}{\partial x_1}. \tag{23.9}$$

Let us take an arbitrary vector $a \in R^{m_1}$ and perturb the initial condition. According to the theorem on the existence of a solution to the Cauchy problem for small perturbations of the initial conditions, for any number $\alpha > 0$, the Cauchy problem

$$\dot{x_1} = f_1(t, x_1, u_1^*(t)),\ x_1(t_1) = x_1^* + \alpha a$$

has a solution $x_1(t, \alpha)$, $t \in [t_1, t_2]$. Then, for any $\alpha > 0$, the inequality

$$\begin{aligned} &\alpha^{-1}(K_0(x_1(t_1,\alpha), x_2(t_2,\alpha), x_3(t_3,\alpha), \cdots, x_m(t_m,\alpha)) \\ &\quad -K_0(x_1(t_1), x_1(t_2), x_2(t_3), \cdots, x_m)) \geqslant 0. \end{aligned} \tag{23.10}$$

Moving to the limit in this inequality as $\alpha \to 0+$, we obtain

$$\begin{aligned} &\langle\frac{\partial K_0}{\partial x_1}, a\rangle + \langle\frac{\partial K_0}{\partial x_2}, \delta_1(t_2)\rangle + \langle\frac{\partial K_0}{\partial x_3}, \delta_2(t_3)\rangle + \cdots \\ &+\langle\frac{\partial K_0}{\partial x_{m-1}}, \delta_{m-2}(t_{m-1})\rangle + \langle\frac{\partial K_0}{\partial x_m}, \delta_{m-1}(t_m)\rangle \geqslant 0, \end{aligned} \tag{23.11}$$

where

$$\delta_{i-1}(t_i) = \frac{\partial x_{i-1}}{\partial \alpha}(t_i, 0), \qquad i = \overline{2, m}.$$

Here and below, the argument p^* is omitted for brevity.

Let us write the equation in variations for the system

$$\dot{x_1} = f_1(t, x_1, u_1),$$

we get that $\delta_1(t)$ is the solution of the linear homogeneous equation

$$\dot{\delta}_1(t) = \frac{\partial f_1}{\partial x_1}(t)\delta_1(t),\ \delta_1(t_1) = a. \tag{23.12}$$

Let us denote the fundamental matrix of equation (23.12) via Φ_1. Then $\Phi_1(t_1)$ is the identity matrix and $\delta_1(t_2) = \Phi_1(t_2)a$.

Since $q_1(x_1(t_2,\alpha)) = x_2(t_2,\alpha)$, then

$$\delta_2(t_2) = \frac{\partial x_2}{\partial \alpha} = \frac{\partial q_1}{\partial x_1}\frac{\partial x_1(t_2,\alpha)}{\partial \alpha} = \frac{\partial q_1}{\partial x_1}\delta_1(t_2).$$

Similarly to the previous case, for the i-th system

$$\dot{x}_i = f_i(t, x_i, u_i),$$

on the time interval $[t_i, t_{i+1}](i = \overline{2, m-1})$, we obtain that $\delta_i(t)$ is the solution of the linear homogeneous equation

$$\dot{\delta}_i(t) = \frac{\partial f_i}{\partial x_i}(t)\delta_i(t),\ \delta_i(t_i) = \frac{\partial q_{i-1}}{\partial x_{i-1}}\delta_{i-1}(t_i). \tag{23.13}$$

Let us denote by Φ_i the fundamental matrix of equation (23.13). Then $\Phi_i(t_i)$ is a identity matrix and

$$\delta_i(t_{i+1}) = \Phi_i(t_{i+1})\frac{\partial q_{i-1}}{\partial x_{i-1}}\delta_{i-1}(t_i) = \Phi_i(t_{i+1})\cdot\ \ldots\ \cdot\frac{\partial q_1}{\partial x_1}\Phi_1(t_2)a.$$

Now, let us consider the conjugate equations for variables $\psi_i(i = \overline{2, m-1})$

$$\dot{\psi}_i(t) = -\frac{\partial f_i}{\partial x_i}(t)\psi_i(t), t \in [t_i, t_{i+1}], \tag{23.14}$$

which have solutions ψ_i on the intervals $[t_i, t_{i+1}]$ with initial conditions determined by the conditions

$$\psi_{i-1}(t_i) + \frac{\partial K_0}{\partial x_i} = (\frac{\partial q_{i-1}}{\partial x_{i-1}})^*\psi_i(t_i), i = \overline{2, m-1}. \tag{23.15}$$

From (23.4) and (23.12) we get

$$\frac{d}{dt}\langle\psi_1(t), \delta_1(t)\rangle \equiv 0.$$

Integrating this identity over the interval $[t_1, t_2]$ we get

$$\langle\psi_1(t_2), \delta_1(t_2)\rangle - \langle\psi_1(t_1), \delta_1(t_1)\rangle = 0,$$

$$\langle\psi_1(t_2), \Phi_1(t_2)a\rangle = \langle\frac{\partial K_0}{\partial x_1}, a\rangle. \tag{23.16}$$

Similarly,

$$\frac{d}{dt}\langle\psi_i(t),\delta_i(t)\rangle \equiv 0, i=\overline{2,m}.$$

Integrating these identities, respectively, on the intervals $[t_i,t_{i+1}]$ $(i=\overline{2,m-1})$ we get

$$\langle\psi_i(t_{i+1}),\delta_i(t_{i+1})\rangle - \langle\psi_i(t_i),\delta_i(t_i)\rangle = 0. \tag{23.17}$$

From (23.16) with $i=2$ and taking into account the relation (23.17) with $i=2$, as well as the transversality condition "at the junction" (23.15) with $i=2$, we get

$$\langle\psi_2(t_3),\delta_2(t_3)\rangle = \langle\psi_2(t_2),\delta_2(t_2)\rangle,$$

$$\langle\psi_2(t_3),\Phi_2(t_3)\frac{\partial q_1}{\partial x_1}\Phi_1(t_2)a\rangle = \langle\psi_2(t_2),$$

$$\frac{\partial q_1}{\partial x_1}\delta_1(t_2)\rangle = \langle(\frac{\partial q_1}{\partial x_1})^*\psi_2(t_2),\delta_1(t_2)\rangle = \langle(\frac{\partial q_1}{\partial x_1})^*\psi_2(t_2),\Phi_1(t_2)a\rangle,$$

$$\langle\frac{\partial K_0}{\partial x_2},\delta_1(t_2)\rangle = \langle-\psi_1(t_2),\Phi_1(t_2)a\rangle + \langle(\frac{\partial q_1}{\partial x_1})^*\psi_2(t_2),\Phi_1(t_2)a\rangle,$$

where from

$$\langle\frac{\partial K_0}{\partial x_2},\delta_1(t_2)\rangle = \langle-\psi_1(t_2),\Phi_1(t_2)a\rangle + \langle\psi_2(t_3),\Phi_2(t_3)\frac{\partial q_1}{\partial x_1}\Phi_1(t_2)a\rangle. \tag{23.18}$$

Similar transformations will be applied to all internal intervals. For the last time interval $[t_{m-1},t_m]$, we have

$$\langle\frac{\partial K_0}{\partial x_{m-1}},\delta_{m-2}(t_{m-1})\rangle = \langle-\psi_{m-2}(t_{m-1}),\Phi_{m-2}(t_{m-1})\frac{\partial q_{m-3}}{\partial x_{m-3}}\cdot\ldots\cdot\Phi_1(t_2)a\rangle$$
$$+\langle\psi_{m-1}(t_m),\Phi_{m-1}(t_m)\frac{\partial q_{m-2}}{\partial x_{m-2}}\Phi_{m-2}(t_{m-1})\cdot\ldots\cdot\Phi_1(t_2)a\rangle. \tag{23.19}$$

Substituting formulas (23.16), (23.18) and (23.19) into inequality (23.11), we get

$$\langle\psi_1(t_2),\Phi_1(t_2)a\rangle + \langle-\psi_1(t_2),\Phi_1(t_2)a\rangle + \langle\psi_2(t_3),\Phi_2(t_3)\frac{\partial q_1}{\partial x_1}\Phi_1(t_2)a\rangle$$
$$+\ldots+\langle-\psi_{m-2}(t_{m-1}),\Phi_{m-2}(t_{m-1})\frac{\partial q_{m-3}}{\partial x_{m-3}}\cdot\ldots\cdot\Phi_1(t_2)a\rangle$$
$$+\langle\psi_{m-1}(t_m),\Phi_{m-1}(t_m)\frac{\partial q_{m-2}}{\partial x_{m-2}}\Phi_{m-2}(t_{m-1})\cdot\ldots\cdot\Phi_1(t_2)a\rangle$$
$$+\langle\frac{\partial K_0}{\partial x_m},\Phi_{m-1}(t_m)\frac{\partial q_{m-2}}{\partial x_{m-2}}\Phi_{m-2}(t_{m-1})\cdot\ldots\cdot\Phi_1(t_2)a\rangle \geqslant 0,$$

where from

$$\langle\psi_{m-1}(t_m)+\frac{\partial K_0}{\partial x_m},\Phi_{m-1}(t_m)\frac{\partial q_{m-2}}{\partial x_{m-2}}\Phi_{m-2}(t_{m-1})\cdot\ldots\cdot\Phi_1(t_2)a\rangle\rangle \geqslant 0.$$

Using the arbitrariness of vector a, from the latter inequality we have

$$(\Phi_1(t_2))^* \cdot \ldots \cdot (\Phi_{m-2}(t_{m-1}))^* (\frac{\partial q_{m-2}}{\partial x_{m-2}})^* (\Phi_{m-1}(t_m))^* (\psi_{m-1}(t_m) + \frac{\partial K_0}{\partial x_m}) = 0.$$

Then, due to the non-degeneracy of the matrices $\Phi_i(\overline{1, m-1})$ and the mappings $q_i(i = \overline{1, m-2})$, we obtain the desired transversality condition at the boundary point t_m

$$\psi_{m-1}(t_m) = -\frac{\partial K_0}{\partial x_m}.$$

Let us now prove that for almost all t the conditions for the maximum of Pontryagin's function are satisfied,

$$H_i(t, u_i^*(t)) = \sup_{u_i \in U_i(t)} H_i(t, u_i), \quad i = \overline{1, m-1} \tag{23.20}$$

where

$$H_i(t, x_i, u_i, \psi_i) = \langle f_i(t, x_i, u_i), \psi_i \rangle, \quad i = \overline{1, m-1}.$$

Hereinafter, the following designations are used

$$H_i(t, u_i) = H_i(t, x_i^*(t), u_i, \psi_i(t)), \quad i = \overline{1, m-1},$$

$$H_i(t) = H_i(t, u_i^*(t)), \quad i = \overline{1, m-1}.$$

The proof is based on the application of needle variations. Consider an arbitrary number k from 1 to m and, accordingly, two adjacent intervals $[t_{k-1}, t_k]$ and $[t_k, t_{k+1}]$. We will prove that the maximum conditions are satisfied on them. First, we obtain the maximum condition on the interval $[t_{k-1}, t_k]$. Let us take a countable set of measurable selectors $\{v_i^{k-1}\}$ of a multivalued mapping U_{k-1} such that the set $\{v_i^{k-1}\}_{i=1}^{\infty}$ is dense in $U_{k-1}(t)$ for almost all t. We will assume that each of the functions v_i^{k-1} is essentially bounded on $[t_{k-1}, t_k]$.

We will prove (23.20) at $i = k-1$ for an arbitrary point $\tau \in (t_{k-1}, t_k)$ for which the following conditions are satisfied: the point τ is a Lebesgue point of the functions

$$f_{k-1}(t, x_{k-1}, u_{k-1}), f_{k-1}(t, x_{k-1}, v_i^{k-1}), \ i = 1, 2, \ldots$$

and the set of points $\{v_i^{k-1}(\tau)\}$ is dense everywhere in $U_{k-1}(\tau)$. Let us fix the number i and the number $\alpha > 0$. Consider the control

$$u_i^{k-1}(t, \alpha) = \begin{cases} u_{k-1}^*(t), & t \notin [\tau - \alpha, \tau], \\ v_i^{k-1}(t), & t \in (\tau - \alpha, \tau), \end{cases}$$

- needle variation control u_{k-1}^*.

Let us denote the solution of the system via $x_{k-1}(t, \alpha)$

$$\dot{x}_{k-1}(t) = f_{k-1}(t, x_{k-1}(t), u_{k-1}(t)),$$

corresponding to the control $u_i^{k-1}(\cdot, \alpha)$ and satisfying the initial condition

$$x_{k-1}(t_{k-1}, \alpha) = x_{k-1}^*.$$

For small α, this solution exists on the entire interval $[t_{k-1}, t_k]$. Since τ is the Lebesgue point of the corresponding functions, we have:

$$x_{k-1}(\tau, \alpha) = x_{k-1}(\tau - \alpha) + \alpha f_{k-1}(\tau, x_{k-1}^*(\tau), v_i^{k-1}(\tau)) + o(\alpha),$$

$$x_{k-1}^*(\tau) = x_{k-1}^*(\tau - \alpha) + \alpha f_{k-1}(\tau, x_{k-1}^*(\tau), u_{k-1}^*(\tau)) + o(\alpha).$$

Consequently

$$\begin{aligned} x_{k-1}(\tau, \alpha) - x_{k-1}^*(\tau) &= x_{k-1}(\tau - \alpha) - x_{k-1}^*(\tau - \alpha) \\ &+ \alpha(f_{k-1}(\tau, x_{k-1}^*(\tau), v_i^{k-1}(\tau)) - f_{k-1}(\tau, x_{k-1}^*(\tau), u_{k-1}^*(\tau))). \end{aligned}$$

Then, the limit

$$y_{k-1}(\tau) = \lim_{\alpha \to 0} \frac{1}{\alpha}(x_{k-1}(\tau, \alpha) - x_{k-1}^*(\tau))$$

exists and is equal to

$$y_{k-1}(\tau) = f_{k-1}(\tau, v_i^{k-1}(\tau)) - f_{k-1}(\tau, u_{k-1}^*(\tau)). \tag{23.21}$$

On the interval $[\tau, t_k]$, both functions x_{k-1}^* and $x_{k-1}(\cdot, \alpha)$ satisfy the same equation

$$\dot{x}_{k-1}(t) = f_{k-1}(t, x_{k-1}(t), u_{k-1}(t)).$$

By virtue of the theorem on the differentiability of the solution of a differential equation under the initial conditions, the functions $x_{k-1}(\cdot, \alpha)$ converge uniformly as $\alpha \to 0$ to x_{k-1}^* and the limit

$$y_{k-1}(t) = \lim_{\alpha \to 0+} \frac{1}{\alpha}(x_{k-1}(t, \alpha) - x_{k-1}^*(t))$$

exists for all $t \geq \tau$. For all $t > \tau$ we have

$$x_{k-1}(\tau, \alpha) = x_{k-1}^*(\tau) + \int_\tau^t f_{k-1}(\theta, x_{k-1}(\theta, \alpha), u_{k-1}^*(\theta))d\theta,$$

$$x_{k-1}^*(t) = x_{k-1}^*(\tau) + \int_\tau^t f_{k-1}(\theta, x_{k-1}(\theta), u_{k-1}^*(\theta))d\theta.$$

From which we obtain

$$y_{k-1}(t) = y_{k-1}(\tau) + \int_\tau^t \frac{\partial f_{k-1}}{\partial x_{k-1}}(\theta) y_{k-1}(\theta) d\theta.$$

Therefore, on $[\tau, t_k]$, the function y_{k-1} satisfies the equation

$$\dot{y}_{k-1} = \frac{\partial f_{k-1}}{\partial x_{k-1}}(t, x^*_{k-1}(t), u^*_{k-1}(t))y_{k-1}. \tag{23.22}$$

For almost all $t > \tau$, we have

$$\frac{d}{dt}\langle\psi_{k-1}(t), y_{k-1}(t)\rangle = \langle\dot{\psi}_{k-1}(t), y_{k-1}(t)\rangle + \langle\psi_{k-1}(t), \dot{y}_{k-1}(t)\rangle$$
$$= \langle -\frac{\partial f^*_{k-1}}{\partial x_{k-1}}(t)\psi_{k-1}(t), y_{k-1}(t)\rangle + \langle\psi_{k-1}(t), \frac{\partial f_{k-1}}{\partial x_{k-1}}(t)y_{k-1}(t)\rangle = 0.$$

thus

$$\frac{d}{dt}\langle\psi_{k-1}(t), y_{k-1}(t)\rangle = 0.$$

We integrate the resulting equality over $[\tau, t_k]$

$$\langle\psi_{k-1}(t_k), y_{k-1}(t_k)\rangle - \langle\psi_{k-1}(\tau), y_{k-1}(\tau)\rangle = 0.$$

Using the transversality condition "at the junction" (23.5), we obtain

$$\langle\psi_{k-1}(\tau), y_{k-1}(\tau)\rangle = \langle -\frac{\partial K_0}{\partial x_k} + (\frac{\partial q_{k-1}}{\partial x_{k-1}})^*\psi_k(t_k), y_{k-1}(t_k)\rangle.$$

From (23.21) we have

$$\begin{aligned}&\langle\psi_{k-1}(\tau), f_{k-1}(\tau, v_i^{k-1}(\tau)) - f_{k-1}(\tau, u^*_{k-1}(\tau))\rangle\\ &\quad= \langle -\frac{\partial K_0}{\partial x_k} + (\frac{\partial q_{k-1}}{\partial x_{k-1}})^*\psi_k(t_k), y_{k-1}(t_k)\rangle.\end{aligned} \tag{23.23}$$

Since in the problem under consideration the minimum is equal to K_0^*, we have the inequality

$$\lim_{\alpha\to 0}(K_0(x_1^*, \ldots, x_{k-1}(t_{k-1}, \alpha), x_k(t_k, \alpha), \cdots, x_m(t_m, \alpha))$$
$$-K_0(x_1^*, \ldots, x^*_{k-1}, x^*_k, \cdots, x^*_m)) \geq 0.$$

Calculating this limit, we obtain

$$\begin{aligned}&\langle\frac{\partial K_0}{\partial x_k}, y_{k-1}(t_k)\rangle + \langle\frac{\partial K_0}{\partial x_{k+1}}, \delta_k(t_{k+1})\rangle + \langle\frac{\partial K_0}{\partial x_{k+2}}, \delta_{k+1}(t_{k+2})\rangle + \ldots\\ &\quad+\langle\frac{\partial K_0}{\partial x_{m-1}}, \delta_{m-2}(t_{m-1})\rangle + \langle\frac{\partial K_0}{\partial x_m}, \delta_{m-1}(t_m)\rangle \geq 0.\end{aligned} \tag{23.24}$$

Here, $\delta_k(t_{k+1})$ is the solution to the equation in variations

$$\begin{cases} \dot{\delta_k}(t) = \dfrac{\partial f_k}{\partial x_k}(t)\delta_k(t), \\ \delta_k(t_k) = \dfrac{\partial q_{k-1}}{\partial x_{k-1}} y_{k-1}(t_k) \end{cases}$$

for the k - th of the systems (23.1). Here, Φ_k is the fundamental matrix of the given system, $\Phi_k(t_k)$ is the identity matrix, and

$$\delta_k(t_{k+1}) = \Phi_k(t_{k+1})\frac{\partial q_{k-1}}{\partial x_{k-1}} y_{k-1}(t_k).$$

Similarly to the case for ψ_{k-1}, for ψ_k the following equality holds

$$\frac{d}{dt}\langle\psi_k(t), \delta_k(t)\rangle = 0,$$

integrating which on $[t_k, t_{k+1}]$, we get $\langle\psi_k(t_{k+1}), \delta_k(t_{k+1})\rangle = \langle\psi_k(t_k), \delta_k(t_k)\rangle$

$$\begin{aligned} \langle\psi_k(t_{k+1}), \Phi_k(t_{k+1})\frac{\partial q_{k-1}}{\partial x_{k-1}} y_{k-1}(t_k)\rangle &= \langle\psi_k(t_k), \frac{\partial q_{k-1}}{\partial x_{k-1}} y_{k-1}(t_k)\rangle \\ &= \langle(\frac{\partial q_{k-1}}{\partial x_{k-1}})^* \psi_k(t_k), y_{k-1}(t_k)\rangle. \end{aligned} \tag{23.25}$$

Taking into the accout the transversality condition at point t_{k+1} and (23.25), we have

$$\begin{aligned} \langle\frac{\partial K_0}{\partial x_{k+1}}, \delta_k(t_{k+1})\rangle &= \langle-\psi_k(t_{k+1}), \Phi_k(t_{k+1})\frac{\partial q_{k-1}}{\partial x_{k-1}} y_{k-1}(t_k)\rangle \\ &+\langle(\frac{\partial q_k}{\partial x_k})^* \psi_{k+1}(t_{k+1}), \Phi_k(t_{k+1})\frac{\partial q_{k-1}}{\partial x_{k-1}} y_{k-1}(t_k)\rangle \\ &= \langle-(\frac{\partial q_{k-1}}{\partial x_{k-1}})^* \psi_k(t_k), y_{k-1}(t_k)\rangle \\ &+\langle(\frac{\partial q_k}{\partial x_k})^* \psi_{k+1}(t_{k+1}), \Phi_k(t_{k+1})\frac{\partial q_{k-1}}{\partial x_{k-1}} y_{k-1}(t_k)\rangle. \end{aligned}$$

Using similar transformations for the corresponding indices, we get

$$\begin{aligned} \langle\frac{\partial K_0}{\partial x_{k+2}}, \delta_{k+1}(t_{k+2})\rangle &= -\langle(\frac{\partial q_k}{\partial x_k})^* \psi_{k+1}(t_{k+1}), \Phi_k(t_{k+1})\frac{\partial q_{k-1}}{\partial x_{k-1}} y_{k-1}(t_k)\rangle \\ &+\langle(\frac{\partial q_{k+1}}{\partial x_{k+1}})^* \psi_{k+1}(t_{k+1}), \Phi_{k+1}(t_{k+2})\frac{\partial q_k}{\partial x_k}\Phi_k(t_{k+1})\frac{\partial q_{k-1}}{\partial x_{k-1}} y_{k-1}(t_k)\rangle \end{aligned}$$

$$\begin{aligned} \langle\frac{\partial K_0}{\partial x_{m-1}}, \delta_{m-2}(t_{m-1})\rangle &= -\langle\psi_{m-2}(t_{m-1}), \delta_{m-2}(t_{m-1})\rangle \\ &+\langle(\frac{\partial q_{m-2}}{\partial x_{m-2}})^* \psi_{m-1}(t_{m-1}), \delta_{m-2}(t_{m-1})\rangle. \end{aligned}$$

Using (23.25) and the transversality condition at the right boundary point, we get

$$\langle \frac{\partial K_0}{\partial x_m}, \delta_{m-1}(t_m)\rangle = -\langle \psi_{m-1}(t_m), \delta_{m-1}(t_m)\rangle$$
$$= \langle -\psi_{m-1}(t_{m-1}), \frac{\partial q_{m-2}}{\partial x_{m-2}} \delta_{m-2}(t_{m-1})\rangle.$$

Substituting the obtained equalities into inequality (23.24), performing pairwise elimination of the corresponding terms and using the transversality condition at point t_{k+1} , we get

$$\langle \frac{\partial K_0}{\partial x_k}, y_{k-1}(t_k)\rangle + \langle -(\frac{\partial q_{k-1}}{\partial x_{k-1}})^* \psi_k(t_k), y_{k-1}(t_k)\rangle$$
$$= \langle \frac{\partial K_0}{\partial x_k} - (\frac{\partial q_{k-1}}{\partial x_{k-1}})^* \psi_k(t_k), y_{k-1}(t_k)\rangle = \langle \psi_{k-1}(t_k), y_{k-1}(t_k)\rangle \geq 0.$$

Using (23.21), we get

$$-\langle \psi_{k-1}(\tau), f_{k-1}(\tau, v_i^{k-1}(\tau)) - f_{k-1}(\tau, u_{k-1}^*(\tau))\rangle \geq 0,$$

where from

$$\langle \psi_{k-1}(\tau), f_{k-1}(\tau, u_{k-1}^*(\tau))\rangle \geq \langle \psi_{k-1}(\tau), f_{k-1}(\tau, v_i^{k-1}(\tau))\rangle,$$
$$H_{k-1}(\tau, u_{k-1}^*(\tau)) \geq H_{k-1}(\tau, v_i^{k-1}(\tau)).$$

The last inequality is proven for any number i. From this, the maximum condition at point τ follows, since, by construction, the sequence $\{v_i^{k-1}\}_{i=1}^{\infty}$ is dense everywhere in the set $U_{k-1}(\tau)$.

Now we will prove the maximum condition to the right of the point t_k, on the interval $[t_k, t_{k+1}]$. If this interval is not the last one, then the proof is similar to the previous case using the conditions of transversality "at the junction" and at the right boundary point. If this interval is the last one, then only the transversality condition at the point t_m is used in the proof. Performing similar arguments to the right of the point t_k, we obtain that the Pontryagin function maximum condition is also satisfied on the interval $[t_k, t_{k+1}]$. The result obtained completes the proof of the theorem.

23.3 Conclusion

This work investigates an optimal control problem involving a switch of phase spaces, which arises in systems with several controlled objects operating sequentially in time. For such a problem, necessary optimality conditions are derived in the form of the Pontryagin maximum principle, supplemented by transversality conditions at the points where the phase spaces switch.

The main results of the work include:

- The formulation of adjoint systems on each time interval corresponding to the different phase spaces.
- Transversality conditions at the initial and final time points, as well as at the "junction" points between consecutive phases, which account for the smooth transition mappings between the spaces.
- A proof of the Pontryagin maximum principle condition on each control interval, based on the method of needle variations and the properties of measurable selectors of set-valued mappings.

The obtained results generalize the classical optimal control theory to the case of systems with a switch of phase spaces and can be applied to physical, technical, and economic models where processes sequentially change their nature or structure.

References

[1] Arutyunov AV (1997) Usloviya ekstremuma. Anormal'nye i vyrozhdennye zadachi (in Russ., Extremum conditions. Anomalous and degenerate problems). Factorial, Moscow

[2] Arutyunov AV (2000) Optimality Conditions. Anomalous and Degenerate Problems. Series Mathematics and Its Applications, Vol 526. Springer, Dordrecht et al, DOI 10.1007/978-94-015-9438-7

[3] Boltansky VG (1983) Zadacha optimizatsii so smenoy fazovogo prostranstva (in Russ., Optimization problem with phase space change). Differentsial'nyye Uravneniya XIX(3):518-521

[4] Medvedev VA, Rozova VN (1972) Optimal'noye upravleniye stupenchatymi sistemami (in Russ., Optimal control of step systems). Avtomatika i Telemekhanika (3):15-23

[5] Maximova I (2023) The problem of controllability with phase space change. Advances in Systems Science and Applications 23(1):61–68, DOI 10.25728/assa.2023.23.01.1364

[6] Maximova IS (2023) Local controllability in the problem with variable structure. 16th International Conference Management of large-scale system development (MLSD), pp 1-3, IEEE, DOI 10.1109/MLSD58227.2023.10303947

[7] Maximova IS (2023) Controllability of triangular systems with phase space change. In: Goldengorin B, Kuznetsov S (eds) Data Analysis and Optimization. Springer Optimization and Its Applications, Vol 202, pp 225–236, DOI 10.1007/978-3-031-31654-8_15

Chapter 24
Supersonic Divergence of a Panel with one Free Edge, Initially Loaded in two Directions: Compressed Along the Gas Flow and Stretched in the Perpendicular Direction

Stella R. Martirosyan

Abstract In the article, in a linear formulation, the problem of the static stability of a rectangular plate in a supersonic gas is investigated, initially loaded in two directions, respectively, compressed along the gas flow and stretched in the direction perpendicular to the gas flow velocity, under the assumption that the gas flow runs on its free edge. An analytical solution to the problem of static stability using the Euler method is obtained. The space of parameters of the «plate-flow» system was partitioned into domains of stability and static instability in the form of Eulerian panel divergence and in the form of localized divergence. It has been established that the boundary between instability subdomains, as a function of the parameter of the ratio of the sides of the rectangle, significantly depends on the stress coefficient of compressive forces, and depends vanishingly little on the coefficient of tensile forces and Poisson's ratio. The critical velocities of the Euler divergence of the panel and localized divergence are found. It is shown that at a certain ratio of compressive and tensile forces, the effect of their mutual compensation takes place.

Key words: Static stability, Rectangular plate, Supersonic gas flow, Combined load, Euler method, Instability, Eulerian panel divergence, Localized divergence, Analytical method

24.1 Introduction

Consideration of problems of aero elastic stability of plates and shells under combined loading has important applied and theoretical significance. Theoretical studies of these problems make it possible to identify various types of loss of stability of the unperturbed equilibrium state of the "plate-flow" system, accompanied by corre-

Stella R. Martirosyan
Institute of Mechanics, National Academy of Sciences of the Republic of Armenia, 24b Marshal Baghramyan Ave, Yerevan, 0019, Republic of Armenia, e-mail: mechinsstella@mail.ru

H. Altenbach and A. V. Sahakyan, *Contact Mechanics and Relevant Problems*,
Advanced Structured Materials 253, https://doi.org/10.1007/978-3-032-20869-9_24

sponding mechanical effects. They also allow us to assess the influence of combined loads on the stability threshold, for the purpose of subsequent analysis of the possibility of controlling it [1–4]. In this article, using the example of a model problem of a rectangular plate flowing around a supersonic gas flow with one free edge, initially compressed along the gas flow and stretched in the perpendicular direction, the influence of the initial stress state on the static stability of the unperturbed equilibrium state of the "plate-flow" system is studied in a linear formulation, when the flow approaches its free edge. An analytical solution to the problem of static stability using the Euler method is found [1, 2]. The space of system parameters is divided into the region of stability and the region of instability, consisting of two subregions: the Euler divergence of the panel and the localized divergence. The critical velocities of panel divergence and localized divergence are found. An accurate assessment is given of the influence of the initial combined loads on the threshold of static stability of the unperturbed equilibrium state of the system. The ratio of the initial compressive and tensile forces has been established, leading either to a decrease in the stability of the system, or to its increase, or to compensation for the action of the combined load on the system. The results of the work can be used in processing data from experimental studies of the loss of stability of skin panels of supersonic aircraft at the design stage and during operation, in order to improve flight safety.

24.2 Statement of the Problem

Consider a thin elastic rectangular plate occupying, in a Cartesian coordinate system $Oxyz$, a domain $0 \le x \le a$, $0 \le y \le b$, $-h \le z \le h$. The Cartesian coordinate system $Oxyz$ is chosen so that the Ox and Oy axes lie in the plane of the undisturbed plate, and the Oz axis is perpendicular to the plate and directed to the side of the supersonic gas flow streamlining it on one side in the direction of the Ox axis with an undisturbed velocity V. It is assumed that the gas flow is plane and potential. Suppose that the edge $x = 0$ of the plate is free, whereas the edges $x = a$, $y = 0$ and $y = b$ are hinged. The hinges are assumed consider to be perfect.

We also assume that the plate initially, even before the gas flow around it, is loaded in two directions, respectively, by compressive $N_x = 2h\sigma_x$ and tensile $N_y = 2h\sigma_y$ forces, which are the result of heating, or some other reason; compressive σ_x and tensile σ_y stresses are assumed to be uniformly distributed over the entire middle surface of the panel, to be constant and unchanged with a change in the deflection $w = w(x, y)$ of the plate [1, 2, 12, 13].

The deflection of the plate $w = w(x, y)$ will cause excessive pressure δp on the upper streamlined surface of the plate from the side of the flowing gas flow, which is taken into account by the approximate formula of the "piston theory" [5, 6]

$$\delta p = -a_0 \rho_0 V \frac{\partial w}{\partial x},$$

where a_0 is the sound velocity in an undisturbed gas medium, and ρ_0 is the density of the undisturbed gas flow. The deflections $w = w(x, y)$ are small as compared with the plate thickness $2h$.

Let us find out the conditions under which, along with the unperturbed form of equilibrium (unbent plate), a curved form of equilibrium (curved plate) is possible in the case in which the bending of the plate is caused by the corresponding aerodynamic loads δp as well as the compressive σ_x and tensile σ_y stresses in the middle surface of the plate. We assume that the compressive stresses σ_x are small as compared to the critical stresses $(\sigma_x)_{\text{cr}}$, that can produce "buckling" of the plate in the absence of the gas flow ($V = 0$) [12], and the tensilee stresses σ_y are small than $(\sigma_y)_{\text{pr}}$, where $(\sigma_y)_{\text{pr}}$ is the force, starting from which the phenomenon of loss of stability of the cylindrical shape of the plate occurs [9]. The values of $(\sigma_x)_{\text{cr}}$ and $(\sigma_y)_{\text{pr}}$ are within the elastic range. Then, under the assumption of the validity of the Kirchhoff hypothesis and the "piston theory" [5, 6], the differential equation of bending of a thin rectangular plate is described by the relation [1, 2]

$$D\Delta^2 w + N_x \frac{\partial^2 w}{\partial x^2} - N_y \frac{\partial^2 w}{\partial y^2} + a_0 \rho_0 V \frac{\partial w}{\partial x} = 0\,, \qquad w = w(x, y), \tag{24.1}$$

where $\Delta^2 w = \Delta(\Delta w)$, Δ is the Laplace's differential operator; D is the cylindrical bending stiffness.

The boundary conditions, under the accepted assumptions regarding the method of securing the edges of the plate, will be of the form [1, 2]

$$\frac{\partial^2 w}{\partial x^2} + \nu \frac{\partial^2 w}{\partial y^2} = 0, \quad D\frac{\partial}{\partial x}\left(\frac{\partial^2 w}{\partial x^2} + (2-\nu)\frac{\partial^2 w}{\partial y^2}\right) + N_x \frac{\partial w}{\partial x} = 0, \quad x = 0, \tag{24.2}$$

$$\begin{aligned} w = 0, \quad & \frac{\partial^2 w}{\partial x^2} = 0 \quad \text{for} \quad x = a, \\ w = 0, \quad & \frac{\partial^2 w}{\partial y^2} = 0 \quad \text{for} \quad y = 0 \quad \text{and} \quad y = b. \end{aligned} \tag{24.3}$$

Here ν is Poisson's ratio.

It is required to find the critical velocity V_{cr} or the lowest velocity value in the interval of supersonic and hypersonic velocities [1, 2]

$$V_{\text{cr}} \in (a_0 M_0,\ a_0 M_{2\text{cosm}}), \quad M_0 = \sqrt{2}, \quad M_{2\text{cosm}} \approx 33.85 \tag{24.4}$$

starting from which the "plate-flow" system loses its static stability at velocities $V \geq V_{\text{cr}}$ under the assumption

$$\sigma_x < (\sigma_x)_{\text{cr}}, \quad \sigma_y < (\sigma_y)_{\text{pr}}. \tag{24.5}$$

Here, M_0 and $M_{2\text{cosm}}$ are the Mach numbers corresponding to the boundary values of the interval of supersonic and hypersonic velocities [1, 2]; the $(\sigma_x)_{\text{cr}}$ is the critical stresses, leading to buckling of the plate in the absence of flow ($V = 0$) and tensile forces $(\sigma_y = 0)$ [12]; the $(\sigma_y)_{\text{pr}}$ is the force, starting from which the phenomenon of loss of stability of the cylindrical shape of the plate occurs. Thus, analysis of the stability of the "plate–flow" system (24.1) – (24.3) is reduced to the study of the differential equation (24.1) with the corresponding boundary conditions (24.2) and (24.3) for the deflection $w = w(x, y)$ under the assumption (24.4) and (24.5).

Let us denote by γ the ratio of the width a of the panel (downstream side) to its length b

$$\gamma = ab^{-1} \in (0, \infty) . \tag{24.6}$$

Obviously, the limiting value of $\gamma = 0$ corresponds to an infinitely elongated plate, and the value of $\gamma = \infty$ corresponds to a semi-infinite plate-strip.

24.3 General Solution of the Stability Problem

To obtain a sufficient sign of loss of stability of the unperturbed equilibrium state of the «plate-flow» system, we will look for a general solution of the differential equation (24.1) that satisfies the boundary conditions (24.2) and (24.3) in the form of harmonic vibrations [1, 2, 12]

$$w(x, y) = \sum_{n=1}^{\infty} C_n \cdot \exp(\mu_n r x) \cdot \sin(\mu_n y), \quad \mu_n = \pi n b^{-1}, \quad \gamma \in (0, \infty) , \tag{24.7}$$

where C_n are the arbitraries constants:

$$\sum_{n=1}^{\infty} C_n^2 \neq 0,$$

n is the number of half-waves along the side b of the plate.

Substituting the solution (24.7) into the differential equation (24.1), we obtain the characteristic equation, which is an algebraic equation of the fourth degree

$$r^4 - 2 \cdot (1 - \beta_x^2) r^2 + \alpha_n^3 \cdot r + (1 + \beta_y^2) = 0, \tag{24.8}$$

$$\begin{aligned} &\alpha_n^3 = a_0 \rho_0 V D^{-1} \mu_n^{-3}, \quad \beta_x^2 = 1/2 \cdot N_{\,x} D^{-1} \mu_n^{-2} = h \sigma_x D^{-1} \mu_n^{-2}, \\ &\beta_y^2 = N_y D^{-1} \mu_n^{-2} = 2h \sigma_y D^{-1} \mu_n^{-2}. \end{aligned} \tag{24.9}$$

In accordance with constraint (24.4) and condition (24.5), we have

$$\alpha_n^3 \in a_0^2 \rho_0 D^{-1} \mu_n^{-3} \left(M_0, \; M_{2\text{cosm}}\right) , \tag{24.10}$$

$$\begin{aligned} &\beta_x^2 = \beta_x^2 \,(n, \gamma, \nu) < \left(\beta_x^2\right)_{\text{cr}} = h(\sigma_x)_{\text{cr}} D^{-1} \mu_n^{-2}, \\ &\beta_y^2 = \beta_y^2 \,(n, \gamma, \nu) < \left(\beta_y^2\right)_{\text{pr}} = 2h(\sigma_y)_{\text{pr}} D^{-1} \mu_n^{-2}. \end{aligned} \tag{24.11}$$

It is clear, that the parameter α_n^3 of the gas flow velocity V characterizes the non–conservative component of the load, and, accordingly, the stress coefficients β_x^2 and β_y^2 characterize the conservative component of the load. In article [12], problem (24.1) – (24.3) was studied in detail under conditions $\beta_x^2 \geq 0$ and $\beta_y^2 = 0$ in the absence of flow around ($V = 0$). The critical values $(\beta_x^2)_{\text{cr}}$ of the stress coefficient β_x^2 of compressive forces σ_x were found (Table 24.1).

From the data in Table 24.1 it follows that in the absence of flow around ($V = 0$) when $\beta_x^2 \geq (\beta_x^2)_{\text{cr}}$, the unperturbed equilibrium state of a wide plate, starting from value $\gamma = \gamma_* \approx 2.9$, loses statistical stability in the form of localized instability in the vicinity of the free edge $x = 0$, in contrast to long plates ($\gamma < \gamma_* \approx 2.9$), the unperturbed equilibrium state of which loses static stability in the form of static instability panels. It is obvious that an infinitely elongated plate ($\gamma = 0$) in the absence of flow ($V = 0$) is unstable for all $\beta_x^2 \geq 0$. In this case we can assume that $(\beta_x^2)_{\text{cr}} = 0$ at all n and ν. According to the algorithm of the Ferrari method, characteristic equation (24.8) can be represented as the multiplication of two square trinomials

$$\begin{aligned}&\left(r^2 + \sqrt{2(q + 1 - \beta_x^2)}r + q - \sqrt{q^2 - 1 - \beta_y^2}\right)\\ &\times \left(r^2 - \sqrt{2(q + 1 - \beta_x^2)}r + q + \sqrt{q^2 - 1 - \beta_y^2}\right) = 0,\end{aligned} \tag{24.12}$$

where $q = q(V) \in R$ is a gas flow velocity parameter, which is the only one real root of the cubic equation

Table 24.1: The values of function $(\beta_x^2)_{cr} = (\beta_x^2)_{cr}(n, \gamma, \nu)$ at $n = 1$.

γ \ ν	0.125	0.25	0.3	0.375	0.5
≤ 0.01	0.8752	0.7501	0.7001	0.6251	0.5000
0.10	0.8911	0.7654	0.7149	0.6391	0.5122
0.15	0.9112	0.7845	0.7332	0.6564	0.5273
0.20	0.9392	0.8108	0.7589	0.6804	0.5480
0.30	1.0178	0.8839	0.8293	0.7459	0.6040
0.50	1.2447	1.0869	1.0215	0.6391	0.5122
0.80	1.4541	1.2744	1.1993	1.0835	0.8828
1.00	1.4039	1.2497	1.1829	1.0774	0.8887
1.50	1.3607	1.2152	1.1524	1.0534	0.8750
1.80	1.3670	1.2180	1.1543	1.0541	0.8747
1.96	1.3671	1.2187	1.1550	1.0547	0.8749
≥ 2.90	1.3672	1.2188	1.1550	1.0547	0.8750

$$8(q + 1 - \beta_x^2)(q^2 - 1 - \beta_y^2) - \alpha_n^6 = 0, \tag{24.13}$$

when

$$q = q\,(V) \in (q_0,\, q\,(a_0 M_{2\text{cosm}})) \subset R$$

due to the obvious limitation (24.4). It is easy to show that the determinant of the cubic equation (24.13) is positive, which implies the uniqueness of its real root.

Using graphical-analytical research methods, it can be shown that the characteristic equation (24.8) in the form (24.12) for all $\gamma \in [0, \infty)$ and $\gamma = \infty$ has two negative roots $r_{1,2}$ and a pair complex conjugate roots $r_{3,4}$ with a positive real part in the interval of values

$$\begin{aligned} &q \in (q_0,\, q\,(M_{2\text{cosm}})), \quad q_0 = \frac{(\beta_x^2 - 1) + 2\sqrt{(\beta_x^2 - 1)^2 + 3(1 + \beta_y^2)}}{3}, \\ &\beta_x^2 < \left(\beta_x^2\right)_{\text{cr}}, \qquad \beta_y^2 < \left(\beta_y^2\right)_{\text{pr}}. \end{aligned} \tag{24.14}$$

The expressions of roots r_i are easily determined from the solution of the corresponding quadratic equations (24.12)

$$\begin{aligned} &r_{1,2} = -0.5\sqrt{2(q + 1 - \beta_x^2)} \pm \sqrt{\sqrt{q^2 - 1 - \beta_y^2} - 0.5(q - 1 + \beta_x^2)}, \\ &r_1 < 0,\ r_2 < 0, \end{aligned} \tag{24.15}$$

$$r_{3,4} = 0.5\sqrt{2(q + 1 - \beta_x^2)} \pm i\sqrt{\sqrt{q^2 - 1 - \beta_y^2} + 0.5(q - 1 + \beta_x^2)}. \tag{24.16}$$

Then, in accordance with expressions (24.9), the general solution (24.7) of Eq. (24.1) will be written in the form of a double series

$$w(x, y) = \sum_{n=1}^{\infty}\sum_{k=1}^{4} C_{nk} \exp(\mu_n r_k x) \sin(\mu_n y) \quad \text{for all} \quad \gamma \in [0, \infty) \tag{24.17}$$

and in the case of a semi-infinite plate-strip one gets

$$w(x, y) = \sum_{n=1}^{\infty}\sum_{k=1}^{2} C_{nk} \exp(\mu_n r_k x) \sin(\mu_n y), \quad \gamma = \infty \tag{24.18}$$

due to the condition of vibration damping at the edge $x = a$ $(a = \infty)$ [7, 8, 10, 12].

Substituting expression (24.10) into the cubic equation (24.13), after simple transformations we obtain the formula for the dependence of the gas flow velocity V on the "essential" parameters of the "plate-flow" system

$$\begin{aligned} &V\left(q, n, \gamma, \nu, \beta_x^2, \beta_y^2\right) \\ &= 2\sqrt{2(q + 1 - \beta_x^2)(q^2 - 1 - \beta_y^2)\pi^3 n^3 \gamma^3 D (a_0 \rho_0 a^3)^{-1}}, \quad \gamma \in [0, \infty) \end{aligned} \tag{24.19}$$

$$V\left(q, n, \nu, \beta_x^2, \beta_y^2\right) = \\ = 2\sqrt{2(q+1-\beta_x^2)(q^2-1-\beta_y^2)\cdot\pi^3 n^3 D(a_0\rho_0 b^3)^{-1}}, \quad \gamma = \infty \tag{24.20}$$

allowing, from a known parameter value $q \in (q_0, q\,(a_0 M_{2cosm}))$, to determine the gas flow velocity $V\,(q)$, or its reduced value: $V\,(q)\cdot D^{-1}(a_0\rho_0 a^3)$ at $\gamma \in [0, \infty)$, or $V\,(q)\cdot D^{-1}(a_0\rho_0 b^3)$ at $\gamma = \infty$. In this case, holds for all "essential" parameters of the stability problem (24.1) – (24.4), as applied to the interval of supersonic velocities (24.4), it is obvious that

$$V(q) \in (V(q_0),\ a_0 M_{2cosm}) \subseteq (a_0 M_0,\ a_0 M_{2cosm}), \gamma \in (0, \infty)\,. \tag{24.21}$$

According to the formula of cylindrical stiffness $D = E(2h)^3/(12\,(1-\nu^2))$, this follows [12]

$$V(q)D^{-1}(a_0\rho_0 a^3) \in (V(q_0)D^{-1}(a_0\rho_0 a^3),\ a_0 M_{2cosm}\Psi_1) \subseteq (a_0 M_0,\ a_0 M_{2cosm})\,\Psi_1,$$
$$V(q)D^{-1}(a_0\rho_0 b^3) \in (V(q_0)D^{-1}(a_0\rho_0 b^3),\ a_0 M_{2cosm}\Psi_2) \subseteq (a_0 M_0,\ a_0 M_{2cosm})\,\Psi_2, \tag{24.22}$$

where

$$\Psi_1 = 12(1-\nu^2)a_0\rho_0 E^{-1}(2ha^{-1})^{-3} \quad \text{for all} \quad \gamma \in (0, \infty)\,,$$
$$\Psi_2 = 12(1-\nu^2)a_0\rho_0 E^{-1}(2hb^{-1})^{-3} \quad \text{when} \quad \gamma = \infty.$$

Substituting the values of Poisson's ratio ν and the relative thickness $2ha^{-1}$ and $2hb^{-1} \in [0.006,\ 0.015]$ of the plate into expressions (24.22), we obtain the intervals $d\left(\nu, 2ha^{-1}\right) = (a_0 M_0,\ a_0 M_{2cosm})\Psi_1$ and $d\left(\nu, 2hb^{-1}\right) = (a_0 M_0,\ a_0 M_{2cosm})\Psi_2$ of permissible values of the reduced flow velocity $VD^{-1}\left(a_0\rho_0 a^3\right)$ and $VD^{-1}\left(a_0\rho_0 b^3\right)$ presented for steel plates in Table 24.2.

Table 24.2: Intervals of permissible values of the reduced flow velocity.

	ν				
$2ha^{-1}, 2hb^{-1}$	0.125	0.25	0.3	0.375	0.5
0.006	(54.8,1311.7)	(52.0,1245.2)	(50.5,1209.0)	(47.7,1141.6)	(41.6,996.3)
0.007	(34.5, 811.1)	(32.7, 769.4)	(32.0, 753.4)	(30.0, 705.3)	(26.2,615.5)
0.008	(23.1, 544.3)	(22.0, 516.4)	(21.5, 505.6)	(20.1, 473.3)	(17.6,413.1)
0.009	(16.2, 381.8)	(15.4, 362.1)	(15.1, 354.6)	(14.1, 332.0)	(12.3, 289.7)
0.010	(11.8, 283.5)	(11.2, 269.1)	(10.9, 261.3)	(10.3, 246.7)	(9.0, 215.3)
0.011	(8.9, 209.4)	(8.4, 198.7)	(8.1, 190.4)	(7.74, 182.1)	(6.75,158.9)
0.012	(6.85, 164.0)	(6.5, 155.7)	(6.3, 151.2)	(5.96, 142.7)	(5.2, 124.6)
0.013	(5.39, 126.9)	(5.11,120.34)	(5.0,117.84)	(4.69,110.32)	(4.09, 96.3)
0.014	(4.3, 101.5)	(4.09, 96.24)	(4.0, 94.24)	(3.75, 88.22)	(3.27, 77.0)
0.015	(3.5, 84.04)	(3.33, 79.73)	(3.23,77.33)	(3.05, 73.10)	(2.67, 63.81)

24.4 Sufficient Sign of Loss of Static Stability of the System "plate–flow" for $\gamma \in [0, \infty]$. Characteristic Determinants

24.4.1 Dispersion Equation for the Rectangular Plates ($\gamma \in (0, \infty)$)

Substituting the general solution (24.17) of the differential equation (24.1) into the boundary conditions (24.2) and (24.3), we obtain a homogeneous system of fourth–order algebraic equations with respect to constants C_{nk}, $k = \overline{1, 4}$. The determinant of this system equated to zero leads to the dispersion equation of the panel divergence for all $\gamma \in (0, \infty)$ in the form

$$\begin{aligned}
F_{\text{div pan}} &= F(q, n, \gamma, \nu, \beta_x^2, \beta_y^2) \\
&= A_1 B_1 B_2 \left\{(q+1-A_2)^2 - 2(q+1)\nu - (1-\nu)^2 - 2\beta_x^2 (q - A_2) \right. \\
&\left. - \left[(q+1+A_2)^2 - 2(q+1)\nu - (1-\nu)^2 - 2\beta_x^2 (q + A_2)\right] \exp(-2A_1 \pi n \gamma)\right\} \\
&+ 2\left\{\Delta_1 \sinh(\pi n \gamma B_1) + A_1^3 A_2 B_1 \cosh(\pi n \gamma B_1)\right\} B_2 \cos(\pi n \gamma B_2) \exp(-A_1 \pi n \gamma) \\
&+ 2\left\{-B_1 \Delta_2 \cosh(\pi n \gamma B_1) - A_1 A_2 \left(2A_2^2 + q A_1^2 - \frac{A_1^4}{4}\right) \sinh(\pi n \gamma B_1)\right\} \\
&\times \sin(\pi n \gamma B_2) \exp(-A_1 \pi n \gamma) = 0.
\end{aligned} \tag{24.23}$$

Here

$$\begin{aligned}
A_1 &= \sqrt{2(q+1-\beta_x^2)}, \qquad A_2 = \sqrt{q^2 - 1 - \beta_y^2}, \\
B_1 &= \sqrt{A_2 - 0.5(q - 1 + \beta_x^2)}, \quad B_2 = \sqrt{A_2 + 0.5(q - 1 + \beta_x^2)}, \\
\Delta_1 &= \frac{1}{2}\left\{2A_2(A_2 - q)^2 + A_1^4 q + A_1^2 (A_2 - q)(A_2 + 3q)\right\} \\
&\quad - (2A_2(A_2 - q) + q A_1^2)\nu + \frac{1}{2}(A_1 + 2A_2)\nu^2, \\
\Delta_2 &= A_2(A_2 + q)^2 - \frac{q}{2} A_1^4 - \frac{1}{2} A_1^2 (A_2 - 3q)(A_2 + 2q) \\
&\quad + 2\left(A_2^2 + q A_2 + \frac{q}{2} A_1^2\right)\nu - \left(\frac{A_1^2}{2} - A_2\right)\nu^2.
\end{aligned}$$

The hypersurface (24.23) with a value of $q = q_{\text{cr div}}$, where $q_{\text{cr div}}$ is the real solution to Eq. (24.23), delimits the domain of stability $\Im_0$ $\left(F_{\text{div pan}} > 0\right)$ and the domain of Eulerian divergence instability $\Im_{11}$ $\left(F_{\text{div pan}} < 0\right)$.

24.4.2 Dispersion Equation of the Localized Divergence for Semi-Infinite Plate-Strip ($\gamma = \infty$)

Substituting the general solution (24.18) of the differential equation (24.1) into the boundary conditions (24.2) and (24.3), we obtain a homogeneous system of second-order algebraic equations with respect to constants C_{nk}, $k = \overline{1,2}$. The determinant of this system equated to zero leads to the dispersion equation of the localized divergence in the form:

$$F_{\text{loc div}} = F(q, \nu, \beta_x^2, \beta_y^2) = \left(q + 1 - \sqrt{q^2 - 1 - \beta_y^2}\right)^2$$
$$- 2(q+1)\nu - (1-\nu)^2 - 2\beta_x^2\left(q - \sqrt{q^2 - 1 - \beta_y^2}\right) = 0, \tag{24.24}$$

where $\beta_x^2 < (\beta_x^2)_{\text{cr}}$(table 1) and $\beta_y^2 < \left(\beta_y^2\right)_{\text{pr}}$.

Likewise, the hypersurface (24.24) with a value of $q = q_{\text{loc div}}$, where $q_{\text{loc div}}$ is the only real solution to the equation (24.24), delimits the domain of stability $\Im_0(F_{\text{loc div}} > 0)$ and the domain of localized divergence $\Im_{12}(F_{\text{loc div}} < 0)$ at $\gamma = \infty$. Using analytical research methods, it is easy to show that for all n, $\beta_x^2 < (\beta_x^2)_{\text{cr}}$ and β_y^2 it is true

$$\lim_{\gamma\to\infty}\left(F_{\text{div pan}} = F(q, n, \gamma, \nu, \beta_x^2, \beta_y^2) = 0\right) = \left(F_{\text{loc div}}\left(q, \nu, \beta_x^2, \beta_y^2\right) = 0\right). \tag{24.25}$$

Numerical research methods have shown the fulfillment of condition (24.25) not only in the case of a semi-infinite plate-strip($\gamma = \infty$), but also in the case of sufficiently wide plates, starting from the value $\gamma = \gamma_{\text{gr}}{=}\gamma_{\text{gr}}\left(\nu, \beta_x^2, \beta_y^2\right)$ (Table 24.3). The boundary value $\gamma_{\text{gr}} = \gamma_{\text{gr}}\left(\nu, \beta_x^2, \beta_y^2\right)$ delimits the domain of static instability $\Im_1$ into the subdomain $\Im_{11}$ of the Eulerian panel divergence and into the subdomain $\Im_{12}$ of localized divergence. Substituting the values $q_{\text{cr div}}$ and $q_{\text{loc div}}$ in expressions (24.19) and (24.20) we obtain the critical panel divergence velocity $V_{\text{cr div}}$ and localized divergence velocity $V_{\text{loc div}}$ respectively. At gas flow velocities of $V \geq V_{\text{cr div}}$ the unperturbed equilibrium state of the "plate-flow" system in the case of rectangular plates when $\gamma \in (0, \gamma_{\text{gr}})$, loses its static stability in the form of Eulerian divergence of the panel; and in the case of sufficiently wide plates $\gamma \in [\gamma_{\text{gr}}, \infty]$ at gas flow velocities of $V \geq V_{\text{loc div}}$ – in the form of localized divergence in the vicinity of the

Table 24.3: Case of sufficiently wide plates.

β_x^2	0	0.1	0.2	≥ 0.3
γ_{gr}	1.96	2.4	2.8	2.9

free edge $x = 0$ of the plate. In both cases, the transitions from the domain of stability $\mathfrak{I}_0$ to the domain of instability $\mathfrak{I}_1$ are "soft".

Accordingly, during the Eulerian panel divergence, the additional stresses arising during the flow lead to the «buckling» of the entire surface of the plate at a limited velocity. In the case of localized divergence, additional stresses lead to "buckling" of only a narrow strip of the plate surface along its free edge $x = 0$. Due to the limited velocity of change in the surface of the plate, these deformations are considered static. In accordance with expressions (24.19) and (24.20), by virtue of condition (24.4), it is obvious that

$$V_{\text{cr div}} = V\left(q_{\text{cr div}}, n, \gamma, \beta_x^2, \beta_y^2\right) \in \left(V\left(q_0, n, \gamma, \beta_x^2, \beta_y^2\right), a_0 M_{2\text{cosm}}\right) \subsetneq \left(a_0 M_0, a_0 M_{2\text{cosm}}\right),$$

$$V_{\text{loc div}} = V\left(q_{\text{loc div}}, n, \beta_x^2, \beta_y^2\right) \in \left(V\left(q_0, n, \beta_x^2, \beta_y^2\right), a_0 M_{2\text{cosm}}\right) \subsetneq \left(a_0 M_0, a_0 M_{2\text{cosm}}\right).$$

24.4.3 *Dispersion Equation for an Infinite Elongated Plate* ($\gamma = 0$)

Tending $b \to \infty$, the differential equation (24.1) and boundary conditions (24.2), (24.3) of the stability problem, take the form

$$D\frac{d^4 w}{d\xi^4} + N_x \frac{d^2 w}{d\xi^2} + a_0 \rho_0 V \frac{dw}{d\xi} = 0, \quad \xi = x a^{-1}, \tag{24.26}$$

$$\frac{d^2 w}{d\xi^2} = 0, \quad D\frac{d^3 w}{d\xi^3} + N_x \frac{dw}{d\xi} = 0, \quad \text{for} \quad \xi = 0,$$
$$w = 0, \quad \frac{d^2 w}{d\xi^2} = 0, \quad \text{for} \quad \xi = 1. \tag{24.27}$$

Hence, it is obvious that in the case of an infinitely elongated panel($b \to \infty$), the description of the stability problem (24.1) – (24.3) for $N_x \neq 0$, $N_y \neq 0$ and $N_x \neq 0$, $N_y = 0$ is the same. The stability problem (24.1) – (24.3) for $N_x \neq 0$, $N_y = 0$ is studied in detail in article [12].

That is why, according to the results of article [12], the dispersion equation in the case of an infinitely elongated plate ($b \to \infty$) has the following description

$$F_\infty(\tilde{q}, \tilde{\beta}_x^2) = \left\{-\exp\left(-1.5\sqrt{2(\tilde{q} - \tilde{\beta}_x^2)}\right) + \cos\sqrt{0.5(3\tilde{q} + \tilde{\beta}_x^2)}\right\}$$
$$\times \sqrt{(3\tilde{q} + \tilde{\beta}_x^2)(\tilde{q} - \tilde{\beta}_x^2)^3} - (3\tilde{q}^2 - \tilde{\beta}_x^4)\sin(\sqrt{0,5(3\tilde{q} + \tilde{\beta}_x^2)}) = 0, \quad \gamma = 0. \tag{24.28}$$

Here $\tilde{q} \in \left(\tilde{q}_0, \tilde{q}\left(a_0 M_{2\text{cosm}}\right)\right)$ is the only real root of the cubic equation

$$
\begin{aligned}
&8\tilde{q}^2\left(\tilde{q}-\tilde{\beta}_x^2\right)-S^6=0, \qquad \tilde{\beta}_x^2=h\sigma_x D^{-1}a^2,\\
&S^3=a_0\rho_0 VD^{-1}a^3, \quad \text{when} \quad \tilde{q}_0>\tilde{\beta}_x^2.
\end{aligned}
\tag{24.29}
$$

It obviously follows from this that

$$
\begin{aligned}
&V\left(\tilde{q},\tilde{\beta}_x^2\right)=2\sqrt{2(\tilde{q}-\tilde{\beta}_x^2)}\cdot\tilde{q}D\left(a_0\rho_0 a^3\right)^{-1},\\
&\tilde{q}\in\left(\tilde{q}_0,\tilde{q}(a_0M_{2\mathrm{cosm}})\right), \quad \tilde{\beta}_x^2=h\sigma_x D^{-1}a^2.
\end{aligned}
\tag{24.30}
$$

As applied to the interval of supersonic velocities (24.4), we obtain

$$
\tilde{q}\in\left(\tilde{q}_0,\tilde{q}(a_0M_{2\mathrm{cosm}})\right)\subseteq\left(\tilde{q}(a_0M_0),\tilde{q}(a_0M_{2\mathrm{cosm}})\right).
\tag{24.31}
$$

In this case, also, the hypersurface (24.28) is the boundary of the stability domain $\Im_0$ $\left(F_\infty(\tilde{q},\tilde{\beta}_x^2)>0\right)$ and the instability domain $\Im_1$ $\left(F_\infty(\tilde{q},\tilde{\beta}_x^2)<0\right)$ in the form of the Eulerian panel divergence $\Im_{11}$. Substituting the first real root $\tilde{q}^{(1)}_{\mathrm{cr\,div}}=\tilde{q}_{\mathrm{cr\,div}}\left(\tilde{\beta}_x^2\right)$ of equation (24.28) into formula (24.30), we obtain the reduced critical panel divergence velocity $\tilde{V}_{\mathrm{cr\,div}}D^{-1}\left(a_0\rho_0 a^3\right)$ of an infinitely elongated plate ($\gamma=0$).

24.4.4 Dispersion Equation for the Sufficiently Elongated Plates

As is known [11, 12], for sufficiently elongated plates ($\gamma\ll 1$) we have

$$
q\in\left(q_0,q\left(a_0M_{2\mathrm{cosm}}\right)\right), q_0\gg 1.
\tag{24.32}
$$

In this case, it is true

$$
\begin{aligned}
&\lim_{\gamma\to\gamma_{**}}\left(F_{\mathrm{div\,pan}}=F(q,n,\gamma,\nu,\beta_x^2,\beta_y^2)=0\right)\\
&=\left(F_{\mathrm{elong}}\left(q,n,\gamma_{**},\beta_x^2\right)=0\right), \quad \gamma_{**}\approx 10^{-2}
\end{aligned}
\tag{24.33}
$$

for all n, $\beta_x^2<\left(\beta_x^2\right)_{\mathrm{cr}}$, $\beta_y^2<\left(\beta_y^2\right)_{\mathrm{pr}}$;

$$
\begin{aligned}
&F_{\mathrm{elong}}(q,n,\gamma,\beta_x^2)=\left\{-\exp\left(-1.5\sqrt{2(q-\beta_x^2)}\pi n\gamma\right)\right.\\
&\left.+\cos\left(\sqrt{0.5(3q+\beta_x^2)}\pi n\gamma\right)\right\}\sqrt{(3q+\beta_x^2)(q-\beta_x^2)}^{\,3}\\
&-(3q^2-\beta_x^4)\sin\left(\sqrt{0,5(3q+\beta_x^2)}\pi n\gamma\right)=0, \quad \gamma\le\gamma_{**}\approx 10^{-2}.
\end{aligned}
\tag{24.34}
$$

The hypersurface (24.34) with the value $q=q_{\mathrm{div\,elong}}$ delimits the domain of stability $\Im_0$ $\left(F_{\mathrm{elong}}>0\right)$ and the subdomain of Eulerian panel divergence

$\Im_{11}\left(F_{\text{elong}} < 0\right)$. In accordance with condition (24.28), expression (24.19) for all $\gamma \le \gamma_{**} \approx 10^{-2}$ can be written in the form

$$V\left(q, n, \gamma, \beta_x^2\right) = 2\sqrt{2(q - \beta_x^2) q \pi^3 n^3 \gamma^3 D (a_0 \rho_0 a^3)^{-1}}. \quad (24.35)$$

Substituting the first real root $q_{\text{div elong}} = q_{\text{div elong}}\left(n, \gamma, \beta_x^2\right)$ of equation (24.34) into formula (24.35), we obtain the reduced critical panel divergence velocity $V_{\text{cr div}} D^{-1}\left(a_0 \rho_0 a^3\right)$ of a sufficiently elongated plate ($\gamma \le \gamma_{**} \approx 10^{-2}$).

Taking into account condition (24.4), it follows that

$$V\left(q, n, \gamma, \beta_x^2\right) \in \left(V\left(q_0, n, \gamma, \beta_x^2\right), a_0 M_{2\text{cosm}}\right) \subseteq \left(a_0 M_0, a_0 M_{2\text{cosm}}\right),$$
$$\gamma \le \gamma_{**} \approx 10^{-2}.$$

In accordance with Eqs. (24.34) and (24.35), we can say that the reduced critical divergence velocity $V D^{-1}\left(a_0 \rho_0 a^3\right)$ of a sufficiently elongated panel ($\gamma \le \gamma_{**} \approx 10^{-2}$) does not depend on the stress coefficient β_y^2 of tensile forces and Poisson's ratio ν, up to an accuracy of the order of a high degree of smallness, as in the case infinitely elongated panel ($\gamma = 0$). Note that the boundary value $\gamma_{**} \approx 10^{-2}$, obtained using analytical methods for analyzing estimates, was also confirmed by numerical methods of analysis.

In addition, we note that the negativity of the roots r_1 and r_2, defined by Eq. (24.15), which is a necessary condition for the loss of static stability of the unperturbed equilibrium state of the "plate-flow" system in the form of localized divergence in the vicinity of edge $x = 0$, determines it in the case of wide plates ($\gamma \ge \gamma_{\text{gr}}$), like a semi-infinite plate-strip ($\gamma = \infty$). Unlike rather wide plates ($\gamma \ge \gamma_{\text{gr}}$), the system in the case rather streamlined elongated plates and plates of moderate dimensions ($\gamma < \gamma_{\text{gr}}$) loses their static stability in the form of the Eulerian panel divergence [11, 12].

24.5 Numerical Results

Using the numerical analysis of dispersion equations (24.23), (24.24), (24.28) and (24.34) families of curves $\left\{q(n, \gamma, \nu, \beta_x^2, \beta_y^2)\right\}$ and $\left\{\tilde{q}(\tilde{\beta}_x^2)\right\}$ were constructed, appropriately parameterized: the most representative of these families of curves were selected and investigated. It has been established that the flow leads to a «decrease» in the critical coefficient β_x^2 of compressive forces σ_x by approximately 1.25 – 2 times.

The static instability domain was divided into the subdomain of panel divergence and localized divergence. The values of function $\gamma_{\text{gr}} = \gamma_{\text{gr}}\left(\nu, \beta_x^2, \beta_y^2\right)$ (Table 24.3), delimiting these domains, are found. As it turned out, γ_{gr} is a increasing function of the stress coefficient of compressive forces β_x^2: as β_x^2 increases from 0 to 0.3, it increases by about 1.48 times. And vanishingly little depends on the stress coefficient of tensile forces β_y^2 and Poisson's ratio ν. This means that with an increase of β_x^2, the domain of panel divergence increases, and the domain of localized divergence, on the contrary, narrows, which indicates a decrease in the stability of the system.

The critical velocities of the panel divergence $V_{\mathrm{cr\,div}}$ and localized divergence $V_{\mathrm{loc\,div}}$ for all values of the parameters of the system (24.1) – (24.4) are found. With $n = 1$ and all $\gamma \in (0, \infty)$, critical velocities $V_{\mathrm{cr\,div}}$ and $V_{\mathrm{loc\,div}}$ take the lowest value with fixed values of the other parameters. And also, they are smaller in plates made of materials with a high Poisson's ratio ν. The critical velocities $V_{\mathrm{cr\,div}}$ and $V_{\mathrm{loc\,div}}$ are a decreasing function of β_x^2 and increasing of β_y^2.

It is shown that elongated plates ($\gamma \in (0, 0.33)$), like an infinitely elongated plate ($\gamma = 0$), of relative thickness $2ha^{-1} \in [0.006, 0.015]$, being unstable near $a_0\sqrt{2}$ for all ν, β_x^2, β_y^2, which is the beginning of the interval of supersonic velocities (24.4), since we have $V_{crdiv} < a_0 M_0 = a_0\sqrt{2}$ [1]. However, at velocities $V \geq V_0$ greater $a_0\sqrt{2}$ than approximately 1.5–7 times depending on γ and β_x^2, the system becomes stable. And at gas flow velocities $V \geq V_{\mathrm{cr\,div}}$ exceeding V_0 by 5 times or more, the system at $2ha^{-1} < 0.009$ loses stability in the form of an Eulerian divergence of the panel. Here $V_{\mathrm{cr\,div}}$ is a slowly decreasing function of β_x^2 and depends vanishingly little on ν and β_y^2 for $\gamma \in (0, 0.33)$. V_{crdiv} does not depend from them, when $\gamma = 0$. Thus, in the case of elongated plates ($\gamma \in (0, 0.33)$), the effect of stabilization of tensile forces σ_y on the stability of the system (24.1) - (24.4) is vanishingly small in comparison with the effect of destabilization of compressive forces σ_x, which is quite consistent with the results of [12].

Tables 24.4 and 24.5 show the values of $V_{\mathrm{cr\,div}} D^{-1} \left(a_0 \rho_0 a^3\right)$ for $\gamma = 0.5$ (Table 24.4), $\gamma = 1$ (Table 24.5) in the case of plates of and the values of $V_{locdiv} D^{-1} \left(a_0 \rho_0 b^3\right)$ (Table 24.6) at $n = 1$ and $\nu = 0.3$. The critical panel divergence velocity $V_{\mathrm{cr\,div}}$ decreases by about 1.5 – 2.2 times with growth $\beta_x^2 \in [0, 0.6)$ and increases by about 2 – 3.5 times with grown $\beta_y^2 \in [0, 3)$ at $\gamma \in \left[0.33, \gamma_{\mathrm{gr}}\right)$ (Table 24.3-24.5). And in the case of sufficiently wide plates $\gamma \in \left[\gamma_{\mathrm{gr}}, \infty\right]$ the critical velocity of localized divergence $V_{\mathrm{loc\,div}}$ decreases by about 1.5 – 2 times with growth$\beta_x^2 \in [0, 0.6)$ and increases by about 2.3 – 2.8 times with grown $\beta_y^2 \in [0, 3)$ (Table 24.3 and 24.6).

Table 24.4: The values $V_{\mathrm{cr\,div}} D^{-1} \left(a_0 \rho_0 a^3\right)$ for $\gamma = 0.5$ and $\nu = 0, 3$.

β_y^2 \ β_x^2	0	0.1	0.2	0.3	0.4	0.5
0.0	11.706	10.459	9.235	8.027	6.843	5.689
0.1	12.128	10.947	9.696	8.461	7.268	6.110
0.5	14.087	12.785	11.510	10.221	8.985	7.758
1.0	16.539	15.151	13.792	12.461	11.152	9.861
1.5	19.043	17.586	16.163	14.770	13.406	12.804
2.0	21.624	20.033	18.515	17.191	15.658	14.234

Table 24.5: The values $V_{\text{cr div}} D^{-1} \left(a_0 \rho_0 a^3\right)$ for $\gamma = 1$ and $\nu = 0, 3$.

β_y^2 \ β_x^2	0	0.1	0.2	0.3	0.4	0.5
0.0	128.476	116.010	104.291	92.395	80.837	69.684
0.1	141.755	128.811	115.285	101.481	89.726	77.364
0.5	207.310	188.417	167.878	148.427	129.062	111.565
1.0	295.415	269.557	241.160	213.117	185.341	160.108
1.5	353.216	326.682	297.972	268.354	239.143	207.683
2.0	398.234	370.577	341.947	312.105	280.542	247.747

Table 24.6: The values $V_{\text{loc div}} D^{-1} \left(a_0 \rho_0 b^3\right)$ for $\nu = 0, 3$.

β_y^2 \ β_x^2	0	0.1	0.2	0.3	0.4	0.5
0.0	143.922	130.123	116.882	103.748	90.442	77.002
0.1	154.717	140.584	126.453	112.163	98.153	84.355
0.5	196.570	179.543	162.591	145.639	129.062	112.042
1.0	245.529	225.670	206.007	185.905	165.935	146.404
1.5	293.345	270.368	247.364	225.579	202.433	179.902.
2.0	338.214	313.963	288.525	263.002	237.828	213.848

At the same time, from a comparison of the data in Table 24.2 and the values of $V_{\text{cr div}} D^{-1} \left(a_0 \rho_0 a^3\right)$ and $V_{\text{loc div}} D^{-1} \left(a_0 \rho_0 b^3\right)$ (Tables 24.4–24.6), the unperturbed form of equilibrium of the system in the interval $\gamma \in [0.84, \infty)$ is stable near $a_0\sqrt{2}$ for plates of relative thickness $2ha^{-1} \in [0.006, 0.015]$ for all values of the remaining parameters. Moreover, the unperturbed form of equilibrium of the system in the case of thick plates $2ha^{-1} \geq 0.01$ for $\gamma \in [0.84, \infty)$ is stable over the entire range of supersonic velocities.

At a certain ratio the initial of compressive forces σ_x and tensile forces σ_y in the middle surface of the plate, the effect of their «mutual compensation» is observed, most clearly manifested in systems at $\gamma \in [0.33, \infty]$ (Table 24.7). In this formulation, contrary to expectations, the initial tensile forces, directed perpendicular to the gas

Table 24.7: The values of coefficients β_x^2 and β_y^2 when the effect "mutual compensation" is observed.

γ	0.4	0.5	0.8	1.0	$\geq \gamma_{\text{gr}}$
$(\beta_{xc}^2; \beta_{yc}^2)$	(0.1; 0.079)	(0.1; 0.060)	(0.1; 0.050)	(0.1; 0.051)	(0.1; 0.063)
$(\beta_{xc}^2; \beta_{yc}^2)$	(0.5; 0.360)	(0.5; 0.286)	(0.5; 0.248)	(0.5; 0.300)	(0.5; 0.380)

flow velocity of the loaded plate in two directions with tensile and compressive forces, generally do not significantly increase the static stability of the "plate-flow" system, in contrast to the case in which they are directed along gas flow [13]. Thus, thanks to the proposed analytical method, it became possible to partition the parametric space of the problem stability (24.1) – (24.4) into domains of dominance of the influence of compressive forces, dominance of the influence of tensile forces, as well as on the domain in which their actions are balanced.

24.6 Conclusions

The graphical-analytical research method presented in this paper can be used to obtain an analytical solution for a wide class of stability problems for rectangular plates in a supersonic gas flow under initial combined loading.

References

[1] Volmir AS (1967) Stability of Elastic Systems. Foreign Technology Division, Air Force Systems Command. Wright-Patterson Air Force Base, Ohio, AD628508

[2] Bolotin VV (1963) Nonconservative Problems of the Theory of Elastic Stability. Pergamon Pess, Oxford et al

[3] Novichkov YN (1967) Flatter plastin i obolochek (in Russ., Flutter of Plates and Shells). Itogi nauki i tekhnologii. Mekhanika deformiruyemykh tverdykh tel, Vol 11, pp 67-122, Nauka, Moscow

[4] Algazin SD, Kijko IA (2006) Flatter plastin i obolochek (in Russ., Flutter of Plates and Shells). Nauka, Moscow

[5] Il'yushin AA (1956) Zakon ploskikh secheniy pri bol'shikh sverkhzvukovykh skorostyakh (in Russ., The law of plane sections at high supersonic speeds). Prikladnaya Matematika i Mekhanika 20(6):733-755

[6] Ashley GH, Zartarian G (1956) Piston theory – a new aerodynamic tool for the aeroelastician. Journal of the Aeronautical Sciences 23(12):1109–1118, DOI 10.2514/8.3740

[7] Ishlinskii AY (1954) Ob odnom predel'nom perekhode v teorii ustoychivosti uprugikh pryamougol'nykh plastin (in Russ., On a limit transition in the theory of stability of elastic rectangular plates). Doklady Akademii Nauk SSSR 95(3):38-46

[8] Konenkov YK (1960) Ob izgibnoy volne "releyevskogo" tipa (in Russ., A Rayleigh–type flexural wave). Akusticheskii Zhurnal 6(1):124–126

[9] Nádai Á (1931) Plasticity, a Mechanics of the Plastic State of Matter. McGraw-Hill, New York

[10] Belubekyan MV, Martirosyan SR (2012) The localized instability of the elastic plate-strip, streamlined by supersonic gas flow. Proceedings of NAS of Armenia. Mechanics 65(1):29–34

[11] Belubekyan MV, Martirosyan SR (2014) O flattere uprugoy pryamougol'noy plastinki, obtekayemoy sverkhzvukovym potokom gaza, nabegayushchim na yeye svobodnyy kray (in Russ., On the problem of the flutter of an elastic rectangular plate, when the supersonic gas flow is in a direction perpendicular to the free edge). Proceedings of NAS of Armenia. Mechanics 67(2):12–42

[12] Belubekyan MV, Martirosyan SR (2018) O sverkhzvukovoy divergentsii paneli, szhatoy po napravleniyu potoka gaza, nabegayushchim na yeye svobodnyy kray (in Russ., On divergence of compressed panel in supersonic gas flow an accumulating on its free edge). Proceedings of NAS of Armenia. Mechanics 71(4):44–68

[13] Martirosyan SR (2022) Supersonic divergence of a panel with a free edge initially loaded in two directions tensile and compressive forces. Journal of Physics: Conference Series 2231:012030, DOI 10.1088/1742âĂŞ6596/2231/1/012030

Chapter 25

Influence of Magnetic Field on Supersonic Flutter Oscillations of Plate, Placed in a Varying Along the Thickness Temperature Field

Marine A. Mikilyan

Abstract This paper investigates the influence of magnetic field on the stability region of an elongated plate under the combined influence of supersonic gas flow and temperature field varying across the thickness. Mathematical modeling of the given problem resulted in obtaining a boundary value problem that characterizes the stable state of the plate under consideration. The constructed model is based on the main assumptions of the theories of magneto-thermo-elasticity of flexible plates and the Piston theory. It should be noted that the extended formula of piston theory is applied here. Using the Routh-Hurwitz theorem, stability regions of the plate under consideration are constructed under the combined action of the temperature field and the flowing stream, and the influence of the present magnetic field on these regions is investigated. Using the previously derived formula to calculate the critical velocity of the flowing stream, in the paper the corresponding values of the critical velocity are calculated for different temperature fields, and also the influence of the magnetic field on these values is studied.

Key words: Supersonic flow, Linear flutter, Magneto-hydro-dynamics, Flutter critical speed, Stability boundary

25.1 Introduction

Panel flutter is stability loss and intense vibrations of aircraft skin panels, excited by interaction with airflow at high flight speeds. When mathematically modeling a gas flowing a solid body, it is generally assumed that the streamlined surface does not deform. If the bending stiffness of the streamlined body is low, this assumption must be abandoned and the combined vibrations of the body and the gas flow

Marine A. Mikilyan
Institute of Mechanics, National Academy of Sciences of the Republic of Armenia, 24b Marshal Baghramyan Ave, Yerevan, 0019, Republic of Armenia, e-mail: mikilyan@rau.am

H. Altenbach and A. V. Sahakyan, *Contact Mechanics and Relevant Problems*,
Advanced Structured Materials 253, https://doi.org/10.1007/978-3-032-20869-9_25

must be considered. Typically, the vibrations initiated by a certain initial impulse quickly dampen due to so-called aerodynamic viscosity, but it is possible for the amplitude of the vibrations to increase sharply over time. This phenomenon, known as flutter, can cause the destruction of streamlined aircraft structural elements - lifting surfaces or skin. Controllability may also be impaired. Flutter has been the cause of many air crashes. Therefore, its study is a priority area of research in mechanics [13, 18, 19]. Supersonic flutter is characteristic of skin elements and high-aspect-ratio load-bearing surfaces. Its occurrence is a transition from damped oscillations to oscillations with increasing amplitude, which is naturally interpreted as a loss of stability [4, 13, 23]. A linear stability analysis allows us to determine the boundary of a stable flow regime, i.e., the critical value of the incident flow velocity [2, 3, 24]. It should be noted that panel flutter can occur on flat or curved panels. Various factors can influence the nature of the flutter boundary in different ways. These factors can be structural or geometric, thermodynamic, magnetic, aerodynamic, arising from flow characteristics, and appearing in the equations of state [5, 8, 11]. The stability of isotropic plates of constant thickness in a supersonic gas flow has been widely studied [2–4, 8, 12, 25], while, as far as the authors know, the problem of aerothermoelastic stability of panels in a temperature field varying in thickness has been considered only to a small extent [5–7, 21]. It should be noted that very few works are devoted to issues of aero-magneto-elastic stability [14, 15, 17, 26]. In particular, this paper studies the issue of loss of stability of an isotropic plate-strip of constant thickness, immersed in a supersonic flow field and subjected to the joint influence of longitudinal magnetic and varying across the thickness thermal fields. Using the basic principles of the theory of aero-thermo-magneto-elasticity of isotropic bodies [10, 20, 22], the theory of flexible plates [27, 28], the theory of magnetoelasticity and the linear law of thermal field changing across the thickness of the plate [5, 6], stability equations and corresponding boundary conditions are obtained. It is shown that the coefficients of these aero-thermo-magneto-elastic governing equations depend on temperature and the magnetic field parameter, and, consequently, the critical flutter velocity of the panel depends on the combined effect of these phenomena. This phenomenon is caused by the fact that, under the influence of a temperature field in the unperturbed state, the plate buckles due to the non-uniformity of the temperature field across its thickness. It is also shown that the applied magnetic field has a significant effect on the critical flutter velocity.

25.2 The Problem Statement

Let us consider a thin isotropic rectangular plate of constant thickness h. Let the plate is placed in a stationery thermal field T and in a longitudinal magnetic field of intensity vector $\vec{H}_0(H_{01}, 0, 0)$. Let us place the plate in a Cartesian coordinate system (x_1, x_2, x_3) in such way, that the plate's middle plane coincide with coordinate plane (x_1, x_2). Let the examined plate is flown at one side by supersonic stream of gas with unperturbed velocity U, which as it shown in Fig. 25.1 is directed along the coordinate axis $0x_1$.

In this paper it is assumed that the plate's edges are simply supported and it long enough in direction perpendicular to flowing stream. That is why we can assume that all quantities do not depend on the coordinate x_2 if coordinate axis $0x_2$ is directed along the long edge of the plate.

25.3 Basics Assumptions of Mathematical Modeling

The following main assumptions are accepted to made mathematical modeling of the examined aero-thermo-magneto-elastic system:

a) Kirchhoff hypotheses [1, 2, 11],
b) model of flexible plates [28, 29],
c) modified formula of the Piston theory [28, 29]

$$q = -\kappa p_\infty \lambda^2 \left(\frac{1}{\sqrt{1+\lambda^2}} \left(\frac{v_3}{a_0} \right) + \frac{5-\kappa+3\lambda^2}{4(1+\lambda^2)^2} \left(\frac{v_3}{a_0} \right)^2 \right.$$
$$\left. + \frac{(2\kappa+3)\lambda^4 + (10-\kappa^2)\lambda^2 + \kappa^2 - 2\kappa + 9}{12(1+\lambda^2)^3\sqrt{1+\lambda^2}} \left(\frac{v_3}{a_0} \right)^3 + \dots \right), \tag{25.1}$$

where p is gas pressure on plate's surface,

$$a_\infty^2 = \kappa p_\infty \rho_\infty^{-1}$$

is the sound speed for unperturbed gas, p_∞ and ρ_∞ are pressure and density of flowing gas in unperturbed state, κ is polytropic index of gas, v_3 is normal component of the velocity on plate's surface, λ is parameter of magnetic field

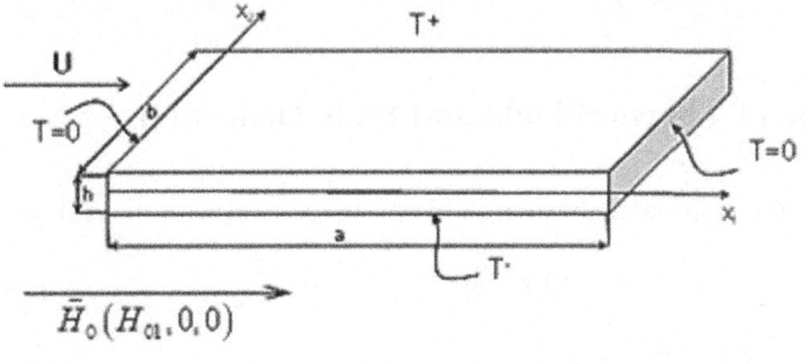

Fig. 25.1: Geometrical representation of the problem.

$$\lambda^2 = \frac{V_A^2}{a_0^2}, \qquad V_A^2 = \frac{B_{01}^2}{\rho_0 \mu_0},$$

V_A is quantity, which is equal to speed of Alfvén electromagnetic waves, $B_{01} = \mu_0 \mu_r H_{01}$, and μ_r is relative magnetic permeability, $\mu_0 = 4\pi 10^7\ \mathrm{N/A^2}$ is magnetic constant;

d) linear law of thermal field [6, 13]

$$T(x_1, x_2, x_3) = T_0(x_1, x_2) + x_3 \Theta(x_1, x_2), \tag{25.2}$$

e) Newman hypothesis on tangential shears [29].

Under the influence of inhomogeneous stationary thermal field ($\Theta \neq 0$) the plate bends (with deflection $w_T(x_1)$) which results an aerodynamic pressure and this pressure is calculated via the formula (1) and in a linear formulation one can cast it in the following representation

$$\kappa p_\infty \sqrt{1 + \lambda^2} a_\infty^{-1} \frac{dw_T}{dx_1}.$$

The goal of this paper is the following: investigate the stability of the noted bending state of the plate under the action of aerodynamic pressure and joint effect of thermal and magnetic fields. For this purpose, we need to make some new assumptions in addition to old ones:

a) deformations of unperturbed state are small enough and we can restrict us by the linear theory of aero-thermo-magnetic elasticity;
b) heat exchange of the plate's surfaces

$$x_3 = \pm \frac{h}{2}$$

occurs according to the Newton-Richmann's law (surfaces have temperatures T^+ and T^-, respectively), and side surfaces ($x_1 = 0$ and $x_1 = a$) are thermally insulated.

25.4 Calculation of Thermal Field and Plate Deflection

As it brought in [6] the stationery thermal field can be defined from the following equation

$$\Delta T = 0 \tag{25.3}$$

with following surface conditions

$$\beta \frac{\partial T}{\partial n} = k\left(T - T^{\pm}\right) \quad \text{for} \quad x_3 = \pm \frac{h}{2}, \tag{25.4}$$

$$\frac{\partial T}{\partial n} = 0 \quad \text{for} \quad x_1 = 0 \text{ and } x_1 = a. \tag{25.5}$$

According to the linear law of thermal field [6, 21], the addressed boundary-value problem (25.3) - (25.5) has the solution

$$T = T_0(x_1) + x_3\Theta(x_1),$$
$$T_0 = \frac{T^+ + T^-}{2}, \quad \Theta = \frac{k\,(T^+ - T^-)}{kh - 2\beta}.$$

In formula (25.4) β is thermal conductivity coefficient, k is heat transfer coefficient, $\vec{n}$ is outward normal to plate surface.

Let us turn to determination of plate's deflection. As noted above, under the action of inhomogeneous stationery thermal field ($\Theta \neq 0$) the plate bends with deflection $w_T(x_1)$ and, therefore, aerodynamic pressure arises. Analogous to the works [6, 21], to determine the deflection w_T, the following ordinary differential equation one can derive with respect to the deflection and displacement u_T

$$\frac{d^2 u_T}{dx_1^2} = 0,$$
$$D\frac{d^4 w_T}{dx_1^4} + \kappa p_\infty M\sqrt{1+\lambda^2}\frac{dw_T}{dx_1} = 0 \tag{25.6}$$

with the following boundary conditions

- the edges are hinge supported and can freely move along the axis $0x_1$

$$w_T = 0, \quad \frac{d^2 w_T}{dx_1^2} + \alpha(1+\mu)\Theta = 0 \quad \text{for} \quad x_1 = 0,\ x_1 = a,$$
$$\frac{du_T}{dx_1} - \alpha(1-\mu)T_0 = 0, \quad \text{for} \quad x_1 = 0,\ x_1 = a, \tag{25.7}$$

- the edges are hinge supported and fixed

$$w_T = 0, \quad \frac{d^2 w_T}{dx_1^2} + \alpha(1+\mu)\Theta = 0 \quad \text{for} \quad x_1 = 0,\ x_1 = a,$$
$$u_T = 0, \quad \text{for} \quad x_1 = 0,\ x_1 = a. \tag{25.8}$$

Solution of brought boundary-value problems (25.6) - (25.7) or (25.6) - (25.8) is obtained via the Bubnov-Galerkin's method and has the form

$$w_T(x_1) = -R\Theta\Bigg(\frac{1}{2}x_1(x_1 - a) + \frac{a^2 K\nu\sqrt{1+\lambda^2}}{36 + \left(K\nu\sqrt{1+\lambda^2}\right)^2}\left[\frac{3\sqrt{1+\lambda^2}}{8\pi}K\nu\sin\frac{\pi x_1}{a} + \frac{9}{16\pi}\sin\frac{2\pi x_1}{a}\right]\Bigg). \tag{25.9}$$

Based on the formula 25.9 the function w_T is calculated and plotted in the Fig. 25.2. In this paper the following initial data is used [27]

$$\alpha = 23.8\,10^{-6}\mathrm{K}^{-1},\ k = 1200\mathrm{Wt}/(\mathrm{m}^2\mathrm{K}),\ \beta = 210\mathrm{Wt}/(\mathrm{m\,K}),\ \mu = 0.34,\\ a = 1\mathrm{m},\ h/a = 1/100.$$

Figure 25.2 shows, that the presence of magnetic field brings to the essential decrease in plate deflection. This means that the magnetic field strengths the plate. Calculations also show that the greater the thermal gradient, the greater the bending of the plate.

25.5 Basic Equations for Rectangular Plate

For rectangular plates the following system of nonlinear differential governing equations of aero-magneto-thermo-elasticity can be cast based on the above addressed assumptions, as it was done in [9, 27]

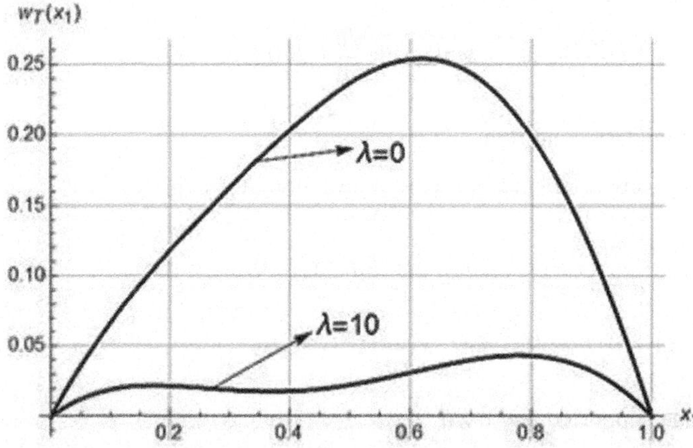

Fig. 25.2: Function w_T for $T_0 = 0$, $\Theta < 0$ and different values of magnetic field parameter λ..

$$D\Delta^2 w + \alpha(1+\mu)D\Delta\Theta + \rho h \frac{\partial^2 w}{\partial t^2} + \left(\rho h \epsilon + \frac{\kappa p_\infty}{a_\infty}\right)\frac{\partial w}{\partial t} + \kappa p_\infty M\sqrt{1+\lambda^2}\frac{\partial w}{\partial x_1}$$
$$= T_{11}\frac{\partial^2 w}{\partial x_1^2} + T_{22}\frac{\partial^2 w}{\partial x_2^2} + 2T_{12}\frac{\partial^2 w}{\partial x_1 \partial x_2}$$
$$- \kappa p_\infty \frac{\kappa+1}{4}\left(M\sqrt{1+\lambda^2}\right)^2\left[\left(\frac{\partial w}{\partial x_1}\right)^2 + \frac{M\sqrt{1+\lambda^2}}{3}\left(\frac{\partial w}{\partial x_1}\right)^3\right],$$
$$\frac{1}{Eh}\left(\frac{\partial^2 T_{11}}{\partial x_2^2} + \frac{\partial^2 T_{22}}{\partial x_1^2} - 2\frac{\partial^2 T_{12}}{\partial x_1 \partial x_2}\right) + \alpha\Delta T_0 = \left(\frac{\partial^2 w}{\partial x_1 \partial x_2}\right)^2 - \frac{\partial^2 w}{\partial x_1^2}\frac{\partial^2 w}{\partial x_2^2}.$$

Here $M = Ua_\infty^{-1}$ is the Mach number of undisturbed flow, $w(x_1, x_2, t)$ is the plate deflection, $D = Eh^3/[12(1-\mu^2)]$, E is the elastic modulus, μ is the Poisson's ratio, α is the coefficient of thermal expansion, ρ is the density of plate's material, ϵ the coefficient of linear attenuation, T_{ij} are internal forces. Let us note that when obtaining Eqs. (**??**) the aerodynamic pressure calculated according to the formula (25.1), expanding it in series according to degrees

$$\frac{v_3}{a_\infty} = \left(\frac{\partial w}{\partial t} + U\frac{\partial w}{\partial x_1}\right)a_\infty^{-1}.$$

In the decomposition, according to the assumptions of the theory of the flexible plates, only first three terms in the equation are considered [4–10, 12, 22–24]. Moreover, from the members, that consider aerodynamic damping, is preserved only the most essential linear term

$$\kappa p_\infty \sqrt{1+\lambda^2}a_\infty^{-1}\frac{\partial w}{\partial t}.$$

Based on the assumption, that tangential components of inertia can be neglected [27], the internal forces T_{ij} satisfy to the following equations

$$\frac{\partial T_{11}}{\partial x_1} + \frac{\partial T_{12}}{\partial x_2} = 0,$$
$$\frac{\partial T_{12}}{\partial x_2} + \frac{\partial T_{22}}{\partial x_1} = 0, \tag{25.10}$$

and according to generalized Hooke's law, they are expressed via the strain components ϵ_{ij} of the middle plane of the plate by the formulas

$$
\begin{aligned}
T_{11} &= \frac{Eh}{1-\mu^2}\left[\epsilon_{11}+\mu\epsilon_{22}-\alpha(1+\mu)T_0\right],\\
T_{22} &= \frac{Eh}{1-\mu^2}\left[\epsilon_{22}+\mu\epsilon_{11}-\alpha(1+\mu)T_0\right],\\
T_{12} &= \frac{Eh}{1-\mu^2}\epsilon_{12},\\
\epsilon_{11} &= \frac{\partial u_1}{\partial x_1}+\frac{1}{2}\left(\frac{\partial w}{\partial x_1}\right)^2,\\
\epsilon_{22} &= \frac{\partial u_2}{\partial x_2}+\frac{1}{2}\left(\frac{\partial w}{\partial x_2}\right)^2,\\
\epsilon_{12} &= \frac{\partial u_1}{\partial x_2}+\frac{\partial u_2}{\partial x_1}+\left(\frac{\partial w}{\partial x_1}\right)\left(\frac{\partial w}{\partial x_2}\right),
\end{aligned}
\tag{25.11}
$$

where $u_1(x_1,x_2,t)$ and $u_2(x_1,x_2,t)$ are tangential displacements of points located on the plate's middle plane.

25.6 Boundary-Value Problems Relative to the Disturbances of Unperturbed State

Finally, from the first equation of the system (25.10) the following main nonlinear differential equation relative to the perturbations $w(x_1,t)$ of the transversal motion of the plate is obtained:

$$
\begin{aligned}
&D\frac{\partial^4 w}{\partial x_1^4}-N_1^T\frac{\partial^2 w}{\partial x_1^2}-N_1\left(\frac{\partial^2 w}{\partial x_1^2}+\frac{d^2 w_T}{dx_1^2}\right)+\rho h\frac{\partial^2 w}{\partial t^2}+\left(\rho h\epsilon+\frac{\kappa p_\infty\sqrt{1+\lambda^2}}{a_\infty}\right)\frac{\partial w}{\partial t}\\
&+\kappa p_\infty M\sqrt{1+\lambda^2}\frac{\partial w}{\partial x_1}+\kappa p_\infty M^2\frac{\kappa+1+(\kappa+4)\lambda^2+3\lambda^4}{4(1+\lambda^2)^2}\left(2\frac{dw_T}{dx_1}+\frac{\partial w}{\partial x_1}\right)\frac{\partial w}{\partial x_1}\\
&+\kappa p_\infty M^3\sqrt{1+\lambda^2}\frac{\kappa+1+(\kappa+1)(\kappa+3)\lambda^2+(\kappa^2-3\kappa+11)\lambda^4+(2\kappa+3)\lambda^6}{12(1+\lambda^2)^4}\\
&\times\left(3\frac{dw_T}{dx_1}+\frac{\partial w}{\partial x_1}\right)\left(\frac{\partial w}{\partial x_1}\right)^2=0
\end{aligned}
\tag{25.12}
$$

Solutions of the Eq. (25.12) should satisfy the fixing conditions of plate's edges $x_1=0$ and $x_1=a$. The following two types of boundary conditions will be considered:

- the edges are simply supported and can freely move along the axes $0x_1$,
- the edges are simply supported and fixed.

The following notations are introduced in Eq. (25.12)

$$N_1^T = \begin{cases} 0, & \text{when plate's edges can move freely,} \\ -\dfrac{Eh}{1-\mu}\alpha T_0, & \text{when plate's edges are fixed,} \end{cases} \tag{25.13}$$

$$N_1 = \begin{cases} 0, & \text{when plate's edges can move freely,} \\ \dfrac{Eh}{2a(1-\mu^2)} \displaystyle\int_0^a \left(2\frac{dw_T}{dx_1}\frac{\partial w}{\partial x_1} + \left(\frac{\partial w}{\partial x_1}\right)^2 \right) dx_1, & \text{when plate's edges are fixed.} \end{cases} \tag{25.14}$$

It is worthy to note, that the solutions of the Eq. (25.12) should satisfy the following simply supporting conditions

$$w = 0, \quad \frac{\partial^2 w}{\partial x_1^2} = 0 \quad \text{for} \quad x_1 = 0,\ x_1 = a. \tag{25.15}$$

25.7 Solution Methodology

Let us solve the obtained non-linear problem (25.12) - (25.15) approximately, having presented it solution in the form [9, 13]

$$w(x_1, t) = \sum_{i=1}^{N} f_i(t) \sin \lambda_i x_1, \quad \lambda_i = \frac{i\pi}{a}, \tag{25.16}$$

where $f_i(t)$ are unknown yet functions. Let us note, that the presentation (25.16) satisfies the boundary conditions (25.15). Having substituted the solution (25.16) into the equation (25.12) and applying the Bubnov-Galerkin's method, one can derive the appropriate non-linear system of ordinary differential equations with respect to the unknown functions $f_i(t)$. The obtained system is huge enough, that is why in this paper we'll restrict us with binomial approximation. Let us note, that binomial approximation gives us an opportunity to investigate analytical behavior of the considered aero-thermo-magnetic system.

So, let us present the solution of the system (25.12) - (25.15) in the form [5–7, 21, 22]

$$w = f_1(t) \sin \lambda_1 x + f_2(t) \sin \lambda_2 x. \tag{25.17}$$

Substituting (25.17) into the equation (25.12), after the well-known orthogonalization method, one can obtain the following non-linear system of ordinary differential equations with respect to the dimensionless functions $y_i = f_i(t)/h$

$$
\begin{aligned}
&\frac{d^2y_1}{d\tau^2}+\chi\frac{dy_1}{d\tau}+(1-\overline{T}\delta)y_1-\frac{2}{3}K\nu y_2+\alpha_{11}^{\lambda}y_1^2+\alpha_{12}^{\lambda}y_2^2+y_2\left[\beta_{11}^{\lambda}y_1^2+\beta_{12}^{\lambda}y_2^2\right]\\
&\quad+3\delta y_1\left[y_1^2+\gamma y_2^2\right]+\Theta\Bigg(\delta_{11}^{(\lambda,\Theta)}y_1+\delta_{12}^{(\lambda,\Theta)}y_2+\alpha_{11}^{(\lambda,\Theta)}y_1^2+\alpha_{12}^{(\lambda,\Theta)}y_2^2\\
&\quad+\alpha_{13}^{(\lambda,\Theta)}y_1y_2+\delta\left[\alpha_{11}^{\Theta}y_1^2+\alpha_{12}^{\Theta}y_2^2+\alpha_{13}^{\Theta}y_1y_2\right]\Bigg)=0,\\
&\frac{d^2y_2}{d\tau^2}+\chi\frac{dy_2}{d\tau}+\frac{2}{3}K\nu y_1+\gamma(\gamma-\overline{T}\delta)y_2+\alpha_{21}^{\lambda}y_1y_2+y_1\left[\beta_{21}^{\lambda}y_1^2+\beta_{22}^{\lambda}y_2^2\right]\\
&\quad+3\gamma\delta y_1\left[y_1^2+\gamma y_2^2\right]+\Theta\Bigg(\delta_{21}^{(\lambda,\Theta)}y_1+\delta_{22}^{(\lambda,\Theta)}y_2+\alpha_{21}^{(\lambda,\Theta)}y_1^2+\alpha_{22}^{(\lambda,\Theta)}y_2^2\\
&\quad+\alpha_{23}^{(\lambda,\Theta)}y_1y_2+\delta\left[\alpha_{21}^{\Theta}y_1^2+\alpha_{22}^{\Theta}y_2^2+\alpha_23^{\Theta}y_1y_2\right]\Bigg)=0,
\end{aligned}
\tag{25.18}
$$

where the following notations are done: $\tau=\omega_1 t$ and

$$
\begin{aligned}
&\overline{T}=RT_0\left(\frac{a}{h}\right)^2\left(\frac{12}{\pi^2}\right),\\
&\overline{\Theta}=R\Theta h\left(\frac{a}{h}\right)^2\left(\frac{1}{36+K^2\overline{\nu}^2}\right),\\
&K=\frac{4\kappa p_\infty}{\rho\omega_1^2h^2},\\
&\overline{\nu}=M\sqrt{1+\lambda^2}\left(\frac{h}{a}\right),\\
&\omega_i=\sqrt{\frac{D}{\rho h}}\left(\frac{i\pi}{a}\right)^2 \quad (i=1,2),\\
&\gamma=\frac{\omega_2}{\omega_1},\\
&\chi=\frac{1}{\omega_1}\left(\epsilon+\frac{a_\infty\rho_\infty\sqrt{1+\lambda^2}}{\rho h}\right)
\end{aligned}
\tag{25.19}
$$

coefficients, taking into the account only the magneto-aerodynamic influence

$$
\begin{aligned}
&\alpha_{11}^{\lambda}=\frac{A}{12},\quad \alpha_{12}^{\lambda}=\frac{7A}{15},\quad \alpha_{21}^{\lambda}=\frac{2A}{15},\\
&\beta_{11}^{\lambda}=\beta_{21}^{\lambda}=\frac{B}{30},\quad \beta_{12}^{\lambda}=-\frac{6B}{35},\quad \beta_{22}^{\lambda}=\frac{22B}{105},
\end{aligned}
\tag{25.20}
$$

coefficients that take into account only thermo-aerodynamic influence

$$\alpha_{11}^{\Theta} = \frac{9C}{8}, \quad \alpha_{12}^{\Theta} = \frac{3C}{2}, \quad \alpha_{13}^{\Theta} = \alpha_{21}^{\Theta} = -\frac{27}{2\pi}K\overline{v},$$
$$\alpha_{22}^{\Theta} = -\frac{81}{\pi}K\overline{v}, \quad \alpha_{22}^{\Theta} = 3C, \tag{25.21}$$

and coefficients taking into account the combined magneto-thermo-aero-dynamic influence

$$\begin{aligned}
\delta_{11}^{(\lambda,\Theta)} &= \frac{9A}{4}, \quad \delta_{12}^{(\lambda,\Theta)} = -\frac{21A}{40}K\overline{v},\\
\delta_{21}^{(\lambda,\Theta)} &= -\frac{3A}{40}K\overline{v}, \quad \delta_{22}^{(\lambda,\Theta)} = \frac{(180 + K^2\overline{v}^2)A}{80},\\
\alpha_{11}^{(\lambda,\Theta)} &= -\frac{3B}{160}K\overline{v}, \quad \alpha_{12}^{(\lambda,\Theta)} = \frac{81B}{280}K\overline{v},\\
\alpha_{13}^{(\lambda,\Theta)} &= \frac{(180 + K^2\overline{v}^2)A}{160},\\
\alpha_{21}^{(\lambda,\Theta)} &= \frac{(900 + K^2\overline{v}^2)A}{640}K\overline{v}, \alpha_{22}^{(\lambda,\Theta)} = \frac{(630 + K^2\overline{v}^2)B}{210}K\overline{v},\\
\alpha_{23}^{(\lambda,\Theta)} &= -\frac{33B}{140}K\overline{v},
\end{aligned} \tag{25.22}$$

where

$$\begin{aligned}
A &= \frac{1 + \kappa + 3\lambda^2}{(1 + \lambda^2)^2}K\overline{v}^2,\\
B &= \frac{\pi\left(\kappa + 1 + (\kappa + 1)(\kappa + 3)\lambda^2 + (\kappa^2 - 3\kappa + 11)\lambda^4 + (2\kappa + 3)\lambda^6\right)}{(1 + \lambda^2)^5}K\overline{v}^3,\\
C &= \frac{1152 + (32 - 3\pi^2)K^2\overline{v}^2}{\pi^3}.
\end{aligned} \tag{25.23}$$

In Eqs. (25.18) and notations (25.19) - (25.23) χ, v, λ, $\overline{T}$, $\overline{\Theta}$ are parameters, characterizing the damping effect, flow velocity and temperature field, respectively, whereas

$$\delta = \begin{cases} 0, & \text{when plate's edges can move freely,} \\ 1, & \text{when plate's edges are fixed.} \end{cases}$$

From the obtained non-linear system one can formulate the appropriate problem for the considered aero-thermo-magnetic stability, considering the corresponding linear problem. In this paper we will restrict us only to considering only linear problem and studying the behavior of solutions depending on supersonic stream velocity (v), temperature ($\overline{T}, \overline{\Theta}$) and magnetic field ($\lambda$). The general nonlinear problem will be investigated in the future.

25.8 Stability Conditions and Stability Boundary

Let us present a linear problem and study it to explore the stability of the examined aero-thermo-magneto-elastic system. The linear one can be obtained from the system (25.18) in the following form

$$\begin{aligned} &\frac{d^2y_1}{d\tau^2}+\chi\frac{dy_1}{d\tau}+(1-\overline{T}\delta)y_1-\frac{2}{3}Kvy_2+\Theta\left[\delta_{11}^{(\lambda,\Theta)}y_1+\delta_{12}^{(\lambda,\Theta)}y_2\right]=0,\\ &\frac{d^2y_2}{d\tau^2}+\chi\frac{dy_2}{d\tau}+\frac{2}{3}Kvy_1+\gamma(\gamma-\overline{T}\delta)y_2+\Theta\left[\delta_{21}^{(\lambda,\Theta)}y_1+\delta_{22}^{(\lambda,\Theta)}y_2\right]=0. \end{aligned} \tag{25.24}$$

Solution of this system one can present in the following form

$$\begin{aligned} y_1&=z_1e^{r\tau},\\ y_2&=z_2e^{r\tau}, \end{aligned} \tag{25.25}$$

and derive he following characteristic equation with respect to the characteristic root r

$$a_0r^4+a_1r^3+a_2r^2+a_3r+a_4=0, \tag{25.26}$$

where

$$\begin{aligned} a_0&=1,\\ a_1&=2\chi,\\ a_2&=\left(1+\gamma^2\right)\left(1-\overline{T}\delta\right)+\Theta\left(\delta_{11}^{(\lambda,\Theta)}+\delta_{22}^{(\lambda,\Theta)}\right)+\chi^2,\\ a_3&=\left[\left(1+\gamma^2\right)\left(1-\overline{T}\delta\right)+\Theta\left(\delta_{11}^{(\lambda,\Theta)}+\delta_{22}^{(\lambda,\Theta)}\right)\right]\chi,\\ a_4&=\frac{4}{9}K^2\overline{v}^2+\frac{2}{3}K\overline{v}\overline{\Theta}\left[\delta_{21}^{(\lambda,\Theta)}-\delta_{12}^{(\lambda,\Theta)}\right]+\gamma^2\left(1-\overline{T}\delta\right)^2\\ &\quad+\overline{\Theta}\left(1-\overline{T}\delta\right)\left[\delta_{11}^{(\lambda,\Theta)}+\delta_{22}^{(\lambda,\Theta)}\right]+\overline{\Theta}^2\left[\delta_{11}^{(\lambda,\Theta)}\delta_{22}^{(\lambda,\Theta)}-\delta_{12}^{(\lambda,\Theta)}\delta_{21}^{(\lambda,\Theta)}\right]. \end{aligned} \tag{25.27}$$

According to the Routh-Hurwitz's theorem stability conditions are those, when all real parts of characteristic roots are negative. Therefore, the stability conditions can be written as follows [16]

$$\begin{aligned} &a_i>0,\quad (i=0,1,2,3,4),\\ &\Delta_2=a_1a_2-a_3a_0>0,\\ &\Delta_3=a_1a_2a_3-a_1^2a_4-a_3^2a_0>0, \end{aligned} \tag{25.28}$$

$$\left(1+\gamma^2\right)\left(1-\overline{T}\delta\right)+\overline{\Theta}\left(\delta_{11}^{(\lambda,\Theta)}+\delta_{22}^{(\lambda,\Theta)}\right)>0,$$

$$\gamma^2\left(1-\overline{T}\delta\right)^2+\frac{4}{9}K^2\overline{v}^2+\frac{2}{3}K\overline{v}\left(\delta_{21}^{(\lambda,\Theta)}-\delta_{12}^{(\lambda,\Theta)}\right)$$
$$+\overline{\Theta}\left(1-\overline{T}\delta\right)\left(\gamma^2\delta_{11}^{(\lambda,\Theta)}+\delta_{22}^{(\lambda,\Theta)}\right)+\overline{\Theta}^2\left(\delta_{11}^{(\lambda,\Theta)}\delta_{22}^{(\lambda,\Theta)}-\delta_{12}^{(\lambda,\Theta)}\delta_{21}^{(\lambda,\Theta)}\right)>0,$$

$$\left(\gamma^2-1\right)^2\left(1+\overline{T}^2\delta^2\right)-\frac{16}{9}K^2\overline{v}^2-2\overline{\Theta}\left(\frac{4}{3}K\overline{v}\left(\delta_{21}^{(\lambda,\Theta)}-\delta_{12}^{(\lambda,\Theta)}\right)\right.$$
$$\left.+\left(\gamma^2-1\right)\left(\delta_{11}^{(\lambda,\Theta)}-\delta_{22}^{(\lambda,\Theta)}\right)\right)+\overline{\Theta}^2\left[\left(\delta_{11}^{(\lambda,\Theta)}-\delta_{22}^{(\lambda,\Theta)}\right)^2+4\delta_{12}^{(\lambda,\Theta)}\delta_{21}^{(\lambda,\Theta)}\right]$$
$$+2\chi^2\left[\left(\gamma^2+1\right)\left(1-\overline{T}\delta\right)+\overline{\Theta}\left(\delta_{11}^{(\lambda,\Theta)}+\delta_{22}^{(\lambda,\Theta)}\right)\right]$$
$$-2\overline{T}\delta\left[\left(\gamma^2-1\right)^2-\overline{\Theta}\left(\gamma^2-1\right)\left(\delta_{11}^{(\lambda,\Theta)}-\delta_{22}^{(\lambda,\Theta)}\right)\right]>0. \tag{25.29}$$

We will study stability boundary and especially the influence of magnetic parameter λ on it having considered the following cases

1. $\overline{T}\neq 0,\ \overline{\Theta}=0$,
2. $\overline{T}=0,\ \overline{\Theta}\neq 0$, and
3. in general, $\overline{T}\neq 0,\ \overline{\Theta}\neq 0$.

25.8.1 Case 1 - Constant Thermal Field ($\overline{T}\neq 0$, $\overline{\Theta}=0$)

In this case, when plate's edges are fixed ($\delta=1$), stability conditions (25.29) have the form

$$1-\overline{T}>0,$$
$$\gamma^2\left(1-\overline{T}\right)^2+\frac{4}{9}k^2\overline{v}^2+\frac{2}{3}K\overline{v}\left(\delta_{21}^{(\lambda,\Theta)}-\delta_{12}^{(\lambda,\Theta)}\right)>0,$$
$$\left(\gamma^2-1\right)^2\left(1+\overline{T}^2\delta^2\right)-\frac{16}{9}K^2\overline{v}^2+2\chi^2\left(\gamma^2+1\right)\left(1-\overline{T}\delta\right)-2\overline{T}\delta\left(\gamma^2-1\right)^2>0. \tag{25.30}$$

On the other hand, when plate's edges can move freely ($\delta=0$), from the inequalities (25.29) the first two are satisfied always and from the third one it is easy to obtain the following formula to define the flutter critical velocity of supersonic stream of flowing gas

$$\overline{v}_{cr}=\frac{3}{4}\frac{\gamma^2-1}{K}\sqrt{1+\frac{2\chi^2\left(\gamma^2+1\right)}{\left(\gamma^2-1\right)^2}} \tag{25.31}$$

or

$$\overline{v}_{cr} = \frac{3}{4}\frac{\gamma^2 - 1}{K\sqrt{1+\lambda^2}}\sqrt{1 + \frac{2\chi^2\left(\gamma^2+1\right)}{\left(\gamma^2-1\right)^2}}. \qquad (25.32)$$

Let us note, that the formula (25.31) coincides with the well-known formula of piston theory [1, 11]. This formula shows, that constant magnetic field does not effect on the stability boundary in that case, when plate's edges can move freely. From the formula (25.31) or (25.32)) it is easy to calculate the critical value of constant thermal field

$$T_0^{cr} = \frac{\pi^2}{12R\delta}\left(\frac{h}{a}\right)^2\left[1 + \frac{\chi^2\left(\gamma^2+1\right)}{\left(\gamma^2-1\right)^2}\right]. \qquad (25.33)$$

Formula (25.32) shows, also, that the value of flutter critical velocity inversely proportional to the value of the applied magnetic field.

To construct stability region on the plane (ν, λ) the inequalities (25.30) will be used. Figure 25.3 shows this region, when $T_0 > 0$, and Fig. 25.4 corresponds to the case of negative temperatures $T_0 < 0$. Numerical influence of magnetic parameter on the values of critical velocity is brought in the Table 25.1. Figures 25.3 and 25.4 show, that

a) when $T_0 < 0$, the presence of a magnetic field significantly narrows the stability region of considered aero-thermo-magneto-elastic system,
b) when $T_0 > 0$, the presence of a magnetic field leads to the bifurcation phenomenon. Namely, an unstable position turns into a stable one and vice versa. The value ν^* we call the upper critical velocity.

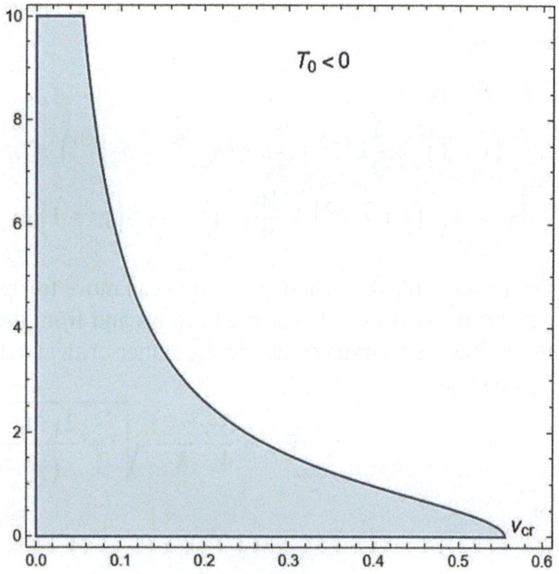

Fig. 25.3 Stability region for $T_0 < 0$.

Table 25.1: Effect of magnetic field on critical velocity at constant temperature field.

λ/T_0, °C	-100	-50	0	5	10
0	0.9904	0.5558	0.1211	0.0777	0.0342
1	0.7003	0.3929	0.0857	0.0549	0.0242
5	0.1942	0.1089	0.0238	0.0152	0.0067
10	0.0985	0.0553	0.0121	0.0077	0.0034

Fig. 25.4 Bifurcation phenomenon at $T_0 \longrightarrow T_0^{cr} > 0$.

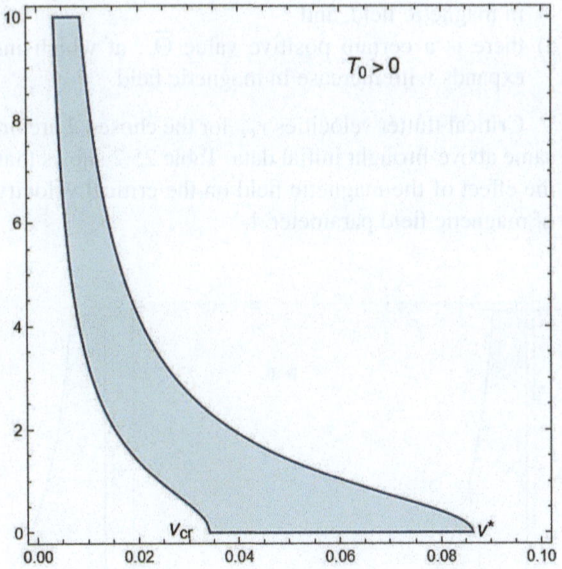

25.8.2 Case 2 - Variable Thermal Field ($\overline{T} = 0,\ \overline{\Theta} \neq 0$)

In this case the stability conditions (25.29) have the form

$$
\begin{aligned}
&1+\gamma^2+\overline{\Theta}\left(\delta_{11}^{(\lambda,\Theta)}+\delta_{22}^{(\lambda,\Theta)}\right)>0,\\
&\gamma^2+\frac{4}{9}K^2\overline{v}^2+\frac{2}{3}K\overline{v}\left(\delta_{21}^{(\lambda,\Theta)}-\delta_{12}^{(\lambda,\Theta)}\right)+\overline{\Theta}\left(\gamma^2\delta_{11}^{(\lambda,\Theta)}+\delta_{22}^{(\lambda,\Theta)}\right)\\
&\qquad+\overline{\Theta}^2\left(\delta_{11}^{(\lambda,\Theta)}\delta_{22}^{(\lambda,\Theta)}-\delta_{12}^{(\lambda,\Theta)}\delta_{21}^{(\lambda,\Theta)}\right)>0,\\
&\left(\gamma^2-1\right)^2-\frac{16}{9}K^2\overline{v}^2-2\overline{\Theta}\left[\frac{4}{3}K\overline{v}\left(\delta_{21}^{(\lambda,\Theta)}-\delta_{12}^{(\lambda,\Theta)}\right)\right.\\
&\qquad\left.+\left(\gamma^2-1\right)\left(\delta_{11}^{(\lambda,\Theta)}-\delta_{22}^{(\lambda,\Theta)}\right)\right]+2\chi^2\left[\gamma^2+1+\overline{\Theta}\left(\delta_{11}^{(\lambda,\Theta)}+\delta_{22}^{(\lambda,\Theta)}\right)\right]\\
&\qquad+\overline{\Theta}^2\left[\left(\delta_{11}^{(\lambda,\Theta)}-\delta_{22}^{(\lambda,\Theta)}\right)^2+4\delta_{12}^{(\lambda,\Theta)}\delta_{21}^{(\lambda,\Theta)}\right]>0.
\end{aligned}
\tag{25.34}
$$

Let us note, that in this case the obtained stability conditions do not depend on plate's edges fixing types. Calculations are done based on the same initial data and shown on the Fig. 25.5 when $\overline{\Theta} < 0$ and on the Fig. 25.6 when $\overline{\Theta} > 0$. Figures 25.5 and 25.6 show, that:

a) if $\overline{\Theta} < 0$, then magnetic field significantly narrows the stability region,
b) there is a certain negative value $\overline{\Theta}_*$, at which magnetic field firstly leads to a noticeable increase in the stability region, after narrows essentially it,
c) if $\overline{\Theta} > 0$, then there is an instability interval $[v_{cr}, v^*]$, which expands with increase in magnetic field, and
d) there is a certain positive value $\overline{\Theta}_*$, at which instability region arises, which expands with increase in magnetic field.

Critical flutter velocities v_{cr} for the chosen λ are brought in the Table 25.2 for the same above-brought initial data. Table 25.2 shows that with a variable thermal field, the effect of the magnetic field on the critical velocity is significant at small values of magnetic field parameter λ.

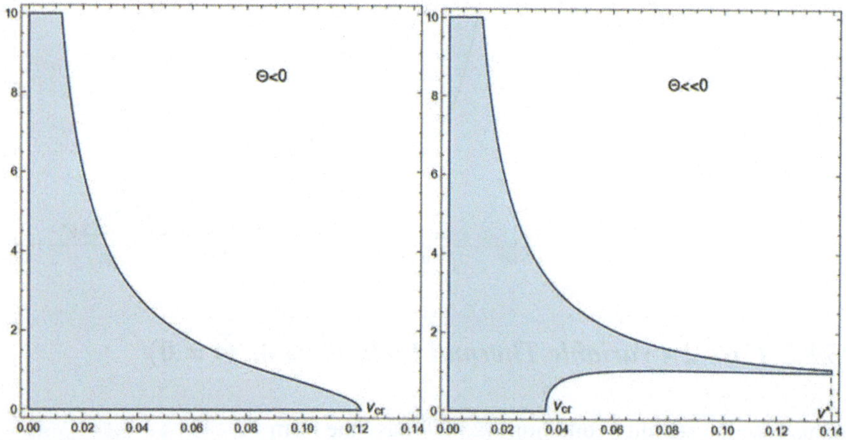

Fig. 25.5: Stability region at $\overline{\Theta} < 0$.

Table 25.2: Effect of magnetic field on critical velocity under moderate temperature fields.

λ/Θ_0, °C/m	-500	-50	0	50	500
0	0.1204	0.1211	0.1211	0.1212	0.1221
1	0.0854	0.0856	0.0857	0.0857	0.0861
5	0.0238	0.02375	0.0238	0.0238	0.0238
10	0.0121	0.0121	0.0121	0.0121	0.0121

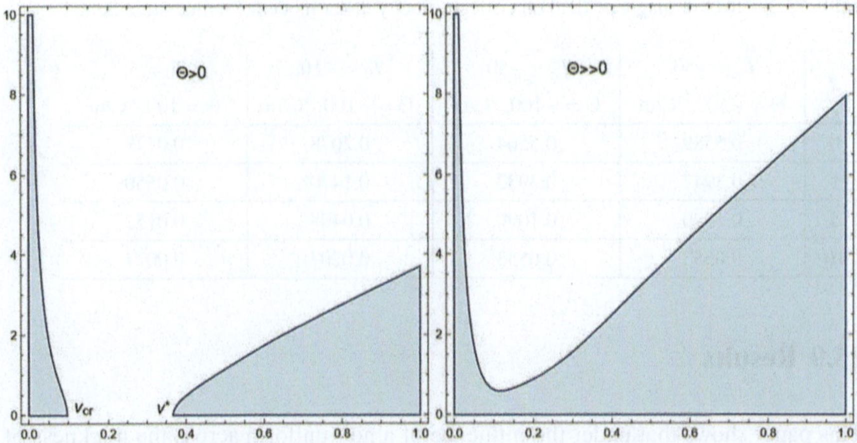

Fig. 25.6: Stability region at $\overline{\Theta} > 0$.

25.8.3 Case 3 - General Case ($\overline{T} \neq 0$, $\overline{\Theta} \neq 0$)

Based on the system (25.29) numerical calculations are also done for general case. The results are brought on the Figs. 25.7. Figures 25.7 and Table 25.3 demonstrate the generalization of the results of particular cases considered above and show that if both components of thermal field are different from zero, then a more emphasized influence of the magnetic field on the stability region is stated.

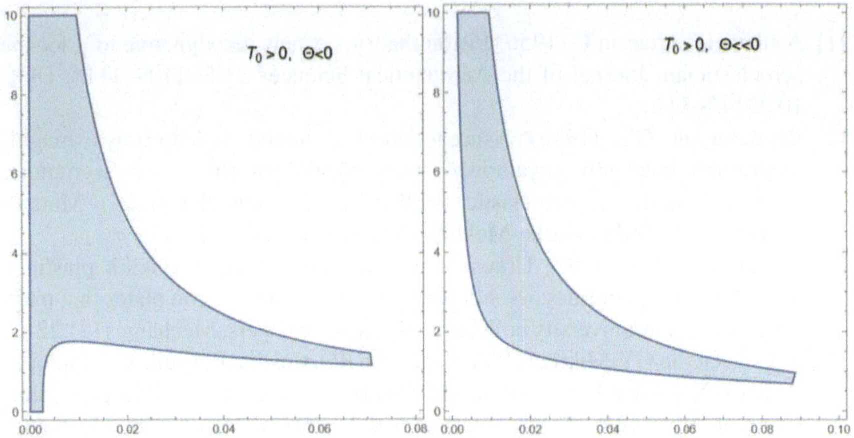

Fig. 25.7: Stability region at $T_0 \longrightarrow T_0^{cr} > 0$, $\overline{\Theta} \ll 0$.

Table 25.3: Effect of magnetic field on critical velocity under moderate temperature fields.

λ	$T_0 = -50$, $\Theta = -500$, °C/m	$T_0 = -50$, $\Theta = -100$, °C/m	$T_0 = -10$, $\Theta = -100$, °C/m	$T_0 = 5$, $\Theta = 100$, °C/m
0	0.5589	0.5564	0.2078	0.0778
1	0.3942	0.3932	0.1470	0.0550
5	0.1090	0.1090	0.0408	0.0152
10	0.0553	0.0553	0.0207	0.0077

25.9 Results

This paper shows that under the influence of a non-uniform across the thickness of the plate stationary thermal field, the plate bends. A formula determining a variable thermal field has been derived and a function determining deflection of the plate has been calculated. Stability conditions of the considered aero-thermo-magneto-elastic system are derived and the stability boundaries are calculated. It is shown that presence of magnetic field has a significant effect on the critical velocity of flutter type oscillations.

Acknowledgements The study was carried out with the financial support of the Science Committee of the Republic of Armenia within the framework of scientific project No. 25T-2C167.

References

[1] Ashley H, Zartarian C (1956) Piston theory - A new aerodynamic tool for the aeroelastician. Journal of the Aeronautical Sciences 23(6):1109–1118, DOI 10.2514/8.3740

[2] Bagdasaryan GY (1986) Asimptoticheskiy metod issledovaniya magnitouprugikh kolebaniy pryamougol'nykh plastin (in Russ., An asymptotic method to study magnetoelastic oscillations of rectangular plates). Matematicheskiye Metody i Fiziko-Mekhanicheskiye Polya (24):72-75

[3] Bagdasaryan GY (1982) Uravneniya nelineynogo flattera tonkikh plastin v magnitnom pole (in Russ., Nonlinear flutter equations for thin plates in a magnetic field). Interuniversity collection of scientific papers, Mechanics (2):22-30

[4] Baghdasaryan GY, Mikilyan MA, Saghoyan RO, Cestino E, Frulla G, Marzocca P (2015) Nonlinear LCO "amplitude–frequency" characteristics for plates fluttering at supersonic speeds. International Journal of Non-Linear Mechanics 77:51–60, DOI 10.1016/j.ijnonlinmec.2015.06.014

[5] Baghdasaryan GY, Mikilyan MA, Vardanyan IA, Danoyan EH, Melikyan K (2020) Influence of boundary conditions on the aero-thermo-elastic stability of

a closed cylindrical shell. IOP Conf. Series: Journal of Physics: Conf. Series 1474:012008, DOI 10.1088/1742-6596/1474/1/012008

[6] Baghdasaryan GY, Mikilyan MA, Vardanyan IA, Melikyan KV, Marzocca P (2021) Thermoelastic non-linear flutter oscillations of rectangular plate. Journal of Thermal Stresses 44(6):731-754, DOI 10.1080/01495739.2021.1914528

[7] Baghdasaryan GY, Mikilyan MA, Vardanyan IA, Panteleev AV, Severina NS (2020) Nonlinear flutter responce of cylindrical shell in thermal field and supersonic gas flow. IOP Conf. Series: Materials Science and Engineering 927(1):012022, DOI 10.1088/1757-899X/927/1/012022

[8] Baghdasaryan GY, Belubekyan MV (1966) Flatter tsilindricheskoy obolochki v potoke szhimayemoy provodyashchey zhidkosti v prisutstvii magnitnogo polya (in Russ., Flutter of a cylindrical shell in a flow of compressible conducting fluid in the presence of a magnetic field). Izvestiya AN SSSR. MTT (6): 50-56

[9] Bagdasaryan GY, Mikilyan MA (2016) Effects of Magnetoelastic Interactions in Conductive Plates and Shells. Springer, Cham , DOI 10.1007/978-3-031-60307-5

[10] Baghdasaryan GY, Mikilyan MA, Marzocca P (2023) Supersonic flutter characteristics of dielectric rectangular plate: The effects of magneto-aero-hydrodynamic interactions. Journal of Fluids and Structures 118:103856, DOI 10.1016/j.jfluidstructs.2023.103856

[11] Baghdasaryan GY, Mikilyan MA (2024) Magnetoelastic Vibrations and Stability of Magnetically Active Plates and Shells. Springer, Cham, DOI 10.1007/978-3-031-60307-5

[12] Baghdasaryan GY, Panteleev AV, Mikilyan MA Mathematical modeling of the problem of magneto-aero-elastic stability of rectangular plate. In: Favorskaya MN, Nikitin IS, Severina NS (eds) Advances in Theory and Practice of Computational Mechanics, Smart Innovation, Systems and Technologies, Vol 274, pp 13–27, DOI 10.1007/978-981-16-8926-0-2

[13] Bolotin VV (1963) Nonconservative Problems of the Theory of Elastic Stability. Pergamon Press, Oxford et al

[14] Borghi CA, Carraro MR, Cristofolini A (2004) Analysis of magnetoplasmadynamic interaction in the boundary layer of a hypersonic vehicle. Journal of Spacecraft and Rockets 42(1):45-50, DOI 10.2514/1.4031

[15] Borghi CA, Carraro MR, Cristofolini A (2006) Magnetohydrodynamics interaction in the shock layer of a wedge in a hypersonic flow. IEEE Transaction on Plasma Science 34(5):2450 - 2463, DOI 10.1109/TPS.2006.883377

[16] Chebotaryev NV, Meyman NS (1949) Problema Rausa-Gurvitsa dlya polinomov i tselykh funktsiy (in Russ., Gauss-Hurwitz Problem for Polynomials and Integer-Valued Functions). Trudy Matematicheskogo instituta imeni V. A. Steklova, Vol 26

[17] Cristofolini A, Borghi CA, Carraro M, Neretti G, Paganucci F, Biagioni L, Fantoni G, Passaro A (2006) Experimental investigation on the MHD interaction around a sharp cone in an ionized argon flow, AIAA-2006-3075, 37th AIAA Plasmadynamics and Lasers Conference, San Francisco, California, June 2006.

[18] Dowell EH (1966) Nonlinear oscillations of a fluttering plate. AIAA Journal 4(7):1267-1275, DOI 10.2514/3.3658

[19] Dowell EH (1967) Nonlinear oscillations of a fluttering plate. II. AIAA Journal 5(10):1856-1862, DOI 10.2514/3.4316

[20] Mikilyan MA (2024) Nature of supersonic flutter of aero-magneto-elastic system at pre-critical flowing speeds. IEEE Aerospace Conference Proceedings pp 529-537, DOI 10.1109/aero58975.2024.10521224

[21] Mikilyan MA (2020) Thermoelastic response of closed cylindrical shells in a supersonic gas flow. Aerospace 7(8):103, DOI 10.3390/aerospace7080103

[22] Mikilyan MA, Vardanyan IA (2024) The effects of magnetic field on supersonic flutter characteristics of dielectric plate: Dependence amplitude-speed. Journal of Fluids and Structures 128:104140, DOI 10.1016/j.jfluidstructs.2024.104140

[23] Miles JW (1957) Supersonic flutter of a cylindrical shell. Journal of Aeronautical Science 24(2):107-118, DOI 10.2514/8.3780

[24] Mikilyan MA, Hambartsumyan HH, Vardanyan IA (2024) Vliyaniye magnitnogo polya na kriticheskuyu skorost' sverkhzvukovogo flattera dielektricheskoy plastinki (in Russ., Effect of magnetic field on the critical speed of supersonic flutter of a dielectric plate). Izvestiya NAS RA, Mechanics 77(2):47-67, DOI 10.54503/0002-3051-2024.77.2-47

[25] Panteleev AV, Mikilyan MA, Vardanyan IA (2023) Stability of a dielectric rectangular plate in a longitudinal magnetic field in the presence of a supersonic flow field. Aerospace Systems 6:605–611, DOI 10.1007/s42401-023-00217-6

[26] Resler RL, Sears WR (1958) The prospects for magneto-aerodynamics. Journal of the Aeronautical Sciences 25(4):235-245, DOI 10.2514/8.7604

[27] Russell R (2011) Zerodur. OPTI521

[28] Vlasov VZ (1949) Obshchaya teoriya obolochek i yeye prilozheniya v tekhnike (in Russ., The General Theory of Shells). Gosudarstvennoye izdatel'stvo tekhniko-teoreticheskoy literatury, Moscow

[29] Volmir AS (1974) Nonlinear Dynamics of Plates and Shells. Foreign Technology Division, Wright-Patterson Air Force Base, Ohio

Chapter 26
On the Stress State of a Piecewise Homogeneous Layer Reinforced by a Stringer with a Collinear System of Cracks and Rigid Inclusions Under Antiplane Deformation

Mher M. Mkrtchyan, Marine S. Grigoryan, and Mushegh S. Mkrtchyan

Abstract In this paper, we consider the problem of determining the components of the stress-strain state of a piecewise homogeneous layer under antiplane deformation. The upper face of which is reinforced by a stringer, the lower face is rigidly clamped, and a collinear system of an arbitrary number of cracks and absolutely rigid inclusions is located along on the horizontal line of the junction of heterogeneous materials. It is assumed that stringers are loaded by tangential forces along their longitudinal edges and by horizontal concentrated forces along their vertical edges. In addition, the crack edges are loaded by distributed tangential forces of different intensities. It is considered that under the action of these loads the composite layer in the direction of one of the coordinate axes is under the conditions of antiplane deformation (longitudinal shear).

Using the Fourier integral transformation, solving the problem is reduced to solving of the system of singular integral equations (SIE). The solution of this system is obtained by the known numerical-analytical method of solving the SIE using Gauss quadrature formulas on Chebyshev nodes. As a result, solving the original system of the SIE is reduced solving the system of linear algebraic equations (SLAE). Various special cases are considered, when the governing SIE and SLAE of the problem are significantly simplified, which make it unable to carry out detailed numerical analyses and reveal regularities of changes in the characteristics of the problems.

Key words: Stresses, Layer, Crack, Stringer, Inclusions, System of singular integral equations, Antiplane deformation

---

Mher M. Mkrtchyan
Institute of Mechanics, National Academy of Sciences of the Republic of Armenia, 24b Marshal Baghramyan Ave, Yerevan, 0019, Republic of Armenia, e-mail: mmher628@gmail.com

Marine S. Grigoryan (corresponding author) · Mushegh S. Mkrtchyan
Institute of Mechanics, National Academy of Sciences of the Republic of Armenia, 24b Marshal Baghramyan Ave, Yerevan, 0019 & National University of Architecture and Construction of Armenia, 105 Teryan Street, Yerevan, 0009, Republic of Armenia e-mail: marinegrigoryan17@gmail.com, mmushegh195@gmail.com

H. Altenbach and A. V. Sahakyan, *Contact Mechanics and Relevant Problems*,
Advanced Structured Materials 253, https://doi.org/10.1007/978-3-032-20869-9_26

26.1 Introduction

It is very important to study the interaction of stress concentrators such as inclusions with massive bodies of different geometric shapes in the mechanics of composites, in the design of various building and aviation structures, in tensometry and in many other fields of applied mechanics and engineering practice. This is explained by the fact that local stress fields with large and intensively changing gradients are formed around these stress concentrators, which significantly reduce the strength of engineering structures and their parts. Therefore, a qualitative and quantitative study of stress concentration problems is of both theoretical and practical interest. Such problems have been studied by many authors. In this direction, we refer to [1, 2, 6, 9, 11].

26.2 Problem Statement and Derivation of the Main Equations

Let a piecewise homogeneous elastic layer, referred to a right rectangular coordinate system $Oxyz$ consist of an upper layer $\Omega_+ = \{-\infty < x, z < \infty, 0 \le y \le h_+\}$ with a shear modulus G_+ and a lower layer $\Omega_- = \{-\infty < x, z < \infty, -h_- \le y \le 0\}$ with a shear modulus G_-, and in the plane $y = 0$ contain a system of through mutually non-intersecting cracks and thin absolutely rigid inclusions. On the plane Oxy we have the system of intervals L_1 and L_2, respectively.

$$L_1 = \bigcup_{k=1}^{N_1} [a_k, b_k]\,; a_k < b_k\,(k = 1, 2, ..., N_1)\,; b_k < a_{k+1}\left(k = \overline{1, N_1 - 1}\right),$$

$$L_2 = \bigcup_{k=1}^{N_2} [c_k, d_k]\,; c_k < d_k\,(k = 1, 2, ..., N_2)\,; d_k < c_{k+1}\left(k = \overline{1, N_2 - 1}\right),$$

$$L = L_1 \cup L_2.$$

Further, let the lower face of the layer be rigidly fixed $u_z^-(x, y)\,|_{y=-h_-} = 0$ and a stringer of height h_1 with a shear modulus G_1 in the form of a strip $\omega_+ = \{e \le x \le f; h_+ \le y \le h_+ + h_1; -\infty < z < \infty\}$ be reinforced on the upper face of the layer $y = h_+ + h_1$, on the interval $[e, f]$.

Assume that uniformly distributed tangential forces of intensities $T_*(x)$ act along the Ox axis (Fig. 26.1), i.e.

$$\tau_{yz}|_{y=h_+ + h_1} = T_*(x) \quad (e < x < f), \tag{26.1}$$

where τ_{yz} is the component of tangential stress. Let tangential concentrated forces F_1, F_2 uniformly distributed along Oz axis act the edges of the stringer $x = e, x = f$. The forces resultant $P_k(k = \overline{1, N_2})$ directed along Oz axis act on the system of inclusions L_2. They cause a longitudinal shear of the elastic layer.

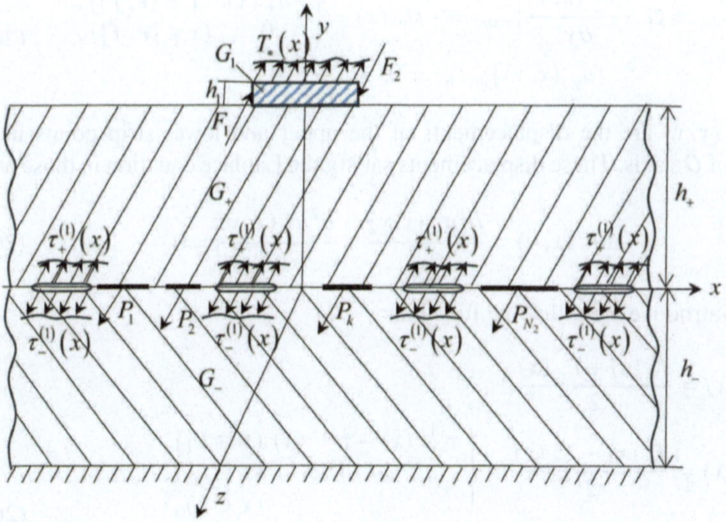

Fig. 26.1: Contact interaction of a stringer with an elastic piecewise homogeneous layer.

It is necessary to determine the continuity of displacement dislocations on the crack edges, crack opening, stress jumps on the inclusion edges, destructive stresses and stress intensity factors (SIF), as well as the tangential contact stresses acting under the stringers. In this case, the one-dimensional elastic continuum model of Melan is adopted for stringers [6, 7].

To derive the governing equations of the problem, we divide the piecewise homogeneous elastic strip along Ox axis into two parts, and then, for the stresses acting on their edges $y = \pm 0$, the following notations are introduced ($L' = R/L$; $R = (-\infty, \infty)$)

$$-\tau_{yz}|_{y=+0} = T_+(x) = \begin{cases} \tau_+^{(1)}(x) & (x \in L_1), \\ \tau_+^{(2)}(x) & (x \in L_2), \\ \tau(x) & (x \in L'), \end{cases}$$
$$-\tau_{yz}|_{y=-0} = T_-(x) = \begin{cases} \tau_-^{(1)}(x) & (x \in L_1), \\ \tau_-^{(2)}(x) & (x \in L_2), \\ \tau(x) & (x \in L'). \end{cases} \tag{26.2}$$

The displacements on the inclusion system L_2 are constant

$$u_z^+(x, y)\,|_{y=+0} = u_z^-(x, y)\,|_{y=-0} = g(x) = \delta_k = \text{const}; x \in (c_k, d_k)\left(k = \overline{1, N_2}\right). \tag{26.3}$$

The boundary conditions for $y = \pm h_\pm$ have the form

$$\tau_{yz}^{+}|_{y=h_+} = G_+ \frac{\partial u_z^+(x,y)}{\partial y}|_{y=h_+} = -H_+(x) = \begin{cases} -h_+(x) & (x \in [e,f]), \\ 0 & (x \notin [e,f]), \end{cases} \tag{26.4}$$
$$u_z^-(x,y)|_{y=-h_-} = 0,$$

where $u_z^{\pm}(x,y)$ are the displacements of the upper and lower strip points in the direction of Oz axis. These displacements satisfy the Laplace equation in those areas

$$\Delta u_z^{\pm}(x,y) = \frac{\partial^2 u_z^{\pm}(x,y)}{\partial x^2} + \frac{\partial^2 u_z^{\pm}(x,y)}{\partial y^2} = 0. \tag{26.5}$$

Next, we introduce the following functions

$$\chi_+(x) = \frac{T_+(x) + T_-(x)}{2},$$
$$\chi_-(x) = \frac{T_+(x) - T_-(x)}{2} = \begin{cases} \tau_+^{(1)}(x) - \tau_-^{(1)}(x) & (x \in L_1), \\ \psi(x) & (x \in L_2), \\ 0 & (x \in L'), \end{cases} \tag{26.6}$$
$$\Phi_+(x) = \frac{1}{2}\left(\frac{du_z^+(x,+0)}{dx} + \frac{du_z^-(x,-0)}{dx}\right),$$
$$\Phi_-(x) = \frac{1}{2}\left(\frac{du_z^+(x,+0)}{dx} - \frac{du_z^-(x,-0)}{dx}\right) = \begin{cases} \varphi(x) & (x \in L_1), \\ 0 & (x \in R \setminus L_1). \end{cases}$$

Using the direct and inverse Fourier transforms of the variable x, we arrive at the key equation of the problem from equation (5) and boundary conditions (26.2) - (26.4) according to the notation (26.6)

$$\begin{aligned} \chi_+(x) = &-\frac{2G_+G_-}{\pi(G_+ + G_-)} \int_{L_1} \frac{\Phi_-(s)\,ds}{s-x} \\ &- \frac{2G_+G_-}{\pi(G_+ + G_-)} \int_{L_1} K(s-x)\,\Phi_-(s)\,ds + \frac{G_+ - G_-}{G_+ + G_-} \\ &\times \chi_-(x) - \frac{2G_+G_-}{\pi(G_+ + G_-)} \int_{L_1 \cup L_2} R(s-x)\,\chi_-(s)\,ds \\ &+ \frac{G_-}{\pi} \int_e^f Q(s-x)\,h_+(s)\,ds \qquad (x \in R), \end{aligned} \tag{26.7}$$

$$\Phi_+(x) = \frac{2}{\pi(G_+ + G_-)} \int\limits_{L_1 \cup L_2} \frac{\chi_-(s)\,ds}{s-x} + \frac{2}{\pi(G_+ + G_-)} \int\limits_{L_1 \cup L_2} K_1(s-x)\,\chi_-(s)\,ds + \frac{2G_+G_-}{\pi(G_+ + G_-)} \int\limits_{L_1} R(s-x)\,\Phi_-(s)\,ds - \frac{1}{\pi}\int\limits_e^f Q_1(s-x)\,h_+(s)\,ds \quad (x \in R), \tag{26.8}$$

with

$$K(x) = \int\limits_0^\infty \frac{G_+ \tanh(\lambda h_+)\left[1-\tanh(\lambda h_-)\right] - G_-\left[1-\tanh(\lambda h_+)\right]}{G_+ \tanh(\lambda h_+)\tanh(\lambda h_-) + G_-} \sin(\lambda x)\,d\lambda,$$

$$K_1(x) = \int\limits_0^\infty \frac{G_+ \tanh(\lambda h_-)\left[1-\tanh(\lambda h_+)\right] - G_-\left[1-\tanh(\lambda h_-)\right]}{G_+ \tanh(\lambda h_+)\tanh(\lambda h_-) + G_-} \sin(\lambda x)\,d\lambda,$$

$$R(x) = \int\limits_0^\infty \frac{1-\tanh(\lambda h_+)\tanh(\lambda h_-)}{G_+ \tanh(\lambda h_+)\tanh(\lambda h_-) + G_-} \cos(\lambda x)\,d\lambda,$$

$$Q(x) = \int\limits_0^\infty \frac{1}{\cosh(\lambda h_+)\left[G_+ \tanh(\lambda h_+)\tanh(\lambda h_-) + G_-\right]} \cos(\lambda x)\,d\lambda,$$

$$Q_1(x) = \int\limits_0^\infty \frac{\tanh(\lambda h_-)}{\cosh(\lambda h_+)\left[G_+ \tanh(\lambda h_+)\tanh(\lambda h_-) + G_-\right]} \sin(\lambda x)\,d\lambda.$$

Now, for the derivatives of the displacements on the edges of the strips $y = \pm h_\pm$, we have

$$\frac{du_z^+(x,h_+)}{dx} = \frac{2G_-}{\pi}\int\limits_{L_1} Q(s-x)\,\Phi_-(s)\,ds + \frac{2}{\pi}\int\limits_{L_1 \cup L_2} Q_1(s-x)\,\chi_-(s)\,ds - \frac{2}{\pi G_+}\int\limits_e^f \frac{h_+(s)\,ds}{s-x} - \frac{1}{\pi G_+}\int\limits_e^f E_0(s-x)\,h_+(s)\,ds, \quad (x \in R) \tag{26.9}$$

with

$$E_0(x) = \int_0^\infty \frac{[1 - \tanh(\lambda h_-)]\,[G_+ \tanh(\lambda h_+) - G_-]}{G_+ \tanh(\lambda h_+) \tanh(\lambda h_-) + G_-} \sin(\lambda x)\, d\lambda.$$

Next, from (26.1) we use for the stringer the differential equation of stringer deformation according to the Melan model under antiplane deformation [7]

$$h_1 G_1 \frac{d^2 w}{dx^2} = h_+(x) - T_*(x) \quad (e < x < f), \tag{26.10}$$

where $w = w(x)$ is the component of the displacements of the stringer points in the direction of Oz axis. In this case, the stringer equilibrium conditions have the form

$$\int_e^f h_+(x)\, dx = F_2 - F_1 + \int_e^f T_*(x)\, dx. \tag{26.11}$$

By integrating (26.11), we easily find

$$h_1 G_1 \frac{dw}{dx} = \frac{1}{2}(F_1 + F_2) + \frac{1}{2} \int_e^f \operatorname{sign}(x - s)\,[h_+(s) - T_*(s)]\, ds. \tag{26.12}$$

Now, we consider the contact condition between the elastic strips and the stringer

$$u_z^+(x, h_+) = w(x) \quad \text{or} \quad \frac{du_z^+(x, h_+)}{dx} = \frac{dw(x)}{dx} \quad (e < x < f).$$

We substitute the expressions from (26.9), (26.10), and (26.12), respectively into the contact condition. After simple transformations concerning to unknown contact stresses, we arrive at the following integral equations

$$\begin{aligned} &\frac{2G_-}{\pi} \int_{L_1} Q(s - x)\,\varphi(s)\, ds + \frac{2}{\pi} \int_{L_1 \cup L_2} Q_1(s - x)\,\chi_-(s)\, ds \\ &- \frac{1}{\pi G_+} \int_e^f \frac{h_+(s)\, ds}{s - x} - \frac{1}{\pi G_+} \int_e^f E_0(s - x)\, h_+(s)\, ds \\ &= \frac{1}{2h_1 G_1}[F_1 + F_2] + \frac{1}{2h_1 G_1} \int_e^f \operatorname{sign}(x - s)\,[h_+(s) - T_*(s)]\, ds, \quad (x \in (e, f)). \end{aligned} \tag{26.13}$$

Now considering equations (26.7) and (26.8) on L_1 and L_2, respectively, as well as equation (26.13) on the interval (e, f), we arrive at the governing SIEs of the problem

$$\begin{cases} \dfrac{1}{\pi}\displaystyle\int_{L_1}\dfrac{\varphi(s)\,ds}{s-x}+\dfrac{1}{\pi}\int_{L_1}K(s-x)\varphi(s)\,ds+\dfrac{1}{\pi}\int_{L_2}R(s-x)\psi(s)\,ds \\ -\dfrac{G_+ + G_-}{2\pi G_+}\displaystyle\int_e^f Q(s-x)h_+(s)\,ds=l_1(x),\quad (x\in L_1), \\ \dfrac{G_+}{\pi}\displaystyle\int_{L_1}R(s-x)\varphi(s)\,ds+\dfrac{1}{\pi G_-}\int_{L_2}\dfrac{\psi(s)\,ds}{s-x}+\dfrac{1}{\pi G_-}\int_{L_2}K_1(s-x)\psi(s)\,ds \\ -\dfrac{G_+ + G_-}{2\pi G_-}\displaystyle\int_e^f Q_1(s-x)h_+(s)\,ds=l_2(x),\quad (x\in L_2), \\ \dfrac{2G_-}{\pi}\displaystyle\int_{L_1}Q(s-x)\varphi(s)\,ds+\dfrac{2}{\pi}\int_{L_2}Q_1(s-x)\psi(s)\,ds-\dfrac{1}{\pi G_+}\int_e^f\dfrac{h_+(s)\,ds}{s-x} \\ -\dfrac{1}{\pi G_+}\displaystyle\int_e^f E_0(s-x)h_+(s)\,ds-\dfrac{1}{2h_1G_1}\int_e^f \operatorname{sign}(x-s)h_+(s)\,ds=l_3(x), \\ (x\in(e,f)), \end{cases} \tag{26.14}$$

$$l_1(x)=-\frac{\tau_+^{(1)}(x)}{G_+}-\frac{\tau_-^{(1)}(x)}{G_-}-\frac{1}{\pi}\int_{L_1}R(s-x)\left[\tau_+^{(1)}(s)-\tau_-^{(1)}(s)\right]ds,$$

$$l_2(x)=-\frac{1}{\pi G_-}\int_{L_1}\frac{\left[\tau_+^{(1)}(s)-\tau_-^{(1)}(s)\right]ds}{s-x}$$

$$-\frac{1}{\pi G_-}\int_{L_1}K_1(s-x)\left[\tau_+^{(1)}(s)-\tau_-^{(1)}(s)\right]ds,$$

$$l_3(x)=\frac{F_1+F_2}{2h_1G_1}-\frac{1}{2h_1G_1}\int_e^f \operatorname{sign}(x-s)T_*(s)\,ds$$

$$-\frac{2}{\pi}\int_{L_1}Q_1(s-x)\left[\tau_+^{(1)}(s)-\tau_-^{(1)}(s)\right]ds.$$

The system (26.14) should be considered under the conditions

$$\int_{a_k}^{b_k}\varphi(x)\,dx=0,\quad \left(k=\overline{1,N_1}\right)\quad \int_{c_k}^{d_k}\psi(x)\,dx=P_k.\quad \left(k=\overline{1,N_2}\right). \tag{26.15}$$

The first of them is equivalent to the conditions of continuity of displacements at the end points of the section, and the second is the conditions of equilibrium of the k-th inclusion; we also add the equilibrium conditions of the stringer (26.11) to them. Considering equation (26.7) on the junction line of the strips outside the inclusion and cracks, we have

$$\begin{aligned}\tau(x) = &-\frac{G_+G_-}{\pi(G_++G_-)}\int_{L_1}\frac{\varphi(s)\,ds}{s-x} - \frac{G_+G_-}{\pi(G_++G_-)}\int_{L_1}K(s-x)\,\varphi(s)\,ds \\ &+\frac{G_-}{2\pi}\int_e^f Q(s-x)\,h_+(s)\,ds - \frac{G_+G_-}{\pi(G_++G_-)}\int_{L_2}R(s-x)\,\psi(s)\,ds \quad (26.16) \\ &-\frac{G_+G_-}{\pi(G_++G_-)}\int_{L_1}R(s-x)\left[\tau_+^{(1)}(s)-\tau_-^{(1)}(s)\right]ds \qquad (x\in L').\end{aligned}$$

Thus, the problem of the stress state of an elastic strip with cracks and stringers is reduced to solving the system (26.14) under conditions (26.11), (26.15). After solving (26.14), (26.11), (26.15), the stress on the junction line outside the inclusion is determined by the formula (26.16).

26.3 Solution of the Governing SIE

To solve the governing SIE (26.14) with the conditions (26.11) and (26.15), we introduce dimensionless coordinates and quantities

$$\begin{aligned}&\left\{\xi,\eta,\overline{h}_+,\overline{h}_-,\alpha_k,\beta_k\right\} = \frac{1}{|a_1|}\{x,s,h_+,h_-,a_k,a_k\}\ (k=\overline{1,N_1}),\ L_{01}=\bigcup_{k=1}^{N_1}(\alpha_k,\beta_k), \\ &\{\gamma_k,\delta_k\} = \frac{1}{|a_1|}\{c_k,d_k\},\ \left(k=\overline{1,N_2}\right),\ \mu=\frac{G_+}{G_-},\ \mu_1=\frac{G_1}{G_+},\ L_{02}=\bigcup_{k=1}^{N_2}(\gamma_k,\delta_k), \\ &\left\{\overline{e},\overline{f},\overline{h}_1\right\} = \frac{1}{|a_1|}\{e,f,h_1\},\ \varphi_0(\xi)=\varphi(|a_1|\xi),\ \psi_0(\xi)=\frac{\psi(|a_1|\xi)}{G_++G_-}, \\ &\overline{h}_+(\xi)=\frac{h_+(|a_1|\xi)}{G_+},\ T_*^{(0)}(\xi)=\frac{T_*(|a_1|\xi)}{G_1}, \\ &F_1^{(0)}=\frac{F_1}{|a_1|G_1},\ F_2^{(0)}=\frac{F_2}{|a_1|G_1},\ \tau_\pm^{(0)}(\xi)=\frac{\tau_\pm^{(1)}(|a_1|\xi)}{G_++G_-}.\end{aligned} \qquad (26.17)$$

Then, the equation system (26.14) is transformed into the following SIEs

$$\begin{cases} \frac{1}{\pi}\int\limits_{L_{01}} \frac{\varphi_0(\eta)\,d\eta}{\eta-\xi} + \frac{1}{\pi}\int\limits_{L_{01}} K_0(\eta-\xi)\,\varphi_0(\eta)\,d\eta + \frac{\mu+1}{\pi}\int\limits_{L_{02}} R_0(\eta-\xi)\,\psi_0(\eta)\,d\eta \\ -\frac{\mu_1(\mu+1)}{2\pi}\int\limits_{\overline{e}}^{\overline{f}} Q_0(\eta-\xi)\,\overline{h}_+(\eta)\,d\eta = l_{01}(\xi) \quad (\xi\in L_{01}), \\ \frac{\mu}{\pi}\int\limits_{L_{01}} R_0(\eta-\xi)\,\varphi_0(\eta)\,d\eta + \frac{\mu+1}{\pi}\int\limits_{L_{02}} \frac{\psi_0(\eta)\,d\eta}{\eta-\xi} + \frac{\mu+1}{\pi}\int\limits_{L_{02}} M_0(\eta-\xi)\,\psi_0(\eta)\,d\eta \\ -\frac{\mu_1(\mu+1)}{2\pi}\int\limits_{\overline{e}}^{\overline{f}} G_0(\eta-\xi)\,\overline{h}_+(\eta)\,d\eta = l_{02}(\xi) \quad (\xi\in L_{02}) \\ \frac{2}{\pi}\int\limits_{L_{01}} Q_0(\eta-\xi)\,\varphi_0(\eta)\,d\eta + \frac{2(\mu+1)}{\pi}\int\limits_{L_{02}} G_0(\eta-\xi)\,\psi_0(\eta)\,d\eta \\ -\frac{\mu_1}{\pi}\int\limits_{\overline{e}}^{\overline{f}} \frac{\overline{h}_+(\eta)\,d\eta}{\eta-\xi} - \frac{\mu_1}{\pi}\int\limits_{\overline{e}}^{\overline{f}} E_0(\eta-\xi)\,\overline{h}_+(\eta)\,d\eta \\ -\frac{1}{2\overline{h}_+}\int\limits_{\overline{e}}^{\overline{f}} \operatorname{sign}(\xi-\eta)\,\overline{h}_+(\eta)\,d\eta = l_{03}(\xi) \quad \left(\xi\in\left(\overline{e},\overline{f}\right)\right), \end{cases} \tag{26.18}$$

where

$$K_0(x) = \int\limits_0^\infty \frac{\mu\tanh\left(\lambda\overline{h}_+\right)\left[1-\tanh\left(\lambda\overline{h}_-\right)\right] - \left[1-\tanh\left(\lambda\overline{h}_+\right)\right]}{\mu\tanh\left(\lambda\overline{h}_+\right)\tanh\left(\lambda\overline{h}_-\right)+1}\sin(\lambda x)\,d\lambda,$$

$$M_0(x) = \int\limits_0^\infty \frac{\mu\tanh\left(\lambda\overline{h}_-\right)\left[1-\tanh\left(\lambda\overline{h}_+\right)\right] - \left[1-\tanh\left(\lambda\overline{h}_-\right)\right]}{\mu\tanh\left(\lambda\overline{h}_+\right)\tanh\left(\lambda\overline{h}_-\right)+1}\sin(\lambda x)\,d\lambda,$$

$$R_0(x) = \int\limits_0^\infty \frac{1-\tanh\left(\lambda\overline{h}_+\right)\tanh\left(\lambda\overline{h}_-\right)}{\mu\tanh\left(\lambda\overline{h}_+\right)\tanh\left(\lambda\overline{h}_-\right)+1}\cos(\lambda x)\,d\lambda,$$

$$E_0(x) = \int\limits_0^\infty \frac{\left(1-\tanh\left(\lambda\overline{h}_-\right)\right)\left(\mu\tanh\left(\lambda\overline{h}_+\right)-1\right)}{\mu\tanh\left(\lambda\overline{h}_+\right)\tanh\left(\lambda\overline{h}_-\right)+1}\sin(\lambda x)\,d\lambda,$$

$$Q_0(x) = \int\limits_0^\infty \frac{1}{\cosh\left(\lambda\overline{h}_+\right)\left[\mu\tanh\left(\lambda\overline{h}_+\right)\tanh\left(\lambda\overline{h}_-\right)+1\right]}\cos(\lambda x)\,d\lambda,$$

$$G_0(x) = \int\limits_0^\infty \frac{\tanh\left(\lambda\overline{h}_-\right)}{\cosh\left(\lambda\overline{h}_+\right)\left[\mu\tanh\left(\lambda\overline{h}_+\right)\tanh\left(\lambda\overline{h}_-\right)+1\right]}\sin(\lambda x)\,d\lambda,$$

$$l_{01}(\xi) = -\frac{\mu+1}{\mu}\tau_{+}^{(0)}(\xi) - (\mu+1)\tau_{+}^{(0)}(\xi)$$
$$-\frac{\mu+1}{\pi}\int_{L_{01}} R_0(\eta-\xi)\left[\tau_{+}^{(0)}(\eta) - \tau_{-}^{(0)}(\eta)\right]d\eta \qquad (\xi \in L_{01},)$$

$$l_{02}(\xi) = -\frac{\mu+1}{\pi}\int_{L_{01}} \frac{\left[\tau_{+}^{(0)}(\eta) - \tau_{-}^{(0)}(\eta)\right]d\eta}{\eta-\xi}$$
$$-\frac{\mu+1}{\pi}\int_{L_{01}} M_0(\eta-\xi)\left[\tau_{+}^{(0)}(\eta) - \tau_{-}^{(0)}(\eta)\right]d\eta \qquad (\xi \in L_{02}),$$

$$l_{03}(\xi) = \frac{F_1^{(0)} + F_2^{(0)}}{2\overline{h}_1} - \frac{1}{2\overline{h}_1}\int_{\overline{e}}^{\overline{f}} sgn(\xi-\eta)\,T_{*}^{(0)}(\eta)\,d\eta$$
$$-\frac{2(\mu+1)}{\pi}\int_{L_{01}} G_0(\eta-\xi)\left[\tau_{+}^{(0)}(\eta) - \tau_{-}^{(0)}(\eta)\right]d\eta \qquad \left(\xi \in \left(\overline{e},\overline{f}\right)\right).$$

Now, each interval $(\alpha_k, \beta_k), (\gamma_k, \delta_k)$ of the system L_{01} and L_{02} and the interval (e, f), is transformed into the interval $(-1, 1)$, assuming

$$\xi = \frac{\beta_k-\alpha_k}{2}t + \frac{\beta_k+\alpha_k}{2}, \eta = \frac{\beta_k-\alpha_k}{2}u + \frac{\beta_k+\alpha_k}{2} \quad \left(k=\overline{1,N_1}\right) \quad (-1<t,u<1),$$
$$\xi = \frac{\delta_k-\gamma_k}{2}t + \frac{\delta_k+\gamma_k}{2}, \eta = \frac{\delta_k-\gamma_k}{2}u + \frac{\delta_k+\gamma_k}{2} \quad \left(k=\overline{1,N_2}\right) \quad (-1<t,u<1),$$
$$\xi = \frac{\overline{f}-\overline{e}}{2}t + \frac{\overline{f}+\overline{e}}{2}, \qquad \eta = \frac{\overline{f}-\overline{e}}{2}u + \frac{\overline{f}+\overline{e}}{2} \qquad (-1<t,u<1).$$

As a result, the SIE (26.18) is transformed into a system of integral equations for the functions

$$\varphi_k(t) = \varphi_0\left(\frac{\beta_k-\alpha_k}{2}t + \frac{\beta_k+\alpha_k}{2}\right) \qquad \left(-1<t<1, k=\overline{1,N_1}\right),$$
$$\psi_k(t) = \psi_0\left(\frac{\delta_k-\gamma_k}{2}t + \frac{\delta_k+\gamma_k}{2}\right) \qquad \left(-1<t<1, k=\overline{1,N_2}\right),$$
$$\psi_*(t) = \overline{h}_{+}\left(\frac{\overline{f}-\overline{e}}{2}t + \frac{\overline{f}+\overline{e}}{2}\right) \qquad (-1<t<1)$$

on the interval $(-1, 1)$

$$
\begin{cases}
\dfrac{1}{\pi}\displaystyle\int_{-1}^{1}\frac{\varphi_k(u)\,du}{u-t}+\frac{1}{\pi}\sum_{\substack{m=1\\ m\neq k}}^{N_1}\int_{-1}^{1}K_{km}^{(1)}(t,u)\,\varphi_m(u)\,du \\
+\dfrac{1}{\pi}\displaystyle\sum_{m=1}^{N_1}\int_{-1}^{1}K_{km}(t,u)\,\varphi_m(u)\,du+\frac{\mu+1}{\pi}\sum_{m=1}^{N_2}\int_{-1}^{1}R_{km}^{(1)}(t,u)\,\psi_m(u)\,du \\
-\dfrac{\mu_1(\mu+1)}{2\pi}\displaystyle\int_{-1}^{1}\widetilde{Q}(t,u)\,\psi_*(u)\,du=l_{1k}(t)\qquad\left(t\in(-1,1)\,;k=\overline{1,N_1}\right) \\
\dfrac{\mu}{\pi}\displaystyle\sum_{m=1}^{N_1}\int_{-1}^{1}R_{km}^{(2)}(t,u)\,\varphi_m(u)\,du+\frac{\mu+1}{\pi}\int_{-1}^{1}\frac{\psi_k(u)\,du}{u-t} \\
+\dfrac{\mu+1}{\pi}\displaystyle\sum_{\substack{m=1\\ m\neq k,}}^{N_2}\int_{-1}^{1}M_{km}^{(1)}(t,u)\,\psi_m(u)\,du+\frac{\mu+1}{\pi}\sum_{m=1}^{N_2}\int_{-1}^{1}M_{km}(t,u)\,\psi_m(u)\,du, \\
-\dfrac{\mu_1(\mu+1)}{2\pi}\displaystyle\int_{-1}^{1}\widetilde{G}(t,u)\,\psi_*(u)\,du=l_{2k}(t)\qquad\left(t\in(-1,1)\,;k=\overline{1,N_2}\right), \\
\dfrac{2}{\pi}\displaystyle\sum_{m=1}^{N_1}\int_{-1}^{1}Q_m(t,u)\,\varphi_m(u)\,du+\frac{2(\mu+1)}{\pi}\sum_{m=1}^{N_2}\int_{-1}^{1}G_m(t,u)\,\psi_m(u)\,du \\
-\dfrac{\mu_1}{\pi}\displaystyle\int_{-1}^{1}\frac{\psi_*(u)\,du}{u-t}-\frac{\mu_1}{\pi}\int_{-1}^{1}\widetilde{E}(t,u)\,\psi_*(u)\,du \\
-\dfrac{\overline{f}-\widetilde{e}}{4\overline{h}_+}\displaystyle\int_{-1}^{1}\operatorname{sign}(t-u)\,\psi_*(u)\,du=l_{30}(t)\qquad(t\in(-1,1)),
\end{cases}
\tag{26.19}
$$

where

$$
K_{km}^{(1)}(t,u)=\left(u-\frac{\beta_k-\alpha_k}{\beta_m-\alpha_m}t+\frac{\beta_m+\alpha_m}{\beta_m-\alpha_m}-\frac{\beta_k+\alpha_k}{\beta_m-\alpha_m}\right)^{-1}\qquad\left(m,k=\overline{1,N_1},m\neq k\right),
$$

$$
M_{km}^{(1)}(t,u)=\left(u-\frac{\delta_k-\gamma_k}{\delta_m-\gamma_m}t+\frac{\delta_m+\gamma_m}{\delta_m-\gamma_m}-\frac{\delta_k+\gamma_k}{\delta_m-\gamma_m}\right)^{-1}\qquad\left(m,k=\overline{1,N_1},m\neq k\right),
$$

$$
R_{km}^{(1)}(t,u)=\lambda_m^{(2)}R_0\left(\lambda_m^{(2)}u+\frac{\delta_m+\gamma_m}{2}-\lambda_k^{(1)}t-\frac{\beta_k+\alpha_k}{2}\right)\ \left(k=\overline{1,N_1},m=\overline{1,N_2}\right),
$$

$$
R_{km}^{(2)}(t,u)=\lambda_m^{(1)}M_0\left(\lambda_m^{(1)}u+\frac{\beta_m+\alpha_m}{2}-\lambda_k^{(2)}t-\frac{\delta_k+\gamma_k}{2}\right)\ \left(k=\overline{1,N_2},m=\overline{1,N_1}\right),
$$

$$K_{km}(t,u) = \lambda_m^{(1)} K_0\left(\lambda_m^{(1)} u + \frac{\beta_m+\alpha_m}{2} - \lambda_k^{(1)} t - \frac{\beta_k+\alpha_k}{2}\right) \quad \left(m,k=\overline{1,N_1}\right),$$

$$M_{km}(t,u) = \lambda_m^{(2)} M_0\left(\lambda_m^{(2)} u + \frac{\delta_m+\gamma_m}{2} - \lambda_k^{(2)} t - \frac{\delta_k+\gamma_k}{2}\right) \quad \left(m,k=\overline{1,N_2}\right),$$

$$Q_m(t,u) = \lambda_m^{(1)} Q_0\left(\lambda_m^{(1)} u + \frac{\beta_m+\alpha_m}{2} - \lambda_0 t - \frac{\overline{f}+\overline{e}}{2}\right) \quad \left(m=\overline{1,N_1}\right),$$

$$G_m(t,u) = \lambda_m^{(2)} G_0\left(\lambda_m^{(2)} u + \frac{\delta_m+\gamma_m}{2} - \lambda_0 t - \frac{\overline{f}+\overline{e}}{2}\right) \quad \left(m=\overline{1,N_2}\right),$$

$$\widetilde{Q}(t,u) = \lambda_0 Q_0(\lambda_0(u-t)),$$

$$\widetilde{G}(t,u) = \lambda_0 G_0(\lambda_0(u-t)),$$

$$\widetilde{E}(t,u) = \lambda_0 E_0(\lambda_0(u-t)),$$

$$l_{1k}(t) = l_{01}\left(\lambda_k^{(1)} t + \frac{\beta_k+\alpha_k}{2}\right) \quad \left(k=\overline{1,N_1}, t\in(-1,1)\right),$$

$$l_{2k}(t) = l_{02}\left(\lambda_k^{(2)} t + \frac{\delta_k+\gamma_k}{2}\right) \quad \left(k=\overline{1,N_2}, t\in(-1,1)\right),$$

$$l_{30}(t) = l_{03}\left(\lambda_0 t + \frac{\overline{f}+\overline{e}}{2}\right) \quad (t\in(-1,1)),$$

$$\lambda_m^{(1)} = (\beta_m-\alpha_m)/2 \quad \left(m=\overline{1,N_1}\right),$$

$$\lambda_k^{(2)} = (\delta_k-\gamma_k)/2 \quad \left(k=\overline{1,N_2}\right), \quad \lambda_0 = \left(\overline{f}-\overline{e}\right)/2,$$

and conditions (26.11) and (26.15) are transformed into the following conditions:

$$\int_{-1}^{1} \varphi_k(u)\,du = 0 \qquad \left(k=\overline{1,N_1}\right),$$

$$\int_{-1}^{1} \psi_k(u)\,du = \widetilde{P}_k, \quad \left(\widetilde{P}_k = \frac{2P_k}{(\delta_k-\gamma_k)\,|a_1|\,(G_+ + G_-)}\right), \quad \left(k=\overline{1,N_2}\right),$$

$$\int_{-1}^{1} \psi_*(u)\,du = \frac{2\left(F_2^{(0)}-F_1^{(0)}\right)}{\overline{f}-\overline{e}} + \int_{-1}^{1} \widetilde{T}_1(u)\,du, \tag{26.20}$$

where

$$\widetilde{T}_1(u) = T_*^{(0)}\left(\lambda_0 u + \frac{\overline{f}+\overline{e}}{2}\right).$$

To the system of SIE (26.19) under conditions (26.20), we apply the method for solving SIE described in [3, 8, 10]. For this purpose, we put

$$\varphi_k(t) = \frac{x_k(t)}{\sqrt{1-t^2}} \quad \left(-1 < t < 1, k = \overline{1, N_1}\right),$$
$$\psi_k(t) = \frac{y_k(t)}{\sqrt{1-t^2}} \quad \left(-1 < t < 1, k = \overline{1, N_2}\right),$$
$$\psi_*(t) = \frac{z(t)}{\sqrt{1-t^2}} \qquad (-1 < t < 1),$$

where $x_k(t)$, $y_k(t)$ and $z(t)$ are functions belonging to the Hölder class of functions on the interval $[-1, 1]$, and then we choose an arbitrary natural number M. Further, following the well-known procedure [6], the system (26.19) under conditions (26.20) is reduced to the SLAE

$$\begin{cases} \dfrac{1}{M}\displaystyle\sum_{n=1}^{M}\frac{x_k(u_n)}{u_n - t_r} + \frac{1}{M}\sum_{m=1, m\neq k}^{N_1}\sum_{n=1}^{M} K_{km}^{(1)}(t_r, u_n)\, x_m(u_n) \\ +\dfrac{1}{M}\displaystyle\sum_{m=1}^{N_1}\sum_{n=1}^{M} K_{km}(t_r, u_n)\, x_m(u) + \frac{\mu+1}{M}\sum_{m=1}^{N_2}\sum_{n=1}^{M} R_{km}^{(1)}(t_r, u_n)\, y_m(u_n) \\ -\dfrac{\mu_1(\mu+1)}{2M}\displaystyle\sum_{n=1}^{M}\widetilde{Q}(t_r, u_n)\, z(u_n) = l_{1k}(t_r) \quad \left(k = \overline{1, N_1}, r = \overline{1, M-1}\right), \\ \displaystyle\sum_{n=1}^{M} x_k(u_n) = 0 \quad \left(k = \overline{1, N_1}\right), \\ \dfrac{\mu}{M}\displaystyle\sum_{m=1}^{N_1}\sum_{n=1}^{M} R_{km}^{(2)}(t_r, u_n)\, x_m(u_n) + \frac{\mu+1}{M}\sum_{n=1}^{M}\frac{y_k(u_n)}{u_n - t_r} \\ +\dfrac{\mu+1}{M}\displaystyle\sum_{m=1, m\neq k}^{N_2}\sum_{n=1}^{M} M_{km}^{(1)}(t_r, u_n)\, y_m(u_n) + \frac{\mu+1}{M}\sum_{m=1}^{N_2}\sum_{n=1}^{M} M_{km}(t_r, u_n)\, y_m(u_n) \\ -\dfrac{\mu_1(\mu+1)}{2M}\displaystyle\sum_{n=1}^{M}\widetilde{G}(t_r, u_n)\, z(u_n) = l_{2k}(t_r) \quad \left(k = \overline{1, N_2}, r = \overline{1, M-1}\right), \\ \dfrac{\pi}{M}\displaystyle\sum_{n=1}^{M} y_k(u_n) = \widetilde{P}_k \quad \left(k = \overline{1, N_2}\right), \\ \dfrac{2}{M}\displaystyle\sum_{m=1}^{N_1}\sum_{n=1}^{M} Q_m(t_r, u_n)\, x_m(u) + \frac{2(\mu+1)}{M}\sum_{m=1}^{N_2}\sum_{n=1}^{M} G_m(t_r, u_n)\, y_m(u_n) \\ -\dfrac{\mu_1}{M}\displaystyle\sum_{n=1}^{M}\frac{z(u_n)}{u_n - t_r} - \frac{\mu_1}{M}\sum_{n=1}^{M}\widetilde{E}(t_r, u_n)\, z(u_n) \\ -\dfrac{\pi\left(\overline{f} - \overline{e}\right)}{4\overline{h}_+ M}\displaystyle\sum_{n=1}^{M}\operatorname{sign}(t_r - u_n)\, z(u_n) = l_{30}(t_r) \quad \left(r = \overline{1, M-1}\right), \\ \dfrac{\pi}{M}\displaystyle\sum_{n=1}^{M} z(u_n) = \frac{2\left(F_2^{(0)} - F_1^{(0)}\right)}{\overline{f} - \overline{e}} + \frac{\pi}{M}\sum_{n=1}^{M}\sqrt{1 - u_n^2}\,\widetilde{T}_1(u_n). \end{cases} \tag{26.21}$$

The unknowns are determined from (26.21)

$$\{x_k(u_n)\}\left(k=\overline{1,N_1}\right), \{y_k(u_n)\}\left(k=\overline{1,N_2}\right), \{z(u_n)\}\ \left(n=\overline{1,M}\right).$$

Here

$$u_n=\cos\left(\frac{2n-1}{2M}\pi\right)\ \left(n=\overline{1,M}\right),\quad t_r=\cos\left(\frac{\pi r}{M}\right)\ \left(r=\overline{1,M-1}\right) \tag{26.22}$$

are Chebyshev nodes, i.e. the roots, respectively, of the equations $T_M(u)=0$, $U_{M-1}(t)=0$, where $T_M(u)$ and $U_{M-1}(t)$ are Chebyshev polynomials of the first and second kind, respectively.

Let us proceed to determining the coefficients of destructive tangential SIF in the antiplane problem of elasticity theory at the end points a_k, b_k of cracks $(a_k, b_k)\left(k=\overline{1,N_1}\right)$. If in the known formulas for a homogeneous body SIFs of under antiplane deformation [8] the shear modulus is replaced by the reduced shear modulus, then for the shear modulus of a piecewise homogeneous body we obtain the following formulas [3]

$$\begin{aligned} K_{\mathrm{III}}(a_k)&=\frac{G_+}{1+\mu}\lim_{x\to a_k+0}\sqrt{2\pi(x-a_k)}\varphi(x),\\ K_{\mathrm{III}}(b_k)&=-\frac{G_+}{1+\mu}\lim_{x\to b_k-0}\sqrt{2\pi(b_k-x)}\varphi(x),\quad (\mu=G_+/G_-). \end{aligned} \tag{26.23}$$

After transforming formula (26.23) in accordance with (26.17) and moving to the variable t, we have

$$K_{\mathrm{III}}^{(0)}(a_k)=\frac{\sqrt{\lambda_k^{(1)}}}{1+\mu}x_k(-1),\quad K_{\mathrm{III}}^{(0)}(b_k)=-\frac{\sqrt{\lambda_k^{(1)}}}{1+\mu}x_k(1)\quad\left(k=\overline{1,N_1}\right), \tag{26.24}$$

where dimensionless SIFs are introduced

$$K_{\mathrm{III}}^{(0)}(a_k)=\frac{K_{\mathrm{III}}(a_k)}{\sqrt{\pi|a_1|}G_+},\quad K_{\mathrm{III}}^{(0)}(b_k)=\frac{K_{\mathrm{III}}(b_k)}{\sqrt{\pi|a_1|}G_+}\quad\left(k=\overline{1,N_1}\right). \tag{26.25}$$

The values $x_k(\pm 1)$ of the function $x_k(t)$ included in (26.24)) are determined using the Lagrange interpolation polynomial at the Chebyshev nodes (26.22) and expressed by the formulas [6]

$$\begin{aligned} x_k(1)&=\frac{1}{M}\sum_{n=1}^{M}(-1)^{n+1}x_k(u_n)\cot\left(\frac{2n-1}{4M}\pi\right),\\ x_k(-1)&=\frac{1}{M}\sum_{n=1}^{M}(-1)^{M+n}x_k(u_n)\tan\left(\frac{2n-1}{4M}\pi\right)\quad\left(k=\overline{1,N_1}\right). \end{aligned} \tag{26.26}$$

For dimensionless crack openings $w_k(t)$, taking into account Eq. (26.17), we write [6]

$$w_k(t) = \int_{-1}^{1} \frac{\chi_k(\eta)\, d\eta}{\sqrt{1-\eta^2}} = \frac{\pi}{M} \sum_{u_n \le t} \chi_k(u_n) \qquad (-1 \le t \le 1), \tag{26.27}$$

$$\hat{w}_k(x) = \frac{b_k - a_k}{2} w_k\left(\frac{2x}{b_k - a_k} - \frac{b_k + a_k}{b_k - a_k}\right) \qquad (a_k \le x \le b_k),$$

where $\hat{w}_k(x)$ is the opening of the kth crack.

26.4 Numerical Results

Let us consider a special case where the connection line between dissimilar bands contains a symmetrically located crack and an absolutely rigid thin inclusion, $L = L_1 \cup L_2, L_1 = (-b, -a), L_2 = (a, b), (0 \le a < b)$. The stringer is located on the upper edge of the band on the interval $[-e, e]$. In this case, we assume that equal stresses of arbitrary intensities act on the crack edges $\tau_+^{(1)}(x) = \tau_-^{(1)}(x) = \tau(x)\ (x \in L_1)$. In this particular case, the system (26.19) can be written in the form

$$\begin{cases} \frac{1}{\pi}\int_{-1}^{1} \frac{\widetilde{\varphi}(u)\, du}{u-t} + \frac{1}{\pi}\int_{-1}^{1} \widetilde{K}(u-t)\widetilde{\varphi}(u)\, du + \frac{\mu+1}{\pi}\int_{-1}^{1} \widetilde{R}_1(u-t)\widetilde{\psi}(u)\, du \\ -\frac{\mu_1(\mu+1)}{2\pi}\int_{-1}^{1} \widetilde{Q}_1(t,u)\psi_*(u)\, du = -\frac{(\mu+1)^2}{\mu}\tau_0(t) \quad (t \in (-1,1)), \\ \frac{\mu}{\pi}\int_{-1}^{1} \widetilde{R}_2(t,u)\widetilde{\varphi}(u)\, du + \frac{\mu+1}{\pi}\int_{-1}^{1} \frac{\widetilde{\psi}(u)\, du}{u-t} + \frac{\mu+1}{\pi}\int_{-1}^{1} \widetilde{M}(u-t)\widetilde{\psi}(u)\, du \\ -\frac{\mu_1(\mu+1)}{2\pi}\int_{-1}^{1} \widetilde{G}_1(t,u)\psi_*(u)\, du = 0 \quad (t \in (-1,1)), \\ \frac{2}{\pi}\int_{-1}^{1} \widetilde{Q}_2(t,u)\widetilde{\varphi}(u)\, du + \frac{2(\mu+1)}{\pi}\int_{-1}^{1} \widetilde{G}_2(t,u)\widetilde{\psi}(u)\, du - \frac{\mu_1}{\pi}\int_{-1}^{1} \frac{\widetilde{\psi}_*(u)\, du}{u-t} \\ -\frac{\mu_1}{\pi}\int_{-1}^{1} \widetilde{E}(u-t)\widetilde{\psi}_*(u)\, du - \frac{\widetilde{e}}{2\overline{h}_1}\int_{-1}^{1} \operatorname{sign}(t-u)\widetilde{\psi}_*(u)\, du = l_{30}(t) \\ (t \in (-1,1)), \end{cases} \tag{26.28}$$

where

$$\widetilde{\varphi}(u) = \varphi_0\left(\frac{1-k}{2}u + \frac{1+k}{2}\right) \qquad (-1 < u < 1),$$

$$\widetilde{\psi}(u) = \psi_0\left(\frac{1-k}{2}u + \frac{1+k}{2}\right) \qquad (-1 < u < 1),$$

$$\psi_*(u) = \overline{h}_+(\overline{e}u) \quad (-1 < u < 1),$$

$$\widetilde{K}(u-t) = \frac{1-k}{2}K_0\left(\frac{1-k}{2}(u-t)\right),$$

$$\widetilde{R}_1(u-t) = \frac{1-k}{2}R_0\left(\frac{1-k}{2}(u-t)\right),$$

$$\widetilde{R}_2(t,u) = \frac{1-k}{2}R_0\left(\frac{1-k}{2}(u-t) - 1 - k\right),$$

$$\widetilde{Q}_1(t,u) = \overline{e}Q_0\left(\overline{e}u - \frac{1-k}{2}t + \frac{1+k}{2}\right),$$

$$\widetilde{Q}_2(t,u) = \frac{1-k}{2}Q_0\left(\frac{1-k}{2}u - \overline{e}t - \frac{1+k}{2}\right),$$

$$\widetilde{G}_1(t,u) = \overline{e}G_0\left(\overline{e}u - \frac{1-k}{2}t - \frac{1+k}{2}\right),$$

$$\widetilde{G}_2(t,u) = \frac{1-k}{2}G_0\left(\frac{1-k}{2}u - \overline{e}t + \frac{1+k}{2}\right), \quad l_{30} = l_{03}(\overline{e}t)\,; (-1 < t < 1),$$

$$\widetilde{M}(u-t) = \frac{1-k}{2}M_0\left(\frac{1-k}{2}(u-t)\right), \quad \widetilde{E}(u-t) = \overline{e}E_0(\overline{e}(u-t)),$$

$$\mu = \frac{G_+}{G_-}, \quad \mu_1 = \frac{G_1}{G_+}, \quad k = \frac{a}{b}, \quad \rho = \frac{\overline{e}}{\overline{h}_1}, \quad F_1^{(0)} = F_2^{(0)} = 0.001.$$

Further, using the well-known numerical-analytical method from system (26.28) and conditions (26.20), we have the following system of SLAE:

$$\begin{cases} \frac{1}{M}\sum_{n=1}^{M}\left[\frac{1}{u_n-t_r}+\widetilde{K}\left(u_n-t_r\right)\right]x\left(u_n\right)+\frac{\mu+1}{M}\sum_{n=1}^{M}\widetilde{R}_1\left(u_n-t_r\right)y\left(u_n\right) \\ -\frac{\mu_1\left(\mu+1\right)}{2M}\sum_{n=1}^{M}\widetilde{Q}_1\left(t_r,u_n\right)z\left(u_n\right)=-\frac{\left(\mu+1\right)^2}{\mu}\tau_0\left(t_r\right)\quad\left(r=\overline{1,M-1}\right), \\ \sum_{n=1}^{M}x\left(u_n\right)=0, \\ \frac{\mu}{M}\sum_{n=1}^{M}\widetilde{R}_2\left(t_r,u_n\right)x\left(u_n\right)+\frac{\mu+1}{M}\sum_{n=1}^{M}\left[\frac{1}{u_n-t_r}+\widetilde{M}\left(u_n-t_r\right)\right]y\left(u_n\right) \\ -\frac{\mu_1\left(\mu+1\right)}{2M}\sum_{n=1}^{M}\widetilde{G}_1\left(t_r,u_n\right)z\left(u_n\right)=0\quad\left(r=\overline{1,M-1}\right), \\ \frac{\pi}{M}\sum_{n=1}^{M}y\left(u_n\right)=\widetilde{P}, \\ \frac{2}{M}\sum_{n=1}^{M}\widetilde{Q}_2\left(t_r,u_n\right)x\left(u_n\right)+\frac{2\left(\mu+1\right)}{M}\sum_{n=1}^{M}\widetilde{G}_2\left(t_r,u_n\right)y\left(u_n\right) \\ -\frac{\mu_1}{M}\sum_{n=1}^{M}\left[\frac{1}{u_n-t_r}+\widetilde{E}\left(u_n-t_r\right)\right]z\left(u_n\right) \\ -\frac{\pi\overline{e}}{2M\overline{h}_1}\sum_{n=1}^{M}\sqrt{1-u_n^2}\,\mathrm{sign}\left(t_r-u_n\right)z\left(u_n\right)=l_{30}\left(t_r\right)\quad\left(r=\overline{1,M-1}\right), \\ \sum_{n=1}^{M}z\left(u_n\right)=\sum_{n=1}^{M}\sqrt{1-u_n^2\widetilde{T}_1\left(u_n\right)}, \end{cases} \tag{26.29}$$

where

$$\widetilde{P}=\frac{2P}{\left(\delta-\gamma\right)\left|a_1\right|\left(G_++G_-\right)},$$

u_n and t_r are the corresponding Chebyshev nodes from (26.22). In this case, the corresponding formulas (26.18) - (26.19) for the dimensionless SIF at the points $-b$ and $-a$ are replaced by the formulas

$$K_{\mathrm{III}}^{(0)}\left(-b\right)=\frac{1}{\mu+1}x\left(-1\right),\qquad K_{\mathrm{III}}^{(0)}\left(-a\right)=-\frac{1}{\mu+1}x\left(1\right), \tag{26.30}$$

$$K_{\mathrm{III}}\left(-b\right)=\sqrt{2\pi b}G_+K_{\mathrm{III}}^{(0)}\left(-b\right),\quad K_{\mathrm{III}}\left(-a\right)=\sqrt{2\pi b}G_+K_{\mathrm{III}}^{(0)}\left(-a\right).$$

Next, for simplicity, assume that $\tau_0\left(t\right)=0.01$ and $\widetilde{P}=0.001$. Then, after solving the system (26.23) for different values of the parameters $\mu,\overline{h}_+,\overline{h}_-,\rho$, we find the corresponding values of the SIF at the crack endpoints using formulas (26.24) and the corresponding crack opening using formula (26.21).

The values of the SIF calculated using formulas (26.24) $K_{\mathrm{III}}^{0}\left(-b\right)$ and $K_{\mathrm{III}}^{0}\left(-a\right)$ depending on the change in the parameter k (when changing this parameter k), it is assumed that the length of symmetrically located cracks and inclusions is constant) and at fixed values of other parameters $\mu, \overline{h}_{+}, \overline{h}_{-}$ are given in the Table 26.1, which are characteristic for these SIF and at other values of the parameters. Here we additionally note that by $K_{\mathrm{III}}^{0}(-a)$ at $a = 0$ ($k = 0$) we should understand the limiting value of the crack's SIF at $-a \to -0$. The values of the SIF $K_{\mathrm{III}}^{(0)}(-b)$, $K_{\mathrm{III}}^{(0)}(-a)$ have a maximum value at $k = 0$, i.e. when the crack and inclusion are connected and $K_{\mathrm{III}}^{(0)}(-a)$ is always greater than $K_{\mathrm{III}}^{(0)}(-b)$. When the parameter k increases to unity, both values decrease and approach each other.

The dimensionless crack openings $w_1(t)$ calculated by the formula (26.27) for different values of the parameter μ at a fixed value $k = 0.5, \overline{h}_{+} = 2, \overline{h}_{-} = 8$ are graphically shown in Fig. 26.2.

For some values of characteristic parameters, at which loads do not exceed the theoretical strength of materials, the crack openings can be larger compared to the values given here. These values are regulated by deformation criteria for crack propagation. For example, according to the well-known Leonov-Panasyuk criterion [4], when the opening reaches a critical value δ_{kr} in the vicinity of the crack tops, the crack spreads. Then, before the opening reaches the calculated large values, the crack is already widespread.

Table 26.1: Values of the SIF $K_{\mathrm{III}}^{(0)}(-b)$, $K_{\mathrm{III}}^{(0)}(-a)$ $\left(\mu = 2, \overline{h}_{+} = 2, \overline{h}_{-} = 8\right), \rho = 5$.

k	0	0.005	0.1	0.2	0.3	0.4
$K_{\mathrm{III}}^{(0)}(-b)$	0.00361	0.00361	0.00359	0.00356	0.00353	0.0035
$K_{\mathrm{III}}^{(0)}(-a)$	0.00396	0.00396	0.00393	0.00389	0.00384	0.00378

k	0.5	0.6	0.7	0.8	0.9	0.95
$K_{\mathrm{III}}^{(0)}(-b)$	0.00347	0.00344	0.00341	0.00338	0.00335	0.00334
$K_{\mathrm{III}}^{(0)}(-a)$	0.00371	0.00364	0.00356	0.00348	0.0034	0.00336

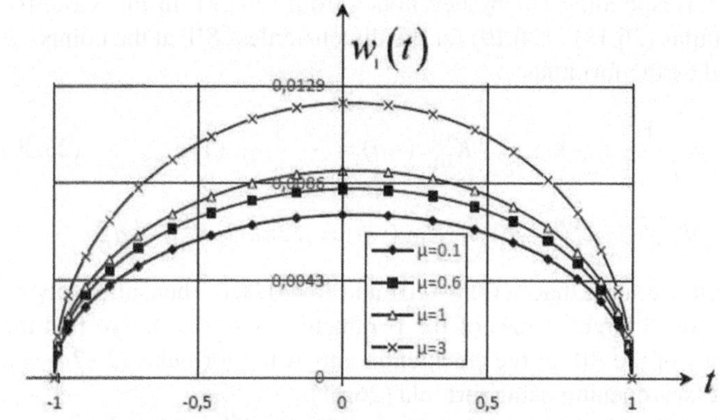

Fig. 26.2: Graphs of $w_1(t)$ for different μ.

26.5 Conclusion

The general formulation and the main governing equations of the problem are obtained and solved. Characteristics of the problems are represented by explicit formulas. For the particular case, the all main characteristics of the problem are represented by numerical-analytical method and given in Table 26.1 and Fig. 26.2.

References

[1] Alexandrov VM, Mkhitaryan SM (1983) Kontaktnyye zadachi dlya tel s tonkimi pokrytiyami i prosloykami (in Russ., Contact Problems for Bodies with Thin Coatings and Interlayers). Nauka, Moscow

[2] Cherepanov GP (1979) Mechanics of Brittle Fracture. McGraw Hill Higher Education

[3] Erdogan FE, Gupta GD, Cook TS (1973) Numerical solution of singular integral equations. In: Sih, G.C. (ed) Methods of Analysis and Solutions of Crack Problems. Mechanics of Fracture, Vol 1. Springer, Dordrecht, DOI 10.1007/978-94-017-2260-5_7

[4] Kachanov LM (1974) Osnovy mekhaniki razrusheniya (in Russ., Fundamentals of Fracture Mechanics). (Nauka, Moscow

[5] Manukyan EA, Mkrtchyan MS (2010) Ob antiploskoy zadache kusochno–odnorodnogo uprugogo sloya, soderzhashchego na linii spaya sistemu shcheley i absolyutno zhestkikh tonkikh vklyucheniy (in Russ., On the antiplane problem of a piecewise homogeneous elastic layer containing a system of cracks and absolutely rigid thin inclusions on the weld line). Proceedings of NAS RA, Mechanics 63(2):21-33

[6] Melan E (1932) Ein Beitrag zur Theorie geschweißter Verbindungen. Ingenieur Archiv 2(3):123-129, DOI 10.1007/BF02079955

[7] Mkhitaryan SM (1993) O dvukh smeshannykh zadachakh, svyazannykh s voprosami vzaimodeystviya kontsentratorov napryazheniy razlichnykh tipov s massivnymi telami pri antiploskoy deformatsii (in Russ., On two mixed problems related to the interaction of stress concentrators of different types with massive bodies under antiplane deformation). In: Mekhanika deformiruyemogo tvordogo tela, Yerevan, Publishing House of the National Academy of Sciences of Armenia, pp. 129-143

[8] Panasyuk VV, Savruk MP, Datsyshin AP (976) Raspredeleniye napryazheniy okolo treshchin v plastinakh i obolochkakh (in Russ., Stress Distribution Around Cracks in Plates and Shells). Naukova Dumka, Kiev

[9] Popov GY (1982) Kontsentratsiya uprugikh napryazheniy vozle shtampov, razrezov, tonkikh vklyucheniy i podklyucheniy (in Russ., Concentration of Elastic Stresses Near Stamps, Cuts, thin Inclusions and Reinforcements). Nauka, Moscow

[10] Theocaris PS, Ioakimidis NI (1977) Numerical integration methods for the solution of singular integral equations. Quarterly of Applied Mathematics XXX(1):173-185

[11] Vorovich II, Alexandrov VM, Babeshko VM (1974) Neklassicheskiye smeshannyye zadachi teorii uprugosti (in Russ., Non-classical Mixed Problems in Elasticity Theory). Nauka, Moscow

Chapter 27
Reliability of Reinforced Concrete Buildings Under Random Seismic Impacts

Oleg V. Mkrtychev, Ekaterina M. Lokhova, Andrey A. Reshetov, and Timur A. Tagirov

Abstract The paper examines the performance characteristics of beam-column junctions in RC frame structures, which are widely used in construction in seismic areas, under seismic impact. Research was conducted using a direct non-linear dynamic method implemented in the ANSYS – LS/DYNA software package. A set of accelerograms with dominant frequencies from 2 to 4 Hz was used as seismic impact.

Isofields of damage accumulation function, isofields of stress intensity and graphs of horizontal displacements of the central part of the junction under random seismic impact are presented as main results.

The results of the study show that the use of devices with increased damping properties in junctions contributes to a significant increase in the overall reliability and seismic resistance of the building's load-bearing system.

Key words: Reliability theory, Direct non-linear dynamic methods, Seismic resistance, Seismic protection

27.1 Introduction

Earthquake is one of the most dangerous and destructive natural phenomena, which can cause numerous human casualties and serious damage to buildings and structures in a matter of minutes. There are approximately 5 000 000 earthquakes worldwide each year. Only 1% of them are felt by people (so-called "felt earthquakes"), because most events occur at great depths or the energy released is relatively small. Strong

---

Oleg V. Mkrtychev · Ekaterina M. Lokhova · Andrey A. Reshetov (corresponding author) · Timur A. Tagirov
National Research Moscow State University of Civil Engineering, Yaroslavskoye Hwy, 26, Moscow, 129337, Russian Federation, e-mail: mkrtychev@yandex.ru, elm97@mail.ru, andrew331@bk.ru, tagirovv_timur@mail.ru

H. Altenbach and A. V. Sahakyan, *Contact Mechanics and Relevant Problems*,
Advanced Structured Materials 253, https://doi.org/10.1007/978-3-032-20869-9_27

earthquakes that lead to catastrophes occur much less frequently - about 10 times a year.

The main damaging factors of earthquakes are displacements, surface ruptures, oscillations or vibrations of soil from seismic waves and tectonic movements of the Earth's crust, accompanied by cracking and compaction, which usually lead to serious damage to buildings and structures and, consequently, to human casualties. Furthermore, earthquakes can cause fires, floods, landslides, avalanches and tsunamis, which pose a serious danger to human life and health.

In seismic areas, reinforced concrete (RC) frame structures are widely used. The seismic resistance and mechanical safety of such frames are determined not only by the load-bearing capacity of the main load-bearing elements (beams, columns), but also by the load-bearing capacity of the connecting joints. Despite their high spatial rigidity and ability to redistribute loads, the connecting joints between elements represent areas with increased stress concentration and, consequently, significant vulnerability to seismic impacts. Therefore, the rational design and calculation of the connecting joints between beams and columns are essential for ensuring the reliability and seismic resistance of the entire structure.

Currently, various seismic protection systems for buildings and structures exist: for example, damping devices, dynamic vibration dampers [16], rubber-metal bearings [1], and others. In many cases, the most effective way to improve seismic resistance is the use of damping devices, which, in turn, can be divided into two groups: systems based on viscous dampers [14] and systems in which energy dissipation occurs through the development of plastic deformations.

Systems capable of developing plastic deformations can be installed directly at the junctions of beams and columns. These systems have been described in a number of modern experimental and theoretical studies [4, 6, 12, 19, 20]. Despite the existence of experimental studies, there is currently a lack of calculation substantiations and developed methods for calculating and assessing the reliability of such systems.

27.2 Problem Statement

The object of the study is a RC beam-column junction of a five-story building (Fig. 27.1) [8]. The floor height of the building under consideration is 3 m; the beam cross-section is 0.4×0.5 m; the column cross-section is 0.4×0.4 m, the building dimensions in plan are 18×24 m. The junction is considered both with and without the use of a seismic protection device.

27.3 Methods

The study is performed using the ANSYS/LS-DYNA software package. The direct nonlinear dynamic method used in this work is based on the equation of motion,

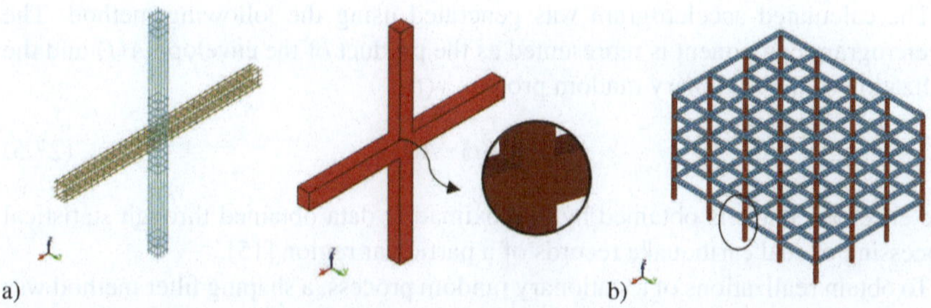

Fig. 27.1: Calculation scheme: a) detailed model of junction; b) RC - frame.

which has the form [7]

$$\mathbf{M}\mathbf{u}_t + \mathbf{C}\mathbf{u}_t + \mathbf{K}\mathbf{u}_t = \mathbf{f}_t^a, \tag{27.1}$$

where: **M** is the mass matrix, **C** is the damping matrix, **K** is the stiffness matrix, **f** is the vector of external loads, and **u** is the vector of nodal displacements.

This paper uses an explicit integration scheme based on the central difference method. This approach allows account the physical, geometric, and structural nonlinearity of the model. Solving the initial differential equation (27.1) lead to a system of nonlinear differential equations solved at each time step. In this method, nodal accelerations and velocities (hereinafter **a** and **v**) are not calculated by numerical differentiation of displacements, but are determined directly (introduced into the number of nodal degrees of freedom as unknowns).

The accelerations that are calculated at the nth layer are expressed through the vectors of external and internal forces (f_i^{ext} and f_i^{int})

$$\mathbf{a}_n = \mathbf{M}^{-1}\left(\mathbf{F}_n{}^{\text{ext}} - \mathbf{F}_n{}^{\text{int}}\right). \tag{27.2}$$

The vector calculating is the most labor-intensive task, as it accounts for all types of nonlinearities. The velocity and displacement vectors at each step are determined as

$$\mathbf{v}_{t+\Delta t/2} = \mathbf{v}_{t-\Delta t/2} + \mathbf{a}_t \Delta t_t, \tag{27.3}$$

$$\mathbf{u}_{t+\Delta t} = \mathbf{u}_t + \mathbf{v}_{t+\Delta t/2} \frac{\Delta t_t + \Delta t_{t+\Delta t}}{2}. \tag{27.4}$$

One of the key tasks in seismic modeling of high-rise buildings is the correct assignment of external impact parameters. The traditional approach, which relies on a single calculated accelerogram, does not always ensure the required reliability and seismic resistance of all load-bearing structural elements. This study employs a methodology based on a representative set of accelerograms [17]. When forming the representative set, the seismic impact was considered as a non-stationary random process. It should be noted that an accelerogram with a certain dominant frequency may be critical for some load-bearing elements of a building, but be harmless for others [15].

The calculated accelerogram was generated using the following method. The accelerogram component is represented as the product of the envelope $A(t)$ and the realization of the stationary random process $y(t)$

$$a(t) = y(t) \cdot A(t). \tag{27.5}$$

The envelope shape is obtained by approximating data obtained through statistical processing of real earthquake records of a particular region [15].

To obtain realizations of a stationary random process, a shaping filter method was used. The basic idea of the method is the linear transformation of an orthonormal sequence of normally distributed random numbers $x(n)$ into a sequence $y(n)$ with a given correlation function

$$y(n) = \sum_{k=0}^{M} c(k) \cdot x(n-k), \quad n = 1, \ldots N, \tag{27.6}$$

where $x(n)$ is the Gaussian white noise and $c(k)$ are weighting factors.

Using the Fourier transform, we translate (27.6) into the frequency domain:

$$Y(k) = C(k) \cdot X(k), \tag{27.7}$$

where $Y(k), C(k), X(k)$ are Fourier transforms of functions $y(n), c(n), x(n)$ respectively (obtained by the fast Fourier transform (FFT))

$$C(k) = |C(k)| \cdot \exp\left(i\phi_C(k)\right), \tag{27.8}$$

$$\mathrm{X}(k) = |X(k)| \cdot \exp\left(i\phi_x(k)\right), \tag{27.9}$$

where $|C(k)|$ is the frequency response of filter, $\phi_C(k)$ is the phase – frequency response of filter, $|X(k)|$ is the frequency response of Gaussian white noise and $\phi_x(k)$ is the phase frequency response of Gaussian white noise (is random and evenly distributed from 0 to 2π for all frequencies). Then

$$Y(k) = |C(k)| \cdot |X(k)| \cdot \exp\left(i\left(\phi_\mathrm{C}(\mathrm{k})+\phi_\mathrm{X}(\mathrm{k})\right)\right). \tag{27.10}$$

Assuming a zero phase-frequency characteristic for the filter ($\phi_C(k) = 0$), we obtain

$$Y(k) = |C(k)| \cdot |X(k)| \cdot \exp\left(i\varphi_X(k)\right). \tag{27.11}$$

In this case, the targeted spectrum function is assumed to be known (like the envelope, it is determined through statistical processing of a large number of real earthquake records). The stationary random process is reconstructed using the inverse Fourier transform

$$y(n) = \mathrm{IFFT}(Y(k)). \tag{27.12}$$

For this study, one accelerogram of a seismic impact with an intensity of 9 ($a_{max} = 4$ m/s^2) with dominant frequencies $f = 2$ Hz was generated using the above-described method (Fig. 27.2). This study utilizes the Karagozian & Case (K&C)

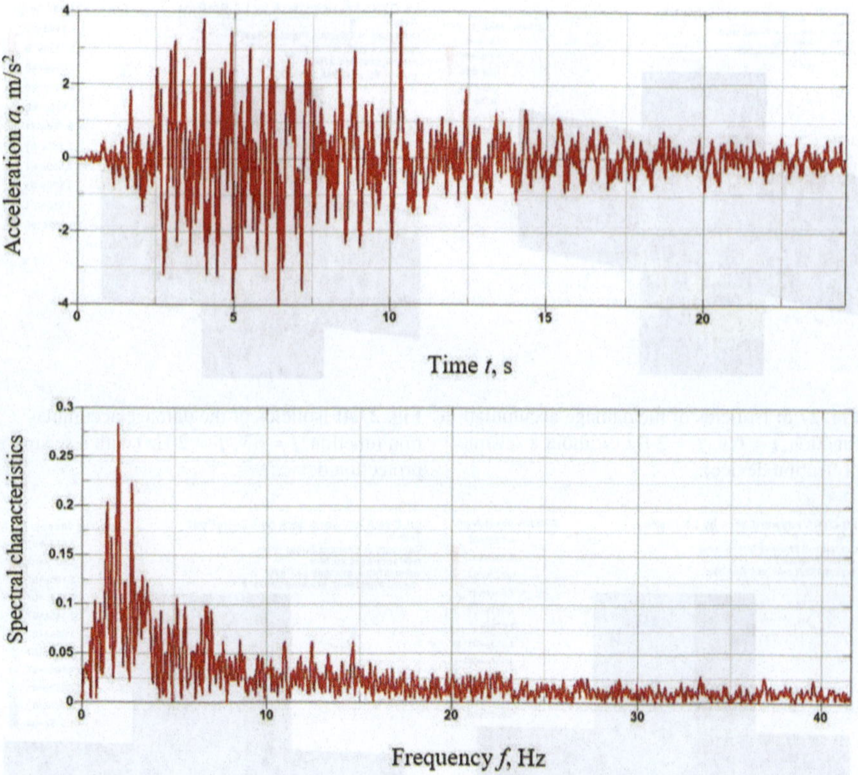

Fig. 27.2: Calculated accelerogram and its spectral characteristics (dominant frequency $f = 2$ Hz).

mathematical model of concrete [5, 18]. This material model is primarily designed to simulate the dynamic responses of RC structures. It takes into account concrete properties such as hardening, stiffness degradation, and strain rate effects. The model is a tri-invariant. In this study, in addition to junction without seismic protection of the device, junction with increased dissipative methods is also considered [8]. The results of the main calculations are presented below.

Figure 27.3 shows the isofields of the damage accumulation function for a junction without a seismic protection device at time $t = 6$ s. Figure 27.4 shows the isofields of the damage accumulation function for a junction with increased dissipative properties at time $t = 6$ s. Damage accumulation for the K&C material model is represented by a function varying from 0 to 2, where 0 is the absence of damage and 2 is the complete exhaustion of the bearing capacity. Figure 27.5 show the stress intensity isofields in concrete for a junction without the use of a seismic protection device at time $t = 6$ s. Figure 27.6 shows the stress intensity isofields in concrete for a junction with increased dissipative properties at time $t = 6$ s. Figure 27.7 shows the mosaics of stress intensity in reinforcing bars for a junction without using a seismic protection device at time $t = 6$ s. Figure 27.8 shows the mosaics of stress intensity in

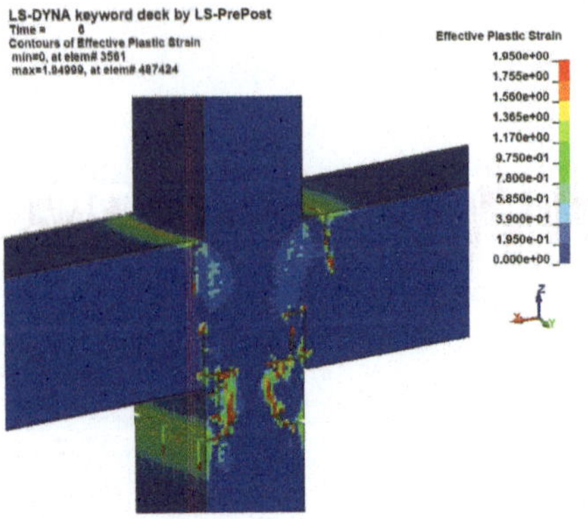

Fig. 27.3: Isofields of the damage accumulation function, $t = 6$ s, $f = 2$ Hz (without a seismic protection device).

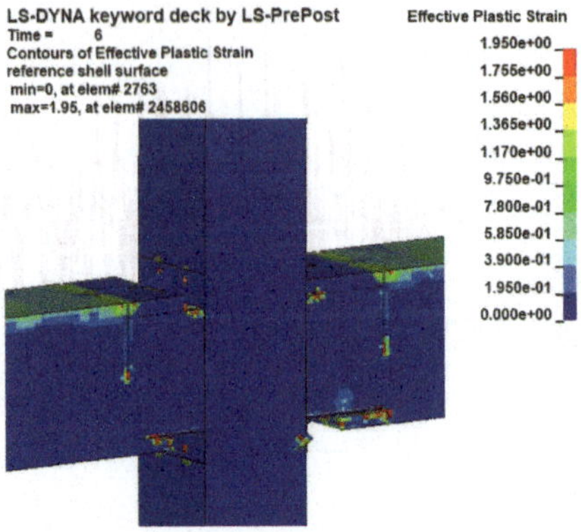

Fig. 27.4: Isofields of the damage accumulation function, $t = 6$ s, $f = 2$ Hz (with a seismic protection device).

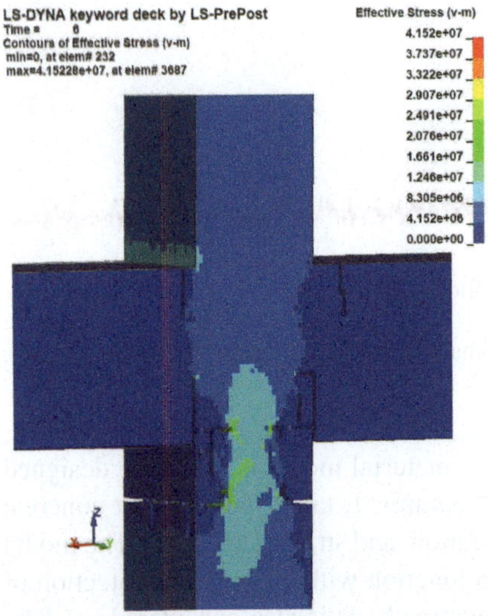

Fig. 27.5: Stress intensity isofields in concrete, MPa, $t = 6$ s, $f = 2$ Hz (without a seismic protection device).

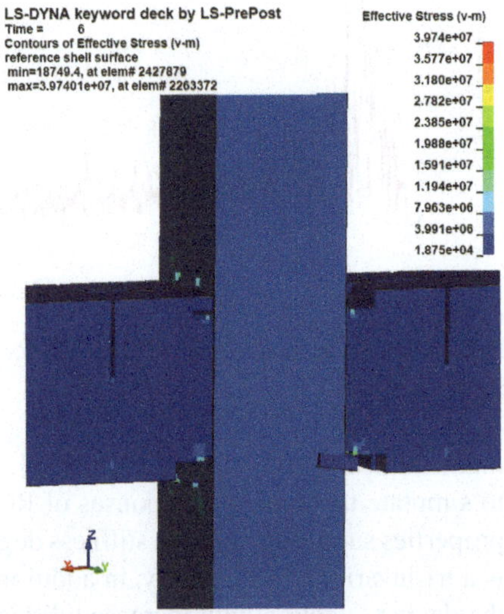

Fig. 27.6: Stress intensity isofields in concrete, MPa, $t = 6$ s, $f = 2$ Hz (with a seismic protection device).

reinforcing bars for a junction with increased dissipative properties at time $t = 6$ s. Figure 27.9 shows a comparison of the horizontal displacements of the central part of the junction for the coupling junction without the use of seismic protection devices. Figure 27.10 shows a comparison of the horizontal displacements of the central part of the junction for a connection with the use of seismic protection devices. According to the research results, the most unfavorable accelerogram for the coupling nodes under consideration is an accelerogram with a dominant frequency of $f = 2$ Hz.

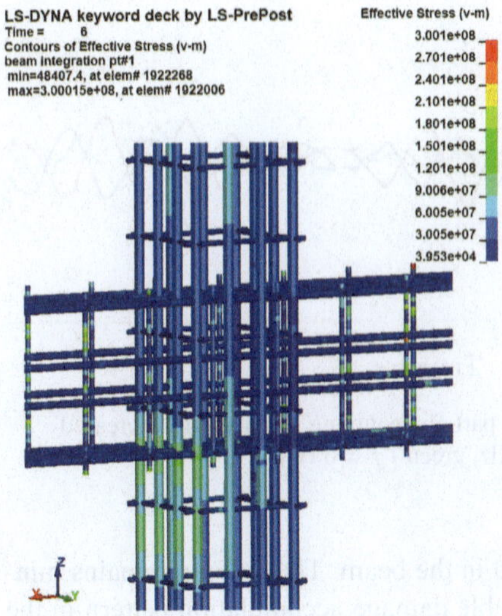

Fig. 27.7: Stress intensity in reinforcing bars, MPa, $t = 6$ s, $f = 2$ Hz (without a seismic protection device).

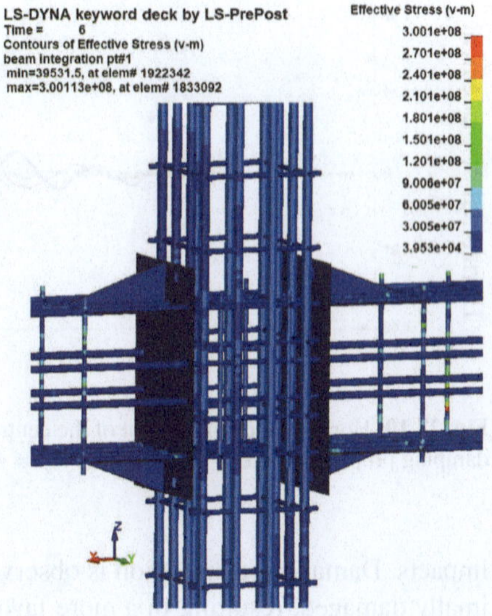

Fig. 27.8: Stress intensity in reinforcing bars, MPa, $t = 6$ s (with a seismic protection device).

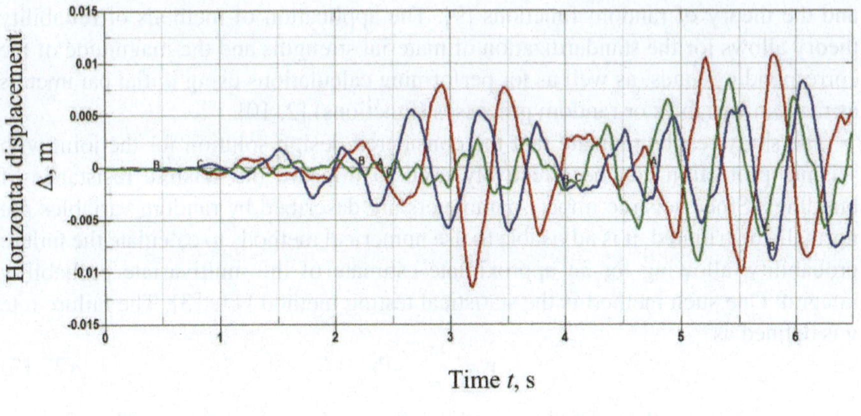

Fig. 27.9: Horizontal displacement of the central part of the coupling junction (red - $f = 2$ Hz, blue - $f = 4$ Hz, green - $f = 6$ Hz).

Based on the results of calculating the volumetric model of the joint without seismic protection, it was concluded that the load-bearing capacity and seismic resistance of the joint with the calculated reinforcement are not ensured. Intensive damage accumulation occurs, with extensive zones of failure where the concrete and reinforcement interact. At $t = 6$ s, through cracks form in the concrete.

When the joint operates with seismic protection, plastic deformations develop in the steel elements, which positively impacts the joint's performance under seismic

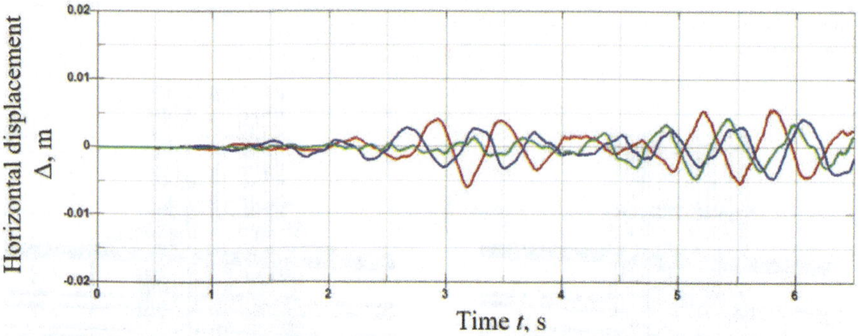

Fig. 27.10: Horizontal displacement of the central part of a coupling junction with increased damping properties (red - f = 2 Hz, blue - f = 4 Hz, green - f = 6 Hz).

impacts. Damage accumulation is observed in the beam. The column remains minimally damaged, resulting in a more favorable damage accumulation pattern in the joint and significantly increasing the structure's maintainability after an earthquake.

The Russian regulatory documents used for the design of buildings and structures are based on methods of reliability theory [3]. The theory of reliability of building structures utilizes the apparatus of probability theory, the theory of random processes, and the theory of random functions [9]. The application of methods of reliability theory allows for the standardization of material strengths and the magnitude of the corresponding loads, as well as for performing calculations using initial parameters as random variables or random processes (functions) [2, 10].

The study results indicate that the considered design solution for the joint with seismic protection can be effectively used to improve the seismic resistance of buildings. Since seismic impact parameters are described by random variables not normally distributed, it is advisable to use numerical methods to calculate the failure probability, allowing for an approximate estimate of the multivariate probability integral. One such method is the statistical testing method [11, 13]. The failure rate ν is defined as

$$\nu = \frac{m}{n} \approx \mathrm{P_f}, \qquad (27.13)$$

where m is the number of failures and n is the total number of tests. The resulting failure rate ν will be an estimate of the desired failure probability $\mathrm{P_f}$.

The statistical testing method is based on a number of theorems:

1. Bernoulli's theorem: as $n \to \infty \quad \nu \to P_f$,
2. Khinchin's theorem (law of large numbers): $n \to \infty, \quad \langle \nu \rangle \to \mu_\nu$, and
3. Lindeber-Levy theorem (central limit theorem): the average failure rate ν has an asymptotically normal distribution.

Acknowledgements This research was supported by Ministry Science and Higher Education of the Russian Federation (project No. FSWG-2023-0004) "A system of territorial seismic protection of critical infrastructure facilities based on granular metamaterials with the properties of wide-range phonon crystals".

References

[1] Bunov AA (2014) Otsenka nadezhnosti zdaniy s sistemoy seysmoizolyatsii v rezinometallicheskikh opor (in Russ., Reliability Assessment of Buildings with Rubber-Metal Bearing Seismic Isolation System). PhD Thesis "Candidate of Technical Sciences, 05.23.17", Moscow

[2] Branstein MF (1960) Prilozheniye veroyatnostnykh metodov k raschetu sooruzheniy na seysmicheskoye vozdeystviye (in Russ., Application of probabilistic methods to calculating structures for seismic impact). Stroitel'naya mekhanika i raschet sooruzheniy (2):25-29

[3] Chirkov VP (2006) Prikladnyye metody teorii nadezhnosti v raschetakh stroitel'nykh konstruktsiy (in Russ., Applied Methods of Reliability Theory in Calculating Building Structures). Route, Moscow

[4] Li W, Fan H-T, Jiang H-M, Chen Y (2024) Seismic behavior of demountable reinforced concrete (RC) beam-to-column joints with damage-control fuses. Buildings 14(1):20, DOI 10.3390/buildings14010020

[5] Malvar LJ, Crawford JE, Wesevich JW, Simons D (1997 A plasticity concrete material model for DYNA3D. International Journal of Impact Engineering 19(9-10):847-873, DOI 10.1016/S0734-743X(97)00023-7

[6] Meas K, Li B, Pham TP (2014) Experimental and numerical studies on the seismic performance of RC interior beam-column joints. Advances in Structural Engineering 17(2):233-247, DOI 10.1260/1369-4332.17.2.233

[7] Mkrtychev OV (2022) Razvitiye pryamykh nelineynykh dinamicheskikh metodov rascheta na seysmicheskiye vozdeystviya (in Russ., Development of direct nonlinear dynamic methods for calculating seismic effects). Promyshlennoe i grazhdanskoe stroitel'stvo (7):12-16, DOI 10.33622/0869-7019.2022.07.12-16

[8] Mkrtychev OV, Lokhova EM (2025) Seysmostoykost' uzlov zhelezobetonnogo ramnogo karkasa s povyshennymi dissipativnymi svoystvami (in Russ., Seismic resistance of reinforced concrete frame units with increased dissipative properties). Izvestiya vuzov. Stroitel'stvo (7)16-26

[9] Pshenichkina VA, Pshenichkin AP (2009) Nadezhnost' zdaniy i osnovaniy v osobykh usloviya (in Russ., Reliability of Buildings and Foundations under Special Conditions). VolgGASU Publishing House, Volgograd

[10] Pshenichkina VA, Voronkova GV, Drozdov VV (2021) Veroyatnostnyye metody stroitel'noy mekhaniki i teoriya nadezhnosti stroitel'nykh konstruktsiy (in Russ., Probabilistic Methods of Structural Mechanics and Reliability Theory of Building Structures), Vol 2. Volgograd State Technical University, Volgograd

[11] Rayzer VD (2010) Teoriya nadezhnosti sooruzheniy (in Russ., Reliability Theory of Structures). ASV, Moscow

[12] Rong X, Zhang X, Zhang J (2021) Seismic behavior of innovation steel-embedded precast concrete beam-to-column joints. Structures 34:4952–4964, DOI 10.1016/j.istruc.2021.10.082

[13] Sargsyan AE, Rayzer VD, Mkrtychev OV (1999) Metod statisticheskikh ispytaniy pri raschete stroitel'nykh konstruktsiy na nadezhnost' (in Russ., Statistical Testing Method in Reliability Analysis of Structural Systems). RGOTUPS, Moscow

[14] Sigaher-Boyle AN, Constantinou MC (2003) Scissor-jack-damper energy dissipation system. Earthquake Spectra 19(1):133–158

[15] STO 01.01.2024 (2024) Seysmostoykost' zdaniy i sooruzheniy. Raschetnyye polozheniyan (in Russ., Seismic Resistance of Buildings and Structures. Design Provisions). Nauchno-issledovatel'skiy institut seysmostoykogo stroitel'stva i nadezhnosti konstruktsiy, Moscow

[16] Uzdin AM, Sandovich TA, Al-Naser-Mohammad SA (1993) Osnovy teorii seysmostoykosti i seysmostoykogo stroitel'stva zdaniy i sooruzheniy (in Russ., Fundamentals of the Theory of Earthquake Resistance and Seismic-Resistant Construction of Buildings and Structures). BE Vedeneev VNIIG, St Petersburg

[17] Wu J, Li L, Du X (2017) Numerical study on the asphalt concrete structure for blast and impact load using the Karagozian and Case concrete model. Applied Sciences 7(2):202, DOI 10.3390/app7020202

[18] Wu Y, Crawford JE (2015) Numerical modeling of concrete using a partially associative plasticity model. Journal of Engineering Mechanics 141(12):04015051, DOI 10.1061/(ASCE)EM.1943-7889.0000952

[19] Wu Z, Lu X, Bao H, Li L, Lu Z (2024) Experimental response of semi-rigid reinforced concrete beam-column joints with bolted angle dissipating connections. Journal of Building Engineering 90:109345, DOI 10.1016/j.jobe.2024.109345

[20] Zhang H, Huang W, Li J, Quan W (2025) Seismic behavior of precast concrete beam-to-column connection with double box bolted couplers. Engineering Structures 326:119499, DOI 10.1016/j.engstruct.2024.119499

Chapter 28
Asymptotic Solutions of Problems with Localized Right-Hand Sides with Application to the Wave Equation and Maxwell's Equations

Vladimir E. Nazaikinskii

Abstract Semiclassical asymptotic solutions of inhomogeneous differential and pseudodifferential equations can be obtained by taking the convolution of the asymptotics of Green's function with the right-hand side. However, if the right-hand side is localized in a neighborhood of a point (fixed source) or a curve (moving source), then it is much more computationally efficient to avoid the laborious operation of taking the convolution of two rapidly varying functions and solve the asymptotic problem directly. We present recent results in this direction based on the use of Maslov's canonical operator in conjunction with an auxiliary Cauchy problem and a formal Duhamel principle.

Key words: Semiclassical asymptotics, Green's function, Localized right-hand side, Maslov canonical operator

28.1 Introduction

We deal with linear evolution equations of the wave type with a localized, possibly, moving source of the form

$$f(x,t) = F\left(t, \frac{x - x(t)}{h}\right) \tag{28.1}$$

on the right-hand side, where the function $F(t, y)$ rapidly decays as $|y| \to \infty$ and $h \to 0$ is a small parameter, so that the source has efficient size $\sim h$. The aim is to construct asymptotic solutions of such equations as $h \to 0$. Examples of interest to us include:

Vladimir E. Nazaikinskii
Ishlinsky Institute for Problems in Mechanics RAS, Vernadsky Ave. 101-1, Moscow, 119526 Russian Federation, e-mail: nazaikinskii@googlemail.com

H. Altenbach and A. V. Sahakyan, *Contact Mechanics and Relevant Problems*, Advanced Structured Materials 253, https://doi.org/10.1007/978-3-032-20869-9_28

- The wave equation
$$\frac{\partial^2 u}{\partial t^2} - c^2(x)\frac{\partial^2 u}{\partial x^2} = f(x,t).$$
- The Maxwell equations
$$\operatorname{div}\mathbf{B} = 0, \qquad \operatorname{rot}\mathbf{E} + \frac{1}{c}\frac{\partial \mathbf{B}}{\partial t} = 0,$$
$$\operatorname{div}\mathbf{D} = 4\pi\rho, \qquad \operatorname{rot}\mathbf{H} - \frac{1}{c}\frac{\partial \mathbf{D}}{\partial t} = \frac{4\pi}{c}\mathbf{J}$$
with regard to inhomogeneity and dispersion: the permittivity ε and the permeability μ in the material relations
$$\mathbf{D} = \varepsilon\mathbf{E}, \qquad \mathbf{B} = \mu\mathbf{H}$$
may depend on x and ω.
- The water wave equation for a basin of finite variable depth $D(x)$
$$\frac{\partial^2 u}{\partial t^2} + G\left(x, -i\frac{\partial}{\partial x}\right)u = f(x,t), \quad G(x,p) = |p|\tanh(|p|D(x)).$$

A solution of an equation with a right-hand side can always be obtained as the convolution of the right-hand side with a Green's function of the equation (plus an appropriate solution of the homogeneous equation), and hence it is natural to obtain asymptotic solutions by convolution with the asymptotics of the Green's function. There are numerous papers on the semiclassical asymptotics of the Green's function, and we do not even attempt to give a complete list; instead, we only cite some of the papers most important to us.

Much attention was paid to the case of the Helmholtz equation. J.B. Keller [11] was apparently the first to consider the asymptotics of the Green's function in the case of variable coefficients. V.M. Babich [3, 4] obtained the asymptotics by sewing the exact Green's function for the equation with frozen coefficients with asymptotic solutions of the homogeneous equation outside the source, but he did not consider caustics. V.V. Kucherenko [12] included the consideration of caustics by using the Maslov canonical operator [13]. Further references can be found in [5].

As to the general theory not limited to the Helmholtz equation, J.J. Duistermaat and L.Hörmander [10] and R.B. Melrose and G.A. Uhlmann [15] developed a construction using a so-called *Lagrangian intersection* to obtain the asymptotics of Green's function with respect to smoothness. In [19], B.R. Vainberg carried out a comprehensive study of the semiclassical (short-wave) asymptotics of the Green's function; in particular, in several cases he provided estimates for the difference between the asymptotic and exact solutions. B.Yu. Sternin and V.E. Shatalov [17] developed a modification (canonical operator on a Lagrangian manifold with boundary) of the construction in [10, 15] for the case of semiclassical asymptotics.

The list could be continued. However, note that the approach using the Green's function is not very efficient for practical computations. The Green's function, as well as its semiclassical asymptotics, is rapidly oscillating, and hence computing the convolution with the right-hand side, which itself may be rapidly varying, is not an easy task. Hence the natural idea to obtain asymptotic solutions of the equations in question directly, without resorting to Green's function. This idea is viable if the right-hand side itself admits an expression via the canonical operator (as is the case for the localized function (28.1). The corresponding construction for equations with localized right-hand side was carried out by A.Yu. Anikin, S.Yu. Dobrokhotov, M. Rouleux, and the present author in [1, 2].

Note that, in contrast with much of the work concerning the asymptotics of Green's function (e.g., Vainberg's paper [19]), we deal exclusively with asymptotic solutions (i.e., functions that, when substituted into the equation, produce small discrepancy) and never even speak of the asymptotics of the solutions. In other words, we do not study whether these asymptotic solutions approximate some real solutions (if any of these exist at all). At first glance, this might seem strange. Nevertheless, there are deep underlying reasons. The equations dealt with in asymptotic theory are often not exact themselves. They might describe the original physical problem only approximately and/or be valid only in specific ranges of parameters occurring in these equations. It might happen that these equations do not have exact solutions at all, and even if they have solutions, these solutions might have little to do with real physical processes. The following example given in [1] provides a perfect illustration of this. In the above-mentioned equation of water waves in a basin of finite depth, for a certain relation between the typical wavelength and the source size, it is natural to expand the symbol in a Taylor series in powers of the momenta and retain only the first few terms. In the first approximation, we obtain the wave equation, and in the second approximation, one has something like the linearized Boussinesq equation of fourth order in the spatial variables. The asymptotic solutions of the Cauchy problem for this equation describe the physical processes fairly well, while the Cauchy problem itself is ill-posed, and it would be meaningless to consider exact solutions. On the other hand, even if the equation to be studied is well posed, asymptotic solutions may fail to approximate exact solutions; for example, in eigenvalue problems semiclassical asymptotic eigenfunctions (also known as quasi-modes) often approximate only linear combinations of true eigenfunctions with slightly different eigenvalues rather than individual eigenfunctions.

The paper is organized as follows. In Sect. 28.2, we recall the main points of the construction of the Maslov canonical operator, and Sect. 28.3 gives a brief survey of the results in [1, 2] and some more recent results in this direction.

28.2 Maslov's Canonical Operator

There is a well-developed tool, known as Maslov's canonical operator [13, 14], for the construction of asymptotic solutions of differential equations with small parameter $h \to 0$ multiplying the derivatives. In this section, we very briefly recall relevant notions, mainly following [7, 8] (where modern computationally efficient version of the canonical operator is developed), but with some deviations and simplifications.

The canonical operator acts on functions on a *Lagrangian manifold* in the phase space $\mathbb{R}^{2n}_{(x,p)}$ with coordinates $(x, p) = (x_1, \dots, x_n, p_1, \dots, p_n)$ and takes them to rapidly oscillating functions on the configuration space $\mathbb{R}^n_x$.

28.2.1 Rapidly Oscillating Functions

Let $H^s_h(\mathbb{R}^n)$, $h > 0$, be the Sobolev space of order $s \in \mathbb{R}$ equipped with the equivalent norm $\|u\|_{s,h}$ defined by

$$\|u\|^2_{s,h} = \int_{\mathbb{R}^n} \overline{u(x)}(1 - h^2\Delta)^s u(x)\, dx,$$

and let $\mathcal{F}^h(\mathbb{R}^n)$ be the space of smooth functions $f(x, h)$, $h \in (0, 1)$, such that the semi-norm

$$\|f\|_{\psi,s} := \sup_{h\in(0,1)} \|\psi f(\cdot, h)\|_{s,h}$$

is finite for any $\psi \in C_0^\infty(\mathbb{R}^n)$ and any $s \in \mathbb{R}$. The elements of $\mathcal{F}^h(\mathbb{R}^n)$ are called *rapidly oscillating functions*; we write $f = \widehat{O}(h^k)$ if $h^{-k} f \in \mathcal{F}^h(\mathbb{R}^n)$ and $f = \widehat{O}(h^\infty)$ if $f = \widehat{O}(h^k)$ for all k. A simple example is given by a WKB function

$$f(x, h) = e^{\frac{i}{h}S(x)} a(x), \tag{28.2}$$

where the phase function $S(x)$ is smooth and real-valued and the amplitude $a(x)$ is smooth.

28.2.2 Lagrangian Manifolds

Let Λ be a smooth n-dimensional manifold. The points of Λ will be denoted by α. Whenever convenient, we also denote by $\alpha = (\alpha_1, \dots, \alpha_n)$ local coordinate systems on Λ. From now on, we assume that Λ is oriented and only consider coordinate systems consistent with the orientation. Now let an embedding

$$\iota\colon \Lambda \to \mathbb{R}^{2n}_{(x,p)}$$

be given. One says that Λ is a *Lagrangian manifold* in $\mathbb{R}^{2n}_{(x,p)}$ if $\iota^*(\omega^2) = 0$, where

$$\omega^2 = dp \wedge dx \equiv dp_1 \wedge dx_1 + \cdots + dp_n \wedge dx_n$$

is the standard symplectic form on the phase space $\mathbb{R}^{2n}_{(x,p)}$. We denote the functions specifying the embedding by

$$x = X(\alpha), \quad p = P(\alpha).$$

Locally, Λ being a Lagrangian manifold is equivalent to the conditions

$$\operatorname{rank}\begin{pmatrix} X_\alpha \\ P_\alpha \end{pmatrix} = n, \qquad X_\alpha^\top P_\alpha = P_\alpha^\top X_\alpha,$$

where (as well as throughout the following) X and P are considered as column vectors, the superscript $^\top$ indicates transposition of matrices, and subscripts stand for derivatives.

If Λ is a Lagrangian manifold, then

- Locally on Λ there always exists a smooth real-valued function $S(\alpha)$ (an *action* on Λ) such that $dS(\alpha) = P(\alpha)\,dX(\alpha) \equiv P_1(\alpha)\,dX_1(\alpha) + \cdots + P_2(\alpha)\,dX_2(\alpha)$.
- The Jacobian
$$J^1(\alpha) = \det(X_\alpha - iP_\alpha)$$
is nowhere zero on Λ.

In the present paper, we only need to consider Lagrangian manifolds Λ such that

- There exists a globally defined action $S(\alpha)$ on Λ.
- On Λ, there exists a continuous branch of the argument $\arg J^1(\alpha) \pmod{4\pi}$.

We say that Λ is *simple* if these conditions are satisfied. From now on, we only deal with simple Lagrangian manifolds. Note that the Bohr-Sommerfeld-Maslov quantization conditions are always satisfied on simple Lagrangian manifolds. Examples of simple Lagrangian manifolds include

- The manifold $\Lambda = \{(x, p) \colon p = S_x(x)\}$ associated with the WKB function (28.2).
- The conormal bundle $N^*\Gamma$ of an arbitrary smooth submanifold Γ of $\mathbb{R}^n_x$.

28.2.3 Canonical Operator as a "Black Box"

Let Λ be a simple Lagrangian manifold in $\mathbb{R}^{2n}_{(x,p)}$. We choose and fix an action $S(\alpha)$, a branch of the argument $\arg J^1(\alpha) \pmod{4\pi}$, and a real-valued volume form (a *measure*) $d\mu(\alpha)$ on Λ. These choices essentially (modulo lower-order terms) fix the definition of the canonical operator

$$K \equiv K^h_{\Lambda,d\mu} \colon C^\infty_{\mathrm{pr}}(\Lambda) \longrightarrow \mathcal{F}^h(\mathbb{R}^n),$$

where $C^\infty_{\mathrm{pr}}(\Lambda)$ is the space of smooth functions $A(\alpha)$ on Λ such that the restriction to $\operatorname{supp} A$ of the natural projection

$$\pi_x : \Lambda \to \mathbb{R}^n_x, \quad \alpha \mapsto X(\alpha),$$

is proper (i.e., the preimage of any compact set is compact). The canonical operator is naturally extended to the space of functions $A(\alpha, h)$, $h \in [0, 1)$, that are smooth functions of $h \in [0, 1)$ with values in $C^\infty_{\mathrm{pr}}(\Lambda)$.

The most important properties of the canonical operator are given by commutation formulas, which show how h-(pseudo)differential operators (i.e., pseudodifferential operators with small parameter h multiplying the derivatives) act on functions of the form KA. Let

$$\widehat{\mathscr{H}} = \mathscr{H}\left(\overset{2}{x}, -ih\overset{1}{\frac{\partial}{\partial x}}, h\right)$$

be an h-pseudodifferential operator with symbol

$$\mathscr{H}(x, p, h) = H_0(x, p) + hH_1(x, p) + \cdots .$$

Then the following commutation formulas hold.

- First commutation formula

$$\widehat{\mathscr{H}} KA = K\left[H_0\big|_\Lambda A + O(h)\right] + \widehat{O}(h^\infty), \tag{28.3}$$

where $H_0\big|_\Lambda = \iota^*(H_0)$ is the restriction of the *principal symbol* $H_0(x, p)$ of the operator $\widehat{\mathscr{H}}$ to the Lagrangian manifold Λ.

- Second commutation formula
Now assume that $H_0\big|_\Lambda = 0$ identically on Λ. Then the Hamiltonian vector field

$$V(H_0) = H_{0p}(x, p)\frac{\partial}{\partial x} - H_{0x}(x, p)\frac{\partial}{\partial p}$$

is everywhere tangent to Λ, its restriction $V(H_0)\big|_\Lambda$ to Λ is well defined, and

$$\widehat{\mathscr{H}} KA = -ihK[\Pi A + O(h)] + \widehat{O}(h^\infty), \tag{28.4}$$

where Π is the *transport operator*,

$$\Pi = V(H_0)\big|_\Lambda + iH_{\mathrm{sub}}\big|_\Lambda + \frac{1}{2}\frac{\mathscr{L}_{V(H_0)|_\Lambda} d\mu}{d\mu}.$$

Here $\mathscr{L}_{V(H_0)|_\Lambda} d\mu$ is the Lie derivative of the measure $d\mu$ along the vector field $V(H_0)\big|_\Lambda$, and

$$H_{\mathrm{sub}}(x, p) = H_1(x, p) + \frac{i}{2}\sum_{j=1}^n \frac{\partial^2 H_0}{\partial x_j \partial p_j}(x, p)$$

is the *subprincipal symbol* of the operator $\widehat{\mathscr{H}}$. The second commutation formula becomes especially simple if the subprincipal symbol of the operator is zero and the measure $d\mu$ is invariant with respect to the Hamiltonian vector field; then Π is just the differentiation along $V(H_0)\big|_\Lambda$.

28.2.4 Solution Scheme for the Homogeneous Equations

The canonical operator has been designed to construct asymptotic solutions of homogeneous equations of the form $\widehat{\mathscr{H}}u = 0$. The general scheme of this construction is as follows.

1. Find a Lagrangian manifold $\Lambda \subset \mathbb{R}^{2n}$ on which the principal symbol of the operator $\widehat{\mathscr{H}}$ vanishes, $H_0(x, p)\big|_\Lambda = 0$.
2. Ensure that the quantization conditions are satisfied on Λ. (In the present paper, these conditions are always satisfied automatically.)
3. Find a measure $d\mu$ on Λ invariant with respect to the Hamiltonian vector field.
4. Seek an asymptotic solution of the equation in the form $u = KA$, where $K \equiv K^h_{\Lambda,d\mu}$ is the canonical operator on the Lagrangian manifold Λ with the measure $d\mu$.
5. The unknown amplitude A is found as a solution of the first-order linear ODE $\Pi A = 0$ (the *transport equation*) along the trajectories of the Hamiltonian system $\dot{x} = H_{0p}, \dot{p} = -H_{0x}$ on the manifold Λ
6. The function u thus obtained satisfies the equation $\widehat{\mathscr{H}}u = \widehat{O}(h^2)$. To ensure higher accuracy where needed, one adds lower-order terms to the amplitude,
$$A = A_0 + hA_1 + h^2 A_2 + \dots,$$
and successively solves higher-order transport equations, which have the form
$$\Pi A_j = F_j, \quad j = 1, 2, \dots,$$
where the right-hand sides F_j are already known functions that can be expressed via
$$A_0, \dots, A_{j-1}.$$

28.2.5 Local Canonical Operators

Now let us present the actual construction of the canonical operator. First, we do this locally, i.e., on functions A supported in a neighborhood of an arbitrary given point of the Lagrangian manifold.

A smooth real function $\Phi(x, \theta)$, $\dim \theta = m \geq 0$, is called a *nondegenerate phase function* if the condition
$$\operatorname{rank}\begin{pmatrix}\Phi_{\theta x} & \Phi_{\theta\theta}\end{pmatrix} = m$$
is satisfied on the set

$$C_\Phi = \{(x,\theta)\colon \Phi_\theta(x,\theta) = 0\},$$

then C_Φ is a smooth n-dimensional manifold, and the mapping

$$j_\Phi\colon (x,\theta) \mapsto (x, \Phi_x(x,\theta))$$

is a diffeomorphism of C_Φ onto its image

$$j_\Phi(C_\Phi) = \Lambda_\Phi,$$

which is a Lagrangian manifold, and which we identify with C_Φ via this diffeomorphism. This Lagrangian manifold is called a Φ-*chart on* Λ if $\Lambda_\Phi \subset \Lambda$ (as a subset of $\mathbb{R}^{2n}$). On

$$\Lambda_\Phi \simeq C_\Phi$$

we define an *action*

$$S(\alpha) = \Phi\big|_{C_\Phi}.$$

In addition, we choose and fix some branch of the argument $\arg J^1(\alpha)$. The local canonical operator on Λ_Φ is defined in terms of *oscillatory integrals*

$$I[\Phi, a] = \frac{e^{i\pi m/4}}{(2\pi h)^{m/2}} \int e^{\frac{i}{h}\Phi(x,\theta)} a(x,\theta)\, d\theta_1 \cdots d\theta_m \qquad (a \in C_0^\infty). \tag{28.5}$$

Since the phase function is nondegenerate, it follows that if $a\big|_{C_\Phi} = 0$, then $a \in \mathcal{J}_\Phi$ (the ideal generated by the derivatives of Φ_θ), which, combined with the formula

$$I[\Phi, b \cdot \Phi_\theta] = ihI[\Phi, \operatorname{tr} b_\theta],$$

where

$$\operatorname{tr} b_\theta = \sum_{j=1}^m \frac{\partial b_j}{\partial \theta_j}$$

is the trace of the martrix b_θ, shows that, up to minor terms, the oscillatory integral is determined by the values of its amplitude a on C_Φ. The local canonical operator $K^h_{\Phi,d\mu}$ in the Φ-chart on Λ is given by the formula

$$K^h_{\Phi,d\mu} A = I[\Phi, \{\sqrt{F}A\}], \qquad A \in C_0^\infty(\Lambda_\Phi),$$

where the curly brackets denote an (arbitrary) extension of a compactly supported function on $\Lambda_\Phi \simeq C_\Phi$ to a compactly supported function on the ambient space of variables (x,θ), and F is a function on C_Φ defined as the ratio of two volume forms defined on the tangent spaces to $\mathbb{R}^{n+m}$ at the points of the set C_Φ

$$F \equiv F_{\Phi,d\mu} = \frac{d\mu \wedge d(-\Phi_\theta)}{dx \wedge d\theta} = \frac{d\mu \wedge d(-\Phi_{\theta_1}) \wedge \cdots \wedge d(-\Phi_{\theta_m})}{dx_1 \wedge \cdots \wedge dx_n \wedge d\theta_1 \wedge \cdots \wedge d\theta_m}.$$

The argument of F for extracting the square root is chosen according to the following rule

$$\arg F = -\frac{\pi m}{2} - \arg J^1 + \sum_{k=1}^{n+m} \arg \lambda_k,$$

where the λ_k are the eigenvalues of the matrix

$$\begin{pmatrix} E - i\Phi_{xx} & -i\Phi_{x\theta} \\ -i\Phi_{\theta x} & -i\Phi_{\theta\theta} \end{pmatrix},$$

and their arguments are chosen in the interval

$$\arg \lambda_k \in \left[-\frac{\pi}{2}, \frac{\pi}{2}\right],$$

which is possible because all these eigenvalues are nonzero and lie in the closed right half-plane. Thus, the canonical operator on Λ_Φ is defined (up to lower-order terms).

The simplest example of the local canonical operator is obtained if $m = 0$ and $\Phi(x, \theta) = S(x)$. Then

$$\Lambda_\Phi = \left\{(x, p) \colon p = \frac{\partial S}{\partial x}(x)\right\}$$

is the graph of the gradient of $S(x)$, one can take the local coordinates $\alpha = x$ on Λ_Φ, and the canonical operator gives the WKB function

$$[K^h_{\Lambda, d\mu} A](x) = e^{\frac{i}{h} S(x)} A(x) \sqrt{\mu(x)},$$

with an appropriate branch of the square root, where $\mu(x)$ is the density of the measure $d\mu$ in the coordinates x

$$d\mu = \mu(x)\, dx_1 \wedge \cdots \wedge dx_n.$$

This is always the case if

$$\pi_x \colon \Lambda_\Phi \to \mathbb{R}^n_x, \qquad (x, p) \mapsto x,$$

is a diffeomorphism onto the image of itself; such a chart with $m = 0$ is called a *nonsingular chart* on Λ.

If Λ is not diffeomorphically projected onto the image of itself in $\mathbb{R}^n_x$, then the canonical operator is given by an oscillatory integral (28.5) of the general form.

28.2.6 Global Canonical Operator

Now we paste the global canonical operator on Λ from local canonical operators in Φ-charts Λ_Φ. First, note that the Lagrangian manifold Λ can be completely covered by Φ-charts; i.e., a neighborhood of each point $\alpha_0 \in \Lambda$ can be represented as a Φ-chart. Consider a locally finite cover $\{\Lambda_{\Phi_j}\}$ of the manifold Λ by Φ-charts. The action in each chart can be chosen to coincide with $S(\alpha)$ (by adding an appropriate constant to Φ_j), and we take the argument $\arg J^1(\alpha) \pmod{4\pi}$ in each chart to be consistent with the specified global choice. Then we can set

$$K^h_{\Lambda,d\mu}A = \sum_j \chi_j(x) K^h_{\Phi_j,d\mu}(e_j A), \qquad A \in C_0^\infty(\Lambda),$$

where $\{e_j\}$ is a smooth partition of unity subordinate to the cover $\{\Lambda_{\Phi_j}\}$ and $\chi_j(x)$ are cutoff functions such that $\chi_j(x) = 1$ on $\pi_x(\operatorname{supp} e_j)$ and for each compact set $W \Subset \mathbb{R}^n_x$ all but finitely many of the functions $\chi_j(x)$ are zero on W.

For the canonical operator thus defined, the commutation formulas (28.3) and (28.4) hold.

28.3 Equations with Localized Right-Hand Side

Consider the equation

$$\widehat{\mathscr{H}} u = f \tag{28.6}$$

with the right-hand side (28.1), where $\widehat{\mathscr{H}}$ is an h-pseudodifferential operator. The canonical operator cannot be applied directly to solve this equation because of the nontrivial right-hand side. Here we describe method by which one can obtain asymptotic solutions.

We are interested in two types of right-hand sides

$$f(x,t,h) = e^{i\omega t} F\left(\frac{x-\xi}{h}\right) \tag{28.7}$$

(a fixed harmonic source of high frequency ω) and

$$f(x,t,h) = F\left(t, \frac{x - x(t)}{h}\right), \tag{28.8}$$

where $F(t, y)$ does not contain rapid oscillations (a moving source such as a ship or a charged body).

28.3.1 Harmonic Source: Helmholtz-Type Equations

Consider the case of the right-hand side (28.7) (using the wave equation as an example). We seek a solution that depends harmonically on time. Using the Fourier transform, from the wave equation we obtain the Helmholtz equation

$$\Delta u + k^2 n^2(x) u = F\left(\frac{x-\xi}{h}\right), \qquad k = \frac{\omega}{\overline{c}},$$

where $\overline{c}$ is a typical velocity. If $k \approx h^{-1}$ (which we will assume), then we obtain the h-differential equation

$$\widehat{\mathcal{H}} u = F\left(\frac{x-\xi}{h}\right),$$

where

$$\widehat{\mathcal{H}} = \mathcal{H}\left(x, -ih\frac{\partial}{\partial x}\right), \qquad \mathcal{H}(x, p) = -p^2 + n^2(x).$$

The right-hand side can be written as

$$F\left(\frac{x-\xi}{h}\right) = \frac{i^{3/2}}{(2\pi)^{3/2}} \int_{\mathbb{R}^3} e^{\frac{i}{h}p(x-\xi)} A(p)\, dp_1\, dp_2\, dp_3,$$

where A is the Fourier transform of the function F. This expression is none other than the canonical operator on the three-dimensional "vertical" Lagrangian manifold $\Lambda_0 = \{(x, p) \in \mathbb{R}^6 : x = \xi,\ p = \alpha \in \mathbb{R}^3\}$ in the space $\mathbb{R}^6$

$$F\left(\frac{x-\xi}{h}\right) = h^{3/2}[K^h_{\Lambda_0, d\mu_0} A](x), \tag{28.9}$$

where the measure $d\mu_0$ is given by the formula

$$d\mu_0 = d\alpha_1 \wedge d\alpha_2 \wedge d\alpha_3.$$

Asymptotic solutions of such problems were constructed in [1, 2]. Let us describe these results. The main idea of the construction of an asymptotic solution is as follows. Consider Eq. (28.6) with the right-hand side (28.9). Let $v(x, \tau)$ be the solution of the Cauchy problem

$$-ih\frac{\partial v}{\partial \tau} + \widehat{\mathcal{H}} v = 0, \quad v(x, 0) = F\left(\frac{x-\xi}{h}\right). \tag{28.10}$$

Then the solution of Eq. (28.6) can be written in the form

$$u(x, h) = \frac{i}{h} \int_0^\infty v(x, \tau)\, d\tau. \tag{28.11}$$

This is of course purely formal, because the integral usually diverges.

However, if we consider asymptotic solutions rather than exact ones, then, under certain conditions, we can ascribe a meaning to this integral. More precisely, this integral gives us a hint as to how the actual asymptotic solution looks like. Let $\zeta(\tau)$ be a compactly supported function equal to unity near $\tau = 0$. Using the cutoff function $\zeta(\tau)$, we can write the integral (28.11) as the sum

$$u(x,h) = \frac{i}{h}\int_0^\infty \zeta(\tau)v(x,\tau)\,d\tau + \frac{i}{h}\int_0^\infty (1-\zeta(\tau))v(x,\tau)\,d\tau \equiv I_1 + I_2.$$

The solution of the Cauchy problem (28.10) is given by the canonical operator. The integral I_1 is over a finite interval and already makes sense. The integral I_2 can under certain conditions be formally computed by the stationary phase method, and the resulting expression is well-defined as the canonical operator on some Lagrangian manifold Λ_+ (see below). Substituting this expression for I_2, we obtain a well-defined function, and a straightforward substitution shows that this function is an asymptotic solution of the original equation.

Let us present conditions under which this construction works. These are conditions on the principal symbol (which we denote by $H(x,p)$, omitting the subscript "0") of the operator $\widehat{\mathscr{H}}$.

We introduce the following notation:

- $L_0 = \{q \in \mathbb{R}^n : H(\xi, q) = 0\}$.
- $(X(q,\tau), P(q,\tau))$ is the solution of the Cauchy problem for the Hamiltonian system (where $\dot{} = d/d\tau$)

$$\begin{cases} \dot{x} = H_p(x,p), \\ \dot{p} = -H_x(x,p), \end{cases} \qquad x|_{\tau=0} = \xi, \quad p|_{\tau=0} = q \in \mathbb{R}^n.$$

The principal symbol $H(x,p)$ must satisfy the following conditions:

1. $H(\xi,p)$ is nonzero for large $|p|$.
2. For $q \in L_0$, the solutions $(X(q,\tau), P(q,\tau))$ are endlessly continuable to the right.
3. For $q \in L_0$, the trajectories $X(q,\tau)$ leave any ball $|x| < R$ in finite time T_R (the absence of traps).
4. $H_p(\xi,q) \neq 0$ for $q \in L_0$ (the *principal type* condition).

Assuming that these conditions are satisfied, let us define Lagrangian manifolds Λ_0, Λ, and Λ_+. They are defined parametrically. Let us describe the parameter ranges.

Let $\Lambda \subset \mathbb{R}^{n+1}_{(q,\tau)}$ be the maximal domain where the solutions $(X(q,\tau), P(q,\tau))$ are defined. We identify $\mathbb{R}^n_q$ with the subset

$$\Lambda_0 = \{\tau = 0\} \subset \Lambda$$

and the $(n-1)$-dimensional manifold L_0 with a subset in Λ_0. Further, let

$$\Lambda_+ = L_0 \times (0,\infty) \subset \Lambda.$$

Now we define the Lagrangian manifolds themselves

$$\begin{aligned}
\iota_0 &: \Lambda_0 \longrightarrow \mathbb{R}^{2n}_{(x,p)}, & \iota_0(q) &= (\xi, q), \\
\iota &: \ \Lambda \longrightarrow \mathbb{R}^{2n+2}_{(x,t,p,E)}, & \iota(q,\tau) &= (X(q,\tau), \tau, P(q,\tau), E(q)), \quad E(q) = -H(\xi,q), \\
\iota_+ &: \Lambda_+ \longrightarrow \mathbb{R}^{2n}_{(x,p)}, & \iota_+(\alpha,\tau) &= (X(q(\alpha),\tau), P(q(\alpha),\tau)).
\end{aligned}$$

We equip these manifolds with the respective measures $d\mu_0$, $d\mu$, and $d\mu_+$, where $d\mu_0$ has been defined above, $d\mu = d\mu_0 \wedge d\tau$, and $d\mu_+ = d\sigma \wedge d\tau$, where $d\sigma$ is the measure on L_0 defined by the equation $d\mu_0 = -dH \wedge d\sigma$ on the tangent spaces to Λ_0 at the points of L_0.

Now we are in a position to write formulas for the asymptotic solutions. Let $\tilde{F}$ be the Fourier transform of F.

Theorem 28.1 ([1, 2]). *If $\tilde{F} \in C_0^\infty(\mathbb{R}^n)$, then the equation*

$$\widehat{\mathscr{H}} u = F\left(\frac{x-\xi}{h}\right)$$

has the asymptotic solution

$$u(x,h) = ih^{\frac{n}{2}-1} \int_0^\infty [K^h_{\Lambda,d\mu}(\chi A)](x,t)\,dt,$$

where χ is an appropriate cutoff function[1] *and the leading term of the amplitude*

$$A = A(q,\tau,h) \in C^\infty(\Lambda \times [0,1])$$

has the form

$$A_0(q,\tau) \equiv A(q,\tau,0) = e^{-i\int_0^\tau H_{\mathrm{sub}}(X(q,\tau'),P(q,\tau'))\,d\tau'}\,\tilde{F}(q).$$

Recall that $H_{\mathrm{sub}}(x,p)$ is the subprincipal symbol of the operator $\widehat{\mathscr{H}}$. Further, this asymptotic solution has a nice representation for $x \neq \xi$.

Theorem 28.2 ([1, 2]). *In the domain $\mathbb{R}^n_x \setminus \{\xi\}$, the asymptotic solution can be represented in the form*

$$\psi(x,h) = (2\pi)^{\frac{1}{2}} e^{\frac{\pi i}{4}} h^{-\frac{n-1}{2}} [K^h_{\Lambda_+,d\mu_+} Q](x,h) + \widehat{O}(h^\infty),$$

where the leading term of the amplitude $Q = Q(\alpha,\tau,h)$ is given by the formula

$$Q_0(\alpha,\tau) \equiv Q(\alpha,\tau,0) = A_0(q(\alpha),\tau).$$

28.3.2 Moving Source

Consider the case of the right-hand side (28.8) (using the Maxwell equations as an example). Since the source is no longer harmonic, reduction to stationary equations by means of the Fourier transform with respect to t does not help, and we must deal with the nonstationary equations, including the variable t into consideration as

[1] See [1, 2] for the exact description of this cutoff function.

yet another coordinate (say, x_0) on the configuration space. As a consequence, one must deal with the change of multiplicity of roots of the characteristic equation at $p = 0$ by extending the construction of asymptotic solutions of h-pseudodifferential equations with localized right-hand sides [1] to the case of asymptotically homogeneous Hamiltonians. More precisely, we use the theory of Maslov's canonical operator on punctured (asymptotically homogeneous) Lagrangian manifolds [6, 16]. An additional difficulty is created by the fact that, under the standard condition of the absence of "traps," the trajectories do not go to infinity uniformly in momenta, since the effective Hamiltonian is of the second order. This problem is solved by reformulating the indicated condition and making suitable changes of variables in the proofs.

Consider a model problem about a charged body of small but finite dimensions $\sim \delta$, which moves along a given trajectory $x = \phi(t)$ in an inhomogeneous dispersive medium filling the entire space $\mathbb{R}^3 \ni x$, with magnetic permeability $\mu = 1$ and permittivity of the form

$$\varepsilon = \varepsilon(x, \omega) = f(x)g(\omega),$$

where ω is the circular frequency. Let us assume that

$$f(x) \ge f_{\min} > 1$$

and

$$g(\omega) = g(-\omega) \ge 1$$

are smooth functions, the derivative of the function $\omega^2 g(\omega)$ is strictly positive for $\omega > 0$, and the velocity of the body satisfies the condition

$$c > \phi'(t) > c f_{\min}^{-1/2},$$

i.e., it is greater than the speed of light in the medium. The mechanical interaction of the body with the medium is not taken into account. The following equation with a source on the right-hand side for the electric field $\mathbf{E}$ can be derived in the standard way from Maxwell's equations and material relations:

$$\nabla \times \nabla \times \mathbf{E} + \frac{1}{c^2} f(x) g\left(-i\frac{\partial}{\partial t}\right) \frac{\partial^2 \mathbf{E}}{\partial t^2} = -\frac{4\pi}{c^2}\frac{\partial \mathbf{J}}{\partial t}, \tag{28.12}$$

where $\mathbf{J}$ is the current density created by the motion of the charged body.

Suppose that $\delta \ll L$, where L is the characteristic distance scale at which the equation is considered. We pose the problem of constructing asymptotic solutions of Eq. (28.12), assuming $h = \delta/L$ to be a small parameter. To formulate the problem mathematically, we switch to dimensionless variables $\tilde{x} = x/L$ and $\tilde{t} = ct/L$ and assume that

$$f(x) = \tilde{f}(\tilde{x}), \qquad \phi(t) = L\tilde{\phi}(\tilde{t}), \qquad g(\omega) = \tilde{g}(\omega\delta/c),$$

and

$$\mathbf{J}(x,t) = \frac{J_0}{4\pi i} V\left(\frac{x-\phi(t)}{\delta}, \frac{ct}{L}\right) = \frac{J_0}{4\pi i} V\left(\frac{\tilde{x}-\tilde{\phi}(\tilde{t})}{h}, \tilde{t}\right),$$

where J_0 is a constant with the dimension of the current density, and $\tilde{f}, \tilde{\phi}, \tilde{g}$, and V are dimensionless functions of dimensionless arguments, uniformly bounded as $h \to 0$ with all derivatives, with $V(y,\tilde{t})$ being compactly supported in y, and its dependence on $\tilde{t}$ taking into account the change in orientation and relativistic distortion of the body's shape during motion. We introduce the dimensionless unknown function $\tilde{\mathbf{E}}(\tilde{x},\tilde{t}) = c\mathbf{E}(x,t)/\delta J_0$. The dimensionless equation (28.12) is written as an h-pseudodifferential equation (here and below, tildes are omitted everywhere)

$$\hat{\mathcal{H}} \equiv \hat{p}\times\hat{p}\times\mathbf{E} + f(x)g(\hat{\omega})\hat{\omega}^2\mathbf{E} = \hat{\omega}V\left(\frac{\tilde{x}-\tilde{\phi}(\tilde{t})}{h}, \tilde{t}\right), \quad \hat{p} = -ih\frac{\partial}{\partial x}, \quad \hat{\omega} = -ih\frac{\partial}{\partial t} \tag{28.13}$$

with the matrix Hamiltonian

$$\mathcal{H}(x,t,p,\omega) = f(x)g(\omega)\omega^2 I - p^2 I + \begin{pmatrix} p_1^2 & p_1p_2 & p_1p_3 \\ p_1p_2 & p_2^2 & p_2p_3 \\ p_1p_3 & p_2p_3 & p_3^2 \end{pmatrix}$$

independent of t, where I is the 3×3 identity matrix and (p,ω) are the momenta dual to the coordinates (x,t). For $p \neq 0$, this Hamiltonian has a single eigenvalue

$$H_0(x,\omega) = f(x)g(\omega)\omega^2$$

with eigenvector $\xi = p$ and a double eigenvalue

$$H_{1,2}(x,p,\omega) \equiv H(x,p,\omega) = f(x)g(\omega)\omega^2 - p^2$$

with a two-dimensional eigenspace consisting of vectors orthogonal to ξ. We denote the orthogonal projection onto this subspace by $\Pi = \Pi(p)$.

The right-hand side of the equation can be written as $K^h_{\Lambda_0,d\mu_0}A$, where $K^h_{\Lambda_0,d\mu_0}$ is the Maslov canonical operator on the Lagrangian manifold

$$\Lambda_0 = \{(x,t,p,\omega) \in \mathbb{R}^8 : x = \phi(t),\ \omega = -\langle p, \phi'(t)\rangle\}$$

with the standard Riemannian volume form $d\mu_0$ and the amplitude A in canonical coordinates (p,t) on Λ_0 has the form $A(p,t) = -\langle p,\phi'(t)\rangle\tilde{V}(p,t) - ih\tilde{V}_t(p,t)$, here $\tilde{V}(p,t)$ is the Fourier transform of the function $V(y,t)$ with respect to the variables y.

Under the above assumptions, the set

$$L_0 = \Lambda_0 \cap \{(x,t,p,\omega) : H(x,p,\omega) = 0,\ p \neq 0\}$$

is a smooth three-dimensional manifold that decomposes into the union of two components

$$L_0 = L_0^+ \cup L_0^-$$

such that $\pm\omega > 0$ on $L_0^\pm$. Let Λ_+ denote the union of positive semitrajectories of the Hamiltonian system corresponding to the Hamiltonian $H(x, p, \omega)$ issuing from the submanifold L_0^+, and let Λ_- denote the union of negative semitrajectories of the Hamiltonian system corresponding to the Hamiltonian $H(x, p, \omega)$ issuing from the submanifold L_0^-.

Theorem 28.3. *The sets $\Lambda_\pm$ are punctured Lagrangian manifolds in $\mathbb{R}^8$ in the sense of [6, 16].*

We denote the Maslov canonical operator on these manifolds by $K_\pm$.

Condition 1 (absence of "traps") *The projections of manifolds with boundary $\overline{\Lambda}_\pm$ onto the space-time $\mathbb{R}^4_{(x,t)}$ are proper (i.e., the preimages of compact sets are compact).*

Theorem 28.4. *In the absence of traps, equation* (28.13) *has an asymptotic solution* $\mathbf{E}(x, t, h)$, *which outside an arbitrarily small neighborhood of the trajectory $x = \phi(t)$ can be represented as*

$$\mathbf{E}(x,t,h) = (2\pi)^{\frac{1}{2}} e^{\frac{\pi i}{4}} h^{\frac{n-1}{2}} \left([\mathcal{K}_+ Q_+](x,t,h) - [\mathcal{K}_- Q_-](x,t,h) \right) + O\left(h^\infty\right),$$

where the leading terms of the amplitudes $Q_\pm$ are the solutions of the transport equation corresponding to the Hamiltonian $H(x, p, \omega)$ with initial conditions ΠA on $L_0^\pm$. This asymptotic solution satisfies the causality principle.

The proofs of Theorems 28.3 and 28.4 will be given in a forthcoming publication.

Remark 28.1. Note that asymptotic solutions of the Cauchy problem for Eq. (28.12) with zero right-hand side and of Eq. (28.12) with right-hand side harmonically depending on t were in particular studied in [9, 18].

Acknowledgements The author is grateful to Sergei Dobrokhotov, Anton Tolchennikov, and Aleksandr Klevin for useful discussions. My special thanks are to Lilit Dashtoyan, Vahram Hakobyan, Avetik Sahakyan, and all the organizers of the IX International Conference on Topical Problems of Continuum Mechanics for their patience and kind hospitality.

This work was financially supported by the Russian Science Foundation, project 24-11-00213 (https://rscf.ru/en/project/24-11-00213/) at the Ishlinsky Institute for Problems in Mechanics RAS.

References

[1] Anikin AY, Dobrokhotov SY, Nazaikinskii VE, Rouleux M (2023) Lagrangian manifolds and the construction of asymptotics for (pseudo)differential equations with localized right-hand sides. Theoretical and Mathematical Physics 214(1):1-23, DOI 10.1134/S0040577923010014

[2] Anikin AY, Dobrokhotov SY, Nazaikinskii VE, Rouleux M (2017) The Maslov canonical operator on a pair of Lagrangian manifolds and asymptotic solutions of stationary equations with localized right-hand sides. Doklady Mathematics 96(1):406-410, DOI 10.1134/S1064562417040275

[3] Babich VM (1964) O korotkovolnovoy asimptotike funktsii Grina dlya uravneniya Gel'mgol'tsa (in Russ., On the short-wave asymptotic behaviour of the Green's function for the Helmholtz equation). Matematicheskii Sbornik 65(107)(4):576-630

[4] Babich VM (1965) The short wave asymptotic form of the solution for the problem of a point source in an inhomogeneous medium. USSR Computational Mathematics and Mathematical Physics 5(5):247-251, DOI 10.1016/0041-5553(65)90021-2

[5] Babich VM, Buldyrev VS (2007) Asymptotic Methods in Short-Wavelength Diffraction Theory (Alpha Science Series on Wave Phenomena). Alpha Science, Oxford

[6] Dobrokhotov SY, Nazaikinskii VE, Shafarevich AI (2021) Canonical operator on punctured Lagrangian manifolds. Russian Journal of Mathematical Physics 28(1):22-36, DOI 10.1134/S1061920821010040

[7] Dobrokhotov SY, Nazaikinskii VE, Shafarevich AI (2021) Efficient asymptotics of solutions to the Cauchy problem with localized initial data for linear systems of differential and pseudodifferential equations. Russian athematical Surveys 76(5):745-819, DOI 10.1070/RM9973

[8] Dobrokhotov SY, Nazaikinskii VE, Shafarevich AI (2017) New integral representations of the Maslov canonical operator in singular charts. Izvestiya: Mathematics 81(2):286-328, DOI 10.1070/IM8470

[9] Dobrokhotov SY, Tolchennikov AA (2024) Uniform formulas for the asymptotic solution near the leading front for Maxwell's equations with temporal dispersion and localized initial data. Mathematical Notes 116(3):458-464, DOI 10.1134/S0001434624090062

[10] Duistermaat JJ, Hörmander L (1972) Fourier integral operators II. Acta Mathematica 128(3-4):183-269, DOI 10.1007/BF02392165

[11] Keller JB (1962) Geometrical theory of diffraction. Journal of the Optical Society of America 52(2):116-130, DOI 10.1364/JOSA.52.000116

[12] Kucherenko VV (1969) Kvaziklassicheskaya asimptotika funktcii tochechnogo istochnika dlya statsionarnogo uravneniya Shredingera (in Russ., Quasiclassical asymptotics of a point-source function for the stationary Schrödinger equation). Theoreticheskaya i Mathematicheskaya Fizika 1(3):294-310

[13] Maslov VP (1965) Teoriya vozmushcheniy i asimptoticheskiye metody (in Russ., Perturbation Theory and Asymptotic Methods). Moscow State University, Moscow

[14] Maslov VP, Fedoriuk MV (1981) Semiclassical Approximation in Quantum Mechanics. Reidel, Dordrecht

[15] Melrose RB, Uhlmann GA (1979) Lagrangian intersection and the Cauchy problem. Communications on Pure and Applied Mathematics 32(4):483-519, DOI 10.1002/cpa.3160320403

[16] Nazaikinskii VE (2022) Canonical operator on punctured Lagrangian manifolds and commutation with pseudodifferential operators: local theory. Mathematical Notes 112(5):709-725, DOI 10.1134/S0001434622110086
[17] Sternin BY, Shatalov VE (1983) On a method of solving equations with simple characteristics. Mathematics of the USSR - Sbornik 44(1):23-59, DOI 10.1070/SM1983v044n01ABEH000950
[18] Tolchennikov AA (2025) Asymptotic solution of a system of Maxwell equations with a time-harmonic spatially localized electromagnetic field source. Mathematical Notes 118(6):1340-1343, DOI 10.4213/mzm14859
[19] Vainberg BR (1975) On the short wave asymptotic behaviour of solutions of stationary problems and the asymptotic behaviour as $t \to \infty$ of solutions of non-stationary problems. Russ. Math. Surveys 30(2):1-58, DOI 10.1070/RM1975v030n02ABEH001406

Chapter 29
Prediction of the Long-Term Strength of Metallic Materials at Different Temperatures

Regina R. Saitova, Alexander R. Arutyunyan, and Holm Altenbach

Abstract The problem of high-temperature creep and long-term strength of metallic materials is demanded in such important fields of modern engineering, as thermal and nuclear power plants, aircraft and spacecraft, etc. Because these effects are observed in elements of many important engineering objects, the problem of brittle fractures became a subject of numerous theoretical and experimental research. The damage conception was introduced in the mechanics of materials to describe long-term strength under conditions of high-temperature creep. This conception has been developed in the fundamental works of Kachanov and Rabotnov. In this work, to describe the creep and long-term strength, a damage conception is used. A system of interconnected kinetic equations for the creep rate and damage parameter is formulated. A compressible medium is considered and the mass conservation law is taking into account. The damage parameter is specifying in the form of the ratio of the current density of the material to the initial one. Exact and approximate analytical solutions of these equations are obtained. Assuming fracture conditions, a long-term strength criterion is formulated, describing the brittle fracture region. Changes in the model parameters depending on the test temperature are studied. An expression for describing the temperature dependence of the one model parameter is proposed. The resulting expression adequately describes the change in the system parameters for the 5Cr-0.5Mo Steel, 0.2C Silicon-Killed Steel and Ni based 15.5Cr-2.5Ti-0.7Al-1Nb-7Fe superalloy at different temperatures. Thus, the proposed system of interrelated kinetic equations for the creep rate and damage parameter allows one to describe long-term strength curves at various temperatures, demonstrating the universality of the proposed system.

Regina R. Saitova (corresponding author) · Alexander R. Arutyunyan
St. Petersburg State University, 7/9 Universitetskaya Emb., St. Petersburg, 199034, Russian Federation, e-mail: rigastr@yandex.ru, a.arutyunyan@spbu.ru

Holm Altenbach
Institut für Werkstoffe, Technologien und Mechanik, Fakultät für Maschinenbau, Otto-von-Guericke-Universität Magdeburg, Universitätsplatz 2, 39106 Magdeburg, Germany, e-mail: holm.altenbach@ovgu.de

H. Altenbach and A. V. Sahakyan, *Contact Mechanics and Relevant Problems*,
Advanced Structured Materials 253, https://doi.org/10.1007/978-3-032-20869-9_29

Key words: Creep, Damage parameter, Long-term strength criterion, Experiments at different temperatures

29.1 Introduction

Metallic materials exhibit elastic and inelastic material behavior under mechanical and thermal stress. Examples of inelastic material behavior are plasticity and creep, whereby plastic behavior generally occurs at elevated mechanical stresses even at normal temperatures. The measure of critical mechanical stress is the yield stress. Creep and relaxation, on the other hand, are associated with elevated temperatures (approximately 0.3 times the melting temperature of the material) but moderate mechanical stresses. Creep also occurs at extremely low temperatures — however, this situation is not taken into account here.

Let us consider creep behavior of metallic materials at elevated temperatures. In addition to the primary and secondary creep, the performance of metals under these conditions is primarily determined by damage processes that contribute to the development of thermal embrittlement at elevated temperatures of the material. Under the influence of relatively low stresses and high temperatures, some metallic materials become brittle and fractured with only small residual strains. Physical research methods have established that thermal embrittlement is inherent to all metallic materials and is associated with the accumulation of porosity along grain boundaries through the mechanisms of vacancy diffusion and grain boundary sliding. This effect is observed in the elements of many important engineering structures, and therefore the problem of brittle fractures has become the subject of numerous theoretical and experimental studies. To address this problem in mechanics of materials [10], the concept of continuity (damage) was introduced in the works of Kachanov [8, 9] (continuity) and Rabotnov [19, 20] (damage). To describe the brittle region of the experimental long-term strength curve, a system of simple kinetic equations for the damage parameter and creep strain is proposed, and a long-term strength criterion is formulated.

29.2 Theoretical Description of the System of Kinetic Equations of Creep and Long-Term Strength

When formulating creep equations and long-term strength criteria in the brittle fracture region, it is necessary to consider the compressibility condition and formulate creep rheological relations and long-term strength criteria based on the mass conservation law. In addition, a physical meaning for the damage parameter should be given. Many authors [3–7, 13, 21] consider the material density to be the most representative characteristic of porosity and damage. The appearance of pores according to the results of density measurements was detected at the very initial stage of creep. Moreover, it was shown [7, 21], that the healing of pores by the application of hydrostatic pressure leads to a sharp deceleration of the creep deformation and a significant increase in the time to fracture. These experiments convincingly showed that loosening [18] (an irreversible change in density) is the main damage factor that determines the performance of metallic materials during creep. Porosity is the main characteristic of damage, and the change in density due to loosening of the material is taken as an integral measure of porosity.

In this case, the irreversible change in density (loosening) of the material should be considered as the damage parameter [3–7, 13, 18, 21]. Based on these principles, interconnected equations for the damage parameter and creep deformation have been formulated [1]

$$\frac{d\varepsilon}{dt} = B\sigma_0^m \psi^{m-\beta} e^{m\varepsilon}, \tag{29.1}$$

$$\frac{d\psi}{dt} = -A\sigma_0^n \psi^{n-\alpha} e^{n\varepsilon}, \tag{29.2}$$

where $\psi = 1 - \omega = \rho/\rho_0$ is the continuity (first introduced by L. Kachanov [9]) and related to the changes of the density (discussed, for example, in [12]), ρ_0 is the initial density, ρ is the current density of the specimen, B, A, m, n, α, β are parameters. Note that Eqs. (29.1) and (29.2) are only one possibility to describe the creep-damage behavior - there are much more suggestions in the literature (see, e.g., [11, 17]).

A modified version of the system of kinetic equations was proposed in [2] and [23]. Solutions of the system of Eqs (29.1) and (29.2) were considered in the following works: for the case of uniaxial tension [1, 23] and for the case of two-stage loading [2]. In the considered works, exact, approximate analytical and numerical solutions of the system of interconnected Eqs. (29.1) and (29.2) were obtained and a comparison was given with the results of experiments to determine the damage accumulation laws under high-temperature creep conditions, obtained for various metals and alloys.

In the case $e^{m\varepsilon} \approx 1$, $e^{n\varepsilon} \approx 1$ (small strains), with initial conditions $t = 0$, $\psi = 1$, from Eq. (29.2) we obtain [22]

$$\psi = \left[1 - (\alpha - n + 1)A\sigma_0^n t\right]^{\frac{1}{\alpha - n + 1}}. \tag{29.3}$$

Taking the fracture conditions in the form $t = t_f$, $\psi = 0$, from Eq. (29.3) one can obtain the following long-term strength criterion [22]

$$t_f = \frac{1}{(\alpha - n + 1)A\sigma_0^n}. \tag{29.4}$$

When $\alpha = 2n$ criterion (29.4) coincides with the Kachanov-Rabotnov long-term strength criterion [8, 19].

29.3 Specification of Model Parameters and Comparison of Theoretical Curves with Experimental Results

The accuracy of strength and service life estimates for structural elements depends on how exactly the mathematical model used describes the deformation and damage processes in the material in dangerous areas of the structural elements as a result of high-temperature creep. It is important to accurately determine the material parameters included in the mathematical model used. Particular attention should be paid to experimental studies of high-temperature creep processes.

A series of experiments on high-temperature creep and the long-term strength of the 5Cr-0.5Mo Steel, 0.2C Silicon-Killed Steel and Ni based 15.5Cr-2.5Ti-0.7Al-1Nb-7Fe superalloy at various temperatures were conducted in [14–16]. The values of coefficients in calculations by formula (29.4) for these experimental data were determined (Table 29.1).

To describe the temperature dependence, we assume that the parameter A is not constant but rather temperature-dependent. We will further specify this parameter as the following relationship

$$A\left(T\right) = k\,10^{aT^2+bT+c}, \tag{29.5}$$

where $k,\ a,\ b,\ c$ are parameters.

Figures 29.1 - 29.3 shows the specification of the parameter $A(T)$ in the form of (29.5) as solid lines and the values of the parameter A obtained by relation (29.4) (Table 29.1) for the STB 410 (0.2C Silicon-Killed Steel), for the STBA 25 (5Cr-0.5Mo Steel) and for the NCF 750-B (Nickel based 15.5Cr-2.5Ti-0.7Al-1Nb-7Fe superalloy) as crosses. The values of the coefficients used in the calculations according to formula (29.5) are given in Table 29.2.

Table 29.1: Values of coefficients in theoretical calculations using formula (29.4).

Brand of the material	Test temperature	α	n	A, $[\text{MPa}]^{-6}[h]^{-1}$
STB 410	400°C	8	6	$2.0 \cdot 10^{-19}$
STB 410	450°C	8	6	$5.0 \cdot 10^{-18}$
STB 410	500°C	8	6	$1.5 \cdot 10^{-16}$
STBA 25	500°C	8	6	$6.0 \cdot 10^{-18}$
STBA 25	550°C	8	6	$1.0 \cdot 10^{-16}$
STBA 25	600°C	8	6	$2.0 \cdot 10^{-15}$
STBA 25	650°C	8	6	$4.0 \cdot 10^{-14}$
NCF 750-B	600°C	8	6	$1.0 \cdot 10^{-20}$
NCF 750-B	700°C	8	6	$4.0 \cdot 10^{-19}$
NCF 750-B	800°C	8	6	$1.0 \cdot 10^{-16}$
NCF 750-B	900°C	8	6	$1.0 \cdot 10^{-13}$

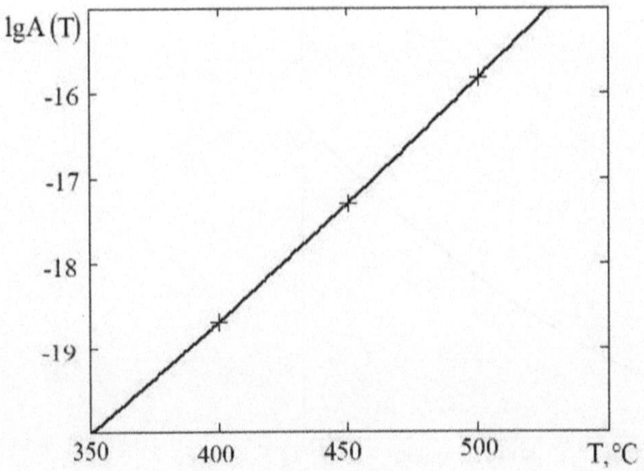

Fig. 29.1: Specification of the parameter $A(T)$ in the form (29.5) is solid line, and the values of the parameter A obtained according to the relation (29.4) for the STB 410 (0.2C Silicon-Killed Steel) are crosses.

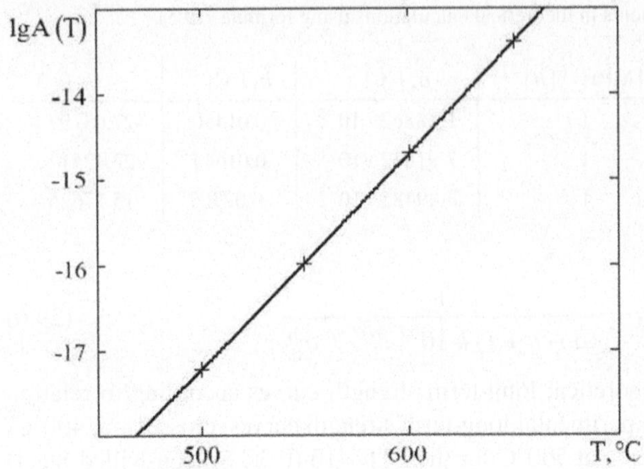

Fig. 29.2: Specification of the parameter $A(T)$ in the form (29.5) is solid line, and the values of the parameter A obtained according to the relation (29.4) for the STBA 25 (5Cr-0.5Mo Steel) are crosses.

There is good agreement between the values of the parameter A that were determined at different temperatures for the 5Cr-0.5Mo Steel, 0.2C Silicon-Killed Steel and Ni based 15.5Cr-2.5Ti-0.7Al-1Nb-7Fe superalloy and the proposed relationship (29.5) for the temperature dependence of the parameter $A(T)$.

Taking into account the ratio (29.5), the long-term strength criterion (29.4) has the following form

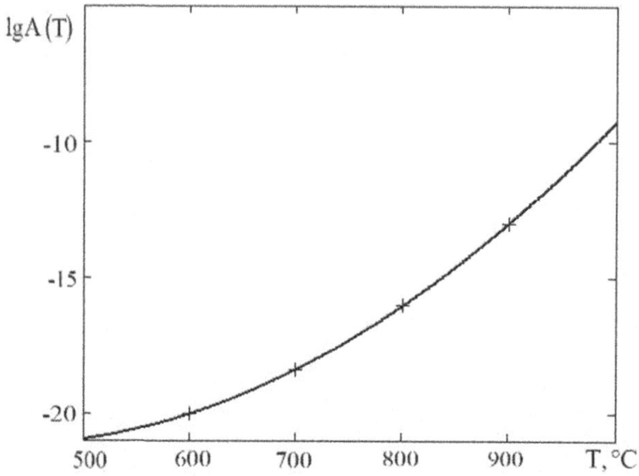

Fig. 29.3: Specification of the parameter $A(T)$ in the form (29.5) is solid line, and the values of the parameter A obtained according to the relation (29.4) for the NCF 750-B (Nickel based 15.5Cr-2.5Ti-0.7Al-1Nb-7Fe superalloy) are crosses.

Table 29.2: Values of coefficients in theoretical calculations using formula (29.5).

Brand of the material	k, $[\mathrm{MPa}]^{-6}[h]^{-1}$	a, $[°\mathrm{C}]^{-2}$	b, $[°\mathrm{C}]^{-1}$	c
STB 410	1	$1.58362 \cdot 10^{-5}$	0.01450	-27.03197
STBA 25	1	$7.91812 \cdot 10^{-6}$	0.01644	-27.42517
NCF 750-B	1	$3.49485 \cdot 10^{-5}$	-0.02902	-15.17626

$$t_{\mathrm{f}} = \frac{1}{(\alpha - n + 1)\, k\, 10^{aT^2+bT+c}\, \sigma_0^n}. \tag{29.6}$$

Figure 29.4 shows the theoretical long-term strength curves according to relation (29.6) (solid lines), and experimental long-term strength curves: crosses – at 400°C, squares – at 450°C, circles – at 500°C for the STB 410 (0.2C Silicon-Killed Steel) [14]. Figure 29.5 shows the theoretical long-term strength curves according to relation (29.6) (solid lines), and experimental long-term strength curves: crosses – at 500°C, squares – at 550°C, circles – at 600°C, triangles – at 650°C for the STBA 25 (5Cr-0.5Mo Steel) [15]. Figure 29.6 shows the theoretical long-term strength curves according to relation (29.6) (solid lines), and the experimental long-term strength curves: crosses – at 600°C, squares – at 700°C, circles – at 800°C, triangles – at 900°C for the NCF 750-B (Nickel based 15.5Cr-2.5Ti-0.7Al-1Nb-7Fe superalloy) [16]. A good agreement between the theoretical and experimental creep curves at different temperatures according to Figs. 29.4 - 29.6 is observed.

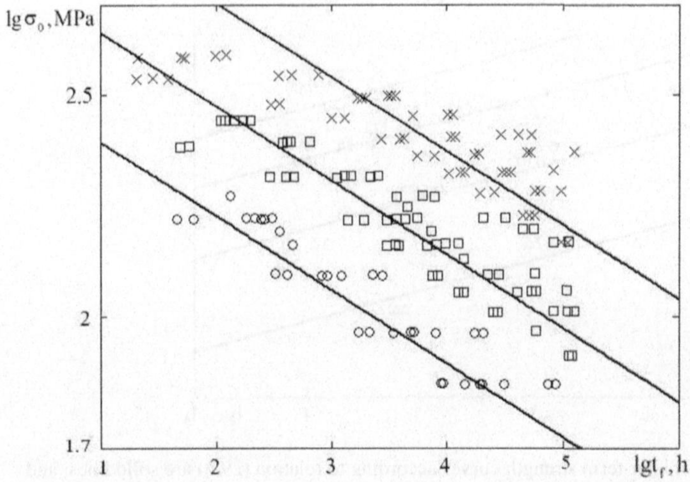

Fig. 29.4: Theoretical long-term strength curves according to relation (29.6) are solid lines, and experimental long-term strength curves: crosses – at 400°C, squares – at 450°C, circles – at 500°C for the STB 410 (0.2C Silicon-Killed Steel) [14].

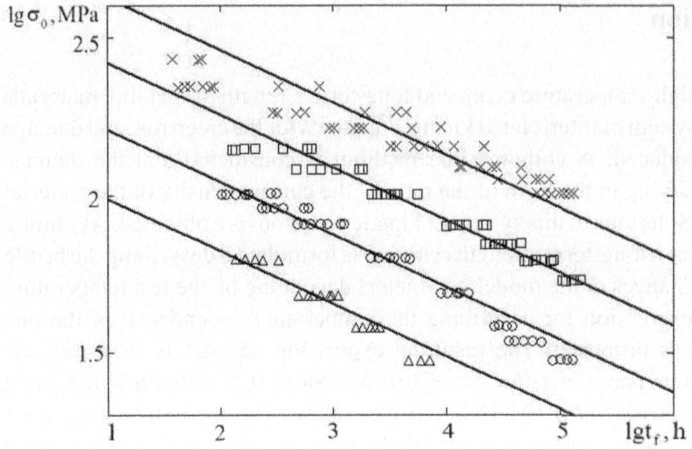

Fig. 29.5: Theoretical long-term strength curves according to relation (29.6) are solid lines, and experimental long-term strength curves: crosses – at 500°C, squares – at 550°C, circles – at 600°C, triangles – at 650°C for the STBA 25 (5Cr-0.5Mo Steel) [15].

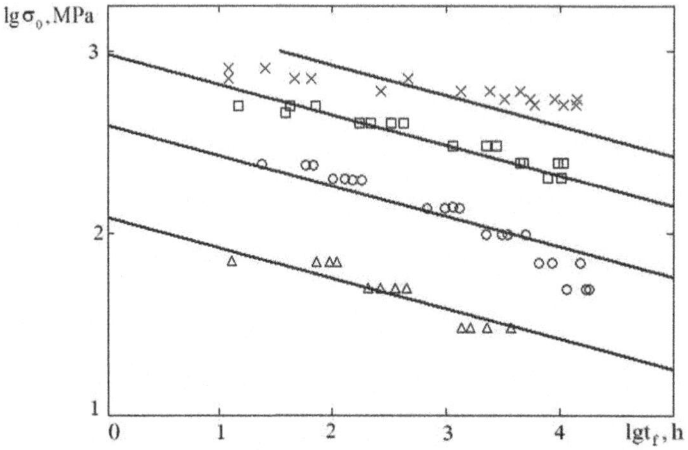

Fig. 29.6: Theoretical long-term strength curves according to relation (29.6) are solid lines, and experimental long-term strength curves: crosses – at 600°C, squares – at 700°C, circles – at 800°C, triangles – at 900°C for the NCF 750-B (Nickel based 15.5Cr-2.5Ti-0.7Al-1Nb-7Fe superalloy) [16].

29.4 Conclusion

The problem of high-temperature creep and long-term strength of metallic materials is considered. A system of interrelated kinetic equations for the creep rate and damage parameter is introduced. A compressible medium is considered and the damage parameter is specifying in the form of the ratio of the current density of the material to the initial one. Solutions to this system of kinetic equations are obtained. Assuming fracture conditions, a long-term strength criterion is formulated, describing the brittle fracture region. Changes in the model parameters depending on the test temperature are studied. An expression for describing the temperature dependence of the one model parameter is proposed. The resulting expression adequately describes the change in the system parameters for the 5Cr-0.5Mo Steel, 0.2C Silicon-Killed Steel and Ni based 15.5Cr-2.5Ti-0.7Al-1Nb-7Fe superalloy. The theoretical long-term strength curves are plotted and compared with the experimental results for considered materials. It is shown, that the experimental results are in good agreement with the theoretical ones. Thus, the proposed system of interrelated kinematic equations allows us to describe the creep and long-term strength behavior of metallic materials at different temperatures, demonstrating the universality of the system.

References

[1] Arutyunyan AR, Saitova RR (2022) Exact and approximate solutions of the system of interrelated equations of the theory of creep and long-term strength. Journal of Physics: Conference Series. 2231(1):012001, DOI 10.1088/1742-6596/2231/1/012001

[2] Arutyunyan AR, Saitova RR (2024) Experimental and theoretical investigations of high-temperature creep of the ENAW 5251 (AMg2) aluminum alloy under step loading. Doklady Physics 68(7):203-206, DOI 10.1134/S1028335823070017

[3] Arutyunyan RA (1997) O roli poristosti v processah vyazkogo techeniya i hrupkogo razrusheniya pri polzuchesti (in Russ., On the role of porosity in the processes of viscous flow and brittle fracture during creep). Doklady RAN 352(2):190-192

[4] Arutyunyan RA (1998) On brittle fracture under the conditions of high-temperature creep. Metal Science and Heat Treatment 39(8):354-357, DOI 10.1007/BF02467636

[5] Arutyunyan RA (2004) Problema deformacionnogo stareniya i dlitel'nogo razrusheniya v mekhanike materialov (in Russ., The Problem of Deformation Aging and Long-Term Fracture in the Mechanics of Materials). St. Petersburg State University Publishing House, St. Petersburg

[6] Arutyunyan RA (2015) High-temperature embrittlement and long-term strength of metallic materials. Mechanics of Solids 50(2):191-197, DOI 10.3103/S0025654415020089

[7] Betekhtin VI (1997) Porosity of Solids. Transactions St.-Petersburg Academy of Sciences for Strength Problems 1:202-210

[8] Kachanov LM (1986) Introduction to Continuum Damage Mechanics. Martinus Nijhoff, Dordrecht

[9] Kachanov LM (1958) O vremeni razrusheniya v usloviyakh polzuchesti (in Russ., On the time to rupture under creep conditions). Izvestiya AN SSSR. Otdelenie Tekhnicheskikh Nauk (8):26–31

[10] Lemaitre L (1996) A Course on Damage Mechanics, 2nd ed. Springer, Berlin, Heidelberg, DOI 10.1007/978-3-642-18255-6

[11] Lemaitre J, Desmorat R (2005) Engineering Damage Mechanics: Ductile, Creep, Fatigue and Brittle Failures, Springer, Berlin, Heidelberg, DOI 10.1007/b138882

[12] Lemaitre J, Dufailly J (1987) Damage measurements. Engineering Fracture Mechanics 2885–6):643 - 661, DOI 10.1016/0013-7944(87)90059-2

[13] Lokoshchenko AM (2017) Creep and Long-Term Strength of Metals. CRC Press, Boca Raton, DOI 10.1201/b22242

[14] National Research Institute for Metals (1988) Data Sheets on the Elevated-Temperature Properties of 0.2 Silicon-Killed steel tubes for boilers and heat exchangers (STB 410). Creep Data Sheet No. 7B

[15] National Research Institute for Metals. (1988) Data Sheets on the Elevated-Temperature Properties of 5Cr-0.5Mo Steel Tubes for boilers and heat exchangers (STBA 25). Creep Data Sheet No. 12B
[16] National Research Institute for Metals. (1988) Data Sheets on the Elevated-Temperature Properties of Nickel based 15.5Cr-2.5Ti-0.7Al-1Nb-7Fe Superalloy Bars for High-Temperature Service. Creep Data Sheet No. 39
[17] Naumenko K, Altenbach H (2016) Modeling High Temperature Materials Behavior for Structural Analysis, Part I: Continuum Mechanics Foundations and Constitutive Models, Advanced Structured Materials, Vol 28, Springer, Cham, DOI 10.1007/978-3-319-31629-1
[18] Novozhilov VV (1965) On plastic cavitation. Journal of Applied Mathematics and Mechanics 29(4):811-819, DOI 10.1016/0021-8928(65)90090-0
[19] Rabotnov YN (1969) Creep Problems in Structural Members. North-Holland, Amsterdam
[20] Rabotnov YN (1959) O mechanizme dlitel'nogo razrusheniya (in Russ., A mechanism of the long term fracture). Voprosy prochnosti materialov i konstruktsii, AN SSSR pp 5–7
[21] Ratcliffe RT Greenwood GW (1965) Mechanism of cavitation in magnesium during creep. The Philosophical Magazine: A Journal of Theoretical Experimental and Applied Physics, Series 8 12(115):59-69, DOI 10.1080/14786436508224947
[22] Saitova RR (2023) High temperature creep and embrittlement of the materials under conditions of the long-term usage. PhD thesis (Candidate of Physical and Mathematical Sciences). St. Petersburg State University, St. Petersburg
[23] Saitova RR, Arutyunyan AR, Altenbach H (2024) High temperature creep and embrittlement in metals and alloys under conditions of the long-term usage. Acta Mechanica 235(10):6033-6055, DOI 10.1007/s00707-024-04026-6

Chapter 30
On Stress Singularities in an Elasticity Problem for a Poroelastic Wedge

Azat M. Sargsyan and Eghine G. Kanetsyan

Abstract Within the framework of the Cowin–Nunziato theory of poroelastic bodies, the stress singularity in the vicinity of the apex of a poroelastic wedge under the plane deformation is studied. For a poroelastic wedge with an arbitrary opening angle, the second and mixed homogeneous singular boundary-value problems are considered. Based on the asymptotic representations of the governing poroelasticity equations near the wedge tip, a local solution to the formulated homogeneous problem is constructed, and the energetically admissible – that is, physically realizable – solutions are identified. Various special cases corresponding to different wedge opening angles are discussed, and their graphical representations are provided.

Key words: Poroelasticity, Poroelastic wedge, Covin-Nunziato model, Singularity of stresses

30.1 Introduction

In modern technology, particularly in technology, in mechanical engineering, both traditional materials, and new-generation materials with tailored physical and mechanical properties are widely used. Regardless of their origin or method of manufacture, such materials inevitably contain various heterogeneities such as pores, grains, microcracks, thin-walled inclusions and geometric irregularities such as cor-

Azat M. Sargsyan
Institute of Mechanics, National Academy of Sciences of the Republic of Armenia, 24b Marshal Baghramyan Ave, Yerevan, 0019, Republic of Armenia, e-mail: azat-sargsyan@mail.ru

Eghine G. Kanetsyan (corresponding author)
Institute of Mechanics, National Academy of Sciences of the Republic of Armenia, 24b Marshal Baghramyan Ave, Yerevan, 0019 & National University of Architecture and Construction of Armenia, 105 Teryan Street, Yerevan, 0009, Republic of Armenia, Republic of Armenia, e-mail: ekanetsyan@gmail.com

H. Altenbach and A. V. Sahakyan, *Contact Mechanics and Relevant Problems*,
Advanced Structured Materials 253, https://doi.org/10.1007/978-3-032-20869-9_30

ner points. These heterogeneities give rise to local stress fields with steep gradients, forming zones of stress concentration. Stress concentration plays a critical role in determining the strength characteristics of engineering structures and their components, often leading to a reduction in their load-bearing capacity and durability. Therefore, qualitative and quantitative analyses of stress concentration are essential, including the identification of the nature of stress singularity and the development of effective approaches for reducing the level of concentration. Based on the actual mechanical properties of materials and the design purposes of their use, physically adequate generalized models of an elastic body are proposed. For the rigorous mathematical investigation of stress concentration phenomena, the methods of homogeneous singular boundary value problems provide a powerful analytical framework.

One of the generalized models of an elastic medium is the Covin-Nunziato theory of poroelasticity [5, 10], which provides a more accurate description of the deformation processes of elastic porous materials. In contrast to the classical theory of elasticity, which involves only two independent material constants, the poroelasticity model introduces five such constants, thereby significantly expanding the possibilities for capturing the mechanical response of heterogeneous media. Within the framework of this theory, many problems were solved: the problem of homogeneous deformations [10], the Saint-Venant problem for beams [15], the Kirsch problem [9], the Lamé problem [11], some contact problems [5, 21], problems of fracture mechanics [8], the plane problem for a half-plane [24], and the others. The Saint-Venant principle for poroelasticity was proved in [2]. The existence and uniqueness theorems were proved in [12]. Thin porous beams were investigated in [3]. A significant contribution to the work of Covin-Nunziato was made in [13, 14], where static and dynamic problems for two and three-dimensional domains were considered using the finite element method.

Since the equations of elasticity theory break down in the vicinity of corner points of elastic bodies, it becomes necessary to conduct a separate analysis of these regions and to identify the characteristic features of the resulting stresses and deformations. The study of such problems originates in the well-known works of Williams [26, 27], where three homogeneous boundary value problems for a wedge-shaped region were thoroughly investigated in different boundary condition combinations. In [26], a new class of boundary conditions associated with contact problems for a smooth contact model was introduced as an extension of the boundary value problems considered in [27]. These studies were then continued in a number of works [4, 6, 7, 17, 20, 25]. In particular, in [6], based on group theory, a rigorous mathematical theory of singular homogeneous boundary value problems in elasticity theory was developed, and, using Kolosov-Muskhelishvili complex potentials, the type of singular solution in the vicinity of corner points was strictly justified. The behavior of solutions to elliptic equations, including the equation of elasticity theory near corner points, was mathematically rigorously studied in [18]. In [22], another new type of boundary condition was proposed, together with various combinations with existing ones, to describe contact interactions of thin-walled elements with massive elastic bodies. In [23], the first homogeneous singular boundary-value problem in the theory of poroelasticity was considered under the condition that elastic stresses are prescribed on the faces of the wedge.

In the present paper, for the first time, the second and the mixed homogeneous singular boundary-value problem of elasticity theory is investigated for a poroelastic wedge with an arbitrary opening angle. The analysis focuses on the stress singularity in the vicinity of the apex of a wedge under a plane deformation state based on the Covina-Nunziato theory of poroelastic bodies.

30.2 Problem Formulation and Derivation of Governing Equations

If the displacements of the points of a poroelastic body $u_x(x, y)$ and $u_y(x, y)$ are taken as the primary unknowns, then their determination is governed by the following system of equations [24]

$$\begin{gathered}
\frac{\partial^2 u_x}{\partial x^2} + c^2 \frac{\partial^2 u_x}{\partial y^2} + \left(1 - c^2\right) \frac{\partial^2 u_y}{\partial x \partial y} + H \frac{\partial \Phi}{\partial x} = 0, \\
\frac{\partial^2 u_y}{\partial y^2} + c^2 \frac{\partial^2 u_y}{\partial x^2} + \left(1 - c^2\right) \frac{\partial^2 u_x}{\partial x \partial y} + H \frac{\partial \Phi}{\partial y} = 0, \\
\alpha \left(\frac{\partial^2 \Phi}{\partial x^2} + \frac{\partial^2 \Phi}{\partial y^2} \right) - \xi \Phi - \beta \left(\frac{\partial u_x}{\partial x} + \frac{\partial u_y}{\partial y} \right) = 0,
\end{gathered} \tag{30.1}$$

where function $\Phi(x, y)$ describes changes in the volume fraction of pores, α is the void diffusion coefficient, β is the parameter of the relationship between micro and macro dilation properties, ξ is the void rigidity, c and H are dimensionless quantities,

$$c^2 = \mu_0 / (\lambda_0 + 2\mu_0)\ ,\ H = \beta / (\lambda_0 + 2\mu_0).$$

λ_0 and μ_0 are Lamé's parameter. In addition to system (30.1), the fundamental set of equations in poroelasticity theory also includes the relations between displacements and strains

$$\varepsilon_x = \partial u_x / \partial x,\ \ \varepsilon_y = \partial u_y / \partial y,\ \ \varepsilon_{xy} = \left(\partial u_y / \partial x + \partial u_x / \partial y\ \right) / 2, \tag{30.2}$$

as well as the generalized Hooke's law

$$\begin{gathered}
\frac{\sigma_x}{\lambda_0 + 2\mu_0} = \frac{\partial u_x}{\partial x} + \left(1 - 2c^2\right) \frac{\partial u_y}{\partial y} + H\Phi, \\
\frac{\sigma_y}{\lambda_0 + 2\mu_0} = \left(1 - 2c^2\right) \frac{\partial u_x}{\partial x} + \frac{\partial u_y}{\partial y} + H\Phi, \\
\frac{\tau_{xy}}{\mu_0} = \frac{\partial u_x}{\partial x} + \frac{\partial u_y}{\partial y}.
\end{gathered} \tag{30.3}$$

These equations must be supplemented by boundary conditions on the body's surface, where either stresses and/or displacements are prescribed. Evidently, Eqs. (30.1)- (30.3) also remain valid when $\beta = 0$, in which case they reduce to the well-known equations of classical elasticity theory. In what follows, we first seek a fundamental solution to system (30.1) and then construct a solution satisfying the boundary conditions by superposing appropriate particular solutions corresponding to $\beta = 0$.

To obtain a fundamental solution to the system of equations (30.1), we introduce, by analogy with thermoelasticity theory, the concept of a poroelastic displacement potential $\Psi(x, y)$. It is assumed that the displacements u_x and u_y can be expressed as

$$\bar{u}_x = \partial\Psi/\partial x\ ,\ \bar{u}_y = \partial\Psi/\partial y\ , \tag{30.4}$$

where $\Psi(x, y)$ is a function that is four times differentiable within the given domain.

From the first two equations in (30.1), it follows that

$$\Delta\Psi = -H\Phi, \tag{30.5}$$

and from the third equation we have

$$\Delta\Phi - k_1\Phi = 0, \tag{30.6}$$

where Δ is the Laplace operator in the Cartesian coordinate system, the parameter $k_1 = (\xi - H\beta)/\alpha\ > 0$ [11].

Thus, by introducing the poroelastic displacement potential, the associated poroelasticity problem is reduced to determining the function $\Phi(x, y)$ from the Helmholtz equation (30.6). Once this function $\Phi(x, y)$ is obtained, the poroelastic displacement potential $\Psi(x, y)$ and the displacements $\bar{u}_x,\ \bar{u}_y$ can be determined.

Equations (30.2) and (30.3) then take the following forms, respectively:

$$\bar{\varepsilon}_x = \partial^2\Psi/\partial x^2,\ \bar{\varepsilon}_y = \partial^2\Psi/\partial y^2\ ,\ \bar{\varepsilon}_{xy} = \partial^2\Psi/\partial x\partial y\ , \tag{30.7}$$

$$\bar{\sigma}_x = -2\mu_0\partial^2\Psi/\partial y^2,\ \bar{\sigma}_y = -2\mu_0\partial^2\Psi/\partial x^2\ ,\ \bar{\tau}_{xy} = 2\mu_0\partial^2\Psi/\partial x\partial y\ . \tag{30.8}$$

These solutions may only incidentally satisfy the prescribed values on the boundary contour of the poroelastic body, since expression (30.4) for $\bar{u}_x$ and $\bar{u}_y$ at $\beta = 0$ does not, in the general case, represent a general solution to the equations of elasticity theory. If the obtained solution does not meet the boundary conditions, it is supplemented by the solution of the equation of elasticity theory for $\beta = 0$ to enforce the prescribed conditions on the boundary contour.

Let us consider the second and the mixed homogeneous singular boundary-value problems in poroelasticity theory. A poroelastic wedge with an opening angle φ_1, as shown in Fig. 30.1, is associated with both rectangular and polar coordinate systems and is assumed to be in a state of plane deformation.

Assuming that zero displacements are prescribed on the portions of the boundary counter adjacent to the corner point, on the cross-sections of the wedge $\varphi = 0$ and $\varphi = \varphi_1$. The boundary conditions can be written as

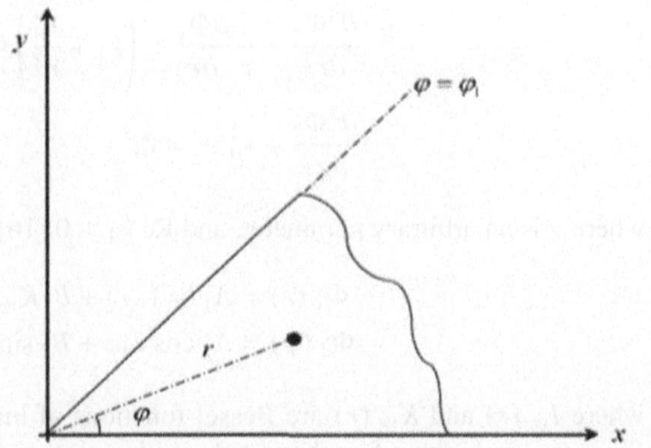

Fig. 30.1 Diagram of a poroelastic isotropic wedge.

$$u_r(r,0) = u_\varphi(r,0) = 0, \tag{30.9}$$

$$u_r(r,\varphi_1) = u_\varphi(r,\varphi_1) = 0 \tag{30.10}$$

for the second problem, and as

$$\sigma_\varphi(r,0) = \tau_{r\varphi}(r,0) = 0, \tag{30.11}$$

$$u_r(r,\varphi_1) = u_\varphi(r,\varphi_1) = 0 \tag{30.12}$$

for the mixed problem.

For function $\Phi(x,y)$, we impose the following boundary conditions [11, 24]

$$\left.\frac{\partial\Phi}{\partial\varphi}\right|_{\varphi=0} = \left.\frac{\partial\Phi}{\partial\varphi}\right|_{\varphi=\varphi_1} = 0, \tag{30.13}$$

$$\Phi(r,0) = \Phi(r,\varphi_1) = 0, \tag{30.14}$$

$$\Phi(r,0) = \left.\frac{\partial\Phi}{\partial\varphi}\right|_{\varphi=\varphi_1} = 0. \tag{30.15}$$

In the polar coordinate system, formulas (30.5) and (30.6) take the form

$$\frac{\partial^2\Psi}{\partial r^2} + \frac{1}{r}\frac{\partial\Psi}{\partial r} + \frac{1}{r^2}\frac{\partial^2\Psi}{\partial\varphi^2} + H\Phi = 0, \tag{30.16}$$

$$\frac{\partial^2\Phi}{\partial r^2} + \frac{1}{r}\frac{\partial\Phi}{\partial r} + \frac{1}{r^2}\frac{\partial^2\Phi}{\partial\varphi^2} - k_1\Phi = 0. \tag{30.17}$$

Representing the solution $\Phi(x,y)$ as

$$\Phi(r,\varphi) = \Phi_1(r)\,\Phi_2(\varphi) \tag{30.18}$$

and substituting into (30.17), we obtain two ordinary differential equations

$$\frac{\partial^2 \Phi_1}{\partial r^2} + \frac{1}{r}\frac{\partial \Phi_1}{\partial r} - \left(k_1 + \frac{\nu_0^2}{r^2}\right)\Phi_1 = 0,$$
$$\frac{\partial^2 \Phi_2}{\partial \varphi^2} + \nu_0^2 \Phi_2 = 0, \tag{30.19}$$

where ν is an arbitrary parameter, and $\operatorname{Re} \nu_0 > 0$ [16]. The solutions of (30.19) are

$$\Phi_1(r) = A_1 I_{\nu_0}(kr) + B_1 K_{\nu_0}(kr),$$
$$\Phi_2(\varphi) = A_2 \cos \nu_0 \varphi + B_2 \sin \nu_0 \varphi, \tag{30.20}$$

where $I_{\nu_0}(r)$ and $K_{\nu_0}(r)$ are Bessel functions of imaginary argument ($k = \sqrt{k_1}$). Since we are looking for a bounded solution as $r \to 0$, then $B_1 = 0$.

The asymptotic behavior of functions $I_\nu(r)$ for small r, which is what we need, is given by the formula [16]

$$I_{\nu_0}(r) \approx \frac{1}{\Gamma(\nu_0 + 1)}\left(\frac{r}{2}\right)^{\nu_0}. \tag{30.21}$$

Therefore, using formula (30.21), the function $\Phi(r, \varphi)$ in the vicinity of the apex of the poroelastic wedge is

$$\Phi(r, \varphi) = \frac{(k/2)^{\nu_0}}{\Gamma(\nu_0 + 1)} r^{\nu_0} \left(\tilde{A} \cos \nu_0 \varphi + \tilde{B} \sin \nu_0 \varphi\right).$$

Now, note that from (30.17) $\Phi = \Delta\Phi/k_1$. Substituting this expression into (30.16), we arrive at the equation

$$\Delta\left(\Psi(r, \varphi) + \frac{H}{k_1}\Phi(r, \varphi)\right) = 0.$$

Let us further assume that

$$\Psi(r, \varphi) + \frac{H}{k_1}\Phi(r, \varphi) = 0 \Rightarrow \Psi(r, \varphi) = -\frac{H}{k_1}\Phi(r, \varphi),$$

this implies that we accept the proportionality between the potentials of change in the volumetric part of the pores and displacements. In other words, we assume the proportionality of the corresponding components of the fields of the volumetric part of the pores and displacements, which are interconnected.

In a polar coordinate system, displacements $\bar{u}_r, \bar{u}_\varphi$ are easily determined using the formula for converting displacements from a Cartesian coordinate system to a polar coordinate system.

$$u_r = u_x \cos \varphi + u_y \sin \varphi = \frac{\partial \psi}{\partial x} \cos \varphi + \frac{\partial \psi}{\partial y} \sin \varphi,$$
$$u_\varphi = -u_x \sin \varphi + u_y \cos \varphi = -\frac{\partial \psi}{\partial x} \sin \varphi + \frac{\partial \psi}{\partial y} \cos \varphi.$$

Taking into account the behavior of the functions $I_{\nu_0}(r)$ as $r \to 0$, these displacements in the vicinity of the wedge apex will have the form

$$\begin{aligned}
\overline{u}_r(r,\varphi) &= \chi(\lambda^*)\left[\tilde{A}_2\cos(\lambda^*+1)\varphi+\tilde{B}_2\sin(\lambda^*+1)\varphi\right]r^{\lambda^*},\\
\overline{u}_\varphi(r,\varphi) &= -\chi(\lambda^*)\left[-\tilde{A}_2\sin(\lambda^*+1)\varphi+\tilde{B}_2\cos(\lambda^*+1)\varphi\right]r^{\lambda^*},\\
\overline{\sigma}_r &= \frac{2\mu H}{k_1}\frac{(k/2)^{\lambda^*+1}}{\Gamma(\lambda^*+2)}r^{\lambda^*-1}\left[a_1(\lambda^*,\varphi)\tilde{A}_2+b_1(\lambda^*,\varphi)\tilde{B}_2\right],\\
\overline{\sigma}_\varphi &= \frac{2\mu H}{k_1}\frac{(k/2)^{\lambda^*+1}}{\Gamma(\lambda^*+2)}r^{\lambda^*-1}\left[a_2(\lambda^*,\varphi)\tilde{A}_2+b_2(\lambda^*,\varphi)\tilde{B}_2\right],\\
\overline{\tau}_{r\varphi} &= \frac{2\mu H}{k_1}\frac{(k/2)^{\lambda^*+1}}{\Gamma(\lambda^*+2)}r^{\lambda^*-1}\left[a_3(\lambda^*,\varphi)\tilde{A}_2+b_3(\lambda^*,\varphi)\tilde{B}_2\right],\\
\chi(\lambda) &= \frac{H}{k_1}\frac{(k/2)^{\lambda-1}}{\Gamma(\lambda+2)}(\lambda+1),\ \lambda=\nu_0-1,\\
\nu &= \lambda^*+1;\ \chi(\lambda^*)=H(\lambda^*+1)(k/2)^{\lambda^*-1}/k_1\Gamma(\lambda^*+2),
\end{aligned} \tag{30.22}$$

where $a_j(\lambda^*,\varphi)$, $b_j(\lambda^*,\varphi)$ are known function of λ^* and φ.

To satisfy the boundary conditions (30.9)–(30.13), we impose on the solution (30.22) the elasticity theory solution $\tilde{u}_r$, $\tilde{u}_\varphi$ corresponding to $\beta=0$

$$\begin{aligned}
u_r &= \tilde{u}_r+\overline{u}_r = \frac{1}{E}r^\lambda\\
&\times\left[-A\lambda^+\nu^+S_\varphi^+ - B\lambda^+\nu^+C_\varphi^+ + C(4-\lambda^+\nu^+)S_\varphi^- + D(4-\lambda^+\nu^+)C_\varphi^-\right]\\
&+\chi(\lambda)r^\lambda\left[\tilde{A}\cos\lambda^+\varphi+\tilde{B}\sin\lambda^+\varphi\right],\\
u_\varphi &= \tilde{u}_\varphi+\overline{u}_\varphi = \frac{1}{E}r^\lambda\\
&\times\left[-A\lambda^+\nu^+C_\varphi^+ + B\lambda^+\nu^+S_\varphi^+ - C(4+\lambda^-\nu^+)C_\varphi^- + D(4+\lambda^-\nu^+)S_\varphi^-\right]\\
&+\chi(\lambda)r^\lambda\left[-\tilde{A}\sin\lambda^+\varphi+\tilde{B}\cos\lambda^+\varphi\right],
\end{aligned} \tag{30.23}$$

where, $\lambda=\nu_0-1$, $\lambda^\pm=\lambda\pm1$, $\nu^\pm=\nu\pm1$, $S^\pm=\sin(\lambda\pm1)\varphi$, $C^\pm=\cos(\lambda\pm1)\varphi$, ν is the Poisson coefficient.

As follows from (30.3) and (30.23), the stresses have a singularity of order $(1-\mathrm{Re}\,\lambda)$ at the wedge apex. By satisfying the boundary conditions (30.9)–(30.13), we obtain a homogeneous system of linear algebraic equations for the unknown coefficients $A, B, C, D, \tilde{A}$, and $\tilde{B}$

$$\sum_{j=1}^{6}P_{ij}V_j=0 \quad (i=1,2,...,6),$$

where $V_j=\left[A,B,C,D,\tilde{A},\tilde{B}\right]$, and the matrix of coefficients P_{ij} is

$$
\begin{bmatrix}
0 & 0 & 0 & 0 & 0 & \dfrac{\left(\frac{k}{2}\right)^{\lambda-1}}{\Gamma(\lambda+2)} \\
0 & 0 & 0 & 0 & -\dfrac{\left(\frac{k}{2}\right)^{\lambda-1} S^{+}_{\varphi_1}}{\Gamma(\lambda+2)} & \dfrac{\left(\frac{k}{2}\right)^{\lambda-1} C^{+}_{\varphi_1}}{\Gamma(\lambda+2)} \\
0 & -\dfrac{\lambda^{+}\nu^{+}}{E} & 0 & \dfrac{4-\lambda^{+}\nu^{+}}{E} & -\chi(\lambda) & 0 \\
-\dfrac{\lambda^{+}\nu^{+}}{E} & 0 & \dfrac{4+\lambda^{-}\nu^{+}}{E} & 0 & -\chi(\lambda) & 0 \\
-\dfrac{\lambda^{+}\nu^{+} S^{+}_{\varphi_1}}{E} & -\dfrac{\lambda^{+}\nu^{+} C^{+}_{\varphi_1}}{E} & \dfrac{4-\lambda^{+}\nu^{+}}{E} S^{-}_{\varphi_1} & \dfrac{4-\lambda^{+}\nu^{+}}{E} C^{-}_{\varphi_1} & -\chi(\lambda)\, C^{+}_{\varphi_1} & -\chi(\lambda)\, S^{+}_{\varphi_1} \\
-\dfrac{\lambda^{+}\nu^{+} C^{+}_{\varphi_1}}{E} & \dfrac{\lambda^{+}\nu^{+} S^{+}_{\varphi_1}}{E} & -\dfrac{4+\lambda^{-}\nu^{+}}{E} C^{-}_{\varphi_1} & \dfrac{4+\lambda^{-}\nu^{+}}{E} S^{-}_{\varphi_1} & -\chi(\lambda)\, S^{+}_{\varphi_1} & -\chi(\lambda)\, C^{+}_{\varphi_1}
\end{bmatrix}.
$$

For boundary conditions (30.14), and (30.15) in the coefficient matrices P_{ij}, only the last two elements of the first two rows change.

30.3 Solutions of Transcendental Equations and Their Analysis

Consider the second homogeneous singular boundary-value problem. The condition for the existence of a non-trivial solution to the above system of linear algebraic equations leads to the equation $\det P_{ij} = \Delta_6 = \Delta_2 \cdot \Delta_4 = 0$.From the condition $\Delta_4 = 0$, the well-known Williams transcendental equation follows [27]

$$\sin^2\lambda\varphi_1 - \lambda^2 \sin\varphi_1 = 0. \tag{30.24}$$

From the condition $\Delta_2 = 0$, for the parameter λ, we obtain

1. in the case of boundary conditions (30.13), (30.14)

$$\tilde{\lambda}_n = n\pi/\varphi_1 - 1, \tag{30.25}$$

2. for the boundary condition (30.15)

$$\lambda_m = (2m-1)\,\pi/2\varphi_1 - 1\,. \tag{30.26}$$

The requirement of finite elastic deformation energy in a small vicinity of the wedge apex (under a finite external load) imposes the condition on the roots (30.24) - (30.26)

$$\operatorname{Re}\lambda > 0. \tag{30.27}$$

Condition (30.27), depending on the wedge opening angle, limits the range of variation for the parameters n and m as follows

for boundary conditions (30.11) and (30.12)

a) $n = 2, 3, 4, \ldots$ when $0 < \varphi_1 < 2\pi$,

b) $n = 1, 2, 3, \ldots$ when $0 < \varphi_1 < \pi$,
c) $n = 2, 3, 4, \ldots$ when $\pi < \varphi_1 < 2\pi$,
for boundary conditions (30.13)
a1) $m = 3, 4, 5, \ldots$ when $0 < \varphi_1 < 2\pi$,
b1) $m = 1, 2, 3, \ldots$ when $0 < \varphi_1 < \pi/2$,
c1) $m = 2, 3, 4, \ldots$ when $\pi/2 < \varphi_1 < 3\pi/2$,
d1) $m = 3, 4, 5, \ldots$ when $3\pi/2 < \varphi_1 < 2\pi$.

Figures 30.2 and 30.3 show graphs of roots (30.25) and (30.26) as a function of the wedge opening angle; at certain values of the angle φ_1, there are infinite stresses in the vicinity of the apex of the poroelastic wedge.

The graph of the real part of the first root of Eq. (30.25), which gives the maximum order of stress singularity $|\mathrm{Re}\,\lambda - 1|$ (curve w), is provided for comparison. Note that the curves $\lambda_1,\ \lambda_2$ and $\tilde{\lambda}_1,\ \tilde{\lambda}_2$ coincide with the corresponding curves of work [23], while the curves w differ. However, in both cases, the order of the stress singularity at $\varphi_1 = \pi$ coincides. The intersection of these graphs with the line $\mathrm{Re}\,\lambda = 1$ determines the limiting wedge opening angles, beyond which the stresses in the vicinity of the sector vertex tend to infinity (for larger angles) or to zero (for smaller angles).

Thus, the stress behavior near the wedge apex should be investigated in different regions of variation of the wedge opening angle φ_1. So, for case a), the limiting angle of the wedge opening is $\varphi_1^* = \pi$. If $\varphi_1 < \pi$, the stresses tend to zero as $r \to 0$ (the vicinity of the wedge apex is in a low-stress state [7]). In the case of $\varphi_1 > \pi$, the stresses exhibit a singularity (a local stress concentration field forms near the wedge apex [7]). The maximum order of the singularity varies within the range $0 < 1 - \tilde{\lambda}_2 < 1$. When $\varphi_1 \to 2\pi$ the maximum order of singularity can reach up to unity, i.e., up to the order of the stress singularity under the concentrated force. Stress singularities of this kind arise in contact problems on the interaction of two unequally loaded absolutely rigid stamps or two pads loaded with horizontal forces

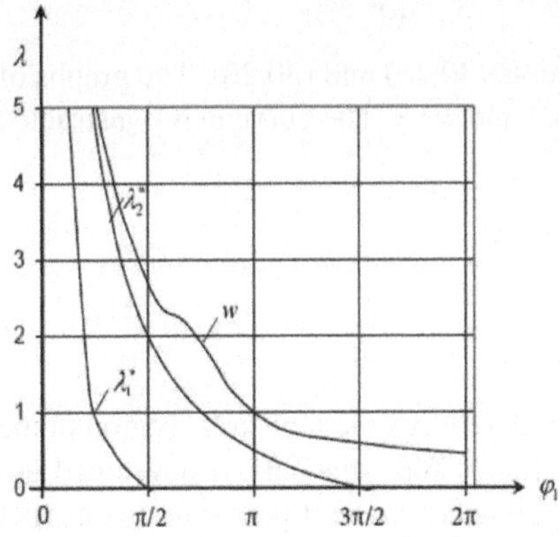

Fig. 30.2: Dependence of the parameter λ on the opening angle of the poroelastic wedge for boundary condition (30.13).

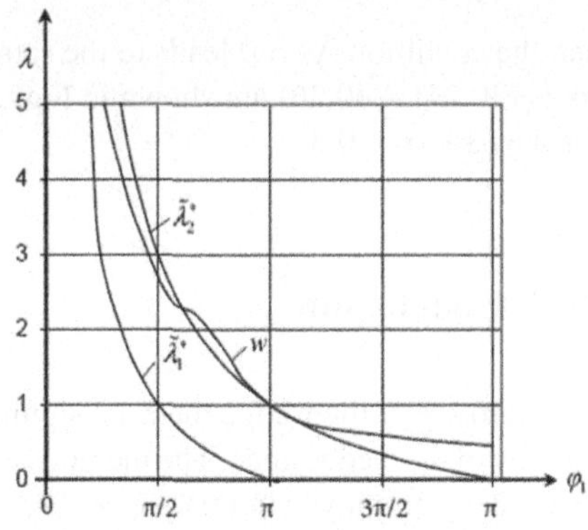

Fig. 30.3: Dependence of the parameter λ on the opening angle of the poroelastic wedge for boundary conditions (30.14) and (30.15).

that are absolutely rigid in tension and absolutely flexible in bending with an elastic half-plane [1]:

- For case b), $\varphi_1^* = \pi/2$, and the maximum order of singularity $1 - \tilde{\lambda}_1$ tends to unity as $\varphi_1 \rightarrow \pi$.
- For case c), the vicinity of the wedge's apex is always in a concentrated state and as $\varphi_1 \rightarrow 2\pi$, $1 - \tilde{\lambda}_2 \rightarrow 1$.
- For case a1), $\varphi_1^* = \pi$, and the maximum order of singularity tends to $1/2$ as $\varphi_1 \rightarrow 2\pi$, (see curve w).
- For case b1), $\varphi_1^* = \pi/4$, and the maximum order of singularity $1 - \lambda_1 \rightarrow 1$ as $\varphi_1 \rightarrow \pi/2$ For case c1), $\varphi_1^* = 3\pi/4$, $1 - \lambda_2 \rightarrow 1$.
- For d1), $\varphi_1^* = \pi$, the maximum order of singularity is $1/2$ as $\varphi_1 \rightarrow 2\pi$.

30.4 The Mixed Homogeneous Singular Boundary-Value Problem

This problem is solved in a similar manner. After satisfying the boundary conditions (30.9) – (30.13), we obtain a new homogeneous system of linear algebraic equations for the unknown coefficients $A, B, C, D, \tilde{A}$ and $\tilde{B}$:

$$\sum_{j=1}^{6} P_{ij}^* V_j = 0 \qquad (i = 1, 2, ..., 6) .$$

The determinant of this system is $\det P_{ij}^* = \Delta_6^* = \Delta_2 \cdot \Delta_4^*$. From the conditions $\Delta_4^* = 0$, we obtain the well-known Williams equation [26]

$$\sin^2 \lambda \varphi_1 = \frac{4(1-\sigma)^2}{3-4\sigma} - \frac{\sin^2 \varphi_1}{3-4\sigma},$$

and the condition $\Delta_2 = 0$ leads to the same roots (30.25) and (30.26). The graphs of roots (30.24) –(30.26) are shown in Figs. 30.4 and 30.5. The curve w is constructed for the case $\sigma = 0.3$.

30.5 Conclusion

The porosity of the wedge material significantly reduces the low-stress region in the vicinity of the wedge apex. The methods and ideas of Williams [26] are developed and extensively employed in this paper. These approaches make it possible to construct and analyze other homogeneous boundary-value problems for a poroelastic wedge, which will subsequently enable the study of contact problems and fracture mechanics problems that take into account the poroelastic effect.

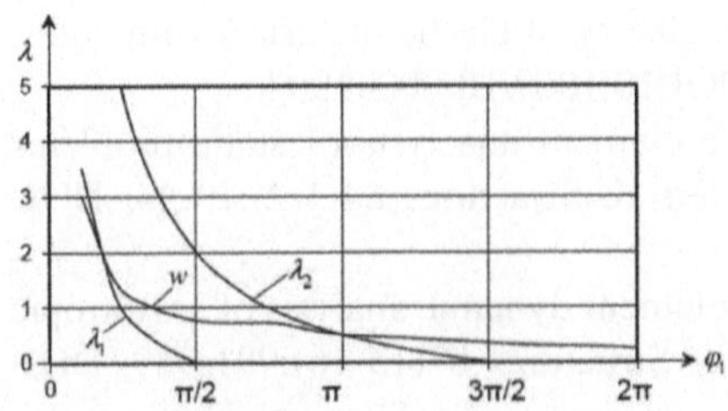

Fig. 30.4: Dependence of the parameter on the opening angle of the poroelastic wedge for boundary condition.

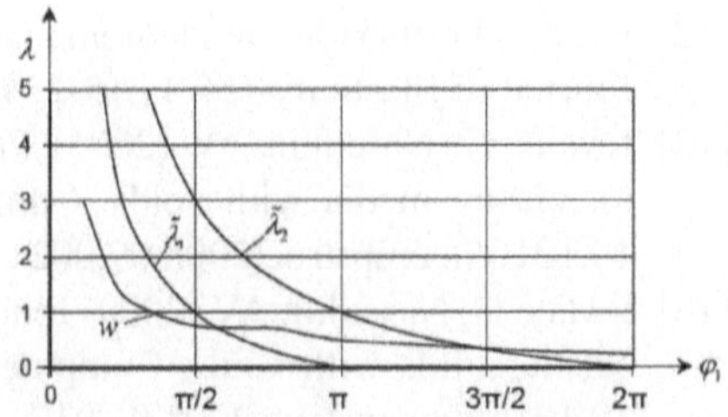

Fig. 30.5: Dependence of the parameter on the opening angle of the poroelastic wedge for boundary conditions.

References

[1] Alexandrov VM, Mkhitaryan SM (1983) Kontaktnyye zadachi dlya tel s tonkimi pokrytiyami i prosloykami (in Russ., Contact problems for bodies with thin coatings and interlayers). Nauka, Moscow

[2] Barta RC, Yang JS (1995) Saint-Venant's principle for linear elastic materials. Journal of Elasticity 39(3):265-271, DOI 10.1007/BF00041841

[3] Bîrsan M, Altenbach H (2011) Theory of thin thermoelastic rods made of porous materials. Archive of Applied Mechanic 81(10):1365-1391, DOI 10.1007/s00419-010-0490-z

[4] Bogy DB (1971) Two edge-bonded elastic wedges of different materials and wedge angles under surface tractions. Journal of Applied Mechanics 38(2):87-96, DOI 10.1115/1.3408786

[5] Chebakov MI, Kolosova EM (2021) Contact interaction of an axisymmetric stamp and elastic layer fixed on a poroelastic base. Mechanics of Composite Materials 56(6):769-778, DOI 10.1007/s11029-021-09922-9

[6] Cherepanov GP (1979) Mechanics of Brittle Fracture. McGraw-Hill, New York

[7] Chobanyan KS (1987) Napryazheniya v sostavnykh uprugikh telakh (in Russ., Stresses in Composite Elastic Bodies). Publishing house of the Academy of Sciences of the ArmSSR, Yerevan

[8] Ciarletta M, Iovane G, Sumbatyan MA (2003) On stress analysis for cracks in elastic materials with voids. International Journal of Engineering Science 41(20):2447-2461, DOI 10.1016/S0020-7225(03)00236-2

[9] Cowin SC (1984) The stresses around a hole in a linear elastic material with voids. The Quarterly Journal of Mechanics and Applied Mathematics 37(3):441-465, DOI 10.1093/qjmam/37.3.441

[10] Cowin SC, Nunziato JW (1983) Elastic materials with voids. Journal of Elasticity 13(2):125-147, DOI 10.1007/BF00041230

[11] Cowin SC, Puri P (1983) The elastic pressure vessel problems for linear elastic materials with voids. Journal of Elasticity 13(2):157-163, DOI 10.1007/BF00041232

[12] Ieaşan D (1985) Some theorems in the theory of elastic materials with voids. Journal of Elasticity 15(2):215-224, DOI 10.1007/BF00041994

[13] Iovane G, Nasedkin AV (2005) Finite element ana lysis of static problems for elastic media with voids. Computers & Stractures 84(1-2):19-24, DOI 10.1016/j.compstruc.2005.09.002

[14] Iovane G, Nasedkin AV (2009) Finite element dynamic analysis of anisotropic elastic solids with voids. Computers & Structures 87(15-16):981-989, DOI 10.1016/j.compstruc.2008.04.017

[15] Dell'Isola F, Gupta A (1997) Saint-Venants problem for porous linear elastic materials. Journal of Elasticity 47(1):73-81, DOI 10.1023/A:1007478322647

[16] Janke E, Emde F (1945) Tables of functions with formulae and curves. Dover, New York

[17] Kalandia AI (1969) Remarks of singularities of elastic solutions near corners. Journal of Applied Mathematics and Mechanics 33(1):132-135, DOI 10.1016/0021-8928(69)90121-X

[18] Kondrat'ev VA (1967) Krayevyye zadachi dlya ellipticheskikh uravneniy v oblastyakh s konicheskimi i uglovymi tochkami (in Russ., Boundary value problems for elliptic equations in domains with conical or angular points). Trudy Moskovskogo Matematicheskogo Obshchestva 16:209-292

[19] Nunziato JW, Cowin SC (1979) Nonlinear theory of elastic materials with voids. Archive for Rational Mechanics and Analysis 72(2):175-201, DOI 10.1007/BF00249363

[20] Parton VZ, Perlin PI (1981) Metody matematicheskoy teorii uprugosti (in Russ., Methods of mathematical theory of elasticity). Nauka, Moscow

[21] Pompei A, Rigano A, Sumbatyan MA (2005) Contact problem for a rectangular punch on the porous elastic half-space. Journal of Elasticity 76(1):1-19, DOI 10.1007/s10659-004-0043-1

[22] Sargsyan AM (2018) On homogeneous singular problems of elasticity theory. Reports of NAS RA 118(2):134-141

[23] Sargsyan AM, Kanetsyan EG (2025) On a homogeneous singular boundary value problem of poroelasticity theory. LJM 46(8):4150-4158 (in press)

[24] Scalia AA, Sumbatyan MA (2000) Contact problem for porous elastic half-plane. Journal of Elasticity 60(2):91-102, DOI 10.1023/A:1010880823544

[25] Sinclair JB (1980) On the singular eigenfunction for plane harmonic problems in composite regions. Journal of Applied Mechanics 47(1):87-92, DOI 10.1115/1.3153644

[26] Williams ML (1957) On the stress distribution at the base of a stationary crack. Journal of Applied Mechanics 14(1):109-114, DOI 10.1115/1.4011454

[27] Williams ML (1952) Stress singularities resulting from various boundary conditions in angular corners of plates in extension. Journal of Applied Mechanics 19(4):526-528, DOI 10.1115/1.4010553

Chapter 31
On the Substantiation of the Modified Method of Timoshenko for the Solution of the Problem of transverse Doubly Antisymmetric Bending of a Rectangular Plate Clamped Along the Contour

Suren P. Seyranyan

Abstract The problem of bending of a rectangular plate clamped on a contour under a transverse load that is doubly antisymmetric with respect to the axes of symmetry of the rectangle is considered. Previously, the author, representing bending moments on the contour of a rectangular plate as supposedly converging trigonometric series containing unknown coefficients, and applying the method of S.P. Timoshenko in a modified way, reduced the solution of this problem for deflection to an infinite, completely regular system of linear algebraic equations. However, when deriving the infinite system, the following assumptions were made: **a)** term-by-term differentiation of the series in the expression for the deflection into convergent series in the direction of the normals to the plate contour, **b)** convergence of the trigonometric series of bending moments on the plate contour, **c)** the admissibility of changing the order of summation in the arising repeated series. In the presented work, it is proven that, for the adopted type, but an arbitrary mechanical load, in the class of bounded solutions there exists a solution of the infinite system tending to zero (simultaneously, unique), which can be obtained by the reduction method. The assumptions made in points **a)** - **c)** are performed. The solution for the deflection is continuously adjacent to the boundary conditions.

Key words: Substantiation, Solution, Bending, Rectangular plate

31.1 Introduction

The solution to the problem of bending a rectangular plate clamped on a contour is the subject of works reflected in the review [4], as well as in the review sections of books [10, 11], etc. One of the well-known and universal methods for solving this

---

Suren P. Seyranyan
Institute of Mechanics, National Academy of Sciences of the Republic of Armenia, 24b Marshal Baghramyan Ave, Yerevan, 0019, Republic of Armenia, e-mail: surenparuir@gmail.com

H. Altenbach and A. V. Sahakyan, *Contact Mechanics and Relevant Problems*,
Advanced Structured Materials 253, https://doi.org/10.1007/978-3-032-20869-9_31

problem is the method of S.P. Timoshenko [11], which, however, as noted in [1], refers to a mathematically non-rigorous method. In this paper, as a continuation of [7], the gap that has arisen is filled.

31.2 Original Known Author's Results

The solution to the problem of bending a rectangular plate clamped on a contour under a transverse load $q(x, y)$ that is antisymmetric relative to the symmetry axes of the rectangle for a deflection $w(x, y)$ in a rectangle $(x, y) \in G = [-a/2, a/2] \times [-b/2, b/2]$, starting from work [11], in work [7] was sought in the form

$$
\begin{aligned}
w(x,y) &= \frac{1}{D} \sum_{m=2,4\ldots}^{\infty} \sum_{n=2,4\ldots}^{\infty} \frac{q_{mn} \sin \lambda_m x \, \sin \mu_n y}{(\lambda_m^2 + \mu_n^2)^2} \\
&+ \frac{1}{2D} \sum_{m=2,4\ldots}^{\infty} E_m \frac{(-1)^{m/2}}{\lambda_m \sinh \alpha_m} \left(\frac{\alpha_m}{\lambda_m} \coth \alpha_m \sinh \lambda_m y - y \cosh \lambda_m y \right) \sin \lambda_m x \\
&+ \frac{1}{2D} \sum_{n=2,4\ldots}^{\infty} F_n \frac{(-1)^{n/2} \sinh \mu_n x}{\mu_n \sinh \beta_n} \left(\frac{\beta_n}{\mu_n} \tanh \beta_n - x \coth \mu_n x \right) \sin \mu_n y.
\end{aligned}
\tag{31.1}
$$

Here D is the classical bending stiffness of the plate, $\lambda_m = m\pi/a, \mu_n = n\pi/b$, $\alpha_m = m\pi b/(2a), \beta_n = n\pi a/(2b)$,

$$
q_{mn} = \frac{4}{a\,b} \int_{-a/2}^{a/2} \int_{-b/2}^{b/2} q\,(x,y) \sin \lambda_m x \, \sin \mu_n y \, dx \, dy\,. \tag{31.2}
$$

Representation (31.1) initially satisfies the Sophie Germaine–Lagrange equation with the right-hand side being the load $q(x, y)$ and the support conditions of the edges, bent by moments

$$
\begin{aligned}
M_y(x, b/2) = -M_y(x, -b/2) &= \sum_{m=2,4\ldots}^{\infty} (-1)^{m/2} E_m \sin \lambda_m x, \\
M_x(a/2, y) = -M_x(-a/2, y) &= \sum_{n=2,4\ldots}^{\infty} (-1)^{n/2} F_n \sin \mu_n y.
\end{aligned}
\tag{31.3}
$$

Further, $w(x, y)$ on the sides of the plate obeys the edge clamping conditions. Assuming the convergence of series (31.3) and the admissibility of changing the order of summation in expansions in repeated trigonometric series, in [7] a completely regular infinite system of linear algebraic equations [3] with respect to unknowns $\widehat{E}_m = (a/b)^{1/2} E_m$, $\widehat{F}_n = (b/a)^{1/2} F_n$, was obtained

$$
\begin{cases}
\widehat{E}_m = -\dfrac{8\,\xi_m^2}{\pi\,\varphi\,(\alpha_m)} \displaystyle\sum_{n=2,\,4,\ldots}^{\infty} \dfrac{n}{(\xi_m^2 + n^2)^2} \widehat{F}_n + b_m^{(1)}, \\
\widehat{F}_n = -\dfrac{8\chi_n^2}{\pi\,\varphi\,(\beta_n)} \displaystyle\sum_{m=2,\,4,\ldots}^{\infty} \dfrac{m}{[\chi_n^2 + m^2]^2} \widehat{E}_m + b_n^{(2)},
\end{cases}
\tag{31.4}
$$

where $m, n = 2, 4, \ldots, \varphi(t) = \coth t - t/\sinh^2 t\,,\ t = \alpha_m, \beta_m,$

$$
b_m^{(1)} = \frac{(-1)^{m/2}}{\varphi_1\,(\alpha_m)}\sqrt{\frac{b}{a}} \int\limits_{-a/2}^{a/2} (\,y/b\,\cosh\lambda_m y - 0.5\,\coth\alpha_m\,\sinh\lambda_m y)
\times \int\limits_{-b/2}^{b/2} q\,(x, y)\,\sin\lambda_m x\,dx\,dy,
\tag{31.5}
$$

$$
b_n^{(2)} = \frac{(-1)^{n/2}}{\varphi_1\,(\beta_n)}\sqrt{\frac{a}{b}} \int\limits_{-a/2}^{a/2} (\,x/a\,\cosh\mu_n x - 0.5\,\coth\beta_n\,\sinh\mu_n x)
\times \int\limits_{-b/2}^{b/2} q\,(x, y)\,\sin\mu_n y\,dx\,dy,
\tag{31.6}
$$

$$
\begin{aligned}
\xi_m &= m\,b/a,\ \chi_n = n\,a/b,\ \varphi_1\,(t) = \varphi\,(t)\,\sinh t \\
&= (1 - 2t/\sinh 2t)\,\cosh t > 0,\ t = \alpha_m, \beta_n.
\end{aligned}
\tag{31.7}
$$

31.3 Analysis of the Limit Behavior of Free Members of an Infinite System

Let the plate be loaded with four concentrated forces of a quantity $P/4$

$$
\begin{aligned}
q\,(x, y) = P/4\,[&\delta(x - \xi)\,\delta(y - \eta) - \delta(x - \xi)\,\delta(y + \eta) \\
&- \delta(x + \xi)\,\delta(y - \eta) + \delta(x + \xi)\,\delta(y + \eta)].
\end{aligned}
\tag{31.8}
$$

Substituting (31.8) into (31.5), (31.6), we arrive at the results

$$
\begin{aligned}
b_m^{(1)} &= P(-1)^{m/2}\sqrt{b/a}\,(\,\eta/b\,\cosh\lambda_m\eta - 0.5\,\coth\alpha_m\,\sinh\lambda_m\eta) \\
&/\varphi_1\,(\alpha_m)\ \sin\lambda_m\xi = O(e^{\lambda_m(\eta - b/2)}),
\end{aligned}
\tag{31.9}
$$

$$
\begin{aligned}
b_n^{(2)} &= P(-1)^{n/2}\sqrt{a/b}\,(\xi/a\,\cosh\mu_n\xi - 0.5\,\coth\beta_n\,\sinh\mu_n\xi) \\
&/\varphi_1\,(\beta_m)\,\sin\mu_n\eta = O(e^{\mu_n(\xi - a/2)}).
\end{aligned}
\tag{31.10}
$$

Now let be $q(x, y)$ a distributed limited load – $|q(x, y)| \leq q* = \text{const} > 0$. Let us show that the integrand in (31.5), containing y, when $0 \leq y \leq b/2$ is not positive.

$$\begin{aligned} f(y) &= y/b \cosh \lambda_m y - 0.5 \coth \alpha_m \sinh \lambda_m y \\ &= \cosh \lambda_m y \, (2y/b - \coth \alpha_m \tanh \lambda_m y)/2 \leq 0 . \end{aligned} \tag{31.11}$$

Let us put $y = t\, b/2$, where $0 \leq t \leq 1$. Then statement (31.11) is reduced to the form

$$f_1(t) = t - \coth \alpha_m \tanh \alpha_m t \leq 0. \tag{31.12}$$

Note that $f_1(0) = f_1(1) = 0$. Moreover

$$f_1{}'(t) = 1 - \alpha_m \coth \alpha_m / \cosh^2 \alpha_m t < \infty, \; 0 \leq t \leq 1. \tag{31.13}$$

Then, according to Rolle's theorem [2], there exists a value $t = t^*$ such that $0 < t^* < 1$, at which $f_1{}'(t^*) = 0$. But from the monotonic increase $\cosh \alpha_m t$, and with it $\cosh^2 \alpha_m t$ when $t \in [0,\ 1]$ in (31.13) it follows that for $t \in [0, t^*)$ $((t^*, 1])f_1{}'(t) < 0\,(f_1{}'(t) > 0)$. Thus the function $f_1(0) = 0$, decreases monotonically when $t \in (0, t^*)$, at $t = t^*$ - takes on a minimum value, at $t \in (t^*, 1)$ – increases monotonically [2], and $f_1(1) = 0$. Hence, inequality (31.12), and with it (31.11), are satisfied.

Let us now move on to estimating the free term $b_m^{(1)}$ for a distributed load satisfying the condition $|q\,(x, y)| \leq q*$. Taking into account inequalities (31.7) and (31.11), we have

$$\begin{aligned} \left|b_m^{(1)}\right| &\leq \frac{4}{\varphi_1(\alpha_m)}\sqrt{\frac{b}{a}} \int_0^{a/2} |\sin \lambda_m x| \, dx \int_0^{b/2} |q\,(x, y)|\,(-f(y))\, dy \\ &\leq \frac{2\sqrt{a\,b}\, q*}{\pi\, \varphi_1(\alpha_m)} \frac{\cosh \alpha_m - 1}{\lambda_m} \left(\frac{1}{\alpha_m} - \frac{1}{\sinh \alpha_m}\right) = O(1/m^2) \end{aligned} \tag{31.14}$$

and similarly $\left|b_n^{(2)}\right| = O(1/n^2)$.

But $q(x, y)$ in the general case consists of the sum of a distributed limited load and a finite number of loads of type (31.8). Therefore, in the general case $q(x, y)$, which follows from (31.9), (31.10) and (31.14),

$$\begin{aligned} \left|b_m^{(1)}\right| &= O(1/m^2) \leq K/m^2, \; m = 2,\ 4, \ldots, \\ \left|b_n^{(2)}\right| &= O(1/n^2) \leq K/n^2, n = 2,\ 4, \ldots , \end{aligned} \tag{31.15}$$

where $K > 0$ is some constant.

31.4 On the Existence of a Solution to an Finite System that Converges to zero

The system of equations (31.4) is a completely regular infinite system of linear algebraic equations [3] with limited, as follows from (31.15), free terms. Therefore, this system has a main limited solution [3]. Let us now show that at infinity the solution $\tilde{E}_m, \tilde{F}_n$ tends to zero, which will simultaneously lead to its uniqueness in the class of limited solutions [3]. Let us consider the series summed in [11, p. 165]

$$I_1(z) = \sum_{n=1}^{\infty} \frac{\cos nz}{(\alpha^2+n^2)^2} = -\frac{1}{2\alpha^4} + \frac{\pi}{4\alpha^3}\frac{\cosh\alpha(\pi-z)}{\sinh\pi\alpha} - \frac{\pi(\pi-z)}{4\alpha^2}\frac{\sinh\alpha(\pi-z)}{\sinh\pi\alpha} + \frac{\pi^2}{4\alpha^2}\frac{\cosh\alpha(\pi-z)\cosh\pi\alpha}{\sinh^2\pi\alpha}, \tag{31.16}$$

where $\alpha \neq 0,\ 0 \leq z \leq 2\pi$. Then

$$\begin{aligned}\frac{I_1(0)+I_1(\pi)}{2} &= \sum_{n=2,4,\ldots}^{\infty} \frac{1}{(\alpha^2+n^2)^2} = -\frac{1}{2\alpha^4} + \frac{\pi}{8\alpha^3}\frac{\cosh\alpha\pi+1}{\sinh\pi\alpha} \\ &- \frac{\pi^2}{8\alpha^2} + \frac{\pi^2}{8\alpha^2}\left(\frac{\cosh^2\alpha\pi}{\sinh^2\pi\alpha} + \frac{\cosh\pi\alpha}{\sinh^2\pi\alpha}\right) = -\frac{1}{2\alpha^4} + \frac{\pi}{8\alpha^3} \\ &\times \frac{2\cosh^2\alpha\pi/2}{2\sinh\alpha\pi/2\,\cosh\alpha\pi/2} + \frac{\pi^2}{8\alpha^2}\frac{1+\cosh\alpha\pi}{\sinh^2\pi\alpha} = -\frac{1}{2\alpha^4} \\ &+ \frac{\pi}{8\alpha^3}\left(\coth\frac{\alpha\pi}{2} + \frac{\alpha\pi}{2\sinh^2\frac{\alpha\pi}{2}}\right) = -\frac{1}{2\alpha^4} + \frac{\pi^2}{8\alpha^2\sinh^2\frac{\alpha\pi}{2}} \\ &+ \frac{\pi}{8\alpha^3}\left(\coth\frac{\alpha\pi}{2} - \frac{\alpha\pi}{2\sinh^2\alpha\pi/2}\right) < -\frac{1}{2\alpha^4} + \frac{\pi^2}{8\alpha^2(\alpha\pi/2)^2} \\ &+ \frac{\pi}{8\alpha^3}\left(\coth\frac{\alpha\pi}{2} - \frac{\alpha\pi}{2\sinh^2\frac{\alpha\pi}{2}}\right) = \frac{\pi}{8\alpha^3}\left(\coth\frac{\alpha\pi}{2} - \frac{\alpha\pi}{2\sinh^2\frac{\alpha\pi}{2}}\right).\end{aligned} \tag{31.17}$$

Here we used the inequality $\sinh t > t$ for t > 0. Hence

$$\frac{8\alpha^3}{\pi\left(\coth(\alpha\pi/2) - \alpha\pi/(2\sinh^2\alpha\pi/2)\right)} \sum_{n=2,4,\ldots}^{\infty} \frac{1}{(\alpha^2+n^2)^2} < 1. \tag{31.18}$$

Next, using (31.18), the Cauchy–Bunyakovsky inequality for infinite series [2] and the complete regularity of system (31.4) [3], proven in [5], [8], we obtain

$$\frac{8\,\alpha^{5/2}}{\pi\,\varphi\,(\pi\,\alpha/2)}\sum_{n=2,4,\ldots}^{\infty}\frac{n^{1/2}}{(\alpha^2+n^2)^2}=\frac{8\,\alpha^{5/2}}{\pi\,\varphi\,(\pi\,\alpha/2)}\sum_{n=2,4,\ldots}^{\infty}\left(\frac{1}{\alpha^2+n^2}\right.$$
$$\left.\times\frac{n^{1/2}}{\alpha^2+n^2}\right)\le\frac{8\,\alpha^{5/2}}{\pi\,\varphi\,(\pi\,\alpha/2)}\sqrt{\sum_{n=2,4,\ldots}^{\infty}\frac{1}{(\alpha^2+n^2)^2}}$$
$$\times\sqrt{\sum_{n=2,4,\ldots}^{\infty}\frac{n}{(\alpha^2+n^2)^2}}=\sqrt{\frac{8\,\alpha^3}{\pi\,\varphi\,(\pi\,\alpha/2)}\sum_{n=2,4,\ldots}^{\infty}\frac{1}{(\alpha^2+n^2)^2}} \tag{31.19}$$
$$\times\sqrt{\frac{8\,\alpha^2}{\pi\,\varphi\,(\pi\,\alpha/2)}\sum_{n=2,4,\ldots}^{\infty}\frac{n}{(\alpha^2+n^2)^2}}<\sqrt{0.871}<0.934.$$

Let us now note that the replacement of unknowns in system (31.4)

$$\widehat{E}_m=\bar{E}_m(a/b)^{1/4}/m^{1/2},\ \widehat{F}_n=\bar{F}_n(b/a)^{1/4}/n^{1/2} \tag{31.20}$$

again for $0<b/a<\infty$ leads to a completely regular infinite system

$$\begin{cases}\bar{E}_m=-\dfrac{8\,\xi_m^{5/2}}{\pi\,\varphi(\alpha_m)}\displaystyle\sum_{n=2,4,\ldots}^{\infty}\frac{n^{1/2}}{(\xi_m^2+n^2)^2}\bar{F}_n+b_m^{(1)}m^{1/2}\left(\frac{b}{a}\right)^{1/4}, \\ \bar{F}_n=-\dfrac{8\chi_n^{5/2}}{\pi\,\varphi(\beta_n)}\displaystyle\sum_{m=2,4,\ldots}^{\infty}\frac{m^{1/2}}{[\chi_n^2+m^2]^2}\bar{E}_m+b_n^{(2)}n^{1/2}\left(\frac{a}{b}\right)^{1/4},\end{cases} \tag{31.21}$$

where $m,n=2,4,\ldots.$

Since (31.19) for the value $\alpha=\xi_m$ and $\alpha=\chi_n$ is the executed condition of complete regularity [3] for the first and second equations of the infinite system (31.21). But the free members of the named system are limited, because taking into account (31.15) $\left|b_m^{(1)}\right|(b/a)^{1/4}m^{1/2}=O\,(1/m^{3/2})$, $\left|b_n^{(2)}\right|(a/b)^{1/4}n^{1/2}=O\,(1/n^{3/2})$. This means that the resulting system has a principal bounded solution $\left|\hat{E}_m\right|\le K_1$, $\left|\hat{F}_n\right|\le K_1$, where $K_1>0$ is some constant [3]. Therefore, the main solution of the original system (31.4), taking into account (31.20), satisfies the

$$\left|\tilde{E}_m\right|\le(a/b)^{1/4}K_1/m^{1/2},\ \left|\tilde{F}_n\right|\le(b/a)^{1/4}K_1/n^{1/2}. \tag{31.22}$$

Therefore $\tilde{E}_m$ and $\tilde{F}_n\to 0$ when m and $n\to\infty$, which simultaneously leads to the uniqueness of the solution in the class of bounded solutions [3].

31.5 On Changing the Order of Summation in Repeated Series

In [7] the derivation of the above system [1] was attended by an assumption similar to S.P. Timoshenko [11] about changing the order of summation in repeated series, which, when replacing the initial unknowns $\widehat{E}_m, \widehat{F}_n$ with solutions $\tilde{E}_m, \tilde{F}_n$ of the system [1], are written in the form

$$\begin{aligned} &\sum_{n=2,4,\ldots}^{\infty} \sum_{m=2,4,\ldots}^{\infty} \frac{(-1)^{m/2}\tilde{F}_n\, n\, m\, \sin\lambda_m x}{[(m\,b\,/\,a)^2+n^2]^{\,2}} \\ &= \sum_{m=2,4,\ldots}^{\infty} \sum_{n=2,4,\ldots}^{\infty} \frac{(-1)^{m/2}\tilde{F}_n\, n\, m\, \sin\lambda_m x}{[(m\,b\,/\,a)^2+n^2]^{\,2}}, \end{aligned} \tag{31.23}$$

$$\begin{aligned} &\sum_{n=2,4,\ldots}^{\infty} \sum_{m=2,4,\ldots}^{\infty} \frac{(-1)^{m/2}\tilde{E}_n n\, m\, \sin\mu_m x}{[(m\,a\,/\,b)^2+n^2]^{\,2}} \\ &= \sum_{m=2,4,\ldots}^{\infty} \sum_{n=2,4,\ldots}^{\infty} \frac{(-1)^{m/2}\tilde{E}_n\, n\, m\, \sin\mu_m x}{[(m\,a\,/\,b)^2+n^2]^{\,2}}. \end{aligned} \tag{31.24}$$

Let us prove that equalities (31.23), (31.24) are true. We will use Young's inequality, transformed in [6] with a generalization to the form

$$\alpha^2+\beta^2 \geq C\,\alpha^\gamma\,\beta^{2-\gamma}, \tag{31.25}$$

where $0 \leq \gamma \leq 2$, $\alpha, \beta \geq 0$ and $C > 0$ is some constant. Then applying (31.25), (31.22) to the evaluation of the common terms in (31.23), (31.24) leads to the results

$$\begin{aligned} &\left|\frac{(-1)^{m/2}\tilde{F}_n\, n\, m}{[(m\,b\,/\,a)^2+n^2]^{\,2}} \sin\lambda_m x\right| \leq \frac{\left|\tilde{F}_n\right|\, n\, m}{[(m\,b\,/\,a)^2+n^2]^{\,2}} \\ &\leq \frac{(b/a)^{1/4}K_1\, n^{1/2} m}{[(m\,b\,/\,a)^2+n^2]^{\,2}} \leq \left.\frac{(b/a)^{1/4}K_1\, n^{1/2} m}{C^2(m\,b\,/\,a)^{2\gamma} n^{4-2\gamma}}\right|_{\gamma=9/8} \\ &= \frac{K_1}{C^2(b/a)^2\, m^{\,5/4} n^{5/4}}. \end{aligned} \tag{31.26}$$

Hence replacing the common term in the repeated series on the left in (31.23), with their moduli leads to convergent series, which, according to theorem 3 [2], establishes equality (31.23). Equality (31.24) is proved similarly.

31.6 On the Convergence of Trigonometric Series of Bending Moments at the Edges of a Plate

A proof of the convergence of series (31.3) is given. Let us first note that when $x = \pm a/2, y = \pm b/2$ the sum of these series is equal to zero. Now let $-a/2 < x < a/2$, $-b/2 < y < b/2$. Let's put

$$c_m = (m\,a\,/b)^2 \sum_{n=2,4,\ldots}^{\infty} \tilde{F}_n n\,/\Lambda_{nm}^2, \text{ where } \Lambda_{nm} = (\,n\,a\,/\,b)^2 + m^2. \tag{31.27}$$

Then using (31.22) and (31.25) we obtain

$$\begin{aligned} |c_m| \le & \left(\frac{m\,a}{b}\right)^2 \sum_{n=2,4,\ldots}^{\infty} \frac{|\tilde{F}_n|\,n}{\Lambda_{nm}^2} \le \left(\frac{m\,a}{b}\right)^2 \frac{K_1}{C^2}\left(\frac{b}{a}\right)^{1/4} \\ & \times \sum_{n=2,4,\ldots}^{\infty} \frac{n^{1/2}}{(\,n\,a\,/\,b)^{2\gamma} m^{4-2\gamma}}\bigg|_{\gamma=7/8} = \frac{K_1}{C^2 m^{1/4}} \sum_{m=2,4,\ldots}^{\infty} \frac{1}{n^{5/4}} \\ & = \frac{K_2}{m^{1/4}}, \quad \lim_{m\to\infty} c_m = 0. \end{aligned} \tag{31.28}$$

Next we have

$$\begin{aligned} & \sum_{m=2,4,\ldots}^{N} (-1)^{m/2} c_m \sin\lambda_m x = \frac{1}{2\cos\lambda_1 x} \sum_{m=2,4,\ldots}^{N} (-1)^{m/2} c_m\,(\sin\lambda_{m+1} x \\ & + \sin\lambda_{m-1} x) = \frac{1}{2\,\cos\lambda_1 x}[-c_2 \sin\lambda_1 x + (-1)^{N/2} c_N \sin\lambda_{N+1} x \\ & - \sum_{m=3,5,\ldots}^{N-1} (-1)^{(m-1)/2}(c_{m+1} - c_{m-1}) \sin\lambda_m x]. \end{aligned} \tag{31.29}$$

Tending to the limit in (31.29) as $N \to \infty$ tends and taking into account (31.28), we arrive at the result

$$\begin{aligned} & \sum_{m=2,\,4,\ldots}^{\infty} (-1)^{m/2} c_m \sin\lambda_m x = -\frac{1}{2\cos\lambda_1 x}[c_2 \sin\lambda_1 x \\ & + \sum_{m=3,\,5,\ldots}^{\infty} (-1)^{(m-1)/2}(c_{m+1} - c_{m-1}) \sin\lambda_m x]\,. \end{aligned} \tag{31.30}$$

Here we used the convergence of the series on the right, which led to the convergence of the series on the left. Indeed, using (31.22), (31.25) and the **Wolfram Mathematica 11** package, we have

where

$$
\begin{aligned}
|c_{m+1} - c_{m-1}| \le & \left(\frac{a}{b}\right)^2 \sum_{n=2,4,\dots}^{\infty} |\tilde{F}_n| n \left| \frac{(m+1)^2}{(\Lambda_{nm} + 2m + 1)^2} \right. \\
& \left. - \frac{(m-1)^2}{(\Lambda_{nm} - 2m + 1)^2} \right| = \left(\frac{a}{b}\right)^2 \sum_{n=2,4,\dots}^{\infty} \left(|\tilde{F}_n| n m \right. \\
& \left. \times \left| \frac{4(\Lambda_{nm}^2 - 2m^2(\Lambda_{nm} - 1) - 1)}{((\Lambda_{nm} + 1)^2 - 4m^2)^2} \right| \right) \le \left(\frac{a}{b}\right)^{7/4} K_1 K_3 \sum_{n=2,4,\dots}^{\infty} \frac{m n^{1/2}}{\Lambda_{nm}^2} \qquad (31.31) \\
& \le \left(\frac{a}{b}\right)^{7/4} \frac{K_1 K_3}{C^2} \sum_{n=2,4,\dots}^{\infty} \left. \frac{m n^{1/2}}{(a/b)^{2\gamma} n^{2\gamma} m^{4-2\gamma}} \right|_{\gamma = 7/8} = \frac{K_1 K_3}{C^2} \frac{1}{m^{5/4}} \\
& \times \sum_{n=2,4,\dots}^{\infty} \frac{1}{n^{5/4}},
\end{aligned}
$$

where

$$
\left| \frac{4(\Lambda_{nm}^2 - 2m^2(\Lambda_{nm} - 1) - 1)}{((\Lambda_{nm} + 1)^2 - 4m^2)^2} \right| = O\left(\frac{1}{\Lambda_{nm}^2}\right) \le \frac{K_3}{\Lambda_{nm}^2},
$$

$K_3 > 0$ is some constant [2].

Estimate (31.31) shows that the series on the right in (31.30) is majorized by a numerical convergent series, and therefore converges [2]. Therefore, in (31.30) the series on the left also converges.

Let us now note that

$$
\begin{aligned}
& \left| (1/\varphi(\alpha_m) - 1)(-1)^{m/2} c_m \sin \lambda_m x \right| = |(-e^{-\alpha_m} + \alpha_m / \sinh \alpha_m) / \\
& / \varphi_1(\alpha_m)| \, |c_m| = O(e^{-2\alpha_m} m^{-1/4}) \le K_4 e^{-2\alpha_m} m^{-1/4},
\end{aligned} \qquad (31.32)
$$

where $K_4 > 0$ - is some constant. Then

$$
\begin{aligned}
\sum_{m=2,4\dots}^{\infty} \frac{(-1)^{m/2} c_m \sin \lambda_m x}{\varphi(\alpha_m)} = & \sum_{m=2,4,\dots}^{\infty} \left(\frac{-e^{-\alpha_m} + \alpha_m / \sinh \alpha_m}{\varphi_1(\alpha_m)} \right. \\
& \left. \times (-1)^{m/2} c_m \sin \lambda_m x \right) + \sum_{m=2,4,\dots}^{\infty} (-1)^{m/2} c_m \sin \lambda_m x < \infty .
\end{aligned} \qquad (31.33)
$$

Since here the second series on the right is the series in (31.30) on the left, it has been proven that it converges. The first series on the right is majorized by a numerical convergent series, which follows from inequalities (31.28), (31.32). Hence, statement (31.33) is true [2]. Then

$$
\sum_{m=2,4,\dots}^{\infty} (-1)^{m/2} b_m^{(1)} \sin \lambda_m x < \infty. \qquad (31.34)
$$

Since taking into account (31.15) and [2]

$$\sum_{m=2,4,\ldots}^{\infty} \left|(-1)^{m/2} b_m^{(1)} \sin \lambda_m x\right| \leq \sum_{m=2,4,\ldots}^{\infty} \left|b_m^{(1)}\right| \leq K \sum_{m=2,4,\ldots}^{\infty} \frac{1}{m^2} < \infty. \quad (31.35)$$

Hence, substituting the solution $\tilde{E}_m, \tilde{F}_n$ of the infinite system (31.4) into the first equation of system (31.4), multiplying this equation by $(-1)^{m/2} \sqrt{b/a} \sin \lambda_m x$, summing over m for values $m = 2, 4, \ldots$ and taking into account (31.3), (31.33) and (31.34), we arrive at the required proof

$$M_y(x, \pm b/2) = \sum_{m=2,4,\ldots}^{\infty} (-1)^{m/2} \sqrt{\frac{b}{a}}\, \tilde{E}_m \sin \lambda_m x = \sqrt{\frac{b}{a}} \left[-\frac{8}{\pi} \right.$$
$$\left. \times \sum_{m=2,4,\ldots}^{\infty} \frac{(-1)^{m/2} \sin \lambda_m x}{\varphi(\alpha_m)} c_m + \sum_{m=2,4,\ldots}^{\infty} (-1)^{m/2} b_m^{(1)} \sin \lambda_m x \right] < \infty. \quad (31.36)$$

The convergence of the series of bending moments $M_x(\pm a/2,\ y)$ (31.3) is established in a similar. Which is what needed to be proven.

Note 31.1. The convergence of bending moments on the contour of the plate, proven for the only solution in the class of bounded solutions $\tilde{E}_m, \tilde{F}_n$ of the infinite system (31.4), will be violated for other possible unbounded solutions of the system (31.4), the substitution of which into the expression for deflection (31.1) leads to the satisfaction of the Sophie Germaine–Lagrange equation within G.

31.7 On Continuous Adjacency to the Boundary Conditions of the Solution for Deflection

Using the substitution of the solution $\tilde{E}_m, \tilde{F}_n$, which is unique in the class of bounded infinite system (31.4), into (31.1), a solution for the deflection of the problem of bending a rectangular plate clamped along the contour is considered under an arbitrary, but antisymmetric relative to the axes of symmetry of the rectangle transverse load

$$\tilde{w}(x, y) = \frac{1}{D} \sum_{m=2,4\ldots}^{\infty} \sum_{n=2,4,\ldots}^{\infty} \frac{q_{mn} \sin \lambda_m x \, \sin \mu_n y}{(\lambda_m^2 + \mu_n^2)^2}$$
$$+ \frac{1}{2D} \sum_{m=2,4,\ldots}^{\infty} \tilde{E}_m \frac{(-1)^{m/2}}{\lambda_m \sinh \alpha_m} \left(\frac{\alpha_m}{\lambda_m} \coth \alpha_m \sinh \lambda_m y - y \cosh \lambda_m y \right) \sin \lambda_m x$$
$$+ \frac{1}{2D} \sum_{n=2,4,\ldots}^{\infty} \tilde{F}_n \frac{(-1)^{n/2} \sinh \mu_n x}{\mu_n \sinh \beta_n} \left(\frac{\beta_n}{\mu_n} \tanh \beta_n - x \coth \mu_n x \right) \sin \mu_n y. \quad (31.37)$$

It has been proven that the partial derivatives of the plate $\tilde{w}(x,\ y)$ with respect to the x and y coordinates $\partial \tilde{w}(x,\ y)/\partial x$ and $\partial \tilde{w}(x,\ y)/\partial y$ (and with them $\tilde{w}(x,\ y)$) are continuous in a closed rectangle of the plate plan as functions of two variables, that is, the solution $\tilde{w}(x, y)$ satisfies the classical condition of continuous adjacency to the boundary conditions [9].

Let us first prove that the ordinary series in (31.1) are continuously differentiated term-by-term with respect to the variables x and y. We have

$$\begin{aligned} &\alpha_m \coth\alpha_m \sinh\lambda_m y \; - \; \lambda_m y \cosh\lambda_m y = [\alpha_m \coth\alpha_m \sinh\lambda_m |y| \\ &- \lambda_m |y| \; \cosh\lambda_m y\,]\,\mathrm{sign}(y) = [\alpha_m \coth\alpha_m \sinh\lambda_m |y| - \lambda_m \\ &\times (|y| - b/2) \cosh\lambda_m y - \lambda_m b/2 \cosh\lambda_m |y|]\,\mathrm{sign}(y) \\ &= [\alpha_m \sinh\lambda_m(|y| - b/2)\,/\sinh\alpha_m \; + \lambda_m(b/2 - |y|) \\ &\times \cosh\lambda_m y]\,\mathrm{sign}(y). \end{aligned} \tag{31.38}$$

Then, using the inequalities

$$\sinh t \le e^t, \quad e^t/2 \le \cosh t \le e^t \quad \text{at } t \ge 0, \tag{31.39}$$

$$1/\sinh\alpha_m = \coth\alpha_m/\cosh\alpha_m \le 2\coth\alpha_2\, e^{-\alpha_m}, \tag{31.40}$$

we consistently obtain

$$\begin{aligned} &\frac{\alpha_m \sinh\lambda_m(b/2 - |y|)}{\lambda_m^2 sh^2\,\alpha_m} \le \frac{b}{2\,\lambda_m}\, e^{\lambda_m(b/2-|y|)} 4\coth^2\alpha_2 e^{-2\alpha_m} \\ &= \frac{2b\coth^2\alpha_2}{\lambda_m}\, e^{-\lambda_m(b/2+|y|)} \le \frac{2b\coth^2\alpha_2}{\lambda_m} e^{-\alpha_m}, \end{aligned} \tag{31.41}$$

$$\begin{aligned} &\frac{\lambda_m(b/2 - |y|)\cosh\lambda_m y}{\lambda_m^2 \sinh\alpha_m} \le \frac{2\coth\alpha_2}{\lambda_m} e^{-\alpha_m}(b/2 - |y|)\, e^{\lambda_m |y|} \\ &= \frac{2\coth\alpha_2}{\lambda_m}(b/2 - |y|)\, e^{-\lambda_m(b/2-|y|)}. \end{aligned} \tag{31.42}$$

We continue the estimate (31.42) as follows. Investigate the function $f(c,t) = t\exp(-ct)$ for $c = \text{const} > 0$, $t \ge 0$ at the extremum. Differentiating $f(c,t)$ with respect to t and equating the derivative to zero, we obtain

$$f_t'(c,t) \; = \; (1 - ct)\exp(-ct) = 0, \tag{31.43}$$

from which

$$\begin{aligned} &f_t'(c, 1/c\,) \; = \; 0, f_t'(c,t) \; > \; 0\,(f_t'(c,t) \; < \; 0), \\ &0 \le t < 1/c \;(t > 1/c\,),\; f_{\max}(c,t) \; = f(c, 1/c\,) = 1/(ec)\,. \end{aligned} \tag{31.44}$$

Therefore, using (31.42), (31.44) we obtain

$$\frac{\lambda_m(b/2 - |y|)\cosh\lambda_m y}{\lambda_m^2 \sinh\alpha_m} \le \frac{2\coth\alpha_2}{\lambda_m} f(\lambda_m,\, b/2 - |y|) \le \frac{2\coth\alpha_2}{e\,\lambda_m^2}. \tag{31.45}$$

Then, taking into account (31.41), (31.45) and (31.22) for the general term of the ordinary series with summation index m and its partial derivative with respect to x, we arrive at the estimates

$$\left|\frac{\partial^s}{\partial x^s}\left[\tilde{E}_m \frac{(-1)^{m/2}}{\lambda_m^2 \sinh \alpha_m}(\alpha_m \coth \alpha_m \sinh \lambda_m y \; - \; \lambda_m y \cosh \lambda_m y)\right.\right.$$
$$\left.\left.\times \sin \lambda_m x\right]\right| \le |\tilde{E}_m| \left[\frac{\alpha_m \sinh \lambda_m (b/2 - |y|)}{\lambda_m^2 \sinh^2 \alpha_m}\right.$$
$$\left. + \frac{(b/2 - |y|) \cosh \lambda_m y}{\lambda_m \sinh \alpha_m}\right] \lambda_m^s \le 2\left(\frac{a}{b}\right)^{1/4} K_1 \coth \alpha_2 \tag{31.46}$$
$$\times \left[b \coth \alpha_2 \, e^{-\alpha_m} + \frac{1}{e\,\lambda_m}\right] \frac{\lambda_m^{s-1}}{m^{1/2}} = O\left(\frac{1}{m^{2.5-s}}\right),$$

where $s = 0, 1$, $(x, y) \in G$.

Further

$$\frac{d}{dy}(\alpha_m \coth \alpha_m \sinh \lambda_m y \; - \; \lambda_m y \cosh \lambda_m y) = \lambda_m \alpha_m$$
$$\times \coth \alpha_m \cosh \lambda_m y - \lambda_m \cosh \lambda_m y - \lambda_m^2 y \sinh \lambda_m y = \lambda_m$$
$$\times \alpha_m \coth \alpha_m \; \cosh \lambda_m |y| - \lambda_m \cosh \lambda_m y - \lambda_m^2 (|y| - b/2) \tag{31.47}$$
$$\times \sinh \lambda_m |y| - \lambda_m \alpha_m \sinh \lambda_m y = \frac{\lambda_m \alpha_m}{\sinh \alpha_m} \cosh \lambda_m (|y| - b/2)$$
$$- \lambda_m \cosh \lambda_m y - \lambda_m^2 (|y| - b/2) \sinh \lambda_m |y| \,.$$

Then, using inequalities (31.39), (31.40), and also (31.44), we obtain

$$\frac{\lambda_m \alpha_m \cosh \lambda_m (|y| - b/2)}{\lambda_m^2 \sinh^2 \alpha_m} \le \frac{b}{2} \coth^2 \alpha_2 \frac{\cosh \lambda_m (|y| - b/2)}{\cosh^2 \alpha_m}$$
$$\le 2b \coth^2 \alpha_2 \, e^{\lambda_m (b/2 - |y| - b)} \le 2b \coth^2 \alpha_2 e^{-\alpha_m}, \tag{31.48}$$

$$\frac{\lambda_m \cosh \lambda_m y}{\lambda_m^2 \sinh \alpha_m} \le \frac{\coth \alpha_2}{\lambda_m} \frac{\cosh \lambda_m |y|}{\cosh \alpha_m} \le \frac{2 \coth \alpha_2}{\lambda_m} e^{\lambda_m (|y| - b/2)} \le \frac{2 \coth \alpha_2}{\lambda_m}, \tag{31.49}$$

$$\frac{\lambda_m^2 (b/2 - |y|)}{\lambda_m^2 \sinh \alpha_m} \sinh \lambda_m |y| = \coth \alpha_m \; (b/2 - |y|) \frac{\sinh \lambda_m |y|}{\cosh \alpha_m}$$
$$\le 2 \coth \alpha_2 \; (b/2 - |y|) \, e^{\lambda_m (|y| - b/2)} = 2 \coth \alpha_2 \, f(\lambda_m, b/2 - |y|) \tag{31.50}$$
$$\le \frac{2 \coth \alpha_2}{e\,\lambda_m}.$$

Then, taking into account (31.48)-(31.50) and (31.22) for the partial derivative with respect to y of the general term of an ordinary series with summation index m, we arrive at the estimate

$$\left|\frac{\partial}{\partial y}\left[\tilde{E}_m \frac{(-1)^{m/2}}{\lambda_m^2 \sinh\alpha_m}(\alpha_m \coth\alpha_m \sinh\lambda_m y - \lambda_m y\cosh\lambda_m y)\sin\lambda_m x\right]\right|$$
$$\leq |\tilde{E}_m|\left[\frac{\alpha_m \cosh\lambda_m(|y|-b/2)}{\lambda_m \sinh^2\alpha_m} + \frac{\cosh\lambda_m y}{\lambda_m \sinh\alpha_m} + \frac{(b/2-|y|)}{\sinh\alpha_m}\right.$$
$$\left.\times \sinh\lambda_m|y|\right] \leq 2K_1\left(\frac{a}{b}\right)^{1/4}\coth\alpha_2\left[b\coth\alpha_2\, e^{-\alpha_m}\right. \tag{31.51}$$
$$\left.+\frac{1}{\lambda_m}\left(1+\frac{1}{e}\right)\right]\frac{1}{m^{1/2}} = O\left(\frac{1}{m^{1.5}}\right),\ (x,\ y)\in G.$$

Estimates (31.46) for $s = 0$ and $s = 1$ (31.51) show that the ordinary series in (31.37) with summation index m and the series obtained by its formal term-by-term partial differentiation with respect to x (y) in the rectangle G are majorized by numerical convergent series. This means that these series converge uniformly with respect to $(x, y) \in G$. [2]. But, the common terms of these series are continuous functions of two variables $(x, y) \in G$. Therefore, the sums of these series are also continuous in G as functions of the variables x and y [2]. Fixing any $y = y_0 \in [-b/2, b/2\ ]$ $(x = x_0 \in [-a/2, a/2\ ])$, we notice that the noted functional series with respect to the variable $x \in [-a/2, a/2\ ]$ $(y \in [-b/2, b/2\ ])$, satisfy the conditions of Theorem 7 [2]. Therefore, the ordinary series in (31.37) with summation index m at y_0 (x_0) is continuously partially term-by-term differentiable with respect to $x \in [-a/2\ , a/2]$; $(y \in [-b, b/2\ ])$. But, by the assumption, y_0 (x_0) is arbitrary. Hence, the same is true for $y \in [-b/2\ , b/2\ ]$ $(x \in [-a/2\ , a/2\ ])$ Which is what needed to be proven. Similarly, with precision of notation, it is proved the continuously term-by-term partial differentiability with respect to x and y of an ordinary series with summation index n , simultaneously with its continuity.

Let us now consider repeated series

$$\sum_{m=2,4\ldots}^{\infty}\frac{d^s \sin\lambda_m x}{dx^s}\sum_{n=2,4\ldots}^{\infty}\frac{q_{mn}}{(\lambda_m^2+\mu_n^2)^2}\sin\mu_n y \tag{31.52}$$

at $(x,\ y) \in G,\ \ s = 0, 1$ with a load of general form

$$q(x,y) = \bar{q}(x,y) + \sum_{n=1}^{N}P_n\sum_{k_1=1}^{2}\sum_{k_2=1}^{2}[(-1)^{k_1+k_2}\delta(x+(-1)^{k_1}\xi_n)$$
$$\times\,\delta(y+(-1)^{k_2}\eta_n)], \tag{31.53}$$

where $\bar{q}(x, y)$ is a doubly antisymmetric distributed limited load

$$|\bar{q}(x,y)| \leq q* = \text{const} > 0,$$

P_n are concentrated forces.

Then

$$|q_{mn}| \leq \frac{4}{a\,b}\left|\int\limits_{-a/2}^{a/2}\int\limits_{-b/2}^{b/2} q\,(x,y)\ \sin\lambda_m x\ \sin\mu_n y\ d\,x\ d\,y\right|$$
$$\leq 4\,q^* + \frac{16}{a\,b}\sum_{n=1}^{N}|P_n| = K_5 = \text{const} > 0\,, \tag{31.54}$$

Therefore, taking into account Young's inequality (31.25)

$$\left|\frac{q_{mn}\ \sin\mu_n y}{\left(\lambda_m^2+\mu_n^2\right)^2}\right| \leq \frac{K_5}{\left(\lambda_m^2+\mu_n^2\right)^2} \leq \left.\frac{K_5}{C^2\lambda_m^{2\gamma}\mu_n^{4-2\gamma}}\right|_{\gamma=5/4} = \frac{K_5}{C^2\lambda_m^{2.5}\mu_n^{1.5}}, \tag{31.55}$$

$$\left|\frac{d^s\sin\lambda_m x}{dx^s}\sum_{n=2,4,\ldots}^{\infty}\frac{q_{mn}\sin\mu_n y}{\left(\lambda_m^2+\mu_n^2\right)^2}\right| \leq \frac{K_5}{C^2\lambda_m^{2.5-s}}\sum_{n=2,4,\ldots}^{\infty}\frac{1}{\mu_n^{1.5}} < \infty, \tag{31.56}$$

$s = 0, 1,\ (x, y) \in G$.

Inequality (31.55) shows that the ordinary series in the repeated series in (31.52) with summation index n with a common term – a continuous function $y \in [-b/2, b/2]$ – is majorized by a numerical convergent series. This means that the named series converges uniformly [2] to the continuous function $y \in [-b/2, b/2]$ [2]. Then the common terms of the ordinary series in (31.52) with summation index m for $s = 0, 1$ are continuous functions of two variables x and y in G, and it follows from inequalities (31.56) that the named series are majorized by numerical convergent series. This implies their uniform convergence [2] to continuous functions x and y in G and the continuous term-by-term partial differentiability with respect x and y in G to $x \in [-a/2, a/2]$ for $s = 0$ of the series (31.52) and any fixed $y = y_0 \in [-b/2, b/2]$ [2], and hence for all $y \in [-b/2, b/2]$.

$$\frac{\partial}{\partial x}\sum_{m=2,4\ldots}^{\infty}\sin\lambda_m x\sum_{n=2,4\ldots}^{\infty}\frac{q_{mn}}{\left(\lambda_m^2+\mu_n^2\right)^2}\sin\mu_n y \;=\; \sum_{m=2,4\ldots}^{\infty}\frac{d\sin\lambda_m x}{dx}$$
$$\times\sum_{n=2,4\ldots}^{\infty}\frac{q_{mn}}{\left(\lambda_m^2+\mu_n^2\right)^2}\sin\mu_n y\ \in C(G)\,,\ s = 0, 1. \tag{31.57}$$

Similarly, with precision of notation, we prove the continuous partial term-by-term differentiability with respect to y for $s = 0$ of the sum of series (31.52), if, taking into account (31.55), the inequalities

$$\left|\frac{q_{mn}\sin\lambda_m x\ \sin\mu_n y}{\left(\lambda_m^2+\mu_n^2\right)^2}\right| \leq \left|\frac{q_{mn}\ \sin\mu_n y}{\left(\lambda_m^2+\mu_n^2\right)^2}\right| \leq \frac{K_5}{C^2\lambda_m^{2.5}\mu_n^{1.5}}. \tag{31.58}$$

with the statement in [2] before changing the order of summation to the inverse

$$\frac{\partial}{\partial y}\sum_{m=2,4...}^{\infty}\sum_{n=2,4...}^{\infty}\frac{q_{mn}}{(\lambda_m^2+\mu_n^2)^2}\sin\lambda_m x\,\sin\mu_n y$$
$$=\frac{\partial}{\partial y}\sum_{n=2,4...}^{\infty}\sum_{m=2,4...}^{\infty}\frac{q_{mn}}{(\lambda_m^2+\mu_n^2)^2}\sin\lambda_m x\,\sin\mu_n y \qquad (31.59)$$
$$=\sum_{n=2,4...}^{\infty}\frac{d\,\sin\mu_n y}{dy}\sum_{m=2,4...}^{\infty}\frac{q_{mn}}{(\lambda_m^2+\mu_n^2)^2}\sin\lambda_m x\in C(G)\,,\; s=0,1.$$

Therefore [2], $\tilde{w}(x,y), \partial\tilde{w}(x,y)/\partial x$ and $\partial\tilde{w}(x,y)/\partial y \in C(G)$.

Note 31.2. Formulas (31.57), (31.59) together with the partial term-by-term differentiability with respect to x and y of the ordinary series in (31.1) into convergent series were previously only supposedly applied in calculating the derivative of deflection along the normal on the contour of the plate when deriving the infinite system (31.4).

31.8 Summary and Conclusions

The author substantiates the previously proposed modified method of S.P. Timoshenko, which is intended for solving the problem of transverse doubly antisymmetric bending of a rectangular plate clamped along the contour, by which the solution of this problem for deflection is reduced to an infinite, completely regular system of linear algebraic equations. It is proven that an infinite system in the class of bounded solutions has a unique solution that simultaneously tends to zero, and the trigonometric series of bending moments on the contour of the plate converge, and the deflection continuously approaches the boundary conditions. All intermediate mathematical operations that were used in deriving the infinite system are substantiated. It is noted that possible unbounded solutions of an infinite system, in which the deflection within the plate satisfies the Sophie Germaine–Lagrange equation, are excluded from consideration, since they violate the convergence of the series of bending moments on the plate contour.

References

[1] Darevsky VM, Sharinov IL (1975) Novoye resheniye zadachi ob izgibe zashchemlennoy po krayam pryamougol'noy plastinki (in Russ., New solution to the problem of bending of a rectangular plate clamped at the edges). In: Nauka, Moscow, pp 183-194

[2] Fikhtengolts GM (2003, 2008) Kurs differentsial'nogo i integral'nogo ischisleniya (in Russ., Course of Differential and Integral Calculus, Vol II, III, Fizmatlit, Moscow

[3] Kantorovich LV, Krylov VI (1958) Approximate Methods of Higher Analysis. Noordhoff Ltd, Groningen
[4] Meleshko VV (2004) Bigarmonicheskaya zadacha dlya pryamougol'nika: istoriya i sovremennost' (in Russ., Biharmonic problem for a rectangle: history and modernity). Matematychni metody ta fizyko-mekhanichni polya 47(3):45-68
[5] Seyranyan S.P. (2020) Ob odnoy beskonechnoy sisteme lineynykh algebraicheskikh uravneniy B.L. Abramyana v pervoy krayevoy ploskoy zadache teorii uprugosti dlya pryamougol'nika (in Russ., On one infinite system of linear algebraic equations of B.L.Abramyan in the first boundary plane problem theory of elasticity for a rectangle). Proceedings of the XIII International Conference on Applied Mathematics and Mechanics in the Aerospace Industry, September 6-13, 2020, Alushta. MAI Publishing House, Moscow, pp 337-340
[6] Seyranyan SP (2009) Ob odnom reshenii Nav'ye dlya chastichno nagruzhennoy pryamougol'noy plastiny (in Russ., On one Navier solution for a partially loaded rectangular plate). Vestnik Tomskogo Gosudarstvennogo Universiteta. Matematika i Mekhanika 1(5):82-95
[7] Seyranyan SP (2020) K resheniyu zadachi ob antisimmetrichnom izgibe pryamougol'noy zashchemlennoy po konturu plastiny modifitsirovannym metodom S.P. Timoshenko (in Russ., On the solution for the problem of antisymmetric bending of a rectangular plate clamped along the contour using the modified method of S.P. Timoshenko). Proceedings of the XIV International Conference on Applied Mathematics and Mechanics in the Aerospace Industry, September 6-13, 2022, Alushta. MAI Publishing House, Moscow, pp 229-232
[8] Seyranyan SP (2019) K resheniyu pervoy krayevoy ploskoy zadachi teorii uprugosti dlya pryamougol'nika pri antisimmetrichnykh otnositel'no osey simmetrii granichnykh usloviyakh (in Russ., On the solution of the first boundary plane problem of the theory of elasticity for the rectangle with antisymmetrical relative axial symmetry of the boundary conditions). Topical problems of continuum mechanics. Proceedings of the VI International Conference, RA, Dilijan, October 1-6, 2019. Publishing House of NUACA RA, IM NAS RA, IPMech. RAS, pp 304-308
[9] Sobolev SL (1964) Equations of Mathematical Physics. Pergamon Press, Oxford et al
[10] Sukhoterin MV (2009) Metod superpozitsiy ispravlyayushchikh funktsiy (in Russ., Method of superpositions of correcting functions). Publishing house of the Polytechnic University, St Petersburg
[11] Timoshenko SP, Woinowsky-Krieger S (1959) Theory of Plates and Shells. McGraw-Hill Book Company, New York et al

Chapter 32
Boundary Integral Equation with Varying Impedance for the Harmonic Wave Equation

Mezhlum A. Sumbatyan

Abstract The work proposes a method to solve a two-dimensional problem of reflecting a wave from a horizontal surface with impedance, which is a given function varying along the boundary. In the general case, the problem is reduced to the boundary integral equation (BIE) for pressure, in which the specific impedance of the boundary is included in the kernel. As an example, we consider as the reflecting surface an alternation of acoustically hard and acoustically soft intervals on the boundary.

Key words: Wave dynamics, Boundary integral equation, Reflection coefficient, Absorption coefficient, Periodic structure

32.1 Introduction

It is known [2, 5] that in reflection of sound waves from flat infinite boundaries, the modulus of the reflection coefficient is equal to unity if the boundary surface has no sound absorption. In this case, the energy of the reflecting wave is equal to the energy of the incident wave. The examples of such boundaries are given by absolutely hard and absolutely soft surfaces. In the case of non-perfect surfaces, there is an absorption of energy in the reflection.

Let us restrict the consideration of this problem by a two-dimensional (2d) case of a normal (i.e. orthogonal) incidence of a plane wave, downward to a straight horizontal surface in the natural Cartesian coordinates (x, y), as well as a time-harmonic wave process, with the factor $e^{-\imath\omega t}$, which reduces the wave equation to the stationary (free of time) Helmholtz partial differential equation

Mezhlum A. Sumbatyan
Vorovich Institute of Mathematics, Mechanics and Computer Science, Southern Federal University, Milchakova Street 8a, 344090, Rostov-on-Don, Russian Federation, e-mail: masumbatyan@sfedu.ru

H. Altenbach and A. V. Sahakyan, *Contact Mechanics and Relevant Problems*,
Advanced Structured Materials 253, https://doi.org/10.1007/978-3-032-20869-9_32

$$\Delta p + k^2 p = 0, \qquad (k = \omega/c). \tag{32.1}$$

For the problem on reflection of the plane wave from an interface boundary between two different media, the acoustic pressure should satisfy Eq. (32.1) both in the first (upper) and in the second (lower) half-planes. The solution to this problem is well known in literature [2], and its form is given by the following expressions

$$p_1(x, y) = \mathrm{e}^{-ik_1 y} + R\mathrm{e}^{ik_1 y}, \ (y \geq 0), \qquad p_2(x, y) = T\mathrm{e}^{-ik_2 y}, \ (y \leq 0), \tag{32.2}$$

in the first and the second media, respectively. Here $k_{1,2} = \omega/c_{1,2}$ are wave numbers, R is the reflection coefficient, T is the transmission (transparency) coefficient. With so doing, one obtains

$$R = \frac{Z_2 - Z_1}{Z_2 + Z_1}, \qquad T = \frac{2Z_2}{Z_2 + Z_1}, \tag{32.3}$$

where $Z_1 = \rho_1 c_1$, $Z_2 = \rho_2 c_2$ are acoustic impedances of the upper and the lower half-planes, respectively [2], ρ and c for each medium are respectively mass densities and wave speeds. It is easily seen from Eq. (32.3) that for acoustically hard ($Z_2 \to \infty$, $R \to 1$) and acoustically soft ($Z_2 \to 0$, $R \to -1$) reflecting surfaces one obtains in the both cases $|R| = 1$, which is discussed above in the first paragraph of the present section, for the two perfectly reflecting types of boundaries.

In the general case, by introducing the coefficient of sound absorption α ($0 \leq \alpha \leq 1$), the energy in the reflected wave, which is proportional to pressure in square, is related with α: $|R|^2 = 1 - \alpha$. This determines the sound absorption coefficient in terms of the reflection coefficient: $\alpha = 1 - |R|^2$. In order to restrict the study by only wave process in the first (upper) medium, where incidence and reflection take place, it is usually introduced the so-called specific impedance

$$Z_s = \frac{p_1}{(-v_{1y})} = \frac{p_2}{(-v_{2y})} \tag{32.4}$$

over the boundary, as a ratio of the acoustic pressure to the normal component of the velocity vector. It should be noted that equality (32.4) is valid since both the pressure and the normal component of the velocity are continuous when crossing the interface boundary $y = 0$, if we speak about non-viscous acoustic media.

Since the linearized equation of motion for the upper medium has the form

$$\rho_1 \frac{\partial \bar{v}_1}{\partial t} + \nabla p_1 = 0 \qquad \sim \qquad -i\omega\rho_1 \bar{v}_1 + \nabla p_1 = 0\,, \tag{32.5}$$

then for the specific impedance one obtains from (32.2)

$$\begin{gathered} v_{1y}\Big|_{y=0} = \frac{1}{i\omega\rho_1} \frac{\partial p_1}{\partial y}\Big|_{y=0} = \frac{ik_1(R-1)}{i\omega\rho_1} = \frac{R-1}{Z_1} \quad \Longrightarrow \\ Z_s = \frac{p_1}{(-v_{1y})}\Big|_{y=0} = \frac{1+R}{1-R} Z_1 = Z_2\,. \end{gathered} \tag{32.6}$$

It follows from relations (32.6) that one can study wave process in the first (upper) medium, if instead of physical parameters in the two considered media one takes for the upper medium the following boundary condition of the 3rd type

$$p_1 + Z_s v_{1y} = 0 \quad (y = 0), \tag{32.7}$$

and accepts the specific impedance Z_s as a known function over the boundary. It is obvious that for the two extremal values of this function one obtains two perfect reflectors: the case $Z_s \equiv 0$ yields the boundary condition $p_1 = 0$, which corresponds to acoustically soft boundary; while the case $Z_s \to \infty$ leads to boundary condition $v_{1y} = 0, \sim \partial p_1/\partial y = 0$, which corresponds to acoustically hard boundary. In general case, we have a boundary condition with a varying specific impedance.

32.2 Reducing the Problem to a BIE

For the wave incidence downwards to a straight infinite horizontal reflector, the boundary values of the physical quantities involved should satisfy the following relation (with $|x| < \infty$) [3, 8]

$$\frac{p(x,0)}{2} - \int_{-\infty}^{\infty} \left[p(\xi,0)\frac{\partial G(\xi-x,\eta)}{\partial \eta} - \frac{\partial p(\xi,\eta)}{\partial \eta} G(\xi-x,0) \right]_{\eta=0} d\xi = p^{\text{inc}}(x,0), \tag{32.8}$$

where $p^{\text{inc}}(x,y) = \mathrm{e}^{-\imath k_1 y}$ is the incident wave, and the Green's function is expressed in terms of the Hankel function [1]

$$G(\xi - x, \eta - y) = \frac{\imath}{4} H_0^{(1)} \left[k\sqrt{(\xi - x)^2 + (\eta - y)^2} \right], \tag{32.9}$$

as a function depending upon the distance between the observation point (x, y) and the integration point (ξ, η). It should also be noted that, when considering wave process only in the upper medium, we omit the subscript 1 for function p_1, for the sake of brevity: $p_1 = p$.

It is obvious that if $y = 0$ then $(\partial G/\partial \eta)_{(\eta=0)} = 0$. Therefore, the first term under the integral in (32.8) is trivial and so Eq. (32.8) can be simplified, as follows:

$$\frac{p(x,0)}{2} + \frac{i}{4} \int_{-\infty}^{\infty} \frac{\partial p(\xi,0)}{\partial \eta} H_0^{(1)}(k|\xi - x|) d\xi = p^{inc}(x,0). \tag{32.10}$$

Equality (32.10) is valid for arbitrary incident wave field p^{inc}. It is interesting that in the case when the boundary is absolutely hard: $v_{1y} = 0, \sim \partial p/\partial y = 0$, then the integral vanishes, and the boundary value of acoustic pressure coincides with that predicted by the Kirchhoff theory [8]: $p(x,0) = 2p^{\text{inc}}(x,0)$.

In general case, when the impedance boundary condition (32.7) is valid, according to the first relation in (32.6), Eq. (32.10) takes the form:

$$\frac{\partial p}{\partial y} = \iota\omega\rho v_y = -\frac{\iota\omega\rho}{Z_s} p \quad \Longrightarrow$$

$$\frac{p(x,0)}{2} + \frac{\omega\rho}{4}\int_{-\infty}^{\infty} p(\xi,0) H_0^{(1)}(k|\xi - x|)\frac{d\xi}{Z_s(\xi)} = p^{\text{inc}}(x,0), \quad (|x| < \infty). \tag{32.11}$$

One thus can conclude that in the most general case, the problem of reflection from the impedance boundary is reduced to the BIE (32.11) of the second kind, whose kernel contains the specific impedance, function Z_s.

32.3 The Case of Periodic Alternation of Hard and Soft Boundary Intervals

Two simplest types of the reflecting boundary are the perfect reflectors, since $|R| = 1$ for both them, therefore the incident wave keeps its energy after reflection. It is interesting to investigate whether alternating these two ideally reflective boundary types can generate energy absorption upon reflection, and if so, to what extent. The problem in this formulation is related to the search for effective sound-absorbing materials in room acoustics [4]. This problem is particularly relevant in the low-frequency range, as it is known that most sound absorbers are effective only in the mid- and high-frequency range. A possible arrangement of these two materials is shown in Fig. 32.1, being periodic.

Obviously, in the problem under consideration, equality (32.10) is valid, and integrals over only acoustically soft segments with the boundary condition $p = 0$ remain non-trivial, since on acoustically hard sections the normal derivative of the pressure is equal to zero, and the integrals over those intervals vanish. Let us designate acoustically soft intervals as $\ell_m : -b+(2a)m < x < b+(2a)m$, $(m = 0, \pm 1, \pm 2, \dots)$. The union of all these segments is a periodic lattice

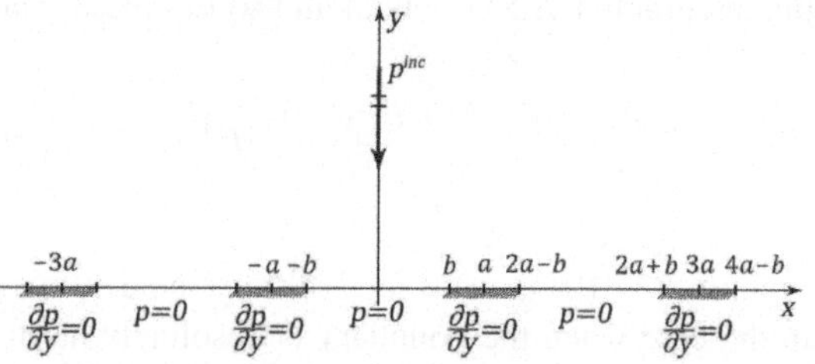

Fig. 32.1 Reflection of the plane wave from a periodic horizontal boundary with the period $2a$.

$$\ell = \bigcup_{m=-\infty}^{\infty} \ell_m$$

of the period $2a$, which consists of intervals – each of length $2b$.

It is obvious that in the case of normal incidence, when $p^{inc} = e^{-iky}$, the solution is locally the same function $g(x) = [\partial p/\partial y]_{y=0}$ of variable x on each interval ℓ_m. Therefore, the integral in (32.10) is

$$\int_{-\infty}^{\infty} \frac{\partial p(\xi,0)}{\partial \eta} H_0^{(1)}(k|\xi - x|)d\xi = \sum_{m=-\infty}^{\infty} \int_{-b}^{b} g(\xi) H_0^{(1)}(k|\xi + 2am - x|)d\xi$$

$$= \frac{1}{\pi \imath} \sum_{m=-\infty}^{\infty} \int_{-b}^{b} g(\xi)d\xi \int_{-\infty}^{\infty} \frac{e^{\imath(\xi+2am-x)\alpha}}{\sqrt{\alpha^2 - k^2}} d\alpha\,, \tag{32.12}$$

where we have used integral representation of the Hankel function [1, 6]. Now, let us apply the well-known result from the theory of generalized functions [8]:

$$\sum_{-\infty}^{\infty} e^{\imath mz} = 2\pi \sum_{-\infty}^{\infty} \delta(z + 2\pi m)\,, \tag{32.13}$$

where δ is Dirac's delta-function. As a consequence, one obtains for the key term in Eq. (32.12), by taking into account the basic property of delta-function:

$$\sum_{m=-\infty}^{\infty} \int_{-\infty}^{\infty} \frac{e^{\imath(\xi+2am-x)\alpha}}{\sqrt{\alpha^2 - k^2}} d\alpha = 2\pi \sum_{m=-\infty}^{\infty} \int_{-\infty}^{\infty} \frac{\delta(2a\alpha + 2\pi m)}{\sqrt{\alpha^2 - k^2}} e^{\imath(\xi-x)\alpha}\, d\alpha$$

$$= 2\pi \sum_{m=-\infty}^{\infty} \int_{-\infty}^{\infty} \frac{\delta(\alpha + 2\pi m)}{\sqrt{\alpha^2 - (2ak)^2}} e^{\imath(\xi-x)\alpha/(2a)}\, d\alpha = \frac{\pi}{a} \sum_{m=-\infty}^{\infty} \frac{e^{\imath(x-\xi)\pi m/a}}{q_m}\,, \tag{32.14}$$

where

$$q_m = \sqrt{\left(\frac{\pi m}{a}\right)^2 - k^2}\,. \tag{32.15}$$

Therefore, in the periodic problem under consideration, if one takes the observation point x over soft part of the boundary ($p = 0$, $|x| \le b$), the BIE (32.10) is reduced to the following form:

$$\sum_{m=-\infty}^{\infty} \int_{-b}^{b} g(\xi) \frac{e^{\imath(x-\xi)\pi m/a}}{4aq_m} d\xi = 1, \quad (|x| \le b) \quad \sim$$

$$\int_{-b}^{b} g(\xi) \left[K(x-\xi) - \frac{1}{2\imath ak} \right] d\xi = 2, \; (|x| \le b), \; K(x) = \sum_{m=1}^{\infty} \frac{\cos(\pi m x/a)}{a q_m} . \tag{32.16}$$

32.4 Low-Frequency Approximation

If $ka < \pi$ then in (32.15) all q_m are positive: $q_m > 0$, $(m = 1, 2, \dots)$. This is a so-called "one-mode regime". If we accept the more strict low-frequency approximation: $ka \ll \pi$, then for all m one obtains: $q_m \approx \pi m/a$, and the series for the kernel in Eq. (32.16) can be calculated explicitly [6]

$$K(x) \approx K_0(x) = \frac{1}{\pi} \sum_{m=1}^{\infty} \frac{\cos(\pi m x/a)}{m} = -\frac{1}{\pi} \ln \left| 2 \sin \frac{\pi x}{2a} \right|, \tag{32.17}$$

being free of wave number k. Let us extract the terms in (32.16), which depend on the wave number, if the kernel in the low-frequency limit is given by Eq. (32.17)

$$\int_{-b}^{b} g(\xi) K_0(x-\xi) d\xi = 2 + \frac{G_0}{2iak}, \quad (|x| \le b), \qquad G_0 = \int_{-b}^{b} g(\xi) d\xi . \tag{32.18}$$

It is now obvious that if a certain function $h(x)$, which does not depend upon parameter k, is defined from the following equation

$$\int_{-b}^{b} h(\xi) K_0(x-\xi) d\xi = 1, \quad (|x| \le b), \tag{32.19}$$

then

$$g(x) = \left(2 + \frac{G_0}{2iak}\right) h(x), \implies G_0 = \left(2 + \frac{G_0}{2iak}\right) H_0, \; H_0 = \int_{-b}^{b} h(\xi) d\xi . \tag{32.20}$$

Moreover, constant G_0, which is included in definition of function $g(x)$ in (32.20), can simply be determined from the second relation of the same Eq. (32.20)

$$G_0 = \frac{2H_0}{1 - \dfrac{H_0}{2iak}}, \tag{32.21}$$

as soon as function $h(x)$ is determined from integral equation (32.19), – so that constant H_0 becomes a certain known quantity.

The solution to integral equation (32.19) with the kernel $K_0(x)$ defined in (32.17) is well known in literature (see, for example [7]), ($|x| < b$)

$$h(x) = -\frac{\pi \cos\left(\dfrac{\pi x}{2a}\right)}{2a \ln\left(\sin\dfrac{\pi b}{2a}\right)\left(\sin^2\dfrac{\pi b}{2a} - \sin^2\dfrac{\pi x}{2a}\right)^{1/2}}, \quad H_0 = -\frac{\pi}{\ln\left(\sin\dfrac{\pi b}{2a}\right)}, \tag{32.22}$$

and the integral for H_0 in (32.20) can easily be obtained by applying the change of variable: $\zeta = \sin[(\pi x)/(2a)]$. After that constant G_0 is directly defined by Eq. (32.21).

32.5 Calculation of Acoustic Pressure over Rigid Intervals of the Boundary

The whole infinite boundary line $y = 0$ (see Fig. 32.1), which we denote E_1 as a one-dimensional Euclidean space, contains two types of intervals

1. $x \in \ell = \bigcup_{m=-\infty}^{\infty} \ell_m$, which are acoustically soft intervals with the boundary condition $p(x, 0) = 0$. Here $(\partial p/\partial y)(x, 0)$ is given by function $g(x)$, see the paragraph just before Eq. (32.12), and function $g(x)$ is already known from a resolved BIE (4.2), in terms of function $h(x)$, from Eq. (32.20). Therefore, on the soft intervals of the boundary there are already known both acoustic pressure and its normal derivative.
2. Acoustically hard intervals: $x \in E_1 \setminus \ell$, as the rest of the boundary, outside the union of soft intervals. The normal derivative there is trivial, due to the boundary condition for the acoustically hard parts, but the acoustic pressure itself is still unknown over these parts of the boundary line. Since equality (32.10) is valid for all $x \in E_1$, let us consider this equality over domain $x \in E_1 \setminus \ell$ for acoustic pressure

$$\frac{p(x,0)}{2} = p^{\text{inc}}(x,0) - \frac{\iota}{4}\int_{\ell} \frac{\partial p(\xi,0)}{\partial \eta} H_0^{(1)}(k|\xi - x|)d\xi, \quad (x \in E_1 \setminus \ell), \tag{32.23}$$

where the integration is restricted by only soft intervals, due to trivial values of the normal derivative on rigid parts. However, the values of normal derivative of the pressure on the union of soft parts, set ℓ, is defined by function $g(x)$, being already known from (32.20)–(32.22), and they can directly be submitted to Eq. (32.23). Indeed, if $b < |x| < a$ then in the low-frequency approximation one obtains, by analogy to what was done above in derivation of Eqs. (32.16) and (32.17)

$$\frac{\imath}{4}\int_{\ell} \frac{\partial p(\xi,0)}{\partial \eta} H_0^{(1)}(k|x-\xi|)d\xi$$

$$= \frac{1}{4a}\int_{-b}^{b} g(\xi)\left[-\frac{1}{\imath k}+\frac{2a}{\pi}\sum_{m=1}^{\infty}\frac{1}{m}\cos\frac{\pi m(x-\xi)}{a}\right]d\xi \tag{32.24}$$

$$= -\frac{G_0}{4\imath ak}-\frac{1}{2\pi}\left(2+\frac{G_0}{2\imath ak}\right)\int_{-b}^{b} h(\xi)\ln\left|2\sin\frac{\pi(x-\xi)}{2a}\right|d\xi\,,$$

where function $h(x)$ is given by Eq. (32.22). By using the evenness of function $h(x)$, the integral in (32.24) can be simplified, as follows

$$J = J(x) = \int_{-b}^{b} h(\xi)\ln\left|2\sin\frac{\pi(x-\xi)}{2a}\right|d\xi$$

$$= \int_{0}^{b} h(\xi)\left[\ln\left|2\sin\frac{\pi(x-\xi)}{2a}\right|+\ln\left|2\sin\frac{\pi(x+\xi)}{2a}\right|\right]d\xi \tag{32.25}$$

$$= \int_{0}^{b} h(\xi)\ln\left|4\sin\frac{\pi(x-\xi)}{2a}\,\sin\frac{\pi(x+\xi)}{2a}\right|d\xi\,.$$

Here, by representing the product of two sine functions as a half of difference between cosine of difference and cosine of sum of their arguments, one obtains:

$$J=\int_0^b h(\xi)\ln\left|2\left(\cos\frac{\pi\xi}{a}-\cos\frac{\pi x}{a}\right)\right|d\xi=\int_0^b h(\xi)\ln\left|4\left(\sin^2\frac{\pi\xi}{2a}-\sin^2\frac{\pi x}{2a}\right)\right|d\xi, \tag{32.26}$$

where we have also used a relation between cosine of an argument and sine in square of half its argument. Now, J can be calculated explicitly by applying the change of variable: $\zeta = \sin[(\pi\xi)/(2a)]$, $\zeta_0 = \sin[(\pi b)/(2a)]$:

$$J=\int_0^b h(\xi)\left[\ln\left|2\left(\sin\frac{\pi\xi}{2a}-\sin\frac{\pi|x|}{2a}\right)\right|+\ln\left|2\left(\sin\frac{\pi\xi}{2a}+\sin\frac{\pi|x|}{2a}\right)\right|\right]d\xi$$

$$=-\int_{-\zeta_0}^{\zeta_0}\frac{\ln\left|2\left(\zeta-\sin\frac{\pi|x|}{2a}\right)\right|}{\ln\zeta_0}\frac{d\zeta}{\sqrt{\zeta_0^2-\zeta^2}}=-\pi\,\frac{\ln\left(\sin\frac{\pi|x|}{2a}+\sqrt{\sin^2\frac{\pi x}{2a}-\zeta_0^2}\right)}{\ln\zeta_0}, \tag{32.27}$$

where we have used a tabulated integral [6].

At last, by substituting (32.27) to (32.24), and after that (32.24) to (32.23), we finally come to the following representation of acoustic pressure on the acoustically hard intervals (when $b < |x| < a$):

$$p(x,0) = 2 - \left[-\frac{G_0}{2\imath ak} + \left(2 + \frac{G_0}{2\imath ak}\right) \frac{\ln\left(\sin\frac{\pi|x|}{2a} + \sqrt{\sin^2\frac{\pi x}{2a} - \zeta_0^2}\right)}{\ln \zeta_0} \right]$$

$$= \left(2 + \frac{G_0}{2\imath ak}\right) \left[1 - \frac{\ln\left(\sin\frac{\pi|x|}{2a} + \sqrt{\sin^2\frac{\pi x}{2a} - \sin^2\frac{\pi b}{2a}}\right)}{\ln\left(\sin\frac{\pi b}{2a}\right)} \right] \tag{32.28}$$

$$= \frac{4\imath ak}{2\imath ak - H_0} \left[1 - \frac{\ln\left(\sin\frac{\pi|x|}{2a} + \sqrt{\sin^2\frac{\pi x}{2a} - \sin^2\frac{\pi b}{2a}}\right)}{\ln\left(\sin\frac{\pi b}{2a}\right)} \right],$$

due to relation (32.21).

It is interesting to notice that if $x \to \pm b$ then it is easily seen from Eq. (32.28) that $p(x,0) \to 0$ in this limiting case, as it should be when approaching to a border of acoustically soft interval moving from a neighbour rigid interval, by continuity of the solution.

32.6 Calculation of the Reflected Wave

The reflected wave field is expressed by formula analogous to (32.8)

$$p^{\text{ref}}(x,y) = \frac{\imath}{4} \int_{-\infty}^{\infty} \left\{ p(\xi,0) \frac{\partial}{\partial \eta} H_0^{(1)} \left[k\sqrt{(\xi - x)^2 + (y - \eta^2)} \right] \Big|_{\eta=0} \right. \tag{32.29}$$

$$\left. - \frac{\partial p(\xi,0)}{\partial \eta} H_0^{(1)} \left[k\sqrt{(\xi - x)^2 + y^2} \right] \right\} d\xi.$$

Passing in (32.29) to a periodic geometry, the summation over "soft" part ℓ of the boundary line (with boundary condition $p(x,0) = 0$) operates with the second term in the integrand (32.29), in the following way

$$S_2=\frac{\imath}{4}\sum_{m=-\infty}^{\infty}H_0^{(1)}\left[k\sqrt{(\xi+2am-x)^2+y^2}\right]=\frac{1}{4a}\sum_{m=-\infty}^{\infty}\frac{e^{\imath(x-\xi)\pi m/a}}{q_m}\,e^{-q_m y} \qquad (32.30)$$

by analogy to (32.12)–(32.15), if one applies an integral representation for Hankel function, then change summation of exponents to summation of Dirac's delta-functions, and then applies the basic property of the delta-function in integration.

In the low-frequency limit ($k \to 0$) and in the far-zone approximation ($y \to \infty$) Eq. (32.30) takes the following form

$$S_2 \overset{(k\to 0)}{\approx} \frac{1}{4a}\left\{\frac{e^{\imath ky}}{-\imath k}+\frac{2a}{\pi}\sum_{m=1}^{\infty}\frac{\cos[\pi m(x-\xi)/a]}{m}\,e^{-\pi m y/a}\right\} \overset{(y\to\infty)}{\approx} \frac{\imath e^{\imath ky}}{4ak}\,. \qquad (32.31)$$

Then the second part of the integral (32.29) takes asymptotically the following form:

$$\begin{aligned}&\frac{\imath}{4}\int_{\ell}\frac{\partial p(\xi,0)}{\partial\eta}H_0^{(1)}\left[k\sqrt{(\xi-x)^2+y^2}\right]d\xi\\ &=\imath\int_{\ell}g(\xi)\frac{e^{\imath ky}}{4ak}d\xi=\imath G_0\frac{e^{\imath ky}}{4ak}=-\frac{H_0 e^{\imath ky}}{2\imath ak-H_0}\end{aligned} \qquad (32.32)$$

By analogy, the first part of the integral (32.29), taken over acoustically hard parts of the boundary line, can be calculated as follows

$$2\int_b^a g(\xi,0)\frac{\partial}{\partial\eta}\frac{\imath e^{\imath k(y-\eta)}}{4ak}\bigg|_{\eta=0}d\xi=\frac{2ke^{\imath ky}}{4ak}\,\frac{4\imath ak}{2\imath ak-H_0}\int_b^a g(\xi,0)d\xi\,, \qquad (32.33)$$

where function $g(\xi,0)$ is given by Eq. (32.28).

Since the complex integral in (32.33) can be reduced, again – by the change of variable $\zeta=\sin[(\pi\xi)/(2a)]$, $\zeta_0=\sin[(\pi b)/(2a)]$, to a tabulated one [6]

$$\begin{aligned}&\int_b^a \ln\left(\sin\frac{\pi\xi}{2a}+\sqrt{\sin^2\frac{\pi\xi}{2a}-\sin^2\frac{\pi b}{2a}}\right)d\xi=\frac{2a}{\pi}\int_{\zeta_0}^{1}\ln\left(\zeta+\sqrt{\zeta^2-\zeta_0^2}\right)\frac{d\zeta}{\sqrt{1-\zeta^2}}\\ &=\frac{2a}{\pi}(-\arcsin\zeta_0)\ln\zeta_0=\frac{2a}{\pi}\left(-\frac{\pi b}{2a}\right)\ln\zeta_0=-b\ln\zeta_0\,,\end{aligned} \qquad (32.34)$$

hence finally the contribution of the first part of the integral in (32.29) takes the following form

$$\frac{2\imath k e^{\imath k y}}{2\imath a k - H_0} \int_b^a \left[1 - \frac{\ln\left(\sin\frac{\pi\xi}{2a} + \sqrt{\sin^2\frac{\pi\xi}{2a} - \sin^2\frac{\pi b}{2a}} \right)}{\ln\left(\sin\frac{\pi b}{2a} \right)} \right] d\xi$$

$$= \frac{2\imath k e^{\imath k y}}{2\imath a k - H_0} \left[(a-b) + \frac{b \ln \zeta_0}{\ln\left(\sin\frac{\pi b}{2a} \right)} \right] = \frac{2\imath a k\, e^{\imath k y}}{2\imath a k - H_0} . \tag{32.35}$$

Now, collecting together the contributions of the first and second parts in the integral (32.29), one obtains for the reflected wave-field in the far-zone the following expression

$$p^{\mathrm{ref}}(x, y) = \frac{2\imath a k + H_0}{2\imath a k - H_0}\, e^{\imath k y}, \quad H_0 = -\frac{\pi}{\ln\left(\sin\frac{\pi b}{2a} \right)}, \qquad (ak \ll 1,\ y/a \gg 1). \tag{32.36}$$

This obtained result can be tested in the two opposite cases:

1. $b = 0$. In this case the whole reflecting infinite boundary is acoustically hard. As noticed in the Introduction, this corresponds to the reflection coefficient $R = 1$. This value is obtained indeed in this particular case, since for $b \to 0$ one obtains $H_0 \to 0$, which leads in Eq. (32.36) to the acoustic pressure in the reflected wave as follows: $p^{\mathrm{ref}}(x, y) \to e^{\imath k y}$. We thus come to a far-field plane reflected wave with the unit reflection coefficient, that is to be proved.
2. $b = a$. In this case the whole reflecting boundary is acoustically soft. It follows from the Introduction that this corresponds to the reflection coefficient $R = -1$. Such a value is really obtained in this particular case, since for $b \to a - 0$ one obtains $H_0 \to +\infty$. As follows from (32.36), in this case the reflected wave is of the following form: $p^{\mathrm{ref}}(x, y) \to -e^{\imath k y}$. Therefore, the reflected wave here is plane again, whose amplitude is equal to -1, that is to be proved.

32.7 Conclusions

The following conclusions may be drawn from the above study:

1. In the case when a straight infinite boundary is made of a combination of two perfectly reflecting materials – acoustically hard and acoustically soft ones distributed periodically, the reflected wave for a plane incident wave is plane again if we speak about a far-field zone.
2. The complex magnitude of the reflected wave is not unit, however as can be seen from Eq. (6.8), its real physical amplitude is a unit value:

$$|R|^2 = |p^{\text{ref}}|^2 = \frac{4a^2k^2 + H_0^2}{4a^2k^2 + H_0^2} = 1. \tag{7.1}$$

Therefore, the considered periodic reflecting boundary is a perfect reflector, the same as its two components. It is interesting to study if a combination of two opposite perfect reflecting materials can yield an absorbing reflector. The present work proves that in the low-frequency limit this is impossible. This question is worthy of a deeper investigation under less strict assumptions.

3. Despite the unit reflecting amplitude, the phase in the reflected wave may vary, depending upon geometry of the structure. As shown in the previous section, in the second of the two considered opposite limiting cases, the phase of the reflected wave changes by 180 degrees upon reflection. Expression (6.8) shows that, with variation of parameter b from 0 to a, the phase of the reflected wave can vary within a wide range.

Acknowledgements The work has been performed under financial support of the Russian Science Foundation (RSF), Project 25-21-00139, in the Southern Federal University.

References

[1] Abramowitz M, Stegun IA (1965) Handbook of Mathematical Functions. Dover Publ, New York
[2] Brekhovskikh L (1976) Waves in Layered Media. Academic Press, New York
[3] Colton D, Kress R (1983) Integral Equation Methods in Scattering Theory. John Wiley, New York
[4] Long M (2006) Architectural Acoustics. Elsevier, San Diego
[5] Pierce AD (1991) Acoustics: An Introduction to Its Physical Principles and Applications. Acoustical Society of America, New York
[6] Prudnikov AP, Brychkov YA, Marichev OI (1986) Integrals and Series, Vol 1. Gordon & Breach, Amsterdam
[7] Scarpetta E, Sumbatyan MA (1995) Explicit analytical results to one-mode normal reflection and transmission by a periodic array of screens. Journal of Mathematical Analysis and Applications 195(3):736-749, DOI 10.1006/jmaa.1995.1386
[8] Sumbatyan MA, Scalia A (2005) Equations of Mathematical Diffraction Theory. CRC Press, Boca Raton, Florida

Chapter 33
Method of Boundary Integral Equation for the Wave Process Inside Canonical Triangles

Evgeny A. Titarenko and Mezhlum A. Sumbatyan

Abstract In frames of a classical problem of wave dynamics for domains with angles, we study the internal problem for a two-dimensional triangular region. The main difficulty is that any numerical method loses its stability near the corners. This paper proposes a method to overcome this obstacle. It is based on the fact that the instability of the algorithm takes place when both the observation point and the integration point simultaneously approach the corner, i.e. when the distance between them becomes extremely small. In this case, it is proposed to use the asymptotics of the Green's function for small argument. Then the integral near the corner can be calculated analytically, which leads to a stable algorithm. The obtained results are compared with the known analytical results for three considered canonical triangular regions.

Key words: Wave dynamics, Boundary integral equation, Green's function, eigen-frequency

33.1 Introduction

The application of the Boundary Integral Equations method (the BIE), as well as of some alternative direct numerical methods, is effective for smooth boundary surfaces with regular boundary conditions. The foundations of the BIE method were laid in the second half of the 20th century, and it has been intensively developed in a wide variety of scientific fields. Sometimes in a discrete form, this is called Boundary Element Method (the BEM) [3, 4]. The method has proven its efficiency, since with numerical discretization it reduces the dimension of the problem by one. For example, for spatial problems it is reduced to integral equations on a two-dimensional boundary surface

Evgeny A. Titarenko · Mezhlum A Sumbatyan (corresponding author)
Southern Federal University, Milchakova Street 8a, Rostov-on-Don, 344090, Russian federation,
e-mail: evtitarenko@sfedu.ru,masumbatyan@sfedu.ru

H. Altenbach and A. V. Sahakyan, *Contact Mechanics and Relevant Problems*,
Advanced Structured Materials 253, https://doi.org/10.1007/978-3-032-20869-9_33

of the region under consideration. Moreover, the boundary can be either closed (for a bounded region) or open (for a semi-infinite region). In two-dimensional (plane) problems, the construction of a solution by this method is generally reduced to a one-dimensional BIE along the boundary curve. In particular, the BIE method has demonstrated effectiveness in problems of wave radiation and scattering [5, 15].

Typically, the BIE method is faced with some obstacles for domains with edges and corners. To overcome this difficulty, some ideas have been proposed in this concern. Thus, in [2], the BIE method is used in combination with the Galerkin method and the energy interpretation of wave propagation. The authors claim the effectiveness of this approach, but there is no numerical example, to confirm this statement. The energy interpretation of the BIE is also discussed in [6]. In [7], an approach is used, in which the volume of the domain is divided into simple polyhedra, each of them being based on the ray propagation of waves. It is clear that this method works only at high frequencies. In [11], the use of the open BEM++ code is mentioned, in which the BIEs are solved not by the generally accepted collocation method on the boundary mesh of nodes, but by the projection Galerkin method with the choice of an appropriate basis, which is allied to [2]. It should also be noted that in [17], when analyzing cells located close to the corners of a domain, the main efforts are directed at how to correctly arrange the ordering of nodes near such a corner. Methods for overcoming the instability of the solution in the vicinity of a corner are not considered.

In the present work an approach developed in our previous works [12–14] is used, to improve the stability of the BIE solution with boundaries containing corners and edges. Specifically, in [12, 13], the asymptotic behavior of the BIE's kernel in the vicinity of a corner is used in the two-dimensional problem of wave diffraction by an infinite wedge. In particular, this made it possible to explicitly obtain the exact value of the pressure in the wave field at the corner point of the wedge. This idea was then transferred to the diffraction of waves by bounded obstacles with corners [14]. To test the stability of the proposed approach in application to the interior problem for domains with angles, we calculate the low-frequency normal modes of three types of triangular domains, where the solution is known in the analytical form.

33.2 Problem Formulation

Let us consider a two-dimensional triangular region (Fig. 33.1). Here $x_0 = (x_1^0, x_2^0)$ is a point wave source. Assuming that the boundary of the region is perfectly rigid, the wave process can be described by the Helmholtz equation with the homogeneous Neumann boundary condition [8]

$$\Delta p + k^2 p = 0, \quad \Delta = \frac{\partial^2}{\partial x_1^2} + \frac{\partial^2}{\partial x_2^2}, \quad \left.\frac{\partial p}{\partial n}\right|_l = 0, \quad p = p_0 + p^{\text{sc}}, \tag{33.1}$$

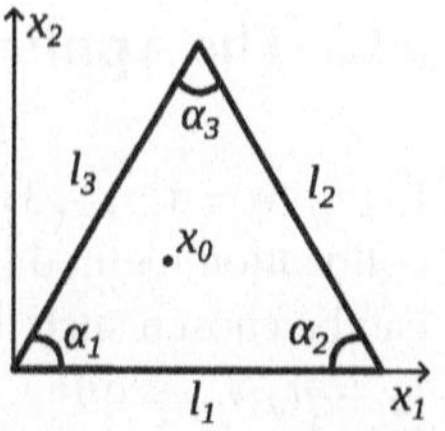

Fig. 33.1 Harmonic wave process inside a triangular region.

where p is the full wave pressure, $k = \omega/c$ is the wave number, c is the wave speed, $\omega = 2\pi f$ is the angular frequency, l is the closed boundary contour, p^{sc} is the scattered wave field, $p_0 = H_0^{(1)}(kr_0)$ is the wave field of the point source x_0,

$$r_0 = |x - x_0| = \sqrt{(x_1 - x_1^0)^2 + (x_2 - x_2^0)^2}.$$

The time dependence is given with the factor $e^{-i\omega t}$. Any point is defined by its two components, e.g. $x = (x_1, x_2)$.

If ω is not an eigenfrequency and the boundary points lie on the smooth part of the contour l, then the solution of problem (33.1) is equivalent to the following boundary integral equation (BIE) [5, 15]

$$\frac{p(x)}{2} - \int_l \frac{\partial G(y,x)}{\partial n_y} p(y) dl_y = p_0(x), \quad x \in l, \quad G(y,x) = \frac{i}{4} H_0^{(1)}(kr), \quad (33.2)$$

where n_y is the unit normal at the boundary point $y \in l$ directed inward the region bounded by contour l, dl_y is the elementary length in the vicinity of the boundary point $y \in l$, $H_0^{(1)}$ is the Hankel function of the first kind of zero's order [1]. In the considered case, there exists a unique solution to Eq. (33.2), which provides a unique solution to problem (33.1), in the following form:

$$p^{sc} = \int_l \frac{\partial G}{\partial n_y} p(y) dl_y, \qquad x \in l. \quad (33.3)$$

It should be noted that when point x in (33.2) coincides with one of the angles of the boundary contour, the coefficient $1/2$ in the first term of the left-hand side in Eq. (33.2) should be replaced by a coefficient, which depends on the magnitude of the internal angle of the boundary contour at the given corner of the boundary. This property may negatively affect the stability of the solution of the BIE (33.2), since a slight shift of the collocation point to the corner of the boundary contour can lead to a significant change in the diagonal elements of the corresponding system of linear algebraic equations (SLAE) resulting from a discretization of the BIE.

33.3 The Applied Method

Let l_q $(q = 1, \ldots, 3)$ be the sides of the considered region, then $l = \sum_{q=1}^{3} l_q$. For the collocation method [15], we assume that some small constant step for each side h_q can be chosen such that side l_q is divided into N_q equal segments of length h_q, i.e., $l_q = h_q N_q$, see Fig. 33.2 where the mesh is presented only schematically since in fact there are many small subintervals over every side. Obviously, the total size of the mesh is $N = \sum_{q=1}^{3} N_q$. Let us place the nodes of the discrete grid $y_j = (y_1^j, y_2^j)$ $(j = 1, \ldots, N)$ in the middle of each small segment, so that none of the nodes fall into any of the corners of the boundary, and each node near the corner is at a distance of $h_q/2$ from the corner. Such a grid allows us to keep the form of the BIE (2) with the coefficient $1/2$ in the first term in the left-hand side.

Taking into account that

$$\frac{\partial G}{\partial n_y} = \frac{\partial G}{\partial r}\frac{\partial r}{\partial n_y}, \quad \frac{\partial G}{\partial r} = -\frac{ik}{4}H_1^{(1)}(kr), \quad \frac{\partial r}{\partial n_y} = \frac{(r \cdot n_y)}{r}, \tag{33.4}$$

one can rewrite (33.2) in the following form

$$\frac{p(x)}{2} - \int_l K(y,x)p(y)dl_y = F(x), \quad x \in l, \tag{33.5}$$

where

$$K(x,y) = \frac{ik}{4}H_1^{(1)}(kr)\frac{(\bar{r} \cdot \bar{n}_y)}{r}, \quad F(x) = p_0(x), \quad \bar{r} = y - x, \quad r = |y - x|$$

Let us set in (33.5) $x = y_m \in l$, $(m = 1, \ldots, N)$. Then the integral (33.5) can be represented as follows

$$\int_l K(y,x)p(y)dl_y = \sum_{i=1}^{N}\int_{a_i}^{b_i} K(y,y_m)p(y)dl_y \approx \sum_{i=1}^{N} b_{mj}p(y_i),$$
$$b_{mj} = \int_{a_j}^{b_j} K(y,y_m)dl_y, \tag{33.6}$$

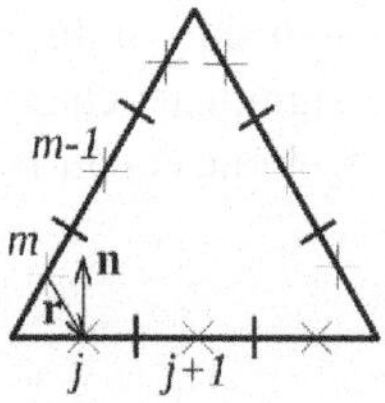

Fig. 33.2 A discrete mesh over the perimeter of the triangular domain.

where, a_j and b_j are the ends of the interval of length h_q with the central node y_j. We assume that the unknown function $p(y)$ is approximately constant on each small interval. It should be noted that a_j, y_j, b_j always belong to the same side of the considered region, so the segment $[a_j, b_j]$ is a straight line of length h_q with central point y_j. Quantities b_{mj} can be calculated by applying a certain quadrature rule, like Simpson's approximation.

As a result of discretization, the BIE (33.5) can be represented in matrix form

$$\mathbf{AP} = \mathbf{F}, \tag{33.7}$$

where

$$\mathbf{A} = (a_{mj}), \quad a_{mj} = \frac{\delta_{mj}}{2} + (1 - \delta_{mj}) b_{mj},$$

$$P = [p(y_j)], \quad F = [F(y_m)],$$

and δ_{mj} is Kronecker's delta. The elements b_{mj} are non-zero only if the nodes y_m and y_j belong to different sides. If they lie on the same side, then $(\bar{r} \cdot \bar{n}_y) = 0$. Therefore, in practice, in (33.7) the expression for b_{mj} from (33.6) should be used only to calculate the off-diagonal elements of the matrix, and for the diagonal elements one should set $b_{mm} = 0$.

To calculate the normal frequencies, the following algorithm is proposed. When the wave number k is different from any eigenvalue of the interior domain, the boundary value problem (33.1) is equivalent to the BIE (33.5) or to the SLAE (33.7) in the discrete form. The solution of the SLAE in turn exists and is unique if and only if its matrix is not degenerate. The normal frequencies correspond only to the cases when this matrix is degenerate, i.e. when its condition number is theoretically infinite but in practice is extremely large. Therefore, it is sufficient to go through all values of the wave number k with a small step, calculating each time the condition number CN (Condition Number) of the matrix and selecting the frequency values for which CN has a sharp local maximum.

The worst-case scenario for the stability of the algorithm takes place when the observation point (y_m) and the integration point (y_j) are both located close to the same corner. In this case, we apply the improvement proposed in the recent work [16]. This is based upon an explicit-form integration over the "closest to the angle α" small interval of length h_j, by using the small-argument representation for the Hankel function in (33.5), see [1]

$$H_1^{(1)}(z) \sim -\frac{2i}{\pi z} + \frac{iz}{\pi} \ln z + \frac{z}{\pi} \left[\frac{\pi}{2} + i \left(\gamma - \ln 2 - \frac{1}{2} \right) \right], \quad z \to 0, \tag{33.8}$$

where $\gamma = 0.57722$ is Euler's number.

Then the coefficients closest to the angle α in (33.7) should be replaced by the following quantities

$$b_{mj}^{\alpha} = \left(-\frac{h_m \sin\alpha}{8}\right) \int_0^{h_j} \left\{ \frac{2}{\pi r^2} - \frac{k^2}{\pi} \ln r + \frac{k^2}{\pi} \left[\frac{\pi i}{2} - \left(\gamma + \ln\frac{k}{2} - \frac{1}{2} \right) \right] \right\} dy_1, \tag{33.9}$$
$$r^2 = \left(y_1 - \frac{h_m \cos\alpha}{2} \right)^2 + \left(\frac{h_m \sin\alpha}{2} \right)^2 .$$

All integrals in (33.9) are tabulated [10]

$$\int_0^{h_j} \frac{dy_1}{r^2} = \frac{2}{h_m \sin\alpha} \left[\arctan\left(\frac{2h_j}{h_m \sin\alpha} - \cot\alpha \right) + \frac{\pi}{2} - \alpha \right] \tag{33.10}$$

and

$$\int_0^{h_j} \ln r \, dy_1 = \left(h_j - \frac{h_m \cos\alpha}{2} \right)$$
$$\times \left(\frac{1}{2} \ln \left[h_j^2 - h_j h_m \cos\alpha + \left(\frac{h_m}{2} \right)^2 \right] - 1 \right) + \frac{h_m \cos\alpha}{2} \left(\ln\frac{h_m}{2} - 1 \right) \tag{33.11}$$
$$+ \left(\frac{h_m \sin\alpha}{2} \left[\arctan\left(\frac{2h_j}{h_m \sin\alpha} - \cot\alpha \right) + \frac{\pi}{2} - \alpha \right] \right).$$

The quantities (33.9) should be substituted instead of general expression for matrix' elements b_{mj}, obtained on the basis of the quadrature rule in (33.6), twice for "closest to a corner" elements, for every corner. Thus, if for the triangular domain there are in common three angles, hence six elements of the matrix should be used in the form (33.9), to provide the stability of the algorithm. With so doing, expressions (33.9)-(33.11) are universal with respect to the value of the angle α, regardless of whether this angle is acute or not.

33.4 Numerical Implementation and Calculation Results

The method was implemented with the authors' program in C++. The condition number is calculated by using Gaussian elimination technique [9]. The calculations are performed for $c = 340\,\text{m/s}$, for the frequencies in the range 20 – 200 Hz with the step 0.001 Hz. For the examples considered below, every geometry requires few minutes of calculations on a personal computer. To test the method, three canonical triangles are selected, see Fig. 33.3, whose normal frequencies can be calculated explicitly [15]:

1. and 2. Equilateral triangle and right triangle with acute angles 30° and 60°:

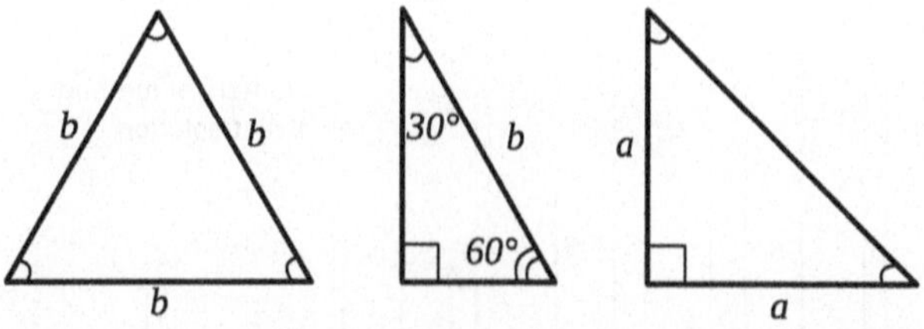

Fig. 33.3: Three canonical triangles. From left to right: 1. equilateral triangle with the side b; 2. right triangle with acute angles 30° and 60° and hypotenuse b; 3. isosceles right triangle with the leg a.

$$k_{mn} = \frac{4\pi}{3b}\sqrt{m^2 + mn + n^2}, \qquad (m, n = 0, 1, 2, \ldots), \tag{33.12}$$

3. Isosceles right triangle:

$$k_{mn} = \frac{\pi}{a}\sqrt{m^2 + n^2}, \qquad (m \geq n), \qquad (m, n = 0, 1, 2, \ldots). \tag{33.13}$$

As a result of the calculations, we demonstrate dependence of the condition number versus frequency for all three triangular domains, see Figs. 33.4 – 33.6. The dotted lines correspond to exact expressions (33.12) and (33.13), where the CN parameter is theoretically infinite. The solid lines show respective values of CN, in frames of the proposed algorithm. It is seen from Figs. 33.4 – 33.6 that parameter CN, as

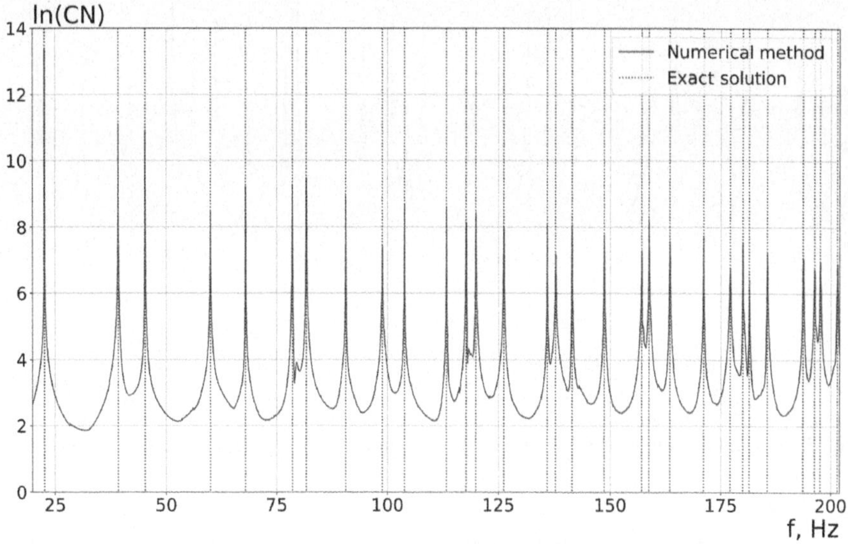

Fig. 33.4: Condition number (CN) versus frequency in Hz: equilateral triangle with the side $b = 10\,\text{m}$.

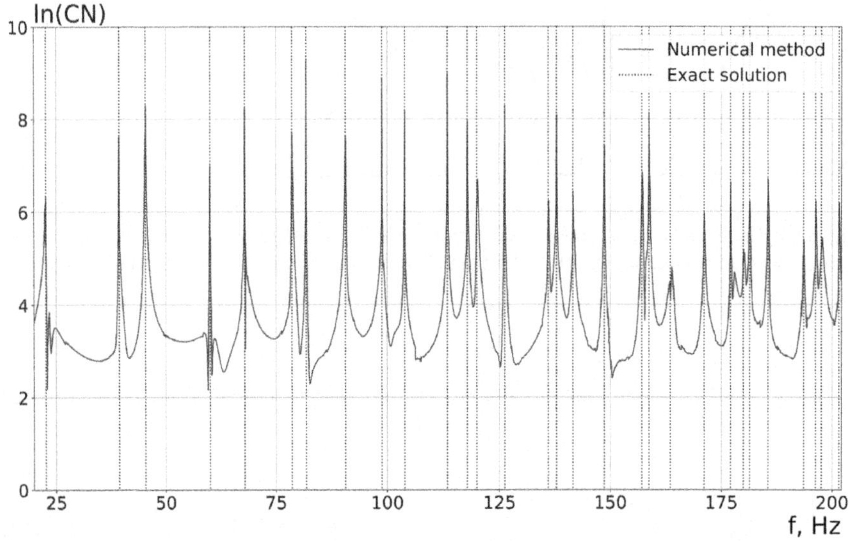

Fig. 33.5: Condition number (CN) versus frequency in Hz: right triangle with acute angles 30° and 60° and hypotenuse $b = 10$ m.

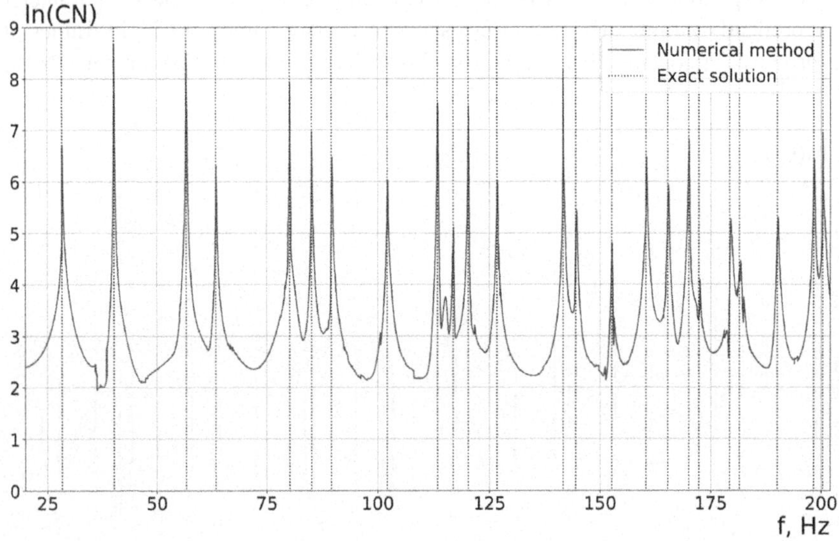

Fig. 33.6: Condition number (CN) versus frequency in Hz: isosceles right triangle with the leg $a = 6$ m.

a function of frequency, has indeed sharp maximum values near the exact values predicted theoretically. The figures confirm a known property of the distribution of modal frequencies – that the distribution becomes denser with frequency increasing. Besides, it is clearly seen from Fig. 33.4 that in some cases, regarding parameter CN, which at modal frequencies should be infinite, in numerical treatment the magnitude of sharp maxima of this parameter may decrease with increasing frequency.

In numbers, the calculated quantities are also presented in the three Tables 33.1 - 33.3, with three significant digits after the decimal point. It is also very interesting to see from Tables 33.1 and 33.2 that for these geometries, where exact values coincide to each others, the numerical values are slightly different, within the accuracy of numerical calculations.

33.5 Conclusions

The proposed method demonstrates high accuracy and stability in solving the wave equation for triangular regions. The following conclusions can be drawn from the obtained results:

- The developed algorithm provides stability near the corners of the domain.

Table 33.1: Comparison of the obtained normal frequencies with exact values for equilateral triangle, $b = 10\,\text{m}$.

Normal Frequency No	1	2	3	4	5	6
Exact value	22.667	39.260	45.333	59.970	68.000	78.520
Proposed method	22.667	39.263	45.336	59.986	68.000	78.530
Normal Frequency No	7	8	9	10	11	12
Exact value	81.723	90.667	98.802	103.872	113.333	117.78
Proposed method	81.741	90.675	98.824	103.873	113.344	117.797
Normal Frequency No	13	14	15	16	17	18
Exact value	119.941	126.203	136.000	137.876	141.553	148.635
Proposed method	119.966	126.227	136.001	137.909	141.554	148.669
Normal Frequency No	19	20	21	22	23	24
Exact value	157.039	158.667	163.452	171.130	177.032	179.911
Proposed method	157.059	158.669	163.489	171.134	177.073	179.912
Normal Frequency No	25	26	27	28	29	30
Exact value	181.333	185.535	193.664	196.299	197.603	201.466
Proposed method	181.367	185.580	193.711	196.322	197.637	201.518

Table 33.2: Comparison of the obtained normal frequencies with exact values for right triangle with acute angles 30° and 60°, $b = 10\,m$.

Normal Frequency No	1	2	3	4	5	6
Exact value	22.667	39.260	45.333	59.970	68.000	78.520
Proposed method	22.679	39.220	45.365	59.924	67.989	78.582
Normal Frequency No	7	8	9	10	11	12
Exact value	81.723	90.667	98.802	103.872	113.333	117.780
Proposed method	81.700	90.701	98.806	103.883	113.336	117.798
Normal Frequency No	13	14	15	16	17	18
Exact value	119.941	126.203	136.000	137.876	141.553	148.635
Proposed method	120.088	126.213	136.083	137.906	141.623	148.654
Normal Frequency No	19	20	21	22	23	24
Exact value	157.039	158.667	163.452	171.130	177.032	179.911
Proposed method	157.132	158.670	163.364	171.208	177.090	180.016
Normal Frequency No	25	26	27	28	29	30
Exact value	181.333	185.535	193.664	196.299	197.603	201.466
Proposed method	181.389	185.590	193.689	196.392	197.793	201.554

Table 33.3: Comparison of the obtained normal frequencies with exact values for isosceles right triangle, $a = 6\,m$.

Normal Frequency No	1	2	3	4	5	6
Exact value	28.333	40.096	56.667	63.355	80.139	85.000
Proposed method	28.382	40.111	56.693	63.426	80.185	85.050
Normal Frequency No	7	8	9	10	11	12
Exact value	89.598	102.157	113.333	116.821	120.208	126.710
Proposed method	89.710	102.244	113.386	116.900	120.277	126.828
Normal Frequency No	13	14	15	16	17	18
Exact value	141.667	144.472	152.580	160.277	165.210	170.000
Proposed method	141.672	144.671	152.685	160.399	165.384	170.074
Normal Frequency No	19	20	21	22	23	24
Exact value	172.345	179.196	181.422	190.066	198.333	200.347
Proposed method	172.583	179.465	181.689	190.235	198.428	200.358

- The key point of the proposed algorithm is the use of the small-argument asymptotics of the Green's function, which corresponds to mesh nodes closest to the corners.
- Comparison with analytical solutions shows good agreement for all tested cases.
- The method is applicable to various types of triangular regions.

The results confirm the effectiveness of the proposed approach and its potential for solving more complex wave problems in polygonal domains. Further research may be directed to extend the method to three-dimensional cases and more complex geometries.

Acknowledgements The work has been performed under financial support of the Russian Science Foundation (RSF), Project 25-21-00139, in the Southern Federal University.

References

[1] Abramowitz M, Stegun IA (1965) Handbook of Mathematical Functions. Dover Publ, New York
[2] Aimi A, Diligenti M, Guardasoni C, Mazzieri I, Panizzi S (2009) An energy approach to space–time Galerkin BEM for wave propagation problems. The International Journal for Numerical Methods in Engineering 80(9):1196-1240, DOI 10.1002/nme.2660
[3] Banerjee PK, Butterfield R (1981) Boundary Element Methods in Engineering Science. McGraw Hill, London
[4] Brebbia CA, Telles JCF, Wrobel LC (1984) Boundary Element Techniques: Theory and Application in Engineering. Springer-Verlag, Berlin Heidelberg, DOI 10.1007/978-3-642-48860-3
[5] Colton D, Kress R (1983) Integral Equation Methods in Scattering Theory. John Wiley, New York
[6] Hargreaves JA, Lam YW (2014) An energy interpretation of the Kirchhoff-Helmholtz boundary integral equation and its application to sound field synthesis. Acta Acustica united with Acustica 100: 912-920, DOI 0.3813/AAA.918770
[7] Hargreaves JA, Rendell LR, Lam YW (2019) A framework for auralization of boundary element method simulations including source and receiver directivity. Journal of the Acoustical Society of America 145(4):2625–2637, DOI 10.1121/1.5096171
[8] Pierce AD (1991) Acoustics: An Introduction to Its Physical Principles and Applications, 3rd ed. Springer, Cham, DOI 10.1007/978-3-030-11214-1
[9] Press WH, Teukolsky SA, Vetterling WT, Flannery BP (2002) Numerical Recipes in C: The Art of Scientific Computing. Cambridge University Press, Cambridge
[10] Prudnikov AP, Brychkov YA, Marichev OI (1986), Integrals and Series, vol 1. Amsterdam, Gordon & Breach

[11] Smigaj W, Betcke T, Arridge S, Phillips J, Schweiger M (2015) Solving boundary integral problems with BEM++. ACM Transactions on Mathematical Software 41(2):1-40, DOI 10.1145/2590830

[12] Sumbatyan MA, Martynova TS, Musatova NK (2021) Boundary element methods in diffraction of a point-source acoustic wave by a rigid infinite wedge. Engineering Analysis with Boundary Elements 127:157–167, DOI 10.1016/j.enganabound.2021.01.017

[13] Sumbatyan MA, Martynova TS, Musatova NK (2022) On diffraction of a point sound source by an infinite wedge. Acoustical Physics 68(4):307-315, DOI 10.1134/S1063771022030149

[14] Sumbatyan M, Musatova NK (2023) An efficient treatment of sound diffraction by arbitrary obstacles with angles. In: Altenbach H, Mkhitaryan SM, Hakobyan V, Sahakyan AV (eds) Solid Mechanics, Theory of Elasticity and Creep. Advanced Structured Materials, vol 185, pp 347-359. Springer, Cham. DOI 10.1007/978-3-031-18564-9_25

[15] Sumbatyan MA, Scalia A (2005) Equations of Mathematical Diffraction Theory. CRC Press, Boca Raton, Florida

[16] Sumbatyan MA, Titarenko EA (2025) Development of the boundary equations method for wave probles in 2D domains with irregularities of the boundary curve. In: Altenbach H (ed) Current Developments in Solid Mechanics and Their Applications. Advanced Structured Materials, vol 223, pp 675-686. Springer, Cham, DOI 10.1007/978-3-031-90022-8_39675-686

[17] Yan ZY (2006) Treatment of sharp edges and corners in the acoustic Boundary Element Method under Neumann boundary condition. Computer Modeling in Engineering & Sciences 13(2):81-90, DOI 10.3970/cmes.2006.013.081

Chapter 34
Investigation of Factors Influencing the Load-Bearing Capacity of Building Structures

Mher S. Torosyan and Mihran G. Stakyan

Abstract The condition for the consideration of factors affecting the load-bearing capacity of the responsible elements and nodes in the gas transportation system is characterized by multiple variations, complex nature, and combinations, with different quantities and time periods. Using the method of system analysis, multi-parameter functional connections and systems of parametric equations have been formulated according to the principle of the priority of the influencing factor, which is dictated by the requirements presented for the structure. To increase the cyclic strength and durability of the elements, the method of plastic deformation of working surfaces has been employed, which significantly improves the load-bearing capacity indicators of the structure when using low and medium carbon steels. In order to reveal the physical and mechanical characteristics of the reinforced surface layer, inclined planes have been developed on the surfaces and grid schemes have been formulated for the study of micro-roughness according to depth. The results have been processed using mathematical statistical methods and obtained from regression equations in order to determine the optimal parameters of plastic reinforcement regimes.

Key words: Gas transportation system, Construction element, Load capacity, Complex influence of factors, Surface plastic deformation, Microhardness network scheme, Regression equations

Mher S. Torosyan (corresponding author) · Mihran G. Stakyan
Institute of Mechanics, National Academy of Sciences of the Republic of Armenia, 24b Marshal Baghramyan Ave, Yerevan, 0019, Republic of Armenia, e-mail: mhertorosyan@list.ru, stakyan.mihran@yandex.ru

H. Altenbach and A. V. Sahakyan, *Contact Mechanics and Relevant Problems*, Advanced Structured Materials 253, https://doi.org/10.1007/978-3-032-20869-9_34

34.1 Introduction

The continuous improvement of construction structures cannot be achieved without the modernization of the entire complex of design, technical, production, and operational measures. Among these, technological factors occupy a special place, particularly highlighting the processes of surface plastic deformation (SPD), which are the main methods for strengthening construction materials. SPD is a technologically simple and efficient method for improving the surface state of elements and is therefore widely used in the production of various constructions. It allows for a high-quality improvement of working surfaces and load-bearing indicators with minimal costs, including fatigue resistance, durability, wear resistance, and more [2, 3].

The factors affecting the change in the physico-mechanical state of surface layers due to the impact of the SPD are considered (material science, computational design, technological, operational, etc.), the cumulative result of which establishes the load-bearing capacity of structural elements and joints according to the criteria of the observed factors. The work mainly presents the condition parameters of the outer surface layers of test samples made of 40X grade structural steel, which were obtained as a result of reinforcement in SPD modes.

In the work, the most common case of SPD has been selected: the reinforcement on the surfaces of elements through a rolling ball device, where a main deforming ball is located, performing a centrifugal rotation with the help of three additional balls. This device has certain advantages: point contact, absence of relative sliding, and self-regulation concerning the processed surface. This excludes the formation of sliding lines, creates favorable conditions for plastic deformation, and ensures a high-quality surface with relatively low normal force application, which prevents the phenomenon of surface superfinishing and the risk of degradation, facilitates the free deformation of surface microelements, and establishes optimal microgeometric characteristics of the surface. The mentioned SPD method is a relatively accessible and high-performance means that does not require additional technological equipment.

Regression equations have been obtained between the indicators of the specified factors [10, 15], and multi-parametric functional connections have been established, which are essentially mathematical models of the structure's carrying capacity. To increase the reliability of the calculations, inclined planes have been developed on the external surfaces of the samples, where a network scheme has been formed, and mass measurements of HV microhardness have been performed. The results have been processed using mathematical statistical methods with standard software tools, which also included a subroutine ensuring the selection of optimal operations in three-dimensional data according to values of the determination coefficient $R^2 > 0,9$ [12, 13].

The purpose of the work is to reveal the variations in the load-bearing capacity of constructions under the combined procedures of the observed factors, with different combinations, quantities, and time periods of influence, which will allow for the application of the optimal parameters of the specified factors in accounting design activities. In this regard, a multi-parameter relationship is proposed

$$\Phi_1\left[(F,\ V,\ S),\ (d,\alpha_\sigma),\ (\Delta h,\ \mathrm{HV},\ \mathrm{HV}_{\max}),\right.$$
$$\left.(R_a,\ R_z,\ R_{\max}),\ \left(\bar{C},\ \bar{m},\ \lg \bar{N}_G,\bar{\sigma}_R\right)\right] = 0, \quad (34.1)$$

where F, V, S are the normal force (N) of the deforming device, the linear speed of the device's rolling body (m/s), and the longitudinal supply (mm/rev), d, α_σ are the diameter (mm) of the specimen and the accumulation factor for stresses, Δh, HV, $\mathrm{HV}_{\max}$ are the depth (mm) of the reinforced layer, microhardness according to Vickers and maximum microhardness, R_a, R_z, $R_{\max}$ are the surface micro-roughness (mm), $\bar{C}$, $\bar{m}$, $\lg \bar{N}_G$, $\bar{\sigma}_R$ are the fatigue median indicators. In computational operations (34.1), the function is replaced by a system of parametric equations

$$\begin{aligned}
&\Delta h = f_1(F,V,S),\ \mathrm{HV} = \varphi_1(\Delta h),\ \bar{C}_{(v)} = \varphi_6\left(\bar{m}_{(v)}\right),\\
&\mathrm{HV}_{\max} = f_2(F,V,\ S),\ \mathrm{HV}_{\max} = \varphi_2(\Delta h),\ \bar{m}_{(v)} = \varphi_7(\mathrm{HV}_{\max}),\\
&R_a = f_3(F,V,\ S),\ \bar{\sigma}_{R(v)} = \varphi_3(\Delta h,\ d,\alpha_\sigma),\ \bar{m}_{(v)} = \varphi_8(\Delta h),\\
&R_z = f_4(F,V,\ S),\ d = \varphi_4\left(\bar{\sigma}_{R(v)},\alpha_\sigma,\Delta h\right),\ \bar{\sigma}_{R(v)} = \varphi_9\left(\lg \bar{N}_{G(v)}\right),\\
&R_{\max} = f_5(F,V,\ S),\bar{m}_{(v)} = \varphi_5\left(\bar{\sigma}_{R(v)}\right),\lg \bar{N}_{G(v)} = \varphi_{10}(\Delta h).
\end{aligned} \quad (34.2)$$

The significant functional relationships mentioned are $\overline{m}_v = \varphi_5\left(\bar{\sigma}_{R(v)}\right)$, $\bar{C} = \varphi_6(\bar{m})$ and $\bar{\sigma}_{R(v)} = \varphi_9\left(\lg \bar{N}_{G(v)}\right)$, which are parametric equations that characterize the variations of cyclic loads and lifetimes under influencing factors. However, these functions are also multi-parameter, where the priority of the parameters is dictated by the dominance of any one of the factors affecting fatigue resistance within the framework of the study. For example, for the function $\bar{\sigma}_{R(v)} = \varphi_9\left(\lg \bar{N}_{G(v)}\right)$, the following combinations of the priority of arguments and parameters are possible

$$\begin{aligned}
&\left(\lg \bar{N}_{G(v)},\alpha_\sigma,d,\Delta h\right),\ \left(\lg \bar{N}_{G(v)},d,\alpha_\sigma,\Delta h\right),\ \left(\lg \bar{N}_{G(v)},\Delta h,\alpha_\sigma,d\right),\\
&\left(\lg \bar{N}_{G(v)},\alpha_\sigma,\Delta h,d\right),\ \left(\lg \bar{N}_{G(v)},d,\Delta h,\alpha_\sigma\right),\ \left(\lg \bar{N}_{G(v)},\Delta h,d,\alpha_\sigma\right).
\end{aligned} \quad (34.3)$$

By performing similar calculations for the cases of $\bar{m}_{(v)} = \varphi_5\left(\bar{\sigma}_{R(v)}\right)$ and $\bar{C}_{(v)} = \varphi_6\left(\bar{m}_{(v)}\right)$, it is possible to obtain the entire system of regression equations describing the characteristics of cyclic strength and durability, and to carry out design calculations for various combinations of influencing factors and principles of precedence, which cannot be realized under the conditions of evaluating the separate effects of these factors.

The equations for the selection of the form of the mathematical model are as follows [4–7, 11, 14]

$$y_x = a_0 + a_1 x, \quad (34.4)$$

$$y_x = a_0 + a_1 x + a_2 x^2, \quad (34.5)$$

$$y_x = a_0 + a_1 x + a_2 x^2 + a_3 x^3, \quad (34.6)$$

$$y_x = a_0 + a_2 x^{-(m+\Delta m)}. \quad (34.7)$$

The selection of optimal mathematical models for functions is carried out using any of the methods below or by combining them:

1. According to the selected data group, the regression equations are calculated using (34.4) - (34.7), their R^2 determination coefficients are evaluated, and the function that ensures the condition $R^2 > 0.9$ is selected. If there are two or more such functions, the option that provides the $R^2_{\max}$ value is chosen; if the difference in $R^2_{\max}$ values varies within the range of $R^2_{\max} = 0.02 \ldots 0.06$, a relatively lower degree function can be chosen if necessary.
2. The regression equations are calculated using the same procedure as in step 1, and for each of them, the sum of the squares of the differences between the taken data and their corresponding functional values is calculated

$$E_k = \sum_{i=1}^{n} (y_i - y_{xi})^2,$$

where k is the order of approximation polynomial in (34.4) - from (34.7). The optimal order of approximation will be the order corresponding to the smallest of E_k

$$E_{\min} = \min E_k. \tag{34.8}$$

If the values of E_k differ from each other within $8\% - 10\%$, then a lower order of approximation can be selected [9].

For the function $\bar{\sigma}_{R(\upsilon)} = \varphi_9 \left(\lg \bar{N}_{G(\upsilon)}\right)$, the method 1 for obtaining an optimal mathematical model has been applied. The coefficient a $\bar{a}_{\sigma i}$ is chosen as the primary parameter, under which the specified function primarily exhibits a linear nature (Fig. 34.1, lines 1,. . . ,4). The statistical calculations performed according to computer programs confirm this version, with linear regression equations achieved with high linearization indicators $R^2 = 0,914 \ldots 0,998$ (see Table 34.1) [16, 17].

The preliminary graphical analysis of the data indicates that linear connections only become curved in the case of stress accumulation. This coincides with the nature of the functional relationships between the parameters of fatigue lines, consistent with previously obtained data [16], but these results have only been derived and validated in the case of the application of reinforcing technologies.

The change in the gradient of the function $\bar{\sigma}_{R(\upsilon)} = \varphi_9 \left(\lg \bar{N}_{G(\upsilon)}\right)$ is considered new in the presence of a sharp energy accumulator. In this case, there is also a complete restoration of the load-bearing capacity of the structural loaded component, with the resistance limits of elements with surface grooves reaching up to the baseline values obtained for flat elements. In other words, reinforcing technologies almost completely eliminate the phenomenon of energy accumulation.

In the calculations of structural elements designed for short-term working mode (3. . . 5 years), the equations derived for the inclined sections of fatigue lines $\left(\bar{\sigma}_{RN(\upsilon)} < \bar{\sigma}_{RG(\upsilon)}, \bar{N}_{(\upsilon)} < \bar{N}_{G(\upsilon)}\right)$ are used, the parameters of which are presented with functional relationships $\bar{m}_{(\upsilon)} = \varphi_5 \left(\bar{\sigma}_{R(\upsilon)}\right)$, $\bar{C}_{(\upsilon)} = \varphi_6 \left(\bar{m}_{(\upsilon)}\right)$. $\bar{m}_{(\upsilon)} = \varphi_5 \left(\bar{\sigma}_{R(\upsilon)}\right)$ is characterized by a relatively dense distribution in different sections of the actual range of stress accumulation ($\alpha_\sigma = 1,0 \ldots 3,5$), while the

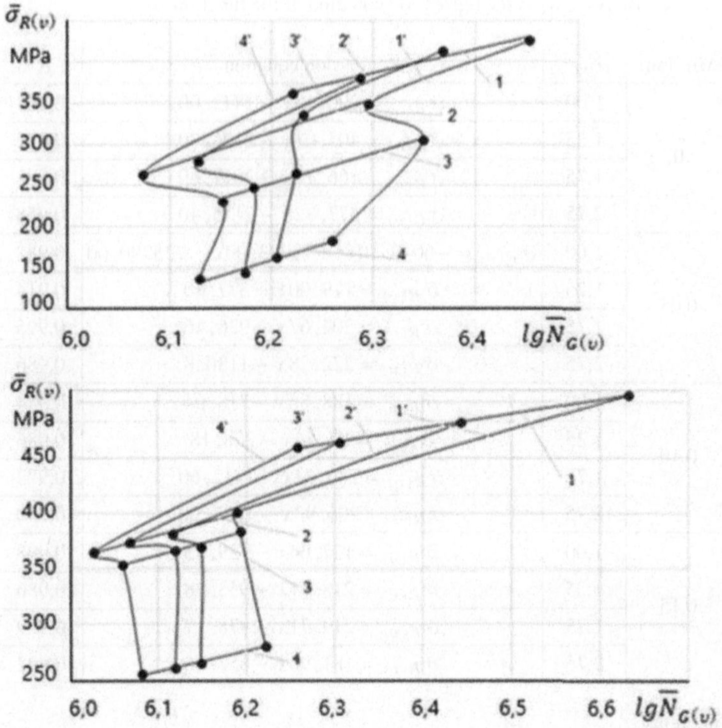

Fig. 34.1: The functions $\bar{\sigma}_{R(\upsilon)} = \varphi_9\left(\lg \bar{N}_{G(\upsilon)}\right)$ for $\Delta h_1 = 0$ (a) and $\Delta h_4 = 0, 15$ (b). The groups of functions 1, 2, 3, 4 and 1', 2', 3', 4' correspond to the variants $\left(\lg \bar{N}_{G(\upsilon)},\ \alpha_\sigma,\ d,\ \Delta h\right)$ and $\left(\lg \bar{N}_{G(\upsilon)},\ \alpha_\sigma,\ \Delta h,\ d\right)$.

parameter of reinforcement Δh exhibits a monotonous increase (Fig. 34.2), which allows to express that functional relationship in the selected ranges of construction (α_σ) and technological (Δh) factors in the form of a general linear regression equation (Table 34.2) [1, 17]:

$\bar{C}_{(\upsilon)} = \varphi_6\left(\bar{m}_{(\upsilon)}\right)$ function groups the parameters of the equations of fatigue median lines for $N = 10^4 \ldots 5 \cdot 10^6$ cycles, which encompasses the real time interval for limited and long-term fatigue processes. In this regard, the mentioned function actually regulates the (2) connection element mapping, while in the case of application of SPD, the phenomena of data linearization and sequential ordering of separate sections of the function present themselves more clearly and prominently (Fig. 34.3), which also proposes the extraction of the general equation (Table 34.2). The results of the calculations confirm the graphical interpretation of the variations of the functions $\bar{\sigma}_R$, $\overline{m}$, $\bar{C}$, $\lg \bar{N}_G$ and justify the primacy of linear regression relationships for the specified functions (Tables 34.1 and 34.2 and Figs. 34.2 and 34.3).

The systematic analysis and classification of data from previously conducted large-scale and various fatigue tests, as well as the mathematical modeling of co-

Table 34.1: $\bar{\sigma}_{R(\upsilon)} = \varphi_9 \left(\lg \bar{N}_{G(\upsilon)}\right)$ is the regression equation of the function.

No.	d_i, mm	Δh_i, mm	$\bar{a}_{\sigma i}$	Regression equation	R^2
1	7,5		1,00	$\bar{\sigma}_{R(\upsilon)} = 290,19x - 1442,00$	0,965
2	10,0	0	1,25	$\bar{\sigma}_{R(\upsilon)} = 405,01x - 2196,80$	0,977
3	15,0		1,75	$\bar{\sigma}_{R(\upsilon)} = 366,12x - 2017,60$	0,985
4	20,0		2,75	$\bar{\sigma}_{R(\upsilon)} = 373,32x - 2155,40$	0,958
5	7,5		1,00	$\bar{\sigma}_{R(\upsilon)} = -6049,10x^2 + 73903,00x - 225290,00$	0,988
6	10,0	0,05	1,25	$\bar{\sigma}_{R(\upsilon)} = 119,00x - 377,80$	0,914
7	15,0		1,75	$\bar{\sigma}_{R(\upsilon)} = 204,67x - 936,46$	0,995
8	20,0		2,75	$\bar{\sigma}_{R(\upsilon)} = 222,28x - 1130,8$	0,986
9	7,5		1,00	$\bar{\sigma}_{R(\upsilon)} = 188,86x - 718,04$	0,974
10	10,0	0,10	1,25	$\bar{\sigma}_{R(\upsilon)} = 95,86x - 228,18$	0,986
11	15,0		1,75	$\bar{\sigma}_{R(\upsilon)} = 286,24x - 1412,60$	0,972
12	20,0		2,75	$\bar{\sigma}_{R(\upsilon)} = 206,93x - 1027,00$	0,950
13	7,5		1,00	$\bar{\sigma}_{R(\upsilon)} = 127,64x - 339,85$	0,998
14	10,0	0,15	1,25	$\bar{\sigma}_{R(\upsilon)} = 218,61x - 953,18$	0,986
15	15,0		1,75	$\bar{\sigma}_{R(\upsilon)} = 219,71x - 978,17$	0,979
16	20,0		2,75	$\bar{\sigma}_{R(\upsilon)} = 181,96x - 852,66$	0,992

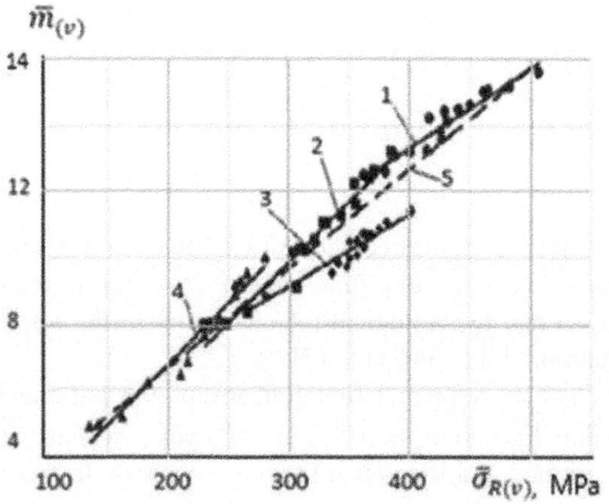

Fig. 34.2 The functions $\bar{m}_{(\upsilon)} = \varphi_5 \left(\bar{\sigma}_{R(\upsilon)}\right)$:
1 - $\bar{\alpha}_{\sigma 1} = 1,00$,
2 - $\bar{\alpha}_{\sigma 2} = 1,25$,
3 - $\bar{\alpha}_{\sigma 3} = 1,75$,
4 - $\bar{\alpha}_{\sigma 4} = 2,75$,
5 - the overall line,
the points correspond to $\bar{\alpha}_{\sigma 1} \ldots \bar{\alpha}_{\sigma 4}$.

influencing factors and phenomena, have revealed the regularities of these factors concerning the strength and durability indicators of the load-bearing structural elements of the specified machines and technological equipment, which allows for determining the optimal parameters of the factors according to the mathematical models obtained in order to ensure the fulfillment of certain technical requirements for the construction.

Table 34.2: $\bar{m}_{(\upsilon)} = \varphi_5\left(\bar{\sigma}_{R(\upsilon)}\right)$ and $\bar{C}_{(\upsilon)} = \varphi_6\left(\bar{m}_{(\upsilon)}\right)$ functions.

No.	d, mm	Δh_i, mm	$\bar{a}_{\sigma i}$	Regression equation	R^2
			$\bar{m}_{(\upsilon)} = \varphi_5\left(\bar{\sigma}_{R(\upsilon)}\right), (x = \bar{\sigma}_{R(\nu)})$		
1	7, 5	0	1,00	$\bar{m}_{(\upsilon)} = 0,024x + 3,754$	0,920
2	...	...	1,25	$\bar{m}_{(\upsilon)} = 0,022x + 2,709$	0,906
3	20, 0	0, 15	1,75	$\bar{m}_{(\upsilon)} = 0,037x - 1,101$	0,953
4			2,75	$\bar{m}_{(\upsilon)} = 0,037x - 0,647$	0,960
5	Equation of a straight line		1,00... 2,75	$\bar{m}_{(\upsilon)} = 0,030x + 0,645$	0,929
			$\bar{m}_{(\upsilon)} = \varphi_6\left(\bar{\sigma}_{R(\upsilon)}\right), (x = \bar{m}_{(\nu)})$		
6	7, 5	0	1,00	$\bar{C}_{(\upsilon)} = 3,192x - 1,597$	0,987
7	...	...	1,25	$\bar{C}_{(\upsilon)} = 3,006x + 1,434$	0,992
8	20, 0	0, 15	1,75	$\bar{C}_{(\upsilon)} = 2,705x + 3,486$	0,999
9			2,75	$\bar{C}_{(\upsilon)} = 2,854x + 2,488$	0,998
10	Equation of a straight line		1,00... 2,75	$\bar{C}_{(\upsilon)} = 2,942x + 1,804$	0,997

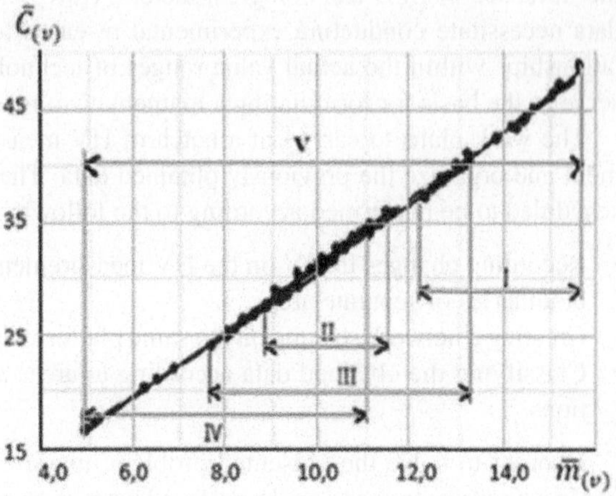

Fig. 34.3 Functions $\bar{C}_{(\upsilon)} = \varphi_6(\bar{m}_{(\upsilon)}, \alpha_\sigma, \Delta h, d)$. Sections I,..., IV correspond to $\bar{a}_{\sigma 1} \ldots \bar{a}_{\sigma 4}$, while the Section V corresponds to the general section.

The choice of a mathematical optimal model for the quantitative assessment of the fatigue failure process of construction elements among such relationships obtained for the selection of an optimal reinforcing technology allows for the formation of an interconnected system of parametric equations for the technological provision of problems of strength, durability, and reliability of constructions.

34.2 Organization of the Scientific Experiment for the Surface Reinforcement of Elements

A significant part of research in the field of SPD is devoted to increasing the bearing capacity of responsible elements, as well as improving the technological processes and tools in the field and proposing new methods. The physicomechanical state of reinforced surface layers, their characteristics, and the regularities of the variation of indicators have been comparatively less studied. Ultimately, these indicators characterize and define the fatigue resistance of elements, resisting the onset, cracking, and development processes of fatigue damage. A significant emphasis is placed on HV microhardness, which is included as a fundamental indicator in the systems of equations (34.1) and (34.2) of the SPD process. In a certain respect, it has a general nature and can be classified as an integral indicator in the SPD process, characterizing the level of strengthening both at the surface of the element (HV_{max}) and at the depth of the deformed layer Δh ($HV \to HV_0$). In this context, there arises a need to derive the functional relationships related to HV microhardness through the analysis of conducted studies and data classification, which are primarily formulated under the influence of SPD technological factors. However, unstructured and small-scale data necessitate conducting experimental research to uncover these functional relationships within the actual value ranges of technological parameters, which will serve as the basis for formulating a mathematical model of the SPD process [5, 6].

The work plans to carry out a batch of HV measurements, which will complement and organize the previously obtained data. The mentioned measurements are scheduled to be performed according to the following procedure:

- Recording changes in HV on the HV measurement platform and designating the boundaries of separate areas,
- Creating a network scheme on the same platform and carrying out measurements,
- Classifying the obtained data according to areas and obtaining regression equations.

In order to solve the presented problem, massive measurements of HV microhardness have been made on the surface layers of 15 samples of steel grade 40X subjected to SPD (approximately 2100 measurements, 34.4, $D = 35,5$ mm, $d = 25$ mm, $l = 26$ mm): The external surfaces of the samples were polished at a circumferential speed of $V = 52.2$ m/min under normal force values $F = 100, 200, 300, 400, 500$ N. For each normal force F, three corresponding long-range values of the fastening tool have been taken: $S_{1,2,3} = 0.050,\ \ 0.100$ and 0.155 mm/rev. After deformation of the external surfaces, mechanical processing was carried out with fine grinding at an angle $\alpha = 2°20'$ relative to the axis of the roller. The resulting surface was subjected to smoothing and a sloped platform with an elliptical outline was obtained to perform HV measurements.

In order to ensure necessary accuracy and take into account the sharp change of the HV gradient at the beginning of the platform ($\Delta h_i \to 0$), the measurement step of HV in that section is reduced ($t = 0.2$ mm) and formed into a grid scheme (Fig. 34.4), as well as to obtain the multi-dimensional coordinate system $HV = \varphi_2\,(\Delta h,\ x, y)$

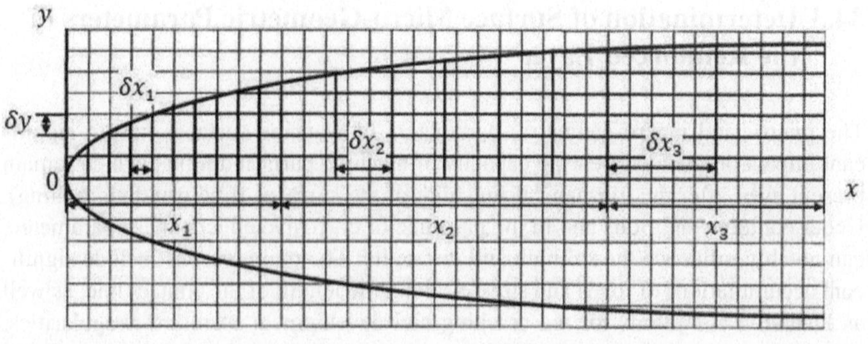

Fig. 34.4: The network scheme for measuring HV on the inclined platform.

function. The grid scheme in the $(\Delta h,\ x, y)$ system is adjusted to the characteristics of HV changes, which allows for a relatively accurate assessment of the relationships of experimental data with an optimal quantity of measurements and to reveal the regularities of HV changes.

It is arranged as follows:

- The interval along the Ox axis from x=0 to x=2.0 mm is divided into three segments, where the grid steps δx_i are different. For $x_1 = 0 \ldots 2,0$ mm $\delta x_1 = 0.2$ mm, $x_2 = 2.0 \ldots 5.0$ mm $\delta x_2 = 0.5$ mm, and for $x_3 = 5.0 \ldots 7.0$ mm $\delta x_3 = 1.0$. The following ranges correspond to the maximum depths of the reinforced surface layers $\Delta h_{i\,\max}$. $\Delta h_{1\,\max} = 0 \ldots 0.081$ mm with a step 0.01 mm size; $\Delta h_{2.3\,\max} = 0.081 \ldots 0.28$ mm with a step of 0.02 mm.
- The interval along the Oy axis $y = 3.17$ mm is divided into 6 parts with the same grid step of $\delta y = 0.5$ mm, where $\Delta h_{1,2,3\,\max} = 0 \ldots 0.285$ mm.

On the external surfaces of the samples subjected to different technological modes of SPD, measurements of HV have been performed according to the points indicated in the network diagram (140. . . 142 measurements for each sample), which are classified in orthogonal two-dimensional coordinate systems as (HV-x) and (HV-y).

The advantage of HV measurements on the inclined elliptical surface formed at angle α is the possibility to determine the main parameters of the reinforced surface layer (HV,Δh) under continuous values of Δh. This method is particularly justified from the perspective of generating reliable information regarding the reinforcement of rotating or curved working surfaces and obtaining precise functional relationships based on that.

34.3 Determination of Surface Micro-Geometric Parameters of the Reinforced Layer

The micro- and macro-geometric parameters of working surfaces have a significant impact on the load-bearing capacity of machine parts and joints, which remain present even after the mechanical finishing of the surfaces (fine grinding, honing). Under contact conditions and in the presence of corrosive effects, these parameters can notably influence the strength and durability. The main reasons include significant accumulations of loads and stresses along the length of the contact line, as well as favorable conditions for the cracking and development of micro-irregularities, which can lead to the formation of fatigue damage. In the case of corrosive effects, which are characteristic of a considerable number of production tools (transport vehicles, construction and agricultural machines and equipment, etc.), there is corrosion fatigue degradation.

The SPD has a double positive impact on the stress-deformational state of working surfaces in the specified cases

- it reduces the heights of surface irregularities from (R_α = 2.5 . . . 5.0 mm) to (R_α = 0.12 . . . 0.25 mm) and creates a relatively dense medium at their bases, significantly reducing the microcracking focal points of stresses,
- forming a directed deformed microstructure in the surface layers, eliminating microstructural defects and non-metallic elements.

The mentioned positive processes are conditioned by the application of optimal technological regimes, in which the amount of F normal force is highlighted, which has a positive effect within certain values, as well as the S longitudinal delivery, where the relatively small values lead to a somewhat repeated plastic reinforcement process of the treated surface after one rotation of the surface. In this regard, it is important to reveal such functional connections between the indicators of the physical and mechanical state of the surface and the parameters of the reinforcement technological process, which is provided for in the scientific experiment program.

34.4 Conclusions

A classification and systematic analysis of the fatigue tests conducted in significant volumes in the past has been carried out regarding the fatigue resistance indicators of parts made of 40X grade steel reinforced with SPD technology, and regression equations have been obtained for the specified indicators and the technological parameters of SPD, taking into account the physical and mechanical properties of the reinforced surface layers.

The mathematical modeling of the SPD technological process has been performed, and multi-parametric functional relationships have been obtained to achieve optimal parameters for the calculation, design, preparation, and operation of constructions.

In order to improve the reliability of mathematical models, an experiment has been organized according to the developed methodology:

- Tilted planes have been created on reinforced surfaces of samples, a network scheme has been established, and *HV* microhardness measurements have been conducted (about 2100 measurements). The macrogeometric parameters R_α, R_z, R_{max} have been measured on the same surfaces to obtain functional connections between the physical and mechanical characteristics of the coated surface layers and the technological parameters of reinforcement.

Based on the results obtained, it is possible to solve the following problems:

- by comprehensively accounting for the material, design, technological, and operational measures of the construction, select the optimal parameters of the reinforcement technology,
- designate a justified service life of the machine while ensuring normal working conditions,
- design a construction that has low material consumption, the necessary level of uninterrupted operation, and safety features, taking into account conflicting and synergistic influencing factors.

References

[1] Chibukhchyan SS, Stakyan MG, Torosyan MS (2013) Otsenka nesushchey sposobnosti valov i soyedineniy transportnykh sredstv, podvergnutykh poverkhnostnomu plasticheskomu deformirovaniyu (in Russ.) Assessment of the load-bearing capacity of shafts and joints of vehicles subjected to surface plastic deformation. Transport: Science, Technology, Management, Scientific and informational collection (4):7-13

[2] Dashchenko AF, Kravchuk VS, Iorgachev VD (2004) Nesushchaya sposobnost' uprochneniya detaley mashin (in Russ., Load-bearing capacity of strengthened machine parts). Astroprint, Odessa

[3] Ezhelev AV, Bobrovsky IN, Lukyanov AA (2012) Analiz sposobov obrabotki poverkhnostno-plasticheskim deformirovaniyem (in Russ., Analysis of methods of surface treatment by plastic deformation). Fundamental Research (6-9):642-646

[4] Guo YB, Liu CR (2002) 3D FEA modeling of hard turning. Journal of Manufacturing Science and Engineering 124(2):189-199, DOI 10.1115/1.1430678

[5] Gurov RV (2011) Razrabotka programmy dlya modelirovaniya otdelochnoy i otdelochno-uprochnyayushchey obrabotki PPD (in Russ., Development of a program for modeling finishing and surface hardening treatment of PVD coatings). Izvestiya Tul'skogo gosudarstvennogo universiteta. Tekhnicheskiye nauki (6/2):481-485

[6] Kirpichev VA, Saushkin MN, Smyslov VA (2009) (in Russ., On mathematical modeling of plastic deformation fields arising during various types of strength-

ening treatment). Mathematical modeling of physical, technical, and social systems and processes, Proceedings of the XVII International Conference, Ulyanovsk, pp. 116-117

[7] Lee CW, Shin YC (2000) Evolutionary modeling and optimization of grinding processes. International Journal of Production Research 38(1-2):2787-2813, DOI 10.1080/002075400411484

[8] Pirumyan NV, Torosyan MS, Stakyan MG (2019) Complex changes in fatigue resistance indicators of structures using strengthening technologies (in Armenian). NUASA Bulletin 1(66):58-66

[9] Sistani SJ, Parikyan FH, Babayan AS (2012) Stress state of elements with stress elements under variable complex loading and application of strengthening technologies (in Armenian). Bulletin of NPUA, Yerevan 1:124-128

[10] Sokolov GA, Gladkikh IM (2007) Matematicheskaya statistika (in Russ., Mathematical Statistics), 2nd ed. Ekzamen Publishing House, Moscow

[11] Stakyan MG (2003) Methods of statistical processing of research results (in Armenian). NPUA, Yerevan

[12] Stakyan MG, Jrbashyan TE, Mnatsakanyan AS (1992) Veroyatnostnyye metody rascheta otnositel'noy prochnosti i dolgovechnosti detaley mashin, Algoritmy i programmy (in Russ., Probabilistic methods for calculating the relative strength and durability of machine parts, Algorithms and programs). Information bulletin of the All-Russian Scientific and Technical Center (3):29-38

[13] Stakyan MG, Oganesyan LG, Manukyan GA (1990) Korrelyatsionnyy i regressionnyy analiz rezul'tatov mekhanicheskikh ispytaniy, Algoritmy i programmy (in Russ., Correlation and regression analysis of mechanical test results, Algorithms and programs). Information bulletin of VINITI Center, State Fund of Algorithms and Programs of the USSR 5:(75-81)

[14] Stakyan MG, Sistani SJ, Aykazyan ME (2013) Modelirovaniye fizikomekhanicheskogo sostoyaniya poverkhnostnykh sloyev detaley s primeneniyem uprochnyayushchikh tekhnologiy (in Russ., Modeling of the physicomechanical state of surface layers of parts using strengthening technologies). Bulletin of NPUA. Series: Mechanics, Machine Science, Mechanical Engineering 16(3):52-58

[15] Stepnov MN, Shavrin AV (2005) Statisticheskiye metody obrabotki rezul'tatov mekhanicheskikh ispytaniy: Spravochnik (in Russ., Statistical Methods for Processing the Results of Mechanical Tests: Handbook), 2nd ed. Mashinostroenie Publishing House, Moscow

[16] Stakyan MG, Sistani SJ, Torosyan MS Otsenka prochnosti i dolgovechnosti detaley mashin pri poverkhnostnom plasticheskom deformirovanii (in Russ., Machine elements strength and longevity Assessment in case of surface plastic deformation). Proceedings of the National Academy of Sciences of Armenia and State Engineering University of Armenia, Serie Technical Sciences 66(2):95-102

[17] Torosyan MS, Stakyan MG (2017) Increasing the bearing capacity of structural elements of construction equipment by surface plastic deformation (in Armenian). NUACA 67(4)68-79

Chapter 35
On the Elastic-Plastic Bending of Axisymmetric Plates of the Föppl - von Kármán Type Under Generalized Elastic Embedment Conditions

Konstantin B. Ustinov and David V. Gandilyan

Abstract The stress-strain state of a circular plate, conjugate along the contour with a base, under a constant normal load, is investigated. The problem is considered within the framework of the Föppl - von Kármán model with two types of edge embedment: rigid embedment and generalized elastic embedment type. The Tresca - Saint-Venant criterion is used as a plasticity criterion, in which plastic deformation begins when the tangential stresses reach a critical value. The solution to the problem is obtained using a numerical method similar to the "shooting method".

Key words: Circular plate, Föppl-von Kármán model, Generalized elastic embedment, Elastic-plastic deformation

35.1 Introduction

Studies of deformation of thin-walled structural elements have been conducted since the end of the 19th century and they remain relevant to this day, especially in micro- and nanoelectronics. When studying thin-walled structural elements associated with massive bodies, these elements are conveniently modeled within the framework of two-dimensional or one-dimensional models: plates, shells, beams, rods. Despite the fact that such structures are usually operated in an elastic mode, when critical stresses are reached, brittle or plastic failure is possible under increasing loading. There are quite a large number of works on elastic-plastic research of rods and plates [7–9, 12–15, 21–23, 34–37, 39] however, this topic is still relevant.

In this paper, we consider the initiation of plastic fracture of an axisymmetrically loaded circular plate, conjugate along the contour to the base, in the approximation

Konstantin B. Ustinov (corresponding author) · David V. Gandilyan
Russian Academy of Sciences, Ishlinsky Institute for Problems in Mechanics RAS, Prospekt Vernadskogo, Moscow, 119526, Russian Federation, e-mail: ustinov@ipmnet.ru,david.ghandilyan@mail.ru

H. Altenbach and A. V. Sahakyan, *Contact Mechanics and Relevant Problems*,
Advanced Structured Materials 253, https://doi.org/10.1007/978-3-032-20869-9_35

of ideal plasticity within the framework of the Föppl–von Karman model (taking into account the Kirchhoff–Love hypotheses). In previous studies (e.g. for cylindrically curved rectangular plates [31] and elastic-plastic stability of rods [13]), the relationship between forces, strain and curvature was used, forming a system of ordinary differential equations. However, for plates, these equations are insufficient: there remains uncertainty in the components of the stress tensor. To complete the definition of the system, we decompose the strain into elastic and plastic parts with an associated flow law [40]. This leads to additional terms in the compatibility equations and modifications of the classical Föppl–von Karman equations [17, 27, 42, 43].

In this case, the condition of rigid restraint is usually used as the boundary conditions, justifying this by the "massiveness" of the bodies with which the thin-walled elements under consideration are conjugated. A more detailed examination shows that the condition of rigid restraint is satisfied only approximately, and that more accurate are the conditions of elastic restraint, when the kinematic parameters — the components of the midplane displacement and the rotation angles are related to the force parameters — the longitudinal and transverse forces and bending moments by means of the compliance matrix [1–4, 6, 11, 16, 18–20, 26, 28–30, 32, 33, 41]. The condition of rigid restraint can be obtained as a special case by zeroing the compliance coefficients of the matrix. In this paper, we consider the deformation of a thin layer, coupled along the contour with a massive base having a cylindrical hole, due to the pressure applied to this layer through the hole. Such problems arise in identifying experimentally determined parameters of ultra-thin MEMS (microelectromechanical systems) elements used, in particular, in projection lithographic systems [5, 24, 25, 38]. The problem is reduced to a system of ordinary differential equations and solved numerically, using a method similar to the "shooting method".

35.2 Statement of the Problem

Consider a thin layer of thickness h, conjugated to a base having a cylindrical hole of radius R, from the side of which a constant pressure q is applied to the layer (Fig. 35.1). To solve the problem, a cylindrical coordinate system (r, α, z) is used, where the z axis coincides with the center of the hole. The considered area of the plate corresponds to the values: $z \in [-h/2, h/2]$, $r \in [0, R]$, $\alpha \in [0, 2\pi)$. Due to axial symmetry, one of the tangential components of the displacement $v = 0$, the remaining components of the displacement and stresses depend on two coordinates: $u(r,z)$, $w(r,z)$, $\sigma(r,z)_{ij}$, $i, j = r, \alpha, z$. In the framework of the plate theory, the Kirchhoff-Love hypothesis is used, consisting of two elements: kinematic ($\varepsilon_{rz} = \varepsilon_{\alpha z} = \varepsilon_{zz} = 0$), and force ($\sigma_{zz}(r,z) = 0$).

The total deformation is assumed to be representable as the sum of elastic and plastic components

$$\begin{aligned} \varepsilon^T_{rr} &= \varepsilon_{rr} + \varepsilon^P_{rr}, \\ \varepsilon^T_{\alpha\alpha} &= \varepsilon_{\alpha\alpha} + \varepsilon^P_{\alpha\alpha}, \end{aligned} \tag{35.1}$$

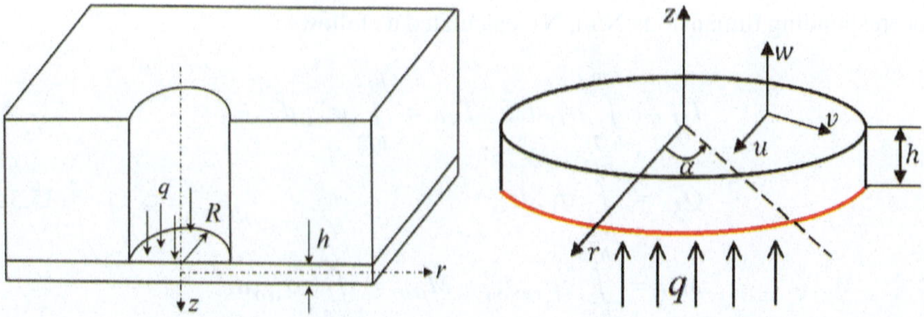

Fig. 35.1: Configuration of the element in question.

where ε_{ij} — components of elastic deformation, ε_{ij}^{P} — components of plastic deformation. Accepting the kinematic hypotheses of Kirchhoff—Love, we write out the dependences of the radial and circumferential components of the total deformation (e.g. [17, 27, 42, 43]) in the median plane ($z = 0$), chosen as the reference

$$\begin{aligned} \varepsilon_{rr}^{T} &= \frac{du}{dr} + \frac{1}{2}\left(\frac{dw}{dr}\right)^2 - z\frac{d^2w}{dr^2}, \\ \varepsilon_{\alpha\alpha}^{T} &= \frac{u}{r} - z\frac{1}{r}\frac{dw}{dr}. \end{aligned} \tag{35.2}$$

It is worth emphasizing that this decomposition (35.1) is not rigorous; to obtain a more accurate result, it is necessary to use a multiplicative decomposition, but when compared with the additive (35.1), it leads to the appearance of additional, insignificant terms that can be discarded within the framework of the current approximation (see, for example, [17]). The compatibility equations must necessarily be satisfied for the total deformation, but are not required for its elastic or plastic component separately.

Expressions for the components of the stress tensor (the second Piola— Kirchhoff tensor) can be written as the following constitutive relations

$$\sigma_{rr} = \bar{E}\left[\frac{du}{dr} + \frac{1}{2}\left(\frac{dw}{dr}\right)^2 + \nu\frac{u}{r} - z\left(\frac{d^2w}{dr^2} + \frac{\nu}{r}\frac{dw}{dr}\right)\right] - \bar{E}\varepsilon_{rr}^{P} - \bar{E}\nu\varepsilon_{\alpha\alpha}^{P}, \tag{35.3}$$

$$\sigma_{\alpha\alpha} = \bar{E}\left[\frac{u}{r} + \nu\frac{du}{dr} + \frac{\nu}{2}\left(\frac{dw}{dr}\right)^2 - z\left(\frac{1}{r}\frac{dw}{dr} + \nu\frac{d^2w}{dr^2}\right)\right] - \bar{E}\nu\varepsilon_{rr}^{P} - \bar{E}\varepsilon_{\alpha\alpha}^{P}, \tag{35.4}$$

where

$$\bar{E} = \frac{E}{1-\nu^2}.$$

E, ν are the Young's modulus and Poisson's ratio of the plate material. In the theory of shells and plates, instead of the usual concepts of "force" and "moment", integral characteristics such as "force" and "bending moment" (are used with the corresponding dimensions N/m, N), calculated as follows

$$
\begin{aligned}
T_{rr} &= \int_{-h/2}^{h/2} \sigma_{rr} dz, \quad T_{\alpha\alpha} = \int_{-h/2}^{h/2} \sigma_{\alpha\alpha} dz, \\
Q_{rz} &= \int_{-h/2}^{h/2} \sigma_{rz} dz, \\
M_{rr} &= \int_{-h/2}^{h/2} z\sigma_{rr} dz, \quad M_{\alpha\alpha} = \int_{-h/2}^{h/2} z\sigma_{\alpha\alpha} dz
\end{aligned}
\tag{35.5}
$$

for which the equilibrium equations are valid (e. g. [27])

$$\frac{dM_{rr}}{dr} + \frac{M_{rr} - M_{\alpha\alpha}}{r} - Q_{rz} = 0, \tag{35.6}$$

$$\frac{dT_{rr}}{dr} + \frac{T_{rr} - T_{\alpha\alpha}}{r} = 0, \tag{35.7}$$

$$\frac{dQ_{rz}}{dr} + \frac{Q_{rz}}{r} + \frac{d}{dr}\left(T_{rr}\frac{dw}{dr}\right) + \frac{1}{r}\left(T_{rr}\frac{dw}{dr}\right) + q = 0. \tag{35.8}$$

Substituting the expressions for stresses (35.3), (35.4) into the expressions for integral forces and moments (35.6) gives

$$T_{rr} = \bar{E}h\left[\frac{du}{dr} + \frac{1}{2}\left(\frac{dw}{dr}\right)^2 + \nu\frac{u}{r}\right] - \bar{E}\int_{-h/2}^{h/2} \varepsilon_{rr}^P dz - \bar{E}\nu\int_{-h/2}^{h/2} \varepsilon_{\alpha\alpha}^P dz, \tag{35.9}$$

$$T_{\alpha\alpha} = \bar{E}h\left[\frac{u}{r} + \nu\frac{du}{dr} + \frac{\nu}{2}\left(\frac{dw}{dr}\right)^2\right] - \bar{E}\nu\int_{-h/2}^{h/2} \varepsilon_{rr}^P dz - \bar{E}\int_{-h/2}^{h/2} \varepsilon_{\alpha\alpha}^P dz, \tag{35.10}$$

$$M_{rr} = -D\left(\frac{d^2w}{dr^2} + \frac{\nu}{r}\frac{dw}{dr}\right) - \bar{E}\int_{-h/2}^{h/2} \varepsilon_{rr}^P z dz - \bar{E}\nu\int_{-h/2}^{h/2} \varepsilon_{\alpha\alpha}^P z dz, \tag{35.11}$$

$$M_{\alpha\alpha} = -D\left(\nu\frac{d^2w}{dr^2} + \frac{1}{r}\frac{dw}{dr}\right) - \bar{E}\nu\int_{-h/2}^{h/2} \varepsilon_{rr}^P z dz - \bar{E}\int_{-h/2}^{h/2} \varepsilon_{\alpha\alpha}^P z dz, \tag{35.12}$$

where

$$D = \frac{Eh^3}{12\left(1 - \nu^2\right)} = \frac{\bar{E}h^3}{12}.$$

The obtained 7 equations (35.6) - (35.12) with respect to 7 unknowns u, w, M_{rr}, $M_{\alpha\alpha}$, $T_{rr}, T_{\alpha\alpha}, Q_{rz}$ form a closed system for given plastic deformations ε_{rr}^P, $\varepsilon_{\alpha\alpha}^P$. For unknowns ε_{rr}^P, $\varepsilon_{\alpha\alpha}^P$, to close the system, it is necessary to specify the governing equations for them.

To specify the boundary conditions at the edge of the circular plate, we use the boundary conditions of generalized elastic embedment [11, 32]:

$$\bar{E}\begin{pmatrix} u(R) \\ -hw'(R) \\ w(R) \end{pmatrix} = \begin{pmatrix} a_{11}\ a_{12}\ a_{13} \\ a_{12}\ a_{22}\ a_{23} \\ a_{13}\ a_{23}\ a_{33} \end{pmatrix}\begin{pmatrix} -T_{rr} \\ h^{-1}M_{rr} \\ Q_{rz} \end{pmatrix}, \quad (35.13)$$

where a_{ij} are the coefficients of the compliance matrix [1–4, 6, 11, 16, 18–20, 26, 28–30, 32, 33, 41]. For $a_{ij} = 0$ we obtain the boundary conditions corresponding to rigid restraint. Note that the system (35.6) - (35.12) is not consistent even in the absence of plastic deformations; the adopted kinematic constraints contradict the stress equilibrium equation. In this paper, an approach is used in which local equilibrium equations are neglected and only the global equilibrium equations (35.6) - (35.8) are required to be satisfied.

35.3 Solution Within the Framework of Elasticity Theory

To solve this problem within the framework of elasticity theory, there are a sufficient number of approaches that allow obtaining analytical, semi-analytical and numerical solutions of varying degrees of rigor [11, 32]:

- Model of a plate without taking into account longitudinal forces (Sophie Germain model).
  The system of equations (35.6) - (35.8) is reduced to a single equation with terms responsible for bending, which allows one to obtain an analytical solution even for the boundary conditions of elastic embedment (35.13).
- Membrane model.
  This model assumes the predominance of the role of longitudinal forces. In Eqs. (35.6) - (35.8), the terms associated with bending are discarded. The solution for taking into account the boundary conditions (35.13) is also obtained analytically, assuming the constancy of forces $T_{rr} = const$. It is worth emphasizing that in this model the assumption of constancy of membrane forces does not allow to satisfy exactly the equations of compatibility of deformations, however, this model is used in approximate calculations, where longitudinal forces are significant.
- Föppl-von Kármán model in the approximation of constant longitudinal forces.
  Here, the constancy of longitudinal forces T_{rr} = const is also assumed, with the help of which it is possible to obtain an analytical representation of the solution of the displacement components through modified Bessel functions and hypergeometric functions, and from the boundary conditions (35.13), using a numerical method for solving a system of nonlinear equations, the necessary unknown quantities are found.
- Complete model of Föppl — von Kármán; solution in series.
  From the problem statement (due to symmetry), we can represent the longitudinal force component T_{rr} and the derivative of the normal displacement component

$\frac{dw}{dr}$ as power series with respect to the radial coordinate r [11, 27]. Substituting these expressions into the main equations (35.6) - (35.8), we obtain a system in which, by grouping terms with the same powers in r, we arrive at recurrence relations, where the unknown coefficients are determined through the boundary conditions (35.13). The required number of terms in the series is determined by the value of the deflection at the center of the plate for n and the $(n + 1)$th term.

The methods described are quite effective under certain loading conditions within the framework of elasticity theory, however, for the elastic-plastic case, to ensure the continuity of the solution, it is necessary to use a numerical method similar to the "shooting method", which for the elastic case is described in detail in [11].

35.4 Plasticity Condition

Plastic deformation will be considered within the framework of the Tresca-Saint-Venant theory, according to which the onset of plastic deformation corresponds to the maximum shear stresses reaching a critical value. Within the framework of the accepted plate theory (namely, neglecting the shear stresses σ_{rz} compared to the normal σ_{rr}, $\sigma_{\alpha\alpha}$) it follows that the condition for the shear stresses to reach a limiting value σ_c becomes equivalent to the condition for the stress components σ_{rr}, or $\sigma_{\alpha\alpha}$, or both, to reach a critical value. When deforming a plate in the considered formulation of the problem, the transition to a plastic state can be expected at the edges of the plate and in its center (Fig. 35.2), areas II and III, respectively (area I corresponds to the elastic state), and, as the analysis of the elastic state [11] shows, in area II, namely at the lower edge of the plate (at $z = -h/2$, $r = R$) the critical stress occurs earlier.

Some potential cases are shown in Fig. 35.3, where case (a) corresponds to the elastic state, cases (b) – (f) — states in which part of the material in the section becomes plastic. Consider the case of transition to plasticity along the lower edge of the plate (Fig. 35.3b), where all quantities related to this case are denoted with index 1. In this case, in region I (Fig. 35.2) there arises a zone $-h/2 < z < z_1(r)$, in which stresses σ_{rr} reach the critical value $\sigma_{rr} = \sigma_c$ and remain equal to it, and the stress component $\sigma_{\alpha\alpha}$ remains less than the critical value. Consequently, plastic deformations $\varepsilon^P_{rr}(r, z)$ appear, while circumferential plastic deformations are absent $\varepsilon^P_{\alpha\alpha}(r, z) = 0$. To satisfy the condition $\sigma_{rr} = \sigma_c$ according to (35.3), the plastic deformations must have the form

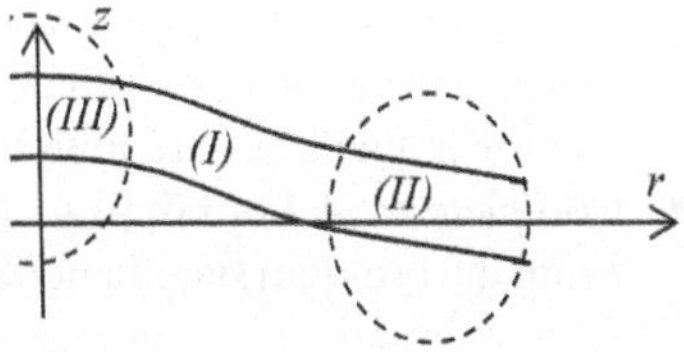

Fig. 35.2 Possible areas of occurrence of plastic deformations.

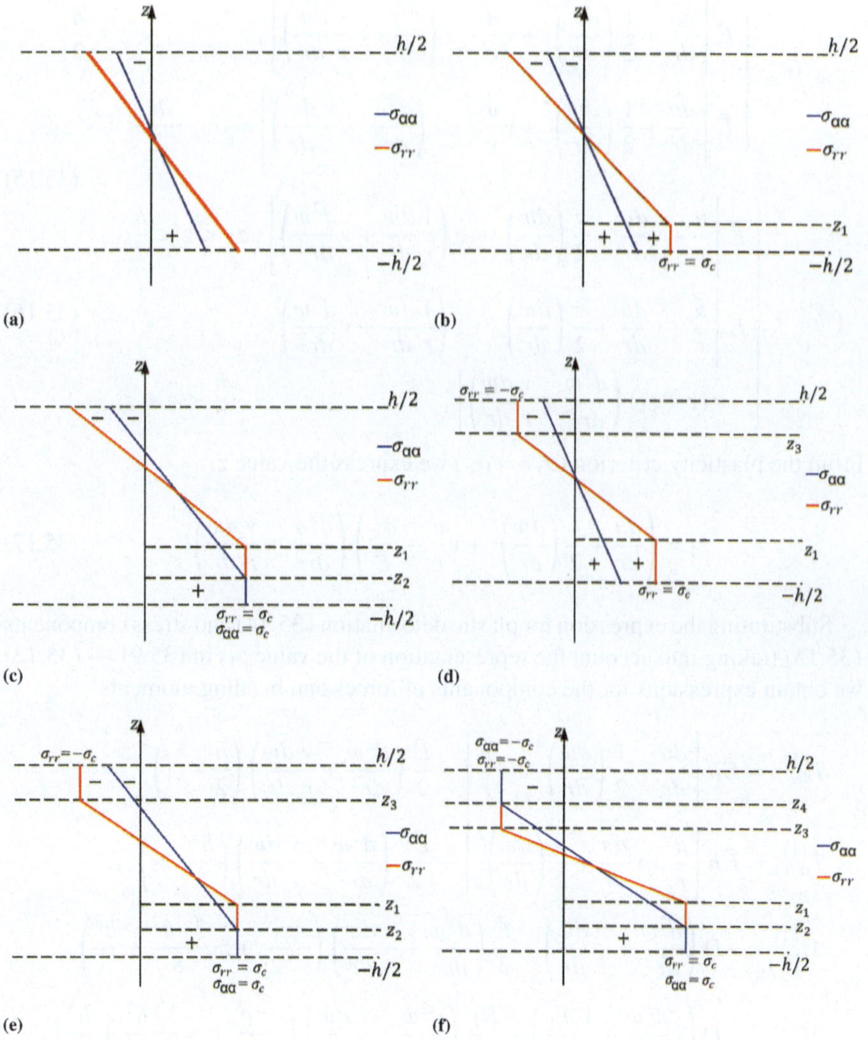

Fig. 35.3: Distribution of stress components σ_{rr}, $\sigma_{\alpha\alpha}$ in the cross-section.

$$
\varepsilon_{rr}^{P(1)} = \begin{cases} 0, & z_1 < z < \dfrac{h}{2}, \\ -(z - z_1)\left(\dfrac{d^2w}{dr^2} + \dfrac{\nu}{r}\dfrac{dw}{dr}\right), & -\dfrac{h}{2} < z < z_1, \end{cases} \tag{35.14}
$$
$$
\varepsilon_{\alpha\alpha}^{P(1)} = 0
$$

accordingly, the stress components will have the form

$$\sigma_{rr}^{(1)} = \begin{cases} \bar{E}\left[\dfrac{du}{dr} + \dfrac{1}{2}\left(\dfrac{dw}{dr}\right)^2 + \nu\dfrac{u}{r} - z\left(\dfrac{d^2w}{dr^2} + \dfrac{\nu}{r}\dfrac{dw}{dr}\right)\right], & z_1 < z < \dfrac{h}{2}, \\ \bar{E}\left[\dfrac{du}{dr} + \dfrac{1}{2}\left(\dfrac{dw}{dr}\right)^2 + \nu\dfrac{u}{r} - z_1\left(\dfrac{d^2w}{dr^2} + \dfrac{\nu}{r}\dfrac{dw}{dr}\right)\right] = \sigma_c, & -\dfrac{h}{2} < z < z_1, \end{cases} \tag{35.15}$$

$$\sigma_{\alpha\alpha}^{(1)} = \begin{cases} \bar{E}\left[\dfrac{u}{r} + \nu\dfrac{du}{dr} + \dfrac{\nu}{2}\left(\dfrac{dw}{dr}\right)^2 - z\left(\dfrac{1}{r}\dfrac{dw}{dr} + \nu\dfrac{d^2w}{dr^2}\right)\right], & z_1 < z < \dfrac{h}{2}, \\ \bar{E}\left[\dfrac{u}{r} + \nu\dfrac{du}{dr} + \dfrac{\nu}{2}\left(\dfrac{dw}{dr}\right)^2 - z\left(\dfrac{1}{r}\dfrac{dw}{dr} + \nu\dfrac{d^2w}{dr^2}\right)\right. & \\ \left. +\nu\left(z - z_1\right)\left(\dfrac{d^2w}{dr^2} + \dfrac{\nu}{r}\dfrac{dw}{dr}\right)\right], & -\dfrac{h}{2} < z < z_1, \end{cases} \tag{35.16}$$

From the plasticity criterion ($\sigma_{rr} = \sigma_c$) we express the value z_1

$$z_1 = \left(\frac{du}{dr} + \frac{1}{2}\left(\frac{dw}{dr}\right)^2 + \nu\frac{u}{r} - \frac{\sigma_c}{\bar{E}}\right)\left(\frac{d^2w}{dr^2} + \frac{\nu}{r}\frac{dw}{dr}\right)^{-1}. \tag{35.17}$$

Substituting the expression for plastic deformation (35.14) and stress components (35.15) (taking into account the representation of the value z_1) in (35.9) — (35.12) we obtain expressions for the components of forces and bending moments

$$\begin{aligned} T_{rr}^{(1)} &= \bar{E}h\left[\frac{du}{dr} + \frac{1}{2}\left(\frac{dw}{dr}\right)^2 + \nu\frac{u}{r}\right] - \frac{\bar{E}}{2}\left(\frac{d^2w}{dr^2} + \frac{\nu}{r}\frac{dw}{dr}\right)\left(\frac{h}{2} + z_1\right)^2, \\ T_{\alpha\alpha}^{(1)} &= \bar{E}h\left[\frac{u}{r} + \nu\frac{du}{dr} + \frac{\nu}{2}\left(\frac{dw}{dr}\right)^2\right] - \frac{\bar{E}\nu}{2}\left(\frac{d^2w}{dr^2} + \frac{\nu}{r}\frac{dw}{dr}\right)\left(\frac{h}{2} + z_1\right)^2, \\ M_{rr}^{(1)} &= -D\left(\frac{d^2w}{dr^2} + \frac{\nu}{r}\frac{dw}{dr}\right) + \frac{\bar{E}}{3}\left(\frac{d^2w}{dr^2} + \frac{\nu}{r}\frac{dw}{dr}\right)\left(-\frac{{z_1}^3}{2} + z_1\frac{3\cdot h^2}{8} + \frac{h^3}{8}\right), \\ M_{\alpha\alpha}^{(1)} &= -D\left(\nu\frac{d^2w}{dr^2} + \frac{1}{r}\frac{dw}{dr}\right) + \frac{\bar{E}\nu}{3}\left(\frac{d^2w}{dr^2} + \frac{\nu}{r}\frac{dw}{dr}\right)\left(-\frac{{z_1}^3}{2} + z_1\frac{3\cdot h^2}{8} + \frac{h^3}{8}\right). \end{aligned} \tag{35.18}$$

Also consider the case shown in Fig. 3d (in this case, all the corresponding quantities are denoted with index 3), when an increase in load can cause the formation of an additional plastic zone on the opposite side of the plate, in the region of $z = h/2$, $r = R$. In this case, a region $z_3 < z < h/2$ arises, in which the stress component σ_{rr} reaches the critical value $\sigma_{rr} = -\sigma_c$ and remains equal to it. In this case, the stress component $\sigma_{\alpha\alpha}$ also remains below the critical value everywhere. According to the considered model (the associated law used), only plastic deformations $\varepsilon_{rr}^P(r, z)$ appear, and plastic deformations $\varepsilon_{\alpha\alpha}^P(r, z)$ are absent $\left(\varepsilon_{\alpha\alpha}^P = 0\right)$. Also, to satisfy the condition $\sigma_{rr} = -\sigma_c$ according to (35.3, plastic deformations must have the form

$$\varepsilon_{rr}^{P(2)} = \begin{cases} -(z - z_3)\left(\dfrac{d^2w}{dr^2} + \dfrac{\nu}{r}\dfrac{dw}{dr}\right), & z_3 < z < \dfrac{h}{2}, \\ 0, & z_1 < z < z_3 \\ -(z - z_1)\left(\dfrac{d^2w}{dr^2} + \dfrac{\nu}{r}\dfrac{dw}{dr}\right), & -\dfrac{h}{2} < z < z_1 \end{cases} \tag{35.19}$$

$$\varepsilon_{\alpha\alpha}^{P(2)} = 0$$

accordingly, the voltage components will have the form

$$,\sigma_{rr}^{(2)} = \begin{cases} \bar{E}\left[\dfrac{du}{dr} + \dfrac{1}{2}\left(\dfrac{dw}{dr}\right)^2 + \nu\dfrac{u}{r} - z_3\left(\dfrac{d^2w}{dr^2} + \dfrac{\nu}{r}\dfrac{dw}{dr}\right)\right] = -\sigma_c, & z_3 < z < \dfrac{h}{2}, \\ \bar{E}\left[\dfrac{du}{dr} + \dfrac{1}{2}\left(\dfrac{dw}{dr}\right)^2 + \nu\dfrac{u}{r} - z\left(\dfrac{d^2w}{dr^2} + \dfrac{\nu}{r}\dfrac{dw}{dr}\right)\right], & z_1 < z < z_3, \\ \bar{E}\left[\frac{du}{dr} + \frac{1}{2}\left(\frac{dw}{dr}\right)^2 + \nu\frac{u}{r} - z_1\left(\frac{d^2w}{dr^2} + \frac{\nu}{r}\frac{dw}{dr}\right)\right] = \sigma_c, & -\dfrac{h}{2} < z < z_1, \end{cases} \tag{35.20}$$

$$\sigma_{\alpha\alpha}^{(2)} = \begin{cases} \bar{E}\left[\dfrac{u}{r} + \nu\dfrac{du}{dr} + \dfrac{\nu}{2}\left(\dfrac{dw}{dr}\right)^2 - z\left(\dfrac{1}{r}\dfrac{dw}{dr} + \nu\dfrac{d^2w}{dr^2}\right)\right. \\ \left. +\nu(z - z_3)\left(\dfrac{d^2w}{dr^2} + \dfrac{\nu}{r}\dfrac{dw}{dr}\right)\right], & z_3 < z < \dfrac{h}{2}, \\ \bar{E}\left[\dfrac{u}{r} + \nu\dfrac{du}{dr} + \dfrac{\nu}{2}\left(\dfrac{dw}{dr}\right)^2 - z\left(\dfrac{1}{r}\dfrac{dw}{dr} + \nu\dfrac{d^2w}{dr^2}\right)\right], & z_1 < z < z_3, \\ \bar{E}\left[\dfrac{u}{r} + \nu\dfrac{du}{dr} + \dfrac{\nu}{2}\left(\dfrac{dw}{dr}\right)^2 - z\left(\dfrac{1}{r}\dfrac{dw}{dr} + \nu\dfrac{d^2w}{dr^2}\right)\right. \\ \left. +\nu(z - z_1)\left(\dfrac{d^2w}{dr^2} + \dfrac{\nu}{r}\dfrac{dw}{dr}\right)\right], & -\dfrac{h}{2} < z < z_1. \end{cases} \tag{35.21}$$

From the plasticity criterion ($\sigma_{rr} = -\sigma_c$) we express the value z_3

$$z_3 = \left(\frac{du}{dr} + \frac{1}{2}\left(\frac{dw}{dr}\right)^2 + \nu\frac{u}{r} + \frac{\sigma_c}{\bar{E}}\right)\left(\frac{d^2w}{dr^2} + \frac{\nu}{r}\frac{dw}{dr}\right)^{-1}. \tag{35.22}$$

Similarly, substituting the expressions for plastic deformation (35.19) and stress components (35.20), (35.21) (taking into account the representation of the quantities z_1 from (35.17) and z_3 from (35.22)) in (35.9) - (35.12) we also obtain expressions for the components of forces and bending moments

$$
\begin{aligned}
T_{rr}^{(2)} &= (z_1 + z_3)\,\sigma_c = 2\left(\frac{du}{dr} + \frac{1}{2}\left(\frac{dw}{dr}\right)^2 + \nu\frac{u}{r}\right)\left(\frac{d^2w}{dr^2} + \frac{\nu}{r}\frac{dw}{dr}\right)^{-1}\sigma_c,\\
T_{\alpha\alpha}^{(2)} &= \bar{E}h\left[\frac{u}{r} + \nu\frac{du}{dr} + \frac{\nu}{2}\left(\frac{dw}{dr}\right)^2\right] - \frac{\bar{E}\nu}{2}\left(\frac{d^2w}{dr^2} + \frac{\nu}{r}\frac{dw}{dr}\right)(h + z_1 - z_3)(z_1 + z_3)\\
&= \bar{E}h\left[\frac{u}{r} + \nu\frac{du}{dr} + \frac{\nu}{2}\left(\frac{dw}{dr}\right)^2\right]\\
&\quad - \bar{E}\nu\left[h - \frac{2\sigma_c}{\bar{E}\left(\frac{d^2w}{dr^2} + \frac{\nu}{r}\frac{dw}{dr}\right)}\right]\left(\frac{du}{dr} + \frac{1}{2}\left(\frac{dw}{dr}\right)^2 + \nu\frac{u}{r}\right),\\
M_{rr}^{(2)} &= -D\left(\frac{d^2w}{dr^2} + \frac{\nu}{r}\frac{dw}{dr}\right) + \frac{\bar{E}}{3}\left(\frac{d^2w}{dr^2} + \frac{\nu}{r}\frac{dw}{dr}\right)\\
&\quad \times\left(-\frac{{z_1}^3}{2} + \frac{{z_3}^3}{2} + \frac{h^3}{4} + z_1\frac{3\cdot h^2}{8} - z_3\frac{3\cdot h^2}{8}\right),\\
M_{\alpha\alpha}^{(2)} &= -D\left(\nu\frac{d^2w}{dr^2} + \frac{1}{r}\frac{dw}{dr}\right) + \frac{\bar{E}\nu}{3}\left(\frac{d^2w}{dr^2} + \frac{\nu}{r}\frac{dw}{dr}\right)\\
&\quad \times\left(-\frac{{z_1}^3}{2} + \frac{{z_3}^3}{2} + \frac{h^3}{4} + z_1\frac{3\cdot h^2}{8} - z_3\frac{3\cdot h^2}{8}\right).
\end{aligned}
\tag{35.23}
$$

The remaining cases marked in Fig. 35.3 can theoretically occur, however, in this formulation of the problem, the most probable outcomes are cases (a), (b), (d) from Fig. 35.3.

35.5 Solution Algorithm

As noted in the section on the solution in a purely elastic state, an algorithm similar to the numerical "shooting method" was used to ensure a continuous solution to the problem. Before its implementation, dimensionless quantities were used for convenience

$$\rho = \frac{r}{R},$$

where

$$
\begin{aligned}
r = 0 &\Rightarrow \rho = 0,\\
r = R &\Rightarrow \rho = 1
\end{aligned}
$$

and

$$\tilde{u}(\rho) = \frac{u(r)}{R}, \quad \tilde{w}(\rho) = \frac{w(r)}{R}, \quad p = \frac{q}{\bar{E}},$$
$$\tilde{T}_{rr} = \frac{T_{rr}}{\bar{E}h}, \quad \tilde{T}_{\alpha\alpha} = \frac{T_{\alpha\alpha}}{\bar{E}h}, \quad \tilde{M}_{rr} = \frac{M_{rr}}{\bar{E}h^2}, \quad \tilde{M}_{\alpha\alpha} = \frac{M_{\alpha\alpha}}{\bar{E}h^2}.$$

To implement the numerical method, the system of differential equations (35.6) - (35.8) was reduced to a system of two differential equations with two unknown quantities $\tilde{u}\,(r)$, $\tilde{w}\,(r)$ with given starting values

$$\begin{cases} \tilde{w}'(\rho = \Delta\rho) \approx \tilde{w}'(\rho = 0) + \Delta\rho \cdot \tilde{w}''(\rho = \Delta\rho) = \Delta\rho \cdot \tilde{w}''(\rho = \Delta\rho), \\ \tilde{w}''(\rho = \Delta\rho), \\ \tilde{u}(\rho = \Delta\rho) \approx \tilde{u}(\rho = 0) + \Delta\rho \cdot \tilde{u}'(\rho = \Delta\rho) = \Delta\rho \cdot \tilde{u}'(\rho = \Delta\rho), \\ \tilde{u}'(\rho = \Delta\rho). \end{cases}$$

where the quantities $\tilde{u}'(\rho = \Delta\rho)$, $\tilde{w}''(\rho = \Delta\rho)$ are generally set randomly, $\Delta\rho$ is the interval of "advancement" from the center to the edge of the circular plate. Then, using any known numerical method [10], for example, the Runge-Kutta method of the 4th order, one can obtain a solution to the problem. In this case, it is important to check at each step of "advancement" from the center to the edge of the circular plate whether the condition for the transition to plasticity from the possible cases indicated in Fig. 35.3 is met, and when this criterion is met (the Tresca - Saint-Venant criterion), it is necessary to solve a similar system of differential equations for the corresponding case. If the obtained solution satisfies the boundary conditions (35.13) with sufficient accuracy, then we consider the solution to be acceptable, otherwise we take other starting values. For a faster solution to the problem under consideration, we can take the starting values corresponding to a similar elastic problem [11] as starting values. Then, by executing the algorithm of the numerical method in the specified neighborhood of the starting values, we obtain a solution to the problem that satisfies the boundary conditions with sufficient accuracy.

35.6 Results

In the calculations, aluminum was taken as the material of the circular plate with the corresponding values of Young's modulus: E = 70 GPa, Poisson's ratio: $\nu = 0.35$ and plastic limit: 175 MPa, and the ratio of the plate thickness to its radius was considered $h/R = 1/150$. The values of the compliance matrix coefficients were taken for the case of an aluminum plate on a silicon base [32]:

$$a_{11} = 10, \; a_{12} = a_{21} = 1.7, \; a_{22} = 12, \; a_{13} = 0, \; a_{23} = 0, \; a_{33} = 0.$$

The choice of such materials is due to the interest in identifying the parameters of ultra-thin MEMS elements used in projection lithographic systems.

The dependence of the main parameters on the dimensionless value of the applied load p is presented in Tables 35.1 and 35.2, Figs. 35.4 - 35.7; the presented parameters include

Table 35.1: Dependence of parameters on the magnitude of the applied normal load. Rigid embedment.

Type of state	p	$\tilde{w}(0)$	$\tilde{T}_{rr}(1)$	$\tilde{M}_{rr}(1)$	ρ_{pl1}	ρ_{pl2}
(a)	1.0E-07	1.519E-02	1.268E-04	-1.385E-04	-	-
(a)	2.0E-07	1.968E-02	2.244E-04	-2.254E-04	-	-
(a)	3.0E-07	2.277E-02	3.091E-04	-2.986E-04	-	-
(b)	4.0E-07	2.521E-02	3.862E-04	-3.636E-04	9.841E-01	-
(d)	5.0E-07	2.733E-02	4.567E-04	-4.198E-04	9.714E-01	9.982E-01
(d)	6.0E-07	2.927E-02	5.241E-04	-4.637E-04	9.636E-01	9.914E-01
(d)	7.0E-07	3.113E-02	5.872E-04	-4.924E-04	9.597E-01	9.880E-01
(d)	7.5E-07	3.204E-02	6.178E-04	-4.993E-04	9.589E-01	9.874E-01

Table 35.2: Dependence of parameters on the magnitude of the applied normal load. Elastic embedment.

Type of state	p	$\tilde{w}(0)$	$\tilde{T}_{rr}(1)$	$\tilde{M}_{rr}(1)$	ρ_{pl1}	ρ_{pl2}
(a)	1.0E-07	1.556E-02	1.235E-04	-1.348E-04	-	-
(a)	2.0E-07	2.014E-02	2.193E-04	-2.171E-04	-	-
(a)	3.0E-07	2.330E-02	3.022E-04	-2.858E-04	-	-
(b)	4.0E-07	2.581E-02	3.773E-04	-3.467E-04	9.885E-01	-
(b)	5.0E-07	2.795E-02	4.472E-04	-3.999E-04	9.754E-01	-
(d)	6.0E-07	2.988E-02	5.125E-04	-4.443E-04	9.670E-01	9.944E-01
(d)	7.0E-07	3.168E-02	5.747E-04	-4.775E-04	9.620E-01	9.899E-01
(d)	8.0E-07	3.344E-02	6.336E-04	-4.959E-04	9.597E-01	9.881E-01

- type of state: elasticity (case (a) from Fig. 35.3), presence of a plastic zone on one side, i.e. when one of the stress components σ_{rr} from below reaches the yield point σ_c (case (b) from Fig. 35.3); presence of a plastic zone on both sides, i.e. when the stress component σ_{rr} from below and above reaches the yield point (case (d) from Fig. 35.3);
- $\tilde{w}(0)$ - dimensionless value of deflection at the center of the plate;
- $\tilde{T}_{rr}(\rho = 1)$ - dimensionless value of the radial component of the force at the edge of the plate;
- $\tilde{M}_{rr}(\rho = 1)$ - dimensionless value of the radial component of the bending moment at the edge of the plate;
- ρ_{pl1} - radial coordinate where the plastic region begins from below;
- ρ_{pl2} - radial coordinate where the plastic region begins from above.

Figures 35.4 and 35.5 show the graphs of the deflection at the center (the buckling value) and the bending moment at the edge of the plate when solving the problem without taking into account the effect of plasticity (purely elastic solution) and

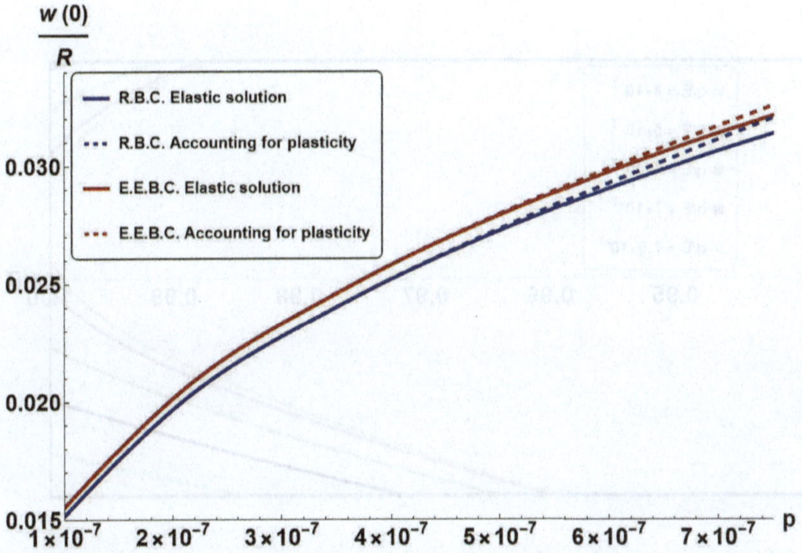

Fig. 35.4: Deflection in the center of the plate depending on the specified load: red curves — in case of elastic embedment, black curves — in case of rigid embedment.

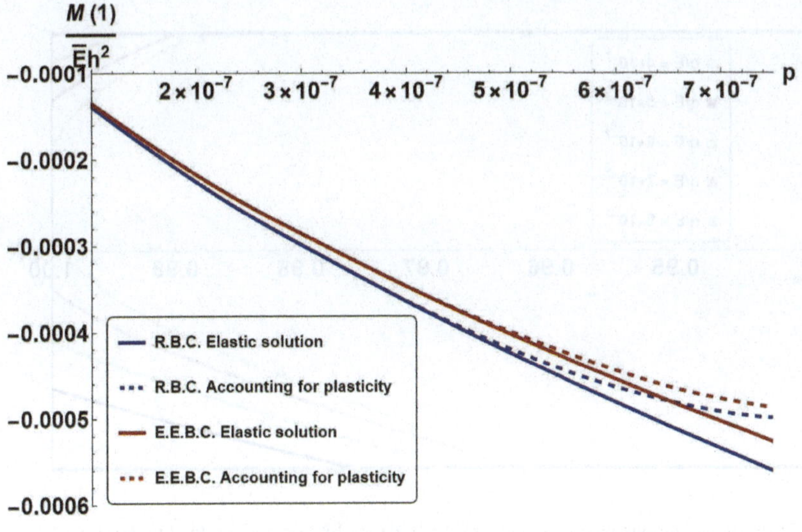

Fig. 35.5: Bending moment at the edge of the plate depending on the specified load: red curves — in case of elastic embedment, black curves — in case of rigid embedment.

taking into account plasticity depending on the specified normal load for the case of rigid and generalized elastic embedment (r.b.c. - rigid type of boundary condition, e.e.b.c. - elastic embedment type of boundary condition). Figures 35.6 and 35.7

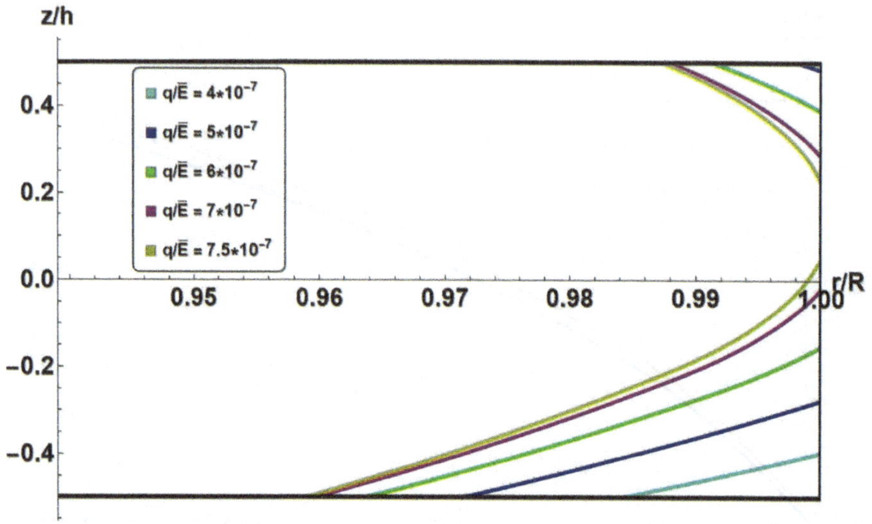

Fig. 35.6: Plasticity regions when solving a problem depending on a given load under the condition of rigid embedment.

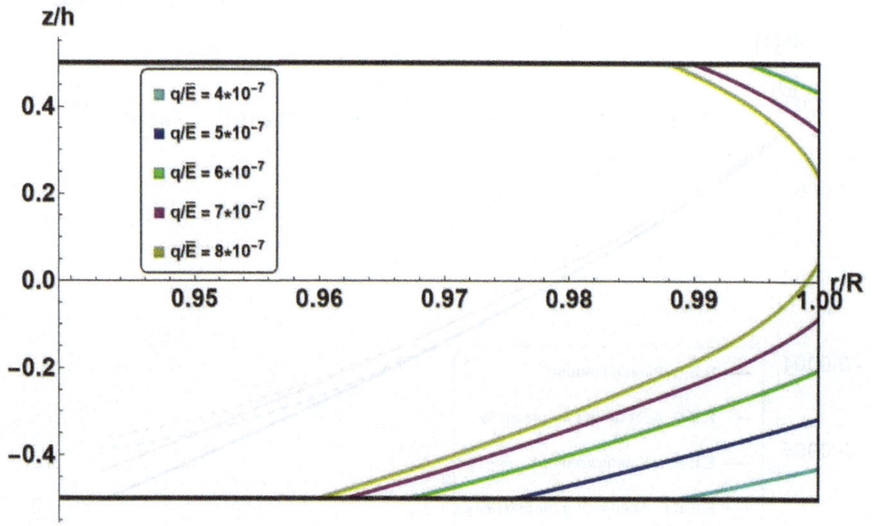

Fig. 35.7: Plasticity regions when solving a problem depending on a given load under the condition of generalized elastic embedment.

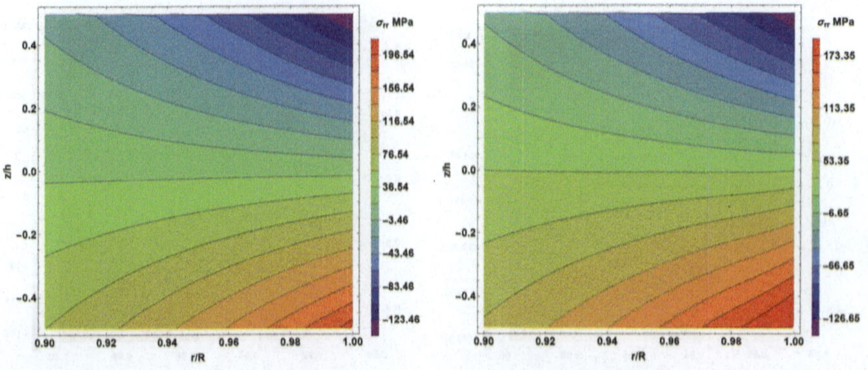

(a) Within the framework of elasticity theory.

(b) Elastic-plastic solution.

Fig. 35.8: Distribution of the stress component σ_{rr} at the edge of the plate for the dimensionless load value $p = 4.0E - 07$ in the case of rigid embedment.

show the location of the plasticity regions for different values of the specified load under the condition of rigid and elastic embedment. Figures 35.8 - 35.11 show the spectral graphs of the stress component σ_{rr} for different values of the transverse load under the condition of rigid and elastic embedment in the case of purely elastic and elastic-plastic states.

From the results it can be concluded that with an increase in the normal load the plastic zone also increases, first in the edge zone from below (tensile zone), then from above (compression zone). Also, with rigid fixation the plastic state is achieved at lower values of the normal load than with a generalized elastic embedment, and from these calculations, at a load value of $p \geq 8.0E - 07$ there is an "intersection" of the plastic zones from above and below in the case of rigid embedment (a different

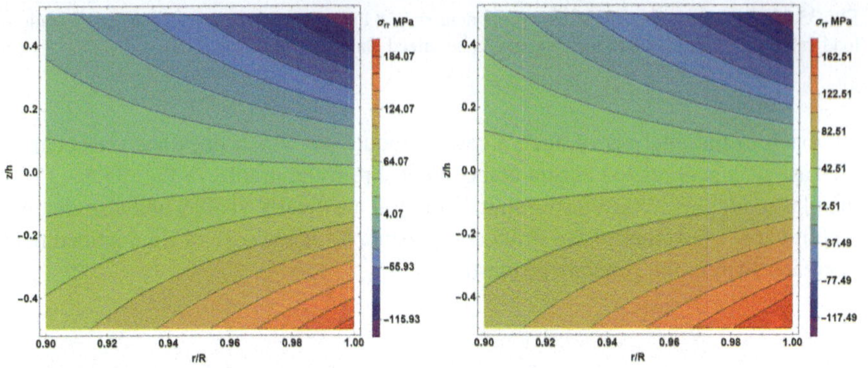

(a) Within the framework of elasticity theory.

(b) Elastic-plastic solution.

Fig. 35.9: Distribution of the stress component σ_{rr} at the edge of the plate for the dimensionless load value $p = 4.0E - 07$ in the case of a generalized elastic embedment.

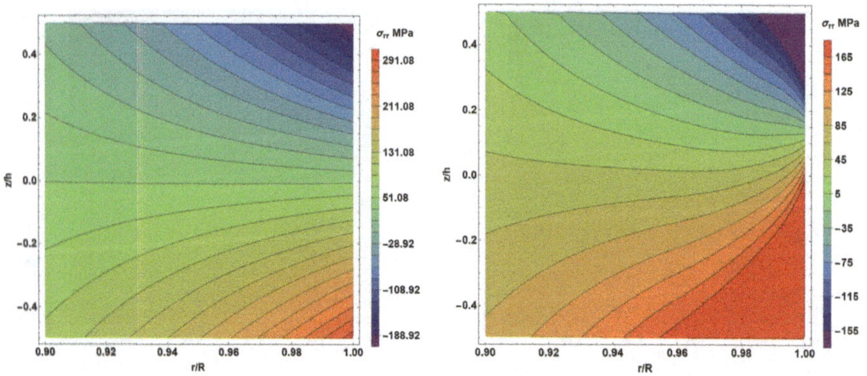

(a) Within the framework of elasticity theory.

(b) Elastic-plastic solution.

Fig. 35.10: Distribution of the stress component σ_{rr} at the edge of the plate for the dimensionless load value $p = 7.0E - 07$ in the case of rigid embedment.

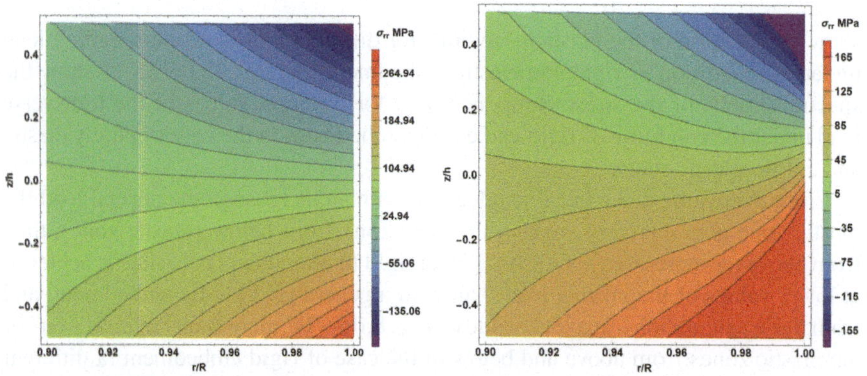

(a) Within the framework of elasticity theory.

(b) Elastic-plastic solution.

Fig. 35.11: Distribution of the stress component σ_{rr} at the edge of the plate for the dimensionless load value $p = 7.0E - 07$ in the case of a generalized elastic embedment.

approach to the analysis of such a case is required), which does not yet occur in the case of a generalized elastic embedment. From the graphs in Fig. 35.8 - 35.11 the plastic regions in the elastic-plastic solution are clearly visible, and also that the amplitude of the values of the stress component σ_{rr} in the case of rigid embedment is greater than in the generalized elastic one.

35.7 Conclusion

Within the framework of plate theory, based on the Föppl - von Kármán model, a solution was obtained for the problem of elastic-plastic deformation of a circular plate, conjugate along the contour with a base and subjected to a uniform normal load under boundary conditions simulating a generalized elastic embedment. The adopted kinematic hypotheses of Kirchhoff - Love allowed us to reduce the problem to a system of ordinary differential equations. The Tresca - Saint-Venant criterion was used as a plasticity criterion. A numerical algorithm similar to the "shooting method" was used to solve it, which allowed us to obtain a solution for both rigid and generalized elastic embedment.

The calculations show that the plastic zone first appears at the fixation points on the tension side of the plate, after which a similar zone appears on the compression side. These plastic zones monotonically increase as the load increases, without forming unloading areas. In addition, the development of plasticity in the edge zones limits the formation of a plastic region in the central part of the plate, which was previously considered potentially dangerous.

Acknowledgements The work was carried out with financial support from the Russian Science Foundation, project No. 23-19-00866 (for K.B. Ustinov).

References

[1] Andrews M, Massabo R (2007) The effects of shear and near tip deformations on energy release rate and mode mixity of edge-cracked orthotropic layers. Engineering Fracture Mechanics 74(17):2700–2720, DOI 10.1016/j.engfracmech.2007.01.013

[2] Andrews M, Massabo R, Cox B (2006) Elastic interaction of multiple delaminations in plates subject to cylindrical bending. International Journal of Solids and Structures 43(5):855–886, DOI 10.1016/j.ijsolstr.2005.04.025

[3] Barbieri L, Massabo R, Berggreen C (2018) The effects of shear and near tip deformations on interface fracture of symmetric sandwich beams. Engineering Fracture Mechanics 201:298–321, DOI 10.1016/j.engfracmech.2018.06.039

[4] Begley MR, Hutchinson JW (2017) The Mechanics and Reliability of Films, Multilayers and Coatings. Cambridge: Cambridge University Press, DOI 10.1017/9781316443606

[5] Cheng YL, Li ML, Lin JH, Lai JH, Ke CT, Huang YC (2005) Development of dynamic mask photolithography system. Proceedings of the IEEE International Conference on Mechatronics (ICM'05), pp. 467-471. DOI 10.1109/ICMECH.2005.1529302

[6] Cotterell B, Chen Z (2000) Buckling and cracking of thin film on compliant substrates under compression. International Journal of Fracture 104(2):169–179, DOI 10.1023/A:1007628800620

[7] Engesser F (1889) Über Knickfestigkeit gerader Stäbe. Zeitschrift des Architekten- und Ingenieur-Vereins zu Hannover 35:455-462
[8] Engesser F (1895) Über Knickfragen. Schweizerische Bauzeitung 26(4):26
[9] Ilyushin AA (1960) Ob uprugoplasticheskoy ustoychivosti konstruktsii, vklyuchayushchey sterzhnevyye elementy (in Russ., On the elastic-plastic stability of a structure including rod elements). Inzhenernyi sbornik 27:87–90
[10] Galanin MP, Savenkov EB (2010) Metody chislennogo analiza matematicheskikh modeley (in Russ., Methods of Numerical Analysis of Mathematical Models). Izdatel'stvo MGTU imeni NE Baumana, Moscow
[11] Gandilyan DV, Ustinov KB (2025) Deformirovaniye tonkoy krugovoy plastiny, sopryazhennoy po konturu s osnovaniyem (in Russ., Deformation of a Thin Circular Plate Adjacent to the Base). Prikladnaya Matematika i Mekhanika 89(1):106-127 DOI 10.31857/S0032823525010089
[12] Jasinski F(1895) Zu den "Knickfragen". Schweizerische Bauzeitung 26(10):63
[13] Ježek K (1937) Die Festigkeit von Druckstäben aus Stahl. Springer, Wien
[14] von Kármán T (1956) Untersuchungen über Knickfestigkeit. Collekted works of Th. von Kármán. Butterworths, London, Vol 1, pp. 1902–1913
[15] Lepik YR (1959) Issledovaniye postkriticheskoy stadii szhatogo uprugoplasticheskogo sterzhnya s uchetom vtorichnykh plasticheskikh deformatsiy (in Russ., Study of the post-critical stage of a compressed elastic-plastic rod taking into account secondary plastic deformations). Uchenye zapisi Tartuskogo universita. Mathematika and Mechanika (73):168–178
[16] Li S, Wang J, Thouless MD (2004) The effects of shear on delamination in layered materials. Journal of the Mechancics and Physics of Solids 52(1):193–214, DOI 10.1016/S0022-5096(03)00070-X
[17] Lychev SA (2023) Nesovmestimye deformacii gibkikh plastin (in Russ., Incompatible deformations of flexible plates). Uchenye Zapiski Kazanskogo Universiteta. Seriya Fiziko-Matematicheskie Nauki 165(4):361–388, DOI 10.26907/2541-7746.2023.4.361-388
[18] Malinin NN (975) Prikladnaya teoriya plastichnosti i polzuchesti (in Russ., Applied Theory of Plasticity and Creep), 2nd ed. Mashinostroenie, Moscow
[19] Massabo R, Ustinov KB, Barbieri L, Berggreen C (2019) Fracture mechanics solutions for interfacial cracks between compressible thin layers and substrates. Coatings 9(3):152, DOI 10.3390/coatings9030152
[20] Monetto I, Massabo R (2020) An analytical beam model for the evaluation of crack tip root rotations and displacements in orthotropic specimens. Frattura ed Integrita Strutturale 14(53):372–393, DOI 10.3221/IGF-ESIS.53.29
[21] Rabotnov YN (1952) O ravnovesii szhatykh sterzhney za predelami predela proportsional'nosti (in Russ., On the equilibrium of compressed rods beyond the proportionality limit). Inzhenernyi sbornik 11:123–126
[22] Shanley FR (1946) The column paradox. Journal of the Aeronautical Sciences 13(12):676, DOI 10.2514/8.11478
[23] Shanley FR (1947) Inelastic column theory. Journal of the Aeronautical Sciences 14(5)261–267, DOI 10.2514/8.1346

[24] Salashchenko NN, Chkhalo NI, Dyuzhev NA (2018) Maskless X-ray lithography based on microoptical electromechanical systems and Microfocus X-ray tubes. Journal of Surface Investigation: X-ray, Synchrotron and Neutron Techniques 12(5):944-952, DOI 10.1134/S0207352818100165
[25] de Teresa, JM (2020) Nanofabrication: Nanolithography Techniques and Their Applications. IOP Publishing Ltd, Bristol
[26] Thouless MD (2018) Shear forces, root rotations, phase angles and delamination of layered materials. Engineering Fracture Mechanics 191:153–167, DOI 10.1016/j.engfracmech.2018.01.033
[27] Timoshenko SP, Woinowsky-Krieger S (1959) Theory of Plates and Shells, 2nd ed. McGraw-Hill, New York
[28] Ustinov KB (2015) On influence of substrate compliance on delamination and buckling of coatings. Engineering Failure Analysis 47, Part B:338–344, DOI 10.1016/j.engfailanal.2013.09.022
[29] Ustinov KB 82019) On semi-infinite interface crack in bi-material elastic layer. European Journal of Mechanics – A/Solids 75:56–69, DOI 10.1016/j.euromechsol.2019.01.013
[30] Ustinov KB (2015) On separation of a layer from the half-plane: Elastic fixation conditions for a plate equivalent to the layer. Mechanics of Solids 50(1):62–80, DOI 10.3103/S0025654415010070
[31] Ustinov KB, Gandilyan DV (2025) On elastic-plastic deformation of a rectangular plate on a cylindrical surface under normal loading. Mechanics of Solids 60(5):3481-3494, DOI 10.1134/S0025654425600394
[32] Ustinov KB, Gandilyan DV (2024) O granichnykh usloviyakh dlya tonkoy krugloy plastiny , sopryazhonnoy s massivnym telom (in Russ., On the boundary conditions for a thin circular plate conjugated to a massive body). Vestnik of Samara University. Natural Science Series 30(1):50–63, DOI 10.18287/2541-7525-2024-30-1-50-63
[33] Ustinov K, Massabo R (2022) On elastic clamping boundary conditions in plate models describing detaching bilayers. International Journal of Solids and Structures 248:111600, DOI 10.1016/j.ijsolstr.2022.111600
[34] Vanko VI (1965) Creep buckling criteria. Journal of Applied Mechanics and Technical Physics 6(1):121-124, DOI 10.1007/BF00914385
[35] Vanko VI (1968) (in Russ., Longitudinal bending of an elastic-plastic rod). Engineering Journal. Mechanics of Solids (4):157–162
[36] Vanko VI, Perelygina ES (2014) Buckling of an elastoplastic bar. Journal of Applied Mechanics and Technical Physics 55(1):52-60, DOI 10.1134/S0021894414010076
[37] Vanko VI, Shesterikov SA (1967) (in Russ., Longitudinal bending and buckling). Engineering Journal. Mechanics of Solids (2):157–163
[38] Vladimirsky Y, Bourdillon A, Vladimirsky O, Jiang W, Leonard Q (1999) Demagnication in proximity X-ray lithography and extensibility to 25 nm by optimizing Fresnel diraction. Journal of Physics D: Applied Physics 32(22):114 - 118, DOI 10.1088/0022-3727/32/22/102

[39] Volmir AS (1967) Stability of Elastic Systems. Foreign Technology Division, Air Force Systems Command. Wright-Patterson Air Force Base, Ohio, AD628508

[40] deWit R (1973) Theory of disclinations: II. Continuous and discrete disclinations in anisotropic elasticity. Journal of Research of the National Bureau of Standards, Section A. 77A(1):49–100, DOI 10.6028/jres.077A.003

[41] Yu H-H, Hutchinson JW (2000) Influence of substrate compliance on buckling delamination of thin films. International Journal of Fracture 113(1):39–55, DOI 10.1023/A:1013790232359

[42] Zubov LM (2007) Von Kármán equations for an elastic plate with dislocations and disclinations. Doklady Physics 52(1)67–70, DOI 10.1134/S102833580701017X

[43] Zubov LM, Stolpovskii AV (2008) A theory of dislocations and disclinations in elastic plates. Journal of Applied Mathematics and Mechanics 72(6):724–737, DOI 10.1016/j.jappmathmech.2009.01.005

Chapter 36
Research of the Mechanical Properties of Aged Glass Textolite Subjected to Unidirectional Tensile Loading and Repeated Static Loading

Sona S. Valesyan[0009-0006-0641-3860]

Abstract Aged glass textolite (GFRP composite-laminates) subjected to unidirectional tensile loading and repeated static loading is studied. Unidirectional tensile curves are plotted, approximated and compared with similar ones obtained about 35 years ago. Hysteresis curves are plotted and approximated. Based on these ones, the dependence between maximal strain and the number of loading/unloading cycles are brought and compared with similar dependencies obtained as a result of research carried our 35 years ago. The energy dissipation coefficient is defined and corresponding curves (dependence between the energy dissipation coefficient and the number of loading/unloading cycles) are charted.

Key words: Aged glass textolite, Unidirectional tensile loading, Repeated static loading, Maximal strain, Coefficient of energy dissipation

36.1 Introduction

Ageing of composite materials affects the mechanical behavior of items subjected to both short-term and long-term force application. With the increasing use of composites in structural applications and in many industries, the study of the aged composite materials mechanical properties is relevant.

In view of physical and chemical nature of components, the orientation of the reinforcement, processing of the material, temperature - moisture conditions of the environment, the numerous research of mechanical properties of composite materials was carried out and the results of research were published in papers and books.

The results of research are necessary for the understanding of the mode of deformation and the evaluation of the reliability of the structure. For this reason, the

Sona S. Valesyan
Institute of Mechanics, National Academy of Sciences of the Republic of Armenia, 24b Marshal Baghramyan Ave, Yerevan, 0019, Republic of Armenia, e-mail: svalesyan@yahoo.com

H. Altenbach and A. V. Sahakyan, *Contact Mechanics and Relevant Problems*,
Advanced Structured Materials 253, https://doi.org/10.1007/978-3-032-20869-9_36

SVAM (fiberglass type) creepage with different time periods after the material production was investigated in [12]. At the reference [15], the mechanical properties of reinforced polymeric plastics after the long-term storage was studied. The research of strength properties and ageing of sheet fiberglasses in the various climatic conditions was carried out in [21]. Research carried out in [4] is dedicated to the effect of natural ageing on fiberglass materials in various climatic areas. The influence of the ageing medium, temperature, and hydrostatic pressure are discussed, then an example of the application of ageing test results to predict long term behavior of the thermal insulation coating of a flow line are presented in reference [3].

The review of processes and modeling of polymeric, glass-ceramic matrix, and silicon carbide composites ageing is brought at the first part of [11]; the second part is dedicated to the ageing of composites in transport applications (aircraft, vehicles and ships); and the ageing of composites applied in medical devices (implants), oil and gas refining, construction, chemical processing and underwater applications is considered in the third part. In [24], the results of the influence of cosmos conditions on the parts of structure subjected to tension, compression, and shear are brought. The research of dissipative properties of plate test pieces cut out from the reinforced composite under $\varphi = 0°; 5°; 15°; 30°; 45°; 60°; 90°$ angles to the direction of reinforcement was carried out in [10].

The different shapes filled composite samples such as plate test pieces of 1.2-4.5 mm of thickness, rings and segments 2-10 mm of thickness exposed on air during spring then kept at temperature of 20-22°C for the stabilizing mass of their were subjected tensile force and bending force [2] The moisture influence on the aging of the plate test pieces subjected tensile force and made of molded polyamide is studied in [6]. The cylindrical sandwich shell stress-strain state caused by an axisymmetric temperature field inhomogeneously distributed across the thickness is investigated in [16]. In [8] the room- and low-temperature influence on the deformation of multilayered fiberglass plastics reinforced with a fabric of sateen weave is studied.

Nowadays, the experimental and theoretical study of a composite beam for bridge structures because of the wide application of composite materials in civil engineering is contemporary issue [18]. Research of deformation behavior, strength and mechanical properties of unidirectional carbon fiber laminates are brought in [9]. The fatigue failure criteria and degradation characteristics of composite laminates subjected to multi-axial loadings are studied in [5]. The load-carrying capacity of circular plates manufactured from the fiber-reinforced composite is considered in [14]. In [1], an experimental study on mechanical properties of GFRP braid-pultruded composite rods subjected to tensile, bending and torsion was carried out. The influence of carbon fibers (CFs) added to polyimide (PI) composite plastics, and the effect of carbon nanotubes (CNTs) on the fiber-matrix interface was studied in [23], as a result was obtained that the mechanical properties of CF/CNT/PI nano-composites are better than those of CF/PI composites. In [22], the mechanical properties of plain-weave composites applying different micro-mechanical models are considered and the expressions of effective material properties for one-, two-, and three-ply woven composites are obtained. Results of study of the influence of possible deviation of the symmetric reinforcement relative to the axis (the disorientation of reinforcement) are occurred during the material technological processing into a product on the deformation behavior and dissipative properties thin-walled glass-plastic tubular elements subjected to repeated-static internal hydrostatic pressure are discussed [7]. In [20], the results of GFRP composite-laminates deformation research under different types of static, repeated-static, low- and multi-cycle loading are brought. For both charts, the curves obtained experimentally are plotted by circles and the approximated curves are plotted as solid lines.

The energy dissipation arisen due to the shear of elastic matrix on the surface of rigid composite particles is discussed in reference [13]. The load for static tests was applied in three directions: axial, radial, and angular (as a moment of force toward the diameter of flywheel). For whole quantity of hysteresis loops, the absence of permanent set is typical. Taking into account the shape of hysteresis loop can be concluded that the shear force arisen in the flywheel depends on the shift, so that the point-to-point friction is present [19]. Hybrid-composite angle-ply laminated plates with an inclined crack, particularly the thermal buckling behavior of their, is considered in [25].

In this paper is studying the results of experimental research of the "age" (ageing) influence on mechanical properties of Glass Fibre Reinforced Polymer composite-laminates (GFRP composite-laminates). GFRP composite-laminates are applicable in various branches of engineering.

36.2 Experimental Part

Research was carried out with the application of glass textolite (GFRP composite-laminate) test pieces. Glass textolite is a Glass Fibre Reinforced Polymer composite-laminate made of glass-cloth saturated by the thermosetting tar. Products of constructional purposes and multi-layered sheets can be produced from glass textolite (GFRP composite-laminate).

Test pieces of glass textolite (GFRP composite-laminate) as double-sided blades, which dimensions satisfy the requirements of GOST (State Standard Specification) 1199-78 (Fig. 36.1), were cut out from the sheet of glass textolite (GFRP composite-laminate) with the angle of the woven fiber orientation of 0° and 90°, and thickness ≈ 9 mm. Glass textolite (GFRP composite-laminate) test pieces were stored in a horizontal position at a distance of not less than 50 mm from the floor. The storage room was closed, dry indoor air temperature was 20-25°C (the temperature interval stated by State Standard Specification from -10°C to +35°C), while the relative humidity did not exceed 80% according to State Standard Specification. The repeated-static tests were performed by the testing machine ZD 10/90 at the loading speed 5 mm/min.

Both cases of research, the unidirectional and repeated-static tensile, test pieces were subjected to loading with the step of loading 0, $1\sigma_u$ (σ_u is the ultimate strength of samples). For unidirectional tensile, the maximal value of the applied loading was

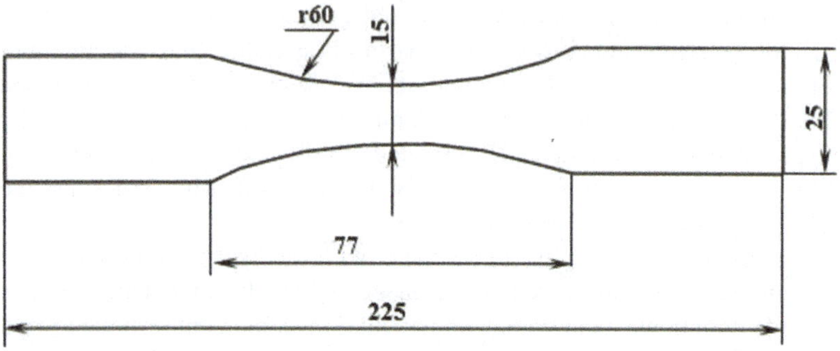

a)

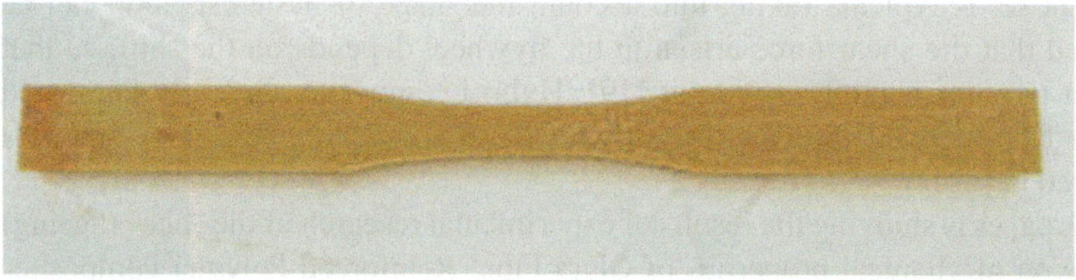

b)

Fig. 36.1: Experimental double-sided spattle-shape GFRP composite-laminate test piece: a) geometry of test piece, b) photo of test piece.

$0,8\sigma_u$, while for repeated-static tensile, the maximal value of the applied loading was $0.7\sigma_u$. The experimental data were written down from the clock-type meter.

After the manufacturing of glass textolite (GFRP composite-laminate) test pieces, experimental researches were carried out by Sargsyan [20] and by the author of this paper in about 35 years later. For unidirectional tensile, the approximation of experimental data (σ and ε) are defined by equation

$$\sigma = \alpha\varepsilon + \beta\varepsilon^n, \tag{36.1}$$

where ε is the value of the strain determined from the value of permanent set, and α, β, n are parameters of the approximation.

For repeated-static tensile, the approximation of experimental data (σ and ε) on the sites of ascending ($\rightarrow$) and descending ($\leftarrow$) branches of hysteresis loop is defined by equation

$$\sigma = \overset{\rightarrow}{\underset{\leftarrow}{\alpha}}\,\varepsilon + \overset{\rightarrow}{\underset{\leftarrow}{\beta}}\,\varepsilon^{\overset{\rightarrow}{\underset{\leftarrow}{n}}}, \tag{36.2}$$

where ε is the value of the strain determined from the value of permanent set, and α, β, n - parameters of approximation defined for every ascending ($\rightarrow$) and descending ($\leftarrow$) branches of hysteresis loop.

Based on the known formula given in [17], the non-elastic deformation of material subjected to the repeated static loading can be estimated through the coefficient of energy dissipation ψ.

$$\psi = \frac{\Delta W}{W}, \tag{36.3}$$

where ΔW is the value of dissipated energy per cycle and W is the value of the total strain energy of the samples.

Applying Eqs. (36.2) and (36.3), and taking into account that values ΔW and W are the area of hysteresis loop and the area restricted between the ascending branch of hysteresis loop and the axis ε, correspondingly, the coefficient of energy dissipation ψ can be defined by the following equation

$$\psi = \frac{\Delta W}{W} = \frac{\int_0^{\varepsilon_{max}} (\vec{\alpha}\,\varepsilon + \vec{\beta}\,\varepsilon^{\vec{n}})d\varepsilon - \int_0^{\varepsilon_{max}} (\overleftarrow{\alpha}\,\varepsilon + \overleftarrow{\beta}\,\varepsilon^{\overleftarrow{n}})d\varepsilon}{\int_0^{\varepsilon_{max}} (\vec{\alpha}\,\varepsilon + \vec{\beta}\,\varepsilon^{\vec{n}})d\varepsilon}. \tag{36.4}$$

Integrating Eq. (36.4), the value of the coefficient of energy dissipation ψ can be obtained by Eq. (36.5)

$$\psi = \frac{\frac{\varepsilon_{max}^2}{2}\left(\vec{\alpha} - \overleftarrow{\alpha}\right) + \vec{\beta}\frac{\varepsilon^{\vec{n}+1}}{\vec{n}+1} - \overleftarrow{\beta}\frac{\varepsilon^{\overleftarrow{n}+1}}{\overleftarrow{n}+1}}{\frac{\vec{\alpha}}{2}\varepsilon_{max}^2 + \vec{\beta}\frac{\varepsilon^{\vec{n}+1}}{\vec{n}+1}}. \tag{36.5}$$

36.3 Results and Discussion

Test pieces made of glass textolite (GFRP composite-laminate) subjected to unidirectional tensile loading were studied in [20] about 35 years ago. Experimental research of glass textolite test pieces subjected to unidirectional tensile loading made at that time was carried out nowadays. On Fig. 36.2, strength σ - strain ε dependencies of unidirectional tensile loading for the aged glass textolite (GFRP composite-laminate) (a) and for the glass textolite tested about 35 years ago (b) with the angle of the woven fiber orientation of 0° (1) and 90° (2) are brought.

Comparing the deformative properties of glass textolite test pieces subjected to unidirectional tensile loading studied 35 years ago and nowadays can be noticed that for ones with the angle of the woven fiber orientation of 0^0, the value of deformation ε is increased on 31.875%, approximately. Meanwhile, for test pieces with the angle of the woven fiber orientation of 90°, the value of deformation ε is increased on $\approx$ 12.12%. From the Fig. 36.2, it is obvious that the value of strength corresponding to the strain $\varepsilon = 1.6\%$ (Fig. 36.2b chart 1) is decreased, in comparison with the value of strength corresponding to the same strain (Fig. 36.2a chart 1), on $\approx$ 29.4%, as well as, the value of strength corresponding to the strain $\varepsilon = 1.65\%$ (36.2b chart 2)

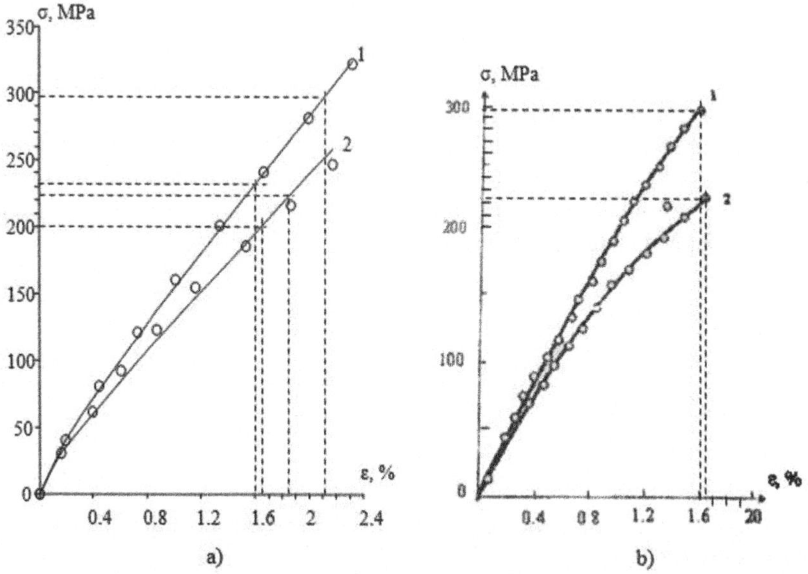

Fig. 36.2: Charts of stress σ and strain ε dependencies: a) for the aged glass textolite (GFRP composite-laminate) and b) for the glass textolite tested about 35 years ago with the angle of the woven fiber orientation of 0°(36.1) and 90°(36.2).

is decreased, in comparison with the value of strength corresponding to the same strain (Fig. 36.2a chart 2), on ≈ 11%.

Research of the mechanical properties of aged glass textolite test pieces subjected to repeated static loading was started with the maximal value of the applied loading $0.8\sigma_u$. However, the test-piece was failed during the third cycle loading/unloading. So that, one was above-mentioned, the tests were carried out with the application of maximal loading equal to $0.7\sigma_u$. Experimental data are evaluated and dependencies are plotted (Fig. 36.3).

Aged glass textolite (GFRP composite-laminate) subjected to $N = 12$ cycles of repeated static loading until $0.7\sigma_u$ tested for both angles of the woven fiber orientation (Fig. 36.3a). Glass textolite subjected to $N = 20$ cycles of repeated static loading until $0.7\sigma_u$ for angle 0°of the woven fiber orientation tested about 35 years ago (Fig. 36.3b) [20] .

For aged glass textolite (GFRP composite-laminate) test pieces with the angle 0° of the woven fiber orientation subjected to repeated static loading, the value of maximal stain ε_{max} is increased by 6.25% in comparison with the glass textolite test pieces tested about 35 years ago.

Dependencies between the coefficient of energy dissipation ψ and the number of loading/unloading cycles N for glass textolite (GFRP composite-laminate) with the angle of the woven fiber orientation of 0°and 90° are plotted as parallel curves (Fig. 36.4). The value of the energy dissipation coefficient ψ is decreased from the first hysteresis curve to the twelfth one for test pieces of both angles of the woven

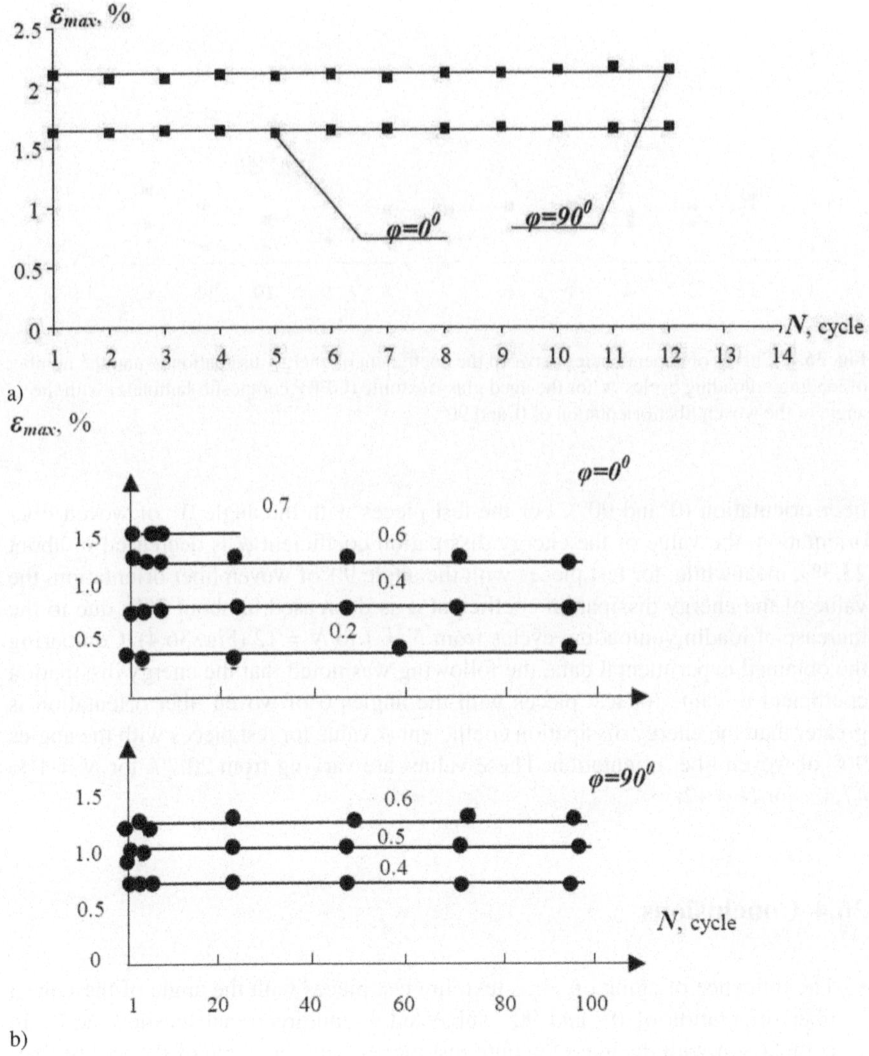

Fig. 36.3: Charts of maximal stain $\varepsilon_{\max}$ and number of loading/unloading cycles N dependencies for the aged glass textolite (GFRP composite-laminate) (a) and for the glass textolite tested about 35 years ago (b) with the angle of the woven fiber orientation of 0° and 90°.

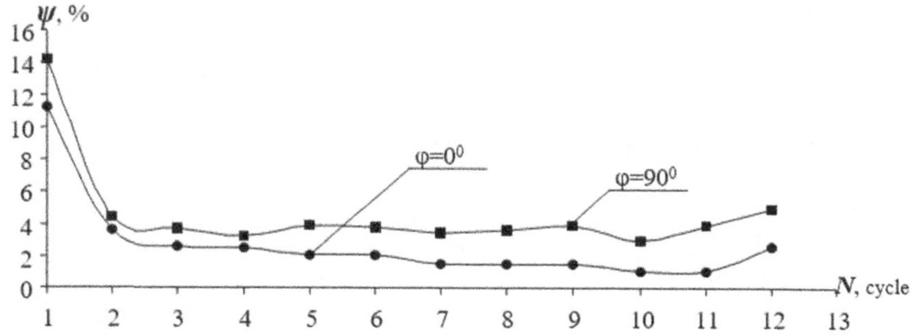

Fig. 36.4: Curves of dependencies between the coefficient of energy dissipation ψ and the number of loading/unloading cycles N for the aged glass textolite (GFRP composite-laminate) with the angle of the woven fiber orientation of 0°and 90°.

fiber orientation (0°and 90°). For the test pieces with the angle 0° of woven fiber orientation, the value of the energy dissipation coefficient ψ is decreased in about 23.3%, meanwhile, for test pieces with the angle 90°of woven fiber orientation, the value of the energy dissipation coefficient ψ is decreased in about 35% due to the increase of loading/unloading cycles from $N = 1$ to $N = 12$ (Fig. 36.4). Comparing the obtained experimental data, the following was noted that the energy dissipation coefficient ψ value for test pieces with the angles 0°of woven fiber orientation is greater than the energy dissipation coefficient ψ value for test pieces with the angles 90° of woven fiber orientation. These values are varying from 20.7% for $N = 1$ to 47.4% for $N = 12$.

36.4 Conclusions

- The influence of aging on glass textolite test pieces with the angle of the woven fiber orientation of 0° and 90° subjected to unidirectional tensile loading in comparison with the glass textolite test pieces with the angle of the woven fiber orientation of 0° and 90° subjected to unidirectional tensile loading studied about 35 years ago is the increase of strain values on ≈ 31.875% (0°) and ≈ 12.12% (90°), correspondingly.
- The maximal strain value increase of aged glass textolite (GFRP composite-laminate) test pieces with the angle of the woven fiber orientation of 0° subjected to repeated static loading by 6.25% is noticed.
- For aged glass textolite (GFRP composite-laminate), the energy dissipation coefficient ψ values for test pieces with the angle of the woven fiber orientation of 90° is greater than the energy dissipation coefficient ψ values for test pieces with the angle of the woven fiber orientation of 0°.

References

[1] Ahmadi MS, Johari MS, Sadighi M, Esfandeh M (2009) An experimental study on mechanical properties of GFRP braid-pultruded composite rods. eXPRESS Polymer Letters 3(9):560–568, DOI 10.3144/expresspolymlett.2009.70

[2] Bulmanis VN, Gunyaev GM, Krivonos VV, Mashinskaya GP, Merkulov VM, Milyutin GI, Gerasimov AA, Kuzmin SA (1991) Atmospheric durability of polymer-fiber composites in cold climates. Mechanics of Composite Materials 27(6):698–705, DOI 10.1007/BF00808081

[3] Le Gac P-Y, Davies P, Choqueuse D (2014) Evaluation of long term behaviour of polymers for offshore oil and gas applications. Oil & Gas Science and Technology - Rev. IFP Energies nouvelles 70(2):279-289, DOI 10.2516/ogst/2013211

[4] Grakova TS, Kiseleva GI, Kochueva LI (1980) Research of natural ageing of fiberglass materials in various climatic areas. Sci.-tech. proceed., Ser. XXIV, 31:554-558

[5] Gude M, Hufenbach W, Koch I, Protz R (2006) Fatigue failure criteria and degradation rules for composites under multiaxial loadings. Mechanics of Composite Materials 42(5):443-450, DOI 10.1007/s11029-006-0054-z

[6] Khramenkov NE, Samoryadov AV, Moiseev JV, Poholov TV, Malikov TP, Nazarov PA (1991) Effect of moisture on aging of glass-filled polyamide-6. Mechanics of Composite Materials 27(3):346-349, DOI 10.1007/BF00616885

[7] Karapetyan KA, Valesyan SS (2023) About the energy dissipation coefficient of thin-walled glass-plastic pipes with the internal reinforcement asymmetry with respect to axis subjected to pulsating internal hydrostatic pressure. In: Altenbach H, Mkhitaryan SM, Hakobyan V, Sahakyan AV (eds) Solid Mechanics, Theory of Elasticity and Creep, Advanced Structured Materials, Vol 185. Springer, Cham, pp. 175-184, DOI 10.1007/978-3-031-18564-9_12

[8] Kucher NK, Dveyrin AZ, Zarazovskii MN, Zemtsov MP (2004) Room- and low-temperature deformation of multilayered fiberglass plastics reinforced with a fabric of sateen weave. Mechanics of Composite Materials 40(3):217-226, DOI 10.1023/B:MOCM.0000033264.89712.ac

[9] Kucher NK, Zemtsov MP, Zarazovskii MN (2006) Deformation behavior and strength of unidirectional carbon fiber laminates. Mechanics of Composite Materials 42(5):407-418, DOI 10.1007/s11029-006-0051-2

[10] Limonov VA, Perevozchikov VG, Tamuzh VP (1989) Fatigue of laminated composites with various reinforcement systems. 1. Experimental results. Mechanics of Composite Materials 24(5):585-594, DOI 10.1007/BF00608842

[11] Martin R. (2008) Aging of Composites, Woodhead Publishing Series in Composites Science and Engineering. Woodhead Publishing, 2008.

[12] Martirosyan MM (1964) Creepage of fiberglass SVAM in the different period after manufacturing of material. Proceedings of Academy of Sciences of Armenian SSR, Seriya of Physical-Mathematical Sciences 17(5):449-463

[13] Mashev VV, Kovrov VN (1995) Experimental study of the contribution of interfacial friction to the deformation resistance of particle-reinforced polymers. Mechanics of Composite Materials 31(5):579-583, DOI 10.1007/BF00617124
[14] Movsumov EA, Shamiev FH (2005) Load-carrying capacity of circular plates made of a fiber-reinforced composite. Mechanics of Composite Materials 41(2):119-130, DOI 10.1007/s11029-005-0039-3
[15] Ogibalov PM, Tuneeva IM (1969) Effect of extended storage on the mechanical properties of reinforced plastics. Polymer Mechanics 5(3):478-480, DOI 10.1007/BF00855177
[16] Paimushin VN, Ivanov VA, Lukankin SA, Bushkov AA (2004) Stress-strain state of a cylindrical sandwich shell in an axisymmetric temperature field inhomogeneous across the thickness. Mechanics of Composite Materials 40(3):203–216, DOI 10.1023/B:MOCM.0000033263.44451.bf
[17] Panovko YG (1960) (in Russ., Internal friction under the waving of elastic systems). State Pub. of Phys.-Math. Literature, Moscow
[18] Piskunov V, Grinevitskii B, Finkelshtein I (2006) Experimental and theoretical investigation of a composite beam for bridge structures. Mechanics of Composite Materials 42(4)315-324, DOI 10.1007/s11029-006-0041-4
[19] Portnov GG, Barinov IN (1992) Experimental study of the natural vibration frequencies and decrements of composite rim flywheels. Mechanics of Composite Materials 28(5):445–454, DOI 10.1007/BF00608614
[20] Sargsyan NE (2012) Anisotropy of Fatigue of Composites. Yerevan: Yerevan State University of Architecture and Construction
[21] Smirnova ZA, Vlasov PV (1971) Strength properties and aging of sheet glass-reinforced plastics under various climatic conditions. Polymer Mechanics 7(3):497-500, DOI 10.1007/BF00854815
[22] Soykasap Ö (2011) Analysis of plain-weave composites. Mechanics of Composite Materials 47(2):161–176, DOI 10.1007/s11029-011-9195-9
[23] Zhang JG (2011) The effect of carbon fibers and carbon nanotubes on the mechanical properties of polyimide composites. Mechanics of Composite Materials 47(4):447-450, DOI 10.1007/s11029-011-9222-x
[24] Zhugun IG, Ivonin YN, Shlitsa RP, Nikishin EF (1988) Effect of prolonged exposure at extreme conditions on the properties of pre-loaded composites subjected to tension, compression, and shear. Mechanics of Composite Materials 23(5):567-572 DOI 10.1007/BF00605680
[25] Yapici A (2005) Thermal buckling behavior of hybrid-composite angle-ply laminated plates with an inclined crack. Mechanics of Composite Materials 41(2):131–138, DOI 10.1007/s11029-005-0040-x

Chapter 37
Study of Mechanical Properties of a Composite Polymer Material Based on Ultra-High Molecular Weight Polyethylene and Modified Aerosil Nanoparticles

Elena G. Zemtsova, Alexander R. Arutyunyan, Vladimir K. Kudymov, Boris N. Semenov, and Sergey O. Kirichenko

Abstract In this paper, we studied the production features of UHMWPE-based composite materials with Aerosil nanofillers surface-modified with SiC nanostructures using molecular layer deposition. To improve the affinity of the inorganic filler to the polymer matrix, the nanoparticle surface was modified with heptyloxysilyl groups by treatment in heptanol vapor. The elastic modulus, hardness, and abrasive wear resistance of nanocomposite samples were assessed using uniaxial tension testing. An analysis of the experimental data showed that the optimal degree of filling with modified aerosil lies in the region of 10 wt.%, where the material has maximum values of elastic modulus and yield strength. For modified UHMWPE, it is shown that the Vicat softening temperature increases linearly depending on the reinforcing phase concentration.

Key words: Ultrahigh molecular weight polyethylene, Nanofillers, Aerosil, Degree of filling, Mechanical properties

37.1 Introduction

Polymer composites are one of the most important and widely used classes of modern functional materials. Their consumption is constantly growing. The development of modern polymer materials emphasizes high performance, both in terms of functional and economic indicators. Therefore, it is necessary to consider not only the material's mechanical, thermal, and other properties, but also its cost, the specifics of its

Elena G. Zemtsova · Alexander R. Arutyunyan (corresponding author) · Vladimir K. Kudymov · Boris N. Semenov · Sergey O. Kirichenko
St. Petersburg State University, 7/9 Universitetskaya Emb., St. Petersburg, 199034, Russian Federation,
e-mail: ezimtsova@mail.ru, a.arutyunyan@spbu.ru, v.k.kudymov@gmail.com, b.semenov@spbu.ru, sergey.kirichenko@spbu.ru

H. Altenbach and A. V. Sahakyan, *Contact Mechanics and Relevant Problems*,
Advanced Structured Materials 253, https://doi.org/10.1007/978-3-032-20869-9_37

processing technology, the reparability of these products, and the possibility of recycling [4].

Ultra-high molecular weight polyethylene (UHMWPE) is increasingly used in various fields of industry, transport engineering, and construction, due to the high performance characteristics of the products obtained from it. The high molecular weight of UHMWPE and its chemical nature determine the entire range of unique properties of this polymer: very high physical and mechanical characteristics, frost resistance, wear resistance, resistance to aggressive environments, low friction coefficient. There is great interest in UHMWPE-based composites containing fillers and functional additives, as they allow achieving even better consumer properties compared to pure polymer [1]. The use of nanodispersed fillers that create assemblies in the polymer matrix and effectively modify the supramolecular structure to form a reinforced polymer system appears promising.

Carbon nanotubes, especially multi-walled ones (MWCNTs), are most often used as reinforcing additives[2, 8, 9]. It is known that even at low content (0.05–1.25 wt.%) MWCNTs significantly improve the hardness, modulus of elasticity and wear resistance of the composite. The second broad class of fillers are nanoceramic particles such as aluminum oxide (Al_2O_3), silicon dioxide (SiO_2), zirconium dioxide (ZrO_2), boron carbide (B_4C), boron nitride (BN) and molybdenum disulfide (MoS_2) [9, 11, 15]. These additives (usually 1-10 wt.%) improve not only mechanical and tribological properties, but also thermal conductivity, radiation protection and biocompatibility. For example, hydroxyapatite and titanium dioxide (TiO_2) are introduced to create bioactive composites designed for bone implants [9]. Two-dimensional nanosized additives, including graphene and graphene oxide, are also actively being investigated [9]. They can significantly increase the strength, rigidity and barrier properties, as well as provide electrical conductivity, which is important for use in electromagnetic screens and sensors.

The advantage of polymer nanocomposite materials lies in the combination of the unique properties of polymers and nanoparticles in a single material. The polymers provide ease of processing, low cost, and viscoelasticity, while the nanoparticles provide hardness, strength, and thermal stability. An analysis of the literature on polymer nanocomposites suggests the potential for such materials to be used in barrier technologies (packaging), electronics, and medical technologies, but only with caution in structural materials. The use of nanodispersed fillers can significantly, and sometimes even orders of magnitude, improve the barrier properties of traditional polymers, while the increase in mechanical properties typically does not exceed 20%. At the same time, under special conditions for the targeted synthesis of polymer nanocomposites, it is possible to obtain materials with truly unique characteristics. Thus, given the current level of nanotechnology development, the use of polymer nanocomposites is more justified in expensive, high-added-value technologies (semiconductor electronics and medical technology and equipment). Where the cost of the material is not a determining factor.

37.2 Methodology for Creating a Polymer Composite and Studying its Mechanical Properties

One possible solution for improving the mechanical properties of a polymer composite may be the homogeneous distribution of reinforcing phase nanoparticles in the polymer matrix, due to increased adhesion between the components of the composite material. To overcome the low affinity of nanoparticles for the polymer due to thermodynamic incompatibility, surface modification of the nanoparticles is sometimes carried out. For this purpose, methods of covalent and non-covalent surface functionalization (chemical and physical methods of surface modification) are used. Chemical modification of the surface of nanoparticles involves changing the nature of the surface functional groups of nanoparticles or the addition of polymer/oligomeric chains [17]. This can result in materials with a unique combination of physical, mechanical, and tribological properties that can be effectively used in the manufacture of friction unit components.

37.2.1 Method for Producing a Bulk Composite Material Based on UHMWPE Reinforced with Nanofillers Based on Aerosil A-300

In this work, we studied the features of obtaining composite materials based on UHMWPE with nanofillers based on Aerosil A-300 (with a particle size of 10 to 20 nm) with grafted silicon-carbide nanostructures (SiO_2/SiC). Carbide nanostructures on the surface of Aerosil were synthesized by molecular layering (ML-ALD) [14, 16]. The molecular layering (ML) method is based on the fact that irreversible chemisorption, occurring with a continuous influx of a reagent in a stream of inert gas, over a certain time leads to the complete population of the substrate surface with the corresponding functional groups.

In our study, we combined nanofiller particles with polymer powder in a ball mill. This mixing method has recently been widely used to produce UHMWPE-based composite materials [10]. Combining the polymer with the nanosized filler is preferable to mixing in a drum mixer or a paddle mixer with a rapidly rotating rotor. The molding of samples of polymeric composite was carried out by thermo-pressing at 180°C and a pressure in the mold of 10 MPa, holding time of 20 minutes.

Initially, experiments were conducted at 140°C (which is 5°C higher than the melting point of the UHMWPE grade GUR 2122 used), as well as at 160°C. However, optical microscopy in transmitted light (Fig. 37.1a) and the surface of the composite using SEM (Fig. 37.2a) revealed the presence of cracks in the resulting composite samples, indicating incomplete fusion of the UHMWPE particles. At 180°C and 10 MPa, there are no cracks in the resulting composite samples, which indicates complete fusion of the UHMWPE particles together (Figs. 37.1b and 37.2b).

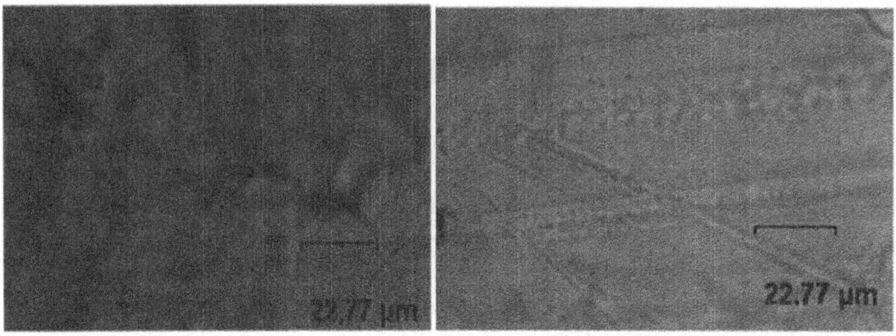

a b

Fig. 37.1: Micrograph of thin section of composite obtained by pressing: **a** - at 140°C, **b** - at 180°C.

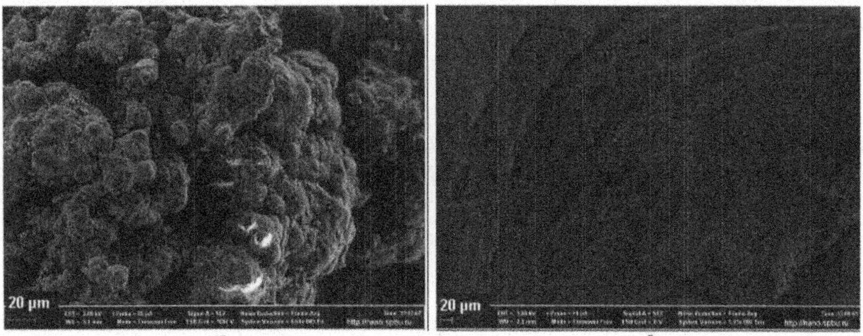

a b

Fig. 37.2: Micrograph of a sample of the composite obtained by pressing: **a** - at 140°C; **b** - at 180°C.

To increase adhesion and activate UHMWPE powder with the introduction of nanomodifiers, we used a surface treatment method in order to increase the surface energy of the polymeric materials. We activated the UHMWPE powder with corona discharge. After corona discharge treatment, carboxyl, carbonyl, peroxide, nitroxide, and other functional groups are formed on the polymeric materials surface, which leads to a surface energy increase. When nanomodifiers are deposited during corona discharge treatment, adhesive bond formation between the nanoparticles and the polymeric surface becomes more efficient.

Experiments on polymer material activation were conducted using a Plazon IVNR 5/50 high-voltage power supply with an adjustable output voltage in the range of 0.3 to 5 kV. The system also allows for the deposition of nanoparticles on a polymer substrate directly during corona discharge treatment. The uniform distribution of nanomodifier particles (SiO_2/SiC) in the polymer matrix bulk is confirmed by microphotography and by the map of the chip surface (Fig. 37.3).

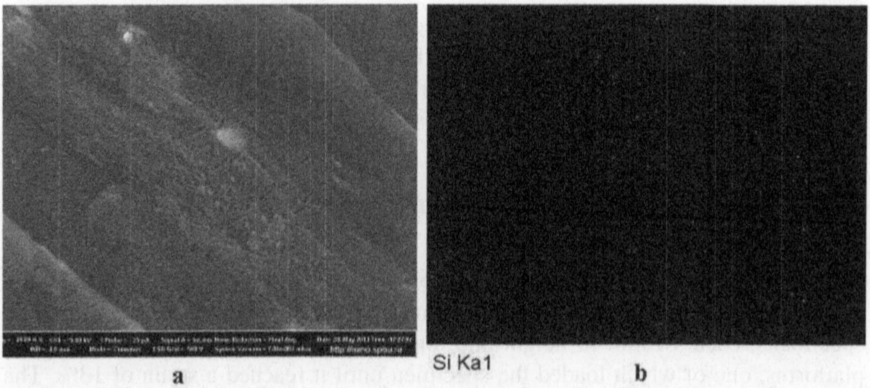

Fig. 37.3: Distribution of nanomodifier particles (SiO_2/SiC) in the polymer matrix bulk: **a** - Micrograph of sintered composite based on UHMWPE reinforced with nanomodifier (5% SiO_2/SiC), **b** - Composite chip surface map.

Thus, based on optical and scanning electron microscopy data, it was concluded that when obtaining a composite based on UHMWPE and aerosil modified with titaniferous nanolayers and surface heptyloxyl groups, under the above-mentioned pressing conditions, a uniform distribution of the modifier is observed in the polyethylene matrix bulk. There are no cracks in the sample, that is, complete fusion of UHMWPE particles and the modifier into a composite material occur.

The resulting composite samples were subjected to mechanical testing to determine their physical and mechanical properties and to identify the optimal filler ratio. The physical and mechanical properties of the resulting composites were examined, their strength and hardness were assessed, and the Vicat softening temperature was measured.

37.2.2 Characterization

The morphology and structure of the composite samples were studied using a Zeiss Merlin scanning electron microscope with additional attachments for X-ray microanalysis Oxford Instruments INCAx-act and an Oxford Instruments CHANNEL5 electron backscatter diffraction (EBSD) recording system.

37.2.3 Mechanical Tests: Conducting Uniaxial Tensile Tests

Tensile testing was conducted on a Shimadzu EZ-1-5-kN testing machine at a temperature of 20°C and a constant tensile rate of 5 mm/min. The machine has a stationary section to which one clamp is attached, and a moving section in which the second clamp is mounted. A dumbbell-shaped specimen with a cylindrical working section 17 mm long and 3 mm in diameter is positioned vertically. The test was continued until failure.

37.2.4 Mechanical Tests: Conducting Compression Tests

Compression tests were conducted on a Shimadzu EZ-1-5-kN testing machine at a temperature of 20°C and a constant compression rate of 5 mm/min. Cylindrical specimens, each 10 mm in height and diameter, were placed between two metal platforms, one of which loaded the specimen until it reached a strain of 13%. The stress in the specimen was measured.

37.2.5 Mechanical tests: Conducting Brinell Tests

To study the material's hardness, we chose the Brinell hardness measurement method, which is an indentation method. An indenter (a polished, hardened steel ball) is pressed into the surface of the test specimen with a specified force. The test is conducted as follows: a small preload is applied to establish the initial position of the indenter on the specimen, then the main load is applied to the specimen, and the penetration depth is measured. Then the main load is removed. Brinell hardness is calculated as the applied load divided by the indentation surface area [6].

Hardness measurements were performed using a Höppler consistometer at 20°C. Cylindrical specimens 15 mm high and 10 mm in diameter were placed on a stand, with a 5 mm indenter inserted into the center. The specimen was then preloaded with 15 N for 10 s. The main load was then applied to the indenter until a total force of 358 N was reached. Based on indenter penetration depth, the Brinell hardness is calculated using the formula

$$HB = \frac{F}{\pi D h}, \tag{37.1}$$

where F is applied load, N; D is indenter diameter, mm; h is indenter penetration depth, mm.

37.2.6 Vicat Softening Temperature Measurement

This test method is used to measure the ability of polymers to retain mechanical properties at high temperatures. It involves determining the temperature at which a standard indenter with a flat bottom penetrates a test specimen heated at a constant rate to a depth of 1 mm under load. The indenter acts perpendicular to the specimen surface. The Vicat softening temperature (VST) is the temperature (°C) measured as close to the specimen surface as possible at which the indenter penetrates to a depth of 1 mm. The temperature at which the specimen compresses by 1 mm under load is determined.

37.3 Results and Discussions

The observed increase in the mechanical properties of composites is associated with a reduction in the particle size of the reinforcing phase to several tens of nanometers. This is due to an increase in the specific area of interaction between the grains of the reinforcing phase and the matrix grains, and a reduction in the size of material defects, which also include the particles of the reinforcing phase [12, 16].

Carbides are of particular interest as reinforcing additives due to their properties: high hardness and elastic modulus, and a high melting point. In this study, SiC nanostructures are a component of Aerosil-based reinforcing particles that are produced through surface chemical reactions. Thanks to the particles of Aerosil A-300, SiC nanostructures can be evenly distributed throughout the bulk polymer.

The hardness of a material is a critical mechanical characteristic. It determines the maximum loads the resulting composite will be able to withstand. Hardness test data (Table 37.1) showed that the hardness of the composite based on UHMWPE and nanofillers increases with the increase of the filling degree and then decreases.

Modified aerosil particles distributed in the matrix structure the polymer near the interface, forming hard and rigid layers that resist the penetration of a foreign body, distributing the load uniformly across the material. However, when the filler content becomes too large, an extensive and continuous interphase boundary is formed, the stress begins to be transferred from the particles to the material, and the particles themselves begin to act as an indenter. Thus, the particle layers pass through the underlying polymer layers [13]. It was also found that the hardness of UHMWPE-based nanocomposites increases with an increase in the filling degree of the reinforcing phase SiO_2/SiC up to 15 wt.%.

Real polymer composites typically undergo compressive deformation during use, so compression tests were conducted. Data for two samples without filler reinforcement (pure UHMWPE) and with maximum nanomodifier reinforcement (30%) are

Table 37.1: Brinell hardness of UHMWPE composite modified with aerosil with Ti-N structures.

Filler amount, wt. %	Hardness, MPa
0	36.1±2.2
5	39.0±2.1
10	43.9±2.0
15	50.0±1.3
30	48.1±2.6

shown in Fig. 37.4. For pure UHMWPE, the stress at a 13% strain is 13±1 MPa, for UHMWPE with 5 % filler it is 17±1 MPa, and for UHMWPE with 30% modified aerosil it is 26 MPa. It is evident that the filled polymer deforms less under loading than the unfilled polymer, and that its deformation requires greater stress, meaning that the material's rigidity increases significantly with filling. However, while the specimen appears to be under uniaxial loading during compression, the resulting stress state is complex due to the simultaneous occurrence of not only compressive deformation but also tension and shear in the direction perpendicular to the compression axis. This is why the mechanical properties of materials under compression differ from those under uniaxial tension. Due to the fundamental nature of the testing, the deformation of compressed specimens is geometrically limited. Consequently, recrystallization and forced high elasticity do not have time to manifest themselves. This explains the low structural and technological sensitivity of this type of testing [3]. Therefore, to analyze the mechanical properties of the samples, it was decided to use the standard, most reliable and widespread test - the uniaxial tensile test.

The results of the evaluation of the elastic modulus in the study of uniaxial tension, yield stress, elongation at break and hardness of samples of UHMWPE nanocomposites with aerosil modified with SiC layers are presented in Table 37.2.

Uniaxial tensile testing of specimens showed that filling with modified aerosil up to 15% (by weight) does not significantly affect the yield strength of the material (26-28 MPa) or the strain corresponding to the yield strength (13-15%), but does lead to a significant increase in tensile strength. The elastic moduli of the material when filled with modified aerosil increase up to a filling level of approximately 10%, after which a decrease is observed (Fig. 37.5).

The data in Table 37.2 and Fig. 37.5 clearly show that the elastic modulus initially increases and then begins to decline. This is primarily due to the cross-linking effect of the reinforcing phase particles. Macromolecules form additional bonds with the

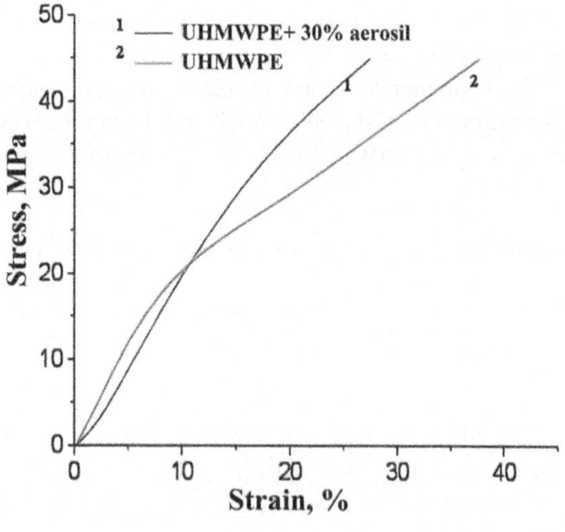

Fig. 37.4 Deformation curves of compression tests

Table 37.2: Mechanical characteristics of composite polymer materials based on UHMWPE matrix.

Specimen	Filler amount	E, MPa	σ_T, MPa	ε_p, %	Hardness, MPa
Pure UHMWPE	0%	570±49	26±41.0	258±49	36.1±42.20
UHMWPE + Aerosil	10%	770±434	28±41.0	210±44	38.0±40.94
UHMWPE + Aerosil with nano SiC	5%	781±455	25±40.5	230±48	39.0±42.10
UHMWPE + Aerosil with nano SiC	10%	840±432	26±40.5	200±44	43.9±42.00
UHMWPE + Aerosil with nano SiC	15%	781±426	26±40.5	168±49	50.0±41.30

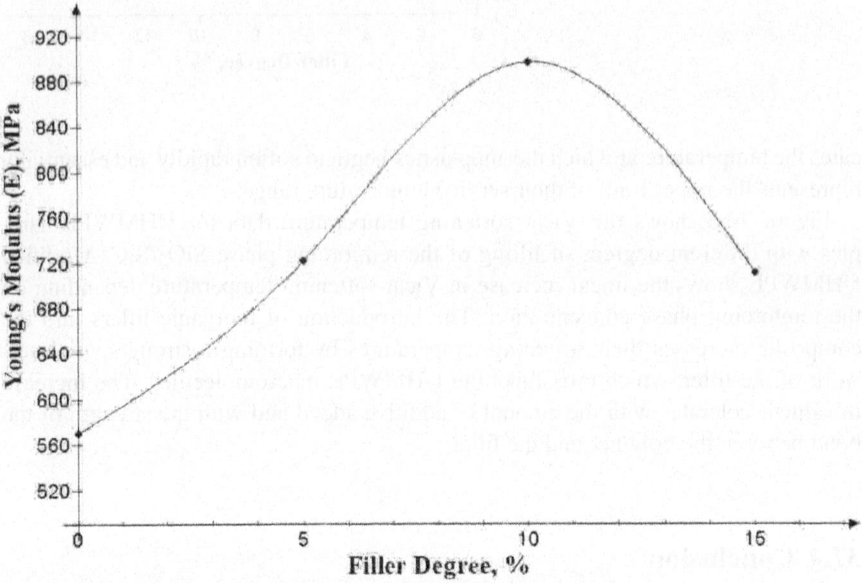

Fig. 37.5: Dependence of the elasticity modulus on the filling degree.

reinforcing phase surface, and the overall material rigidity increases. However, after the formation of a large polymer-filler interface, adhesion decreases, the contribution of weaker filler-filler bonds increases, and the elastic modulus decreases. At the same time, macromolecular coils are able to unfold over a greater length before irreversible separation, resulting in an increase in strain at break.

Changes in yield strength and yield strain are minor and within the error margin. Based on the data obtained, it can be concluded that the optimal degree of UHMWPE reinforcement to improve the composite's strength properties is 10 wt. %.

The study focused on measuring the Vicat softening point, which is used to determine the ability of polymers to retain mechanical properties at high temperatures. This type of test is becoming increasingly popular as plastics continue to replace more traditional materials in various applications. The measurement result indi-

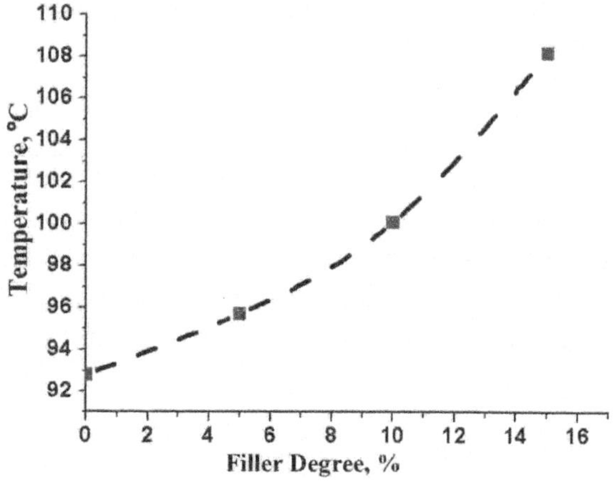

Fig. 37.6 Vicat softening temperature for UHMWPE samples with varying degrees of filling of the reinforcing phase SiO_2/SiC.

cates the temperature at which thermoplastics begin to soften rapidly and essentially represents the upper limit of their service temperature range.

Figure 37.6 shows the Vicat softening temperature data for UHMWPE samples with different degrees of filling of the reinforcing phase SiO_2/SiC. Modified UHMWPE shows the linear increase in Vicat softening temperature depending on the reinforcing phase concentration. The introduction of inorganic fillers into the composite increases their softening temperatures by forming a strong spatial network of the filler, which crosslinks the UHMWPE macromolecules. The increase in value accelerates with the amount of additive added and with the strength of the bond between the polymer and the filler.

37.4 Conclusion

A comparison of uniaxial tensile test data for UHMWPE filled with modified aerosil and modified aerosil with carbide layers allows us to conclude that the optimal degree of filling is approximately 10 wt %, where the material has maximal elastic modulus, yield strength, and tensile strength. This is consistent with known data on other polymer matrix fillers [5, 7]. Increasing the reinforcing phase in the polymer composite by more than 10% leads to a decrease in the elastic modulus and strength characteristics of the material, but, as expected, increases the material hardness.

Structuring a polyethylene matrix with aerosil nanoparticles and SiC nanolayers improves some of the key structural characteristics of polymeric materials (such as elastic modulus, strength, and hardness), thereby improving the mechanical properties of UHMWPE. However, high mechanical properties are only achievable with the correct selection of the reinforcing phase, the method of surface treatment of the reinforcing phase, and the method of combining it with the polymer matrix. The resulting nanocomposite can be used as a structural material for the production of wear-resistant, lightly loaded parts, for example, for railways.

Acknowledgements The work was carried out with the financial support of the Russian Science Foundation (grant No. 25-21-00376). The authors express gratitude to the Science Park of St. Petersburg State University, the Research Center for Extreme States of Materials and Structures, and the Research Center for Nanotechnology.

References

[1] Arutyunyan AR, Arutyunyan RA (2020) Damage and fracture of viscous-plastic compressible materials. Material Physics and Mechanics 44(2):250-255, DOI 10.18720/MPM.4422020_9
[2] Camacho N, Stafford SW, Garza KM, Suro R, Barron KI (2015) Ultra-high molecular weight polyethylene reinforced with multiwall carbon nanotubes: In vitro biocompatibility study using macrophage-like cells. Lubricants 3(3):597–610, DOI 10.3390/lubricants3030597
[3] Danilova SN, Okoneshnikova AV, Okhlopkova AA (2022) Polimernyye materialy na osnove sverkhvysokomolekulyarnogo polietilena: struktura i svoystva (in Russ., Polymer materials based on ultrahigh molecular weight polyethylene: structure and properties). Arctic and Subarctic Natural Resources 27(4):631-642, DOI 10.31242/2618-9712-2022-27-4-631-642
[4] Grashchenkov DV, Chursova LV (2012) Strategiya razvitiya kompozitsionnykh i funktsional'nykh materialov (in Russ., Strategy for the development of composite and functional materials). Aviation Materials and Technologies 231-242
[5] Karami P, Salkhi Khasraghi S. Hashemi M, Rabiei S, Shojaei A (2019) Polymer/nanodiamond composites - a comprehensive review from synthesis and fabrication to properties and applications. Advances in Colloid and Interface Science 269:122-151, DOI 10.1016/j.cis.2019.04.006
[6] Kryzhanovsky VK, Burlov VV, Panimatchenko AD, Kryzhanovskaya YV (2007) Tekhnicheskie svoystva polimernykh materialov (in Russ., Technical Properties of Polymeric Materials). St. Petersburg: Profession
[7] Lebedev OV, Bogdanova OI, Goncharuk GP, Ozerin AN (2020) Tribological and percolation properties of polypropylene/nanodiamond soot composites. Polymers and Polymer Composites 28(6):369-377, DOI 10.1177/0967391119879280
[8] Manoj Kumar R, Sharma SK, Manoj Kumar BV, Lahiri D (2015) Effects of carbon nanotube aspect ratio on strengthening and tribological behaviour of ultra high molecular weight polyethylene composite. Composites Part A: Applied Science and Manufacturing 76:62-72, DOI 10.1016/j.compositesa.2015.05.007

[9] Padhy V, Kandasubramanian B (2024) Polymer nanocomposites of ultra-high molecular weight polyethylene. Polymer Bulletin 81(17):15259–15292, DOI 10.1007/s00289-024-05439-y

[10] Panin SV, Kornienko LA, Vannasri S, Ivanova LR, Shilko SV (2010) Effect of mechanical activation of ultra-high- molecular-weight polyethylene on its mechanical and triboengineering properties. Friction and Wear 31(2):168-176, DOI 10.3103/S1068366610020054

[11] Ravi K, Ichikawa Y, Ogawa K, Deplancke T, Lame O, Cavaille J-Y (2016) Mechanistic study and characterization of cold-sprayed ultra-high molecular weight polyethylene-nano-ceramic composite coating. Journal of Thermal Spray Technology 25(1):160–169, DOI 10.1007/s11666-015-0332-1

[12] Shah V (2009) Spravochnoye rukovodstvo po ispytaniyam plastmass i analizu prichin ikh razrusheniya (in Russ., Reference Guide to Plastics Testing and Analysis of the Causes of Their Failure). St. Petersburg, Nauchnyye Osnovy i Tekhnologii

[13] Shimasaki C, Watanabe N, Fukushima K, Rengakuji S, Nakamura Y, Ono S, Yoshimura T, Morita H, Takakura M, Shiroishi A (1997) Effect of the fire-retardant, melamine, on the combustion and the thermal decomposition of polyamide-6, polypropylene and low-density polyethylene. Polymer Degradation and Stability 58(1):171-180, DOI 10.1016/S0141-3910(97)00043-8

[14] Smirnov VM (2002) Nanoscaled structuring as a way to constructing new solid substances and materials. Russian Journal of General Chemistry 72(4):633-650, DOI 10.1023/A:1016396500708

[15] Suresha B, Harshavardhan B, Rao AM, Koushik UR, Hemanth R (2020) Role of nano-fillers on tribological behaviour of ultra high molecular weight polyethylene composites. Materials Today: Proceedings 27:2169-2173, DOI 10.1016/j.matpr.2019.09.089

[16] Zemtsova EG, Semenov BN, Morozov NF, Sidorov YV, Kudymov VK, Smirnov VM (2025) Creation of aluminum matrix composites reinforced with micro- and nanoparticles of titanium carbide with nickel coating and description of their hardening mechanisms. Materials Physics and Mechanics 53(1):68-77, DOI 10.18149/MPM.5312025_6

[17] Zhang Q, Naito K, Tanaka Y, Kagawa Y (2008) Grafting polyimides from nanodiamonds. Macromolecules 41(3):536-538, DOI 10.1021/ma702268x

Chapter 38
The Effect of Nanostructuring of the Surface of Particles of the Reinforcing Phase of the Nickel Matrix on the Mechanical Properties of the Composite

Elena G. Zemtsova, Pavel E. Morozov, Vladimir K. Kudymov,
Boris N. Semenov, and Yuri V. Sidorov

Abstract The optimal conditions for the production of a non-porous composite are investigated, namely, the temperature and duration of sintering, intermediate heating, as well as the sintering medium (argon and hydrogen) to achieve the best mechanical properties. In order to obtain a non-porous material, we experimentally studied pressing and sintering a nickel-based metal matrix composite reinforced with a TiC frame (1-3 nm). Using micromechanical modeling, the contributions of known hardening mechanisms to the total metal composite hardening are estimated, depending on the concentration, particle volume distribution, and size of reinforcing particles. This study can help to predict the hardening degree of the resulting composites by theoretically assessing the contribution of hardening mechanisms.

Key words: Composite, Nickel matrix, Titanium carbide nanostructures, Mechanical properties

38.1 Introduction

The work relates to the nanotechnology of composites consisted of a metal matrix and nanoscale reinforcing structures. Modern technologies for creating metal matrix composites (MMC) are based on the property management through directional structuring and nanostructuring. The possibility of changing the number and size of reinforcing particles, as well as the variety of matrix alloys, opens up broad prospects for the MMC use in various applications. The observed increase in the metal-matrix composites strength is directly related to the reduction of the reinforcing phase size

Elena G. Zemtsova (corresponding author) · Pavel E. Morozov · Vladimir K. Kudymov · Boris N. Semenov · Yuri V. Sidorov
St. Petersburg State University, 7/9 Universitetskaya Emb., St. Petersburg, 199034, Russian Federation, e-mail: ezimtsova@yandex.ru, comitcont@yandex.ru, v.k.kudymov@gmail.com, b.semenov@spbu.ru, arutalr@rambler.ru

H. Altenbach and A. V. Sahakyan, *Contact Mechanics and Relevant Problems*,
Advanced Structured Materials 253, https://doi.org/10.1007/978-3-032-20869-9_38

to several tens of nanometers. This is due to the increase in the specific area of interaction between the reinforcing phase grains and the matrix grains, and a decrease in the size of the material defects, which are also particles of the reinforcing phase [8]. The processes of strengthening of materials occurring in this case are in good agreement with the Hall-Petch-Orowan mechanisms. There are many calculation works in which the effect of strengthening in composite materials is theoretically substantiated. A number of authors agree that active influence on the morphology, dispersion and nature of the distribution of the dispersed phase makes it possible to obtain a combination of properties that is unattainable even in alloys [5]. The use of metal-insoluble solid inclusions as a reinforcing phase also makes a significant contribution, ensuring the stability of the resulting material's properties over long periods of use. Carbides are widely used as a reinforcing phase in the manufacture of nickel-matrix composites. The popularity of carbides is due to their high hardness, which strengthens the material via the Orowan mechanism (preventing dislocation propagation) and transfers loads to the reinforcing particles.

However, despite the advantages of MMC over traditional materials, their production and use in some areas of mechanical engineering is limited. Along with the development of compositions and the improvement of manufacturing technology, further development of general principles and criteria for the MMC design for machine-building purposes and the accumulation of experience in their operation are required [3]. One of the important criteria when MMC creating is the selection of the reinforcing phase amount introduced into the matrix. With an increase in the reinforcement amount, the material becomes embrittled, which is associated with the formation of local zones with an increased content of particles of the reinforcing phase that are not connected to each other or to the grains of the matrix. The content of carbide reinforcing nanoparticles often does not exceed 1 %, since nanoparticles often have poor adhesion to the matrix material and aggregate, which leads to a decrease in strength and ductility [2, 8].

Traditional methods for consolidating metallic and ceramic particles – high-temperature sintering and hot pressing – have significant limitations, such as the inability to retain the original grain size in the matrix due to excessive grain growth during high-temperature processing [9]. Also, the combination of a matrix and a reinforcing component in one material inevitably encounters interphase interaction. Poor adhesion leads to a loss of strength in the final composite. Compaction of metal-matrix composites (MMCs) is difficult from an adhesion standpoint. Poor wetting leads to sedimentation and agglomeration of the reinforcing component, which clearly negatively impacts mechanical properties. Our research proposed an approach to producing a metal-matrix composite material using surface structuring, ALD (atomic layer epitaxy), and powder metallurgy. The developed approach allowed us to produce a composite with titanium carbide nanostructures measuring 1-5 nm uniformly distributed in the nickel matrix. The absence of interphase boundaries between the nickel matrix particles and the carbide nanostructures minimized the sample's internal porosity. This allows for the most effective manipulation of not only the mechanical properties of the composite material but also the preservation of its ductility, which positively impacts its performance characteristics and processability.

The work is relevant for developing a synthetic approach to the production of a nickel-matrix composite using surface structuring of the reinforcing phase with the possibility of directional control of mechanical properties [7].

38.2 Production of a Nickel-Matrix Composite and its Mechanical Properties

The optimal conditions for the production of a non-porous composite are investigated, namely, the temperature and duration of sintering, intermediate heating, as well as the sintering medium (argon and hydrogen) to achieve the best mechanical properties. Pressing and sintering a nickel-based MMC reinforced with a TiC frame (1-3 nm) are studied in order to evaluate the mechanical properties of the materials.

38.2.1 Production of a Bulk Ni-Based Composite with Reinforcing Phase Ni@TiC

The first step in obtaining a nickel-based composite is the introduction of Ni@TiC particles into the initial matrix in a ratio of 1:99, 3:97 and 5:95 wt.%. Two powders (modified Ni@TiC particles and reinforcing phase) were mixed in a planetary ball mill RM 100 CM.

Samples with a working part length of 5 mm were produced using the pressing method. For pressing, a pneumohydraulic press Sorokin 7.75 was used. The samples pressed at room temperature in 3 loading stages, 15 minutes each: stage 1 – 15 tons, stage 2 – 35 tons, stage 3 – 50 tons.

Due to the stepwise pressing load increase, the resulting stresses in the structure are not enough to destroy the sample matrix under the reinforcing phase action. This pressing makes it possible to evenly distribute the reinforcing phase in the sample and guarantee the formation of a massive matrix without cracks, bubbles and other mechanical defects.

After the pressing stage, the sample was sintered under inert gas at 950°C or 1100°C for 1 h. When the temperature increases from 950 to 1100°C in inert atmosphere, the sample remains brittle, and the tensile strength of the nickel matrix sample increases from 210 to 225 MPa. The fracture curve for a sample heated at 1100°C indicates that as the calcination temperature increases, the brittle fracture model begins to change slightly towards the plastic one (Fig. 38.1).

Subsequently, the samples were sintered at 950°C in inert atmosphere to study the behavior of a nickel matrix reinforced with TiC nanostructures (1-3 nm). TiC nanoparticles in the nickel matrix bulk (Ni/TiC) were synthesized by the method of molecular layering (ALD-ML). This synthesis method makes it possible to obtain 1-3 nm TiC nanostructures on the Ni surface. The study confirms the homogeneous distribution of carbide particles in the metal matrix (Fig. 38.2).

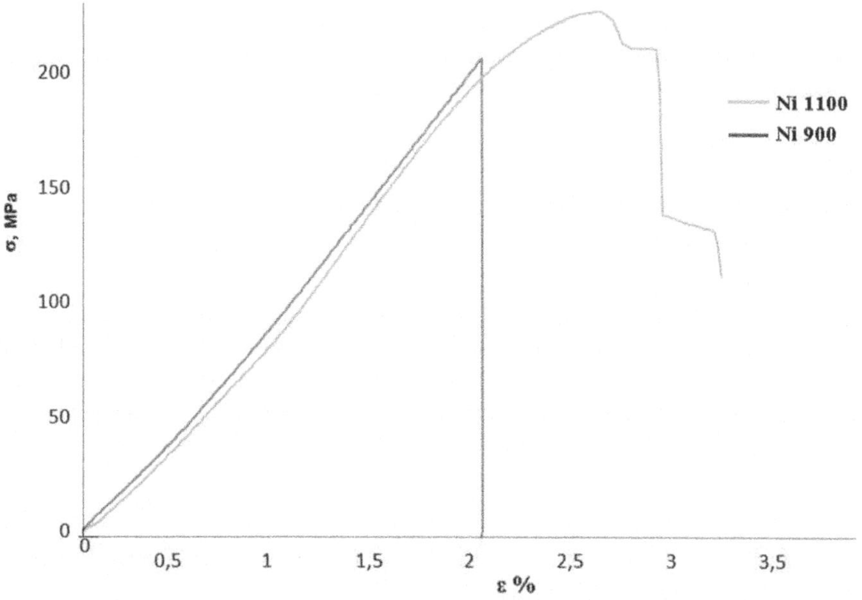

Fig. 38.1: Deformation curves of Ni samples in inert atmosphere.

The absence of interfacial boundaries between nickel matrix particles and carbide nanostructures due to surface chemical reactions (ALD-ML method) made it possible to obtain samples with the absence of external and internal porosity. The presence of oxygen indicates the formation of the nickel oxide phase even at lower sintering temperature (950°C). The oxide phase affects the plasticity of the sample and leads to embrittlement. Peaks at 455 eV (Ti2p)) in X-ray photoelectron spectrum correspond to the C-Ti bond, which confirms the presence of TiC in the samples.

38.2.2 Results of Mechanical Tests of Nickel-Based Composites with Reinforcing Phase Ni@TiC.

In the course of the work, mechanical tests of the obtained composites containing 1% of the reinforcing phase of nickel particles (Ni@TiC) with different ratios of carbide nanostructures in a nickel matrix were carried out. The revealed mechanical characteristics are presented in Table 38.1. A 10 % increase in the MMC strength properties simultaneously with a 2-fold improvement in the plasticity can be explained by creating a high density of dislocations in the Ni matrix.

According to the dislocation theory of plasticity, the movement of dislocations of reinforcing particles (in our case, TiC nanostructures) with high shear resistance limit the movement of dislocations in the nickel matrix, fixing them around themselves.

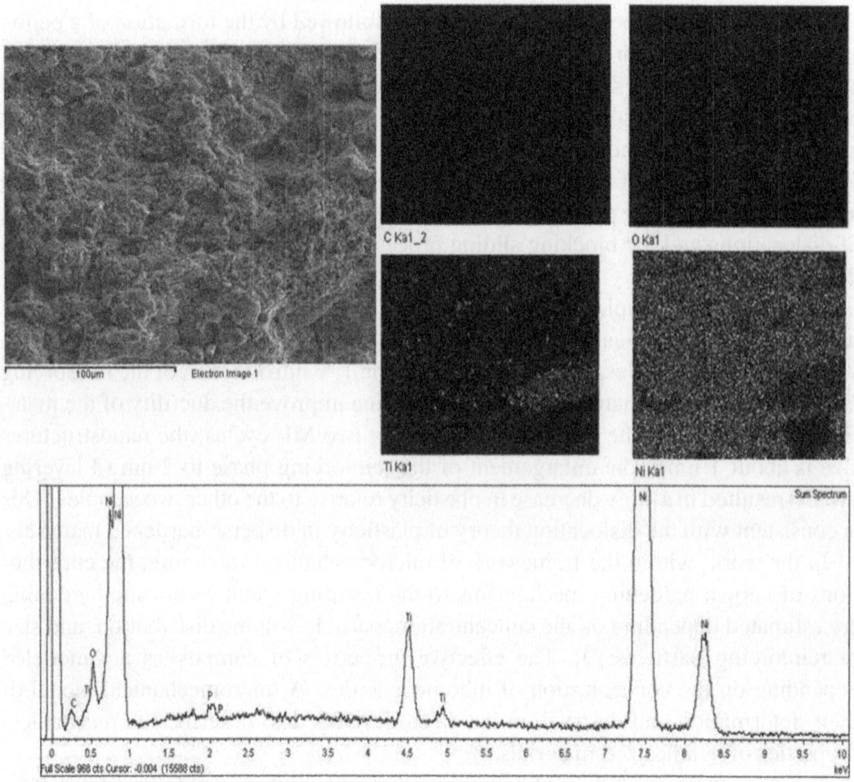

Fig. 38.2: Micrograph and elemental analysis of the chip of a composite based on a nickel matrix and a Ni/TiC reinforcing phase.

Table 38.1: Characteristics of a nickel composite with 1% Ni@TiC.

Samples	σ_B - ultimate strength/ $\sigma_y^{(u)}$ - upper yield strength, MPa	$\sigma_y^{(l)}$ - lower yield strength, MPa	ε_{pl} - length of the flow area, %	ε - elongation, %
TiC, 0.5 nm	223	130	0.645	9.14
TiC, 1 nm	239	198	0.448	10.76
TiC, up to 2 nm	245	170	0.490	6.80
Ni	221	-	-	2.29

First, dislocations are bent (Orowan bending), followed by the formation of a pelion around the nanostructures. This makes it difficult for dislocations to spread further and increases the yield strength. In addition, the interaction between the stress fields created by the dislocation loop makes it difficult for the dislocations to bend further around the carbide nanostructures. Further dislocation movement increases the effective yield strength of the composite [1, 6]. In dispersed-hardened materials, when loaded, dispersed filler particles strengthen the material by resisting the movement of dislocations and, by blocking sliding processes in the matrix, complicate plastic deformation.

In the obtained samples, the hardening phase is not particles, but TiC nanostructures up to 2 nm chemically bonded to the substrate. This made it possible to form a yield point in all samples, taking into account the 1 % introduction of the reinforcing phase into the nickel matrix, and at the same time improve the ductility of the material. The sample had the highest plasticity after two ML cycles (the nanostructures size is about 1 nm). The enlargement of the reinforcing phase to 2 nm (3 layering cycles) resulted in a 40 % decrease in plasticity relative to the other two samples. This is consistent with the dislocation theory of plasticity in disperse-hardened materials.

In the work, within the framework of micromechanical modeling, the contributions of known hardening mechanisms to the resulting metal composite hardening are estimated depending on the concentration, particle volume distribution, and size of reinforcing particles [9]. The effective properties of composites are modeled depending on the concentration of inhomogeneities. A micromechanical model of their deformation and destruction has been developed to describe the mechanical properties of synthesized materials.

To assess the contribution of hardening, the contributions due to the Hall-Petch-Zener mechanism and the mechanism of dispersion hardening were considered [3, 5]. The effect of reinforcing nanoparticles on grain size during recrystallization was taken into account, as well as the fact that the diameter of the reinforcing particles is equal to the thickness of the carbide coating on dispersed particles, i.e. during recrystallization, they do not merge, and the particles are evenly distributed over the volume [4]. The experimentally obtained hardening of the composite corresponds to the theoretically calculated one. This study can help to predict the hardening degree of the resulting composite materials by theoretically assessing the contribution of hardening mechanisms.

38.3 Conclusion

Despite the small TiC volume fraction in the nickel composite structure, carbide affect the material strength and significantly improve its ductility. This fact confirms that the hardening depends not only on the physico-mechanical properties of the reinforcing particles, their shape and concentration and distribution over the matrix, as well as on the presence of a bond at the particle-matrix interface.

The absence of interfacial boundaries between the nickel matrix particles and the carbide nanostructures made it possible to minimize the internal porosity of the sample. It was possible to significantly affect the uniformity of the distribution of the reinforcing phase in the nickel matrix and improve the wettability of the reinforcing phase by the matrix material by introducing TiC nanostructures rather than TiC particles into the matrix and avoiding defects and pores between the particles in the bulk sample [2].

An important feature of the developed composite is the absence of explicit interface boundaries between the metal matrix and the reinforcing element. This ensures the bonding of the matrix and the reinforcing phase into a single whole. This aspect is achieved due to the fact that the nanostructures of the reinforcing phase are connected to the metal matrix by a chemical bond. The atoms of the reinforcing material and the matrix material, connected to each other by chemical bonds through a conventional interface, belong simultaneously to two coordinated structures, as a result of which the resulting conventional interface is characterized by low energy and, consequently, high strength.

The increase in the MMC strength is directly related to a decrease in the particle size of the reinforcing phase to several nm. This is due to an increase in the specific area of interaction between the grains of the reinforcing phase and the grains of the matrix, and a decrease in the size of material defects, which are simultaneously the particles of the reinforcing phase. To achieve the maximum possible effect of strengthening the metal matrix, it is necessary to introduce reinforcing particles with the smallest possible size into the matrix, as well as to achieve the maximum possible uniformity of reinforcement throughout the entire volume of the material.

Acknowledgements This work was supported by the Russian Science Foundation (no. grant 25-21-00376). The authors express their gratitude to the Scientific Park of St. Petersburg State University, RC "Center for Research of Extreme Conditions of Materials and Structures", RC "Nanotechnology".

References

[1] Ferguson JB, Lopez H, Kongshaug D, Schultz B, Rohatgi P (2012) Revised Orowan strengthening: Effective interparticle spacing and strain field considerations. Metallurgical and Materials Transactions A 43(6):2110–2115, DOI 10.1007/s11661-011-1029-9

[2] Gu D, Zhang H, Dai D, Xia M, Hong C, Poprawe R (2019) Laser additive manufacturing of nano-TiC reinforced Ni-based nanocomposites with tailored microstructure and performance. Composites Part B: Engineering 163:585–597, DOI 10.1016/j.compositesb.2018.12.146

[3] Kumaraswamy J, Kumar V, Purushotham G (2021) A review on mechanical and wear properties of ASTM a 494 M grade nickel-based alloy metal matrix

composites. Materials Today: Proceedings 37(2):2027-2032, DOI 10.1016/j.matpr.2020.07.499

[4] Lee GH, Kang S (2006) Sintering of nano-sized WC–Co powders produced by a gas reduction–carburization process. Journal of Alloys and Compounds 419(1-2):281-289, DOI 10.1016/j.jallcom.2005.09.060

[5] Popov VA, Marmulev AV, Kondratenkov MY (2005) Teoreticheskaya otsenka vozmozhnosti polucheniya metallomatrichnykh kompozitov s malym razmerom uprochnyayushchikh chastits (in Russ., Theoretical assessment of the possibility of obtaining metal matrix composites with a small size of reinforcing particle). Izvestiya vysshikh uchebnykh zavedeniy. Tsvetnaya metallurgiya (1):52-56

[6] Somashekar HM, Gowthama K, Suresha V (2020) Microstructure, mechanical properties and sliding wear Behavior of silicon carbide reinforced Al 6026 composites. International Journal of Mechanical and Manufacturing Engineering Research and Development 10(1):259–270

[7] Zemtsova EG, Monin AV, Smirnov VM, Semenov BN, Morozov NF (2017) Synthesis of ceramic composites using three-dimensional nanostructuring (reinforcement) of aluminum matrix with TiN and SiC nanostructures and study of their mechanical properties. Physical Mesomechanics 20(4):438-446, DOI 10.1134/S1029959917040099

[8] Zemtsova EG, Morozov PE, Semenov BN, Morozov NF, Smirnov VM (2019) Chemical preparation of the surface of a porous Nickel matrix for conducting the synthesis of TiC nanolayer. Russian Journal of Genaral Chemistry 89:169-171, DOI 10.1134/S1070363219010328

[9] Zemtsova EG, Yurchuk DV, Morozov PE, Kudymov VK, Smirnov VM (2021) Features of the synthesis of the dispersed tic phase with nickel nanostructures on the surface to create an aluminum-based metal composite. Nanomaterials 11(10):2499, DOI 10.3390/nano11102499

Chapter 39
Extended Hamiltonian Formalism: Geometric Description of Dissipative Systems, Field Theories, and Quantum Deformations

Boris A. Zimin, Yuri V. Sudenkov, and Elizaveta S. Yalych

Abstract This paper presents a unified geometric framework for extending Classical Hamiltonian Mechanics to incorporate dissipative and irreversible processes. We demonstrate that dissipation can be rigorously encoded through an algebraic deformation of the symplectic structure, specifically by augmenting the canonical anti-symmetric Poisson bracket with a symmetric dissipative tensor. This approach, which aligns with the Metriplectic formalism, allows for the derivation of equations of motion for damped linear and nonlinear (Duffing) oscillators while preserving the geometric integrity of the phase space. The framework is further extended to infinite-dimensional continuum mechanics, successfully recovering the dissipative Klein-Gordon (Telegrapher's) equation from first principles. Finally, we explore the profound structural isomorphism between this classical dissipative deformation and the deformed commutation relations characteristic of the Generalized Uncertainty Principle and non-commutative geometry, suggesting a fundamental link between macroscopic irreversibility and the quantum structure of spacetime.

Key words: Extended Hamiltonian formalism, Dissipative systems, Metriplectic formalism, Continuum mechanics

39.1 Introduction

Classical mechanics, standing as the immutable bedrock of the physical description of the macroscopic world, has historically evolved through a process of sequen-

Boris A. Zimin · Yuri V. Sudenkov
St. Petersburg State University, 7/9 Universitetskaya Emb., St. Petersburg, 199034, Russian Federation, e-mail: bazimin@mail.ru, y.sudenkov@yandex.ru

Elizaveta S. Yalych (corresponding author)
Baltic State Technical University, 1st Krasnoarmeyskaya St., Building 1, St. Petersburg, 190005, Russian Federation, e-mail: ialych_es@voenmeh.ru

H. Altenbach and A. V. Sahakyan, *Contact Mechanics and Relevant Problems*, Advanced Structured Materials 253, https://doi.org/10.1007/978-3-032-20869-9_39

tial sophistication and the abstraction of mathematical structures. The traditional exposition of classical mechanics typically begins with the Lagrangian formalism, which is presented as a natural and logical continuation of Newtonian vector mechanics. Where the Newtonian approach operates with vector quantities—forces and accelerations—within the tangible realm of Euclidean space, the Lagrangian method elevates the description to the manifold of configuration space, utilizing scalar functions of energy to determine the system's evolution. These theories, by their very nature, contain second-order derivatives with respect to time and, consequently, lead to systems of second-order differential equations.

However, theoretical inquiry did not arrest its development at the level of configuration space. From the general theory of differential equations, a fundamental property is well understood: a system of equations of the second order can be formally reduced to a system of the first order if the number of independent functions is increased. This mathematical technique, known in mechanics as the Legendre transformation, engenders the mechanics of Hamilton. This formalism does not operate with coordinates q and velocities $\dot{q}$, but with coordinates q and canonical momenta p.

Hamiltonian Mechanics (HM) has played an exceptional and transformative role in the construction of Quantum Mechanics (QM) [20]. The algebraic structure of Poisson brackets became the direct prototype for quantum commutators, while the Hamiltonian function itself was identified as the operator of evolution. Nevertheless, in its canonical form, HM describes conservative systems, where the energy is strictly conserved, and the phase volume remains incompressible in accordance with Liouville's theorem. This "incompressibility" is the hallmark of the symplectic geometry that underpins the standard theory.

However, the physical world is replete with processes of dissipation—friction, viscosity, and heat transfer. It has emerged that HM admits non-trivial generalizations, in distinct contrast to Lagrangian Mechanics (LM), which manifests clearly in the problems of modern physics, such as the nature of the quantum description and quantum gravity [3, 20].

39.2 Fundamental Differences of Variational Principles

39.2.1 Dimension of the Variational Task

The deep distinction between the Lagrangian and Hamiltonian approaches lies in the structure of the calculus of variations. Primarily, in Lagrangian theory, the functional of action S is varied with respect to n functions of time $q_i(t)$ (where n is the number of degrees of freedom), whereas in HM, the variation occurs with respect to $2n$ independent functions $(q_i(t), p_i(t))$.

This doubling of the number of variables fundamentally alters the topology of the space of paths. In the Lagrangian framework, the variational problem is posed on the tangent bundle TM of the configuration manifold. The velocities $\dot{q}$ are not independent variables, but are rigidly locked to the time derivatives of the coordinates. The variation δq propagates to $\delta\dot{q}$ through differentiation. Consequently, if in LM any two points of the configuration space $q(t_1)$ and $q(t_2)$, taken at moments t_1 and t_2, can be connected by a curve simulating the action (the Hamilton-Ostrogradsky principle), the situation in HM is topologically distinct [19].

The phase space of dimensions $2n$ imposes far more rigid restrictions on the structure of boundary conditions. In the Hamiltonian formulation, momenta p are independent variables in the functional action

$$S_H = \int (p\dot{q} - H)dt.$$

They are not defined a priori as partial derivatives of a Lagrangian; rather, their relationship to velocity is a consequence of the equations of motion derived from the variation. This independence implies that the "path" in phase space is a curve in a $2n$-dimensional manifold, and the boundary conditions must be compatible with the symplectic structure that pairs q and p [22].

39.2.2 Boundary Conditions and the Problem of Outside-Integral Terms

In Lagrangian mechanics, at the moment t_1, $2n$ constants are specified (initial coordinates and velocities), which fully determines the trajectory. In the variational principle, instead of this, one can typically fix n initial coordinates $q(t_1)$ and n final coordinates $q(t_2)$. This allows one to turn the variations of the coordinates at the boundaries of integration to zero, destroying the outside-integral terms (boundary terms) of the form

$$\left.\frac{\partial L}{\partial \dot{q}}\delta q\right|_{t_1}^{t_2}.$$

Because $\delta q(t_1) = 0$ and $\delta q(t_2) = 0$, these terms vanish, leaving the Euler-Lagrange equations as the sole condition for stationarity [1]. In Hamiltonian mechanics, the situation regarding boundary conditions is more subtle and reveals the "rigidity" of the phase space geometry. The constants $2n$ (coordinates and momenta) necessarily fix the state of the system at moment t_1 (the Cauchy problem). By virtue of the fact that Hamilton's equations are equations of the first order

$$\dot{q} = \frac{\partial H}{\partial p},$$
$$\dot{p} = -\frac{\partial H}{\partial q}.$$

the specification of the initial point in the phase space uniquely determines the entire trajectory. Consequently, the final point in phase space (PS) can no longer be fixed arbitrarily - it is the strictly deterministic result of the evolution.

This circumstance leads to the fact that the formulation of the variational principle in phase space required substantial clarification. The functional of action in phase space has the form

$$S_H[q,p] = \int_{t_1}^{t_2} \left(p_i \dot{q}^i - H(q,p,t)\right) dt.$$

The variation of this functional yields

$$\delta S_H = \int_{t_1}^{t_2} \left[\left(\dot{q}^i - \frac{\partial H}{\partial p_i}\right)\delta p_i - \left(\dot{p}i + \frac{\partial H}{\partial q^i}\right)\delta q^i\right] dt + \left[p_i \delta q^i\right] {t_1}^{t_2}.$$

For the integral to vanish (yielding the equations of motion), the integral term must be zero. However, the presence of the boundary term (or "outside-integral member")

$$\left[p_i \delta q^i\right]_{t_1}^{t_2}$$

requires careful handling. To make the boundary term vanish, one must either:

1. Fix the coordinates at the ends ($\delta q = 0$), which effectively returns the problem to the Lagrangian approach where endpoints in the configuration space are pinned, but momenta at the boundaries are left free to vary.
2. Add special boundary terms to the action itself. This approach is particularly relevant for theories with boundaries and in the context of holography, where the "boundary physics" encodes the bulk dynamics [14].

This problem of "outside-integral terms" is not merely a technicality; it reflects the deep symplectic topology of Hamiltonian systems. Unlike Riemannian geometry, where lengths are local, symplectic geometry encodes global area-preserving constraints. The boundary term $[p\delta q]$ represents the symplectic flux through the boundary of the time interval. In extended formalisms dealing with dissipation, where the symplectic structure itself may degenerate or deform, the handling of these terms becomes central to ensuring the variational principle remains well-defined [14].

39.2.3 Dual Nature of Laws of Motion in HM

Information about the laws of motion in HM is contained not in a single function (the Lagrangian), but is distributed between two independent geometric structures: the Hamiltonian function H (the generator of dynamics) and the symplectic 2-form ω (the geometry of the space). The equations of motion can be written in a coordinate-free geometric form as

$$\dot{x} = \{x, H\}_\omega$$

or

$$i_X\omega = -dH.$$

This separation allows for the modification of the equations of motion in two distinct ways:

1. Changing the Hamiltonian: this is the standard physics approach - adding potentials, interaction terms, or external fields. This modifies the energy landscape H but leaves the geometry ω (and thus the Poisson brackets) untouched.
2. Deforming the Symplectic Structure: one can deform the symplectic structure itself (changing the Poisson brackets).

The second path opens the route to the description of dissipation. In standard mechanics, dissipation is often added as an external non-conservative force

$$F_{\text{diss}} = -\gamma v.$$

By introducing a symmetric component to the symplectic form (or rather, to the Poisson tensor), one can generate irreversible flows that contract phase volume, a feature impossible in standard symplectic geometry due to Liouville's theorem. This "dual" capability - modifying geometry rather than just energy - is the cornerstone of the Extended Hamiltonian Formalism.

39.3 Generalizations of Hamiltonian Mechanics: from Symplectic to Dissipation

39.3.1 Standard Hamiltonian Structure

In standard HM, three of its main components are specified:

1. Phase Space: an even-dimensional differentiable manifold M^{2n}. The even dimensionality is a topological necessity for the existence of a non-degenerate symplectic form.
2. Symplectic Structure: a closed, non-degenerate 2-form

$$\omega^2 = \omega_{ij}(x)dx^i \wedge dx^j$$

on it, where $x \in M^{2n}$. The non-degeneracy condition ensures the existence of an inverse tensor - the Poisson tensor J^{ij}. The closure condition

$$d\omega = 0$$

is equivalent to the Jacobi identity for the Poisson brackets.

3. Hamiltonian Function: a scalar function

$$H \equiv H(p, q),$$

representing the total energy of the system.

The presence of a non-negative parameter t (time), on which all functions $f(x)$ depend, determines the evolution via the Poisson bracket

$$\dot{f} = \{f, H\} = \omega^{ij} \partial_i f \partial_j H,$$

where

$$\omega^{ik} \omega_{kj} = \delta_i^j,$$

and the bracket satisfies the identity of Jacobi, antisymmetry, and the Leibniz rule.

The triplet (PS, ω, H) forms the classical Hamiltonian structure. This structure is incredibly robust but rigid; it enforces conservation laws that idealize real physical systems. Real systems, especially in continuum mechanics and thermodynamics, are rarely conservative.

39.3.2 Modifications and the Problem of Irreversibility

Any change in the elements of the Hamiltonian structure leads to a modification of the mechanics. These modifications are widely used in solving fundamental problems: the nature of the quantum state, quantum gravity, and superstring theory. However, of particular interest are modifications of HM in the mechanics of complex media and the theory of fracture.

The refusal of the symplecticity of the manifold (phase space) or the introduction of degeneracy into the Poisson tensor allows one to describe dissipative systems with nonlinear behavior within the framework of HM. In classical symplectic geometry, the phase volume is conserved (Liouville's theorem), which prohibits the existence of asymptotically stable points (attractors), which are characteristic of dissipative systems. To include dissipation, it is necessary to violate the symplectic symmetry. This can be done in several ways:

1. Metriplectic Formalism (Metriplectic Formalism): adding a symmetric, metric bracket to the antisymmetric Poisson bracket. This formalism explicitly couples the reversible Hamiltonian dynamics with irreversible entropy-driven dynamics.
2. Contact Geometry: using odd-dimensional manifolds, where Liouville's theorem in its classical form does not work, which allows one to describe systems with friction [10]. Contact manifolds (M^{2n+1}) possess a contact structure that is the odd-dimensional analogue of the symplectic structure. They are naturally suited for thermodynamics and dissipative mechanics because the flow of the Reeb vector field (the contact analogue of the Hamiltonian vector field) does not necessarily preserve the volume form, allowing for contraction to attractors.

3. Deformation of the Symplectic Matrix: direct introduction of diagonal members into the matrix of the Poisson structure, which we will analyze in detail below. This method is algebraic and allows for a smooth transition from conservative to dissipative dynamics via a deformation parameter.

39.3.3 *Linear Oscillator with Friction: Deformation of the Symplectic Matrix*

Let us consider a classical example demonstrating the power of the deformation method. Within the framework of HM, we take the standard Hamiltonian of a harmonic oscillator

$$H = \frac{\omega}{2}(p^2 + q^2).$$

Then the standard Hamilton equations can be written in matrix form

$$\frac{d}{dt}\begin{pmatrix} q \\ p \end{pmatrix} = \begin{pmatrix} 0 & 1 \\ -1 & 0 \end{pmatrix}\begin{pmatrix} \partial_q H \\ \partial_p H \end{pmatrix},$$

where

$$\partial_q H = \omega q$$

and

$$\partial_p H = \omega p.$$

This leads to the standard equation of oscillation without damping

$$\ddot{q} + \omega^2 q = 0.$$

We deform the symplectic matrix Ω, adding to it a symmetric part proportional to a small parameter α:

$$\Omega_{\text{def}} = \begin{pmatrix} 0 & 1 \\ -1 & 0 \end{pmatrix} - \alpha \begin{pmatrix} 1 & 0 \\ 0 & 1 \end{pmatrix} = \begin{pmatrix} -\alpha & 1 \\ -1 & -\alpha \end{pmatrix}.$$

Here α plays the role of the deformation parameter, violating the conservation of phase volume. The matrix ceases to be skew-symmetric, which takes us beyond the limits of classical Poisson geometry. Using HM with the deformed matrix, we write the equations

$$\frac{d}{dt}\begin{pmatrix} q \\ p \end{pmatrix} = \Omega_{\text{def}}\begin{pmatrix} \partial_q H \\ \partial_p H \end{pmatrix}.$$

Hamilton's equations of motion with the generalized matrix Ω represent the system

$$\dot{q} = -\alpha(\omega q) + 1(\omega p) = \omega p - \alpha\omega q,$$

$$\dot{p} = -1(\omega q) - \alpha(\omega p) = -\omega q - \alpha\omega p.$$

Excluding the momentum variable p from the first equation

$$p = \frac{1}{\omega}(\dot{q} + \alpha\omega q)$$

and substituting into the second, we obtain an equation of the second order

$$\ddot{q} + 2\alpha\omega\dot{q} + \omega^2(1 + \alpha^2)q = 0.$$

For $\alpha > 0$, the obtained equation describes damped oscillations. The parameter $2\alpha\omega$ acts as the coefficient of friction. For $\alpha < 0$, HM describes non-equilibrium processes with an influx of energy (negative friction/pumping). It can be noted that in this case the natural frequency of oscillations changes

$$\omega_{\text{eff}} = \omega\sqrt{1 + \alpha^2},$$

which is important when considering the deformation of the oscillatory cell and frequency shifts in resonators.

This frequency shift is a critical prediction of the geometric deformation model that distinguishes it from ad-hoc friction models. In standard Newtonian mechanics, adding a damping term typically lowers the resonance frequency

$$\omega_{\text{d}} = \omega\sqrt{1 - \zeta^2}.$$

However, the geometric deformation of the symplectic structure predicts an increase (or different modification depending on the sign of α^2) in the effective frequency. This theoretical insight finds relevance in the physics of oscillatory cells and resonators [4].

The geometric formalism suggests that the "friction" is not just a drag force but a modification of the time-evolution operator itself, leading to a fundamental shift in the system's spectrum. For example, in high-Q resonators used in MEMS or optical cavities [17], frequency shifts are precise indicators of loss mechanisms (dissipation). The "deformed" Hamiltonian predicts that the measured frequency shift is a signature of the specific geometric nature of the dissipation mechanism. This method of matrix deformation is closely related to conformal Hamiltonian dynamics [8]. In conformal dynamics, a Liouville field Z is added to the Hamiltonian vector field X_H, which pulls the phase flow to an attractor, modeling linear friction. The deformed matrix Ω_{def} effectively combines the symplectic tensor J (rotation) with a metric tensor G (contraction).

39.3.4 Nonlinear Systems: Duffing Equation.

The method of deformation of the symplectic matrix is universal and applicable to nonlinear systems. Similarly, one can build a model of a nonlinear oscillator with a solution. We choose the Hamiltonian in the form

$$H = \frac{\omega}{2}(p^2 + q^2) + \frac{\varepsilon}{4}q^4.$$

We deform the symplectic matrix in the same way

$$\Omega = \begin{pmatrix} -\alpha & 1 \\ -1 & -\alpha \end{pmatrix}.$$

Then the Hamiltonian equations of motion with the matrix Ω will be:

$$\frac{d}{dt}\begin{pmatrix} q \\ p \end{pmatrix} = \Omega \nabla H.$$

Expanding this, we get

$$\dot{q} = \omega p - \alpha(\omega q + \varepsilon q^3),$$

$$\dot{p} = -(\omega q + \varepsilon q^3) - \alpha \omega p.$$

Excluding p and $\dot{p}$, we find the equation of motion

$$\ddot{q} + \alpha\omega(2 + \ldots)\dot{q} + \omega^2 q + \varepsilon\omega q^3 + \cdots = 0.$$

The final result

$$\ddot{q} + (\alpha\omega + 3\alpha\varepsilon q^2)\dot{q} + \omega^2 q + \varepsilon\omega q^3 = 0.$$

This equation represents the nonlinear Duffing equation with friction (including nonlinear friction). Thus, a simple algebraic change in the structure of the brackets (matrix Ω) allows generating complex dissipative dynamics, while maintaining the quasi-Hamiltonian form of the record.

The appearance of the nonlinear friction term $3\alpha\varepsilon q^2\dot{q}$ is particularly interesting. In standard phenomenological modeling, one might simply add a linear term $-\gamma\dot{q}$ to the Duffing equation. However, the geometric deformation dictates that the dissipation must share the symmetry of the potential (via the ε term). This suggests that in systems where dissipation arises from the same microscopic coupling as the nonlinearity, the "geometric" friction model may be more physically accurate than the standard additive model. This has implications for multi-scroll chaotic systems and the analysis of attractors in nonlinear dynamics [18].

39.4 Metriplectic Formalism: Thermodynamic Consistency

Although the matrix deformation described above is effective, it is a special case of a more fundamental theory known as the Metriplectic formalism. This approach was developed by Morrison, Kaufman, and Grmela to combine Hamiltonian mechanics and thermodynamics.

39.4.1 Structure of the Metriplectic Bracket

At the heart of the metriplectic approach lies the idea of splitting dynamics into two parts: conservative (reversible) and dissipative (irreversible). The equation of motion for an arbitrary observable F is written as the sum of two brackets [9]

$$\frac{dF}{dt} = \{F, H\} + (F, S).$$

Here $\{F, H\}$ is the classical Poisson bracket. It is antisymmetric ($\{A, B\} = -\{B, A\}$), satisfies the Jacobi identity, and generates reversible dynamics preserving the energy H. (F, S) is the metric (dissipative) bracket. It is symmetric ($(A, B) = (B, A)$), often semidefinite, and generates irreversible relaxation to equilibrium. The generator of this part acts not as energy, but as entropy S [12].

39.4.2 Casimir Invariants and Laws of Thermodynamics

To ensure thermodynamic consistency (First and Second Laws of Thermodynamics), rigid degeneracy conditions, called conditions of metriplectic completeness, are imposed on the brackets [12]:

1. Conservation of Energy: the Hamiltonian H must be a Casimir invariant for the dissipative bracket
$$(F, H) = 0 \quad \forall F.$$
This guarantees that the dissipative process does not change the total energy of the system (energy passes from mechanical form to thermal internal energy, but H_{total} is conserved). This is crucial for modeling closed systems where friction heats the medium rather than disappearing from the universe.
2. Conservation of Entropy in Reversible Dynamics: entropy S must be a Casimir invariant for the Poisson bracket
$$\{F, S\} = 0 \quad \forall F.$$
This means that adiabatic (Hamiltonian) processes do not change entropy (isentropicity of an ideal fluid).
3. Growth of Entropy: the dissipative bracket must be positively semidefinite on entropy:
$$\frac{dS}{dt} = \{S, H\} + (S, S) \geq 0.$$

This ensures the fulfillment of the Second Law of Thermodynamics (H-theorem).

The example with the matrix

$$\Omega = J - \alpha I$$

can be interpreted as a simplified metriplectic system where H is not conserved (open system, energy dissipates into the external environment), and entropy is not explicitly introduced. The Metriplectic formalism is more general, as it describes closed systems where "lost" mechanical energy goes to heating.

39.4.3 Table of Comparison of Formalisms

In Table 39.1, a comparison of three approaches to the description of dynamical systems is presented. This comparison highlights that the Metriplectic framework is the only one that rigorously conserves total energy while mandating entropy production, making it the suitable geometric language for non-equilibrium thermodynamics [12].

39.5 Foundations of the Hamiltonian Approach in Continuum Mechanics

39.5.1 Functional Description of Fields

At the basis of the Hamiltonian approach to continuous media lies the assignment of the Hamiltonian of the system as a functional of dynamic variables (fields). This is a transition from a finite-dimensional phase space to an infinite-dimensional

Table 39.1: Approaches to the description of dynamic systems.

Characteristic	Hamiltonian Mechanics	Dissipative (Rayleigh)	Metriplectic
Generator	H (Energy)	H (Energy) and $\mathcal{R}$ (Dissipation func.)	$F = H - TS$ (Free energy)
Bracket Structure	Antisymmetric $\{,\}$	No unified algebra	Mixed $\{,\} + (,)$ (Poisson + metric)
Geometry	Symplectic	Riemannian (partially)	Symplectic + Metric
Energy dynamics	Conserved ($\dot{H} = 0$)	Decreases ($\dot{H} < 0$)	Conserved ($\dot{H}_{tot} = 0$)
Entropy Dynamics	Constant (isentropic)	Not defined explicitly	Increases ($\dot{S} \geq 0$)
Application	Ideal systems, QM	Engineering, mechanics	Plasma, hydrodynamics, polymers

functional space. The total Hamiltonian H of the system is defined as an integral over the volume

$$H \equiv \int d^3x\, \mathcal{H}(x).$$

The specific Hamiltonian $\mathcal{H}$ will be a function of the generalized coordinates of the field $\eta_s(x_k, t)$, their spatial derivatives $\partial\eta_s/\partial x_i$, and specific canonical momenta $\pi_s(x_k, t)$. Explicit dependence on time t is also possible [19, 20]:

$$h = h\left(\eta_s, \pi_s, \frac{\partial \eta_s}{\partial x_i}, t\right).$$

Then the Hamilton equations are written using functional (variational) derivatives

$$\frac{\partial}{\partial t}\begin{pmatrix}\eta_s\\ \pi_s\end{pmatrix} = \begin{pmatrix}0 & 1\\ -1 & 0\end{pmatrix}\begin{pmatrix}\delta H/\delta\eta_s\\ \delta H/\delta\pi_s\end{pmatrix},$$

where the variational derivative is defined as [3]

$$\frac{\delta H}{\delta \eta_s} = \frac{\partial h}{\partial \eta_s} - \sum_{j=1}^{n} \frac{d}{dx_j}\frac{\partial h}{\partial(\partial\eta_s/\partial x_j)}.$$

39.5.2 Dissipative Klein-Gordon Equation

We demonstrate the Hamiltonian approach on the example of a relativistic wave equation. We choose the Hamiltonian in the form

$$H = \int \left(\frac{c^2\pi^2}{2} + \frac{1}{2}\left(\frac{\partial u}{\partial x}\right)^2 + \frac{au^2}{2}\right) dx,$$

where u is the field of displacements in a continuous medium (or a scalar field), c is the wave velocity, a is the mass/potential parameter. We use the deformed symplectic matrix Ω (with dissipative parameter α) to find the equation of motion of the continuous medium (one-dimensional case)

$$\frac{d}{dt}\begin{pmatrix}u\\ \pi\end{pmatrix} = \Omega\left(\delta H/\delta u\ \delta H/\delta\pi\right).$$

Calculating the variational derivatives

$$\frac{\delta H}{\delta u} = au - \frac{\partial^2 u}{\partial x^2}, \qquad \frac{\delta H}{\delta \pi} = c^2\pi$$

and substituting them into the equation with the matrix

$$\Omega = \begin{pmatrix} -\alpha & 1 \\ -1 & -\alpha \end{pmatrix},$$

we obtain the system

$$\dot{u} = -\alpha(au - u_{xx}) + c^2\pi,$$
$$\dot{\pi} = -(au - u_{xx}) - \alpha c^2\pi.$$

With small α and certain assumptions (or with redefinition of variables, as is often done in telegraph equations), one can exclude the impulse π. Result of the derivation:

$$\frac{\partial^2 u}{\partial t^2} - c^2\frac{\partial^2 u}{\partial x^2} + \alpha c\frac{\partial u}{\partial t} + ac^2 u = 0.$$

This is the Klein-Gordon equation with friction (or telegraph equation). It describes the propagation of waves in a dissipative medium and has fundamental importance not only in mechanics but also in quantum field theory and cosmology.

The fact that such equations can be obtained from the geometric deformation of the Hamiltonian structure indicates a deep connection between the geometry of the phase space and the physics of irreversible processes. The derivation of the Telegrapher's Equation via this geometric route is significant because it provides a consistent Hamiltonian for a system that is usually considered "non-Lagrangian" due to the first-order time derivative term (friction) [6]. In standard treatments, one must resort to time-dependent Lagrangians ($e^{\gamma t}L$) or non-variational approaches. The extended formalism allows the system to remain autonomous (no explicit time dependence in H) while capturing the dissipation via the algebra Ω.

39.6 Connection with Quantum Mechanics and String Theory

39.6.1 Quantization of Dissipative Systems

Hamiltonian mechanics serves as a bridge to quantization. To quantize a system with friction, the classical Caldirola-Kanai approach is used, which uses an explicitly time-dependent Hamiltonian

$$H(t) = e^{\gamma t}\frac{p^2}{2m} + e^{-\gamma t}\frac{m\omega^2 x^2}{2}.$$

This allows one to obtain equations for a damped oscillator, but leads to problems with the interpretation of quantum states (disappearing uncertainty or violation of canonical commutators).

The Metriplectic approach and deformation of Poisson brackets offer an alternative. The deformed Poisson bracket $\{x, p\}_\alpha$ in classics can correspond to a deformed commutator in quantum mechanics

$$[x, p] = i\hbar(1 + \beta p^2).$$

This relation is known as the Generalized Uncertainty Principle (GUP) [2]. GUP arises in theories of quantum gravity and string theory as a consequence of the existence of a minimal length. The mathematical structure of GUP (deformation of algebra) is analogous to the structure introducing classical dissipation.

39.6.2 Non-Commutative Geometry

In high-energy physics and string theory, non-commutative geometries are widely used, where coordinates do not commute

$$[x, y] = i\theta.$$

This is equivalent to introducing a constant magnetic field

$$B \sim 1/\theta$$

into the phase space, which deforms the symplectic structure [11]. In fluid mechanics, this corresponds to vorticity. Metriplectic extensions of such systems allow describing viscous fluids with non-trivial topology (for example, vortices in superfluid helium or plasma). The deformation of the symplectic matrix creates a "magnetic-like" term in the equations of motion that is mathematically isomorphic to the Berry curvature effects in quantum systems [21].

39.7 Conclusion

The presented analysis demonstrates that Hamiltonian mechanics is a flexible geometric language capable of describing the most complex dissipative processes:

1. Hamiltonian mechanics includes Lagrangian mechanics as a special case, but thanks to the doubling of variables (q, p), it possesses greater flexibility for introducing deformations.
2. Dissipation is no longer an "engineering additive" in the equations of motion. Through the metriplectic formalism and deformation of symplectic matrices, it acquires a strict geometric status, equal to conservative dynamics. The corresponding deformed matrix can be represented as a standard symplectic and some additional (metric) matrix, which is equivalent to the inclusion of an additional force or additional field in a continuous medium.
3. The method scales to infinite-dimensional systems, allowing the derivation of equations of the Klein-Gordon type with friction and the Telegraph equation from first principles [6]. In this theory, there is no standard Darboux transformation, but the absence of triviality allows one to include "non-Lagrangian" systems

in the consideration, opening the way to the description of non-equilibrium thermodynamics, quantum decoherence, and, possibly, quantum gravity.

Acknowledgements The work was carried out with the financial support of the Russian Science Foundation (grant No. 25-29-00218).

References

[1] Banados M, Reyes I (2016) A short review on Noether's theorems, gauge symmetries and boundary terms. International Journal of Modern Physics D 25(10):1630021, DOI 10.1142/S0218271816300214

[2] Bhandari G, Pathak S, Ghotra H (2025) Unified framework for geodesic dynamics with conservative, dissipative, and GUP effects, DOI 10.48550/arXiv.2509.14663

[3] Bogush AA, Moroz LG (2004) Vvedeniye v teoriyu klassicheskikh poley (in Russ., Introduction to the Theory of Classical Fields, 2nd ed). Editorial URSS, Moscow

[4] Cariñena JF, Guha P (2023) Geometry of non-standard Hamiltonian structures of Liénard equations and contact structure. International Journal of Geometric Methods in Modern Physics 21(10):2440005-819, DOI 10.1142/S021988782440005X

[5] Carlier V, Campos Pinto M (2025) A least action principle for visco-resistive Hall magnetohydrodynamic with metriplectic reformulation, DOI 10.48550/arXiv.2502.06840

[6] Gambár K, Márkus F (2023) Generalized Hertz vector in the dissipative electrodynamics. World Journal of Physics 1(1):35-41, DOI 10.56439/WJP/2023.1103

[7] Ghosh A (2025) Invariant measures for some dissipative systems from the Jacobi last multiplier. Physica Scripta 100(3):035220, DOI 10.1088/1402-4896/adb10b

[8] Guha P, Ghose-Choudhury A (2018) Generalized conformal Hamiltonian dynamics and the pattern formation equations. Journal of Geometry and Physics 134:195-208, DOI 10.1016/j.geomphys.2018.07.019

[9] Jing C, Mudiyanselage U, Cho W, Jo, M, Gruber A, Lee K (2025) Meta-learning structure-preserving dynamics, DOI 10.48550/arXiv.2508.11205

[10] de Léon M, Valcázar ML (2019) Contact hamiltonian systems. Journal of Mathematical Physics 60(10):102902, DOI 10.1063/1.50964750

[11] Liang S-D, Lake M (2023) An introduction to noncommutative physics. Physics 5(2)436-460, DOI 10.3390/physics5020031

[12] Materassi M (2016) Entropy as a metric generator of dissipation in complete metriplectic systems. Entropy 18(8):304, DOI 10.3390/e18080304

[13] Materassi M (2015) Metriplectic algebra for dissipative fluids in Lagrangian formulation. Entropy 17(3):1329-1346, DOI 10.3390/e17031329

[14] Materassi M, Tassi E (2012) Algebrizing friction: a brief look at the metriplectic formalism. Intellectual Archive 1(3):45-52, DOI 10.32370/ia_2012_07_3
[15] Morrison PJ, Updike MH (2024) Inclusive curvaturelike framework for describing dissipation: Metriplectic 4-bracket dynamics. Physical Review E 109(4):045202, DOI 10.1103/PhysRevE.109.045202
[16] Nandi P (2022) Some aspects of quantum mechanics and quantum field theory on quantum space-time, DOI 10.48550/arXiv.2209.04758
[17] PAC 2003: Proceedings of the 2003 Particle Accelerator Conference: Portland, Oregon, U.S.A., May 12-16, 2003, American Physical Society. Division of Physics of Beams
[18] Patra T, Ganguli B (2025) Emergence of multi-scroll attractors, DOI 10.48550/arXiv.2511.13332
[19] Prokhorov LV (2008) Gamil'tonova mekhanika i ee obobshchenie (in Russ., Hamiltonian mechanics and its generalizations). Fizika Elementarnykh Chastic i Atomnogo Yadrya 39(5):1565-1569
[20] Prokhorov LV, Shabanov SV (2011) Hamiltonian Mechanics of Gauge Systems, Cambridge Monographs on Mathematical Physics. Cambridge University Press, Cambridge
[21] Resta R (2012) Geometry and Topology in Electronic Structure Theory by Raffaele Resta. University of Trieste, Trieste
[22] Zimin BA, Zorin IS, Sventitskaya VE (2018) On the Hamilton approach of the dissipative systems. 8th Polyakhov's Reading: : Proceedings of the International Scientific Conference on Mechanics, 29 January–2 February 2018, Saint Petersburg, Russia. AIP Conference Proceedings 1959(1):070041, DOI 10.1063/1.5034716

Chapter 40
Modeling of a Shock-Free Actuator made of Shape Memory Material for Transformable Space Structures

Vladimir N. Zimin, Alexey V. Krylov, and Artur O. Shakhverdov

Abstract In the course of developing space technologies, there has been a need to enhance the power of energy sources. This has led to an increase in the size of solar panels, which have become one of the most extensive elements of spacecraft design. Addressing practical issues in the fields of astronomy, communication, and navigation has highlighted the challenge of constructing large-scale radio antennas in space. Research and development in the field of designing large-scale structures for space applications is of significant interest to the scientific community, engineers and designers. These structures must be transported in a tightly packed transport state, enabling transformation from a folded to a deployed state in orbit. This transformation process occurs under the influence of force actuators. In order to prevent dynamic loads that cause shock-like effects on the elements of a transformable system, it is essential to use shock-free actuators with a relatively low mass. It seems advisable to use an actuator with an active element made of titanium nickelide material with a shape memory effect. The temperature field is the controlling factor for these active elements. The phase transformation of the titanium nickelide material are very sensitive to temperature changes, requiring high accuracy in order to ensure it. Experimental and theoretical studies of the active elements of the force actuator made of titanium nickelide material (1.5 mm diameter wire) have confirmed the possibility of generating significant force with a sufficiently low mass and power consumption, as well as shock-free nature of the process of opening large space structures.

Key words: Transformable space structures, Active elements, Shape memory effect

Vladimir N. Zimin (corresponding author) · Alexey V. Krylov · Artur O. Shakhverdov
Bauman Moscow State Technical University, 2nd Baumanskaya str., 5, building 1, Moscow, 105005, Russian Federation, e-mail: `zimin@bmstu.ru`, `kav1982@bmstu.ru`, `shah@bmstu.ru`

H. Altenbach and A. V. Sahakyan, *Contact Mechanics and Relevant Problems*,
Advanced Structured Materials 253, https://doi.org/10.1007/978-3-032-20869-9_40

40.1 Introduction

The development of large-scale space structures involves addressing a range of technical challenges due to the unique nature of these systems, which are characterized by a combination of conflicting requirements, such as increasing geometric dimensions while ensuring sufficient rigidity and minimizing mass [1]. A specific category of such space structures includes transformable systems [4, 10, 11, 16], which have a design that enables automatic configuration changes. These structures are launched into low-Earth orbit in a compact, folded state. Once the required orbit is reached, the system is transformed into working state, as illustrated in Fig. 40.1.

Significant progress has been made in the development of large-scale, transformable space structures. Many designs have seen an increase in the transformation coefficient, which is the ratio of volumes when the structure is in the open, working state compared to when it is in a tightly packed, transport state. The transformation coefficient has increased to a maximum of 30 in some designs [5, 13]. For transformable space antenna systems, the ratio of mass of the working mirror relative to its area has decreased to between 0.5 and 1.5 kg/m^2, and the manufacturing accuracy of parabolic mirrors (reflectors) has improved to within 2×10^{-5} of the diameter.

Further improvement of transformable large-scale space structures may involve optimizing the mass properties of the actuators used to deploy them from their transport state to their working state in orbit. Therefore, the use of an actuator with active elements based on materials with a shape memory effect (SME) represents a rational technical solution for achieving a smooth transformation of the structure with minimal dynamic loading [3, 6, 9, 17]. These actuators have high reliability, simple design, small size and low weight. Titanium nickelide wire is an example of the application of materials with SME as an active element of a force actuator [2, 15]. Due to its unique physical and mechanical characteristics, titanium nickelide has

(a) in the transport state. **(b)** in the working state.

Fig. 40.1: Large-scale transformable space structure.

found widespread application in various areas of mechanical engineering, making it a practical choice for this application [7, 8].

40.2 Experimental Studies of the Active Element made of a Material with Shape Memory Effect

For the design of active elements of force actuators using SME, it is necessary to establish mathematical relationships that describe the deformation of the material under various operational loads. These relationships, known as stress-strain dependencies, can be determined through experimental testing. An experimental study was conducted to investigate the stress-strain characteristics of an active element made of a wire of titanium nickelide material. Two types of loading conditions were applied: in one scenario, the applied force varied during testing, while in the other scenario, it remained constant.

To develop a model of the operation of the active elements of a force actuator with SME, experimental studies were conducted to evaluate the stability of its main characteristics, such as the working stroke (reduction in the relative elongation of the active elements), the generated force (the force generated by the active elements) and the response time. During the testing process, the operating conditions of the active elements of the force actuator under a uniaxial stress-strain conditions was simulated. For this purpose, specially designed experimental setups were used (Fig. 40.2). One setup was designed to prepare the active element of the force actuator for operation. In this setup, the active element was subjected to longitudinal stretching. The change in length was monitored using a wire displacement sensor. The load applied to the active element of the force actuator was measured by a digital dynamometer. At the second setup, the operating characteristics of the active elements of the force actuator made of titanium nickelide (NiTi) wire were studied, including: developed

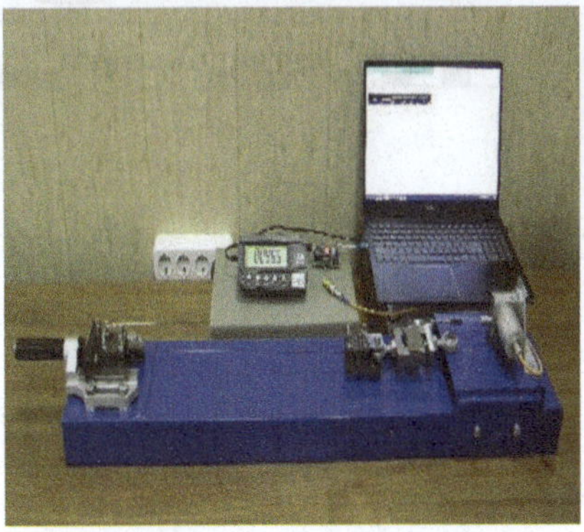

(a) setup for deforming the active element.

(b) setup for determining the main characteristics of the active element.

Fig. 40.2: Experimental setups.

force, maximum displacement, response time and power supply voltage. A series of tests were conducted with various lengths and diameters (0.5 mm, 1.5 mm, 2.1 mm) of active elements.

The controlling factor for the material with SME is the temperature field. Therefore, during the tests, the heating of the active element of the force actuator was carried out by applying a constant electric voltage to the ends of the wire. The value of the current flowing through the active element was recorded using a digital multimeter connected to the electrical circuit, acting as an ammeter. The resulting heating dynamics of the active element was measured in two ways:

1. contact measurement: a thermocouple sensor was attached to the wire and recorded the temperature changes directly,
2. non-contact measurement: a thermal imager was used to measure the radiation emitted by the wire in the infrared range.

Additionally, experiments were conducted to assess the behavior of the wire during repeated sequential cycle of the stretching and actuation. Figure 40.3 shows active elements of various diameters.

Figure 40.4 shows the results of a series of tests with active elements of different diameters. By analyzing the obtained graphs, it can be concluded that the maximum relative decrease in the length (ε_{max}) of the wire active element of the force actuator almost linearly dependent on the applied load. During the experiments, the load on the space structure's force actuator during deployment was simulated using weights suspended from a previously stretched active element.

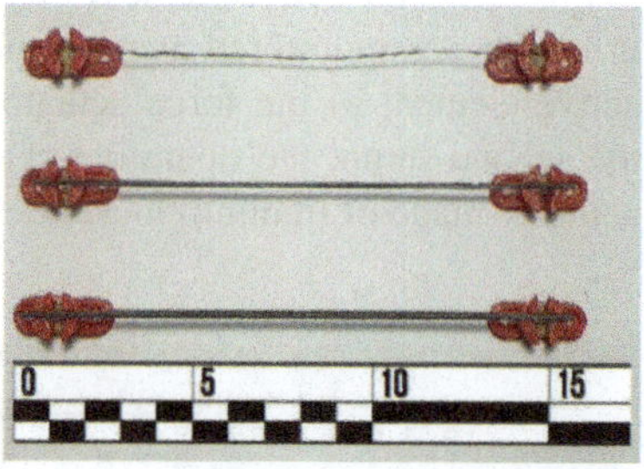

Fig. 40.3 Active elements made of titanium nickelide material with SME: diameter values(from top to bottom): 0.5 mm, 1.5 mm, and 2.1 mm.

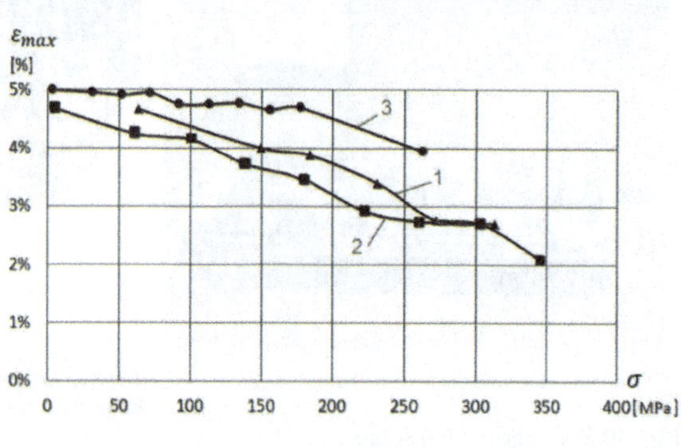

Fig. 40.4 Test results of active elements made of titanium nickelide material:
1 – wire diameter 0.5 mm,
2 – wire diameter 1.5 mm,
3 – wire diameter 2.1 mm.

Based on the experimentally obtained results, a formula can be derived to calculate the relative change in the length of the active element of a force actuator based on the load applied to it

$$\varepsilon_{max} = a\sigma + b, \tag{40.1}$$

where σ is the stress in the cross-section of the active element from the load applied to it; a, b are the coefficients that can be determined from the graphs in Fig. 40.4 for active elements of different diameters.

Therefore, by using formula (40.1), it is possible to estimate the maximum displacement (ε_{max}) that can be achieved using the active element under the required load (developed force) without conducting additional experiments.

Table 40.1 shows the results of a series of experiments conducted to evaluate the performance of active element of a force actuator made of titanium nickelide wire with a diameter of 1.5 mm, through repeated sequential cycle of stretching and actuation. The initial stretch was performed to a relative elongation of 10%. All subsequent stretches were conducted to the same maximum load as during the first stretch: between 0.97 and 0.99 kN. During actuation, the active elements were subjected to a constant load of 0.4 kN. Based on the analysis of the experimental data presented in the table, it can be concluded that by repeating the stretching-actuating cycle within a specific range, the performance characteristics of the active element of the force actuator could be improved.

Table 40.1: Test results of repeated cycle of stretching-actuating.

No.	Initial length, mm	Elongation, mm	Elongation, %	Working length, mm	Working stroke, mm	Working stroke,%
1	160.2	16.13	10.1	173.25	5.33	3.08
2	169.7	11.88	7.0	178.30	6.01	3.37
3	175.3	10.95	6.2	182.40	6.40	3.51
4	179.5	11.34	6.3	187.03	6.79	3.63
5	182.7	11.39	6.2	190.08	7.04	3.70
6	186.4	11.73	6.3	193.78	7.18	3.71
7	189.1	12.71	6.7	197.17	7.43	3.77
8	191.5	13.10	6.8	199.27	7.72	3.87
9	194.8	12.90	6.6	202.38	8.01	3.96
10	197.8	13.29	6.7	205.72	8.21	3.99
11	201.1	12.51	6.2	208.53	7.82	3.75
12	204.0	12.51	6.1	211.53	8.65	4.09
13	205.8	13.44	6.5	214.01	8.51	3.98
14	209.5	13.59	6.5	217.61	8.60	3.95
15	212.3	13.44	6.3	220.17	8.60	3.91
16	215.1	13.39	6.2	222.48	8.26	3.71
17	217.7	12.91	5.9	224.06	8.16	3.64
18	220.4	13.05	5.9	226.80	8.31	3.66
19	221.4	14.13	6.4	228.59	8.70	3.81
20	225.7	11.24	5.0	230.90	8.51	3.69
21	225.6	14.46	6.4	232.93	8.50	3.65

40.3 Mathematical Models of Actuator Operation

Since the controlling factor for an active element made of a material with an SME of a force actuator is a change in its temperature, its operation in outer space can be initiated by two methods: by passing an electric current through it or by radiation heating from solar radiation [14, 18]. The heating process of active elements can be described using a heat balance equation based on the law of conservation of energy.

As a first approximation, we can assume that the change in energy of the active element of a force actuator made of a material with SME is equal to the amount of heat received due to electrical energy, due to the solar radiation, minus the heat losses from emission radiation in orbit and minus the energy spent on phase transformations

$$cm\frac{dT}{dt} = RI^2 + A_r Q_s - \varepsilon\sigma_0 T^4 S - Q, \tag{40.2}$$

where c is the specific heat capacity of the active element; m is the mass of the active element; T is the temperature of the active element; t is the time; R is the electrical resistance of the active element; I is the electrical current strength; A_r is the solar radiation absorption coefficient of the active element; Q_s is the heat flux from solar radiation incident on the active element; ε is the emission coefficient of the active element; σ_0 is the Stefan-Boltzmann constant; S is the radiating surface area; Q is the energy spent on phase transformations.

During ground testing, it can be assumed that the change in the thermal energy of the active element of the force actuator is equal to the amount of heat received due to electrical energy, minus heat losses due to natural convection, and minus the energy spent on phase transformations

$$cm\frac{dT}{dt} = RI^2 - \alpha S(T - T_a) - Q, \tag{40.3}$$

where α is the heat transfer coefficient, T_a is the ambient temperature.

40.3.1 Linear Model

The calculation is carried out using time intervals. Although the terms in Eq. (40.3), that represent heating due to flow of electrical current and heat loss due to natural convection may be considered to be approximately constant throughout the entire operating time of the actuator, the energy consumed for phase transformations changes significantly during the actuator's operation. Therefore, at each time interval, the amount of heat expended on phase transformations (Q') is calculated as

$$Q' = n(RI^2\Delta t - \alpha S(T - T_a)\Delta t). \tag{40.4}$$

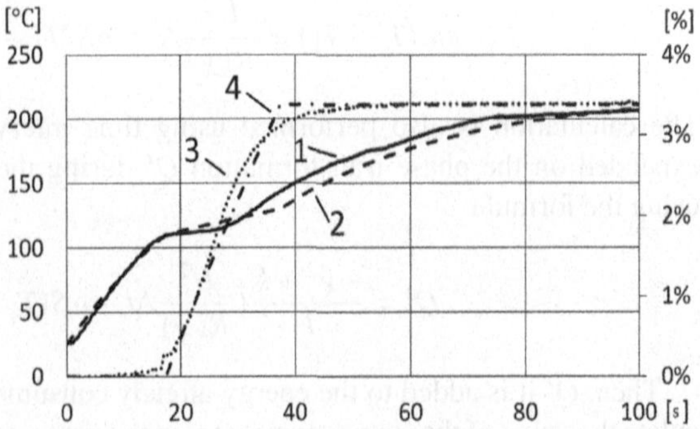

Fig. 40.5 Comparison of the linear model with experimental data:
1 – experimental temperature data,
2 – calculated temperature data,
3 – experimental displacement data,
4 – calculated displacement data.

Additionally, depending on the amount of energy expended during a given time period for phase transformations, it is possible to calculate the displacement generated by the actuator

$$h = \frac{Q'}{Q} h_{\max}. \tag{40.5}$$

Figure 40.5 shows a comparison between calculation made using the linear model with experimental data. By analyzing the graphs, it can be concluded that the results from the linear model calculations closely match the experimental data at the start of actuator operation. However, towards the end of the phase transition process, there is a significant difference due to the non-linear nature of changes in both temperature and displacement.

40.3.2 Nonlinear Model

When modeling, such non-linear behavior during the operation of an active element of an actuator with a shape memory effect can be accounted for using a state parameter, denoted as χ. This parameter represents the fraction of material that has already undergone a phase transition, relative to the total amount of material. As the amount of energy required for the phase transformation is known [12], the state parameter can be determined using the total energy required for the entire material to undergo phase transformation (Q) and the amount of energy already consumed for phase transformation up to the current time point (Q')

$$\chi = \frac{Q'}{Q}, \quad 0 \le Q' \le Q, \quad 0 \le \chi \le 1. \tag{40.6}$$

Fundamentally, the equation that describes the heating of the active element remains unchanged. However, by incorporating the state parameter, it is now possible to consider the variation in the electrical resistance of the actuator, which influences the heating dynamics

$$cm(T_2 - T_1) = \frac{U^2}{R(\chi)}\Delta t - \alpha S(T_1 - T_a)\Delta t - Q. \tag{40.7}$$

The calculation is also performed using time intervals. At each step the energy expended on the phase transformation Q' during the time period Δt is calculated using the formula:

$$Q' = \frac{-\chi^2 + k}{k}\left(\frac{U^2}{R(\chi)}\Delta t - \alpha S(T_1 - T_a)\Delta t\right). \tag{40.8}$$

Then, Q' it is added to the energy already consumed in the previous steps to calculate the value of the state parameter χ in the next step. The multiplier with the state parameter determines how much thermal energy is used for phase transformations, and how much is used for heating. Energy consumption for phase transformation occurs only after the active element reaches the temperature of the beginning of phase transformations and continues until all material has undergone the complete phase transformation [18]. Therefore, calculation of the consumed energy according to Eq. (40.8) only occurs after the active element has reached a given temperature and continues until the energy consumed equals the energy required for complete phase transformation of all material.

Figure 40.6 demonstrates the experimentally determined variation in the electrical resistance of the active element. As the electrical resistivity varies for different phases of a material, it is necessary to calculate its value at each time step. This should be done by taking into account the varying phase ratio of the conductor using formula (40.9). It is important to do so, as a change in electrical resistance will influence the heating dynamics of the active element

$$R(\chi) = (\rho_M(1 - \chi) + \rho_A\chi)\frac{l}{F}, \tag{40.9}$$

where ρ_A is the resistivity of the austenitic phase; ρ_M is the resistivity of the martensitic phase; l is the length of the active element; F is the cross-sectional area of the active element.

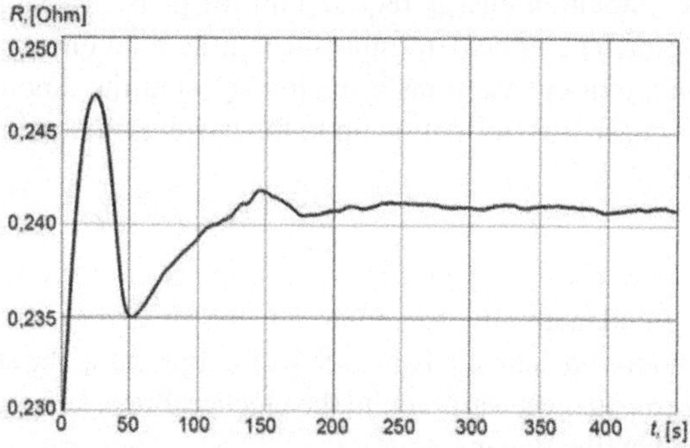

Fig. 40.6 Change in the electrical resistance of the active element.

The displacement generated by the active element of the actuator can be calculated using the state parameter by the following formula

$$h = \chi h_{\max}. \tag{40.10}$$

The results of the calculations for this nonlinear model are presented in Fig. 40.7. Based on the graphs presented in this figure, it can be observed that in addition to the fairly accurate agreement between the calculated and experimental data in the linear region, this model describes well the pronounced nonlinearities in the changes in both temperature and displacement at the end of the phase transformation.

Since the primary parameter of the actuator performance is the change in length, and the temperature field is the controlling factor for the active element made of a shape memory material, the most significant characteristic will be the displacement versus temperature relationship. Therefore, based on the graphs presented in Fig. 40.8, we can conclude that the nonlinear model more accurately describes the final stage of operation of an actuator made of a shape memory material. This model accurately reflects the smoothness of the actuator's operation, which can ensure a stress-free opening of a large-scale space structure.

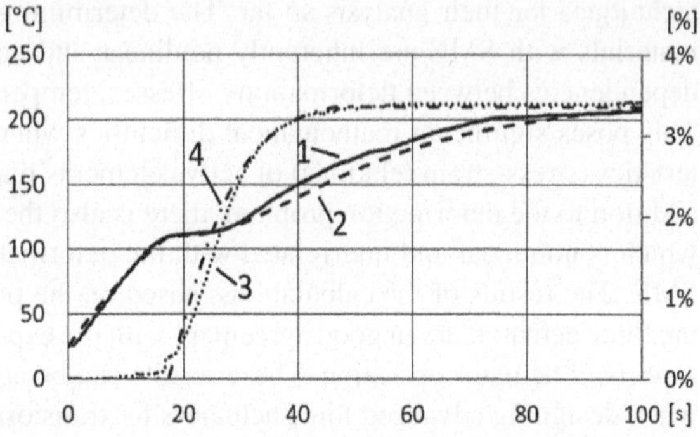

Fig. 40.7 Comparison of the nonlinear model with experimental data:
1 – experimental temperature data;
2 – calculated temperature data;
3 – experimental displacement data;
4 – calculated displacement data.

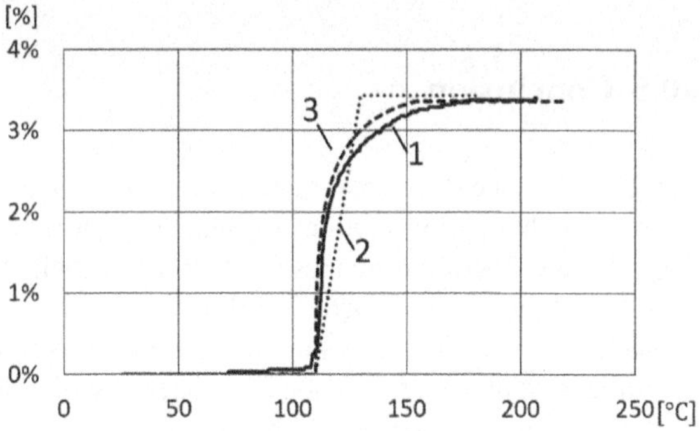

Fig. 40.8 Comparison of displacement dependencies on temperature:
1 – experimental data;
2 – linear model;
3 – nonlinear model.

40.4 Discussion of the Results

The design of promising large-scale space structures is primarily based on mathematical (numerical) experimentation, which aims to validate design solutions from the perspective of their conformity with the given task. Prior to conducting such experiments, it is essential to develop a functional model of the actuation system that will ensure the successful and reliable deployment of the space structure from its tightly packed transport configuration to its operational state in space. Firstly, it is essential to be guided by the results that can be achieved through the use of the model in a practical design environment, while giving its mathematical advantages a secondary consideration. Namely: completeness, versatility, originality, etc. In addition, it is necessary to pay attention to the consumer properties of actuator operation models, which generally are formed in close connection with the design requirements, available computing resources and the time frame for developing a full-scale space structure. Currently, experimental methods play a significant role in the development of models for the operation of actuators utilizing materials with SME. Due to the mathematical challenges associated with describing the behavior of materials with SME and the need for a substantial amount of actual experimental data of their functioning, it has not been possible to develop reliable computational techniques for their analysis so far. The determining relations (state equations) of materials with SME are inherently nonlinear and ambiguous, reflecting the interdependencies between deformations, stresses, temperature, and phase composition. This poses significant mathematical difficulties when calculating the force characteristics (stress-strain relations) of active elements made of a material with SME. In addition to the deformation problem, there is also the issue of thermal conductivity, which is nonlinear and interrelated with the deformation problem in materials with SME. The results of the calculations, based on the proposed model of operation of the force actuator, are in good agreement with the experimentally obtained data in all aspects of actuator operation. These results can be applied in practical applications when designing advanced force actuators for transformable space structures.

40.5 Conclusion

Large-scale space structures are highly complex technical systems that require continuous research for new solutions to enhance performance. One of the most important stages in the operation of these structures is their deployment from the transport state to the operational state in orbit, which occurs under the influence of force actuators. One of the promising areas of development in deployment techniques is the utilization of actuators made of a material with a shape memory effect. However, the fundamental performance characteristics of these materials are quite complex and exhibit significant nonlinearity. To broaden the scope of applications of such materials in large-scale space structures, it is essential to create precise mathematical models that describe the behavior of these actuators during operation. A mathematical model for describing the change in temperature and length of an active element of an actuator is proposed. The findings from calculations based on the proposed mathematical model align with the experimental data. The developed mathematical model has the potential to serve as a foundation for future research in this field, as well as for the design of increasingly advanced large-scale space structures.

Acknowledgements This work was supported by the Ministry of Science and Higher Education of the Russian Federation (project No. FSFN-2024-0004).

References

[1] Banichuk NV, Karpov II, Klimov DM, Markeev AP, Sokolov BN, Sharanyuk AV (1997) Mekhanika bol'shikh kosmicheskikh konstruktsii (in Russ., Mechanics of Large Space Structures). Publishing house "Factorial", Moscow

[2] Barvinok VA, Bogdanovich VI, Lomovsky OV, Vishnyakov MA, Groshev AA (2011) Razrabotka reversivnykh silovykh privodov iz materialov s effektom pamyati formy dlya ustroystv, primenyayemykh v uzlakh raschekovki kosmicheskikh apparatov (in Russ., Development of reversible power drives from materials with a shape memory effect for devices used in the nodes of spacecraft checkout). Izvestiya of Samara Scientific Center of the Russian Academy of Sciences 13(4(2)):301-306

[3] Barvinok VA, Bogdanovich VI, Groshev AA, Plotnikov AN, Lomovskoy OV (2013) Metodika proyektirovaniya silovykh privodov iz materiala s effektom pamyati formy dlya raketno-kosmicheskoy tekhniki (in Russ., Methodology for designing power drives from material with shape memory effect for rocket and space technology). Izvestiya of Samara Scientific Center of the Russian Academy of Sciences 15(6):272-277

[4] Hasanzade V, Sedighy SH, Shahravi M (2017) Compact deployable umbrella antenna design with optimum communication properties. Journal of Spacecraft and Rockets 54(3):782-788, DOI 10.2514/1.A33710

[5] Im E, Thomson M, Fang H, Pearson J, Moore J, Lin J (2007) Prospects of large deployable reflector antennas for a new generation of geostationary Doppler weather radar satellites. In: AIAA SPACE 2007 Conference and Exposition 9917, DOI 10.2514/6.2007-9917

[6] Karla S, Bhattacharya B, Munjal BS (2017) Design of shape memory alloy actuated intelligent parabolic antenna for space applications. Smart Materials and Structures 26(9):095015, DOI 10.1088/1361-665X/aa7468

[7] De Laurentis KJ, Fisch A, Nikitczuk J, Mavroidis C (2002) Optimal design of shape memory alloy wire bundle actuators. In: Proceedings 2002 IEEE International Conference on Robotics and Automation, Vol 3, pp 2363-2368, DOI 10.1109/ROBOT.2002.1013585

[8] Liang C, Rogers CA (1997) Design of shape memory alloy actuators. Journal of Intelligent Material Systems and Structures 8(4), 303-313, DOI 10.1177/1045389X9700800403

[9] Likhachev VA, Razov AI, Cherniavsky AG, Kravchenko Y, Trusov SN (1994) Truss mounting in space by shape memory alloys. In: Proceedings of the First International Conference on Shape Memory and Superelastic Technologies, California, USA, pp. 245-248

[10] Lopatin AV, Rutkovskaya MA (2007) Obzor konstruktsiy sovremennykh transformiruyemykh kosmicheskikh antenn, Chast' 1 (in Russ., Review of the designs of modern transformable space antennas, Part 1). Vestnik of the Reshetnev Siberian State University of Space and Technology (2):78-81

[11] Lopatin AV, Rutkovskaya MA (2007) Obzor konstruktsiy sovremennykh transformiruyemykh kosmicheskikh antenn, Chast' 2 (in Russ., Obzor konstruktsiy sovremennykh transformiruyemykh kosmicheskikh antenn (Chast' 1) Review of the designs of modern transformable space antennas, Part 2). Vestnik of the Reshetnev Siberian State University of Space and Technology (3):51-57

[12] Mishustin IV (2012) Trokhurovnevaya model' splava s pamyat'yu formy i yeyo primeneniye k opisaniyu effekta reversivnoy pamyati formy Trokhurovnevaya model' splava s pamyat'yu formy i yeyo primeneniye k opisaniyu effekta reversivnoy pamyati formy(in Russ., Three-level model of shape memory alloy with application to reversing shape memory effect simulation). Mekhanika Kompozitsionnykh Materialov i Konstruktsiy 18(4):463-474

[13] Ponomarev SV (2011) Transformiruemye reflektory antenn kosmicheskikh apparatov (in Russ., Transformable reflectors of spacecraft antenna). Vestnik Tomskogo Gosudarstvennogo Universiteta. Matematika i Mekhanika 4(16):110-119

[14] Riad A, Ainamany A, Benzohra M (2017) The shape memory alloy actuator controlled by the Sun's radiation. Materials Research Express 4(7):075701, DOI 10.1088/2053-1591/aa75bb

[15] Schiedeck F, Hemsel T, Wallaschek J (2006) The use of shape memory alloy wires in actuators. Solid State Phenomena 113:195-198, DOI 10.4028/www.scientific.net/SSP.113.195

[16] Zheng F (2015) Affordable System Conceptual Structure Design of New Deployable Spaceborne Antenna. In 33rd AIAA International Communications Satellite Systems Conference and Exhibition, DOI 10.2514/6.2015-4343

[17] Zimin VN, Krylov AV, Shakhverdov AO (2021) Development of the mathematical model of the force actuator for deployment of large-sized space structures Journal of Physics: Conference Series 1902:012115, DOI 10.1088/1742-6596/1902/1/012115

[18] Zimin VN, Krylov AV, Kuvyrkin GY, Shakhverdov AO (2022) Development of the mathematical model of the force actuator for the opening of a space structure with transformable configuration. Journal of Physics: conference Series 2231:012029, DOI 10.1088/1742-6596/2231/1/012029

The manufacturer's authorised representative in the EU is Springer Nature Customer Service Centre GmbH, Europaplatz 3, 69115 Heidelberg, Germany. If you have any concerns regarding our products, please contact ProductSafety@springernature.com

Printed and bound by CPI Group (UK) Ltd, Croydon, CR0 4YY
05/10/2026
02228057-0014